Mechanics of Materials

Mechanics of Materials

T. Krishna Rao
B.E., MS (IIT-M), Ph.D., MBA
Professor and Head
Department of Mechanical Engineering
Global Academy of Technology
Bangalore

I.K. International Pvt. Ltd.
NEW DELHI

Published by
I.K. International Pvt. Ltd.
4435-36/7, Ansari Road, Daryaganj
New Delhi-110 002 (India)
E-mail: info@ikinternational.com
Website: www.ikbooks.com

ISBN: 978-93-90620-55-5

Published by Krishan Makhijani for I.K. International Pvt. Ltd., 4435-36/7, Ansari Road, Daryaganj, New Delhi-110 002

Printed at Shree Maitrey Printech Pvt. Ltd., Noida

Preface

I am happy to bring out the book of "Mechanics of Materials" for the use of engineering, polytechnic and AMIE students. The chapters are so chosen that the general requirement of all engineering students are fulfilled.

During my teaching career for more than four decades, I have noticed that students face difficulty in understanding the fundamentals and concepts and find it difficult to solve problem in Mechanics of Materials. An attempt is made to present the concepts in a simple and understandable way with variety of problems solved in each topic, so that students can easily understand and solve any type of problems in examination or competitive examinations. The exercise problems are thought provoking and help them learn the problem solving techniques. MCQs are provided for practice.

The contents of this are carefully prepared in a simple and easy to follow language to help even the average students who can easily study and self-learn the subject.

I express my sincere thanks to Sri Lepaksha S., Hon. Secretary, Sri Venkatappa, Hon. Additional Secretary and the Management of National Education Foundation. My special thanks to our beloved Principal Dr. N. Rana Pratap Reddy, Principal, Global Academy of Technology, for his motivation, encouragement and support for writing this book.

I express my thanks to Smt. Sita Maha Lakshmi, Smt. Meghana and Mr. Pranav for their support and encouragement.

Though every care has been taken in presenting the concepts, solving problems and checking the manuscript and proofreading several times, there may be some mistakes or errors. I shall be very grateful to the readers and teachers who are the users of this book for informing any mistakes that might have crept in. Also suggestions for improvement are very much solicited. The same will be incorporated in the next edition.

T. Krishna Rao

Contents

CHAPTER 1

Simple Stresses and Strains

1.1 INTRODUCTION

When a body is subjected to any external force, the body undergoes deformation. Due to internal bonding, the body tends to resist the deformation. The study of this phenomenon is the Mechanics of Materials. The resistance offered by the body is strength of materials. Study of Mechanics of Materials provides the means to analyse and design various machine elements and structures.

In this subject, we will try to understand the effects of external forces acting on a body. As assumed earlier in Engineering Mechanics, the bodies are no more perfectly rigid. But bodies tend to deform under external loads. We will study and calculate the deformation of bodies under varieties of loads.

1.2 ENGINEERING MATERIALS

Engineering materials are mainly classified as ferrous and non-ferrous materials. Another way of classification of materials is metallic and non-metallic materials. All varieties of steels, and cast iron are examples of ferrous materials. Copper, aluminium, brass, plastics, etc., are examples some non-ferrous materials.

Examples of metals are steels, cost iron, copper, aluminium, brass, etc., and examples of non-metals are plastics, grass, rubber, wood, etc.

The choice of selection of engineering materials depends on their availability, suitability for a given application, cost of material, machinability, etc.

1.3 SOME IMPORTANT PROPERTIES OF MATERIALS

The properties that determine the use of a particular material are mechanical, physical and chemical properties. Some of the important physical properties are, size, electric and thermal conductivities, density and weight, melting point, colour, etc. Some important chemical properties are composition, corrosion resistance, chemical stability, etc.

Important mechanical properties are:

- Elasticity
- Plasticity
- Ductility
- Brittleness
- Malleability
- Resilience
- Toughness
- Stiffness
- Strength
- Hardness
- Creep

1.3.1 Elasticity

Whenever a member is loaded, it will deform. When the load is removed, it may go back to its original shape. The ability of a deformed material to "regain" its original shape and size when external force is removed is called elasticity. This is an important property of all machine members.

1.3.2 Plasticity

The body will regain its previous shape and size only when the deformation caused by the external force is within certain value. There is a limiting value of this external force up to which the deformation disappears completely after the removal of external force. The value of stress corresponding to this limit is known as elastic limit of that material. If the stress produced by the external force is greater than its elastic limit, then there will be a permanent deformation in the member. Now even if the load is completely removed, the material does not regain its original shape and size. The property of material to "retain" the new shape even after the load is removed is called plasticity. This is a desirable property. in the manufacturing processes like forging, shaping, punching, etc.

1.3.3 Ductility

The physical property of a metal that can be drawn into thin wires is called ductility. A ductile material is one having a relatively large tensile strain up to the point of rupture. Examples are structural steel, aluminium, copper, etc. Ductility is usually measured in terms of percentage elongation or percentage reduction in area of cross section of the test specimen.

1.3.4 Brittleness

Brittleness is opposite of ductility. We call a material brittle if it cannot be drawn into thin wires. Brittle material has relatively very small strain up to the rupture. An arbitrary strain of about 0.05 is generally taken as the threshold between ductile and brittleness. Cast iron, glass, brick, stone, ceramics, etc., are examples of ductile materials.

1.3.5 Malleability

Malleability is the physical property of a solid to bend or to be hammered into another shape without breaking. This property permits the material to flow or expand in all directions when subjected to compressive force, without rupture. If malleable, a material may be flattened into thin sheets or plates by hammering or rolling. Examples of malleable metals are gold, silver, aluminium, copper, some carbon steals, etc.

1.3.6 Resilience

Resilience is defined as the ability to absorb energy when deformed elastically and release the energy when unloaded. Modulus of resilience is the measure or indication of resilience property of a material. Modulus of resilience is defined as energy per unit volume, required to deform a particular solid material up to its elastic point in tension. Modulus of resilience is the area below engineering stress-strain diagram up to elastic point.

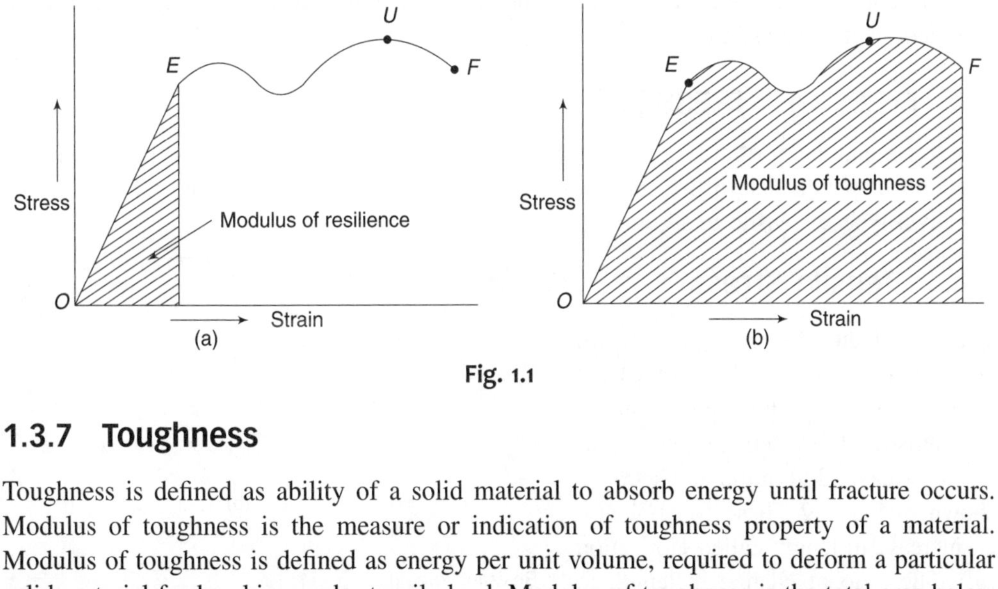

Fig. 1.1

1.3.7 Toughness

Toughness is defined as ability of a solid material to absorb energy until fracture occurs. Modulus of toughness is the measure or indication of toughness property of a material. Modulus of toughness is defined as energy per unit volume, required to deform a particular solid material for breaking under tensile load. Modulus of toughness is the total area below engineering stress-strain curve.

Figures 1.1(a) and (b) show the difference between modulus of resilience and modulus of toughness.

1.3.8 Stiffness

Stiffness is defined as the ability of a material to resist deformation under the action of external load. It is the load required to cause unit deformation of a member.

1.3.9 Strength

Strength is a measure of the amount of stress a material can withstand without breaking or deforming. This is the ability of the material to support maximum load before it deforms or breaks. Based on type of external loads applied, the strength is expressed as tensile strength, compressive strength, bending strength, shear strength, etc. The stress beyond which plastic deformation is caused is called yield stress and the maximum stress a material can carry before breaking is called ultimate stress.

1.3.10 Hardness

Hardness is defined as the resistance offered by a material for scratch or indentation. Hardness is an indirect measure of wear resistance, scratch resistance, deformation, machinability, etc. If hardness increases the strength will also increase.

1.3.11 Creep

It is the tendency of a solid material to move slowly or deform permanently under the influence of persistent mechanical stresses. It can occur as a result of long-term exposure to continuous stresses that are still below the yield strength of a material, e.g., boiler shell, cylinders, belts, chains, etc.

1.4 TYPES OF STRESSES

The stress may be normal stress or shear stress. Normal stress is the stress that acts in the direction perpendicular to the area. It is represented by symbol σ (sigma). Normal stress is of two types – tensile and compressive.

1.4.1 Tensile Stress

The stress induced in a body when two equal and opposite pulling forces act along its axis is called tensile stress. The tensile stress acts perpendicular to the area and it pulls the area as shown in Fig. 1.2 These forces cause an increase in length called elongation (δl). The ratio of change in length to original length is known as tensile strain.

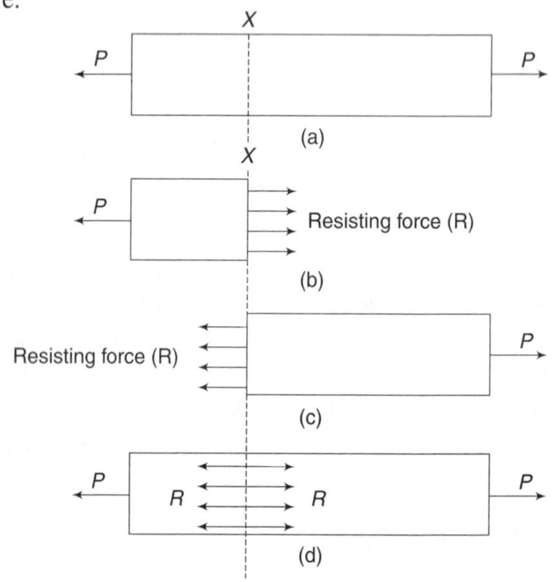

Fig. 1.2

1.4.2. Compressive Stress

The stress induced in a body when two equal and opposite pushing forces act along its axis is called compressive stress.

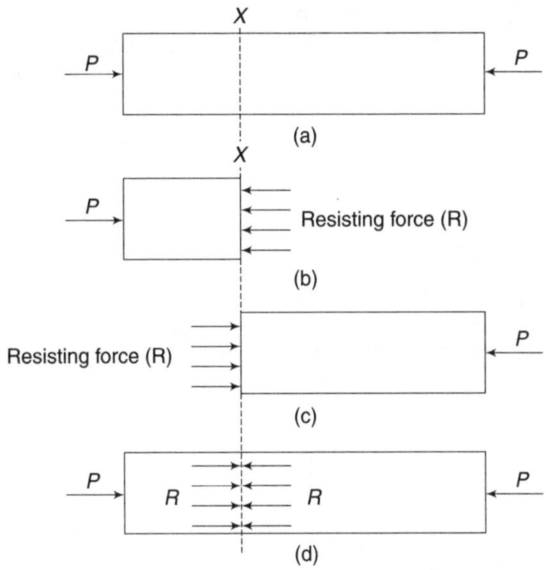

Fig. 1.3

Similar to tensile stress, compressive stress also acts perpendicular to the area of cross section and it pushes the area as shown in Fig. 1.3. These forces cause a decrease in length, δl. The ratio of decrease in length to its original length is called compressive strain.

1.4.3 Shear Stress

The stress induced in a body when subjected to two equal and opposite forces that act tangentially across the resisting area as shown in Fig. 1.4. As a result the body tends to shear off across the cross section. The stress produced due to this is known as shear stress.

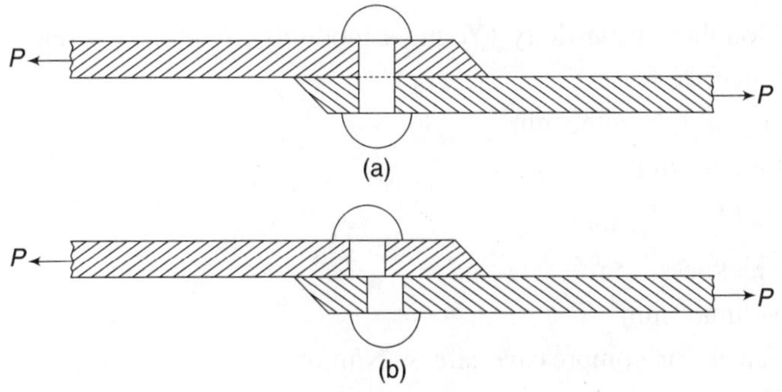

Fig. 1.4

Consider a rectangular block *ABCD* of height '*h*' and length '*L*' subjected to a tangential force *P* as shown in Fig. 1.5. Such force acting tangentially along a surface is known as shear force and the stress induced is called shear stress. As the bottom face (*AB*) is fixed, the top face (*CD*) will be distorted to *C*, *D*, through an angle, ϕ. ϕ is called shear strain.

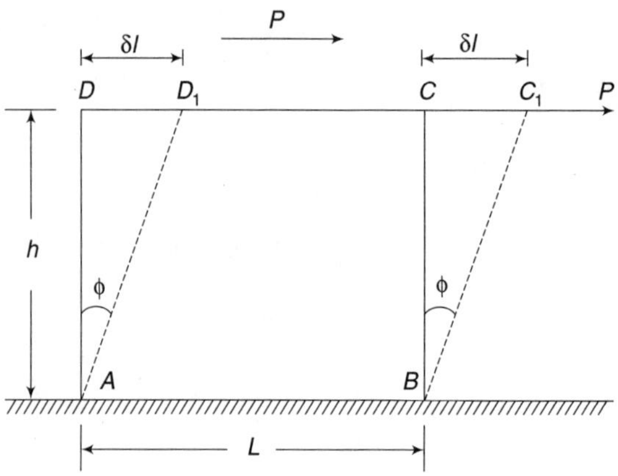

Fig. 1.5

From Δl ADD_1, $\tan \phi = \dfrac{DD_1}{AD} = \dfrac{\delta l}{L}$

For small angles, $\tan \phi = \phi$.

or $\phi = \dfrac{\delta l}{L}$ is called shear strain.

1.5 SYMBOLS

A	=	Area of cross section, mm^2
B, b	=	Width, mm
D, d	=	Diameter, mm
di	=	Inside diameter, mm
do	=	Outside diameter, mm
E	=	Modulus of elasticity (Young's modulus), N/mm^2 or MPa
$F = P$	=	Force, N
h	=	Height or depth, mm
l	=	Length, mm
r	=	Radius $= \dfrac{d}{2}$, mm
t	=	Thickness, mm
V	=	Volume, mm^3
σ	=	Tensile or compressive stress, N/mm^2
τ	=	Shear stress, N/mm^2
μ	=	Poisson's ratio
$\in$	=	Strain
δl	=	Change in length, mm
δb	=	Change in width, mm
δt	=	Change in thickness, mm

δd	=	Change in diameter, mm
δV	=	Change in volume, mm^3
σ_y	=	Yield stress, N/mm^2
σ_U	=	Ultimate stress, N/mm^2
FoS	=	Factor of safety
ΔT	=	Change in temperature, °C
T	=	Temperature, °C

1.6 STRESS AND STRAIN

Stress. Whenever a member is subjected to an axial load, the member develops an internal resistance which is equal and opposite to the external force. This resistance offered by unit area is defined as stress. Symbolically,

$$\text{Stress} = \frac{\text{Resisting force}}{\text{Unit area}}$$

$$\sigma = \frac{P}{A}$$

($\because$ Resisting force is equal to the external force)

The units of stress are N/m^2 = Pa or N/mm^2 = MPa = 10^6 Pa

Note:

$$\frac{N}{m^2} = \frac{N}{(1000\text{ mm})^2} = \frac{N}{10^6 \times mm^2}$$

$$\therefore \quad \frac{N}{mm^2} = 10^6\,\frac{N}{m^2} = 10^6\text{ Pa} = \text{MPa}$$

The other units of stress are: GPa = giga Pa = 10^9 Pa = $10^3 \times 10^6$ Pa = 10^3 MPa.

Strain. Whenever a member is subjected to an axial load, the length of the member will change.

Strain is defined as the ratio of change in length to its original length.

$$\text{Strain } \in = \frac{\delta l}{l} \quad \text{(This is a unitless quantity)}$$

Modulus of Elasticity and Hooke's Law

Robert Hooke proposed that, within elastic limit, the stress is proportional to the strain, i.e., the ratio of stress by strain is constant within the elastic limit. This constant of proportionality is defined as Young's modulus or modulus of elasticity (E).

Within elastic limit, stress $\propto$ strain

$$\sigma \propto \in$$

or

$$\frac{\sigma}{\in} = \text{Const} = E$$

$$\therefore \quad E = \frac{\sigma}{\in}.$$

But stress $\sigma = \dfrac{P}{A}$ and strain $\in\ = \dfrac{\delta l}{l}$

$$\therefore \qquad E = \frac{(P/A)}{(\delta l/l)} = \frac{Pl}{A\delta l}$$

E has same units as the stress.

$$\therefore \qquad \text{Change in length } \delta l = \frac{Pl}{AE}.$$

1.7 STRESS–STRAIN DIAGRAM FOR MILD STEEL SPECIMEN

Figure 1.6 shows a typical stress-strain curve for a mild steel specimen subjected to an axial load on a universal testing machine (UTM).

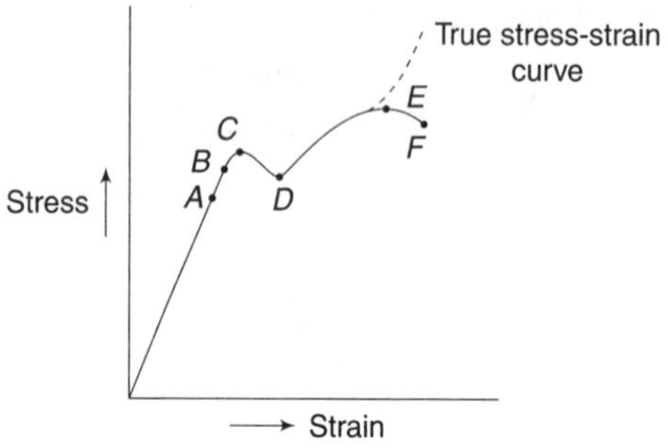

Fig. 1.6

Point A. *Proportionality limit*: Up to this point, the stress-strain curve is a straight line and the stress is proportional to strain. Hooke's law is obeyed up to this point.

Point B. *Elastic limit:* When load is applied, the member deforms. If the member regains its original shape after removal of the applied load, then it is said to be within elastic limit. Beyond this point the member yields. This is very close to proportionality limit.

Point C. *Upper yield point:* Beyond elatic limit, there is sudden increase in strain without appreciable increase in load. Point *C* is known as upper yield point.

Point D. *Lower yield point:* Due to yielding, there is a slight drop in the stress and the strain increases without any increase in load. This point *D* is known as lower yield point. Yield stress is important for the designers.

Point E. *Ultimate stress:* If the load is further increased beyond the yield point, the member goes to plastic zone. Even if the load is removed, the member does not regain its original shape during this stage. This is due to the fact that the member has yielded. The curve beyond elastic limit is parabolic (nonlinear). The maximum stress experienced by the member is known as ultimate stress. Beyond this point, the stress drops and fracture occurs.

Point F. *Breaking stress:* Finally, the member breaks at F and the corresponding stress is known as breaking stress or stress at fracture.

True stress: For drawing the stress-strain curve, the stress is calculated taking original area of cross section. But beyond ultimate stress, necking forms and the diameter of specimen decreases. If stress is calculated by taking actual area at neck then it is called true stress. The corresponding curve is shown in dotted line and is called ture stress-strain curve. Stress-strain curves for other materials are shown in Fig. 1.7.

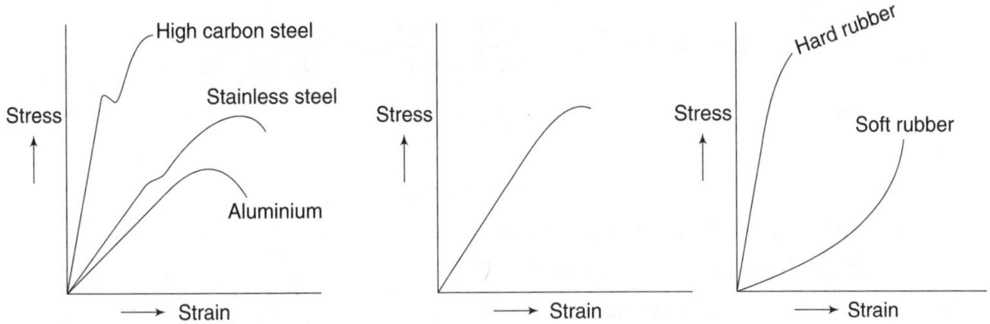

(a) For stainless steel, aluminium and high carbon steels.

(b) For cast iron

(c) For rubber

Fig. 1.7

1.8 PROOF STRESS

For brittle materials like cast iron, there will not be well defined yield stress, a line parallel to the stress-strain curve is drawn at 0.2% strain (AB) and the stress corresponding to this at point B is known as proof stress. For practical calculations this is treated as yield stress.

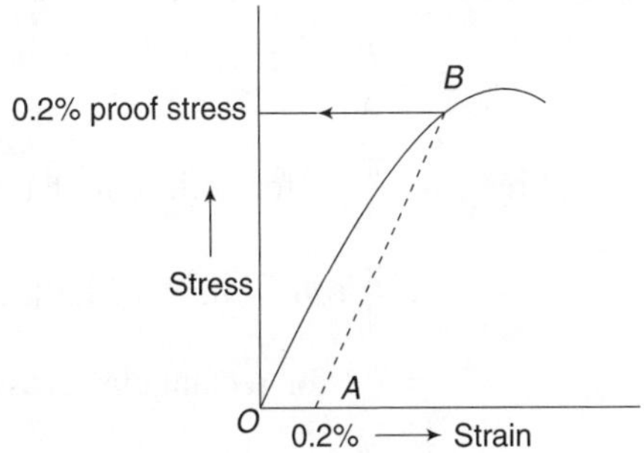

Fig. 1.8 Proof stress

1.9 FACTOR OF SAFETY (FoS)

The yield stress or ultimate stress is taken into consideration while designing the size of a member. This may vary from sample to the material actually used. It is customary to scale down these values and use them for design purpose. This reduced or scaled down value of stress is known as allowable stress, design stress, working stress, limiting stress, permissible stress or safe stress.

The scaling factor is known as factor of safety.

$$\text{Factor of safety} = \frac{\text{Yield stress}}{\text{Design stress}} \text{ or } \frac{\text{Ultimate stress}}{\text{Design stress}}$$

Thus, factor of safety is defined as the ratio of yield stress to design stress or ultimate stress to design stress.

The various reasons for selection of factor of safety are:

(i) Uncertainty of load and working conditions

(ii) Uncertainty of material properties

(iii) Non-homogeneity of material

(iv) Human life at risk

(v) Severity of damage due to its failure etc.

$$\text{FoS} = \frac{\sigma_u}{\sigma} \text{ or } \frac{\sigma_y}{\sigma}$$

Generally, a FoS of 4 to 6 is used based on ultimate stress and 1.5 to 3 is used based on yield stress.

$$\therefore \quad \text{Allowable stress } \sigma = \frac{\sigma_u}{\text{FoS}} \text{ or } \frac{\sigma_y}{\text{FoS}}$$

1.10 IMPORTANT RELATIONS

$$\text{Stress } \sigma = \frac{P}{A}$$

$$\text{Area } A = \frac{\pi}{4} d^2 \text{ for solid circular rod}$$

$$= \frac{\pi}{4} (do^2 - di^2) \text{ for hollow circular rod}$$

$$= b \cdot t \text{ for rectangular cross section}$$

$$\text{Strain } \in = \frac{\delta l}{l}$$

$$\text{Modulus of elasticity} = E = \frac{\sigma}{\in} = \frac{Pl}{A\delta l}$$

$$\text{Change in length } \delta l = \frac{Pl}{AE}$$

$$\% \text{ change in length} = \frac{\text{Change in length}}{\text{Original length}} \times 100 = \frac{\delta l}{l} \times 100$$

$$\% \text{ reduction in area} = \frac{\text{Original area} - \text{Final area}}{\text{Original area}} \times 100$$

$$\text{Yield stress } \sigma_y = \frac{\text{Load at yield point}}{\text{Original area of cross section}}$$

$$\text{Ultimate stress } \sigma_U = \frac{\text{Load at ultimate point (max. load)}}{\text{Original area of cross section}}$$

$$\text{True stress} = \frac{\text{Load}}{\text{Actual area of cross section}}$$

SOLVED PROBLEMS

Problem 1 A steel rod of 30 mm diameter and 1.8 m long is subjected to a tensile load of 81 kN. The change in length was observed to be 1 mm. Find: (a) stress induced (b) strain, and (c) the modulus of elasticity.

Solution: **Given data:**

$$\text{Load } P = 81 \text{ kN} = 81000 \text{ N}$$

$$\text{Diameter } d = 30 \text{ mm}$$

$$\text{Change in length } \delta l = 1 \text{ mm}$$

$$\text{Length } l = 1.8 \text{ m} = 1800 \text{ mm}$$

(a) $$\text{Stress } \sigma = \frac{\text{Load}}{\text{Area}} = \frac{P}{A} = \frac{81000}{\frac{\pi}{4} \times 30^2} = 114.59 \text{ MPa.}$$

(b) $$\text{Strain } \in = \frac{\delta l}{l} = \frac{1}{1800} = 5.556^{-4} \times 10^{-4}$$

(c) $$\text{Modulus of Elasticity } E = \frac{\text{Stress}}{\text{Strain}} = \frac{\sigma}{\in} = \frac{114.59}{5.556 \times 10^{-4}}$$

$$E = 206262 \text{ MPa} = 206.262 \times 10^3 \text{ MPa.}$$

Problem 2 A steel rod of 45 mm diameter and 3 m long is subjected to an axial pull of 90 kN. Taking modulus of elasticity as 200 GPa. Find:
(a) the stress induced, (b) the stain, and (c) the change in length.

Solution: **Given data;**

$$\text{Load } P = 90 \text{ kN} = 90000 \text{ N}$$

$$\text{Diameter } d = 45 \text{ mm}$$

$$\text{Length } l = 3 \text{ m} = 3000 \text{ mm}$$

$$\text{Modulus of elasticity } E = 200 \text{ GPa} = 200 \times 10^3 \text{ MPa}$$

(a) Stress $\sigma = \dfrac{P}{A} = \dfrac{P}{\left(\frac{\pi}{4}d^2\right)} = \dfrac{90000}{\frac{\pi}{4} \times 45^2} = 56.59$ MPa

(b) Strain $\in$

We know that (WKT) Modulus of Elasticity $= \dfrac{\text{stress}}{\text{strain}}$

$$E = \frac{\sigma}{\in}$$

$$\therefore \quad \text{Strain } \in = \frac{\sigma}{E} = \frac{56.59}{200 \times 10^3} = 2.8294 \times 10^{-4}$$

(c) Change in length δl

$$\text{WKT strain } \in = \frac{\delta l}{l}$$

$$\therefore \quad \text{Change in length } \delta l = l \cdot \in = 3000 \times 2.8924 \times 10^{-4}$$

$$\delta l = 0.849 \text{ mm.}$$

Problem 3 A bar of 24 mm diameter and length 270 mm extends by a length of 0.6 mm by the application of a load *P*. What is the value of load *P* if the Young's modulus of the material is 206 kN/mm²? If the extension of this bar is to be limited to 0.3 mm under the same load *P*, what should be the diameter of the bar?

Solution: **Given data:**

$$\text{Diameter } d = 24 \text{ mm}$$

$$\text{Length } l = 270 \text{ mm}$$

$$\text{Extension in length } \delta l = 0.6 \text{ mm}$$

$$\text{Young's modulus } E = 206 \text{ kN/mm}^2$$

$$= 206 \times 10^3 \text{ N/mm}^2 \text{ or MPa}$$

(a) To find Load *P*

$$\text{WKT modulus of elasticity } E = \frac{\text{Stress}}{\text{Strain}} = \frac{Pl}{A\delta l}$$

$$\therefore \quad \text{Load } P = \frac{E \cdot A \cdot \delta l}{l}$$

$$P = \frac{206 \times 10^3 \times \left(\frac{\pi}{4} \times 24^2\right) \times 0.6}{270} = 207.094 \times 10^3 \text{ N}$$

(b) If $\delta l = 0.3$ mm, find d for same load P

Again $$E = \frac{Pl}{A\delta l}$$

$$\therefore \qquad A = \frac{Pl}{E\delta l} = \frac{207.094 \times 10^3 \times 270}{206 \times 10^3 \times 0.3} = 904.779 \text{ mm}^2$$

$$\text{But} \qquad A = \frac{\pi}{4} d^2 = 904.779$$

$\therefore$ Required diameter of the bar $d = 33.94$ mm.

Problem 4 A rod of 21 mm diameter and length 120 mm extends by 0.18 mm. If modulus of elasticity of the material of the rod is 2.1×10^5 N/mm^2, what load and type of load is applied to the rod? If an extension of 20% greater is required for the same load, find the diameter of the rod?

Solution: **Given data:**

Diameter $d = 21$ mm

Length $l = 120$ mm

Extension $\delta l = 0.18$ mm

Modulus of elasticity $E = 2.1 \times 10^5$ N/mm^2 or MPa

(a) To find load, P

WKT modulus of elasticity $E = \dfrac{Pl}{A\delta l}$

$$\therefore \qquad \text{Load, } P = \frac{EA\delta l}{l} = \frac{2.1 \times 10^5 \frac{\pi}{4} \times 21^2 \times 0.18}{120}$$

$$P = 109.104 \times 10^3 \text{ N}$$

Since the rod length is extended, the type of load is tensile in nature.

(b) For 20% greater extension,

$$\delta l = 0.18 + 0.18 \times 20\%$$

$$\delta l = 1.2\,(0.18) = 0.216$$

$$\text{WKT} \qquad E = \frac{Pl}{A\delta l}$$

$$\therefore \qquad A = \frac{Pl}{E\delta l} = \frac{109.104 \times 10^3 \times 120}{2.1 \times 10^5 \times 0.216} = 288.634 \text{ mm}^2$$

$$\text{But} \qquad A = \frac{\pi}{4} d^2 = 288.634$$

$\therefore$ Required diameter $d = 19.17$ mm.

Problem 5 A short piece of steel pipe is to carry an axial compressive load of 1 MN. The factor of safety required is 2 against yielding. Thickness of the pipe is $\frac{1}{5}$th of its outside diameter. Find the diameter of the pipe taking yield stress as 300 N/mm^2.

Solution: **Given data:** **Find:** *di & do*

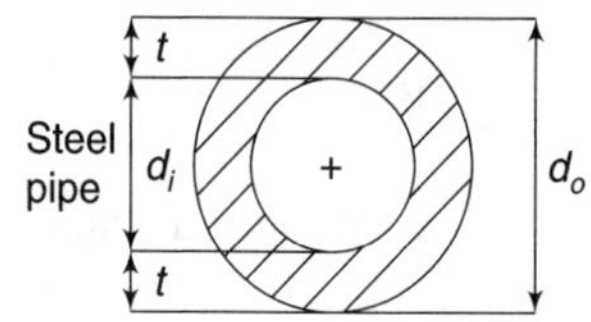

$$\text{Axial load, } P = 1 \text{ MN} = 1 \times 10^6 \text{ N}$$

$$\text{Factor of safety (FoS)} = 2$$

$$\text{Yield stress, } \sigma_y = 300 \text{ MPa or N/mm}^2$$

$$\therefore \quad \text{Allowable stress, } \sigma = \frac{\text{Yield stress}}{\text{FoS}} = \frac{\sigma_y}{\text{FoS}}$$

$$\sigma = \frac{300}{2} = 150 \text{ MPa}$$

$$\text{Thickness of pipe, } t = \frac{d_o - d_i}{2}$$

(From the figure, $t + d_i + t = d_o$)

$$\therefore \quad d_i = d_o - 2t$$

But

$$t = \frac{1}{5} \text{ of outside diameter} = \frac{1}{5} \cdot d_o$$

$$\therefore \quad d_i = d_o - 2\left(\frac{d_o}{5}\right) = 0.6\ d_o$$

$$\text{Area of cross section, } A = \frac{\pi}{4}(d_o^2 - d_i^2)$$

$$A = \frac{\pi}{4}[d_o^2 - (0.6\ d_o)^2] = \frac{\pi}{4} d_o^2 (1 - 0.6^2)$$

$$A = \frac{\pi}{4} d_o^2 (0.64)$$

$$\text{WKT stress, } \sigma = \frac{P}{A}$$

$$\therefore \quad A = \frac{P}{\sigma} = \frac{1 \times 10^6}{150} = 6666.667 \text{ mm}^2$$

$$\frac{\pi}{4} d_o^2 (0.64) = 6666.667$$

$$\therefore \quad \text{Outside diameter} = d_o = 115.165 \text{ mm}$$

$$\text{Inside diameter } d_i = 0.6\ d_o$$

$$d_i = 0.6 \times 115.165 = 69.099 \text{ mm}$$

$$\text{Thickness } t = \frac{d_o}{5} = \frac{115.165}{5} = 234.034 \text{ mm}$$

Problem 6 A hollow steel pipe having inside and outside diameters as 150 mm and 180 mm has to carry a tensile load. If the allowable stress for the steel used is 90 MPa in tension, find the load that can be safely carried by the pipe.

Solution: **Given data:**

$$\text{Hollow pipe inside diameter } d_i = 150 \text{ mm}$$

$$\text{Outside diameter } d_o = 180 \text{ mm.}$$

$$\text{Allowable stress } \sigma = 90 \text{ MPa.}$$

WKT

$$\text{Stress } \sigma = \frac{\text{Load}}{\text{Area}} = \frac{P}{A}$$

$$A = \text{Area of cross section} = \frac{\pi}{4}(d_o^2 - d_i^2)$$

$\therefore$

$$\text{Load } P = \sigma \cdot A$$

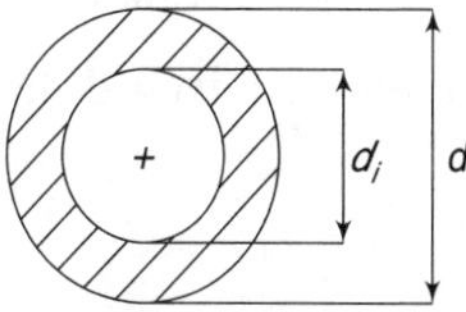

$$P = 90 \times \frac{\pi}{4}(d_o^2 - d_i^2)$$

$$P = 90 \times \frac{\pi}{4}(180^2 - 150^2)$$

$$\text{Load } P = 699.789 \times 10^3 \text{ N.}$$

Problem 7 A hollow C.I. cylinder of wall thickness 30 mm is subjected to a compressive load of 600 kN. Determine the required diameters of the cylinder if the allowable compressive stress for C.I. material is 120 N/mm^2.

Solution: **Given data:**

$$\text{Wall thickness } t = 30 \text{ mm}$$

$$\text{Load } F = 600 \text{ kN} = 600 \times 10^3 \text{ N}$$

$$\text{Allowable stress } \sigma = 120 \text{ N/mm}^2 \text{ or MPa.}$$

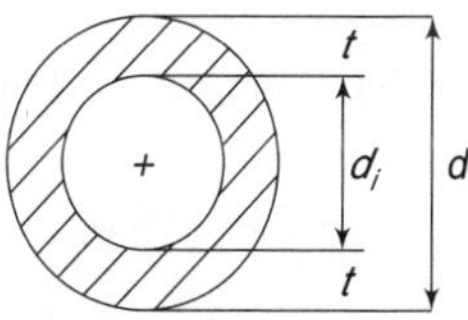

$$\text{Area of cross section } A = \frac{\pi}{4}(d_o^2 - d_i^2)$$

WKT

$$\text{Stress } \sigma = \frac{P}{A}$$

$\therefore$

$$A = \frac{P}{\sigma} = \frac{600 \times 10^3}{120} = 5 \times 10^3 \text{ mm}^2$$

But

$$A = \frac{\pi}{4}(d_o^2 - d_i^2) = 5000$$

$$= \frac{\pi}{4}(d_o + d_i)(d_o - d_i) = 5000$$

But

$$d_o - d_i = 2t = 2 \times 30 = 60$$

$$d_o = d_i + 2t = d_i + 60$$

$$\frac{\pi}{4}(d_i + 60 + d_i)(60) = 5000$$

$\therefore$ Inside diameter $d_i = 23.05$ mm

Outside diameter $d_o = d_i + 2t = 23.05 + 60 = 93.05$ mm.

Problem 8 A steel rod of 30 mm diameter and 300 mm long is tested on UTM for distruction. It showed an extension of 0.185 mm under a load of 90 kN. The load at the elastic limit is 240 kN. The breaking load is 420 kN. The total extension at the fracture is 48 mm. The diameter at the neck is 24 mm.

Find: (i) The Young's modulus

(ii) Percentage elongation

(iii) Stress at elastic point (yield stress)

(iv) Percentage reduction in area

(v) Ultimate stress

Solution: **Given data:**

Diameter $d = 30$ mm

Length $l = 300$ mm

Extension $\delta l = 0.185$ mm at load $P = 90$ kN

Load at elastic limit $P_y = 240$ kN

Breaking load $P_u = 420$ kN

Total elongation $\delta l = 48$ mm

Diameter at the neck $d_n = 24$ mm

$$\text{Area of cross section } A = \frac{\pi}{4} d^2 = \frac{\pi}{4} 30^2 = 706.86 \text{ mm}^2$$

(i) Young's modulus or modulus of elasticity

$$E = \frac{\text{Stress}}{\text{Strain}} = \frac{Pl}{A\delta l}$$

At $P = 90$ kN, $\delta l = 0.185$ mm

$$\therefore \quad E = \frac{90 \times 10^3 \times 300}{706.86 \times 0.185} = 206.47 \times 10^3 \text{ N/mm}^2$$

(ii) $$\text{Percentage elongation} = \frac{\text{Final length} - \text{Original length}}{\text{Original length}} \times 100$$

$$\text{Final length} - \text{Original length} = \text{Total elongation} = 48 \text{ mm}$$

$$\therefore \quad \% \text{ elongation} = \frac{48}{300} \times 100 = 16\%$$

(iii) $$\text{Stress at elastic point} = \text{Yield stress} = \frac{P_{\text{elastic}}}{\text{Area}}$$

$$\sigma_y = \frac{P_y}{A} = \frac{240 \times 10^3}{706.86} = 339.53 \text{ N/mm}^2$$

(iv) Percentage reduction in Area $= \dfrac{\text{Original area} - \text{Final area}}{\text{Original area}} \times 100$

where final area = Area at neck $= \dfrac{\pi}{4} d_n^2 = \dfrac{\left(\frac{\pi}{4} d^2\right) - \left(\frac{\pi}{4} d_n^{\ 2}\right)}{\left(\frac{\pi}{4} \cdot d^2\right)} \times 100$

$$= \frac{706.86 - \left(\frac{\pi}{4} \times 24^2\right)}{706.86} \times 100 = 36\%$$

(v) Ultimate stress $\sigma_u = \dfrac{P_{\text{Ultimate}}}{A} = \dfrac{420 \times 10^2}{706.86}$

$$\sigma_u = 594.177 \text{ N/mm}^2.$$

Problem 9 A tension test was conducted on a specimen and the following readings were noted:

Diameter of specimen = 24 mm, gauge length = 200 mm
Extension at a load of 27 kN = 0.06 mm
Load at yield point = 120 kN
Diameter at neck = 18 mm
Breaking load = 210 kN
Final length = 221 mm.

Find: (i) Modulus of elasticity
(ii) % elongation
(iii) Yield stress
(iv) Ultimate stress
(v) % reaction in area and
(vi) True ultimate stress

***Solution:* Given data:**

Diameter d = 24 mm

Gauge length l = 200 mm

Extension δl = 0.06 mm, load P = 27 kN = 27×10^3 N

Load at yield point P_y = 120 kN = 120×10^3 N

Breaking load P_u = 210 kN = 210×10^3 N

Final length L_F = 221 mm

Diameter at neck d_n = 18 mm

$$\text{Area of cross section } A = \frac{\pi}{4} d^2 = \frac{\pi}{4} 24^2 = 452.39 \text{ mm}^2$$

$$\text{Actual or true area at neck } A_1 = \frac{\pi}{4} d_n^2 = \frac{\pi}{4} \times 18^2 = 254.47 \text{ mm}^2$$

(i) $$\text{Modulus of elasticity } E = \frac{Pl}{A\delta l}$$

$$E = \frac{27 \times 10^3 \times 200}{452.39 \times 0.06} = 198.94 \times 10^3 \text{ MPa}$$

(ii) $$\% \text{ elongation} = \frac{\text{Final length} - \text{Original length}}{\text{Original length}} \times 100$$

$$= \frac{221 - 200}{200} \times 100 = 10.5\%$$

(iii) $$\text{Yield stress } \sigma_y = \frac{P_{\text{yield}}}{\text{Area of cross section}} = \frac{120 \times 10^3}{452.39}$$

$$\sigma_y = 265.26 \text{ MPa.}$$

(iv) $$\text{Ultimate stress } \sigma_u = \frac{P_u}{\text{Area of cross section}}$$

$$\sigma_u = \frac{210 \times 10^3}{452.39} = 464.20 \text{ MPa}$$

(v) $$\% \text{ Reduction area} = \frac{\text{Original area} - \text{Final area}}{\text{Original area}} \times 100$$

$$\text{Final area} = \text{Area at neck} = \frac{\pi}{4} d_n^2 = \frac{\pi}{4} 18^2 = 254.47 \text{ mm}^2$$

$$= \frac{452.39 - 254.47}{452.39} \times 100 = 43.75\ \%$$

(vi) $$\text{True ultimate stress} = \frac{P_u}{\text{True area of cross section}}$$

$$= \frac{210 \times 10^3}{254.47} = 825.25 \text{ N/mm}^2.$$

Problem 10 A steel tape 150 m long has a cross section 20 mm wide and is 0.5 mm thick. A length is measured as 120 m. What will be the actual length if it is measured by another surveyor who applies a force of 150 N. Modulus of elasticity for tape material may be taken as 206 GPa.

***Solution:* Given data:** Length $l = 120 \text{ m} = 120 \times 10^3 \text{ mm}$

Width $b = 20 \text{ mm}$

$$\text{Thickness } t = 0.5 \text{ mm}$$

$$\therefore \quad \text{Area of cross section } A = bt = 20 \times 0.5 = 10 \text{ mm}^2$$

$$\text{Force applied } P = 150 \text{ N}$$

$$\text{Modulus of elasticity } E = 206 \text{ GPa} = 206 \times 10^3 \frac{\text{N}}{\text{mm}^2}$$

$$\text{Increase in length due to stretching } \delta l = \frac{Pl}{AE}$$

$$\delta l = \frac{150 \times 120 \times 10^3}{10 \times 206 \times 10^3} = 8.738 \text{ mm}$$

$$\therefore \quad \text{Actual length measured} = l + \delta l$$

$$= 120 \times 10^3 + 8.738 = 12000\ 8.738 \text{ mm}$$

$$\text{Actual length} = 120.008738 \text{ m.}$$

Problem 11 Find the stress induced in a stepped bar loaded as shown in Fig. 1.9.

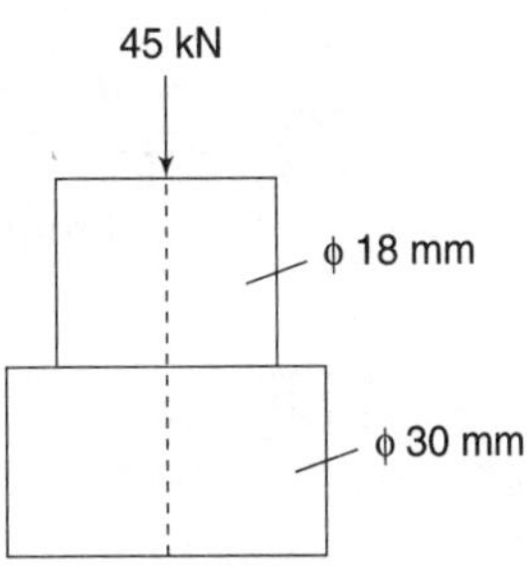

Fig. 1.9

Solution: **Given data:**

$$\text{Load } P = 45 \text{ kN} \times 10^3 \text{ N}$$

$$\text{Area of top block } A_1 = \frac{\pi}{4} d_1^2 = \frac{\pi}{4} 18^2$$

$$= 254.47 \text{ mm}^2$$

$$\therefore \quad \text{Stress in top block } \sigma_1 = \frac{P}{A_1} = \frac{45000}{254.47}$$

$$= 176.84 \text{ N/mm}^2$$

$$\text{Area of bottom block } A_2 = \frac{\pi}{4} d_2^2 = \frac{\pi}{4} \times 30^2 = 706.86 \text{ mm}^2$$

$$\text{Stress in bottom block } \sigma_2 = \frac{P}{A_2} = \frac{45000}{706.86} = 63.66 \text{ N/mm}^2.$$

1.11 ELONGATION IN ROUND TAPER BAR

Figure 1.10 shows a taper round rod of length l that tapers uniformly from bigger diameter d_2 to a smaller diameter d_1 and subjected to tensile load of P.

Consider an elemental length of dx at a distance x from bigger diameter end AB. Let d be the diameter of the element LM.

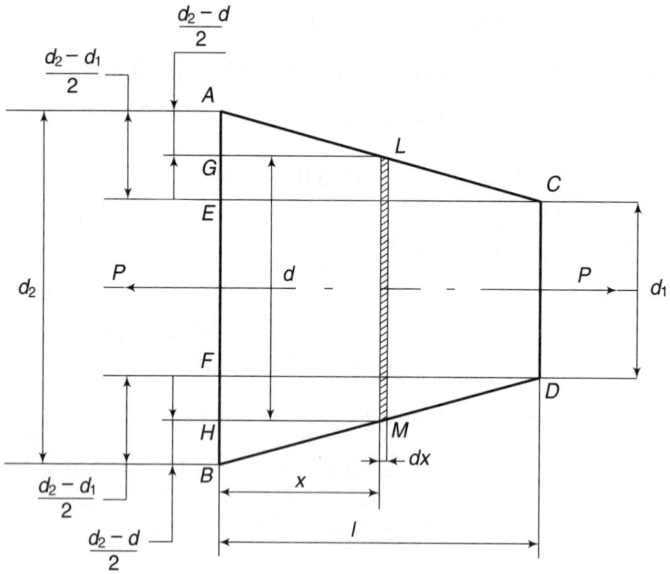

Fig. 1.10

From Fig. 1.10, $EF = CD = d_1$

$$AB = d_2$$

$$GH = LM = d$$

$$AG = BH = \frac{AB - GH}{2} = \frac{d_2 - d}{2}$$

$$AE = BF = \frac{AB - EF}{2} = \frac{d_2 - d_1}{2}$$

Triangles AGL and AEC are similar

$$\therefore \quad \frac{AG}{AE} = \frac{GL}{EC}$$

$$\frac{\left(\frac{d_2 - d}{2}\right)}{\left(\frac{d_2 - d_1}{2}\right)} = \frac{x}{l}$$

$$\therefore \quad d_2 - d = \frac{x}{l}(d_2 - d_1)$$

Put $$\frac{d_2 - d_1}{l} = k \text{ (some constant)}$$

$$\therefore \quad d_2 - d = x.k$$

$$d = d_2 - x\,k$$

Area of cross section of this element (circular)

$$A_x = \frac{\pi}{4}(d_2 - kx)^2$$

$$\text{Stress } \sigma = \frac{P}{A_x} = \frac{P}{\frac{\pi}{4}(d_2 - kx)^2}$$

$$\text{Strain in the element } \in_x = \frac{\delta(dx)}{dx}$$

WKT

$$E = \frac{\text{Stress}}{\text{Strain}}$$

$$E = \left(\frac{P}{\frac{\pi}{4}(d_2 - kx)^2}\right) \bigg/ \left(\frac{\delta(dx)}{dx}\right)$$

$$\therefore \qquad \text{Elongation } \delta\,(dx) = \frac{4F\,dx}{E\pi(d_2 - kx)^2}$$

For total elongation in the bar, integrate this between $x = 0$ to l

$$\delta l = \int \delta(dx) = \int_{x=0}^{l} \frac{4Pdx}{\pi E(d_2 - kx)^2}$$

$$\delta l = \frac{4P}{\pi E}\int_{o}^{l} \frac{dx}{(d_2 - kx)^2} = \frac{4F}{\pi E}(d_2 - kx)^{-2}\,dx$$

$$\delta l = \frac{4P}{\pi E}\left[\frac{(d_2 - kx)^{-1}}{(-1)(-k)}\right]_o^l$$

$$= \frac{4P}{\pi Ek}\left[\frac{1}{(d_2 - kx)}\right]_o^l$$

$$= \frac{4P}{\pi Ek}\left[\frac{1}{d_2 - kl} - \frac{1}{d_2}\right]$$

$$\text{But } k = \left(\frac{d_2 - d_1}{l}\right)$$

Substituting

$$\delta l = \frac{4P}{\pi E\left[\frac{d_2 - d_1}{l}\right]}\left[\frac{1}{d_2 - \left(\frac{d_2 - d_1}{l}\right)l} - \frac{1}{d_2}\right]$$

$$= \frac{4Pl}{\pi E\,(d_2 - d_1)}\left[\frac{1}{d_1} - \frac{1}{d_1}\right]$$

$$= \frac{4Pl}{\pi E(d_2 - d_1)}\left(\frac{d_2 - d_1}{d_1 d_2}\right)$$

$$\delta l = \frac{4Pl}{E\pi d_1 d_2}$$

Note: For round bar of same dia, $d_1 = d_2 = d$

$$\delta l = \frac{4Pl}{E\pi d^2} = \frac{Pl}{E\left(\frac{\pi}{4}d^2\right)} = \frac{Pl}{EA}$$

Same as derived earlier.

1.12 ELONGATION OF TAPER PLATE OF RECTANGULAR CROSS SECTION

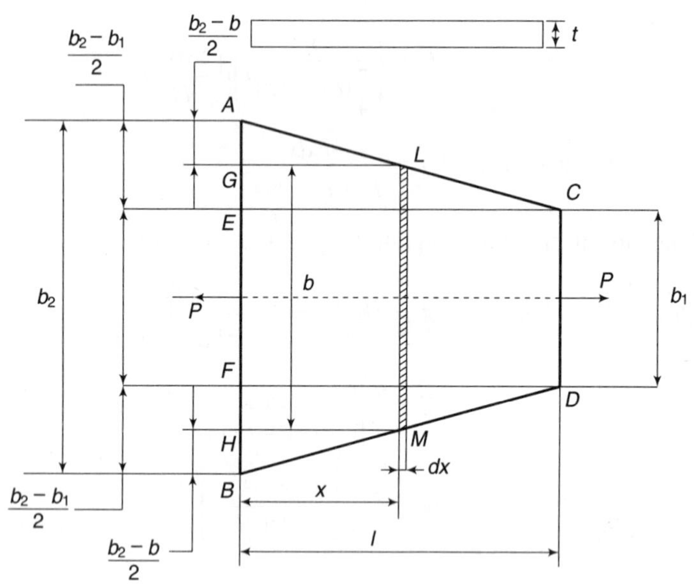

Fig. 1.11

Figure 1.11 shows a taper plate of rectangular cross section of length l subjected to a tensile force of P

b_2 = Width at the bigger end

b_1 = Width at the smaller end

Consider an elemental length of dx at a distance x from bigger width end AB. Let b be the width of the element LM.

From Fig. 1.11,

$$EF = CD = b_1$$

$$AB = b_2$$

$$GH = LM = b$$

$$AG = BH = \frac{AB - GH}{2} = \frac{b_2 - b}{2}$$

$$AE = BF = \frac{AB - EF}{2} = \frac{b_2 - b_1}{2}$$

Triangles *AGL* and *AEC* are similar.

$$\therefore \qquad \frac{AG}{AE} = \frac{GL}{EC}$$

$$\frac{\left(\frac{b_2 - b}{2}\right)}{\left(\frac{b_2 - b_1}{2}\right)} = \frac{x}{l}$$

$$\therefore \qquad b_2 - b = \frac{x}{l}(b_2 - b_1)$$

Put $$\frac{b_2 - b_1}{l} = k \text{ (some constant)}$$

$$\therefore \qquad b_2 - b = x \cdot k$$

or $$b = b_2 - kx$$

Area of cross section of element $= b \cdot t$ (Rectangular cross section)

$$A_x = (b_2 - kx)t$$

We know that elongation $\delta l = \dfrac{Pl}{AE}$

For the element δx,

$$\delta(dx) = \frac{P \cdot dx}{Ax \cdot E}$$

$$\delta(dx) = \frac{P\,dx}{(b_2 - kx)tE}$$

For total elongation in the bar, integrate this between $x = o$ to l

$$\delta l = \int \delta(dx) = \int_o^l \frac{P\,dx}{(b_2 - kx)tE}$$

$$= \frac{P}{tE}\int_o^l \frac{dx}{(b_2 - kx)}$$

This is of the form $\int \frac{dx}{x}$ which is equal to $\log_e x$

$$\delta l = \frac{P}{tE}\left[\frac{\log_e (b_2 - kx)}{(-k)}\right]_o^l$$

$$= \frac{P}{ktE}\left[-\log_e (b_2 - kx)\right]_o^l$$

$$= \frac{P}{ktE}\left[-\log_e (b_2 - kl) - (-\log_e b_2)\right]$$

$$= \frac{P}{ktE}[\log_e b_2 - \log_e (b_2 - kl)]$$

Substituting $$k = \frac{b_2 - b_1}{l}$$

$$\delta l = \frac{P}{\frac{(b_2 - b_1)}{l}\cdot tE}\left[\log_e b_2 - \log_e\left(b_2 - \frac{b_2 - b_1}{l}\times l\right)\right]$$

$$\delta l = \frac{Pl}{(b_2 - b_1)\, tE}[\log_e b_2 - \log_e b_1]$$

$$= \frac{Pl}{(b_2 - b_1)tE}\left[\log_e\left(\frac{b_2}{b_1}\right)\right]$$

$$\delta l = \frac{Pl}{(b_2 - b_1)tE}\log_e\left(\frac{b_2}{b_1}\right)$$

or $$\delta l = \frac{Pl}{(b_2 - b_1)tE}\, l_n\left(\frac{b_2}{b_1}\right).$$

Problem 12 A steel rod 0.6 m long taper uniformly from 60 mm diameter to 40 mm diameter. It is subjected to a tensile load of 72 kN Modulus of elasticity is 210 GPa. Find the elongation in the rod.

Solution: **Given data:**

Length $l = 0.6$ m
$= 600$ mm
Bigger diameter $d_2 = 60$ mm
Smaller diameter $d_1 = 40$ mm
$E = 210$ GPa
$= 210 \times 10^3$ MPa
Load $P = 72\,\text{kN} = 72 \times 10^3\,\text{N}$

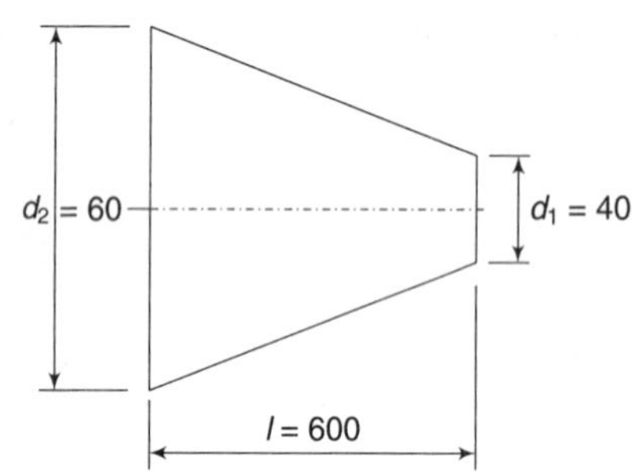

Fig. 1.12

For taper round rod, change in length $$\delta l = \frac{4Pl}{E\pi d_1 d_2}$$

$$= \frac{4 \times 72 \times 10^3 \times 600}{210 \times 10^3 \times \pi \times 60 \times 40} = 0.109 \text{ mm.}$$

Problem 13 A taper rod of length 1 m tapers uniformly from dia 30 mm to diameter *d*. It has to withstand an axial load of 120 kN limiting the elongation to 2 mm. Modulus of elasticity for the rod material may be taken as 1.25×10^5 N/mm^2. Find the diameter at the other end.

Solution: **Given data:**

$$\text{Length } l = 1 \text{ m} = 1000 \text{ mm}$$

$$\text{Diameter } d_2 = 30 \text{ mm}$$

$$d_1 = d$$

$$\text{Load } P = 120 \text{ kN} = 120 \times 10^3 \text{ N}$$

$$\text{Elongation } \delta l = 2 \text{ mm}$$

$$E = 1.25 \times 10^5 \text{ N/mm}^2$$

$$\text{For taper round bar, } \delta l = \frac{4Pl}{E\pi\, d_1 d_2}$$

$$2 = \frac{4 \times 120 \times 10^3 \times 1000}{1.25 \times 10^5 \times \pi \times 30 \times d}$$

$$\therefore \quad \text{Received diameter } d = 20.37 \text{ mm.}$$

Problem 14 The diameter of a circular rod uniformly varies from ($D + a$) to ($D - a$). Show that the percentage error in finding Young's modulus of the rod by considering it an uniform rod of diameter D is $\left(\frac{10\,a}{D}\right)^2$.

Solution: **Given data:**

$$d_2 = D + a$$

$$d_1 = D - a$$

$$\text{For taper bar, elongation } \delta l_T = \frac{4Pl}{E\pi d_1 d_2}$$

$$E_T = \frac{4Pl}{\delta l\pi(D + a)\,(D - a)} = \frac{4Pl}{\delta l\pi(D^2 - a^2)} \quad \text{...(1)}$$

$$\text{Considering it as round rod of } \delta l_R = \frac{4Pl}{E\pi d^2}$$

$$\therefore \quad E_R = \frac{4Pl}{\delta l\,\pi\,(D)^2} \quad \text{...(2)}$$

$$\%\text{ error in calculating } E = \frac{E_T - E_R}{E_T} \times 100$$

$$= \frac{\left(\dfrac{4Pl}{\delta l\pi\,(D^2 - a^2)} - \dfrac{4Pl}{\delta l\,\pi\,D^2}\right)}{\left(\dfrac{4Pl}{\delta l\pi\,(D^2 - a^2)}\right)} \times 100$$

$$= \frac{\left(\frac{1}{D^2 - a^2} - \frac{1}{D^2}\right)}{\left(\frac{1}{D^2 - a^2}\right)} \times 100$$

$$= \frac{\left(\frac{D^2 - (D^2 - a^2)}{(D^2 - a^2)D^2}\right)}{\left(\frac{1}{D^2 - a^2}\right)} \times 100$$

$$\% \text{ Error} = \left(\frac{a^2}{D^2}\right) \times 100 = \left(\frac{10a}{D}\right)^2.$$

Problem 15 A steel plate of trapezoidal shape is 900 mm long and tapers uniformly from 120 mm to 72 mm. Its thickness is 15 mm. Determine the elongation in the plate if it carries an axial load of 45 kN. Take E = 200 GPa.

Solution: **Given data:**

Taper plate, length $l = 900$ mm

Bigger width $b_2 = 120$ mm

Smaller width $b_1 = 72$ mm

Thickness $t = 15$ mm

Load $P = 45$ kN

$= 45 \times 10^3$ N

$E = 200$ GPa

$= 200 \times 10^3$ MPa

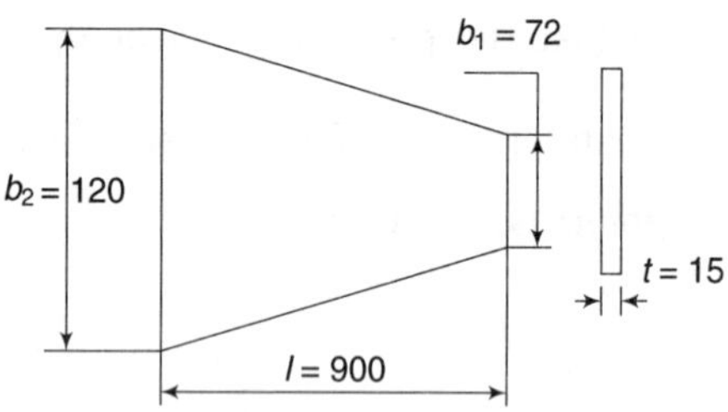

Fig. 1.13

$$\text{For taper plate, elongation } \delta l = \frac{Pl}{tE(b_2 - b_1)} l_n\left(\frac{b_2}{b_1}\right)$$

$$\delta l = \frac{45 \times 10^3 \times 900}{15 \times 200 \times 10^3 (120 - 72)} \log_e \left(\frac{120}{72}\right)$$

$$= 0.1437 \text{ mm.}$$

Problem 16 A tapered steel plate of thickness 10 mm is 750 mm long. It tapers uniformly from 90 mm to 40 mm. The modulus of elasticity for the plate material is 1.2×10^5 N/mm^2. Find the safe load that can be carried if the elongation is not to exceed 1.8 mm.

Solution: **Given data:**

Thickness $t = 10$ mm

$$\text{Length } l = 750 \text{ mm}$$
$$\text{Bigger width } b_2 = 90 \text{ mm}$$
$$\text{Smaller width } b_1 = 40 \text{ mm}$$
$$\text{Modulus of elasticity } E = 1.2 \times 10^5 \text{ N/mm}^2$$
$$\text{Elongation } \delta l = 1.8 \text{ mm}$$
$$\text{For taper plate, elongation } \delta l = \frac{Pl}{tE\,(b_2 - b_1)} \log_e \left(\frac{b_2}{b_1}\right)$$
$$1.8 = \frac{P \times 750}{10 \times 1.2 \times 10^5\,(90 - 40)}\, l_n\left(\frac{90}{40}\right)$$
$$\therefore \quad \text{Load } P = 177.574 \times 10^3 \text{ N.}$$

Problem 17 A taper plate of thickness 10 mm and 2 m long is subjected to a tensile load of 60 kN. Modulus of elasticity is 200 GPa. It is 120 mm wide at one end. Find the width at the other end if the elongation is limited to 0.6 mm.

Solution:

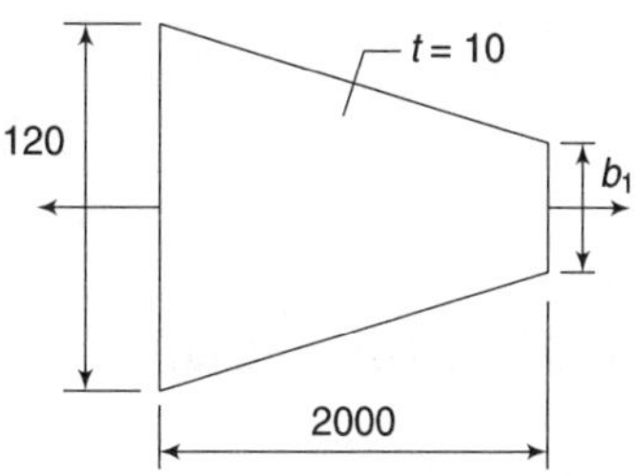

Fig. 1.14

$$\text{Thickness } t = 10 \text{ mm}$$
$$\text{Length } l = 2 \text{ m}$$
$$= 2000 \text{ mm}$$
$$\text{Load } P = 60 \text{ kN}$$
$$= 60 \times 10^3 \text{ N}$$
$$E = 200 \text{ GPa}$$
$$= 200 \times 10^3 \text{ MPa}$$
$$b_2 = 120 \text{ mm}$$
$$\delta l = 0.6 \text{ mm. Find } b_1 = ?$$
$$\text{For taper plate, elongation } \delta l = \frac{Pl}{tE\,(b_2 - b_1)} \log_e \left(\frac{b_2}{b_1}\right)$$
$$0.6 = \frac{60 \times 10^3 \times 2000}{10 \times 200 \times 10^3\,(120 - b_1)}\, l_n\left(\frac{120}{b_1}\right)$$
$$\frac{0.6 \times 10 \times 200 \times 10^3}{60 \times 10^3 \times 2000} = \frac{l_n\left(\frac{120}{b_1}\right)}{120 - b_1}$$
$$l_n\left(\frac{120}{b_1}\right) = \frac{(120 - b_1) \times 0.6 \times 10 \times 200 \times 10^3}{60 \times 10^3 \times 2000}$$

$$
\begin{aligned}
l_n\ 120 - l_n\ b_1 &= (120 - b_1) \times 0.01 \\
4.7875 - l_n\ b_1 &= 1.2 - 0.01\ b_1 \\
l_n\ b_1 - 0.01\ b_1 &= 4.7875 - 1.2 = 3.5875 \\
l_n\ b_1 - 0.01\ b_1 &= 3.5875 \\
\text{Solving by trial and error, } b_1 &= 82.5 \text{ mm.}
\end{aligned}
$$

Problem 18 A steel plate (Fig. 1.15) has uniform thickness of 10 mm and is subjected to a tenside load of 90 kN. The extension in the plate is observed as 0.43 mm. Find the modulus of elasticity.

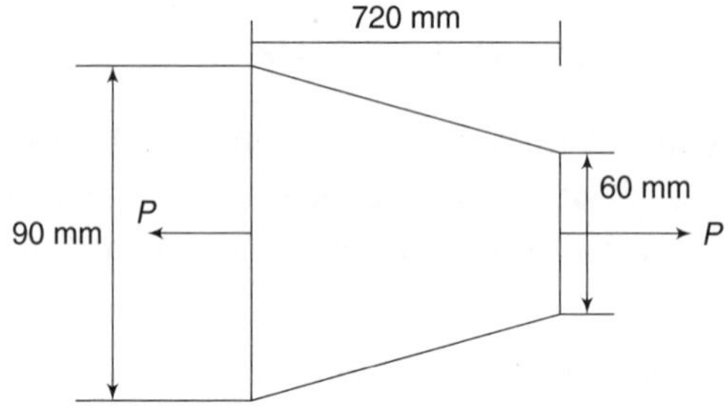

Fig. 1.15

Solution: **Given data:**

$$
\begin{aligned}
b_2 &= 90 \text{ mm} \\
b_1 &= 60 \text{ mm} \\
l &= 720 \text{ mm} \\
P &= 90 \text{ kN} = 90000 \text{ N} \\
t &= 10 \text{ mm} \\
\delta l &= 0.43 \text{ mm.}
\end{aligned}
$$

WKT for taper plate elongation

$$\delta l = \frac{Pl}{tE\ (b_2 - b_1)} \log_e \left(\frac{b_2}{b_1}\right)$$

$$0.43 = \frac{90000 \times 720}{10 \times E\ (90 - 60)} \log_e \left(\frac{90}{60}\right)$$

$$\therefore \quad E = 203.675 \times 10^3 \text{ N/mm}^2.$$

Problem 19 A steel plate of 12 mm thick and 1.2 m long tapers uniformly from a width of 120 mm to a width of 60 mm. The plate is subjected to a tensile load of 240 kN. Modulus of elasticity is 200 GPa. Find the extension in the plate. What will be the extension of the bar if calculated using the average width. What is the percentage error?

Solution: **Given data:**

$$b_2 = 120 \text{ mm} \quad b_1 = 60 \text{ mm}$$
$$P = 240 \text{ kN} = 240 \times 10^3 \text{ N}$$
$$t = 12 \text{ mm} \quad l = 1.2 \text{ m} = 1200 \text{ mm}$$
$$E = 200 \text{ GPa} = 200 \times 10^3 \text{ MPa}$$

Regarding as taper plate, elongation

$$\delta L_T = \frac{Pl}{tE\,(b_2 - b_1)} \log_e \left(\frac{b_2}{b_1}\right)$$
$$= \frac{240 \times 10^3 \times 1200}{12 \times 200 \times 10^3\,(120 - 60)}\; l_n \left(\frac{120}{60}\right)$$
$$= 1.3863 \text{ mm}$$

Regarding as uniform plate of average width

$$b = \frac{b_1 + b_2}{2} = \frac{60 + 120}{2} = 90 \text{ mm}$$
$$A = bt = 90 \times 12 = 1080 \text{ mm}^2$$
$$\delta l_u = \frac{Pl}{AE} = \frac{240 \times 10^3 \times 1200}{1080 \times 200 \times 10^3} = 1.3333 \text{ mm.}$$
$$\% \text{ error} = \frac{\delta L_T - \delta l_u}{\delta L_T} \times 100 = \frac{1.3863 - 1.3333}{1.3863} \times 100$$
$$= 3.82\%.$$

Problem 20 Determine the elongation of uniform taper rod of square cross section tapers from side B to side b over length l subjected to an axial load P.

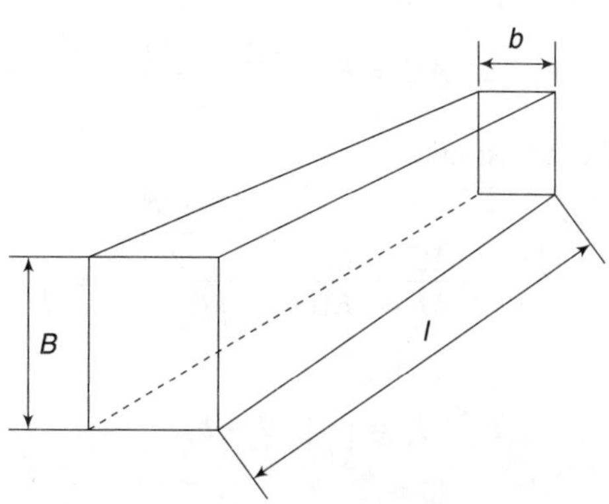

Fig. 1.16 (a)

Solution:

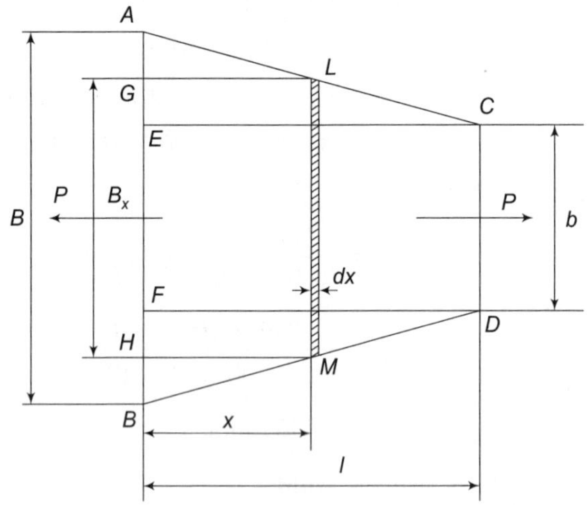

Fig. 1.16 (b)

Figure 1.16 shows a taper square bar a length l taper uniformly from square of side B to square of side b over a length l.

Let P be the load, and E the modulus of elasticity.

Consider an elemental length of dx at a distance x from bigger end AB. Let B_x be the side of the element LM.

From Fig. 1.16, $EF = CD = D$

$$AB = B$$

$$GH = LM = B_x$$

$$AG = BH = \frac{AB - GH}{2} = \frac{B - B_x}{2}$$

$$AE = BF = \frac{AB - EF}{2} = \frac{B - b}{2}$$

Triangles AGL and AEC are similar.

$$\therefore \quad \frac{AG}{AE} = \frac{GL}{EC} \Rightarrow \frac{\left(\dfrac{B - B_x}{2}\right)}{\left(\dfrac{B - b}{2}\right)} = \frac{x}{l}$$

$$\therefore \quad B - B_x = \left(\frac{x}{l}\right)(B - b)$$

or

$$B_x = B - \left(\frac{B - b}{l}\right)x$$

Put $$\frac{B-b}{l} = k$$

$\therefore$ $$B_x = B - kx.$$

Area of cross section of the strip $A_x = B_x^2$ (Square in cross section)

$$\text{But elongation } \delta l = \frac{Pl}{AE}$$

Elongation of the elemental length dx,

$$\delta(dx) = \frac{Pdx}{A_x . E} = \frac{Pdx}{B_x^2 E}$$

For total elongation in the bar, integrate this between $x = 0$ to l.

$$\delta l = \int_o^l \delta(dx) = \int_o^l \frac{Pdx}{B_x^2 E}$$

$$= \frac{P}{E}\int_o^l \frac{dx}{B_x^2}$$

$$= \frac{P}{E}\int_o^l \frac{dx}{(B-kx)^2} = \frac{P}{E}\int_o^l (B-kx)^{-2}\, dx$$

$$= \frac{P}{E}\left[\frac{-(B-kx)^{-1}}{(-1)(-k)}\right]_o^l$$

$$= \frac{P}{EK}\left[\frac{1}{B-kx}\right]_o^l$$

$$= \frac{P}{EK}\left[\frac{1}{B-kl} - \frac{1}{B}\right]$$

$$\text{Put } k = \frac{B-b}{l}$$

$\therefore$ $$\frac{1}{B-kl} = \frac{1}{B-\left(\frac{B-b}{l}\right)l} = \frac{1}{B-B+b} = \frac{1}{b}$$

$$\delta l = \frac{P}{E}\left(\frac{B-b}{l}\right)\left[\frac{1}{b} - \frac{1}{B}\right]$$

$$= \frac{Pl}{E(B-b)}\left[\frac{B-b}{Bb}\right]$$

$$\delta l = \frac{Pl}{EBb}$$

Problem 21 Find the elongation of a square cross section taper bar of length 2 m tapering uniformly from a side of 60 mm to 30 mm. The load on the bar is 90 kN and modulus of elasticity is 2×10^5 N/mm^2.

Solution:

$$\text{Side of bigger square } B = 60 \text{ mm}$$
$$\text{Side of smaller square } b = 30 \text{ mm}$$
$$\text{Length } l = 2 \text{ m} = 2000 \text{ mm}$$
$$\text{Load } P = 90 \text{ kN} = 90000 \text{ N}$$
$$\text{Modulus of elasticity } E = 2 \times 10^5 \text{ N/mn}^2$$

Elongation of a taper square rod,

$$\delta l = \frac{Pl}{BbE}$$

$$\delta l = \frac{90 \times 10^3 \times 2000}{60 \times 30 \times 2 \times 10^5} = 0.5 \text{ mm.}$$

Problem 22 A square taper bar made of copper is 1.2 m long and loaded as shown in Fig. 1.17. Find side of square face B if the elongation is not to exceed 0.3 mm. Take E = 120 GPa.

What will be the elongation if the load is increased to 60 kN.

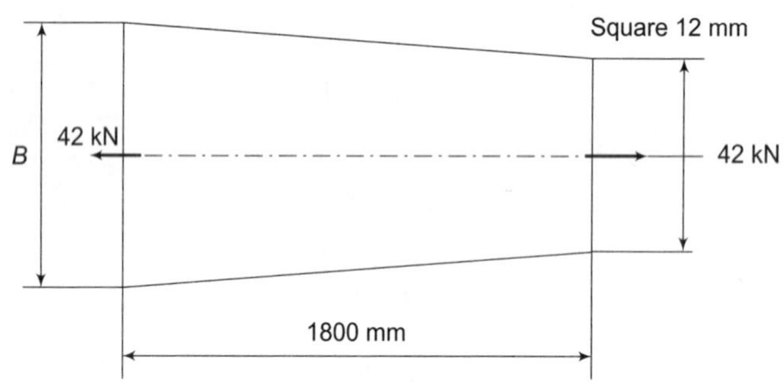

Fig. 1.17

Solution: **Given data:**

(i)

$$\text{Bigger side of square} = B = B$$
$$\text{Smaller side of square } b = 12 \text{ mm}$$
$$\text{Length } l = 1800 \text{ mm}$$
$$\text{Modulus of elasticity } E = 120 \text{ GPa}$$
$$= 120 \times 10^3 \text{ MPa.}$$
$$\text{Load } P = 42 \text{ kN} = 42 \times 10^3 \text{ N}$$

Elongation in taper square bar

$$\delta l = \frac{Pl}{BbE}$$

$$0.3 = \frac{42000 \times 1800}{B \times 12 \times 120 \times 10^3}$$

$\therefore$ Side of bigger square $B = 175$ mm

(ii) If the load is increased to 60 kN

$$P = 60 \times 10^3 \text{ N}$$

$$\delta l = \frac{Pl}{BbE}$$

$$= \frac{60 \times 10^3 \times 1800}{175 \times 12 \times 120 \times 10^3}$$

$$\delta l = 0.4287 \text{ mm.}$$

1.13 BARS OF VARYING CROSS SECTION

Figure 1.18 shows a bar of varying cross section subject to an axial load P.

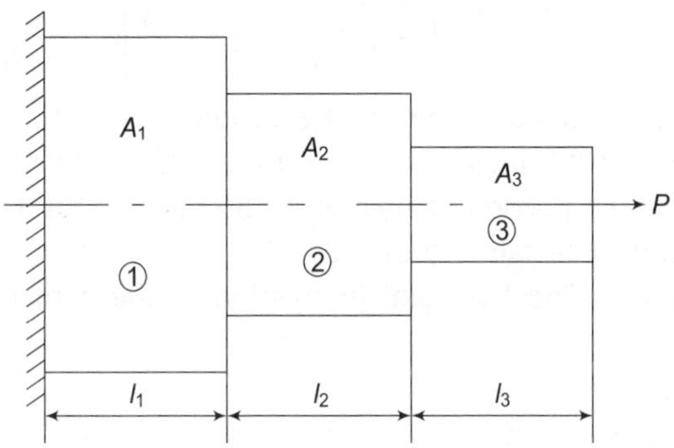

Fig. 1.18

P = Load acting on the compound bar

A_1 = Area of cross section of portion 1

A_2 = Area of cross section of portion 2

A_3 = Area of cross section of portion 3

l_1 = Length of bar in portion 1

l_2 = Length of bar in portion 2

l_3 = Length of bar in portion 3

E_1 = Modulus of elasticity of material of portion 1

E_2 = Modulus of elasticity of material of portion 2

E_3 = Modulus of elasticity of material of portion 3

δl_1 = Elongation in portion 1

δl_2 = Elongation in portion 2

δl_3 = Elongation in portion 3

δl = Total elongation in the bar

Total elongation = Sum of individual elongations in 1, 2 and 3

$$\delta l = \delta l_1 + \delta l_2 + \delta l_3$$

$$\text{But } \delta l = \frac{Pl}{AE}$$

Substituting for portions 1, 2 and 3

$$\delta l = \frac{Pl_1}{A_1E_1} + \frac{Pl_2}{A_2E_2} + \frac{Pl_3}{A_3E_3}$$

If the material is same for all 3 portions,

$$E_1 = E_2 = E_3 = E$$

$$\delta l = \frac{P}{E}\left[\frac{l_1}{A_1} + \frac{l_2}{A_2} + \frac{l_3}{A_3}\right].$$

Problem 23 A stepped compound bar of 1.8 m long has a steel portion of diameter 300 mm over a length of 0.9 m, a copper portion of diameter 210 mm over a length of 0.6 m and a aluminium portion of diameter 150 mm for the remaining length. The Young's moduli for the materials are $E_S = 2 \times 10^5$ N/mm^2, $E_c = 1.1 \times 10^5$ N/mm^2 and $E_{Al} = 0.75 \times 10^5$ N/mm^2. Find the total elongation in the compound bar if it carries a tensile load of 90 kN.

Solution: **Given data:**

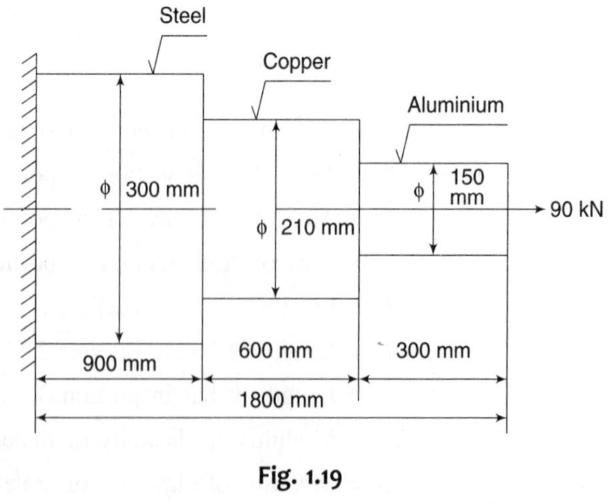

Fig. 1.19

$$d_s = 300 \text{ mm}, \quad d_c = 210 \text{ mm} \quad d_{Al} = 150 \text{ mm}$$

$$l_s = 0.9 \text{ m} = 900 \text{ mm } l_c = 0.6 \text{ m} = 600 \text{ mm}$$

$$\text{Total length } l = 1.8 \text{ m} = 1800 \text{ mm.}$$

$$l_{Al} = \text{Remaining length} = 1800 - 900 - 600 = 300 \text{ mm}$$

$$\text{Modulus of elasticity } E_s = 2 \times 10^5 \text{ N/mm}^2$$

$$E_c = 1.1 \times 10^5 \text{ N/mm}^2$$

$$E_{Al} = 0.75 \times 10^5 \text{ N/mm}^2$$

$$\text{Area of cross section, for steel } A_s = \frac{\pi}{4} d_s^2$$

$$A_s = \frac{\pi}{4} \times 300^2 = 70685.83 \text{ mm}^2$$

$$\text{For copper, } A_c = \frac{\pi}{4} d_c^2 = \frac{\pi}{4} \times 210^2 = 34636.06 \text{ mm}^2$$

$$\text{For aluminium } A_{Al} = \frac{\pi}{4} d_{Al}^2 = \frac{\pi}{4} \times 150^2 = 17671.46 \text{ mm}^2$$

$$\text{Load } P = 90 \text{ kN} = 90 \times 10^3 \text{ N}$$

$$\text{Total elongation in the bar } \delta l = \delta l_S + \delta l_C + \delta l_{Al}$$

$$\delta l = \frac{Pl_s}{A_s E_s} + \frac{Pl_c}{A_c E_c} + \frac{Pl_{Al}}{A_{Al} E_{Al}}$$

$$= 90 \times 10^3 \left[\frac{900}{70685.83 \times 2 \times 10^5} + \frac{600}{34636.06 \times 1.1 \times 10^5} + \frac{300}{17671.46 \times 0.75 \times 10^5} \right]$$

$$= 90 \times 10^3 \, [6.366 \times 10^{-8} + 1.575 \times 10^{-7} + 2.264 \times 10^{-7}]$$

$$\delta l = 0.0403 \text{ mm.}$$

Problem 24 Find the decrease in the length of a steel bar loaded as shown in Fig. 1.20. Take E for steel as 210 kN/mm^2.

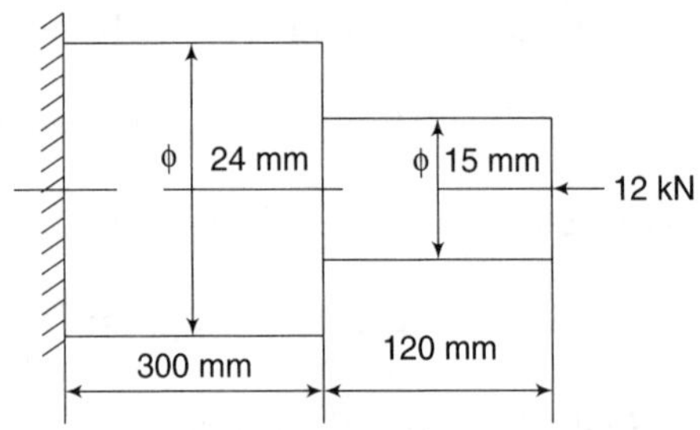

Fig. 1.20

Solution: **Given data:**

$$d_1 = 24 \text{ mm} \quad l_1 = 300 \text{ mm}$$
$$d_2 = 15 \text{ mm} \quad l_2 = 120 \text{ mm}$$
$$P = 12 \text{ kN} = 12 \times 10^3 \text{ N}$$
$$E = 210 \text{ kN/mm}^2 = 210 \times 10^3 \text{ N/mm}^2$$

$$\text{Area of cross section } A_1 = \frac{\pi}{4} d_1^2 = \frac{\pi}{4} \times 24^2 = 452.39 \text{ mm}^2$$

$$A_2 = \frac{\pi}{4} d_2^2 = \frac{\pi}{4} \times 15^2 = 176.71 \text{ mm}^2$$

$$\text{Total decrease in length } \delta l = \delta l_1 + \delta l_2$$

$$= \frac{Pl_1}{A_1E} + \frac{Pl_2}{A_2E} = \frac{P}{E}\left[\frac{l_1}{A_1} + \frac{l_2}{A_2}\right]$$

$$\delta l = \frac{12 \times 10^3}{210 \times 10^3}\left[\frac{300}{452.39} + \frac{120}{176.71}\right] = 0.0767 \text{ mm.}$$

Problem 25 Determine the total change in length of the stepped bar dimensioned in mm as shown in Fig. 1.21. Also find the stress induced in each section. Take E = 200 GPa.

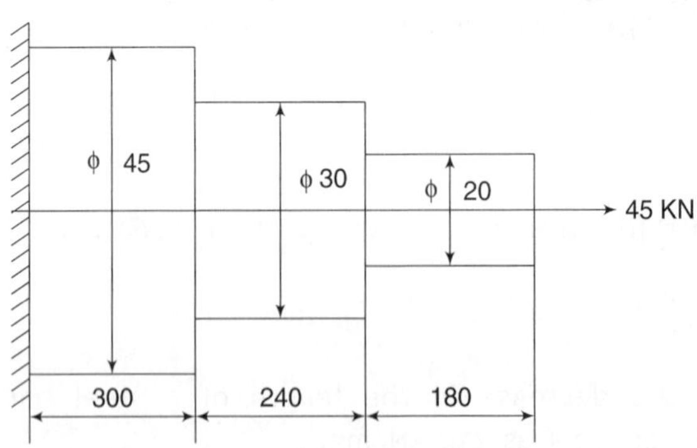

Fig. 1.21

Solution: **Given data:**

$$\text{Diameter } d_1 = 45 \text{ mm}$$

$$d_2 = 30 \text{ mm}$$

$$d_3 = 20 \text{ mm}$$

$$\text{Lengths } l_1 = 300 \text{ mm } l_2 = 240 \text{ mm, } l_3 = 180 \text{ mm}$$

$$\text{Load } P = 45 \text{ kN} = 45 \times 10^3 \text{ N.}$$

$$E = 200 \text{ GPa} = 200 \times 10^3 \text{ MPa or } \frac{\text{N}}{\text{mm}^2}$$

$$\text{Area of cross section } A_1 = \frac{\pi}{4} d_1^2 = \frac{\pi}{4} \times 45^2 = 1590.43 \text{ mm}^2$$

$$A_2 = \frac{\pi}{4} d_2^2 = \frac{\pi}{4} \times 30^2 = 706.86 \text{ mm}^2$$

$$A_3 = \frac{\pi}{4} d_3^2 = \frac{\pi}{4} \times 20^2 = 314.16 \text{ mm}^2$$

$$\text{Total change in length } \delta l = \delta l_1 + \delta l_2 + \delta l_3$$

$$= \frac{P}{E}\left[\frac{l_1}{A_1} + \frac{l_2}{A_2} + \frac{l_3}{A_3}\right]$$

$$= \frac{45 \times 10^3}{200 \times 10^3}\left[\frac{300}{1590.43} + \frac{240}{706.86} + \frac{180}{314.16}\right]$$

$$\delta l = 0.2478 \text{ mm}$$

$$\text{Stress induced } \sigma = \frac{P}{A}$$

$$\sigma_1 = \frac{P}{A_1} = \frac{45000}{1590.43} = 28.29 \text{ N/mm}^2$$

$$\sigma_2 = \frac{P}{A_2} = \frac{45000}{706.86} = 63.66 \text{ N/mm}^2$$

$$\sigma_3 = \frac{P}{A_3} = \frac{45000}{314.16} = 143.24 \text{ N/mm}^2.$$

Problem 26 Determine the total elongation in the rod shown in Fig. 1.22. It is subjected a pull of 60 kN. Modulus of elasticity for the material may be taken as 1.2×10^5 N/mm^2.

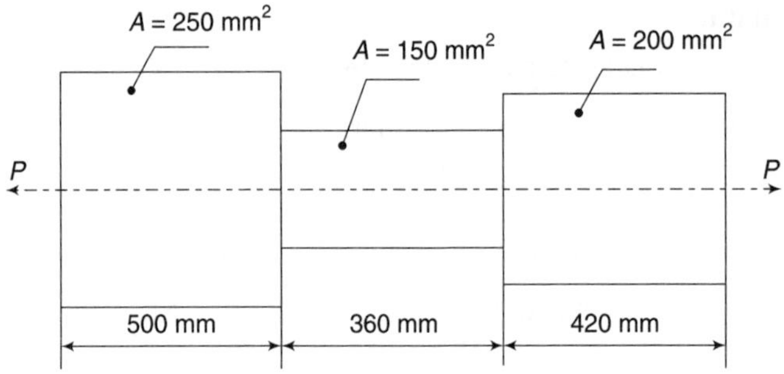

Fig. 1.22

Solution: **Given data:**

$$\text{Load } P = 60 \text{ kN} = 60 \times 10^3 \text{ N}$$

$$\text{Modulus of elasticity } E = 1.2 \times 10^5 \text{ N/mm}^2$$

$$A_1 = 250 \text{ mm}^2 \qquad l_1 = 500 \text{ mm}$$

$$A_2 = 150 \text{ mm}^2 \qquad l_2 = 360 \text{ mm}$$

$$A_3 = 200 \text{ mm}^2 \qquad l_3 = 420 \text{ mm}$$

$$\text{Total elongation } \delta l = \delta l_1 + \delta l_2 + \delta l_3$$

$$= \frac{P}{E}\left[\frac{l_1}{A_1} + \frac{l_2}{A_2} + \frac{l_3}{A_3}\right]$$

$$= \frac{60 \times 10^3}{1.2 \times 10^5}\left[\frac{500}{250} + \frac{360}{150} + \frac{420}{200}\right]$$

$$\delta l = 3.25 \text{ mm.}$$

Problem 27 Determine the safe load that can be carried by a stepped bar dimensioned in mm as shown in Fig. 1.23, limiting the elongation to 3 mm. Take $E = 2 \times 10^5$ MPa.

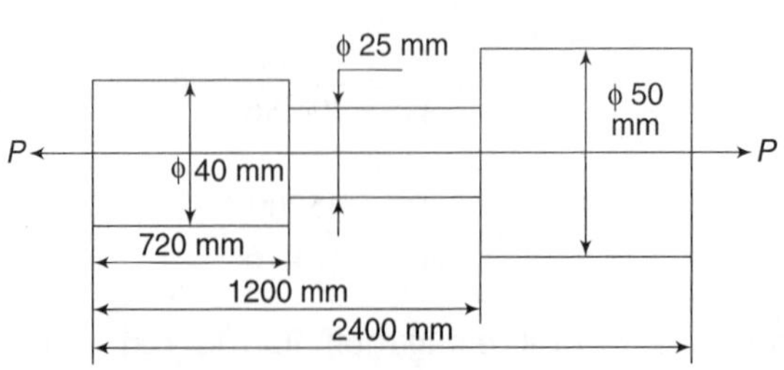

Fig. 1.23

Solution: **Given data:**

Total elongation $\delta l = 3$ mm

$E = 2 \times 10^5$ N/mm^2

$d_1 = 40$ mm $\quad \therefore A_1 = \frac{\pi}{4} d_1^2 = \frac{\pi}{4} \times 40^2 = 1256.64$ mm^2

$d_2 = 25$ mm $\quad \therefore A_2 = \frac{\pi}{4} d_2^2 = \frac{\pi}{4} \times 25^2 = 490.87$ mm^2

$d_3 = 50$ mm $\quad \therefore A_3 = \frac{\pi}{4} d_3^2 = \frac{\pi}{4} \times 50^2 = 1963.5$ mm^2

$l_1 = 720$ mm $\quad l_2 = 1200 - 720 = 480$ mm

$l_3 = 2400 - 1200 = 1200$ mm.

$$\text{WKT } \delta l = \frac{P}{E}\left[\frac{l_1}{A_1} + \frac{l_2}{A_2} + \frac{l_3}{A_3}\right]$$

$$3 = \frac{P}{2 \times 10^5}\left[\frac{720}{1256.64} + \frac{480}{490.87} + \frac{1200}{1963.5}\right]$$

$\therefore$ Load $P = 277.525 \times 10^3$ N.

Problem 28 Figure 1.24 shows a stepped bar made by steel and subjected to a tensile load of 120 kN. Find the diameter *d* if the extension of the bar is not to exceed 1.5 mm. Take $E = 2 \times 10^5$ N/mm^2.

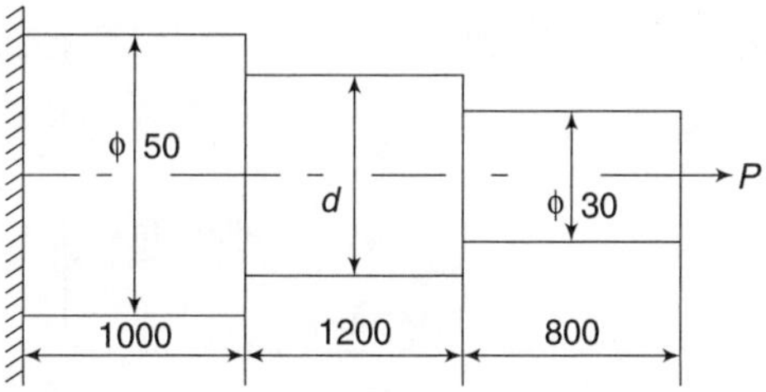

Fig. 1.24

Solution: **Given data:**

$d_1 = 50$ mm, $d_2 = d$, $d_3 = 30$ mm

$l_1 = 1000$ mm $l_2 = 1200$ mm, $l_3 = 800$ mm

$P = 120$ kN $= 120 \times 10^3$ N, $\delta l = 1.5$ mm

$E = 2 \times 10^5$ N/mm^2

$A_1 = \frac{\pi}{4} d_1^2 = \frac{\pi}{4} \times 50^2 = 1963.5$ mm^2

$A_2 = \frac{\pi}{4} d^2$

$$A_3 = \frac{\pi}{4} d_3^2 = \frac{\pi}{4} \times 30^2 = 706.86 \text{ mm}^2$$

$$\text{WKT } \delta l = \delta l_1 + \delta l_2 + \delta l_3$$

$$1.5 = \frac{P}{E}\left[\frac{l_1}{A_1} + \frac{l_2}{A_2} + \frac{l_3}{A_3}\right]$$

$$1.5 = \frac{120 \times 10^3}{2 \times 10^5}\left[\frac{1000}{1963.5} + \frac{1200}{A_2} + \frac{800}{706.86}\right]$$

$$\frac{1.5 \times 2 \times 10^5}{120 \times 10^3} = 0.509 + \frac{1200}{A_2} + 1.1318$$

$$\frac{1200}{A_2} = 2.5 - 0.509 - 1.1318 = 0.8592$$

$$\therefore \quad A_2 = \frac{1200}{0.8592} = 1396.65 \text{ mm}^2$$

$$A_2 = \frac{\pi d^2}{4} = 1396.65$$

$$\therefore \quad \text{Diameter } d = 42.17 \text{ mm.}$$

Problem 29 A CI rod shown in Fig. 1.25 is subjected to a tensile load of 75 kN. Find the elongation in the rod taking Young's modulus as 1.2×10^5 N/mm^2.

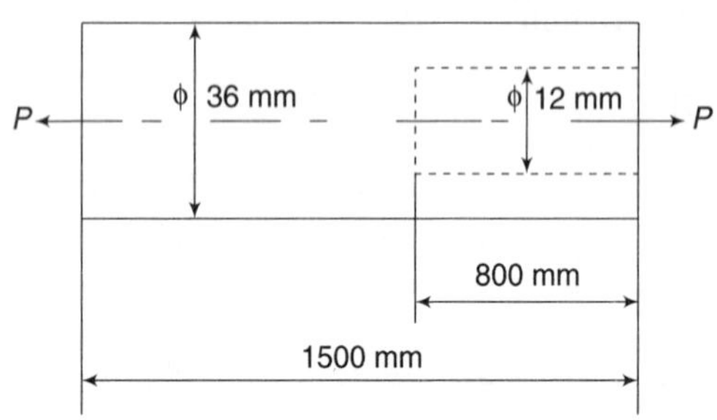

Fig. 1.25

Solution:

This is treated like a stepped bar with varying cross section.

For the solid portion

$$d_1 = 36 \text{ mm } l_1 = 1500 - 800 = 700 \text{ mm}$$

$$A_1 = \frac{\pi}{4} d_1^2 = \frac{\pi}{4} \times 36^2 = 1017.88 \text{ mm}^2$$

For the hollow portion

$$d_o = 36 \text{ mm } d_i = 12 \text{ mm } l_2 = 800 \text{ mm}$$

$$A_2 = \frac{\pi}{4}(d_o^2 - d_i^2) = \frac{\pi}{4}(36^2 - 12^2) = 904.78 \text{ mm}^2$$

$$E = 1.2 \times 10^5 \text{ N/mm}^2,\ P = 75 \text{ kN} = 75 \times 10^3 \text{ N}$$

$$\delta l = \delta l_1 + \delta l_2$$

$$= \frac{P}{E}\left[\frac{l_1}{A_1} + \frac{l_2}{A_2}\right]$$

$$= \frac{75 \times 10^3}{1.2 \times 10^5}\left[\frac{700}{1017.88} + \frac{800}{904.78}\right]$$

$$\delta l = 0.9824 \text{ mm.}$$

Problem 30 Find the total elongation in a compound bar shown in Fig. 1.26. Take E = 200 kN/mm^2.

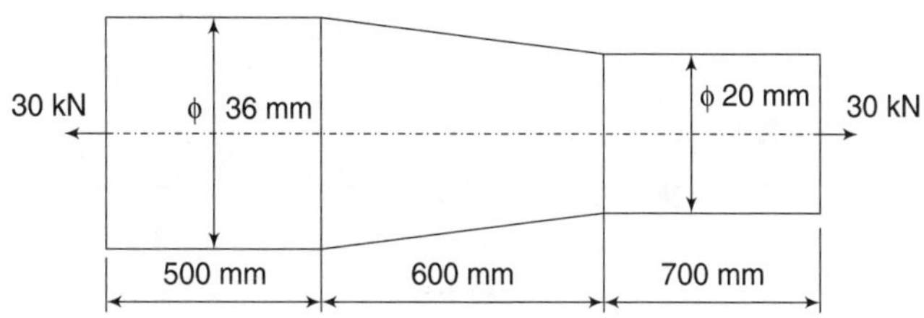

Fig. 1.26

Solution: **Given data:**

$$d_2 = 36 \text{ mm}$$

$$d_1 = 20 \text{ mm}$$

$$l_1 = 500 \text{ mm}$$

$$l_2 = 600 \text{ mm}$$

$$l_3 = 700 \text{ mm}$$

$$P = 30 \text{ kN} = 30 \times 10^3 \text{ N}$$

$$E = 200 \text{ kN/mm}^2 = 200 \times 10^3 \text{ N/mm}^2$$

For composite bar $\delta l = \delta l_1 + \delta l_2 + \delta l_3$

$$= \frac{Pl_1}{EA_1} + \frac{4Fl_2}{E\pi d_1 d_2} + \frac{Fl_3}{EA_3}$$

$$= \frac{P}{E}\left[\frac{l_1}{\left(\frac{\pi}{4}d_2^2\right)} + \frac{4l_2}{\pi d_1 d_2} + \frac{l_3}{\left(\frac{\pi}{4}d_1^2\right)}\right]$$

$$= \frac{P}{E}\left[\frac{4l_1}{\pi d_2^{\,2}} + \frac{4l_2}{\pi d_1 d_2} + \frac{4l_3}{\pi d_1^{\,2}}\right]$$

$$= \frac{4P}{\pi E}\left[\frac{l_1}{d_2^{\,2}} + \frac{l_2}{d_1 d_2} + \frac{l_3}{d_1^{\,2}}\right]$$

$$= \frac{4 \times 30 \times 10^3}{\pi \times 200 \times 10^3}\left[\frac{500}{36^2} + \frac{600}{36 \times 20} + \frac{700}{20^2}\right]$$

$$\delta l = 0.567 \text{ mm.}$$

Problem 31 A composite bar dimensioned in mm is shown in Fig. 1.27. Find the load required to elongate the bar by 1.2 mm. Modulus of elasticity may be taken as 2×10^{11} Pa.

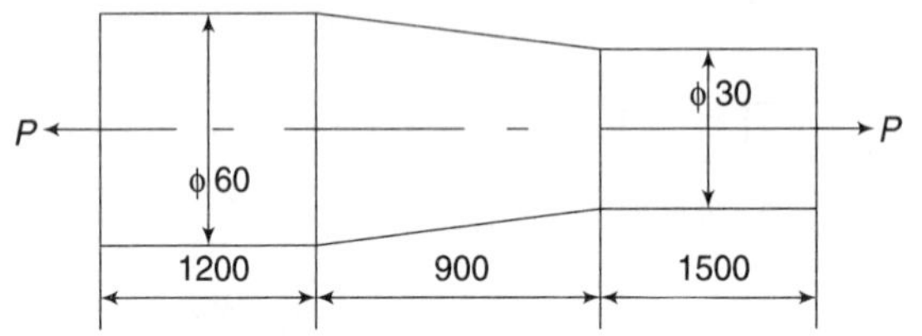

Fig. 1.27

Solution: **Given data:**

$$\text{Bigger diameter } d_2 = 60 \text{ mm}$$

$$\text{Smaller diameter } d_1 = 30 \text{ mm}$$

$$\text{Length } l_1 = 1200 \text{ mm}$$

$$l_2 = 900 \text{ mm}$$

$$l_3 = 1500 \text{ mm}$$

$$E = 2 \times 10^{11} \text{ Pa} = 2 \times 10^5 \text{ MPa or N/mm}^2$$

$$\text{Area of cross section } A_1 = \frac{\pi}{4} d_2^{\,2} = \frac{\pi}{4} \times 60^2 = 2827.43 \text{ mm}^2$$

$$A_3 = \frac{\pi}{4} d_1^{\,2} = \frac{\pi}{4} \times 30^2 = 706.86 \text{ mm}^2$$

$$\text{Elongation } \delta_l = 1.2 \text{ mm}$$

$$\text{WKT} = \delta l = \delta l_1 + \delta l_2 + \delta l_3$$

$$\delta l = \frac{Pl_1}{A_1 E} + \frac{4Pl_2}{E\pi d_1 d_2} + \frac{Pl_3}{A_3 E}$$

$$1.2 = \frac{P}{E}\left[\frac{l_1}{A_1} + \frac{l_2}{\pi d_1 d_2} + \frac{l_3}{A_3}\right]$$

$$1.2 = \frac{P}{2 \times 10^5}\left[\frac{1200}{2827.43} + \frac{900}{\pi \times 60 \times 30} + \frac{1500}{706.86}\right]$$

$$\therefore \quad 1.2 = \frac{P}{2 \times 10^5}[0.4244 + 0.1592 + 2.1221]$$

$$P = \frac{1.2 \times 2 \times 10^5}{(0.4244 + 0.1592 + 2.1221)} = 88703.96 \text{ N.}$$

Problem 32 Figure 1.28 shows a flat plate of thickness 12 mm subjected to a tensile load of 18 kN. Taking Young's modulus as 1.25×10^5 N/mm^2. Find the elongation in the plate.

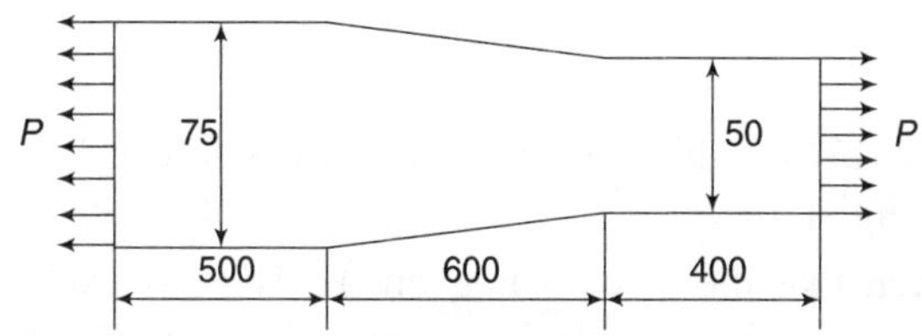

Fig. 1.28

Solution: **Given data:**

$$\text{Load } P = 18 \text{ kN} = 18000 \text{ N}$$

$$\text{Plate thickness } t = 12 \text{ mm}$$

$$E = 1.25 \times 10^5 \text{ N/mm}^2$$

$$\text{Width } b_2 = 75 \text{ mm}$$

$$b_1 = 50 \text{ mm}$$

$$\text{Length } l_1 = 500 \text{ mm}, l_2 = 600 \text{ mm and } l_3 = 400 \text{ mm}$$

$$A_1 = b_2\, t = 75 \times 12 = 900 \text{ mm}^2$$

$$A_3 = b_1\, t = 50 \times 12 = 600 \text{ mmm}^2$$

$$\text{Total elongation } \delta l = \delta l_1 + \delta l_2 + \delta l_3$$

$$\delta l = \frac{Pl_1}{A_1E} + \frac{Pl_2}{tE(b_2 - b_1)} l_n\left(\frac{b_2}{b_1}\right) + \frac{Pl_3}{A_3E}$$

$$= \frac{P}{E}\left[\frac{l_1}{A_1} + \frac{l_2}{t(b_2 - b_1)} l_n\left(\frac{b_2}{b_1}\right) + \frac{l_3}{A_3}\right]$$

$$= \frac{18000}{1.25 \times 10^5}\left[\frac{500}{900} + \frac{600}{12(75 - 50)} l_n\left(\frac{75}{50}\right) + \frac{400}{600}\right]$$

$$\delta l = 0.144\,[0.5555 + 2 + 0,6666] = 0.464 \text{ mm.}$$

1.14 PRINCIPLE OF SUPERPOSITIONING

Principle of superpositioning states that if a body is acted upon by a number of loads, then the net effect is the sum of effect of each load acting independently on the body.

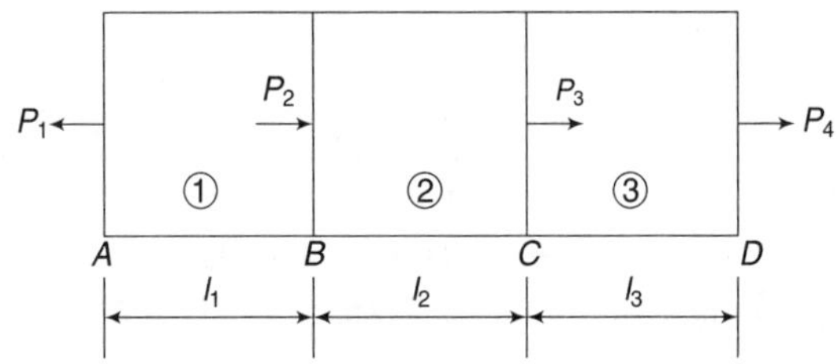

Fig. 1.29 (a)

Consider a member *ABCD* which is acted upon by 4 forces P_1, P_2, P_3 and P_4 at *AB*, *BC*, *CD* as shown in Fig. 1.29.

It can be easily seen that the force acting on *AB* is P_1 (tenside in nature) and force acting on *CD* is P_4 (tensile) since single forces are acting whereas face *B* experiences the algebraic sum of P_1 and P_2 and face *C* experiences algebraic sum of P_3 and P_4.

$$\text{For equilibrium, } \Sigma P = 0$$

Sum of forces acting towards left = Sum of forces acting towards right

$$P_1 = P_2 + P_3 + P_4$$

Also algebraic sum of forces acting $B = P_1 - P_2 = P_3 + P_4$ algebraic sum of forces acting on *C*.

This can be shown diagrammatically as follows:

Free-body diagram

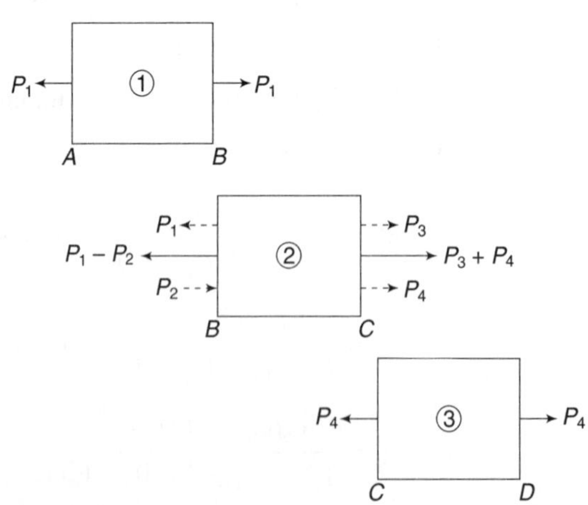

Fig. 1.29 (b)

When a number of forces act on a body, according to the principle of superpositioning, the resulting strain is equal to the algebraic sum of the strains caused by the individual body.

$$\delta l = \pm\, \delta l_1 \pm \delta l_2 \pm \delta l_3$$

+ is used for tensile force

– is used for compressive force

Problem 33 Find the change in length for a composite bar as shown in Fig. 1.30. The area of cross section is 600 mm^2 and modulus of elasticity is 200 GPa.

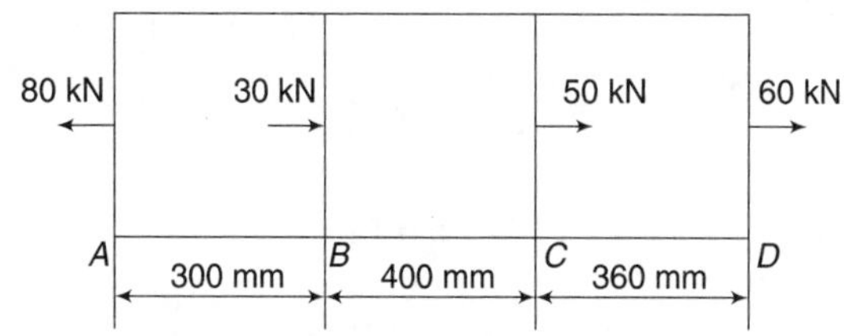

Fig. 1.30 (a)

Solution: **Given data:**

$$A = 600 \text{ mm}^2$$
$$E = 200 \text{ GPa} = 200 \times 10^3 \text{ MPa.}$$

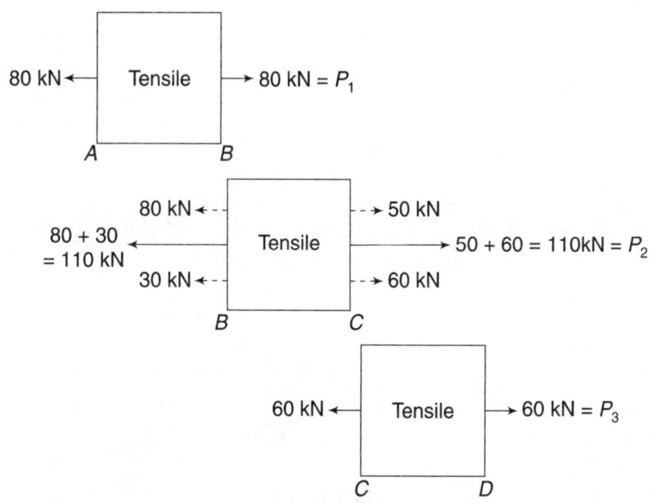

Fig. 1.30 (b)

$$\text{Total elongation} = \delta l = \delta l_1 \pm \delta l_2 \pm \delta l_3$$

$$\delta l = \frac{P_1 l_1}{AE} + \frac{P_2 l_2}{AE} + \frac{P_3 l_3}{AE}$$

$$\delta l = \frac{80 \times 10^3 \times 300}{600 \times 200 \times 10^3} + \frac{110 \times 10^3 \times 400}{600 \times 200 \times 10^3} + \frac{60 \times 10^3 \times 360}{600 \times 200 \times 10^3}$$

$$\delta l = 0.2 + 0.3667 + 0.18 = 0.7467 \text{ mm.}$$

Problem 34 Find the elongation in the circular bar of 36 mm diameter loaded as shown in Fig. 1.31. The modulus of elasticity may be taken as 80 GPa.

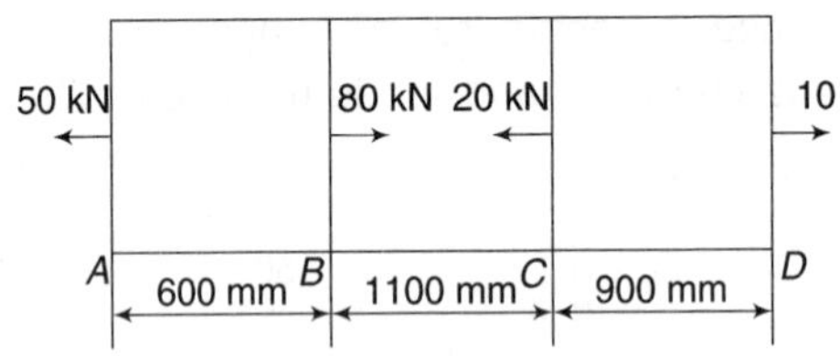

Fig. 1.31 (a)

Solution: **Given data:**

$$\text{Diameter } d = 36 \text{ mm}$$

$$\therefore \quad \text{Area of cross section } A = \frac{\pi}{4} 36^2 = 1017.88 \text{ mm}^2$$

$$\text{Modulus of elasticity } E = 80 \text{ GPa} = 80 \times 10^3 \text{ MPa.}$$

Free-body diagram:

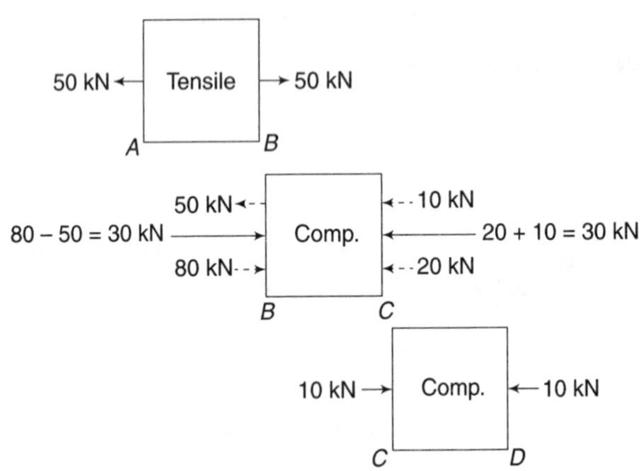

Fig 1.31 (b)

$$\text{Total elongation } \delta l = \delta l \pm \delta l_2 \pm \delta l_3$$

For BC and CD, the load is compressive and hence use –ve sign (length will decrease)

$$\delta l = \delta l_1 - \delta l_2 - \delta l_3$$

$$= \frac{P_1 l_1}{AE} - \frac{P_2 l_2}{AE} - \frac{P_3 l_3}{AE}$$

where, $P_1 = 50 \times 10^3$ N, $l_1 = 600$ mm

$P_2 = 30 \times 10^3$ N, $l_2 = 1100$ mm

$P_3 = 10 \times 10^3$ N, $l_3 = 900$ mm

$A = 1017.88$ mm^2, $E = 80 \times 10^3$ N/mm^2

$$\therefore \quad \delta l = \frac{50 \times 10^3 \times 600}{1017.88 \times 80 \times 10^3} - \frac{30 \times 10^3 \times 1100}{1017.88 \times 80 \times 10^3} - \frac{10 \times 10^3 \times 900}{1017.88 \times 80 \times 10^3}$$

$$= 0.368 - 0.405 - 0.111 = -0.148 \text{ mm}$$

–ve indicates that the length is decreasing

$\therefore$ Decrease in length or change in length $\delta l = 0.148$ mm.

Problem 35 Figure 1.32 shows a bar of 30 mm diameter subjected to forces at *A*, *B*, *C* and *D*. Taking modulus of elasticity for the material as 200 kN/mm^2, find the elongation in the bar. Also find the force acting at *C* for equilibrium.

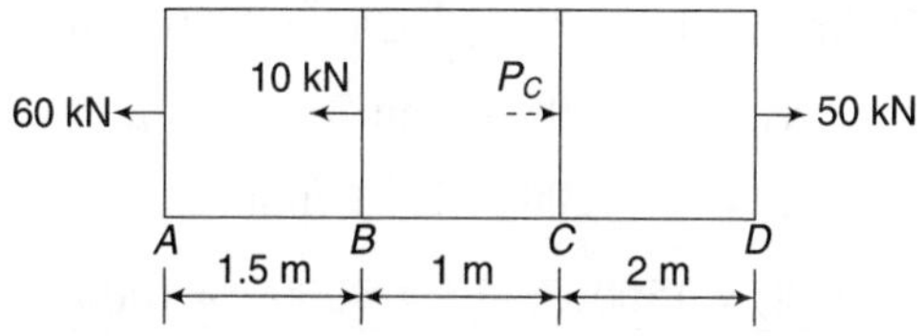

Fig. 1.32 (a)

Solution: **Given data:**

$$\text{Diameter } d = 30 \text{ mm}$$

$$\therefore \quad \text{Area of cross section } A = \frac{\pi}{4} d^2 = \frac{\pi}{4} 30^2$$

$$= 706.86 \text{ mm}^2$$

$$E = 200 \text{ kN/mm}^2 = 200 \times 10^3 \text{ N/mm}^2$$

For equilibrium, $\Sigma F = 0$

$$-60 - 10 + P_c + 50 = 0$$

$$\therefore \quad \text{Force at } C,\ P_c = 20 \text{ kN}$$

Free-body diagram;

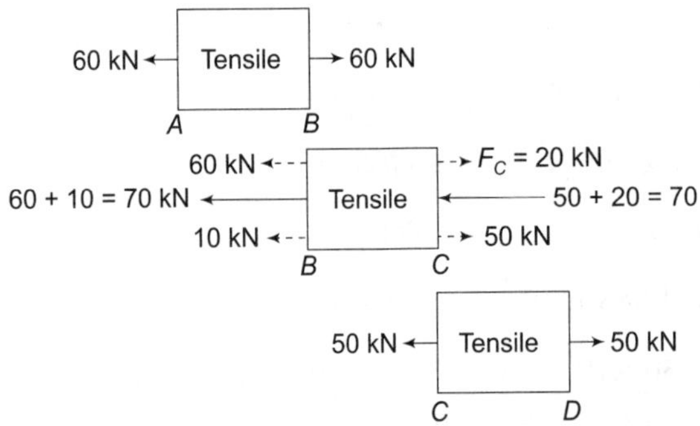

Fig. 1.32 (b)

Total elongation $\delta l = \delta l_1 + \delta l_2 + \delta l_3$

(All are tensile, hence all are +ve)

$$= \frac{P_1 l_1}{A_1 E_1} + \frac{P_2 l_2}{A_2 E_2} + \frac{P_3 l_3}{A_3 E_3}$$

$P_1 = 60 \text{ kN} = 60{,}000 \text{ N}, \; l_1 = 1.5 \text{ m} = 1500 \text{ mm}$

$P_2 = 70 \text{ kN} = 70{,}000 \text{ N}, \; l_2 = 1\text{m} = 1000 \text{ mm}$

$P_3 = 50 \text{ kN} = 50{,}000 \text{ N}, \; l_3 = 2\text{m} = 2000 \text{ mm}$

$A_1 = A_2 = A_3 = A = 706.86 \text{ mm}^2$

$E_1 = E_2 = E_3 = E = 200 \times 10^3 \text{ N/mm}^2$

$$\therefore \quad \delta l = \frac{60000 \times 1500}{706.86 \times 200 \times 10^3} + \frac{70000 \times 1000}{706.86 \times 200 \times 10^3} + \frac{50000 \times 2000}{706.86 \times 200 \times 10^3}$$

$$= 0.6366 + 0.4951 + 0.7074$$

$$\delta l = 1.8391 \text{ mm (elongation).}$$

Problem 36 A MS bar as shown in Fig. 1.33 is subjected to points loads. Calculate the load P for equilibrium of bar. Also find the change in length of bar taking E as 200 kN/mm^2.

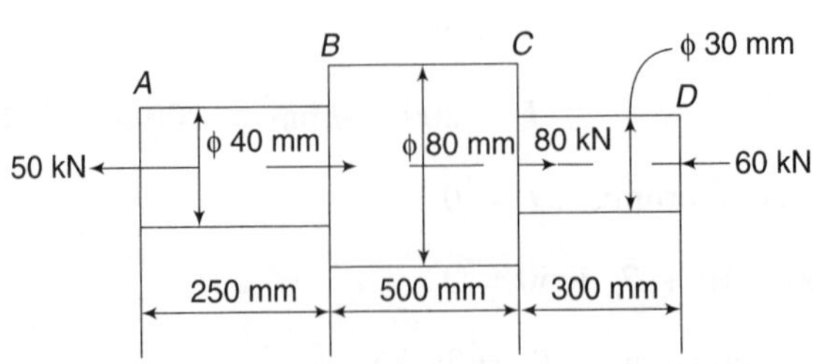

Fig. 1.33 (a)

Solution: **Given data:**

$$d_1 = 40 \text{ mm } d_2 = 80 \text{ mm } d_3 = 30 \text{ mm}$$

$$l_1 = 250 \text{ mm } l_2 = 500 \text{ mm } l_3 = 300 \text{ mm}$$

$$E = 200 \text{ kN/mm}^2 = 200 \times 10^3 \text{ N/mm}^2$$

For equilibrium, $\Sigma P = 0$

$$-50 + P + 80 - 60 = 0$$

$$\therefore \quad P = 30 \text{ kN}$$

Free-body diagram

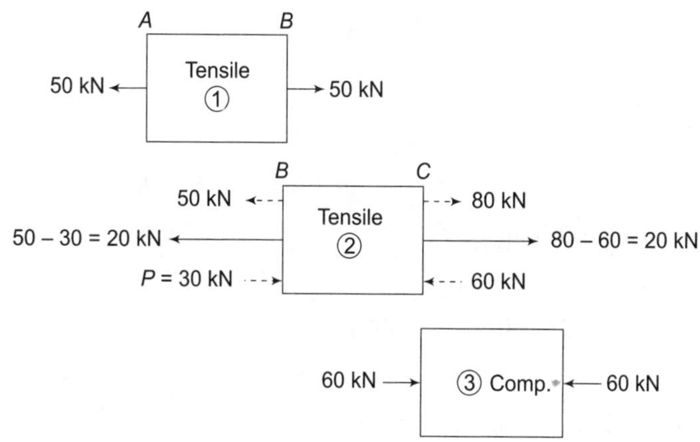

Fig. 1.33 (b)

$$\text{Total elongation } \delta l = \delta l_1 + \delta l_2 + \delta l_3$$

$$\delta l = \frac{P_1 l_1}{A_1 E} + \frac{P_2 l_2}{A_2 E} - \frac{P_3 l_3}{A_3 E}$$

{ 1 and 2 are tensile, hence +ve, 3 is compressive, hence –ve }

$$\delta l = \frac{50 \times 10^3 \times 250}{\frac{\pi}{4} \times 40^2 \times 200 \times 10^3} + \frac{20 \times 10^3 \times 500}{\frac{\pi}{4} \times 80^2 \times 200 \times 10^3} - \frac{60 \times 10^3 \times 300}{\frac{\pi}{4} \times 30^2 \times 200 \times 10^3}$$

$$= 0.796 + 0.0099 - 0.127$$

$$\delta l = 0.6776 \text{ mm.}$$

Problem 37 A bar 1 m long is fixed rigidly at ends *A* and *B* as shown in Fig. 1.34 Two forces 40 kN and 80 kN act at *B* and *C*. Taking Young's modulus as 210 GPa, determine the reactions at ends *A* and *B*. Also find the change in lengths in the three portions.

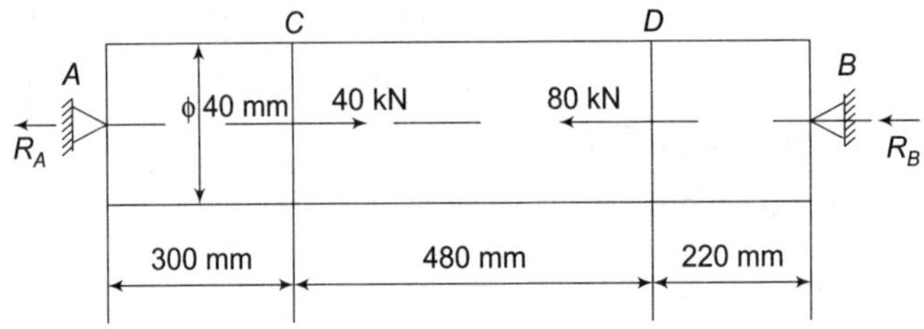

Fig. 1.34 (a)

Solution: **Given data:**

$$d = 40 \text{ mm } E = 210 \text{ GPa} = 210 \times 10^3 \text{ MPa}$$

$$\text{Let reaction } A = R_A \text{ and reaction } B = R_B$$

$$\text{For equilibrium, } \Sigma F = 0$$

$$-R_A + 40 - 80 + R_B = 0$$

$$-R_A + R_B = 80 - 40 = 40 \text{ kN}$$

$$\therefore \quad R_B = R_A + 40 \quad \ldots(1)$$

Since the ends are rigid, change in length $\delta l = 0$

$$\delta l = \delta l_1 + \delta l_2 + \delta l_3 = 0$$

Free-body diagram.

Note: When the reaction at *A* is assured to act towards right, the force will be in the opposite direction, acting towards left. Similarly, the force at *B* will be opposite to reaction, i.e., towards right.

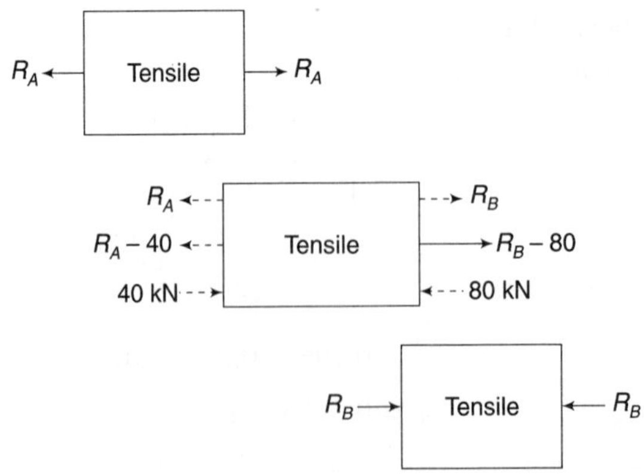

Fig. 1.34 (b)

$$\delta l = \frac{R_A \times l_1}{A_1 E} + \frac{(R_A - 40)l_2}{A_2 E} + \frac{R_B l_3}{A_3 E} = 0$$

$$= \frac{R_A \times 300}{AE} + \frac{(R_A - 40)480}{AE} + \frac{(R_A + 40) \times 220}{AE} = 0$$

$$300\ R_A + 480\ R_A - 19200 + 220\ R_A + 8800 = 0$$

$$\therefore \quad R_A = 10.4 \text{ kN and } R_B = R_A + 40 = 50.4 \text{ kN.}$$

Individual change in length:

$$\text{For } AC\text{: } \delta l_1 = \frac{R_A l_1}{AE} = \frac{10.4 \times 10^3 \times 300}{\frac{\pi}{4} \times 40^2 \times 210 \times 10^3}$$

$$\delta l_1 = \delta l_{AC} = +\ 0.0118$$

$$\text{For } CD\text{: } \delta l_2 = \frac{(R_A - 40)\ l_2}{AE} = \frac{(10.4 - 40) \times 10^3 \times 480}{\frac{\pi}{4} \times 40^2 \times 210 \times 10^3}$$

$$\delta l_2 = \delta l_{CD} = -\ 0.0538 \text{ mm}$$

$$\text{For } DA\text{: } \delta l_3 = \frac{R_B l_3}{AE} = \frac{50.4 \times 10^3 \times 220}{\frac{\pi}{4} \times 40^2 \times 210 \times 10^3}$$

$$\delta l_3 = \delta l_{DB} = +\ 0.0420.$$

Problem 38 Find the total elongation of bar shown in Fig. 1.35. Take $E = 1.05 \times 10^5$ MPa.

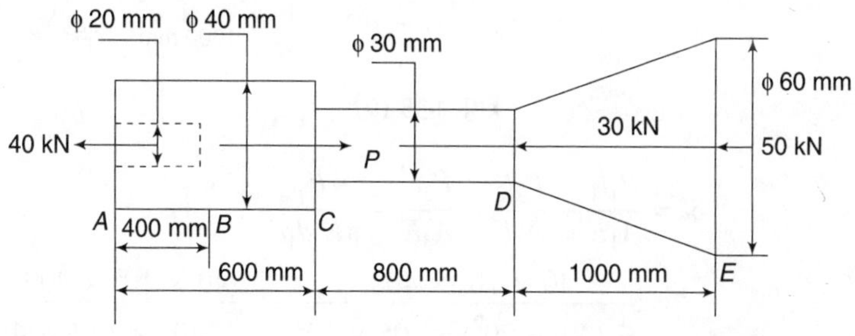

Fig. 1.35 (a)

Solution: **Given data:**

$$E = 1.05 \times 10^5 \text{ MPa}$$

$$\text{For equilibrium, } \Sigma F = 0$$

$$-40 + P - 30 - 50 = 0$$

$$\therefore \quad P = 120 \text{ kN}$$

$$\text{Total change in length } \delta l = \delta l_1 + \delta l_2 + \delta l_3 + \delta l_4$$

Free body diagram

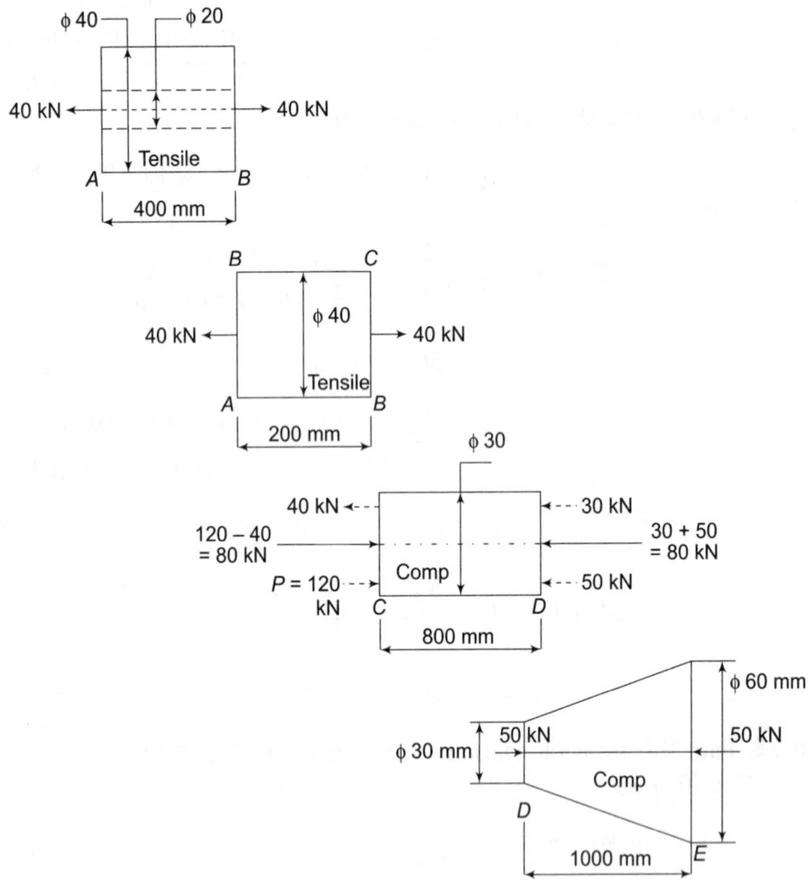

Fig. 1.35 (b)

$$\delta l = \frac{P_1 l_1}{A_1 E} + \frac{P_2 l_2}{A_2 E} - \frac{P_3 l_3}{A_3 E} - \frac{4 P_4 l_4}{\pi E\, d_1 d_2}$$

$$= \frac{40 \times 10^3 \times 400}{\frac{\pi}{4}(40^2 - 20^2) \times 1.05 \times 10^5} + \frac{40 \times 10^3 \times 200}{\frac{\pi}{4} 40^2 \times 1.05 \times 10^5}$$

$$- \frac{80 \times 10^3 \times 800}{\frac{\pi}{4} \times 30^2 \times 1.05 \times 10^5} - \frac{4 \times 50 \times 10^3 \times 1000}{\pi \times 30 \times 60 \times 1.05 \times 10^5}$$

$$\delta l = 0.1617 + 0.0606 - 0.8623 - 0.3368$$

$$\delta l = -0.9768 \text{ mm (Compression)}$$

Problem 39 Determine the stresses in various segments of the circular bar shown in Fig. 1.36. Compute the total elongation taking Young's modulus as 195 GPa.

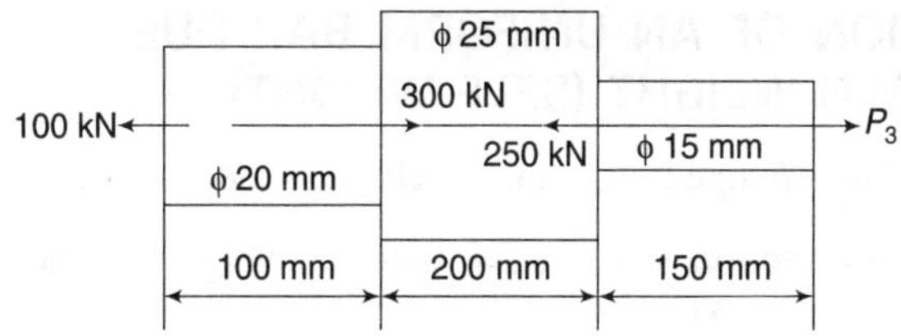

Fig. 1.36 (a)

Solution: Given data $E = 195 \text{ GPa} = 195 \times 10^3 \text{ MPa}$

For equilibrium, $\Sigma P = 0$

$$-100 + 300 - 250 + P_3 = 0$$

$\therefore$ $P_3 = 50 \text{ kN}$

Free-body diagram

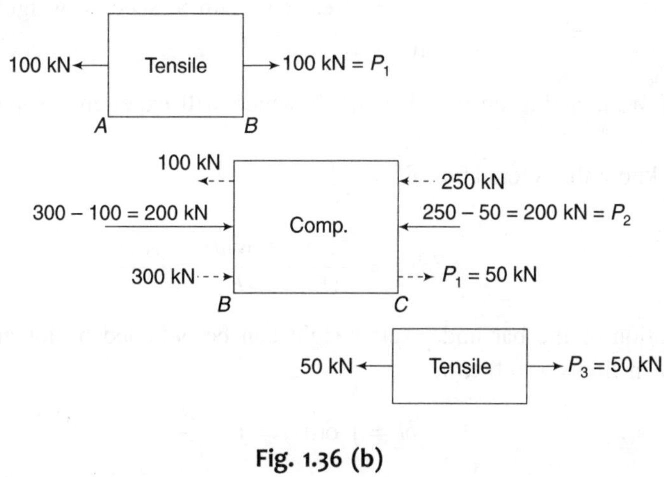

Fig. 1.36 (b)

Total change in length $\delta l = \delta l_1 + \delta l_2 + \delta l_3$

$$\delta l = \frac{P_1 l_1}{A_1 E} - \frac{P_2 l_2}{A_2 E} + \frac{P_3 l_3}{A_3 E}$$

$$= \frac{100 \times 10^3 \times 100}{\left(\frac{\pi}{4} \times 20^2\right) \times 195 \times 10^3} - \frac{200 \times 10^3 \times 200}{\left(\frac{\pi}{4} \times 25^2\right) \times 195 \times 10^3} + \frac{50 \times 10^3 \times 150}{\left(\frac{\pi}{4}\, 15^2 \times 195 \times 10^3\right)}$$

$$= 0.163 - 0.418 + 0.218 = -0.037 \text{ mm}$$

1.15 ELONGATION OF AN UNIFORM BAR DUE TO ITS OWN WEIGHT (SELF WEIGHT)

Figure 1.37 shows a bar *AB* fixed at *A* and freely suspended under its own weight.

A = Area of cross section, mm^2 w = specific weight of the material of the bar $\frac{N}{mm^3}$

W = Total weight, N

l = Length of bar, mm

E = Modulus of elasticity, N/mm^2

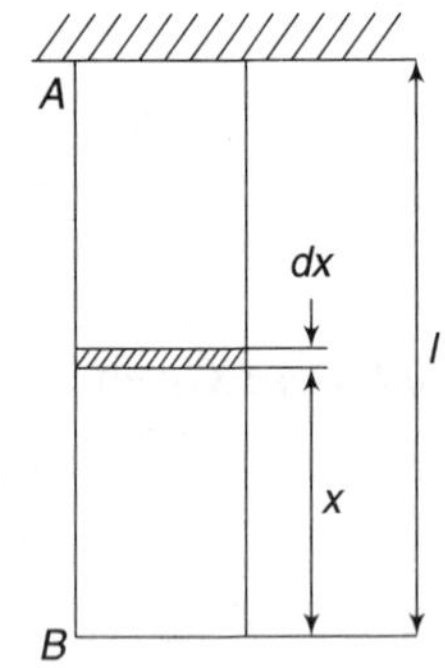

Fig. 1.37

Consider an elemental strip of length dx at a distance x from free end B. Weight of bar over this length x,

$$W_x = \text{Vol} \times \text{Specific weight}$$

$$= \text{Area} \times \text{Length} \times \text{Specific weight}$$

$$W_x = A.x.w$$

This weight W_x is acting on the element dx which will experience some elongation.

$$\text{We know that elongation } \delta l = \frac{Wl}{AE}$$

$$\delta\,(dx) = \frac{W_x dx}{AE}\;\frac{(Axw)dx}{AE} = \frac{xwdx}{E}$$

Total elongation of the bar under self weight can be obtained by integrating the above equation between limits $x = 0$ to l.

$$\delta l = \int_o^l \delta(dx) = \int_o^l \frac{wxdx}{E}$$

$$\delta l = \frac{w}{E}\int_o^l xdx = \frac{w}{E}\left(\frac{x^2}{2}\right)_o^l$$

$$\delta l = \frac{wl^2}{2E}$$

Note: The elongation is independent of area of cross section.

$$\text{Multiply and divide by } A,\ \delta l = \frac{wl^2 A}{2AE}$$

$$\text{But } wAl = W = \text{total weight of bar.}$$

$$\delta l = \frac{Wl}{2AE}$$

The elongation under self weight is half of the elongation with same load.

$\left(\text{For a load of } W,\ \delta l = \dfrac{Wl}{AE}\right)$

1.16 ELONGATION OF CONICAL BAR UNDER SELF WEIGHT

Figure 1.38 shows a conical bar *AB* fixed at end *A* and is freely suspended under its own weight.

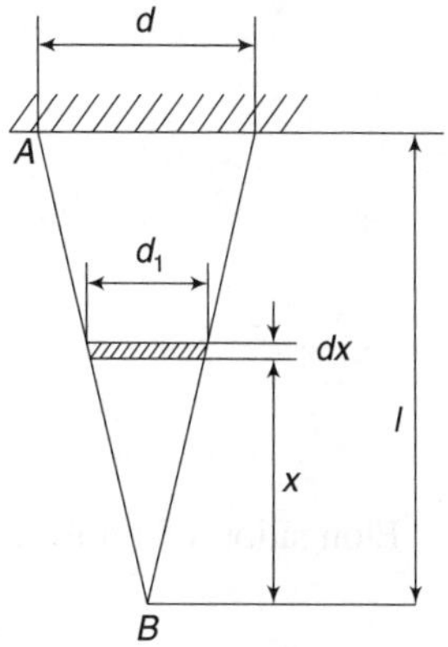

Fig. 1.38

V = Volume of cone $= \dfrac{1}{3}\pi r^2 l$

w = Specific weight of the material of the bar N/mm^3

W = Total weight, N

l = Length of bar

E = Modulus of elasticity, N/mm^2

Consider an elemental strip of length dx at a distance of x form free end B.

Weight of bar over this length x,

$$W_x = V \times w = \left(\frac{1}{3}\pi r_1^2 x\right) w$$

Weight (W_x) is acting on the element dx which will experience some elongation.

We know that elongation $\delta l = \dfrac{Wl}{AE}$

$$\delta(dx) = \frac{W_x . dx}{(\pi r_1^2)\ E}$$

But $$W_x = \frac{1}{3}\pi r_1^2\ x\ w$$

$$\delta(dx) = \frac{\left(\frac{1}{3}\pi r_1^2\ x \cdot w\right)\cdot dx}{(\pi r_1^2)E} = \frac{1}{3}\frac{wxdx}{E}$$

Total elongation of the bar under self weight can be obtained by integrating the above equation between limits $x = 0$ to l.

$$\delta l = \int_o^l \delta\ (dx) = \int_o^l \frac{1}{3}\frac{Wxdx}{E}$$

$$= \frac{w}{3E}\int_o^l xdx = \frac{w}{3E}\left(\frac{x^2}{2}\right)_o^l$$

$$\delta l = \frac{wl^2}{6E}$$

Multiply and divide by Area A

$$\delta l = \frac{wAl^2}{6AE} = \frac{1}{3} wAl.\frac{l}{2EA} = \frac{Wl}{2AE}$$

Problem 40 A steel rod of diameter 60 mm and length 3 m is fixed at the top to a roof and hangs freely. Find the elongation under self weight if specific weight of rod is 77 kN/m^3 and modulus of elasticity is 206 GPa.

***Solution:* Given data:**

Diameter d = 60 mm

Length l = 3 m = 3000 mm

Specific weight w = 77 kN/m^3

$$w = \frac{77 \times 10^3}{(1000)^3} \text{ N/mm}^3 = 77 \times 10^{-6} \text{ N/mm}^3$$

$$E = 206 \text{ GPa} = 206 \times 10^3 \text{ MPa.}$$

Elongation of uniform bar under self weight is $\delta l = \frac{Wl}{2AE} = \frac{Wl^2}{2E}$

$$= \frac{77 \times 10^{-6} \times 3000^2}{2 \times 206 \times 10^3} = 1.68 \times 10^{-3} \text{ mm.}$$

Problem 41 A CI rod of rectangular cross section 30 mm wide, 20 mm thick and 6 m long is fixed at the top and hangs freely. Taking $E = 1.1 \times 10^5$ N/mm^2 and weight of rod as 250 N. Find the elongation in the rod.

***Solution:* Given data:**

Width b = 30 mm, thickness t = 20 mm

Length l = 6 m = 6000 mm

$$E = 1.1 \times 10^5 \text{ N/mm}^2$$

Area $A = bt = 30 \times 20 = 600$ mm^2

W = weight of rod = 250 N

Elongation under self weight $\delta l = \frac{wl^2}{2E} = \frac{Wl}{2AE}$

$$\delta l = \frac{250 \times 6000}{2(30 \times 20) \times 1.1 \times 10^5} = 0.01136 \text{ mm.}$$

Problem 42 A steel conical member of base diameter 120 mm and axis height 2.4 m is freely suspended from its base. Find the elongation due to self weight taking $E = 200$ kN/mm^2 and specific weight as 76 kN/m^3.

Solution: **Given data:**

Conical member

$$\text{Base diameter } d = 120 \text{ mm}$$

$$\text{Height} = \text{Length } l = 2.4 \text{ m} = 2400 \text{ mm.}$$

$$E = 200 \text{ kN/mm}^2 = 200 \times 10^3 \text{ N/mm}^2$$

$$\text{Specific weight } w = 76 \text{ kN/m}^3 = 76 \times 10^{-6} \text{ N/mm}^3$$

Elongation of conical bar under self-weight

$$\delta l = \frac{wl^2}{6E} = \frac{76 \times 10^{-6} \times 2400^2}{6 \times 200 \times 10^3} = 0.0003648 \text{ mm.}$$

1.17 COMPOSITE BARS (COMPOUND BARS)

A bar made up of two or more members of different materials rigidly fixed at the ends and behaving as one unit for extension or compression when subjected to axial load is called composite bar. Figure 1.39 shows a composite bar. For a composite bar, change in length is same for all the members. The total external load applied on the composite bar is equal to the sum of loads carried by each member.

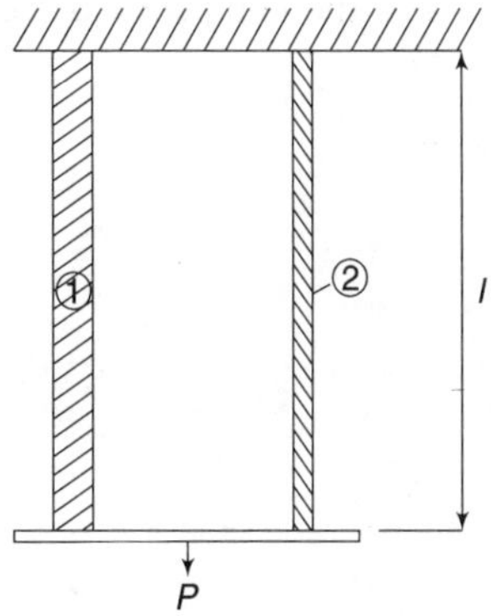

Fig. 1.39

$$\delta l = \delta l_1 = \delta l_2$$

But $$\delta l = \frac{\sigma l}{E}$$

$$\frac{\sigma_1 l}{E_1} = \frac{\sigma_2 l}{E_2} \quad \ldots(1)$$

Load $$P = P_1 + P_2$$

$$P = \sigma_1 A_1 + \sigma_2 A_2 \quad \ldots(2)$$

From Eqs. (1) and (2) we can find the load shared by each member and the stresses induced in them.

Problem 43 A steel rod of 60 mm diameter is placed centrally inside a hollow CI tube of diameters 90 mm and 120 mm. The ends are rigidly fixed and the composite bar is subjected to an axial compressive load of 120 kN. The length of the composite bar is 120 mm.

Find: (i) The load carried by each member

(ii) The stress induced in them.

Take E_S = 206 GPa and E_{CI} = 120 GPa.

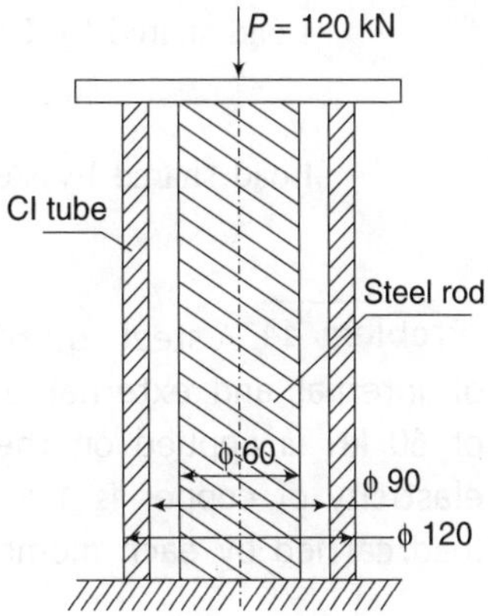

Fig. 1.40

Solution: Given data

$$\text{Diameter of steel rod } d_S = 60 \text{ mm}$$

$$\text{Diameter of CI tube Inside } d_{CI} = 90 \text{ mm}$$

$$\text{Outside dia } d_{CO} = 120 \text{ mm}$$

$$\text{Length } l = l_S = l_{CI} = 120 \text{ mm}$$

$$\text{Load } P = 120 \text{ kN} = 120 \times 10^3 \text{ N}$$

$$E_S = 206 \text{ GPa} = 206 \times 10^3 \text{ N/mm}^2$$

$$E_{CI} = 120 \text{ GPa} = 120 \times 10^3 \text{ N/mm}^2$$

Area of cross section of steel rod $A_S = \frac{\pi}{4} 60^2 = 2827.43 \text{ mm}^2$

Area of cross section of CI tube $A_{CI} = \frac{\pi}{4} (120^2 - 90^2) = 4948.0/\text{mm}^2$

For the composite bar change in length is same.

$$\delta l = \delta l_S = \delta l_{CI}$$

$$\frac{\sigma_S \, l}{E_S} = \frac{\sigma_{CI} l}{E_{CI}}$$

$$\frac{\sigma_S \times 120}{206 \times 10^3} = \frac{\sigma_{CI}.120}{120 \times 10^3}$$

$$\therefore \quad \sigma_S = 1.717 \, \sigma_{CI} \quad \text{...(1)}$$

$$\text{Total load } P = P_1 + P_2$$

$$P = P_S + P_{CI}$$

$$120 \times 10^3 = \sigma_S A_S + \sigma_{CI} A_{CI}$$

$$120 \times 10^3 = (1.717 \, \sigma_{CI}) \times 2827.43 + \sigma_{CI} \, 4948.01$$

$$\therefore \quad \sigma_{C7} = 12.2415 \text{ N/mm}^2$$

$$\therefore \quad \sigma_S = 1.717 \times 12.2415 = 21.018 \text{ N/mm}^2$$

$$\text{Load shared by CI tube } P_{CI} = \sigma_{CI} A_{CI}$$

$$P_{CI} = 12.2415 \times 4948.01 = 60571 \text{ N}$$

$$\text{Load shared by steel rod } P_S = P - P_{CI}$$

$$P_S = 120 \times 10^3 - 60571 = 59429 \text{ N.}$$

Problem 44 A steel rod of 30 mm diameter is placed centrally inside a copper tube of internal and external diameters 45 mm and 75 mm respectively. A tensile load of 60 kN is applied on the composite bar. The length of bar is 0.3 m. Modulus of elasticity of copper is 1×10^5 N/mm^2 and for steel 2×10^5 N/mm^2. Determine the load carried by each member and the stresses induced.

Solution: **Given data:**

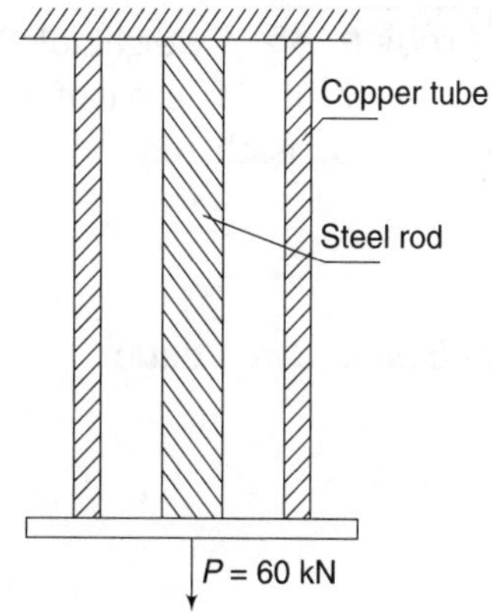

Fig. 1.41

Diameter of steel rod $d_S = 30$ mm

Diameter of copper tube

Inside diameter $d_{CI} = 45$ mm

Outside diameter $d_{CO} = 75$ mm

Length $l_S = l_C = 0.3$ m

$= 300$ mm

$E_S = 2 \times 10^5$ MPa,

$E_C = 1 \times 10^5$ MPa

Area of cross section of steel rod $A_S = \frac{\pi}{4} 30^2$

$= 706.86 \text{ mm}^2$

Area of cross section of copper tube $A_C = \frac{\pi}{4} (75^2 - 45^2)$

$A_C = 2827.43 \text{ mm}^2$

For the composite bar change in length is same $\delta l = \delta l_S = \delta l_C$

$$\text{But } \delta l = \frac{\sigma l}{E}$$

$$\frac{\sigma_S l}{E_S} = \frac{\sigma_C l}{E_C}$$

$$\frac{\sigma_S \times 300}{2 \times 10^5} = \frac{\sigma_C \times 300}{1 \times 10^5} \Rightarrow \sigma_S = 2\,\sigma_C \qquad \ldots(1)$$

$$\text{Load } P = P_S + P_C$$

$$60 \times 10^3 = \sigma_S A_S + \sigma_C A_C$$

$$60 \times 10^3 = (2\,\sigma_C) \times 706.56 + \sigma_C \times 2827.43$$

$\therefore$ $\sigma_C = 14.1471$ MPa

$\therefore$ Stress in steel rod $\sigma_S = 2\,\sigma_C = 28.2942$ MPa

Load shared by copper tube $P_C = \sigma_C A_C$

$P_C = 14.1471 \times 2827.43 = 39999.95$ N

Load shared by steel rod $P_S = P - P_C = 20000.05$ N

Problem 45 A weight of 90 kN is suspended by 3 rods of equal length as shown in Fig. 1.42 of the area of cross section of each rod is 420 mm^2, determine the load carried by each rod.

Take E_C = 120 GPa

E_S = 200 GPa.

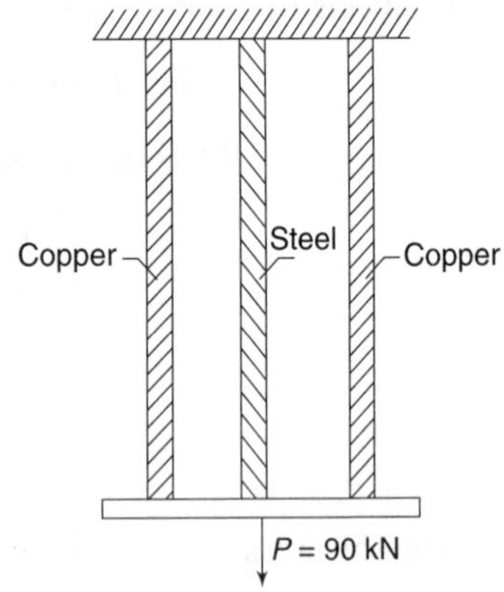

Fig. 1.42

Solution: **Given data:**

$$P = P_1 + P_2 + P_3$$

$$= 90 \times 10^3 \text{ N}$$

$$P_C + P_S + P_C = 90 \times 10^3$$

$$2P_C + P_S = 90 \times 10^3 \text{ N}$$

But $$P = \sigma A$$

A = Area of cross section $A_S = A_C = 420 \text{ mm}^2$

$$2\sigma_C A_C + \sigma_S A_S = 90 \times 10^3$$

$$2\sigma_C \times 420 + \sigma_S \times 420 = 90 \times 10^3$$

$$2\sigma_C + \sigma_S = 214.286 \quad ...(1)$$

Since the length of all rods is same, the elongations are same $\delta l_1 = \delta l_2 = \delta l_3$

But $$\delta l = \frac{\sigma l}{E}$$

$$\frac{\sigma_C l}{E_C} = \frac{\sigma_S l}{E_S}$$

$$\frac{\sigma_C \times l}{120 \times 10^3} = \frac{\sigma_S \times 1}{200 \times 10^3}$$

$\therefore$ $$\sigma_C = 0.6\,\sigma_S \quad ...(2)$$

Substituting in Eq. (1)

$$2(0.6\,\sigma_S) + \sigma_S = 214.286$$

$$\sigma_S = 97.403 \text{ MPa}$$

$$\sigma_C = 0.6\,\sigma_S = 58.442 \text{ MPa.}$$

Load shared by steel rod $P_S = \sigma_S A_S = 97.403 \times 420 = 40909.1$ N

Load shared by each copper rod $P_C = 0.6\,P_S = P_C = 0.6 \times 40909.1 = 24545.45$ N

Problem 46 A steel rod of diameter 30 mm and two brass rods of each 18 mm diameter together support a load of 30 KN as shown in Fig. 1.43. Find the stresses in the rods. Take E_S= 207 kN/mm^2 and E_B = 90 kN/mm^2

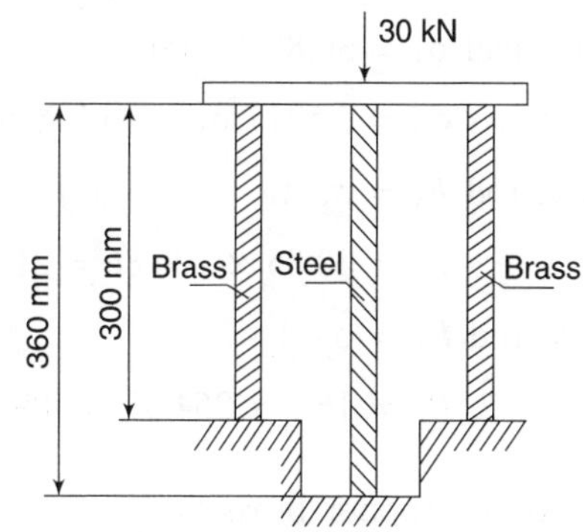

Fig. 1.43

Solution: **Given data:**

$$\text{Load } P = 30 \text{ kN} = 30 \times 10^3 \text{ N}$$

$$\text{Diameter of steel rod } d_S = 30 \text{ mm}$$

∴ Area of cross section of steel rod

$$A_S = \frac{\pi}{4} d_S^2 = \frac{\pi}{4} 30^2 = 706.86 \text{ mm}^2$$

$$\text{Diameter of brass rod } d_B = 18 \text{ mm}$$

∴ Area of cross section of each brass rod

$$A_B = \frac{\pi}{4} d_B^2 = \frac{\pi}{4} 18^2 = 254.47 \text{ mm}^2$$

$$E_S = 207 \text{ kN/mm}^2 = 207 \times 10^3 \text{ N/mm}^2,$$

$$E_B = 90 \times 10^3 \text{ N/mm}^2$$

$$\text{Total load } P = 2\,P_B + P_S = 2(\sigma_B\,A_B) + \sigma_S\,A_S$$

$$30000 = 2\,\sigma_B\,(254.47) + \sigma_S\,(706.86) \quad \ldots(1)$$

Since change in length is same for all the three rods

$$\delta l_B = \delta l_S$$

$$\frac{\sigma_B l_B}{E_B} = \frac{\sigma l_S}{E_S}$$

$$\frac{\sigma_B . 300}{90 \times 10^3} = \frac{\sigma_S \times 360}{207 \times 10^3}$$

$$\therefore \quad \sigma_B = 0.5217\sigma_S \quad \ldots(2)$$

Substituting in Eq. (1)

$$30000 = 2 \times (0.5217\ \sigma_S) \times 254.47 + \sigma_S \times 706.86$$

$$\therefore \quad \text{Stress in steel } \sigma_S = 30.85 \text{ N/mm}^2$$

$$\text{Stress in brass } \sigma_B = 0.5217\ \sigma_S = 16.1 \text{ N/mm}^2$$

$$\text{Load shared by steel rod } P_S = \sigma_S\ A_S$$

$$P_S = 30.85 \times 706.86 = 21{,}806.6 \text{ N}$$

$$\text{Load shared by each brass rod } P_B = \sigma_B\ A_B$$

$$P_B = 16.1 \times 254.47 = 4096.9 \text{ N.}$$

Problem 47 Three bars made of copper brass and aluminium are of equal length and have area of cross section as 600 mm², 500 mm² and 900 mm² respectively. They are rigidly fixed at ends. If this compound member is subjected to a longitudinal pull of 180 KN, determine the loads carried by each member and the stresses induced. Take E_C = 125 GPa, E_B = 98 GPa and E_A = 78 GPa.

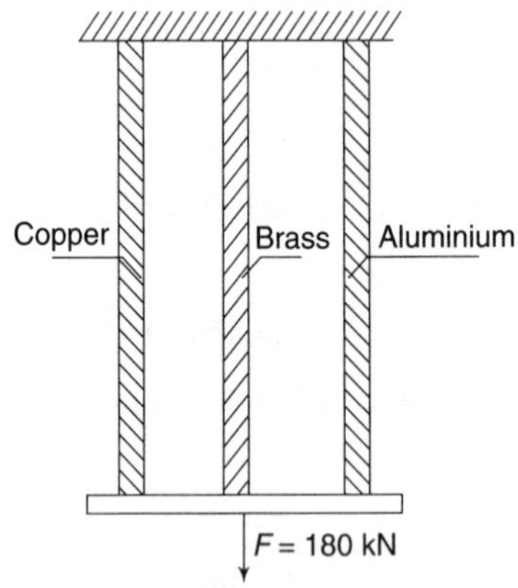

Fig. 1.44

Solution: **Given data:**

$$\text{Area of copper bar } A_C = 600 \text{ mm}^2$$

$$\text{Area of brass bar } A_B = 500 \text{ mm}^2$$

$$\text{Area of aluminium bar } A_A = 900 \text{ mm}^2$$

$$\text{Total load } P = 180 \times 10^3 \text{ N}$$

$$E_C = 125 \times 10^3 \text{ MPa}$$

$$E_B = 98 \times 10^3 \text{ MPa}$$

$$E_A = 78 \times 10^3 \text{ MPa}$$

$$\text{Total load } P = P_C + P_B + P_A$$

$$180 \times 10^3 = \sigma_C\ A_C + \sigma_B\ A_B + \sigma_A\ A_A$$

$$180 \times 10^3 = \sigma_C \times 600 + \sigma_B \times 500 + \sigma_A \times 900 \quad \ldots(1)$$

Since the lengths are same and rigidly fixed, the elongations will be same

$$\delta l_C = \delta l_B = \delta l_A$$

$$\frac{\sigma_C\ l}{E_C} = \frac{\sigma_B\ l}{E_B} = \frac{\sigma_A\ l}{E_A}$$

$$\text{Consider } \frac{\sigma_C\ l}{E_C} = \frac{\sigma_B\ l}{E_B}$$

$$\frac{\sigma_C \times l}{125 \times 10^3} = \frac{\sigma_B \times l}{98 \times 10^3}$$

$$\therefore \quad \sigma_B = 0.784\ \sigma_C \qquad \ldots(2)$$

$$\text{Consider } \frac{\sigma_C\, l}{E_C} = \frac{\sigma_A\, l}{E_A}$$

$$\frac{\sigma_C . l}{125 \times 10^3} = \frac{\sigma_A l}{78 \times 10^3}$$

$$\sigma_A = 0.624\ \sigma_C \qquad \ldots(3)$$

Substituting Eqs. (2) and (3) in Eq. (1)

$$180 \times 10^3 = \sigma_C \times 600 + (0.784\ \sigma_C) \times 500 + (0.624\ \sigma_C) \times 900$$

$$\therefore \quad \sigma_C = 115.86 \text{ N/mm}^2$$

Stress in copper rod $\sigma_C = 115.86$ N/mm^2

Stress in brass rod $\sigma_B = 0.784\ \sigma_C$

$$\sigma_B = 0.784 \times 115.86 = 90.83 \text{ N/mm}^2$$

Stress in aluminium rod $\sigma_A = 0.62\ \sigma_C$

$$\sigma_A = 0.624 \times 115.86 = 72.297 \text{ N/mm}^2$$

Load carried by copper rod $P_C = \sigma_C \cdot A_C$

$$P_C = 115.86 \times 600 = 69519 \text{ N}$$

Load carried by brass rod $P_B = \sigma_B \cdot A_B$

$$P_B = 90.83 \times 500 = 45415 \text{ N}$$

Load carried by aluminium rod $P_A = \sigma_A \cdot A_A$

$$P_A = 72.297 \times 900 = 65067 \text{ N.}$$

Problem 48 Two vertical rods one of steel and the other of copper are fixed rigidly at the top and are 600 mm apart. The diameters of the rods are 30 mm and 36 mm respectively. The length of rods is 3 m. At the bottom ends of these rods, a cross bar is fixed and a load of 6 kN is attached such that the cross bar remains horizontal even after loading.

(i) Find the position of the load on the bar and stresses induced in each rod

(ii) If the load is placed in the middle of the cross bar what will be the slope of the bar and stresses induce in the rods.

Take E_{steel}= 2 × 10^5 N/m^2 and E_{copper} = 1 × 10^5 N/mm^2.

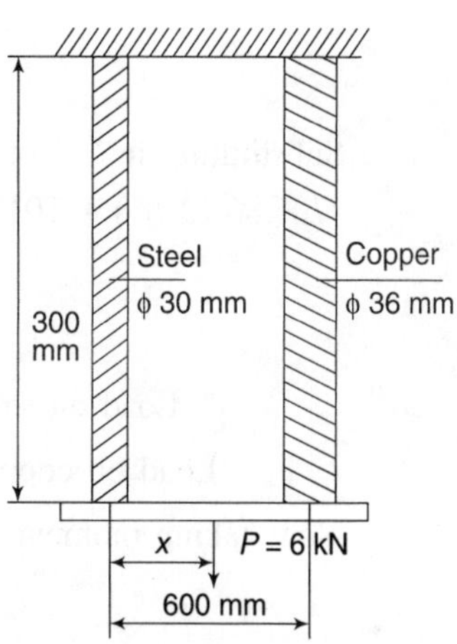

Fig. 1.45

***Solution:* Given data:**

$$\text{Diameter of steel rod } d_S = 30 \text{ mm}$$

$$\text{Diameter of copper rod } d_C = 36 \text{ mm}$$

$$\text{Length } l = 3 \text{ m} = 3000 \text{ mm}$$

$$E_S = 2 \times 10^5 \text{ MPa}$$

$$E_C = 1 \times 10^5 \text{ MPa}$$

(i) For the cross bar to be horizontal under load

Let x be the distance to the load from left end (from steel rod).

$$\text{Area of steel rod } A_S = \frac{\pi}{4} d_S^2 = \frac{\pi}{4} 30^2$$

$$= 706.86 \text{ mm}^2$$

$$\text{Area of copper rod } A_C = \frac{\pi}{4} d_C^2 = \frac{\pi}{4} 36^2 = 1017.88 \text{ mm}^2$$

$$\text{Load } P = P_S + P_C = 6000 \text{ N}$$

$$\text{Force } P = \text{Area} \times \text{Stress}$$

$$P_S = A_S \, \sigma_S \text{ and } P_C = A_C \, \sigma_C$$

$$706.86 \times \sigma_S + 1017.88 \, \sigma_C = 6000 \qquad \text{...(1)}$$

$$\text{Strain} = \frac{\sigma \, l}{E}$$

For the cross bar to be horizontal, the elongation is same $\dfrac{\sigma_S \, l}{E_S} = \dfrac{\sigma_C \, l}{E_C}$

$$\frac{\sigma_S}{2 \times 10^5} = \frac{\sigma_C}{1 \times 10^5} \qquad (l \text{ is same})$$

$$\sigma_S = 2 \times \sigma_C \qquad \text{...(2)}$$

Substituting in Eq. (1),

$$706.86 \, (2 \, \sigma_C) + 1017.88 \times \sigma_C = 6000$$

$$\sigma_C = 2.4675 \text{ MPa}$$

$$\therefore \quad \sigma_S = 2\sigma_C = 4.935 \text{ MPa}$$

$$\therefore \quad \text{Load on steel rod } P_S = A_S \, \sigma_S = 706.86 \times 4.935 = 3488 \text{ N}$$

$$\text{Load on copper rod } P_C = P - P_S = 6000 - 3488 = 2512 \text{ N}$$

Now, taking moments about steel rod (left end)

$$P.x = P_C \times 600$$

$$6000 \, x = 2512 \times 600$$

$$\therefore \quad x = 251.2 \text{ mm}$$

The load is to be placed at a distance of 251.2 mm from steel rod for the cross beam to be horizontal.

$$\text{Stress in steel rod } \sigma_S = 4.935 \text{ MPa}$$

$$\text{Stress in copper rod } \sigma_C = 2.4675 \text{ MPa.}$$

(ii) If the load placed in the middle of cross bar

$$\text{Then } P_S = P_C = \frac{P}{2} = \frac{6000}{2} = 3000 \text{ N.}$$

$$\text{Elongation in the steel rod } \delta l_S = \frac{P_S l}{A_S E_S}$$

$$\delta l_S = \frac{3000 \times 3000}{706.86 \times 2 \times 10^5} = 0.06366 \text{ mm}$$

$$\text{Elongation in the copper rod } \delta l_C = \frac{P_C\, l}{A_C E_C}$$

$$\delta l_C = \frac{3000 \times 3000}{1017.88 \times 1 \times 10^5} = 0.088\ 42 \text{ mm}$$

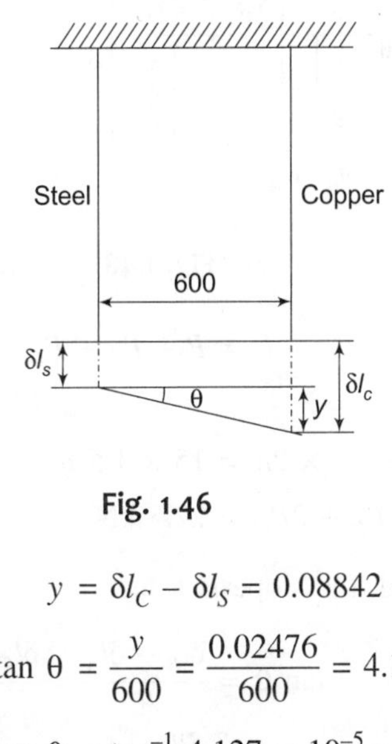

Fig. 1.46

$$y = \delta l_C - \delta l_S = 0.08842 - 0.06366 = 0.02476 \text{ mm}$$

$$\text{Slope } \tan\theta = \frac{y}{600} = \frac{0.02476}{600} = 4.127 \times 10^{-5}$$

$$\text{Angle of inclination } \theta = \tan^{-1} 4.127 \times 10^{-5} = 2.364 \times 10^{-3} \text{ deg}$$

Stresses:

$$\text{Stress in steel rod } \sigma_S = \frac{P_S}{A_S} = \frac{3000}{706.86} = 4.244 \text{ MPa}$$

$$\text{Stress in copper rod } \sigma_C = \frac{P_C}{A_C} = \frac{3000}{1017.88} = 2.947 \text{ MPa}$$

Problem 49 Three identical wires support a rigid bar *ABC* as shown in Fig. 1.47. Determine the force developed in each wire when a load of 15 kN is applied at *D*.

Solution: **Given data:** Identical wires

Hence, *A* and *E* are same

$$\text{Load } P = 15 \text{ kN}$$

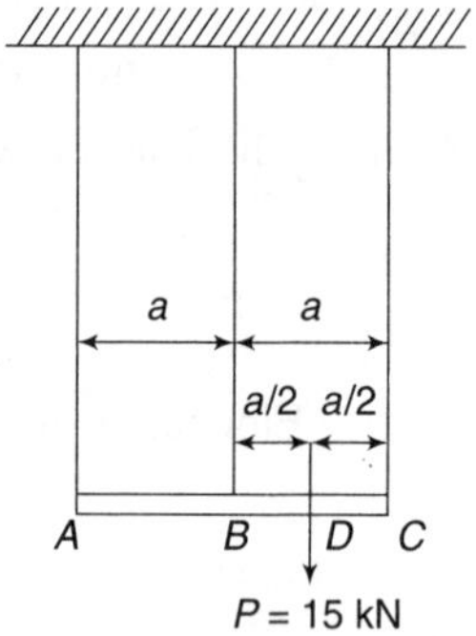

Fig. 1.47

Let F_A, F_B and F_C be the loads shared by wires at *A*, *B* and *C*

$$\delta l_A = \text{Elongation at } A$$

$$\delta l_B = \text{Elongation at } B$$

$$\delta l_C = \text{Elongation at } C$$

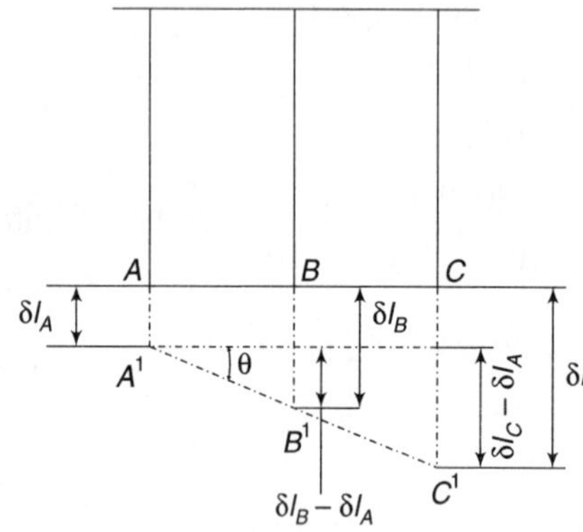

Fig. 1.48

$$P = P_A + P_B + P_C = 15 \text{ kN} \quad \text{...(1)}$$

Taking moments about *A*,

$$P_B \times a + P_C \times 2a = 15 \times 1.5\, a$$

$$P_B + 2P_C = 22.5 \text{ kN} \quad \text{...(2)}$$

Since the bar *ABC* is rigid, the slope is same

$$\tan\theta = \frac{\delta l_B - \delta l_A}{a} = \frac{\delta l_C - \delta l_A}{2a} \quad \text{...(3)}$$

But

$$\delta l = \frac{Pl}{AE}$$

For the wires *l*, *A* and *E* are same

$$\tan\theta = \frac{P_B\, l}{AE} - \frac{P_A\, l}{AE} = \frac{\left(\frac{P_C\, l}{AE} - \frac{P_A l}{AE}\right)}{2}$$

$$P_B - P_A = \frac{P_C - P_A}{2}$$

$$2P_B - 2P_A = P_C - P_A$$

$$P_C = 2P_B - P_A \qquad ...(4)$$

Substituting in Eq. (1)

$$P_A + P_B + (2P_B - P_A) = 15$$

$$3F_B = 15 \text{ or } P_B = 5 \text{ kN}$$

Substituting in Eq. (2)

$$5 + 2P_C = 22.5$$

$$P_C = 8.75 \text{ kN}$$

$$\therefore \quad P_A = 15 - P_B - P_C$$

$$= 15 - 5 - 8.75 = 1.25 \text{ kN}$$

$$\text{Load on } A \; P_A = 1.25 \text{ kN}$$

$$\text{Load on } B \; P_B = 5 \text{ kN}$$

$$\text{Load on } C \; P_C = 8.75 \text{ kN.}$$

Problem 50 A reinforced concrete column 500 mm diameter has four steel rods of each 30 mm diameter embedded in it and carries a load of 900 kN. Find the stresses in the steel and concrete. Take E_S = 204 GPa and E_C = 13.6 GPa.

Solution: **Given data:**

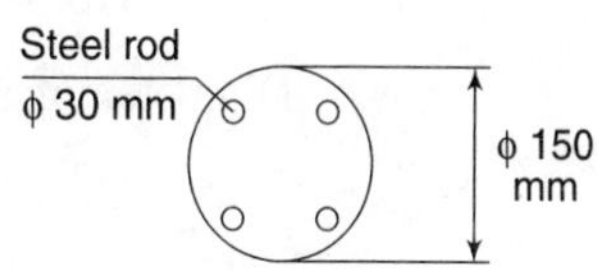

Fig. 1.49

$$\text{Diameter of column} = 500 \text{ mm}$$

$$\text{Diameter of steel rod} = 30 \text{ mm}$$

$$\text{No. of steel rods} = 4$$

$$E_S = 204 \times 10^3 \text{ MPa}$$

$$E_C = 13.6 \times 10^3 \text{ MPa}$$

$$\text{Load } P = 900 \text{ kN} = 900 \times 10^3 \text{ N}$$

$$\text{Area of steel } A_S = 4 \times \frac{\pi}{4} \times 30^2 = 2827.43 \text{ mm}^2$$

$$\text{Total area of column } A = \frac{\pi}{4} \times 500^2 = 196349.54 \text{ mm}^2$$

$$\therefore \quad \text{Area of concrete } A_C = A - A_S = 193522.11 \text{ mm}^2$$

$$P = P_S + P_C = 900 \times 10^3$$

$$\sigma_S A_S + \sigma_C A_C = 900 \times 10^3$$

$$\sigma_S \times 2827.43 + \sigma_C \times 193522.11 = 900 \times 10^3 \qquad ...(1)$$

Since the length is same for the composite bar, the strains are equal.

$$\frac{\sigma_S}{E_S} = \frac{\sigma_C}{E_C}$$

$$\frac{\sigma_S}{204 \times 10^3} = \frac{\sigma_C}{13.6 \times 10^3}$$

$$\therefore \quad \sigma_S = 15\ \sigma_C \qquad \text{...(2)}$$

Substituting in Eq. (1)

$$15\ \sigma_C \times 2827.43 + \sigma_C \times 193522.11 = 900 \times 10^3$$

$$\sigma_C = 3.8146 \text{ N/mm}^2$$

$$\therefore \quad \sigma_S = 15 \times \sigma_C = 15 \times 3.8146 = 57.2195 \text{ N/mm}^2$$

Load shared by steel rods $P_S = \sigma_S A_S = 57.2195 \times 2827.43 = 161.78 \times 10^3$ N

Load shared by concrete $P_C = F - F_S = 738.22 \times 10^3$ N

Problem 51 A RCC column of rectangular section 450 mm wide and 300 mm deep has 6 steel rods each of diameter 25 mm. The column is subjected to a load of 1200 kN. Find the stresses induced in the steel rod and concrete taking $E_S = 2 \times 10^5$ MPa and $E_C = 14 \times 10^3$ MPa.

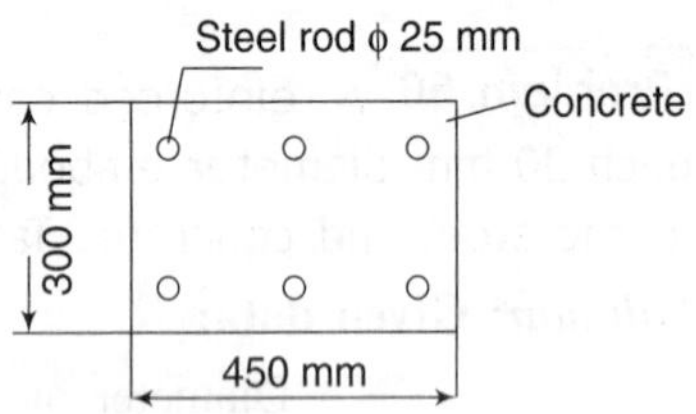

Fig. 1.50

Solution: Given data

Size of column = 450 × 300

No. of steel rods = 6

Diameter of steel rod d_S = 25 mm

Load P = 1200 kN

$$E_S = 2 \times 10^5 \text{ N/mm}^2$$

$$E_C = 14 \times 10^3 \text{ N/mm}^2$$

Area of cross section of column $A = 450 \times 300 = 135 \times 10^3$ mm^2

Area of 6 steel rods $A_S = 6 \times \frac{\pi}{4} \times 25^2 = 2945.24$ mm^2

∴ Area of cross section of concrete $A_C = A - A_S$

$$= 135 \times 10^3 - 2945.24$$

$$A_C = 132.055 \times 10^3 \text{ mm}^2$$

Load $P = P_S + P_C$

$$1200 \times 10^3 = \sigma_S A_S + \sigma_C A_C$$

$$1200 \times 10^3 = \sigma_S \times 2945.24 + \sigma_C \times 132.055 \times 10^3 \qquad \text{...(1)}$$

Since the strains are same, $\dfrac{\sigma_S}{E_S} = \dfrac{\sigma_C}{E_C}$

$$\frac{\sigma_S}{2 \times 10^5} = \frac{\sigma_C}{14 \times 10^3}$$

$$\therefore \qquad \sigma_S = 14.286\ \sigma_C \qquad \ldots(2)$$

Substituting in Eq. (1)

$$14.286\ \sigma_C \times 2945.24 + \sigma_C \times 132.055 \times 10^3 = 1200 \times 10^3$$

$$\sigma_C = 6.89 \text{ N/mm}^2$$

$$\sigma_S = 14.286 \times 6.89 = 98.45 \text{ N/mm}^2$$

Load carried by steel rods $P_S = A_S\ \sigma_S$

$$= 2945.24 \times 98.45 = 289.96 \times 10^3 \text{ N}$$

Load carried by concrete $P_C = P - P_S = 1200 \times 10^3 - 289.96 \times 10^3$

$$F_C = 910.04 \times 10^3 \text{ N.}$$

Problem 52 A weight of 300 kN is supported by a short concrete column of 250 mm square section. The column is reinforced with 4 steel rods of cross-sectional area 5500 mm^2. Find the stress in steel and concrete if E_S = 15 E_C. If the stress in concrete is not to exceed 4.5 MN/m^2 what area of steel is required in order to support a load of 500 kN.

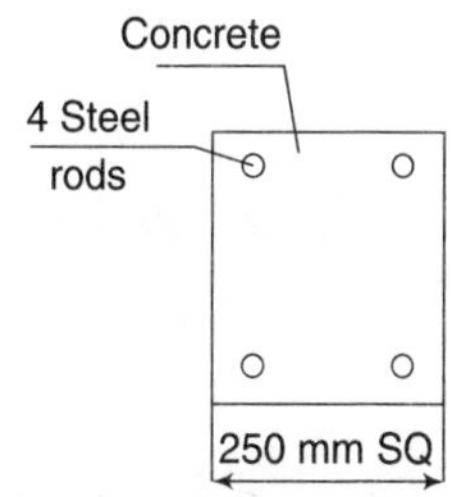

Fig. 1.51

Solution: **Given data:**

$$\text{Load } P = 300 \text{ kN} = 300 \times 10^3 \text{ N}$$

$$\text{Square section} = 250 \times 250$$

$$\text{Area of steel } A_S = 5500 \text{ mm}^2$$

$$E_S = 15\ E_C$$

$$\text{Total Area } A = 250 \times 250$$

$$A = 62500 \text{ mm}^2$$

$\therefore$ Area of concrete

$$A_C = A - A_S = 62500 - 5500 = 57000 \text{ mm}^2$$

$$\text{Total load } P = P_C + P_S = 300 \times 10^3 \text{ N}$$

$$\text{But load } P = \sigma A$$

$$\sigma_C\ A_C + \sigma_S\ A_S = 300 \times 10^3$$

$$\sigma_C \times 57000 + \sigma_S \times 5500 = 300 \times 10^3 \qquad \ldots(1)$$

The strain is same in both, $\epsilon_C = \epsilon_S$

$$\frac{\sigma_C}{E_C} = \frac{\sigma_S}{E_S}$$

$$\frac{\sigma_C}{E_C} = \frac{\sigma_S}{15\ E_C}$$

$$\therefore \qquad \sigma_S = 15\ \sigma_C \qquad ...(2)$$

Substituting in Eq. (1)

$$\sigma_C \times 57000 + (15\ \sigma_C)\ 5500 = 300 \times 10^3$$

$$\therefore \qquad \sigma_C = 2.15 \text{ MPa}$$

$$\sigma_S = 15\ \sigma_C = 32.25 \text{ MPa.}$$

(ii) $\text{Load } P = 500 \text{ kN} = 500 \times 10^3 \text{ N}$

$$\text{Stress is concrete } \sigma_C = 4.5 \text{ MPa}$$

$$\text{Total area } A = 250 \times 250 = 62500 \text{ mm}^2$$

Let A_C = Area of concrete and A_S = Area of steel

$$P = P_C + P_S = 500 \times 10^3$$

$$\sigma_C\ A_C + \sigma_S\ A_S = 500 \times 10^3 \qquad ...(3)$$

For the strain is being same

$$\frac{\sigma_C}{E_C} = \frac{\sigma_S}{E_S}$$

But $E_S = 15E_C$ and $\sigma_C = 4.5$,

$$\frac{4.5}{E_C} = \frac{\sigma_S}{15\ E_C}$$

$$\therefore \qquad \sigma_S = 15 \times 4.5 = 67.5 \text{ MPa}$$

$$A = A_C + A_S = 62500$$

$$A_C = 62500 - A_S$$

Substituting in Eq. (3)

$$4.5\ (62500 - A_S) + 67.5 \times A_S = 500 \times 10^3$$

$$\therefore \qquad \text{Area of steel required } A_S = 3472.22 \text{ mm}^2$$

Problem 53 A solid steel bar 500 mm long and 70 mm diameter is placed inside in aluminium tube having 75 mm inside diameter and 100 mm outside diameter. The aluminium cylinder is 0.15 mm longer than steel bar. An axial load of 600 kN is applied to the bar and the cylinder through rigid cover plates as shown in Fig. 1.52. Find the stresses developed in the steel bar and aluminium tube. Assume E_S = 210 kN/mm^2 and E_A = 70 kN/mm^2.

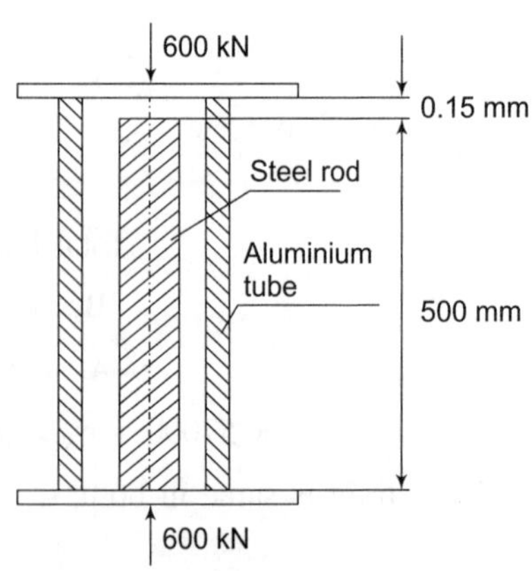

Fig. 1.52

Solution: **Given data:**

$$P = 600 \text{ kN}$$

$$= 600 \times 10^3 \text{ N}$$

Diameter of steel rod $d_S = 70$ mm

Length of steel rod $l_S = 500$ mm

$$\therefore \quad A_S = \frac{\pi}{4} \times 70^2$$

Inside diameter of aluminium tube $d_i = 75$ mm

Outside diameter of aluminium tube $d_o = 100$ mm

$$\therefore \quad A_A = \frac{\pi}{4}(100^2 - 75^2)$$

Length of aluminium tube $= l_S + 0.15 = 500 + 0.15 = 500.15$ mm

$$E_S = 210 \text{ kN/mm}^2 = 210 \times 10^3 \text{ N/mm}^2$$

$$E_A = 70 \text{ kN/mm}^2 = 70 \times 10^3 \text{ N/mm}^2$$

First let us compress the longer aluminium tube by 0.15 mm so that the lengths will be equal. Let P_{A1} be the load required to compress the aluminium tube by 0,15 mm.

We know that $\delta l = \dfrac{Pl}{AE}$

$$\therefore \quad P_{A1} = \left(\frac{\delta l\ AE}{l}\right) = \frac{0.15 \times \frac{\pi}{4}(100^2 - 75^2) \times 70 \times 10^3}{500.15}$$

$$P_{A1} = 72{,}136.8 \text{ N}$$

Remaining load $P_R = P - P_{A1} = 600 \times 103 - 72136.8$

$$P_R = 527863.2 \text{ N}$$

The composite bar will compress by same value under this load (now the lengths are equal)

$$P_R = P_{A2} + P_S = 527863.2 \text{ N} \quad ...(1)$$

For the strains being same $\left(\dfrac{P}{AE}\right)_A = \left(\dfrac{P}{AE}\right)_S$

$$\frac{P_{A2}}{A_A E_A} = \frac{P_S}{A_S E_S}$$

$$\frac{P_{A2}}{\frac{\pi}{4}(100^2 - 75^2) \times 70 \times 10^3} = \frac{P_S}{\frac{\pi}{4}(70)^2 \times 210 \times 10^3}$$

$$P_{A2} = 0.2976\ P_S \quad ...(2)$$

Substituting in Eq. (1)

$$0.2976\, P_S + P_S = 527863.2$$

$$P_S = 406793.6 \text{ N}$$

$$\therefore \quad P_{A2} = P_R - P_S = 52786.2 - 406793.6 = 121069.6 \text{ N}$$

∴ Total load shared by aluminium tube

$$P_A = P_{A1} + P_{A2} = 72136.8 + 121069.6 = 193206.4 \text{ N}$$

$$\text{Load shared by steel rod} = P_S = 406793.6 \text{ N}$$

$$\text{Stress developed in aluminium tube } \sigma_A = \frac{P_A}{A_A}$$

$$\sigma_A = \frac{193206.4}{\frac{\pi}{4}(100^2 - 75^2)} = 56.23 \text{ MPa}$$

$$\text{Stress developed in steel rod } \sigma_S = \frac{P_S}{A_S}$$

$$\sigma_S = \frac{406793.6}{\frac{\pi}{4} \times 70^2} = 105.70 \text{ MPa.}$$

Problem 54 A 30 mm diameter steel rod passes through a copper tube of 63 mm external diameter and 40 mm internal diameter and 1.3 m long. The tube is closed by rigid washers and nuts. If one nut is tightened by 1/2 turn relative to the other, what are the stresses developed in the steel rod and copper tube? There are 5 threads per 10 mm. E_S = 200 GPa and E_C = 100 GPa.

Solution: **Given data:**

$$\text{Diameter of the steel rod } d_S = 30 \text{ mm}$$

$$\therefore \text{ Area of cross section of steel rod } A_S = \frac{\pi}{4} \times 30^2 = 706.86 \text{ mm}^2$$

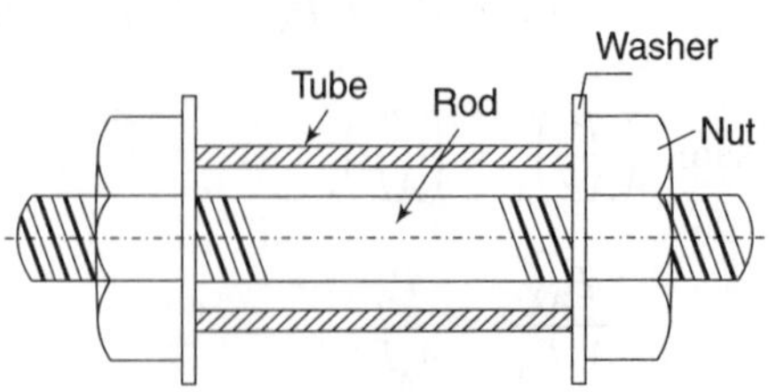

Fig. 1.53

External diameter of copper tube d_o = 63 mm

Internal diameter of copper tube d_i = 40 mm

Area of cross section of copper tube $A_C = \frac{\pi}{4}(63^2 - 40^2)$

$$A_C = 1860.61 \text{ mm}^2$$
$$E_S = 200 \times 10^3 \text{ N/mm}^2$$
$$E_C = 100 \times 10^3 \text{ N/mm}^2$$
$$\text{Length } l = l_S = l_C = 1.3 \text{ m} = 1300 \text{ mm}$$
$$\text{No. of threads} = 5 \text{ in } 10 \text{ mm length}$$
$$\therefore \quad \text{Pitch} = \text{Length/No. of threads} = \frac{10}{5} = 2 \text{ mm}$$
$$\text{Pitch} = \text{Distance moved per revolution}$$
$$1 \text{ Revolution} = 2 \text{ mm}$$
$$\frac{1}{2} \text{ Revolution} = \frac{1}{2} \times P = \frac{1}{2} \times 2 = 1 \text{ mm} = \delta l$$

When the nut is tightened, the rod will elongate and the tube will decrease in length. Total advancement of nut δl = Increase in length of rod + Decrease in length of tube

$$\delta l = \delta l_R + \delta l_T = 1 \text{ mm} \quad \ldots(1)$$

We know that change in length $\delta l = \dfrac{\sigma l}{E}$

$\therefore$ Increase in length of steel rod $\delta l_R = \dfrac{\sigma_S\, l}{E_S}$

$$\delta l_R = \frac{\sigma_S \times 1300}{200 \times 10^3} = 6.5 \times 10^{-3}\, \sigma_S$$

Decrease in length of copper tube $\delta l_T = \dfrac{\sigma_C\, l}{E_C}$

$$\delta l_T = \frac{\sigma_C \times 1300}{100 \times 10^3} = 13 \times 10^{-3}\, \sigma_C$$

For equilibrium, tenside load on rod = Compressive load on tube.

$$P_R = P_T$$
$$\sigma_S A_S = \sigma_C A_C$$
$$\sigma_S \times 706.86 = \sigma_C \times 1860.61$$
$$\therefore \quad \sigma_S = 2.63\, \sigma_C$$

Substituting in Eq. (1)

$$6.5 \times 10^{-3}\, \sigma_S + 13 \times 10^{-3}\, \sigma_C = 1$$
$$6.5 \times 10^{-3} \times 2.63\, \sigma_C + 13 \times 10^{-3}\, \sigma_C = 1$$
$$\therefore \quad \sigma_C = 33.21 \text{ MPa}$$
$$\sigma_S = 2.63 \times \sigma_C = 2.63 \times 33.21 = 87.35 \text{ MPa.}$$

Problem 55 A steel rod of 25 mm in diameter passes centrally through a steel tube of 30 mm internal dia and 40 mm external diameter. The tube is 1 m long and is closed by rigid washers at the ends by nuts. The nuts are tightened until the compressive load on the tube is 30 kN. Calculate the stresses in the rod and tube. The nut is further tightened by one quarter revolution. There are 4 threads in 10 mm. Taking E = 200 GPa, find the increase in stresses in the rod and tube.

Solution: **Given data:**

$$\text{Diameter of steel rod } d_R = 25 \text{ mm}$$

$$\therefore \text{ Area of cross section of rod } A_R = \frac{\pi}{4} \times 25^2 = 490.87 \text{ mm}^2$$

$$\text{Inside diameter of tube } d_i = 30 \text{ mm}$$

$$\text{Outside diameter of tube } d_o = 40 \text{ mm}$$

$$\therefore \text{ Area of cross section of tube } A_T = \frac{\pi}{4}(40^2 - 30^2) = 549.78 \text{ mm}^2$$

$$\text{Length } l = l_R = L_T = 1 \text{ m} = 1000 \text{ mm}$$

(i)

$$\text{Load } P = 30 \text{ kN} = 30 \times 10^3 \text{ N}$$

$$\text{Compressive load on tube} = \text{Tensile load on rod}$$

$$P_T = P_R = 30 \times 10^3 \text{ N}$$

$$\text{Stress in rod } \sigma_{R1} = \frac{P_R}{A_R} = \frac{30 \times 10^3}{490.87} = 61.12 \text{ MPa}$$

$$\text{Stress in tube } \sigma_{T1} = \frac{P_T}{A_T} = \frac{30 \times 10^3}{549.78} = 54.57 \text{ MPa}$$

(ii) The nut is further tightened by quarter revolution $= \frac{1}{4}$ of revolution

There are 4 threads in 10 mm

$$\text{Pitch of thread} = \frac{10}{4} = 2.5 \text{ mm}$$

For $\frac{1}{4}$ of revolution, distance advanced by nut

$$\delta l = \frac{1}{4} \times \text{Pitch} = \frac{1}{4} \times 2.5 = 0.625 \text{ mm}$$

Due to tightening, the screw rod will elongate and the tube will decrease in length

$$\delta l = \delta l_R + \delta l_T = 0.625 \text{ mm} \quad \ldots(1)$$

$$\text{Load on tube} = \text{Load on rod}$$

$$P_T = P_R$$

$$\sigma_T A_T = \sigma_R \cdot A_R$$

$$\sigma_T \times 549.78 = \sigma_R \times 490.87$$

$$\sigma_T = \frac{490.87}{549.78} \times \sigma_R = \sigma_T = 0.8928 \quad \text{...(2)}$$

$$\delta l_R = \frac{\sigma_R \cdot l}{E} = \frac{\sigma_R \times 1000}{200 \times 10^3} = 5 \times 10^{-3}\ \sigma_R$$

$$\delta l_T = \frac{\sigma_T \cdot l}{E} = \frac{\sigma_T \times 1000}{200 \times 10^3} = 5 \times 10^{-3}\ \sigma_T$$

Substituting in Eq. (1)

$$5 \times 10^{-3}\ \sigma_R + 5 \times 10^{-3}\ \sigma_T = 0.625$$

$$5 \times 10^{-3}\ \sigma_R + 5 \times 10^{-3} \times 0.8928\ \sigma_R = 0.625$$

$$\sigma_R = 66.04 \text{ MPa}$$

$$\sigma_T = 58.96 \text{ MPa}$$

1.18 TEMPERATURE STRESSES OR THERMAL STRESSES

Thermal stresses are the stresses induced in a body due to change in temperature. When the temperature of a body is raised or lowered and if the body is not allowed to expand or contract then thermal stresses set in the body. However, if the body is allowed to expand or contract freely, there will be no stress in the body due to change in temperature. Figure 1.54 shows a body of length l heated to a certain temperature.

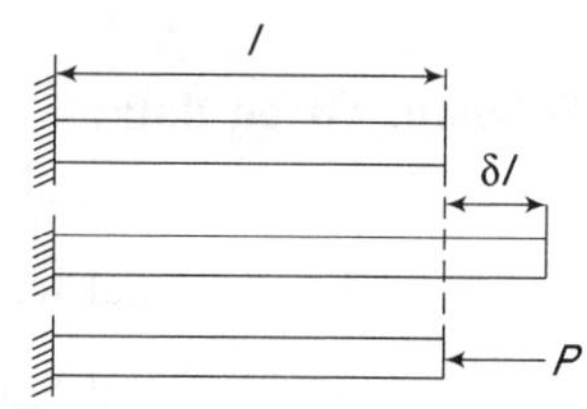

Fig. 1.54

l = Length of the body

δl = Change in length due to change in temperature

ΔT = Change in temperature

E = Modulus of elasticity

α = Coefficient of linear expansion of the material.

If the body is free to expand due to change in temperature, then the change in length is given by

$$\delta l = \alpha\ l\ \Delta T \quad \text{...(1)}$$

Now suppose an external compressive force P is applied on this body to bring it back to original length or compress this by δl,

$$\delta l = \frac{P(l + \delta l)}{AE} \simeq \frac{Pl}{AE}$$

$$\delta l = \frac{\sigma l}{E} \quad \text{...(2)}$$

Equating Eqs. (1) and (2)

$$\alpha \, l \, \Delta T = \frac{\sigma l}{E}$$

$\therefore$ Stress $\sigma = \alpha \, E \, \Delta T$

If the end supports are rigid and if they yield by small value δ, then the net expansion causing the stress $= \delta l - \delta$

$$= \alpha \, l \, \Delta T - \delta \qquad \ldots(3)$$

Equate this to Eq. (2)

$$\alpha \, l \, \Delta T - \delta = \frac{\sigma l}{E}$$

$$\text{Stress } \sigma = \frac{(\alpha \, l \, \Delta T - \delta)E}{l}.$$

Problem 56 A 9 m long steel rod at 30°C is heated to 90°C. Find the free expansion of the rod. Also find the temperature stress when

(i) The rod is prevented from expansion

(ii) The rod is allowed to expand by 2 mm

Take $\alpha = 12 \times 10^{-6}$/°C and $E = 2 \times 10^5$ N/mm^2

Solution: **Given data:**

Length $l = 9$ m $= 9000$ mm

Initial temperature $T_1 = 30$°C

Final temperature $T_2 = 90$°C

Increase in temperature $\Delta T = T_2 - T_1 = 90 - 30 = 60$°C

Coefficient of linear expansion $\alpha = 12 \times 10^{-6}$/°C

$E = 2 \times 10^5$ N/mm^2

(i) Free expansion $\delta l = \alpha \, l \, \Delta T$

$= 12 \times 10^{-6} \times 9000 \times 60 = 6.48$ mm

(ii) Stress in the rod, if prevented to expand

$$\sigma = \alpha \, E \, \Delta T = 12 \times 10^{-6} \times 2 \times 10^5 \times 60 = 144 \text{ MPa}$$

(iii) Stress in the rod, if allowed to expand by 2 mm, $\delta = 2$ mm

$$\sigma = \frac{(\alpha \, l \, \Delta T - \delta)}{l} \times E$$

$$\sigma = \frac{(6.48 - 2)}{9000} \times 2 \times 10^5 = 99.56 \text{ MPa.}$$

Problem 57 A car wheel of diameter 1.2 m is to be filled with a thin steel tyre. Assuming that the wheel is to be forged and the stress in the steel is not to exceed

140 MPa, calculate the minimum diameter of the tyre and the minimum temperature to which it should be heated before slipping on to the wheel Take $E = 2 \times 10^{11}$ N/m^2 and $\alpha = 12.6 \times 10^{-6}$/°C.

Solution: **Given data:**

$$\text{Diameter } d = 1.2 \text{ m} = 1200 \text{ mm}$$

$$\text{Allowable stress } \sigma = 140 \text{ MPa}$$

$$E = 2 \times 10^{11} \text{ N/m}^2 = 2 \times 10^5 \text{ N/m}^2$$

$$\alpha = 12.6 \times 10^{-6} \text{ /°C}$$

$$\text{Length of steel wheel} = \text{circumference} = \pi d = \pi \times 1200 \text{ mm}$$

$$\text{Thermal stress } \sigma = \alpha E \Delta T$$

$$140 = 12.6 \times 10^{-6} \times 2 \times 10^5 \times \Delta T$$

$$\therefore \quad \text{Raise in temperature } \Delta T = 55.56°\text{C}$$

$$\text{Change in length } \delta l = \alpha l \Delta T$$

$$\delta l = 12.6 \times 10^{-6} \times 1200 \times \pi \times 55.56$$

$$\delta l = 2.639 \text{ mm}$$

$$\therefore \quad \text{Length of steel tyre} = 1200\pi - 2.639 = 3767.27 \text{ mm}$$

$$\therefore \quad \text{Diameter } d_1 = \frac{3767.27}{\pi} = 1199.16 \text{ mm}$$

The steel tyre is to made of diameter 1199.16 mm and is to be heated to 55.56°C to limit the stress to 140 MPa.

Problem 58 A 3 m long rod is at a temperature of 12°C. Find the elongation of the rod it the temperature is raised to 100°C. If the expansion is prevented, find the stress induced in the rod. Take $E = 1.2 \times 10^5$ N/mm^2 and $\alpha = 12 \times 10^{-6}$/°C.

Solution: **Given data:**

$$\text{Length of rod} = 3 \text{ m} = 3000 \text{ mm}$$

$$\text{Rise in temperature } \Delta T = T_2 - T_1 = 100 - 12 = 88°\text{C}$$

$$E = 1.2 \times 10^5 \text{ N/mm}^2$$

$$\alpha = 12 \times 10^{-6} \text{ /°C}$$

Elongation due to rise in temperature $\delta l = \alpha l \Delta T$

$$\delta l = 12 \times 10^{-6} \times 3000 \times 88 = 3.168 \text{ mm}$$

The stress induced, if the expansion is prevented

$$\sigma = \alpha E \Delta T$$

$$\sigma = 12 \times 10^{-6} \times 1.2 \times 10^5 \times 88 = 126.72 \text{ N/mm}^2$$

1.19 TEMPERATURE STRESSES IN COMPOSITE BARS

Figure 1.55 shows a composite bar consisting of two members made of different materials. Since the coefficient of linear expansion is different for different materials, the expansion due to rise in temperature will be different. If these members are free to expand, there will not be any stresses induced in them. But for some reason, if these members are rigidly fixed at the ends, then they will expand by same amount. Since the individual expansions are different, the member with more expansion will be pulled back and the one with less expansion will be pulled out as shown in Fig. 1.55(c) But the load acting on both will be same. Member (1) will have tensile stress and (2) will have compressive stress.

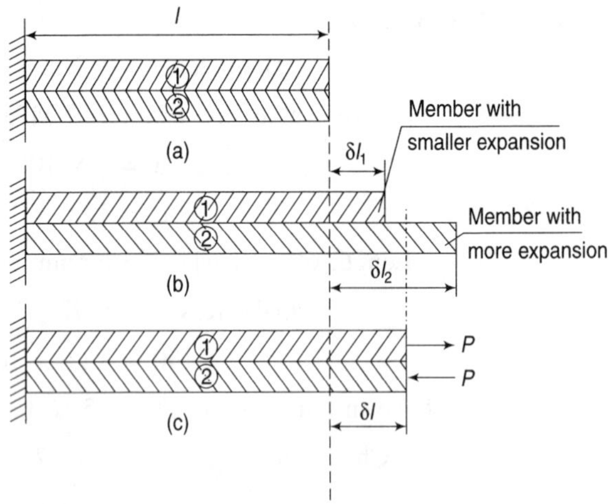

Fig. 1.55

A_1 = Area of cross section of member (1)

A_2 = Area of cross section of member (2)

σ_1 = Stress in member (1)

σ_2 = Stress in member (2)

δl_1 = Elongation of member (1)

δl_2 = Elongation of member (2)

l = Length of each member

ΔT = Rise in temperature

E_1 = Modulus of elasticity of (1)

E_2 = Modulus of elasticity of (2)

Load on member (1) = Load on Member (2)

$$P_1 = P_2$$

$$\sigma_1 A_1 = \sigma_2 A_2 \qquad ...(1)$$

Free expansion of member (1)

$$\delta l_1 = \alpha_1 \, l \, \Delta T$$

Actual expansion of member (1)

$$\delta l = \delta l_1 + \text{Elongation due to load } P$$

$$\delta l = \alpha_1 \, l \, \Delta T + \frac{\sigma_1}{E_1} l \quad \text{...(2)}$$

Similarly, free expansion of member (2)

$$\delta l_2 = \alpha_2 \, l \, \Delta T$$

Actual expansion of member (2)

$$\delta l = \delta l_2 - \text{Compression due to load } P$$

$$= \alpha_2 \, l \, \Delta T - \frac{\sigma_2}{E_2} l \quad \text{...(3)}$$

Since actual expansion is same for the composite bar, equate Eqs. (2) and (3)

$$\delta l = \alpha_1 \, l \, \Delta T + \frac{\sigma_1}{E_1} l = \alpha_2 \, l \, \Delta T - \frac{\sigma_2}{E_2} l \quad \text{...(4)}$$

From Eqs. (1) and (4), we can find stresses σ_1 and σ_2.

Problem 59 A steel tube of 50 mm outside diameter and 10 mm thick is fitted into a copper tube of inner diameter 50 mm and 10 mm thick. They are rigidly connected at the ends and the compound bar is free of stresses at normal temperature. If the length of the compound bar is 600 mm find the stresses produced in the tubes when temperature is raised by 25°C.

Take $\alpha_S = 12 \times 10^{-6}/°C \qquad E_S = 2 \times 10^5 \text{ N/mm}^2$

$\alpha_C = 17.5 \times 10^{-6}/°C \qquad E_C = 1.2 \times 10^5 \text{ N/mm}^2$

Solution: **Given data:**

Outside diameter of steel tube $d_{SO} = 50$ mm

Wall thickness = 10 mm

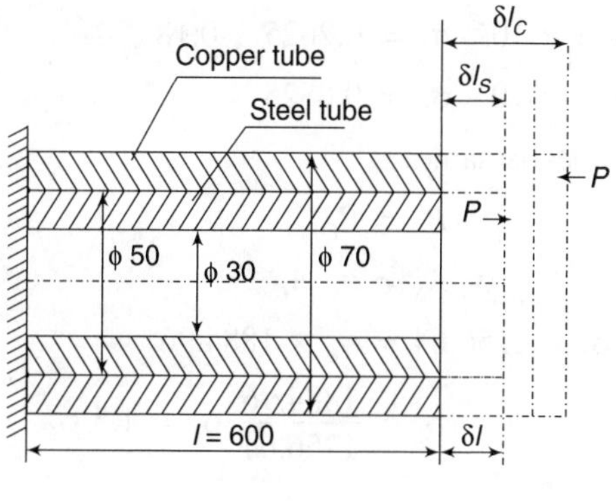

Fig. 1.56

$\therefore$ Inside diameter of steel tube $d_{Si} = 50 - 2 \times 10 = 30$ mm

Inside diameter of copper tube $d_{Ci} = 50$ mm

$$\text{Wall thickness} = 10 \text{ mm}$$

$$\therefore \text{ side diameter of copper tube } d_{Co} = 50 + 2 \times 10 = 70 \text{ mm}$$

$$\text{Length } l = 600 \text{ mm}$$

$$\text{Rise in temperature } \Delta T = 25°\text{C}$$

$$\text{Coefficient of linear expansion } \alpha_S = 12 \times 10^{-6}/°\text{C}$$

$$\alpha_C = 17.5 \times 10^{-6}/°\text{C}$$

$$\text{Modulus of elasticity } E_S = 2 \times 10^5 \text{ N/mm}^2$$

$$E_C = 1.2 \times 10^5 \text{ N/mm}^2$$

$$\text{Area of cross section of steel tube } A_S = \frac{\pi}{4}(50^2 - 30^2)$$

$$A_S = 1256.64 \text{ mm}^2$$

$$\text{Area of cross section of copper tube } A_C = \frac{\pi}{4}(70^2 - 50^2)$$

$$A_C = 1884.96 \text{ mm}^2$$

Since the coefficient linear expansion of copper is more, if allowed freely, copper tube will expand more than steel tube. But as the ends are rigidly fixed, they expand together.

As a result, the steal tube will be subjected to tenside stress and copper tube to compressive stress.

$$\delta l = \alpha_S \, l \, \Delta T + \frac{\sigma_S}{E_S} l = \alpha_C \, l \, \Delta T - \frac{\sigma_C}{E_C} l$$

$$12 \times 10^{-6} \times 600 \times 25 + \frac{\sigma_S}{2 \times 10^5} \times 600 = 17.5 \times 10^6 \times 600 \times 25 - \frac{\sigma_C}{1.2 \times 10^5} \times 600$$

$$0.18 + (3 \times 10^{-3} \times \sigma_S) = 0.2625 - (5 \times 10^{-3} \times \sigma_C)$$

$$3 \times 10^{-3} \, \sigma_S + 5 \times 10^{-3} \, \sigma_C = 0.2625 - 0.18$$

$$3 \times 10^{-3} \, \sigma_S + 5 \times 10^{-3} \, \sigma_C = 0.0825 \quad \text{...(1)}$$

But the load on both tubes is same

$$P_S = P_C$$

$$\sigma_S \, A_S = \sigma_C \, A_C$$

$$\sigma_S \times 1256.64 = \sigma_C \times 1884.96$$

$$\therefore \quad \sigma_S = \frac{1884.96}{1256.64} . \sigma_C = 1.5 \, \sigma_C \quad \text{...(2)}$$

Substituting in Eq. (1)

$$3 \times 10^{-3} \times 1.5 \, \sigma_C + 5 \times 10^{-3} \, \sigma_C = 0.0825$$

$$\sigma_C = 8.684 \text{ N/mm}^2$$

$$\sigma_S = 1.5 \, \sigma_C = 1.5 \times 8.684 = 13.026 \text{ N/mm}^2$$

Problem 60 A steel tube of 25 mm external diameter and 18 mm internal diameter encloses a copper rod of 15 mm diameter. The ends are rigidly fastened to each other. Calculate the stresses in the rod and tube when the temperature is raised from 5° to 200°C. Take $\alpha_S = 11 \times 10^{-6}/°C$, $\alpha_C = 18 \times 10^{-6}/°C$, E_S = 200 GPa and E_C = 100 GPa.

Solution: **Given data:**

Outside diameter of steel tube $d_{So} = 25$ mm

Inside diameter of steel tube $d_{Si} = 18$ mm

Diameter of copper rod $d_C = 15$ mm

Rise in temperature $\Delta T = T_2 - T_1 = 200 - 5 = 195°C \quad \alpha_S = 11 \times 10^{-6}/°C$

$\alpha_C = 18 \times 10^{-6}\ /°C$

$E_S = 200 \times 10^3$ MPa

$E_C = 100 \times 10^3$ MPa

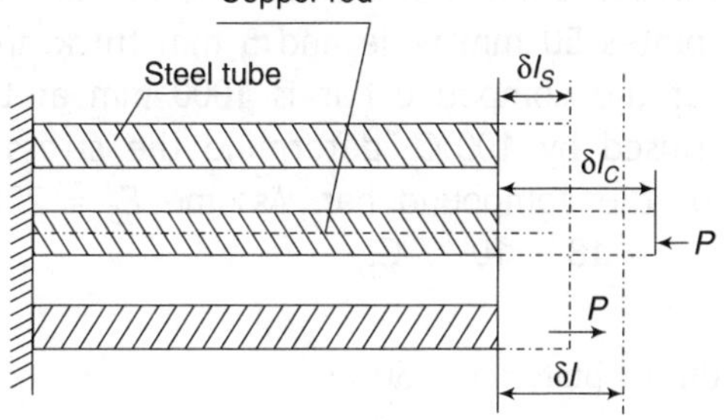

Fig. 1.57

$$\text{Area of steel tube } A_S = \frac{\pi}{4}(25^2 - 18^2) = 236.405 \text{ mm}^2$$

$$\text{Area of copper rod } A_C = \frac{\pi}{4}15^2 = 176.715 \text{ mm}^2$$

Since the coefficient of linear expansion of copper is more than that of steel, copper rod will expand more compared to steel tube.

Since the ends are fixed rigidly, their expansion will be equal. The steel tube will be pulled and subjected to tensile stress. The copper rod will be subjected to compressive stress as the length will reduce.

$$\delta l = \alpha_S\, l\, \Delta T + \frac{\sigma_S l}{E_S} = \alpha_C\, l\, \Delta T - \frac{\sigma_C}{E_C}\, l$$

Since the length is same,

$$\alpha_S\, \Delta T + \frac{\sigma_S}{E_S} = \alpha_C\, \Delta T - \frac{\sigma_C}{E_C}$$

$$11 \times 10^{-6} \times 195 + \frac{\sigma_S}{200 \times 10^3} = 18 \times 10^{-6} \times 195 - \frac{\sigma_C}{100 \times 10^3}$$

$$5 \times 10^{-6}\, \sigma_S + 10 \times 10^{-6}\, \sigma_C = 1.365 \times 10^{-3} \quad \ldots(1)$$

For the load being same, $P_S = P_C$

$$\sigma_S A_S = \sigma_C A_C$$

$$\sigma_S \times 236.405 = \sigma_C \times 176.715$$

$$\sigma_S = 0.7475\, \sigma_C \quad \ldots(2)$$

Substituting in Eq. (1)

$$5 \times 10^{-6} \times 0.7475\, \sigma_C + 10 \times 10^{-6}\, \sigma_C = 1.365 \times 10^{-3}$$

Stress in copper rod $\sigma_C = 99.36\ \text{N/mm}^2$

Stress in steel tube $\sigma_S = 0.7475\, \sigma_C = 0.7475 \times 99.36 = 74.27\ \text{mm}^2$

Problem 61 A compound bar is made of a central steel plate 50 mm wide and 10 mm thick to which copper plates 50 mm wide and 5 mm thick are connected rigidly on each side. The length of the compound bar is 1000 mm at the room temperature. If the temperature is raised by 100°C, determine the stress in each material and the change in length of the compound bar. Assume E_S = 200 GPa, E_C = 100 GPa, $\alpha_S = 12 \times 10^{-6}$/°C and $\alpha_C = 18 \times 10^{-6}$/°C.

Solution: **Given data:**

Width of plates b = 50 mm

Thickness of steel plate t_S = 10 mm

Thickness of each copper plate t_C = 5 mm

Length l = 1000 mm

Increase in temperature ΔT = 100°C

E_S = 200 GPa = 200×10^3 MPa

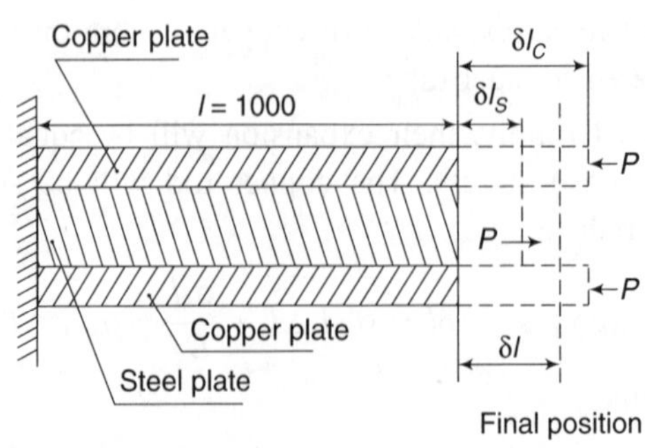

Fig. 1.58

$$E_C = 100 \text{ GPa} = 100 \times 10^3 \text{ MPa}$$

$$\alpha_S = 12 \times 10^{-6}/°\text{C}$$

$$\alpha_C = 18 \times 10^{-6}/°\text{C}$$

Area of cross section of steel plate $A_S = bt_S = 50 \times 10 = 500 \text{ mm}^2$

Area of cross section of two copper plates $= (bt_C) = 2 \times 50 \times 5 = 500 \text{ mm}^2$

Since the coefficient of linear expansion for copper is more than that of steel, copper plate will expand more compared to steel. As the ends are rigidly fixed, for common elongation, the steel plate is expanded and copper plates are compressed. The load is same on the plates.

For the same elongation, $\delta l_S = \delta l_C = \delta l$

$$\alpha_S l \, \Delta T + \frac{\sigma_S}{E_S} l = \alpha_C \, l \, \Delta T - \frac{\sigma_C}{E_C} \cdot l$$

$$12 \times 10^{-6} \times 1000 \times 100 + \frac{\sigma_S}{200 \times 10^3} \times 1000$$

$$= 18 \times 10^{-6} \times 1000 \times 100 - \frac{\sigma_C}{100 \times 10^3} \times 1000$$

$$1.2 + 5 \times 10^{-3} \, \sigma_S = 1.8 - 10 \times 10^{-3} \, \sigma_C$$

$$5 \times 10^{-3} \, \sigma_S + 10 \times 10^{-3} \, \sigma_C = 1.8 - 1.2 = 0.6 \qquad \ldots(1)$$

Since the load is same, $P_S = P_C$

$$\sigma_S \, A_S = \sigma_C \, A_C$$

$$\sigma_S \times 500 = \sigma_C \times 500$$

$$\sigma_S = \sigma_C \qquad \ldots(2)$$

Substituting in Eq. (1)

$$5 \times 10^{-3} \, \sigma_C + 10 \times 10^{-3} \, \sigma_C = 0.6$$

$$\therefore \qquad \sigma_C = 40$$

Stress in copper $= \sigma_C = 40$ MPa

Stress in steel $\sigma_S = \sigma_C = 40$ MPa

Change in length $\delta l = \delta l_C$ or δl_S (Anyone)

Consider $\delta l_S = \alpha_S \, l \, \Delta T + \frac{\sigma_S}{E_S} \cdot l$

$$= 12 \times 10^{-6} \times 1000 \times 100 + \frac{40}{200 \times 10^3} \times 1000$$

$$\delta l = 1.2 + 0.2 = 1.4 \text{ mm.}$$

Problem 62 A rigid block weighing 200 kN is symmetrically supported by two steel bars and one copper bar. There is an increase in temperature by 30°C. Compute

(i) Stress in two materials due to load.

(ii) Stress in two materials due to thermal change.

(iii) Stress in two materials due to combined effect of load and thermal change. The following data is given.

Materials	*Area of cross section mm²*	*Length, m*	*E, Young's Modulus, GPa*	*Coeff. of linear expansion, /°C*
Steel	600	0.6	200	12×10^{-6}
Copper	800	0.6	100	18×10^{-6}

Solution: **Given data:**

$$\text{Load } P_1 = 200 \text{ kN}$$
$$= 200 \times 10^3 \text{ N}$$
$$\text{Area of steel } A_S = 600\text{, two bars}$$
$$\text{Area of copper } A_C = 800$$
$$\text{Length } l = 600 \text{ mm}$$
$$E_S = 200 \times 10^3 \text{ MPa}$$
$$E_C = 100 \times 10^3 \text{ MPa}$$
$$\alpha_S = 12 \times 10^{-6} \text{ /°C}$$
$$\alpha_C = 18 \times 10^{-6} \text{ /°C}$$

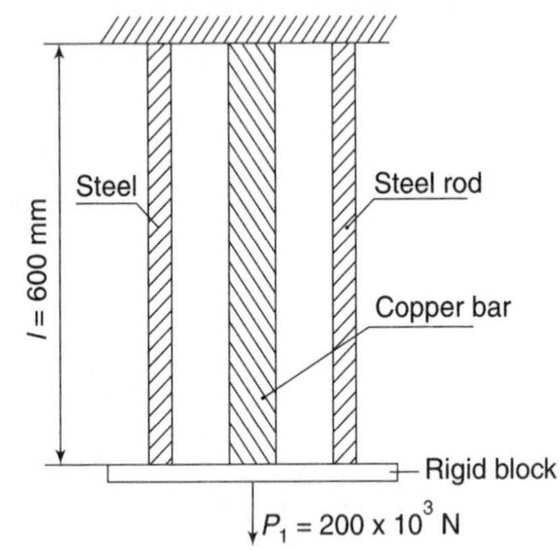

Fig. 1.59

Rise in temperature ΔT = 30°C.

(i) Stresses under static load $P_1 = 200 \times 10^3$ N

Since the ends are attached to the right block, the elongations under load P_1 will be same.

$$P_S + P_C + P_S = P_1 = 20 \times 10^3 \text{ N}$$
$$2P_S + P_C = 200 \times 10^3$$
$$\text{But } P = \sigma A$$
$$2\,\sigma_S A_S + \sigma_C A_C = 200 \times 10^3$$
$$2\,\sigma_S \times 600 + \sigma_C \times 800 = 200 \times 10^3$$
$$12\,\sigma_S + 8\,\sigma_C = 2000 \quad \ldots(1)$$

Since the strain is same (lengths are same)

$$\frac{\sigma_S}{E_S} = \frac{\sigma_C}{E_C}$$

$$\frac{\sigma_S}{200 \times 10^3} = \frac{\sigma_C}{100 \times 10^3}$$

$$\sigma_S = 2\sigma_C \quad \text{...(2)}$$

Substituting in Eq. (1)

$$12\,(2\sigma_C) + 8\,\sigma_C = 2000$$

$$\sigma_C = 62.5 \text{ N/mm}^2$$

$$\sigma_S = 2\sigma_C = 125 \text{ N/mm}^2$$

$$\text{Initital stress in steel bar} = \sigma_{S1} = 125 \text{ N/mm}^2$$

$$\text{In copper bar} = \sigma_{C1} = 62.5 \text{ N/mm}^2$$

(ii) Stress due to thermal change

Since the coefficient of linear expansion for copper is more than steel, copper will elongate more. For equilibrium, copper will compress and steel will elongate.

$$\alpha_S\, l\, \Delta T + \frac{\alpha_S}{E_S}\, l = \alpha_C\, l\, \Delta T - \frac{\sigma_C}{E_C}\, l$$

$$12 \times 10^{-6} \times 600 \times 30 + \frac{\sigma_S}{200 \times 10^3} \times 600$$

$$= 18 \times 10^{-6} \times 600 \times 30 - \frac{\sigma_C}{100 \times 10^3} \times 600$$

$$0.216 + 3 \times 10^{-3}\, \sigma_S = 0.324 - 6 \times 10^{-3}\, \sigma_C$$

$$3 \times 10^{-3}\, \sigma_S + 6 \times 10^{-3}\, \sigma_C = 0.108 \quad \text{...(3)}$$

Since the load is same, $P_S = P_C$

$$\sigma_S\, A_S = \sigma_C\, A_C$$

$$\sigma_S \times 2 \times 600 = \sigma_C \times 800 \qquad \text{(two steel rods)}$$

$$\sigma_S = 0.667\, \sigma_C \quad \text{...(4)}$$

Substituting in Eq. (3)

$$3 \times 10^{-3} \times 0.667\, \sigma_C + 6 \times 10^{-3}\, \sigma_C = 0.108$$

$$\sigma_C = 13.498 \text{ N/mm}^2 = \sigma_{C2}$$

$$\sigma_S = 0.667\, \sigma_C = 9 \text{ N/mm}^2 = \sigma_{S2}$$

(iii) Stresses under combined load are thermal expansion

$$\sigma = \sigma_{\text{Static}} + \sigma_{\text{Thermal}}$$

$$\text{For steel rod } \sigma_S = \sigma_{S1} + \sigma_{S2}$$

$$\sigma_S = 125 + 9.00 = 134 \text{ N/mm}^2$$

$$\text{For copper rod } \sigma_C = \sigma_{C1} + \sigma_{C2}$$

$$\sigma_C = 62.5 + 13.498 = 75.998 \text{ N/mm}^2.$$

Problem 63 A composite member made of steel and aluminium is held between two unyielding supports as shown in Fig. 1.60. The composite member is stress free at 38°C. What will be the stresses in two bars when the temperature reduces to 21°C. Also compute the stresses if the supports yield by 0.1 mm.

Given $E_S = 210 \text{ kN/mm}^2$, $E_A = 74 \text{ kN/mm}^2$

Coefficient of expansion of steel = 11.7×10^{-6} /°C

Coefficient of expansion of aluminium = 23.4×10^{-6} /°C

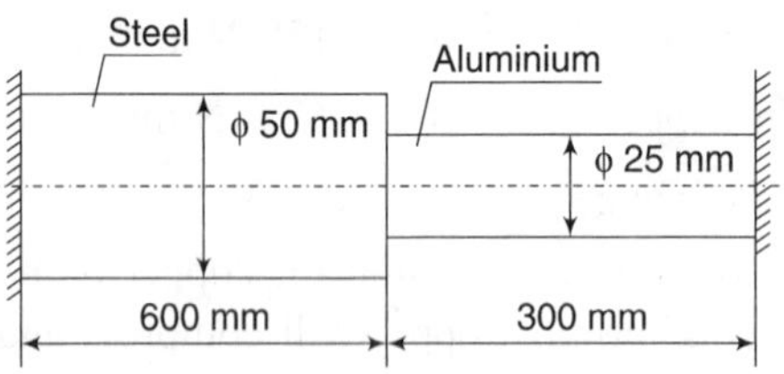

Fig. 1.60

Solution: **Given data:**

$$\text{Diameter of steel rod } d_S = 50 \text{ mm}$$

$$\therefore \quad \text{Area of cross section } A_S = \frac{\pi}{4} 50^2 = 1963.5 \text{ mm}^2$$

$$\text{Diameter of aluminium } d_A = 25 \text{ mm}$$

$$\therefore \quad \text{Area of cross section } A_A = \frac{\pi}{4} \times 25^2 = 490.87 \text{ mm}^2$$

$$\text{Length of steel rod } l_S = 600 \text{ mm}$$

$$\text{Length of aluminium rod } l_A = 300 \text{ mm}$$

$$E_S = 210 \text{ kN/mm}^2 = 210 \times 10^3 \text{ N/mm}^2$$

$$E_A = 74 \text{ kN/mm}^2 = 74 \times 10^3 \text{ N/mm}^2$$

$$\alpha_S = 11.7 \times 10^{-6}/°\text{C}$$

$$\alpha_A = 23.4 \times 10^{-6}/°\text{C}$$

Reduction in temperature $\Delta T = 38 - 21 = 17°\text{C}$. Due to reduction in temperature, the member will experience reduction in length.

Free contraction in both bars = $\alpha_S\, l_S\, \Delta T + \alpha_A\, l_A\, \Delta T$

$$\delta l = 11.7 \times 10^{-6} \times 600 \times 17 + 23.4 \times 10^{-6} \times 300 \times 17$$

$$= 0.23868 \text{ mm}$$

Let σ_S = Stress in steel bar and σ_A = Stress in aluminium bar.

(i) Since the ends are unyielding,

$$\delta l = \delta l_S + \delta l_A = \frac{\sigma_S}{E_S} l_S + \frac{\sigma_A}{E_A} l_A$$

$$0.23868 = \frac{\sigma_S}{210 \times 10^3} \times 600 + \frac{\sigma_A}{74 \times 10^3} \times 300$$

$$0.23868 = 2.857 \times 10^{-3} \sigma_S + 4.05405 \times 10^{-3} \sigma_A \quad \ldots(1)$$

The same load will act on the aluminium and steel bars

$$P_A = P_S$$

$$\sigma_A A_A = \sigma_S A_S$$

$$\sigma_A \times 490.87 = \sigma_S \times 1963.5$$

$$\sigma_A = 4 \sigma_S \quad \ldots(2)$$

Substituting in Eq. (1)

$$0.23868 = 2.857 \times 10^{-3} \sigma_S + 4.05405 \times 10^{-3} \times 4 \sigma_S$$

$$\therefore \quad \sigma_S = 12.514 \text{ N/mm}^2$$

$$\sigma_A = 4 \sigma_S = 4 \times 12.514 = 50.056 \text{ N/mm}^2$$

(ii) If the supports yield by $\delta = 0.1$ mm

$$\delta l = \delta l_S + \delta l_A - 0.1$$

$$= 0.23868 - 0.1 = 0.13868 \text{ mm}$$

Equation (1) will be

$$0.13868 = 2.857 \times 10^{-3} \sigma_S + 4.05405 \times 10^{-3} \sigma_A \quad \ldots(3)$$

From Eq. (2) $\quad \sigma_A = 4 \sigma_S$

Substituting in Eq. (3)

$$0.13868 = 2.857 \times 10^{-3} \sigma_S + 4.05405 \times 10^{-3} \times 4 \sigma_S$$

$$\therefore \quad \sigma_S = 7.27 \text{ N/mm}^2$$

$$\sigma_A = 4 \sigma_S = 29.08 \text{ N/mm}^2.$$

Problem 64 A rigid bar *AB* is hinged at *C* and has steel and aluminium bar supports at *A* and *B* respectively as shown in Fig. 1.61. The bars are stress free at room temperature. What stresses are induced in steel and aluminium bars if the temperature is raised by 40°C. Area of cross section of steel and aluminium bar are A_S = 1200 mm^2 and A_A = 800 mm^2. $E_S = 2 \times 10^5$ N/mm^2 and $E_A = 1 \times 10^5$ N/mm^2. Coefficient of linear expansion of steel and aluminium are $\alpha_S = 12 \times 10^{-6}$/°C and $\alpha_A = 23 \times 10^{-6}$/°C. *AC* = 1000 mm *CB* = 800 mm *AD* = 1000 mm. *BE* = 500 mm.

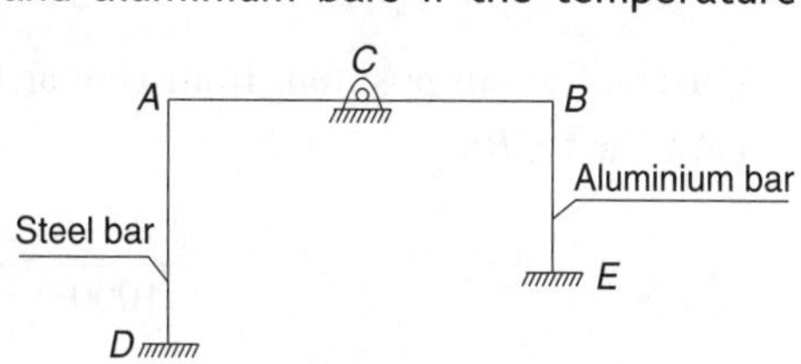

Fig. 1.61

Solution: **Given data:**

$$AD = 1000 \text{ mm} \qquad AC = 1000 \text{ mm}$$
$$BC = 800 \text{ mm} \qquad BE = 500 \text{ mm}$$
$$A_S = 1200 \text{ mm}^2 \qquad A_A = 800 \text{ mm}^2 \quad \Delta T = 40°\text{C}$$
$$E_S = 2 \times 10^5 \text{ N/mm}^2 \qquad E_A = 1 \times 10^5 \text{ N/mm}^2$$
$$\alpha_S = 12 \times 10^{-6}/°\text{C} \qquad \alpha_A = 23 \times 10^{-6}/°\text{C}.$$

Free expansion of steel rod

$$\delta l_S = \alpha_S \, l_S \, \Delta T$$
$$= 12 \times 10^{-6} \times 1000 \times 40 = 0.48 \text{ mm}$$

Free expansion of aluminium rod $\delta l_A = \alpha_A \, l_A \, \Delta T$

$$\delta l_A = 23 \times 10^{-6} \times 500 \times 40 = 0.46 \text{ mm.}$$

Since ABC is a rigid bar, A and B are not allowed to expand freely. Let δ_S be the compression in steel rod and δ_A in aluminium rod. Since the pivot C is not in the middle, the loads will not be equal. Only the moments of loads about C will be equal. Let P_S be the load on steel bar and P_A be the load on aluminium bar.

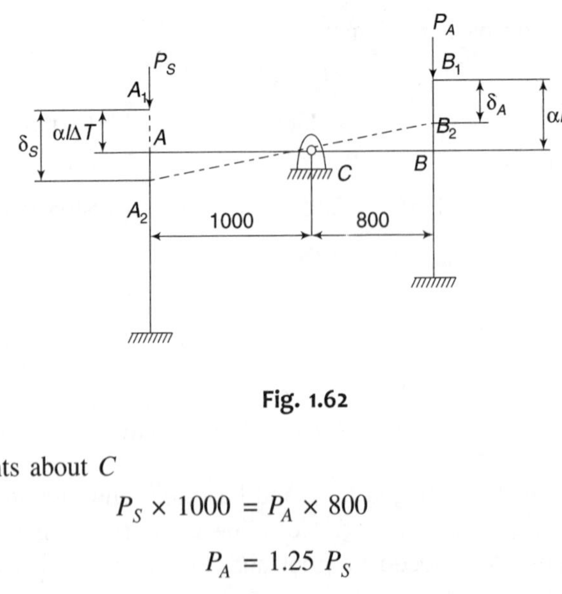

Fig. 1.62

Taking moments about C

$$P_S \times 1000 = P_A \times 800$$

$$\therefore \qquad P_A = 1.25 \, P_S \qquad \text{...(1)}$$

For equilibrium position, from similar triangles CAA_2 and CBB_2

$$\frac{AA_2}{1000} = \frac{BB_2}{800}$$

For equilibrium, the compatibility equation is $\dfrac{\delta_S - \alpha_S \, l_S \, \Delta T}{1000} = \dfrac{\alpha_A \, l_A \, \Delta T - \delta_A}{800}$

$$\frac{\left(\frac{P_{SS}}{A_S F_S} - 0.48\right)}{100} = \frac{\left(0.46 - \frac{P_A L_A}{A_A E_A}\right)}{800}$$

From Eq. (1) $\quad P_A = 1.25\ P_S$

$$\frac{P_S \times 1000}{1200 \times 2 \times 10^5} - 0.48 = \frac{1000}{800}\left[0.46 - \frac{1.25\ P_S \times 500}{800 \times 1 \times 10^5}\right]$$

$$4.167 \times 10^{-6}\ P_S - 0.48 = 0.575 - 9.766 \times 10^{-6}\ P_S$$

$$(4.167 \times 10^{-6} + 9.766 \times 10^{-6})\ P_S = 0.575 + 0.48$$

$$\therefore \quad P_S = 75.721.6 \text{ N}$$

$$P_A = 1.25\ P_S = 94.651.94 \text{ N}$$

$$\text{Stress in steel bar } \sigma_S = \frac{P_S}{A_S} = \frac{75.721.6}{1200} = 63.101 \text{ N/mm}^2$$

$$\text{Stress in aluminium bar } \sigma_A = \frac{P_A}{A_A} = \frac{94651.94}{800} = 118.315 \text{ N/mm}^2.$$

Problem 65 A rigid bar *ABC* is hinged at *A*. It is supported at *B* by a steel bar *BD* and at *C* by a copper bar *CE*. Find the stresses in the steel and copper bars if the temperature of the entire structure is increased by 50°C. $E_S = 2.1 \times 10^5$ N/mm², $E_C = 1.1 \times 10^5$ N/mm². Coefficient of linear expansion of steel $\alpha_S = 12 \times 10^{-6}$/°C and for copper is 18×10^{-6}/°C.

AB = 1000 mm, *BC* = 800 mm, *BD* = 1000 mm, *CE* = 500 mm, A_C = 720 mm², A_S = 450 mm².

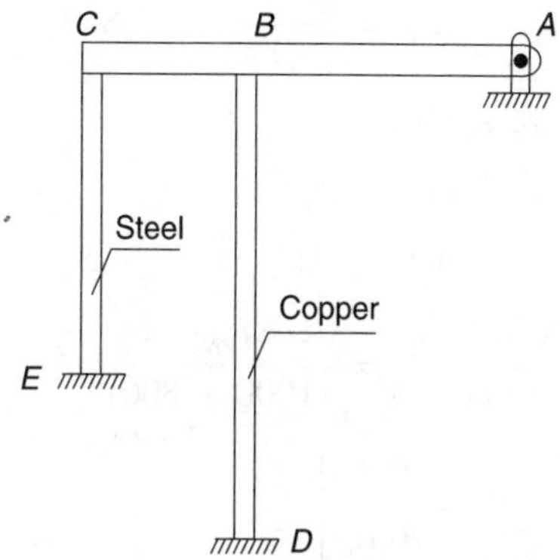

Fig. 1.63

Solution: **Given data:**

$$AB = 1000 \text{ mm} \qquad BC = 800 \text{ mm} \qquad BD = 1000 \text{ mm}$$
$$CE = 500 \text{ mm} \qquad \Delta T = 50°\text{C}$$
$$E_S = 2.1 \times 10^5 \text{ N/mm}^2 \qquad E_C = 1.1 \times 10^5 \text{ N/mm}^2$$
$$\alpha_S = 12 \times 10^{-6}/°\text{C} \qquad \alpha_C = 18 \times 10^{-6} \text{ N/mm}^2$$
$$\text{Area } A_S = 450 \text{ mm}^2 \qquad A_C = 720 \text{ mm}^2$$

Since α_C is greater than α_S, the copper bar *BD* will elongate more compared to steel rod *CE.* Since *ABC* is a rigid bar, the longer elongated copper bar will be subjected to compression (δ_C) and the steel will be subjected to tension (δ_s). The final position of *ABC* is shown in Fig. 1.64.

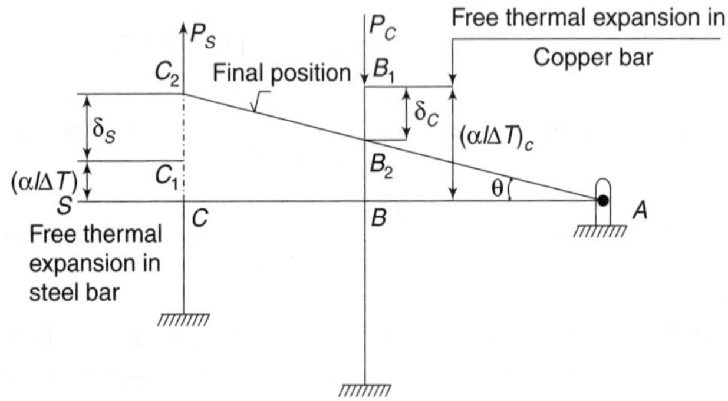

Fig. 1.64

Taking moments about *A* for equilibrium

$$P_S\ (1000 + 800) = P_C \times 1000$$

$$\therefore \qquad P_C = \frac{1800\ P_S}{1000} = 1.8\ P_S \qquad \ldots(1)$$

Triangle ABB_2 and ACC_2 are similar

$$\frac{BB_2}{AB} = \frac{CC_2}{AC} \Rightarrow \frac{BB_1 - B_1B_2}{AB} = \frac{CC_1 + C_1C_2}{AC}$$

$$\frac{(\alpha l \Delta T)_C - \delta_C}{1000} = \frac{(\alpha l \Delta T)_S + \delta_S}{(1000 + 800)}$$

$$\frac{1800}{100}\left[(18 \times 10^{-6} \times 1000 \times 50) - \frac{P_C l_C}{A_C E_C}\right]$$

$$= \left[(12 \times 10^{-6} \times 500 \times 50) + \frac{P_S l_S}{A_S E_S}\right]$$

$$1.8\left[0.9 - \frac{P_C l_C}{A_C E_C}\right] = \left[0.3 + \frac{P_S l_S}{A_S E_S}\right]$$

$$1.8 \times 0.9 - 0.3 = \frac{P_C l_C}{A_C E_C} + \frac{P_S l_S}{A_S E_S}$$

Substitute $P_C = 1.8\ P_S$ from Eq. (1)

$$1.32 = \frac{1.8\ P_S \times 1000}{720 \times 1.1 \times 10^5} + \frac{P_S \times 500}{450 \times 2.1 \times 10^5}$$

$$\therefore \quad P_S = 47\ 112.1\ \text{N}$$

$$P_C = 1.8\ P_S = 1.8 \times 47112.1 = 84801.8\ \text{N}$$

$$\text{Stress in steel bar } \sigma_S = \frac{P_S}{A_S} = \frac{47112.1}{450} = 104.69\ \text{N/mm}^2$$

$$\text{Stress in copper bar } \sigma_C = \frac{P_C}{A_C} = \frac{84801.8}{720} = 117.78\ \text{N/mm}^2.$$

Problem 66 A steel bar of uniformly varying diameter shown in Fig. 1.65 is held between two unyielding supports at room temperature. What is the maximum stress induced in the bar, if temperature rises by 40°C. Take $E_S = 2 \times 10^5$ N/mm^2 and $\alpha_S = 12 \times 10^{-6}$/°C.

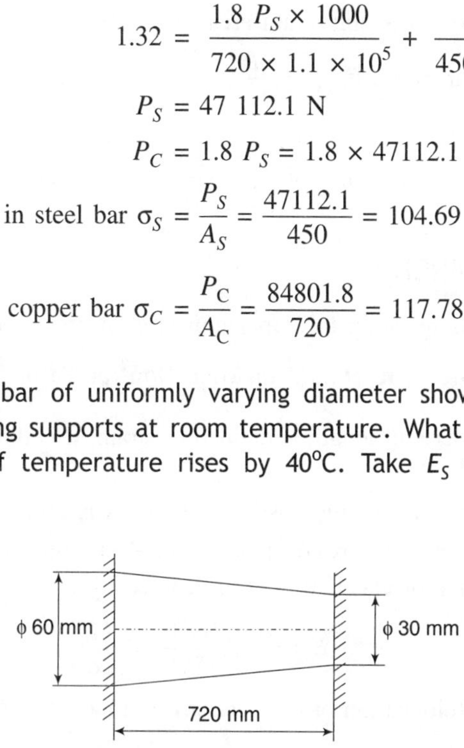

Fig. 1.65

Solution: **Given data:**

Length of taper bar $l = 720$ mm

Bigger diameter $d_2 = 60$ mm

Smaller diameter $d_1 = 30$ mm

$\alpha = 12 \times 10^{-6}$ /°C and $E = 2 \times 10^5$ N/mm^2

Increase in temperature $\Delta T = 40$°C

The ends are unyielding,

$$\text{For taper } \delta l = \frac{4Pl}{\pi d_1 d_2 E}$$

$$\text{Thermal expansion } \delta l = \alpha\, l\, \Delta T$$

$$\delta l = 12 \times 10^{-6} \times 720 \times 40 = 0.3456 \text{ mm}$$

$$\text{Equating, } 0.3456 = \frac{4 \times P \times 720}{\pi \times 30 \times 60 \times 2 \times 10^5}$$

$$\therefore \quad \text{Load } P = 13516.8 \text{ N}$$

Maximum stress will occur at minimum area

$$\therefore \quad \text{Maximum stress } \sigma_{max} = \frac{P}{A_{min}}$$

$$\sigma_{max} = \frac{135716.8}{\left(\frac{\pi}{4} \times 30^2\right)} = 192\,\text{MPa.}$$

THINGS TO REMEMBER

- Stress is defined as resisting force per unit area of cross section. It is expressed as $\sigma = \frac{P}{A}$. Units of stress are N/mm^2 or MPa, N/m^2 or *Pa* or GPa = 10^9 Pa.
- Change in length per unit length is defined as strain. It is dimensionless and denoted by $\in$.
- Hooke's law states that, within elastic limit, stress is proportional to strain.
- The ratio of stress by strain is constant within elastic limit and is known as Young's modulus or modulus of elasticity and is denoted by *E*

$$E = \frac{\sigma}{\in} = \frac{(P/A)}{(\delta l/l)} = \frac{Pl}{A.\delta l}.$$

- Change in length (elongation or contraction) of a bar of length *l*, under load *P*, having area of cross section *A* is given by $\delta l = \frac{Pl}{AE}$.
- The total change in the length of a bar of different length with different area of cross section subjected to an axial load *P* is given by

$$\delta l = \delta l_1 + \delta l_2 + \delta l_3$$

$$= P\left[\frac{L_1}{A_1E_1} + \frac{L_2}{A_2E_2} + \frac{L_3}{A_3E_3}\right]$$

If the material is same, then *E* will be same, then $\delta l = \frac{P}{E}\left[\frac{L_1}{A_1} + \frac{L_2}{A_2} + \frac{L_3}{A_3}\right]$.

- Total elongation of uniformly tapering round rod of diameter d_1 and d_2, length *l* subjected to an axial load *P* is given by

$$\delta l = \frac{4Pl}{\pi d_1 d_2 E}$$

- Total elongation of uniformly tapering plate of widths b_1 and b_2, length l and thickness t subjected to an axial load P is given by

$$\delta l = \frac{Pl}{tE(b_2 - b_1)} l_n \left(\frac{b_2}{b_1}\right).$$

- Principle of superpositioning states that if a body is acted upon by a number of loads, then the net effect is the sum of effects of each load acting independently on the body.
- Elongation of uniform bar due to its own weight is given by $\delta l = \frac{Wl}{2AE} = \frac{wl^2}{2E}$.
- Elongation of conical bar due to its own weight is given by $\delta l = \frac{Wl}{2AE} = \frac{wl^2}{6E}$

where, W = Total self-weight N and w = specific weight of material, N/mm^3.

- For compound bar, $\delta l_1 = \delta l_2$ or $\frac{P_1 l_1}{A_1 E} = \frac{P_2 l_2}{A_2 E_2}$ and total load $P = P_1 + P_2$.
- Elongation of a rod due to change in temperature

$$\delta l = \alpha\, l\, \Delta T$$

where α = Coefficient of linear expansion,
l = Length and ΔT = change in temperature.

- Thermal stress is induced if the elongation is prevented: $\sigma = \alpha\, E\, \Delta T$.
- A composite bar with ends rigidly fixed will expand by same length

$$\delta l_1 = \delta l_2$$

$$\alpha_1\, l\, \Delta T + \frac{\sigma_1}{E_1}\, l = \alpha_2\, l\, \Delta T - \frac{\alpha_2}{E_2}\, l$$

$$(\alpha_2 > \alpha_1)$$

Load on different members of a composite bar is same $P_1 = P_2$ or $\sigma_1 A_1 = \sigma_2 A_2$.

- Increase in length of a square taper bar of sides B and b of length l under a load P is $\delta l = \frac{Pl}{BbE}$.

EXERCISES

1. Define stress and strain and give their units.
2. Draw the stress-strain diagram for mild steel and explain salient points on that.
3. Derive the expression for elongation of a round taper bar.
4. Derive the expression for elongation of a rectangular taper plate.

5. Derive the expression for elongation of uniform rod under self weight.

6. Derive the expression for elongation of conical bar under its own weight.

7. Define the principle of superpositioning.

8. Explain the concept of thermal stress and give expression for elongation and thermal stress.

9. What is a composite bar? How do you calculate the stresses in each member?

10. A round rod of diameter 45 mm and length 600 mm is subjected to a tensile load of 60 kN. Modulus of elasticity is 2×10^5 N/mm^2. Find the stress, strain and change in length
(***Ans.*** 37.73 N/mm^2, 0.0001886, 0.1132 mm)

11. A round rod of diameter 60 mm and length 450 mm carries an axial pull of 90 kN. The elongation is found to be 0.12 mm. Find the stress, strain and Young's modulus of the material. (***Ans.*** 31.83 N/mm^2, 2.67×10^{-4}, 119.36×10^3 N/mm^2)

12. Find suitable diameter of a round rod of length 3 m subjected to an axial load of 12 kN if the elongation is to be limited to 0.09 mm. Take E = 210 GPa. (***Ans.*** 49.25 mm)

13. A round rod of diameters 90 mm and 60 mm and length 1.8 m is subjected to a compressive load of 150 kN. Find the stress, strain and change in length if Young's modulus is 1×10^5 N/mm^2. (***Ans.*** 42.44 N/mm^2, 0.0004244, 0.7639 mm)

14. A steel rod of circular cross section 1.5 m long having diameter of 50 mm is tested on UTM for distruction. It showed an extension of 0.345 mm under a load of 90 kN. The load at elastic limit is 540 kN and the breaking load is 900 kN. The total extension at the fracture is 142 mm. The diameter at the neck is 42 mm.

Find (i) The Young's modulus (ii) Percentage elongation
(iii) Stress at elastic point (iv) Ultimate stress
(v) True stress at fracture (vi) Percentage reduction in area.

(***Ans.*** 199.29×10^3 N/mm^2, 9.47%, 275.02 N/mm^2, 458.37 N/mm^2, 649.61 N/mm^2, 29.44%)

15. Find the elongation of a stepped bar subjected to a tensile load of 180 kN as shown in Fig. 1.66. Take E = 200 kN/mm^2.

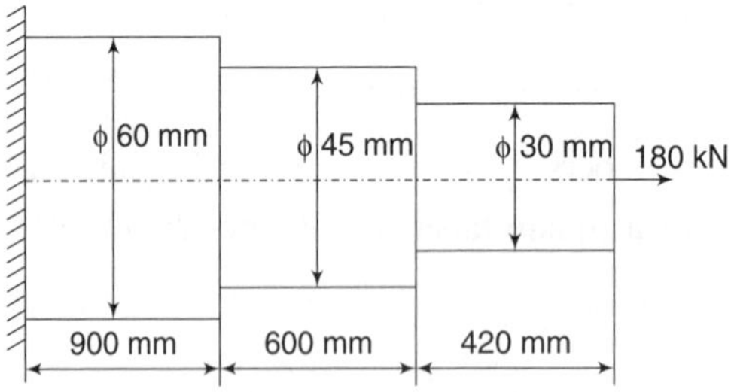

Fig. 1.66

(***Ans.*** 1.1608 mm)

16. Find the safe load that can be applied on to a stepped bar shown in Fig. 1.67 if the total elongation is limited to 1.2 mm. Take $E_S = 2 \times 10^5$ N/mm^2 $E_C = 1.2 \times 10^5$ N/mm^2 and $E_A = 0.8 \times 10^5$ N/mm^2.

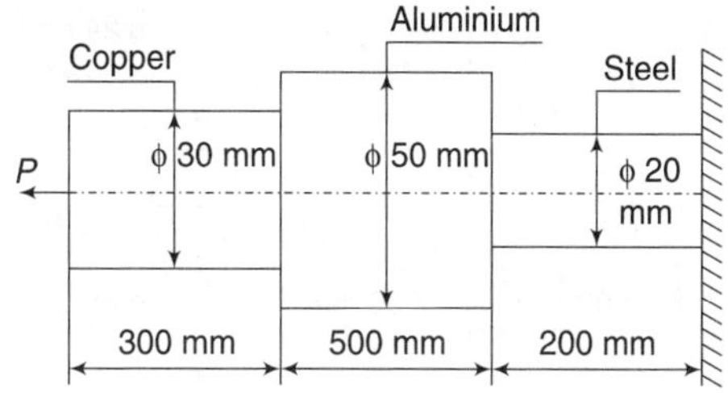

Fig. 1.67

(***Ans.*** $P = 121.176 \times 10^3$ N)

17. A bar of varying cross section made of steel, brass and zinc as shown in Fig. 1.68. Find the length of steel portion if the total elongation is to be united to 0.19 mm when subjected to a load of 150 kN. Take

$$E_S = 206 \text{ GPa}, E_B = 100 \text{ GPa and } E_t = 95 \text{ GPa}.$$

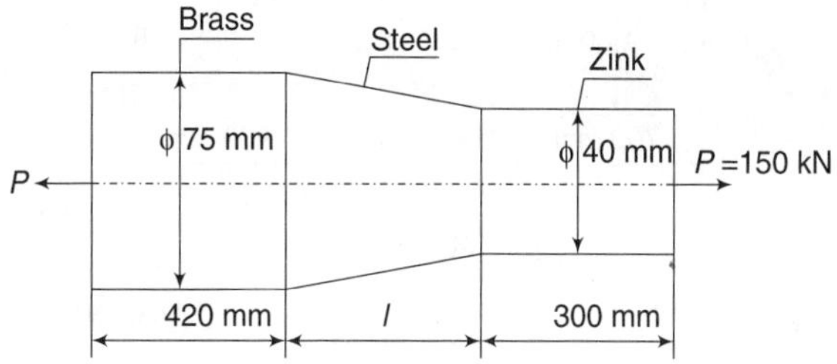

Fig. 1.68

(***Ans.*** l = 879.2 mm)

18. Find the change in length for a bar loaded as shown in Fig. 1.69. Modulus of elasticity for the material of rod is 2×10^5 N/mm^2.

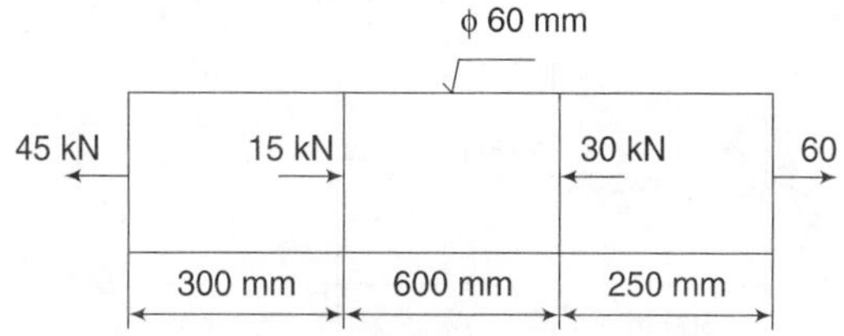

Fig. 1.69

(***Ans.*** 0.0822 mm)

19. Find the load P and change in length for a bar loaded as shown in Fig. 1.70. Modulus of elasticity for the material of rod is 210 GPa.

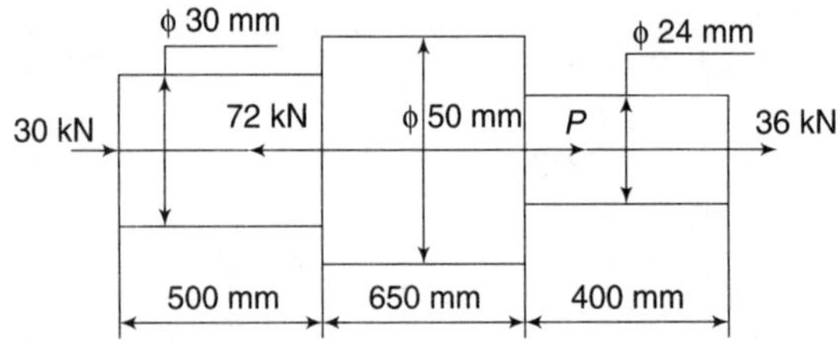

Fig. 1.70

(***Ans.*** $P = 6$ kN, $\delta l = 0.460$ mm)

20. Find the change in length for a composite bar shown in Fig. 1.71. Take $E_S = 2 \times 10^5$ MPa, $E_C = 1.2 \times 10^5$ MPa and $E_A = 0.8 \times 10^5$ MPa.

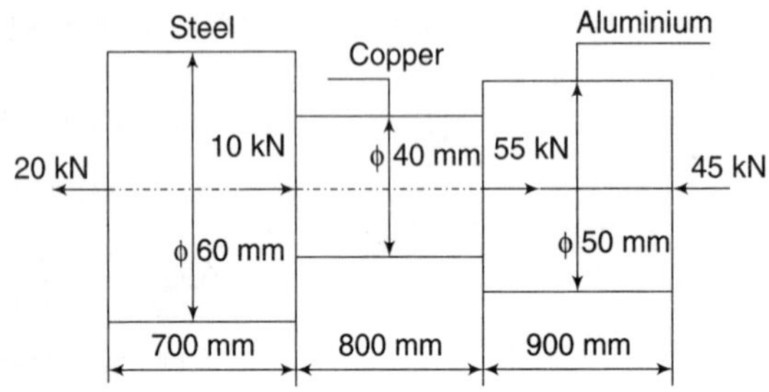

Fig. 1.71

[(***Ans.*** $\delta l = -\ 0.18$ mm contraction)]

21. A bar of 1 m length is held rigidly at A and B as shown in Fig. 1.72. A force of 45 kN is applied at C and 90 kN at D. Determine taking $E = 2 \times 10^5$ N/mm^2:

(i) Reactions at A and B

(ii) Change in length of each portion and

(iii) Stresses induced in each portion.

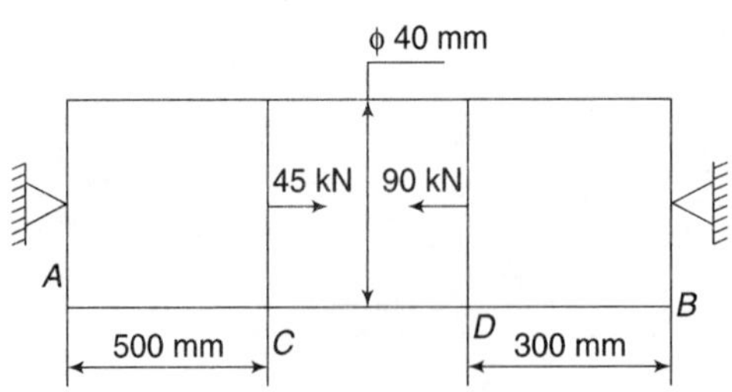

Fig. 1.72

(***Ans.*** $R_A = 4.5$ kN, $R_B = 40.5$ kN, $\delta l_1 = -\ 0.00895$ mm, $\delta l_2 = -\ 0.03939$ mm, $\delta l_3 = 0.04834$ mm)

22. Find the elongation of a rectangular bar 60 mm wide, 40 mm thicks and 3 m long under self weight. The specific weight of the material is 76 kN/m^3. Take $E = 2 \times 10^5$ MPa.

(***Ans.*** $\delta l = 0.00171$ mm)

23. Find the elongation of a conical bar 2.5 m long under self weight when it is suspended on its circular bars. The material weights 7800 kg/m^3. $E = 210$ GPa.

(***Ans.*** 0.00038 mm)

24. A compound bar of length 900 mm has a strip of aluminium 60 mm wide and 30 mm thick and a copper strip of 60 mm wide and 30 mm thick. The two strips are rigidly joined at its ends and carry a load of 60 kN. The modulus of elasticity of aluminium and copper are 0.9×10^5 MPa and 1.2×10^5 MPa respectively. Find stresses developed in strip and the elongation of composite bar. (***Ans.*** $\sigma_A = 17.65$ N/mm^2, $\sigma_C = 23.53$ N/mm^2, $\delta l = 0.1765$ mm)

25. A load of 30 kN is supported by two brass rods and one steel rod as shown in Fig. 1.73. The diameter of steel rod is 30 mm and that of each brass rod is 20 mm. Young's modulus of steel and brass are 2×10^5 N/mm^2 and 0.8×10^5 N/mm^2. Find the stresses developed in each rod.

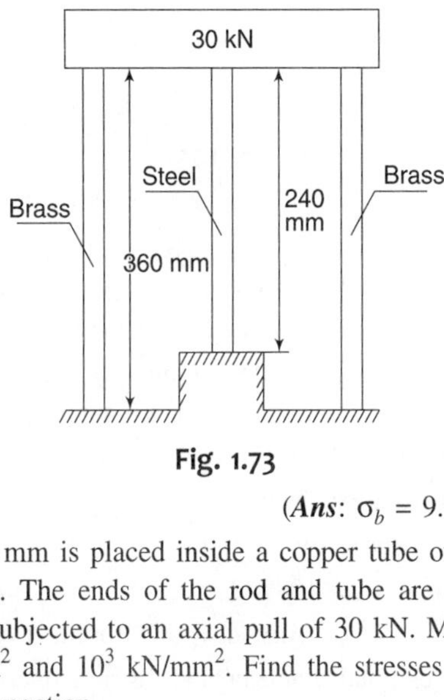

Fig. 1.73

(***Ans***: $\sigma_b = 9.15$ N/mm^2, $\sigma_S = 34.31$ N/mm^2)

26. A mild steel rod of 16 mm is placed inside a copper tube of diameters 20 mm and 30 mm and length of 500 mm. The ends of the rod and tube are rigidly joined together and the compound member is subjected to an axial pull of 30 kN. Moduli of elasticity for steel and copper are 206 kN/mm^2 and 10^3 kN/mm^2. Find the stresses developed in the steel rod and copper tube and the elongation.

(***Ans.*** $\sigma_S = 97.06$ N/mm^2, $\sigma_C = 48.53$ N/mm^2, $\delta l = 0.2356$ mm)

27. A rigid horizontal bar *AB* 1.5 m long is hinged at end *A*. It is supported by a steel and a copper rod and carries a load of 36 kN at the free end *B* as shown in Fig. 1.74. The diameter of steel and copper rods are 30 mm and 25 mm respectively. Find the reaction at *A*, load shared by each rod and the stress induced. Take $E_S = 210 \times 10^3$ N/mm^2, $E_C = 105 \times 10^3$ N/mm^2.

(***Ans:*** $R_B = 20.32$ kN, $P_C = 5.22$ kN, $P_S = 45.1$ kN, $\sigma_C = 10.638$ N/mm^2, $\sigma_S = 63.803$ N/mm^2)

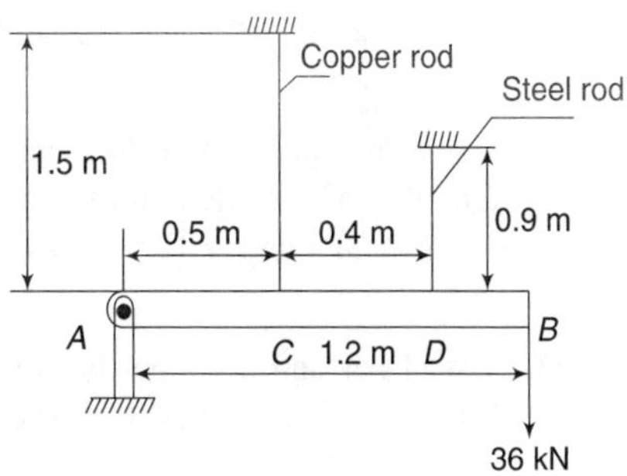

Fig. 1.74

28. A rigid plate is attached to the ends of three rods of equal length as shown in Fig. 1.75. A load of 18 kN is applied initially and the lengths are so adjusted that this initial load is equally shared by these three rods. The outer rods are of copper and the middle one is steel rod having cross-sectional areas as each 300 mm^2. A further load of 30 kN is applied to this assembly. Find the final stresses experienced by each rod. Take E_S = 200 kN/mm^2 and E_C = 100 kN/mm^2.

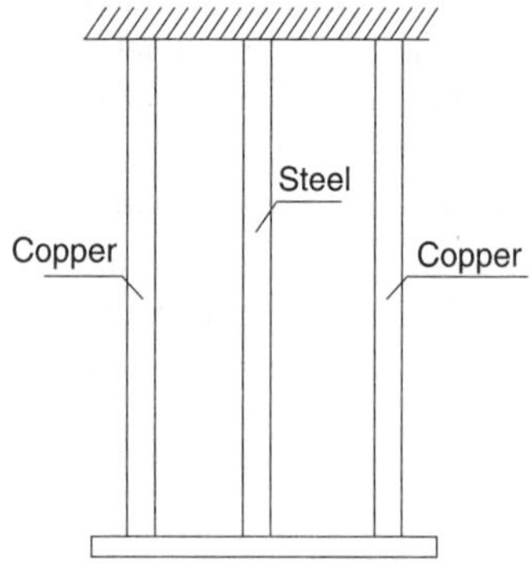

Fig. 1.75

(***Ans.*** σ_C = 45 N/mm^2, σ_S = 70 N/mm^2)

29. A 400 mm × 400 mm square RCC column supports a load of 480 kN. The column has 6 steel rods each 25 mm diameter. Find the stress in steel and concrete if E_S = 14 E_C. If the stress in the concrete is not to exceed 1.8 N/mm^2, what load can be applied the RCC column.

(***Ans.*** σ_S = 33.89 N/mm^2, σ_C = 2.42 N/mm^2, and P = 356.919 kN)

30. A copper rod 6 m long at 32°C is heated to 81°C. Find the free expansion of the rod. Also find the thermal stress when

(i) the rod is prevented from expansion, and

(ii) the rod is allowed to expand by 3.6 mm.

Take E = 120 GPa and α = 18 × 10^{-6}/°C.

(***Ans.*** δl = 5.292 mm, σ = 105.84 MPa, σ = 33.84 MPa)

31. A steel tube of 25 mm external diameter and 18 mm internal diameter encloses a copper rod of 15 mm diameter. The ends are rigidly fastened to each other. Calculate the stress in the rod and tube when the temperature is raised from 15° to 200°C. Take α_S = 12 × 10^{-6}/°C α_C = 18 × 10^{-6}/°C, E_S = 206 GPa, E_C = 103 GPa.

(***Ans.*** σ_S = 62.21 N/mm^2, σ_C = 83.22 N/mm^2)

32. A steel rod of 60 mm diameter is rigidly held in two fixtures at the end at 120°C. Determine the load exerted and the tensile stress developed if the temperature falls to 30°C. Take E = 207 GPa and α = 12 × 10^{-6}/°C.

(***Ans.*** P = 632.1 × 10^3 N, σ = 223.56 N/mm^2)

33. A steel bar and a copper bar of length 1 m each have their ends firmly joined together so as to form a compound bar. The cross sectional area of steel and copper bar are 1200 mm^2

and 1000 mm^2, respectively. Find the stress in each bar when the temperature of the unit is raised by 100°C. Also find the increase in length of the compound bar. What axial force must be applied to prevent this increase in length. Take E_S = 200 GPa, E_C = 120 GPa, $\alpha_S = 12 \times 10^{-6}$/°C and $\alpha_C = 18 \times 10^{-6}$/°C.

(***Ans.*** σ_C = 48 N/mm^2, σ_S = 40 N/mm^2, δl = 1.4 mm, $P = 504 \times 10^3$ N)

34. A steel flat of thickness 16 mm tapers uniformly from 80 mm to 50 mm over a length of 500 mm. If the flat is subjected to a load of 120 kN. Find the extension of the flat. Take $E = 2 \times 10^5$ MPa. Also calculate the percentage error, if average area is used for calculating its extension.

(***Ans:*** 0.47 mm, 1.8%)

35. A short piece of steel pipe is to carry an axial load of 900 kN. If the thickness of pipe is $\frac{1}{5}$th of its outside diameter, find the required diameter of the pipe. Take factor of safety as 2 and yield stress as 360 MPa.

(***Ans.*** d_i = 59.84 mm, d_o = 99.736 mm)

Objective Type Questions

1. Within elastic limit for a loaded member, the stress is

(a) equal to strain (b) inversely proportional to strain

(c) directly proportional to strain (d) twice the strain

2. Hooke's law is valid up to

(a) proportionality limit (b) yield point

(c) elastic limit (d) All the above

3. The change in length of a tensile specimen of length l, subjected to a load P, having area of cross section A and Young's modulus as E is

(a) $\frac{PA}{lE}$ (b) $\frac{PE}{lA}$

(c) $\frac{El}{PE}$ (d) $\frac{Pl}{AE}$

4. The change in length of a circular rod of diameter d, length l subjected to a load P, having Young's modulus E is

(a) $\frac{4Pl}{\pi d^2 E}$ (b) $\frac{4PE}{\pi d^2 l}$

(c) $\frac{\pi Pl}{4d^2 E}$ (d) $\frac{\pi PE}{4d^2 l}$

5. The change in length of a round taper rod of diameters d_1 and d_2 and length l, subjected to an axial load of P having Young's modulus as E is

(a) $\frac{4PE}{\pi l d_1 d_2}$ (b) $\frac{4Pl}{\pi d_1 d_2 E}$

(c) $\frac{\pi Pl}{4d_1d_2E}$ (d) $\frac{\pi PE}{4d_1d_2l}$

6. The change in length of a rectangular cross section rod of length l, width b and depth h, having Young's modulus as E carrying an axial load P is

(a) $\frac{4Pl}{bhE}$ (b) $\frac{4Pl}{\pi b^2E}$

(c) $\frac{4Pl}{\pi h^2E}$ (d) $\frac{Pl}{bhE}$

7. Elongation of uniform bar under self weight is

(a) $\frac{Wl}{AE}$ (b) $\frac{Wl}{2AE}$

(c) $\frac{2Wl}{AE}$ (d) $\frac{WA}{2El}$

8. Elongation of conical bar under self weight is

(a) $\frac{Wl}{AE}$ (b) $\frac{Wl}{6AE}$

(c) $\frac{Wl}{2AE}$ (d) $\frac{2Wl}{AE}$

9. Elongation due to change in temperature T a bar of length l, and coefficient of linear expansion α, Young's modulus E is given by

(a) αlT (b) αTE

(c) αlE (4) lTE

10. Thermal stress induced in a bar of length l, area of cross section A, modulus of elasticity E, and coefficient of linear expansion α is

(a) αlT (b) αAT

(c) αTE (d) αAl

11. A rigidly attached composite bar subjected to thermal stress, the one with higher coefficient of linear expansion will experience

(a) tensile stress. (b) compressive stress.

(c) shear stress. (d) bending stress

12. A rigidly attached composite bar subjected to thermal stress, the one with smaller coefficient of linear expansion will experience

(a) tensile stress (b) compressive stress

(c) shear stress (d) bending stress

13. Unit for stress in SI unit is

(a) kgf/m^2 (b) N/mm^2

(c) kg/m^3 (d) N/m^3

14. Unit for strain is

(a) kgf/m^2 (b) N/mm^2

(c) no units (d) N/mm

15. The resisting force per unit area of cross section is known as

(a) resultant force (b) stress

(c) strain (d) Young's modulus

16. Ratio of longitudinal stress to strain is

(a) rigidity modulus (b) bulk modulus

(c) Young's modulus (d) Poisson's ratio

17. The modulus of elasticity of steels is about

(a) 100 GPa (b) 120 GPa

(c) 200 GPa (d) 75 GPa

18. One Pa is equal to

(a) N/mm^2 (b) N/m^2

(c) kgf/m^2 (d) kgf/mm^2

19. One MPa is same as

(a) N/mm^2 (b) N/m^2

(c) N/cm^2 (d) N/m

20. Strain is defined as deformation per

(a) unit volume (b) unit force

(c) unit area (d) unit length

21. The coefficient of linear expansion of copper is 50% more than that of steel. The elongation of steel rod of same length under same raise in temperature with respect to copper bar is

(a) same (b) 50% more

(c) 50% less (d) $33\frac{1}{3}\%$ less

22. Two copper rods of cross-sectional area *A* and 2*A* of same length are heated to same temperature. The increase in length of rod of cross section *A* compound to 2*A* is

(a) double (b) half

(c) same (d) 50% more

23. The elongation of square taper rod of length *l* tapering uniformly from side *B* to side *b* under a load *P* is

(a) $\frac{4Pl}{BbE}$ (b) $\frac{Pl}{BbE}$

(c) $\frac{PE}{Bbl}$ (d) $\frac{4PE}{Bbl}$

24. A round rod of diameter *d* and a square rod of side *d* made of same material area length and carry same load. The ratio of change in lengths of round rod to square rod is

(a) same (b) $\frac{\pi}{4}$

(c) $\frac{4}{\pi}$ (d) None

Key To Answers

1. (c)	2. (a)	3. (d)	4. (a)	5. (b)	6. (d)	7. (b)	8. (d)
9. (a)	10. (c)	11. (b)	12. (a)	13. (b)	14. (c)	15. (b)	16. (c)
17. (c)	18. (b)	19. (a)	20. (d)	21. (d)	22. (c)	23. (b)	24. (c)

CHAPTER 2

Elastic Constants and Poisson's Ratio

2.1 INTRODUCTION

When a member is subjected to axial tensile load, there is an increase in the length of that member. At the same time there will be decrease in the lateral dimensions that at 90° to the line of action of load. Hence, the member will experience deformation in longitudinal direction as well as lateral directions. If the length in increasing as in the case of tensile load, there will be decrease in lateral dimensions. The lateral dimensions that are perpendicular to axis are width and thickness for a rectangular cross-sectional bar or diameter in case of a circular cross-sectional bar. There will also be change in volume of the bar.

In this chapter, we will study the lateral and longitudinal deformations, volumetric change of members under unidirectional stress and stresses on all three planes, Poisson's ratio, volumetric strain in prismatic bars, bulk modulus and shear modulus or modulus of rigidity. We will also study the relationship between the three elastic constants, viz., Young's modulus E, rigidity modulus G, bulk modulus K and Poisson's ratio μ.

2.2 LATERAL STRAIN AND POISSON'S RATIO

It is seen that in case of a tensile force, the length increases and in a compressive force, the length decreases. This axial increase or decrease results in change of lateral dimensions of the body at right angles to the line of action of the applied force. Hence, any longitudinal deformation will cause a corresponding lateral deformation. If an axial tensile force is applied, then the length will increase resulting in decrease in its lateral dimensions like width and thickness or diameter. For an axial compressive force, the length will reduce and the lateral dimension will increase. This is known as lateral deformation.

The lateral strain is defined as the strain produced in the lateral direction perpendicular to the direction of applied force.

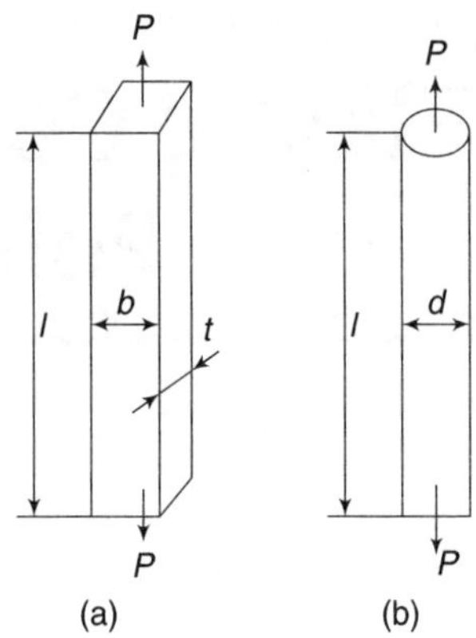

Fig. 2.1

In other words, for an applied axial load, the ratio of change in width to its original width, change in thickness to its original thickness or change in diameter to its original diameter is defined as lateral strains.

For a member of rectangular cross section of width b and thickness t (Fig. 2.1a).

Lateral strain $$\epsilon_b = \frac{\delta b}{b}$$

$$\epsilon_t = \frac{\delta t}{t}$$

For a member of circular cross section of diameter d (Fig. 2.1b).

$$\text{Lateral strain } \epsilon_d = \frac{\delta d}{d}$$

Poisson's Ratio. The ratio of lateral strain to the longitudinal strain is defined as Poisson's ratio. It is denoted by μ. (It is also denoted by $\frac{1}{m}$ or v). For elastic materials the Poisson's ratio is constant.

$$\text{Poisson's ratio } \mu = \frac{\text{Lateral strain}}{\text{Longitudinal strain}}$$

For most of the metals the value of Poisson's ratio varies between 0.25 to 0.33. It is generally taken as 0.3 for steels and 0.15 for concrete. For rubber, Poisson's ratio will be in the range of 0.45 to 0.5. Poisson's ratio has a maximum value of less than or equal to 0.5.

2.3 SIMPLE SHEAR STRESS, SHEAR STRAIN AND SHEAR MODULUS

2.3.1 Shear Stress

When two equal and opposite forces act on a body, not in the same line then shear takes place.

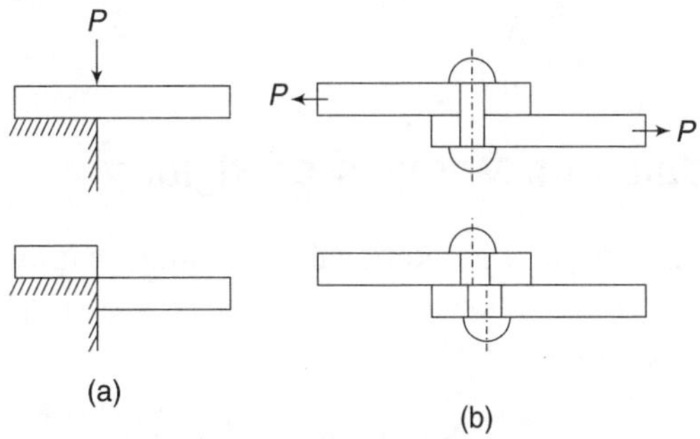

Fig. 2.2 Simple shear

Figure 2.2(a) shows a simple shear of a member subjected to a transverse force. Figure 2.2(b) shows a riveted joint subjected to a simple shear. It is seen that one plate slides over another due to two equal and opposite forces that are applied causing the rivet to experience a stress known as shear stress.

This force tends to shear the rivet into two pieces as shown. The stress induced under this condition is termed as shear stress and is denoted by τ.

$$\text{Shear stress } \tau = \frac{\text{Shear resistance}}{\text{Shear area}}$$

$$\text{Shear resistance} = \text{Applied shear force} = P$$

$$\therefore \qquad \tau = \frac{P}{A}$$

2.3.2 Shear Strain

Consider a rectangular body of length l and height h subjected to a shear force P as shown in Fig. 2.3.

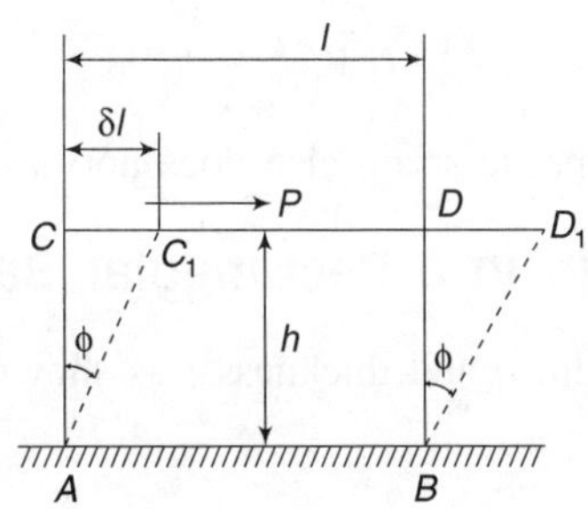

Fig. 2.3

Assuming the bottom of the block is fixed, the upper surface slides relative to the lower surface and the corner angles will distort by a small angle ϕ.

Shear strain is defined as this change in angle measured in radians ϕ.

$$\text{Shear strain } \phi = \frac{\delta l}{h}$$

$$\left(\text{For small angles } \tan\phi = \phi = \frac{\delta l}{h}\right)$$

2.3.3 Shear Modulus or Modulus of Rigidity

For elastic materials, shear stress is proportional shear strain within elastic limit. The ratio of shear stress to shear strain is known as shear modulus or modulus of rigidity and is denoted by G (also denoted by C or N).

$$\text{Rigidity modulus } G \text{ (or } C \text{ or } N) = \frac{\text{Shear stress}}{\text{Shear strain}}$$

$$G = \frac{\tau}{\phi}$$

This has same units as stress since shear strain has no units.

2.4 VOLUMETRIC STRAIN AND BULK MODULUS

2.4.1 Volumetric Strain

When a member is subjected to forces either in one, two or all the three mutually perpendicular directions, it undergoes deformation in all the three directions, resulting in change in volume. The ratio of change in volume to original volume is defined as volumetric stain.

$$\text{Volumetric strain } e_V = \frac{\text{Change in volume}}{\text{Original volume}}$$

$$e_V = \frac{\delta V}{V}$$

Like the other strains, volumetric strain also does not have any unit.

2.4.2 Volumetric Strain in a Rectangular Bar

Consider a bar of length l, width b and thickness t as shown in Fig. 2.4.

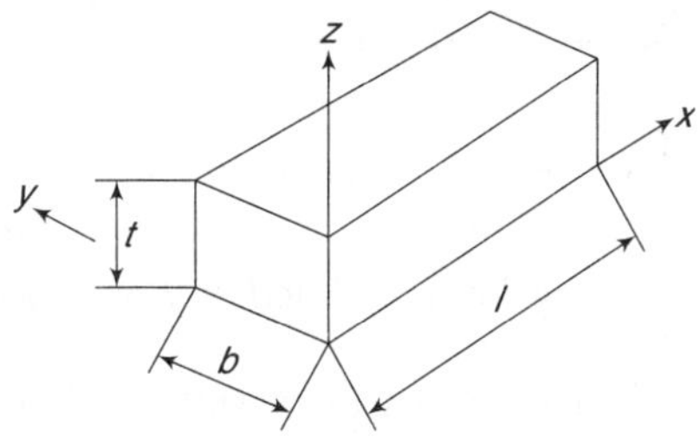

Fig. 2.4

$$\text{Volume } V = lbt$$

Let δl be the change in length,

δb be the change in width, and

δt be the change in thickness.

$$\text{New volume} \quad V^1 = (l + \delta l)(b + \delta b)(t + \delta t)$$

$$V^1 = lbt + lb\delta t + bt\delta l + lt\delta b + l\delta b\delta t + b\delta l\delta t + t\delta l\delta b + \delta l.\delta b.\delta t$$

Since the change in dimension δl, δb and δt are small, their products will be very small and are neglected.

$$V^1 = lbt + lb\delta t + bt\delta t + lt\delta b$$

$\therefore$ Change in volume

$$\delta V = V^1 - V$$

$$= lbt + lb\delta t + bt + \delta l + lt\delta b - lbt$$

$$\delta V = lb\delta t + bt\delta l + lt\delta b$$

$\therefore$ Volumetric strain

$$\in_V = \frac{\delta V}{V} = \frac{lb\delta t + bt\delta l + lt\delta b}{lbt}$$

$$\delta_V = \frac{\delta l}{l} + \frac{\delta b}{b} + \frac{\delta t}{t}$$

$$e_V = \in_l + \in_b + \in_t$$

$$e_V = \in_x + \in_y + \in_z$$

Thus, volumetric strain is the algebraic sum of strains in three mutually perpendicular directions.

2.4.3 Volumetric Strain in a Circular Bar

Consider a circular rod of diameter d and length l as shown in Fig. 2.5.

$$\text{Volume } V = \frac{\pi}{4} d^2 l$$

Let δl be the change in length

δd be the change in diameter

New volume $$V^1 = \frac{\pi}{4}(d + \delta d)^2 \, (l + \delta l)$$

$$= \frac{\pi}{4}[(d^2 + (\delta d)^2 + 2d\delta d) \, (l + \delta l]$$

$$V^1 = \frac{\pi}{4} \, [ld^2 + l(\delta d)^2 + 2ld\delta d + d^2\delta l + \delta e(\delta d)^2 + 2d\delta l\delta d]$$

Since δl and δd are small, their product will be very small and are neglected

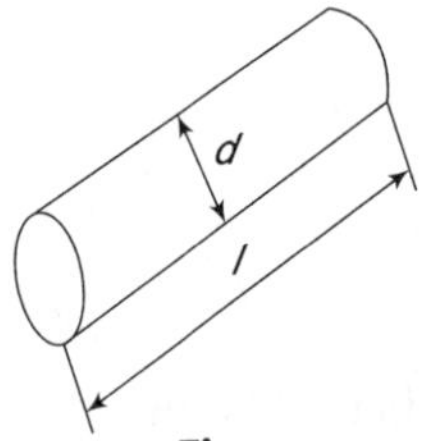

Fig. 2.5

$$V^1 = \frac{\pi}{4} \, (ld^2 + 2ld\delta d + d^2\delta l)$$

Change in volume $\delta V = V^1 - V$

$$\delta V = \frac{\pi}{4} \, ld^2 + \frac{\pi}{4} \, 2ld\delta d + \frac{\pi}{4} \, d^2\delta l - \frac{\pi}{4}d^2 l$$

$$\delta V = \frac{\pi}{4} \, 2ld\delta d + \frac{\pi}{4} \, d^2\delta l$$

∴ Volumetric strain $$\in_V = \frac{\delta V}{V}$$

$$= \frac{\frac{\pi}{4}d^2 l\left(\frac{2\delta d}{d}\right) + \frac{\pi}{4}d^2 l\left(\frac{\delta l}{l}\right)}{\frac{\pi}{4} \, d^2 l}$$

$$\in_V = \frac{\delta l}{l} + \frac{2\delta d}{d}$$

$$\in_V = \in_l + 2\in_d$$

$$\in_V = \in_x + \in_y + \in_z$$

Since $\in_y = \in_z = \dfrac{\delta d}{d}$.

Tensile strain is taken as positive and compressive strain as negative.

Consider a rod subjected to tensile stress then the length will increase and the lateral dimension (width and thickness for rectangular cross reaction and diameter for round cross section) will decrease.

∴ $$\in_V = \in_x - \in_y - \in_z$$

$$= \in_l - \in_b - \in_t$$

$$= \frac{\delta l}{I} - \frac{\delta b}{b} - \frac{\delta t}{t} \qquad \text{...(1)}$$

We know that positions ratio $\mu = \dfrac{\text{Lateral strain}}{\text{Longitudinal strain}}$

For rectangular cross section rod,

$$\mu = \frac{\in_b}{\in_l} \text{ or } = \frac{\in_t}{\in_l}$$

$$= \frac{(\delta l/b)}{(\delta l/l)} \text{ or } \frac{(\delta t/t)}{(\delta l/l)}$$

$$\therefore \quad \frac{\delta b}{b} = \frac{\mu \cdot \delta l}{l} \text{ and } \frac{\delta t}{t} = \frac{\mu \cdot \delta l}{l}$$

Substituting in Eq. (1)

$$\in_V = \frac{\delta l}{l} - \frac{\mu \cdot \delta l}{l} - \frac{\mu \delta l}{l}$$

$$\in_V = \frac{\delta l}{l}(1 - 2\mu) \text{ or } \in_V = \frac{\sigma}{E}(1 - 2\mu)$$

$$\left(\text{since } \frac{\delta l}{l} = \frac{\sigma}{E}\right)$$

For round rod:

$$\mu = \frac{\in_d}{\in_l} = \frac{(\delta d/d)}{(\delta l/l)}$$

$$\therefore \quad \frac{\delta d}{d} = \frac{\mu \cdot \delta l}{l}$$

$$\in_V = \in_l - \in_d - \in_d$$

$$= \in_l - 2\in_d$$

$$= \frac{\delta l}{l} - 2\mu \frac{\delta l}{l}$$

$$\in_V = \frac{\delta l}{l}(1 - 2\mu)$$

But strain $\frac{\delta l}{l} = \frac{\sigma}{E}$

$$\in_V = \frac{\sigma}{E}(1 - 2\mu).$$

2.4.4 Maximum Value of Poisson's Ratio

For Poisson's ratio μ to be maximum, volumetric strain $\in_V$ should be minimum.

$$\in_V = 0 = \frac{\sigma}{E}(1 - 2\mu)$$

Since for a stressed material, stress $\sigma \neq 0$, and Young's modulus E is also $\neq 0$.

$$1 - 2\mu = 0$$

or maximum $\mu = \frac{1}{2} = 0.5$

Hence, for any material, Poisson's ratio cannot be greater than 0.5.

2.4.5 Volumetric Strain in a Rectangular Bar Subjected to the Mutually Perpendicular Stresses

Figure 2.6 shows a block of rectangular cross section of width y and thickness z, having length x, subjected to direct stresses σ_x, σ_y and σ_z along three mutually perpendicular axes.

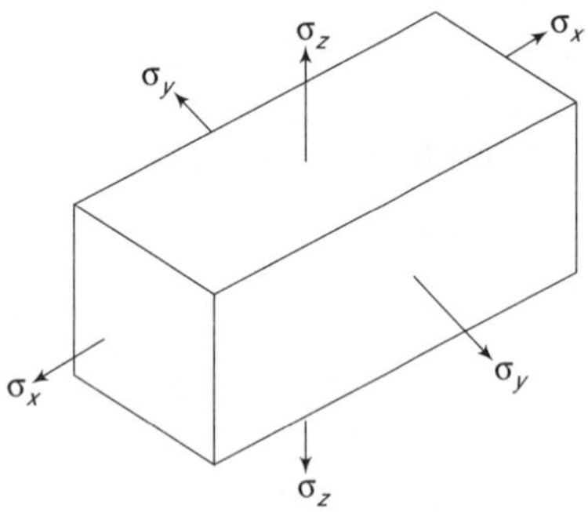

Fig. 2.6

Volume of block $\quad V = x \cdot y \cdot z$

Differentiating, $\quad \delta V = \delta x \cdot yz + \delta y \cdot zx + \delta z \cdot xy$

$$\therefore \text{ Volumetric strain } \quad \frac{\delta V}{V} = \frac{\delta x \cdot yz + \delta y \cdot zx + \delta z \cdot xy}{xyz}$$

$$\in_V = \frac{\delta x}{x} + \frac{\delta y}{y} + \frac{\delta z}{z}$$

$$\in_V = \in_x + \in_y + \in_z$$

where $\in_x$ = Total strain in x direction, $\in_y$ = Total strain in y-direction and $\in_z$ = Total strain in z-direction.

Strain $\in_x$ = Tensile strain due to σ_x + compressive strain due to σ_y in x direction + compressive strain due to σ_z in x-direction,

We know that strain $\in = \dfrac{\sigma}{E}$

Tensile strain in x direction due to $\sigma_x = \dfrac{\sigma_x}{E}$

Compressive strain in x direction due to $\sigma_y = -\dfrac{\mu.\sigma_y}{E}$

Compressive strain in x direction due to $\sigma_z = -\dfrac{\mu\sigma_z}{E}$

$$\therefore \quad \in_x = \frac{\sigma_x}{E} - \frac{\mu\sigma_y}{E} - \frac{\mu\sigma_z}{E}$$

Similarly, $\quad \in_y = \dfrac{\sigma_y}{E} - \dfrac{\mu\sigma_x}{E} - \dfrac{\mu\sigma_y}{E}$ and $\in_x = \dfrac{\sigma_z}{E} - \dfrac{\mu\sigma_x}{E} - \dfrac{\mu\sigma_y}{E}$

Adding all these, $\quad \in_V = \in_x + \in_y + \in_z$

$$= \left(\frac{\sigma_x}{E} - \frac{\mu\sigma_y}{E} - \frac{\mu\sigma_z}{E}\right) + \left(\frac{\sigma_y}{E} - \frac{\mu\sigma_x}{E} - \frac{\mu\sigma_z}{E}\right) + \left(\frac{\sigma_z}{E} - \frac{\mu\sigma_x}{E} - \frac{\mu\sigma_y}{E}\right)$$

$$= \frac{\sigma_x}{E}(1 - 2\mu) + \frac{\sigma_y}{E}(1 - 2\mu) + \frac{\sigma_z}{E}(1 - 2\mu)$$

$$\in_V = \frac{(\sigma_x + \sigma_y + \sigma_z)}{E}(1 - 2\mu)$$

Note: If any of three stresses σ_x, σ_y or σ_z is compressive, it is to be taken as –ve in the above equation.

$\therefore$ Change in volume $\delta V = \in_V \cdot V$

For rectangular block volume $V = lbt$

For circular rod volume $v = \dfrac{\pi}{4} d^2 l$

If δv is positive, then there is increase in volume

If δv is negative, then there is decease in volume.

2.4.6 Bulk Modulus

When a body is subjected to mutually perpendicular like and equal direct stresses (σ), then the ratio of direct equal stress that is acting on three mutually perpendicular directions to volumetric strain is defined as bulk modulus (K).

This direct like stress is proportional to volumetric strain width in elastic limit.

$$\text{Bulk modulus } K = \frac{\sigma}{\left(\dfrac{\delta V}{V}\right)}$$

Bulk modulus is more appropriate to fluid problems where the fluid or hydraulic pressure is equal in all directions.

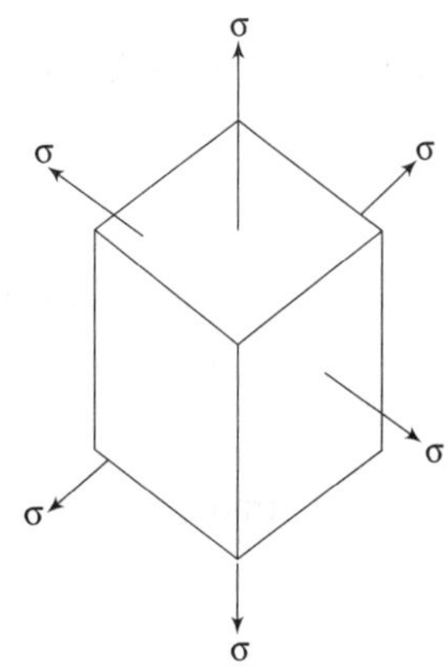

Fig. 2.7

2.5 ELASTIC CONSTANTS

Young's modulus or modulus of elasticity (E), rigidity modulus or shear modulus (G) and bulk modulus (K) are known elastic constants.

2.5.1 Relation between Young's Modulus and Bulk Modulus

Consider a cube of side l subjected to equal direct stress σ in the three mutually perpendicular directions x, y and z as shown in Fig. 2.8.

The stress σ in x direction causes tensile strain $\dfrac{\sigma}{E}$ in x direction and compressive strain of $-\mu\dfrac{\sigma}{E}$ in y and z directions.

Similarly, the stress σ in y direction induces tensile strain $\dfrac{\sigma}{E}$ in y direction and compressive strain $-\mu\dfrac{\sigma}{E}$ in x and z directions and stress σ in z direction induces tensile strain $\dfrac{\sigma}{E}$ in z direction and compressive strain $-\mu\dfrac{\sigma}{E}$ in x and y directions. The same is tabulated below.

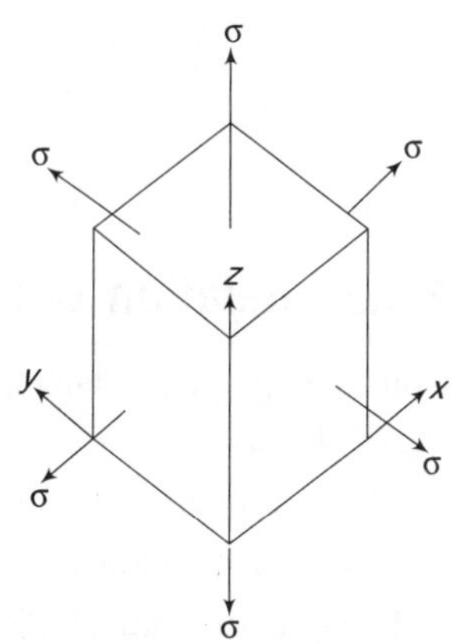

Fig. 2.8

Stress	Strain in		
	Direction x	Direction y	Direction z
σ in x direction	$\frac{\sigma}{E}$	$-\mu\frac{\sigma}{E}$	$-\frac{\mu\sigma}{E}$
σ in y direction	$-\mu\frac{\sigma}{E}$	$\frac{\sigma}{E}$	$-\mu\frac{\sigma}{E}$
σ in z direction	$-\mu\frac{\sigma}{E}$	$-\mu\frac{\sigma}{E}$	$\frac{\sigma}{E}$

$\therefore$ Total strain in x direction $\in_x = \frac{\sigma}{E} - \frac{\mu\sigma}{E} - \frac{\mu\sigma}{E}$

$$\in_x = \frac{\sigma}{E}(1 - 2\mu)$$

Total strain in y direction $\in_y = \frac{\sigma}{E} - \frac{\mu\sigma}{E} - \frac{\mu\sigma}{E}$

$$\in_y = \frac{\sigma}{E}(1-2\mu)$$

Total strain in z direction $\in_z = \frac{\sigma}{E} - \frac{\mu\sigma}{E} - \frac{\mu\sigma}{E}$

$$\in_z = \frac{\sigma}{E}(1 - 2\mu)$$

$\therefore$ Volumetric strain $\in_V = \in_x + \in_y + \in_z$

$$\in_V = \frac{\delta v}{v} = \frac{\sigma}{E}(1 - 2\mu) + \frac{\sigma}{E}(1 - 2\mu) + \frac{\sigma}{E}(1 - 2\mu)$$

$$\in_V = 3\frac{\sigma}{E}(1 - 2\mu)$$

Bulk modulus $K = \frac{\sigma}{\in_v} = \frac{\sigma}{\left[\frac{3\sigma}{E}(1 - 2\mu)\right]}$

$$K = \frac{E}{3(1 - 2\mu)}$$

$\therefore$ $E = 3K(1 - 2\mu)$.

2.5.2 Relation between Young's Modulus and Rigidity Modulus

Consider a square element $ABCD$ of side l subjected to pure shear stress τ as shown in Fig. 2.9.

The square $ABCD$ deforms to ABC_1D_1 under the shear stress τ.

Diagonal AC elongates and BD compresses.

Draw a perpendicular CC_2 from C to AC_1.

Let ϕ be the shear strain and G be the rigidity modulus.

Increase in length of diagonal $AC = AC_1 - AC$

For small angles of ϕ, $AC = AC_2$

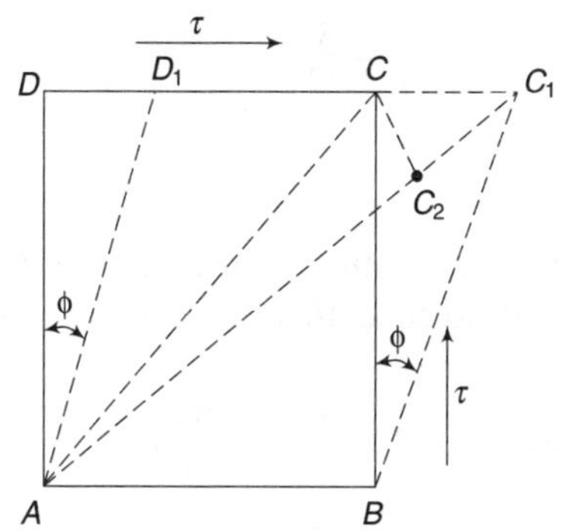

Fig. 2.9

Substituting, increase in length of $AC = AC_1 - AC_2 = C_1C_2$

Since the angle ϕ is very small, angle C_2C_1C may be assumed as 45°

From triangle CC_1C_2, $\quad C_1C_2 = CC_1 \cos 45 = \dfrac{CC_1}{\sqrt{2}}$

From triangle ABC, $\quad AC = \sqrt{AB^2 + BC^2}$

Since $\quad AB = BC, \quad AC = \sqrt{AB^2 + AB^2} = \sqrt{2} \cdot AB$

$\therefore$ Strain in diagonal $\quad AC = \dfrac{C_1C_2}{AC} = \dfrac{C_1C_2}{\sqrt{2}AB}$

From triangle BCC_1, $\quad \tan\phi = \dfrac{CC_1}{BC}$

$$CC_1 = BC \tan\phi$$

But $\quad C_1C_2 = \dfrac{CC_1}{\sqrt{2}} = \dfrac{BC \tan\phi}{\sqrt{2}}$

$\therefore$ Strain in diagonal $\quad AC = \dfrac{BC \tan\phi}{\sqrt{2} \cdot \sqrt{2} \cdot AB}$

Since, $AB = BC$, $\quad \text{Strain} = \dfrac{\tan\phi}{2}$

For small angles, $\quad \tan\phi = \phi$

$\therefore$ Strain in $\quad AC = \dfrac{\phi}{2}$

From definition, rigidity modulus $G = \dfrac{\tau}{\phi}$

$\therefore \qquad \phi = \dfrac{\tau}{G}$

Substituting, $\qquad$ strain in $AC = \dfrac{\tau}{2G}$ $\qquad$ ...(1)

But pure shear induces axial tensile stress τ in diagonal AC and axial compressive stress τ in diagonal BC which B perpendicular to it. These two stresses induce tensile strain in diagonal AC.

$\therefore$ Total tensile strain in diagonal $AC = \dfrac{\tau}{E} - \left(-\dfrac{\mu\tau}{E}\right)$

Strain in $AC = \dfrac{\tau}{E}(1 + \mu)$ $\qquad$ ...(2)

Equating Eqs. (1) and (2), $\dfrac{\tau}{2G} = \dfrac{\tau}{E}(1 + \mu)$

$\therefore \qquad E = 2G(1 + \mu).$

2.5.3 Relation between the Three Moduli: *E*, *G* and *K*

It is seen that relation between Young's modulus and bulk modulus

as $\qquad E = 3K(1 - 2\mu)$ $\qquad$ ...(1)

and relation between Young's modulus and rigidity

as $\qquad E = 2G(1 + \mu)$ $\qquad$ ...(2)

From Eq. (2) $\qquad \mu = \dfrac{E}{2G} - 1$

Substituting in Eq. (1)

$$E = 3K\left[1 - 2\left(\frac{E}{2G} - 1\right)\right] = 3K\left[1 - \frac{2E}{2G} + 2\right]$$

$$E = 3K\left[3 - \frac{E}{G}\right]$$

$$E = 9K - \frac{3KE}{G}$$

$$9K = E + \frac{3KE}{G}$$

$$= E\left(1 + \frac{3K}{G}\right)$$

$$9K = \frac{E(G + 3K)}{G}$$

$$E = \frac{9KG}{3K + G}$$

$$\frac{9}{E} = \frac{3K + G}{KG} \Rightarrow \frac{9}{E} = \frac{3}{G} + \frac{1}{K}$$

Important Relations:

Longitudinal strain $\in_l = \frac{\delta l}{l}$

Lateral strain $\in_b = \frac{\delta b}{b}$

$$\in_t = \frac{\delta t}{t}$$

$$\in_d = \frac{\delta d}{d}$$

$$\text{Poisson's ratio } \mu = \frac{\text{Lateral strain}}{\text{Longitudinal strain}}$$

Volumetric strain in a rectangular cross section or round cross section bar subjected to axial load

$$\in_V = \frac{\sigma}{E}(1 - 2\mu) = \frac{\delta l}{l}(1 - 2\mu)$$

Total volumetric strain $\in_v = \in_x + \in_y + \in_z$

Shear stain $\phi = \frac{\tau}{G}$

Volumetric strain for a body acted upon by stresses σ_x, σ_y and σ_z along three mutually

Perpendicular axes $\in_V = \frac{(\sigma_x + \sigma_y + \sigma_z)}{E}(1 - 2\mu)$

Bulk modulus $K = \frac{\sigma}{\in_V}$

Rigidity modulus $G = \frac{\tau}{\phi}$

Young's modulus $E = \frac{\text{Longitudinal stress}}{\text{Longitudinal strain}} = \frac{\sigma}{\left(\frac{\delta l}{l}\right)}$

Relation between the three modulus

$$E = 2G(1 + \mu)$$

$$E = 3K(1 - 2\mu)$$

$$E = \frac{9KG}{3K + G}$$

$$\frac{9}{E} = \frac{3}{G} + \frac{1}{K}$$

SOLVED PROBLEMS

Problem 1 A bar of diameter 30 mm and length 900 mm was subjected to an axial load of 48 kN. The increase is length and decrease in diameter observed were: 0.315 mm and 0.0035 m. Find the Poisson ratio, Young's modulus, rigidity modulus and bulk modulus of the material.

Solution: **Given data:** Diameter $d = 30$ mm

$$\text{Length } l = 900 \text{ mm}$$

$$\text{Load } P = 48 \text{ kN} = 48{,}000 \text{ N}$$

$$\text{Increase in length } \delta l = 0.315 \text{ mm}$$

$$\text{Change in diameter } \delta d = 0.0035 \text{ mm}$$

$$\text{Longitudinal strain } \in_l = \frac{\delta l}{l} = \frac{0.315}{900} = 0.35 \times 10^{-3}$$

$$\text{Lateral strain } \in_d = \frac{\delta d}{d} = \frac{0.0035}{30} = 1.167 \times 10^{-4}$$

$$\therefore \quad \text{Poisson's ratio } \mu = \frac{\text{Lateral strain}}{\text{Longitudinal strain}} = \frac{\in_l}{\in_d}$$

$$\mu = \frac{1.167 \times 10^{-4}}{0.35 \times 10^{-3}} = 0.3334$$

$$\text{Young's modulus } E = \frac{\text{Stress}}{\text{Strain}} = \frac{P/A}{\in_l}$$

$$E = \frac{\left(\dfrac{48000}{\dfrac{\pi}{4} \times 30^2}\right)}{0.35 \times 10^{-3}} = 190.017 \times 10^3 \text{ N/mm}^2$$

We know that

$$E = 2G(1 + \mu)$$

$$190.017 \times 10^3 = 2G(1 + 0.3334)$$

$$\therefore \quad \text{Rigidity modulus } G = 72.753 \times 10^3 \text{ N/mm}^2$$

$$\text{Again } E = 3K(1 - 2\mu)$$

$$190.017 \times 10^3 = 3K(1 - 2 \times 0.3334)$$

$$\therefore \quad \text{Bulk modulus } K = 194.095 \times 10^3 \text{ N/mm}^2.$$

Problem 2 Determine the change in length, breadth and thickness of a rectangular cross section steel bar 3 m along, 45 mm wide and 12 mm thick subjected to a tensile load of 36 kN along its length. The Poisson's ratio is 0.296 and modulus of elasticity is 207 GPa.

Solution: **Given data:**

$$\text{Length } l = 3 \text{ m} = 3000 \text{ mm}$$

$$\text{Width } b = 45 \text{ mm}$$

$$\text{Thickness } t = 12 \text{ mm}$$

$$\text{Load } P = 36 \text{ kW} = 36000 \text{ N}$$

$$\text{Poisson's ratio } \mu = 0.296$$

$$\text{Modulus of elasticity } E = 207 \text{ GPa} = 207 \times 10^3 \text{ MPa}$$

$$\text{Change in length } \delta l = \frac{Pl}{AE}$$

$$= \frac{36000 \times 3000}{(45 \times 12) \times 207 \times 10^3}$$

$$\delta l = 0.9662 \text{ mm}$$

$$\text{Poisson's ratio } \mu = \frac{\text{Lateral strain}}{\text{Longitudinal strain}}$$

$$\text{Lateral strain} = \frac{\delta b}{b} = \frac{\delta t}{t}$$

$$\text{Longitudinal strain} = \frac{\delta l}{l} = \frac{0.9662}{3000} = 3.221 \times 10^{-4}$$

$\therefore$

$$\text{Lateral strain} = \mu \times \text{Longitudinal strain}$$

$$= 0.296 \times 3.221 \times 10^{-4} = 9.5532 \times 10^{-5}$$

$$\text{But lateral strain} = \frac{\delta b}{b} \text{ or } \frac{\delta t}{t}$$

$$\text{Change in width } \delta b = 9.5532 \times 10^{-5} \times b$$

$$\delta b = 9.5532 \times 10^{-5} \times 45 = 0.00429 \text{ mm}$$

$$\text{Similarly, change in thickness } \delta t = 9.5532 \times 10^{-5} \times t$$

$$\delta t = 9.5532 \times 10^{-5} \times 12 = 0.001144 \text{ mm.}$$

Problem 3 A metal bar 60 mm × 40 mm is subjected to an axial compressive load of 500 kN. The contraction in a gauge length of 200 mm is found to be 0.2 mm. Increase in width of 60 mm side is 0.018 mm.

Find (i) Young's modulus

(ii) Poisson's ratio

(iii) Increase in thickness

(iv) Bulk modulus

(v) Rigidity modulus

Solution: **Given data:**

$$\text{Width } b = 60 \text{ mm}$$

$$\text{Thickness } t = 40 \text{ mm}$$

$$\text{Length } l = 200 \text{ mm}$$

$$\text{Load } P = 500 \text{ kN} = 500 \times 10^3 \text{ N}$$

$$\text{Decrease in length } \delta l = 0.2 \text{ mm}$$

$$\text{Increase in width } \delta b = 0.018 \text{ mm}$$

$$\text{Area of cross section } A = bt = 60 \times 40 = 2400 \text{ mm}^2$$

(i) Young's modulus

$$E = \frac{\text{Stress}}{\text{Strain}} = \frac{Pl}{A\delta l}$$

$$E = \frac{500 \times 10^3 \times 200}{(60 \times 40) \times 0.2} = 208.333 \times 10^3 \text{ MPa}$$

(ii) Poisson's ratio

$$\mu = \frac{\text{Lateral strain}}{\text{Longitudinal strain}}$$

$$\text{Lateral strain} = \frac{\delta b}{b} = \frac{0.018}{60} = 0.0003$$

$$\text{Longitudinal strain} = \frac{\delta l}{l} = \frac{0.2}{200} = 0.001$$

$$\therefore \quad \mu = \frac{0.0003}{0.001} = 0.3$$

(iii) Increase in thickness

$$\text{We know that lateral strain } \frac{\delta t}{t} = 0.0003$$

$$\therefore \quad \text{Increase in thickness } \delta t = t \times 0.0003$$

$$= 40 \times 0.0003$$

$$\delta t = 0.012 \text{ mm}$$

(iv) Bulk modulus

$$\text{WKT } E = 3K(1 - 2\mu)$$

$$208.333 \times 10^3 = 3K(1 - 2 \times 0.3)$$

$$\therefore \quad \text{Bulk modulus } K = 173.608 \times 10^3 \text{ MPa}$$

(v) Rigidity modulus

$$\text{Again } E = 2G(1 + \mu)$$

$$208.333 \times 10^3 = 2G(1 + 0.3)$$

$$\therefore \quad \text{Rigidity modulus } G = 80.128 \times 10^3 \text{ MPa.}$$

Problem 4 A round steel rod 30 mm diameter and 1.8 m along carries an axial tensile load of 240 kN. Find the change in length and change in diameter taking Poisson's ratio as 0.3 and Young's modulus as 200 kN/mm^2.

Solution:

Given data:

$$\text{Diameter } d = 30 \text{ mm}$$

$$\text{Length } l = 1.8 \text{ m} = 1800 \text{ mm}$$

$$\text{Load } P = 240 \text{ kN} = 240 \times 10^3 \text{ N}$$

$$\text{Poisson's ratio } \mu = 0.3$$

$$\text{Young's modulus } E = 200 \text{ kN/mm}^2 = 200 \times 10^3 \text{ N/mm}^2$$

(i)

$$\text{Change in length } \delta l = \frac{Pl}{AE} = \frac{240 \times 10^3 \times 1800}{\left(\frac{\pi}{4} \times 30^2\right) \times 200 \times 10^3}$$

$$\delta l = 3.05577 \text{ mm}$$

(ii) Change in diameter

$$\text{Poisson's ratio } \mu = \frac{\text{Lateral strain}}{\text{Longitudinal strain}}$$

$$\text{Longitudinal strain, } \frac{\delta l}{l} = \frac{3.05577}{1800}$$

$$= 1.69765 \times 10^{-3}$$

$$\therefore \quad \text{Lateral strain } \left(\frac{\delta d}{d}\right) = \mu \times \text{Longitudinal strain}$$

$$= 0.3 \times 1.69765 \times 10^{-3}$$

$$\frac{\delta d}{d} = 5.09296 \times 10^{-4}$$

$$\therefore \quad \text{Change in diameter } \delta d = 5.09296 \times 10^{-4}\, d$$

$$= 5.09296 \times 10^{-4} \times 30$$

$$\delta d = 0.\,01528 \text{ mm.}$$

Problem 5 Under what axial tensile load, the diameter of a steel bar will reduce from 80 mm to 79.95 mm. Take $E = 2 \times 10^5$ MPa and Poisson's ratio = 0.3.

Solution:

Given data:

$$\text{Original diameter } d = 80 \text{ mm}$$

$$\text{Reduced diameter } d_1 = 79.95 \text{ mm}$$

$$\therefore \quad \text{Change in diameter } \delta d = d - d_1$$

$$\delta d = 80 - 79.95 = 0.05 \text{ mm}$$

$$\text{Poisson's ratio } \mu = 0.3$$

$$\text{Young's modulus } E = 2 \times 10^5 \text{ MPa}$$

$$\text{Lateral strain } \in_d = \frac{\delta d}{d} = \frac{0.05}{80} = 6.25 \times 10^{-4}$$

We know that Poisson's ratio

$$\mu = \frac{\text{Lateral strain}}{\text{Longitudinal strain}}$$

$$\therefore \quad \text{Longitudinal strain } \in_l = \frac{\text{Lateral strain}}{\mu}$$

$$\in_l = \frac{\delta l}{l} = \frac{6.25 \times 10^{-4}}{0.3}$$

$$= 2.083 \times 10^{-3}$$

$$\text{But Young's modulus } E = \frac{\text{Stess}}{\text{Strain}} = \frac{P/A}{\in_l}$$

$$\text{Axial load } P = E.A. \in_l$$

$$= 2 \times 10^5 \times \left(\frac{\pi}{4} 80^2\right) \times 2.083 \times 10^{-3}$$

$$\text{Load } P = 2.09 \times 10^6 \text{ N.}$$

Problem 6 A bar of diameter 20 mm and length 200 mm extends by a length of 0.4 mm upon application of load *P*. Young's modulus of the material is known to be 2×10^5 N/mm^2. What is the change in the cross sectional area of the bar and the value of load *P* at this stage. The Poisson's ratio is 0.28. If the extension of the bar is limited to 0.3 mm under the same load *P*, what should be the diameter of the bar.

Solution: **Give data:** Diameter $d = 20$ mm

$$\text{Length } l = 200 \text{ mm}$$

$$\text{Extension in length } \delta l = 0.4 \text{ mm}$$

$$\text{Young's modulus } E = 2 \times 10^5 \text{ N/mm}^2$$

$$\text{Poisson's ratio } \mu = 0.28$$

$$\text{Longitudinal strain } \in_l = \frac{\delta l}{l} = \frac{0.4}{200} = 0.002$$

$$\text{Poisson's ratio } \mu = \frac{\text{Lateral strain}}{\text{Longitudinal strain}}$$

$$\therefore \quad \text{Lateral strain } \in_d = \mu \times \text{Longitudinal strain}$$

$$\in_d = \frac{\delta d}{d} = 0.28 \times 0.002 = 0.00056$$

$$\therefore \quad \text{Decrease in diameter } \delta_d = \in_d \times d$$

$$\delta d = 0.00056 \times 20 = 0.0112 \text{ mm}$$

$$\text{Final diameter } d_1 = d - \delta d = 20 - 0.0112 = 19.9888 \text{ mm}$$

Change in cross-sectional area

$$\delta A = \text{original area} - \text{Final area}$$

$$= \frac{\pi}{4} \times 20^2 - \frac{\pi}{4} \times 19.9888^2$$

$$\delta A = 0.35176 \text{ mm}^2$$

To final load P:

We know that change in length $\delta l = \frac{Pl}{AE}$

$$\therefore \quad P = \frac{\delta l AE}{l} = \frac{0.4 \times \left(\frac{\pi}{4} \times 20^2\right) \times 2 \times 10^5}{200}$$

$$= 125.664 \times 10^3 \text{ N}$$

If the extension of the bar $\delta l = 0.3$ under the same load $P = 125.664 \times 10^3$ N.

$$\delta l = \frac{Pl}{AE}$$

$$\delta A = \frac{Pl}{E.\delta l} = \frac{125.664 \times 10^3 \times 200}{2 \times 10^5 \times 0.3} = 418.879 \text{ mm}^2$$

$$\frac{\pi}{4} d_2^2 = 418.879 \text{ mm}^2$$

$$\text{Required diameter } d_2 = 23.094 \text{ mm}$$

Problem 7 Calculate the modulus of rigidity and bulk modulus of a cylindrical bar of diameter 30 mm and length 1.5 m, if the longitudinal strain in the bar during tension is four times the lateral strain. Find the change in volume, when the bar is subjected to hydrostatic pressure of 100 N/mm^2. Take E = 100 kN/mm^2.

Solution:

Given data:

$$\text{Diameter } d = 30 \text{ mm}$$

$$\text{Length } l = 1.5 \text{ m} = 1500 \text{ mm.}$$

$$\text{Longitudinal strain} = 4 \times \text{Lateral strain}$$

$$\therefore \quad \frac{\text{Lateral strain}}{\text{Longitudinal strain}} = \frac{1}{4} = \text{Poisson's ratio}$$

$$\text{Poisson's ratio } \mu = \frac{1}{4} = 0.25$$

$$\text{Hydrostatic pressure } \sigma = 100 \text{ N/mm}^2$$

$$E = 100 \text{ kN/mm}^2 = 100 \times 10^3 \text{ N/mm}^2$$

We know that $E = 2G(1 + \mu)$

$$\therefore \quad \text{Rigidity modulus } G = \frac{E}{2(1+\mu)}$$

$$G = \frac{100 \times 10^3}{2(1+0.25)} = 40 \times 10^3 \text{ N/mm}^2$$

$$\text{Again } E = 3K(1-2\mu)$$

$$\therefore \quad \text{Bulk modulus } K = \frac{E}{3(1-2\mu)}$$

$$K = \frac{100 \times 10^3}{3(1-2\times 0.25)} = 66.667 \times 10^3 \text{ N/mm}^2$$

$$\text{But bulk modulus} = \frac{\sigma}{\text{Volumetric strain}}$$

$$\therefore \quad \text{Volumetric strain } \in_V = \frac{\delta V}{V} = \frac{\sigma}{K} = \frac{100}{66.667 \times 10^3} = 0.0015$$

$$\therefore \quad \text{Change in volume } \delta V = \in_V \times V$$

$$= \in_V \left(\frac{\pi}{4} d^2 \times l\right)$$

$$= 0.0015 \times \left(\frac{\pi}{4} \times 30^2 \times 1500\right)$$

$$\text{Decrease in volume } \delta V = 159.431 \text{ mm}^3.$$

Problem 8 A round rod of diameter 60 mm and length 1.2 m long is subjected to an axial load of 300 kN. Determine the modulus of rigidity, bulk modulus and change in volume taking Young's modulus as 2.1×10^5 N/mm^2 and Poisson's ratio as 0.3.

Solution:

Given data:

$$\text{Diameter } d = 60 \text{ mm}$$

$$\text{Length } l = 1.2 \text{ m} = 1200 \text{ mm}$$

$$\text{Load } P = 300 \text{ kN} = 300 \times 10^3 \text{ N}$$

$$\text{Young's modulus } E = 2.1 \times 10^5 \text{ N/mm}^2$$

$$\text{Poisson's ratio } \mu = 0.3$$

$$\text{We know that } E = 2G(1+\mu)$$

$$2.1 \times 10^5 = 2G(1+0.3)$$

$$\therefore \quad \text{Rigidity modulus } G = 80.769 \times 10^3 \text{ N/mm}^2$$

$$\text{Again } E = 3\text{K}(1-2\mu)$$

$$\therefore \quad \text{Bulk modulus } K = \frac{E}{3(1-2\mu)} = \frac{2.1 \times 10^5}{3(1-2\times 0.3)}$$

$$K = 175 \times 10^3 \text{ N/mm}^2$$

Volumetric strain for a rod subjected to axial load

$$\in_V = \in_x (1 - 2\mu)$$

$$\text{Linear strain } \in_x = \frac{\delta l}{l} = \frac{\sigma}{E} = \frac{P}{AE}$$

$$= \frac{300 \times 10^3}{\left(\frac{\pi}{4} \times 60^2\right) \times 2.1 \times 10^5}$$

$$\in_x = 0.00050525$$

$$\therefore \quad \in_V = 0.00050525 \, (1 - 2 \times 0.3)$$

$$\in_V = 2.021 \times 10^{-4}$$

$$\text{But volumetric strain } \in_V = \frac{\delta V}{V}$$

$$\therefore \quad \text{Change in volume } \delta V = \in_V \times V$$

$$= \in_V \times \left(\frac{\pi}{4} d^2.l\right)$$

$$= 2.021 \times 10^{-4} \left(\frac{\pi}{4} 60^2 \times 1200\right)$$

$$\delta V = 685.714 \text{ mm}^3.$$

Problem 9 A round bar of diameter 15 mm carries an axial tensile load of 24 kN. The modulus of rigidity of the material is 80 GPa. The diameter of the bar measures 14.9970 mm under load. Determine the Poisson's ratio, Young's modulus and bulk modulus of the material.

Solution: **Given data:** Diameter $d = 15$ mm

$$\text{Load } P = 24 \text{ kN} = 24 \times 10^3 \text{ N}$$

$$\text{Rigidity modulus } G = 80 \text{ GPa} = 80 \times 10^3 \text{ MPa}$$

$$\text{New diameter } d_1 = 14.997 \text{ mm}$$

$$\therefore \quad \text{Decrease in diameter } \delta d = d - d_1$$

$$\delta d = 15 - 14.9970 = 0.0030 \text{ mm}$$

Area of cross section $$A = \frac{\pi}{4} d^2 = \frac{\pi}{4} 15^2 = 176.715 \text{ mm}^2$$

Axial stress $$\sigma = \frac{P}{A} = \frac{24000}{176.715} = 135.812 \text{ N/mm}^2$$

$$\text{WKT Poisson's ratio } \mu = \frac{\text{Lateral strain } \in_d}{\text{Longitudinal strain } \in_l}$$

$$\text{Lateral strain } \in_d = \frac{\delta d}{d} = \frac{0.0030}{15} = 2 \times 10^{-4}$$

$$\therefore \quad \text{Longitudinal strain } \in_l = \frac{\in_d}{\mu}$$

$$\in_l = \frac{2 \times 10^{-4}}{\mu}$$

$$\text{But longitudinal strain } \in_l = \frac{\sigma}{E}$$

$$\therefore \quad E = \frac{\sigma}{\in_l} = \frac{135.812}{\left(\frac{2 \times 10^{-4}}{\mu}\right)} = 679.059\ \mu \qquad \ldots(1)$$

Also

$$E = 2G(1 + \mu)$$

$$E = 2 \times 80 \times 10^3\ (1 + \mu) = 160 \times 10^3(1 + \mu) \quad \ldots(2)$$

Equating Eqs. (1) and (2)

$$679059\ \mu = 160 \times 10^3\ (1 + \mu)$$

$$(679059 - 160 \times 10^3)\ \mu = 160 \times 10^3$$

$$\therefore \quad \text{Poisson's ratio } \mu = 0.308$$

$$\therefore \quad \text{Young's modulus } E = 679059\ \mu$$

$$= 679059 \times 0.308$$

$$E = 209.32 \times 10^3\ \text{N/mm}^2$$

$$\text{Bulk modulus } K = \frac{E}{3(1 - 2\mu)}$$

$$= \frac{209.32 \times 10^3}{3(1 - 2 \times 0.308)}$$

$$K = 181.70 \times 10^3\ \text{N/mm}^2.$$

Problem 10 A round rod of length 1 m and 24 mm diameter carries a tensile load of 90 kN along the longitudinal axis. Taking Young's modulus as 2×10^5 MPa and Poisson's ratio as 0.3, find the percentage change in volume of the rod.

Solution:

Given data: Diameter $d = 24$ mm

Length $l = l$ m $= 1000$ mm

Load $P = 90$ kN $= 90 \times 10^3$ N

Young's modulus $E = 2 \times 10^5$ MPa

Poisson's ratio $\mu = 0.3$

$$\text{Stress } \sigma = \frac{P}{A} = \frac{P}{\left(\frac{\pi}{4} d^2\right)}$$

$$= \frac{90 \times 10^3}{\left(\frac{\pi}{4} \times 24^2\right)} = 198.944 \text{ MPa}$$

For a rod subjected to axial stress, the volumetric strain,

$$\in_V = \frac{\sigma}{E}(1 - 2\mu)$$

$$= \frac{198.944}{2 \times 10^5}(1 - 2 \times 0.3) = 3.9798 \times 10^{-4}$$

$$\text{But volumetric strain } \in_v = \frac{\delta V}{V}$$

$$\therefore \quad \text{Change in volume } \delta V = \in_V \times V$$

$$\text{Volume } V = \frac{\pi}{4} d^2 l = \frac{\pi}{4} 24^2 \times 1000$$

$$V = 452389.34 \text{ mm}^3$$

$$\therefore \quad \delta V = 3.9789 \times 10^{-4} \times 452389.34$$

$$\delta V = 180.0012 \text{ mm}^3$$

$$\therefore \quad \text{Percentage change in volume} = \frac{\text{Change in volume } (\delta V)}{\text{Volume } (V)} \times 100$$

$$= \frac{80.0012}{452389.34} \times 100 = 0.03979\%.$$

Problem 11 A certain material has modulus of elasticity as 206 GPa and modulus of rigidity as 80 GPa. Find the Poisson's ratio and bulk modulus of that material.

Solution:

Given data: Young's modulus $E = 206 \text{ GPa} = 200 \times 10^3$ MPa

$$\text{Rigidity modulus } G = 80 \text{ GPa} = 80 \times 10^3 \text{ MPa}$$

$$\text{WKT } E = 2G(1 + \mu)$$

$$206 \times 10^3 = 2 \times 80 \times 10^3 (1 + \mu)$$

$$\therefore \quad \text{Poisson's ratio } \mu = 0.2875$$

$$\text{Bulk modulus } K = \frac{E}{3(1 - 2\mu)} = \frac{206 \times 10^3}{3(1 - 2 \times 0.2875)}$$

$$K = 161.569 \times 10^3 \text{ MPa.}$$

Problem 12 A 600 mm long steel bar has a rectangular cross section of width 30 mm and thickness 10 mm and is subjected to a pull of 90 kN along the length. Taking Poisson's ratios as 0.3 and E = 200 kN/mm^2, find the change in volume and final volume of the bar.

Solution:

Given data:

$$\text{Length } l = 600 \text{ mm}$$

$$\text{Width } b = 30 \text{ mm}$$

$$\text{Thickness } t = 10 \text{ mm}$$

$$\text{Load } P = 90 \text{ kN} = 90000 \text{ N.}$$

$$\text{Poisson's ratio } \mu = 0.3$$

$$\text{Young's modulus } E = 200 \text{ kN/mm}^2 = 200 \times 10^3 \text{ N/mm}^2$$

$$\text{Stress } \sigma = \frac{P}{A} = \frac{P}{b \times t} = \frac{90000}{30 \times 10} = 300 \text{ MPa}$$

The volumetric strain for a rod subjected to axial stress, $\in_V = \frac{\sigma}{E}(1 - 2\mu)$

$$= \frac{300}{200 \times 10^3}(1 - 2 \times 0.3) = 6 \times 10^{-4}$$

$$\text{But volumetric strain } \in_V = \frac{\delta V}{V}$$

$$\text{Change in volume } V = \in_V \times \delta V$$

$$\text{Volume } V = lbt = 600 \times 30 \times 10 = 180000 \text{ mm}^3$$

$$\delta V = 6 \times 10^{-4} \times 180000 = 108 \text{ mm}^3$$

$$\text{Final volume } V_1 = V + \delta V = 180000 + 108$$

$$= 180108 \text{ mm}^3.$$

Problem 13 A square section bar of side 100 mm and length 1.8 m is subjected to a compressive load of 945 kN. Determine the percentage change in volume and final volume if modulus of elasticity is 2 × 10^5 N/mm^2 and modulus of rigidity is 7.8 × 10^4 N/mm^2.

Solution:

Given data: Side of square section a = 100 mm

$$\text{Length } l = 1.8 \text{ m} = 1800 \text{ mm.}$$

$$\text{Load } P = 945 \text{ kN} = 945 \times 10^3 \text{ N}$$

$$\text{Young's modulus } E = 2 \times 10^5 \text{ N/mm}^2$$

$$\text{Rigidity modulus } G = 7.8 \times 10^4 \text{ N/mm}^2$$

$$\text{Compressive stress } \sigma = \frac{P}{A} = \frac{P}{a^2} = \frac{-945000}{100 \times 100} = -94.5 \text{ MPa}$$

(–ve because compressive)

$$E = 2G(1 + \mu)$$

$$2 \times 10^5 = 2 \times 7.8 \times 10^4 (1 + \mu)$$

$$\therefore \quad \text{Poisson's ratio } \mu = 0.28205$$

For a rod subjected to axial stress σ, the volumetric strain $\in_V = \frac{\sigma}{E}(1 - 2\mu)$

$$= \frac{-94.5}{2 \times 10^5}(1 - 2 \times 0.28205)$$

$$\in_V = -2.05963 \times 10^{-4}$$

(– ve indicates reduction in volume)

$$\text{But } \in_V = \frac{\delta V}{V}$$

$$\text{Change in volume } \delta V = \in_V \times V$$

$$\text{Volume } V = l \times a^2 = 1800 \times 100^2 = 18 \times 10^6$$

$$\therefore \quad \delta V = -2.05963 \times 10^{-4} \times 18 \times 10^6$$

$$= -3707.3295 \text{ mm}^3$$

$$\%\text{ change in volume} = \frac{\delta V}{V} \times 100$$

$$= \frac{3707.3295 \times 100}{18 \times 10^6} = 0.0206\%$$

$$\text{Final volume } V_1 = V + \delta V$$

$$= 18 \times 10^6 - 370.73295 \qquad \text{(Here } \delta V \text{ is –ve)}$$

$$V_1 = 17996292.67 \text{ mm}^3.$$

Problem 14 Figure 2.10 shows a stepped round bar subjected to a tensile load that produces a total elongation of 0.75 mm. The Poisson's ratio is 0.28. Determine the individual change in length, change in diameter and change in volume for the two sections.

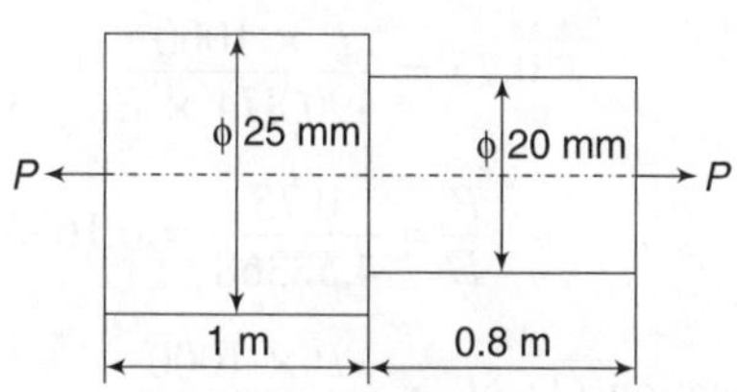

Fig. 2.10

Solution:

Given data: For bigger section (1) diameter $d_1 = 25$ mm

Length $l_1 = 1$ m $= 1000$ mm

For smaller section (2) diameter $d_2 = 20$ mm

Length $l_2 = 0.8$ m $= 800$ mm

Total elongation for both $\delta l = 0.75$ mm

$$\delta l = \delta l_1 + \delta l_2 = 0.75 \text{ mm}$$

Poisson's ratio $\mu = 0.28$

$$\text{Area of section (1), } A_1 = \frac{\pi}{4} d_1^2 = \frac{\pi}{4} 25^2$$

$$= 490.874 \text{ mm}^2$$

$$\therefore \quad \text{Volume of section (1) } V_1 = A_1 \, l_1 = 490.874 \times 1000$$

$$= 490.874 \times 10^3 \text{ mm}^3$$

$$\text{Stress in section (1) } \sigma_1 = \frac{P}{A_1} = \frac{P}{490.874} \text{ MPa}$$

$$\text{Area of cross section of section (2) } A_2 = \frac{\pi}{4} d_2^2$$

$$= \frac{\pi}{4} \times 20^2 = 314.159 \text{ mm}^2$$

$$\text{Volume of section (2) } V_2 = A_2 \, l_2$$

$$= 314.159 \times 800$$

$$= 251.327 \times 10^3 \text{ mm}^3$$

$$\text{Stress is section (2) } \sigma_2 = \frac{P}{A_2} = \frac{P}{314.159} \text{ MPa}$$

$$\text{We know that increase in length } \delta l = \frac{\sigma l}{E}$$

$$\therefore \quad \delta l_1 = \frac{\sigma_1 \, l_1}{E} \text{ and } \delta l_2 = \frac{\sigma_2 \, l_2}{E}$$

$$\delta l = \delta l_1 + \delta l_2$$

$$0.75 = \frac{P \times 1000}{490.874 \times E} + \frac{P \times 800}{314.159 \times E} = \frac{P}{E}(4.58368)$$

$$\therefore \quad \frac{P}{E} = \frac{0.75}{4.58368} = 0.16362$$

$$\therefore \text{ Change in length of section (1) } \delta l_1 = \frac{P \times 1000}{490.874 \times E}$$

$$\delta l_1 = \frac{P}{E} \times \frac{1000}{490.874} = 0.16362 \times \frac{1000}{490.874}$$

$$= 0.3333 \text{ mm}$$

Change in length of section (2) $\delta l_2 = \dfrac{P}{E} \times \dfrac{800}{314.159}$

$$\delta l_2 = 0.16362 \times \frac{800}{314.159} = 0.4167 \text{ mm}$$

To find change in diameter:

Longitudinal strain in section (1) $\in_1 = \dfrac{\delta l_1}{l_1} = \dfrac{0.3333}{1000} = 0.3333 \times 10^{-3}$

$\therefore$ Lateral strain $\in d_1 = \mu \times$ Longitudinal strain $(\in_1)$

$$= 0.28 \times 0.3333 \times 10^{-3} = 9.3324 \times 10^{-5}$$

But lateral strain $\in d_1 = \dfrac{\delta d_1}{d_1}$

$\therefore$ Change in diameter in section (1) $\delta d_1 = \in d_1 \times d_1$

$$\delta d_1 = 9.3324 \times 10^{-5} \times 25 = 0.002333 \text{ mm}$$

Similarly, longitudinal strain in section (2)

$$\in_2 = \frac{\delta l_2}{l_2} = \frac{0.4167}{800} = 5.20875 \times 10^{-4}$$

Lateral strain in section (2) $\in d_2 = \in_2 \times \mu$

$$= 5.20875 \times 10^{-4} \times 0.28$$

$$\in d_2 = \frac{\delta d_2}{d_2} = 1.45845 \times 10^{-4}$$

$\therefore$ Change in diameter in section (2) $\delta d_2 = \in d_2 \times d_2$

$$\delta d_2 = 1.45845 \times 10^{-4} \times 20 = 0.002917 \text{ mm}$$

To find change in volume:

WKT volumetric strain for a member loaded axially is $\in_V = \dfrac{\delta V}{V} = \dfrac{\sigma}{E}(1 - 2\mu)$

Also $\delta l = \dfrac{\sigma l}{E}$ or $\dfrac{\sigma}{E} = \dfrac{\delta l}{l}$

$\therefore$ $$\frac{\delta V}{V} = \frac{\delta l}{l}(1 - 2\mu)$$

For section (1), $\in V_1 = \dfrac{\delta V_1}{V} = \dfrac{\delta l_1}{l_1}(1 - 2\mu)$

$$= \frac{0.3333}{1000}(1 - 2 \times 0.28) = 1.46652 \times 10^{-4}$$

$\therefore$ Change in volume $\delta V_1 = \in V_1 \times V_1$

$$= 1.46652 \times 10^{-4} \times (490.874 \times 10^3)$$

$$= 71.9877 \text{ mm}^3$$

For section (2) $\in V_2 = \dfrac{\delta V_2}{V_2} = \dfrac{\delta l_2}{l_2}(1 - 2\mu)$

$$= \frac{0.4167}{800}(1 - 2 \times 0.28) = 2.29185 \times 10^{-4}$$

$\therefore$ Change in volume $\delta V_2 = \in V_z \times V_2 = 2.29185 \times 10^{-4} \times 251.327 \times 10^3$

$$= 57.60 \text{ mm}^3.$$

Problem 15 A sphere of diameter 1 m is immersed in sea at a depth of 0.9 km. The modulus of rigidity and modulus of elasticity of sphere material are 84 GPa and 210 GPa respectively. The specific weight of seawater may be taken as 10.06 kN/m^3. Find the change in volume of the sphere.

Solution:

Given data: Diameter of the sphere $d = 1\text{m} = 1000$ mm

Depth of sea $H = 0.9$ km $= 900$ m

Modulus of rigidity $G = 84$ GPa $= 84 \times 10^3$ MPa

Modulus of elasticity $E = 210$ GPa $= 210 \times 10^3$ MPa

Specific weight of water $w = 10.06$ kN/m^3

$= 10060$ N/m^3

Intensity of pressure $P = \sigma = w \times H$

$10060 \times 900 = 9.054 \times 10^6$ N/m^2

$p = \sigma = 9.054$ N/mm^2

Volume of sphere $V = \dfrac{\pi d^3}{6} = \dfrac{\pi}{6} \times 1000^3 \text{ mm}^3$

For a member subjected to stress in all three directions, volumetric strain

$$\in_V = \left(\frac{\sigma_x + \sigma_y + \sigma_z}{E}\right)(1 - 2\mu)$$

$$\in_V = \frac{3\sigma}{E}(1 - 2\mu)$$

Young's modulus $E = 2G(1 + \mu)$

$210 \times 10^3 = 2 \times 84 \times 10^3 (1 + \mu)$

$\therefore$ Poisson's ratio $\mu = 0.25$

$$\therefore \qquad \in_V = \frac{3 \times 9.054}{210 \times 10^3}(1 - 2 \times 0.25)$$

$$\in_V = \frac{\delta V}{V} = 6.4671 \times 10^{-5}$$

$$\therefore \qquad \text{Change in volume } \delta V = \in_V \times V = \in_V \times \frac{\pi}{6} d^3$$

$$\in_V = 6.467 \times 10^{-5} \times \left(\frac{\pi}{6} \times 1000^3\right) = 33.862 \text{ mm}^3.$$

Problem 16 A bar of steel 40 mm diameter and 1 m long stretches by 1 mm under simple tension. Find the change in volume if the Poisson's ratio is 0.25. Also what hydrostatic pressure would cause the same change in volume numerically? Take E = 210 GPa.

***Solution:* Given data:** Diameter d = 40 mm

Length l = 1 m = 1000 mm

Increase in length δl = 1 mm

Poisson's ratio μ = 0.25

Young's modulus E = 210 GPa = 210×10^3 MPa

For a bar subjected to axial stren,

$$\text{Volumetric strain } \in_V = \frac{\sigma}{E}(1 - 2\mu)$$

$$\text{But } \frac{\sigma}{E} = \frac{\delta l}{l} = \frac{1}{1000} = 0.001$$

$$\therefore \qquad \in_V = \frac{\delta V}{V} = 0.001\ (1 - 2 \times 0.25) = 0.0005$$

$$\text{Volume } V = \frac{\pi d^2}{4} l = \left(\frac{\pi}{4} \times 40^2 \times 1000\right)$$

$$\therefore \qquad \text{Change in volume } \delta V = \in_V \times V$$

$$\delta V = 0.0005 \times \left(\frac{\pi}{4} \times 40^2 \times 1000\right) = 628.319 \text{ mm}^3$$

$$\text{WKT} \quad E = 3K\ (1 - 2\mu)$$

$$\therefore \qquad \text{Bulk modulus } K = \frac{E}{3\ (1 - 2\mu)} = \frac{210 \times 10^3}{3(1 - 2 \times 0.5)}$$

$$= 140 \times 10^3 \text{ MPa}$$

$$\text{But bulk modulus } K = \frac{\text{Hydrostatic pressure } (p)}{\text{Volumetric strain } (\in_v)}$$

$\therefore$ Hydrostatic pressure $p = K.\in_v = 140 \times 10^3 \times 0.0005$

$$p = 70 \text{ MPa.}$$

Problem 17 A bar of rectangular cross section 30 mm thick, 40 mm wide and 500 mm long carries an axial tensile load of 90 kN. Modulus of elasticity and modulus of rigidity are 2×10^5 MPa and 78×10^3 MPa, respectively.

Find (i) bulk modulus

(ii) change in length, width and thickness, and

(iii) change in volume.

Solution:

Given data:

Width $b = 40$ mm

Thickness $t = 30$ mm

Length $l = 500$ mm

Load $P = 90 \text{ kN} = 90 \times 10^3$ N

Modulus of elasticity $E = 2 \times 10^5$ MPa

Modulus of rigidity $G = 78 \times 10^3$ MPa

WKT $E = 2G(1 + \mu)$

$$2 \times 10^5 = 2 \times 78 \times 10^3 (1 + \mu)$$

$\therefore$ Poisson's ratio $\mu = 0.2821$

(i) $\therefore$ Bulk modulus $K = \dfrac{E}{3(1 - 2\mu)}$

(Since $E = 3K(1 - 2\mu)$)

$$K = \frac{2 \times 10^5}{3(1 - 2 \times 0.2821)} = 152.975 \times 10^3 \text{ MPa}$$

(ii) Change in length $\delta l = \dfrac{Pl}{AE} = \dfrac{90000 \times 500}{(40 \times 30) \times 2 \times 10^5}$

$$= 0.1875 \text{ mm}$$

Poisson's ratio $\mu = \dfrac{\text{Lateral strain}}{\text{Longitudinal strain}}$

Longitudinal strain $\in_l = \dfrac{\delta l}{l} = \dfrac{0.1875}{500} = 3.75 \times 10^{-4}$

$\therefore$ Lateral strain $\in_b = \in_t = \mu \times \in_l$

$$\in_b = \in_t = 0.2821 \times 3.75 \times 10^{-4} = 1.0579 \times 10^{-4}$$

Consider $\in_b = \dfrac{\delta b}{b} = 1.0579 \times 10^{-4}$

$$\therefore \quad \text{Change in width } \delta b = \in_b \times b = 1.0579 \times 10^{-4} \times 40$$

$$\delta b = 0.00423 \text{ mm}$$

$$\text{Again consider } \in_t = \frac{\delta t}{t} = 1.0579 \times 10^{-4}$$

$$\therefore \quad \text{Change in thickness } \delta t = \in_t \times t = 1.0579 \times 10^{-4} \times 30$$

$$\delta t = 0.003174 \text{ mm}$$

(iii) Change in volume:

For a rod subjected to axial stress,

$$\text{Volumetric strain } \in_V = \frac{\delta V}{V} = \frac{\sigma}{E}(1 - 2\mu)$$

But

$$\frac{\sigma}{E} = \frac{\delta l}{l}$$

$$\therefore \quad \in_V = \frac{\delta l}{l}(1 - 2\mu)$$

$$\in_V = \in_l (1 - 2\mu) = 3.75 \times 10^{-4} (1 - 2 \times 0.2821)$$

$$\in_V = \frac{\delta V}{V} = 1.63425 \times 10^{-4}$$

$$\therefore \quad \text{Change in volume } \delta V = \in_V \times V$$

$$\delta V = 1.63425 \times 10^{-4} \times (lbt)$$

$$= 1.63425 \times 10^{-4} \times (500 \times 40 \times 30)$$

$$\delta V = 98.0555 \text{ mm}^3.$$

Problem 18 A metallic block 300 mm × 60 mm × 30 mm is subjected to a tensile force of 45 kN on 60 × 30 face, 84 kN tensile on 300 × 60 face and 60 kN tensile on 300 × 30 face. Determine the change in volume of the block.

Take modulus of elasticity as 206 GPa and Poisson's ratio as 0.3.

***Solution:* Given data:**

$$\text{Length } l = 300 \text{ mm}$$

$$\text{Width } b = 60 \text{ mm}$$

$$\text{Thickness } t = 30 \text{ mm}$$

$$\text{Force in } x\text{-direction } P_x = 45 \text{ kN} = 45 \times 10^3 \text{ N}$$

$$\text{Force in } y\text{-direction } P_y = 84 \text{ kN} = 84 \times 10^3 \text{ N}$$

$$\text{Force in } z\text{-direction } P_z = 60 \text{ kN} = 60 \times 10^3 \text{ N}$$

$$\text{Modulus of elasticity } E = 206 \text{ GPa} = 206 \times 10^3 \text{ MPa}$$

$$\text{Poisson's ratio } \mu = 0.3$$

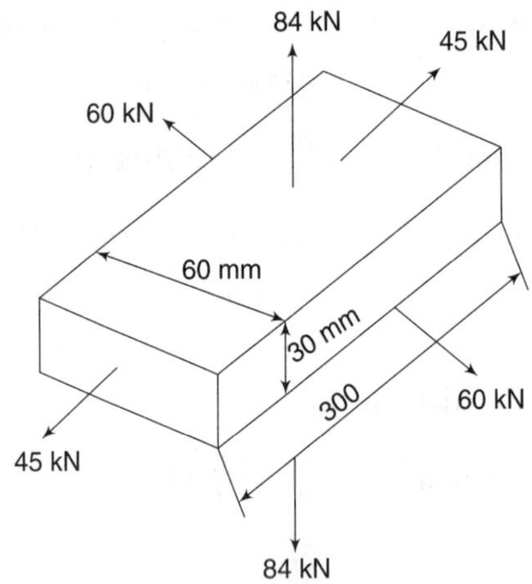

Fig. 2.11

$$\text{Stress in } x\text{-direction } \sigma_x = \frac{P_x}{A} = \frac{45 \times 10^3}{60 \times 30} = 25 \text{ MPa}$$

$$\text{Stress in } y\text{-direction } \sigma_y = \frac{P_y}{A} = \frac{84 \times 10^3}{60 \times 300} = 4.67 \text{ MPa}$$

$$\text{Stress in } z\text{-direction } \sigma_z = \frac{P_z}{A} = \frac{60 \times 10^3}{30 \times 300} = 6.67 \text{ MPa}$$

$$\text{Volume of block } V = lbt = 300 \times 60 \times 30$$

$$V = 540000 \text{ mm}^3$$

For a member subjected to stresses along three mutually perpendicular directions,

$$\text{Volumetric strain } \in_V = \left(\frac{\sigma_x + \sigma_y + \sigma_z}{E}\right)(1 - 2\mu)$$

$$\in_V = \frac{\delta V}{V} = \frac{(25 + 4.67 + 6.67)}{206 \times 10^3}(1 - 2 \times 0.3)$$

$$= 7.0563 \times 10^{-5}$$

$$\therefore \quad \text{Change in volume } \delta V = \in_V \times V$$

$$= 7.0563 \times 10^{-5} \times 540000$$

$$\delta V = 38.104 \text{ mm}^3.$$

Problem 19 A rectangular block is 250 mm long, 75 mm wide and 25 mm thick is loaded with a 250 kN tensile load along the axis of the member. A compressive force of 2 MN acts on the face 250 mm × 75 mm and tensile force of 400 kN acts on the face 250 mm × 25 mm. If μ = 0.3 and E = 200 GPa. Determine the change in length, change in width, change in thickness and change in volume of the bar.

What must be load that should be applied in place of 2 MN so that the change in volume of the bar will be zero.

Solution:

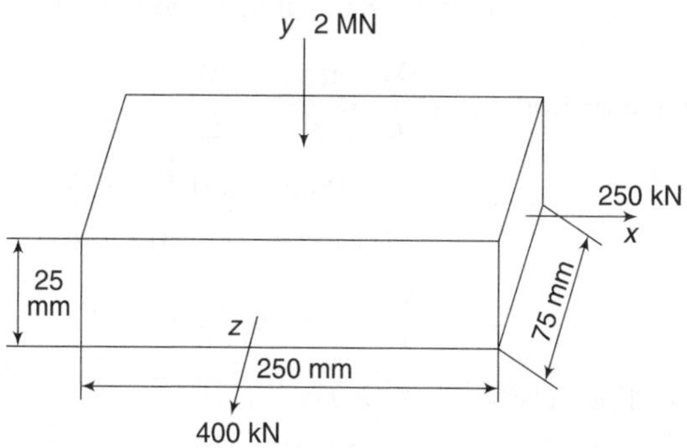

Fig. 2.12

Given data:

$$\text{Length } l = 250 \text{ mm}$$

$$\text{Width } b = 75 \text{ mm}$$

$$\text{Thickness } t = 25 \text{ mm}$$

$$\text{Load in } x\text{-direction } P_x = 250 \text{ kN} = 250 \times 10^3 \text{ N (Tensile)}$$

$$\text{Load in } y\text{-direction } P_y = -2 \text{ MN} = -2 \times 10^6 \text{ N (Compressive)}$$

$$\text{Load in } z\text{-direction } P_z = 400 \text{ kN} = 400 \times 10^3 \text{ N (Tensile)}$$

$$\text{Young's modulus } E = 200 \text{ GPa} = 200 \times 10^3 \text{ MPa}$$

$$\text{Poisson's ratio } \mu = 0.3$$

$$\text{Stress in } x\text{-direction } \sigma_x = \frac{P_x}{b} = \frac{250 \times 10^3}{75 \times 25} = 133.33 \text{ MPa}$$

$$\text{Stress in } y\text{-direction } \sigma_y = \frac{P_y}{lb} = \frac{-2 \times 10^6}{250 \times 75} = -106.67 \text{ MPa}$$

$$\text{Stress in } z\text{-direction } \sigma_z = \frac{Pz}{lt} = \frac{400 \times 10^3}{250 \times 25} = 64 \text{ MPa}$$

$$\text{Strain along } x\text{-direction } \in_x = \frac{\sigma_x}{E} - \frac{\mu}{E}\sigma_y - \frac{\mu}{E}\sigma_z$$

$$\in_x = \frac{133.33}{200 \times 10^3} - \frac{0.3(-106.67)}{200 \times 10^3} - \frac{0.3 \times 64}{200 \times 10^3}$$

$$\in_x = \frac{\delta l}{l} = 7.30655 \times 10^{-4}$$

$\therefore$ Change in length $\delta l = \in_x \times l = 7.30655 \times 10^{-4} \times 250$

$$\delta l = 0.18266 \text{ mm (elongation)}$$

Strain along y-direction $\in_y = \frac{\sigma_y}{E} - \frac{\mu\sigma_x}{E} - \frac{\mu\sigma_z}{E}$

$$\in_y = \frac{-0.106.67}{200 \times 10^3} - \frac{0.3 \times 133.33}{200 \times 10^3} - \frac{0.3 \times 64}{200 \times 10^3}$$

$$\in_y = \frac{\delta t}{t} = -8.29345 \times 10^{-4}$$

Change in thickness $\delta t = \in_y \times t$

$$= -8.29345 \times 10^{-4} \times 25$$

$$\delta t = -0.0207 \text{ mm}$$

–ve indicates reduction in thickness.

Strain along z-direction $\in_z = \frac{\sigma_z}{E} - \frac{\mu\sigma_x}{E} - \frac{\mu\sigma_y}{E}$

$$\in_z = \frac{64}{200 \times 10^3} - \frac{0.3 \times 133.33}{200 \times 10^3} - \frac{0.3\,(-166.67)}{200 \times 10^3}$$

$$\in_z = 2.8001 \times 10^{-4} = \frac{\delta b}{b}$$

$\therefore$ Change in width $\delta b = b \times \in_b$

$$= 75 \times 2.8001 \times 10^{-4}$$

$$\delta b = 0.021 \text{ mm (Elongation)}$$

Volumetric strain $\in_V = \in_x + \in_y + \in_z$

$$\in_V = 7.30655 \times 10^{-4} + (-8.29345 \times 10^{-4}) + 2.8001 \times 10^{-4}$$

$$\in_V = \frac{\delta V}{V} = 1.8132 \times 10^{-4}$$

$\therefore$ Change in volume $\delta V = \in_V \times V$

$$= 1.8132 \times 10^{-4}(lbt)$$

$$= 1.8132 \times 10^{-4}\,(250 \times 75 \times 25)$$

$$= 84.994 \text{ mm}^3$$

Change in 2 MN load so that change in volume is zero.

$$\text{WKT volumetric strain } \in_V = \left(\frac{\sigma_x + \sigma_y + \sigma_z}{E}\right)(1 - 2\mu)$$

$$\text{But change in volume} = \delta V = 0$$

$$\therefore \quad \in_V = 0 = \left(\frac{\sigma_x + \sigma_y + \sigma_z}{E}\right)(1 - 2\mu)$$

$$\therefore \quad \sigma_x + \sigma_y + \sigma_z = 0$$

But $\sigma_x = 133.33$ and $\sigma_z = 64$ MPa.

$$\therefore \quad \sigma_y = -\sigma_x - \sigma_z = -133.33 - 64 = -187.33 \text{ MPa}$$

$\therefore$ Load to be applied in y-direction in place of 2 MN

$$P_y = A \times \sigma_y$$

$$P_y = (250 \times 75)(-197.33) = -3699937.5 \text{ N}$$

$$= -3.6999 \times 10^6 \text{ N}$$

A compressive load of 3.6999×10^6 N is to be applied along 250 × 75 face for the volume to be zero.

Problem 20 A cube of 100 mm side is subjected to 10 N/mm^2 (tensile), 8 N/mm^2 (compressive) and 6 N/mm^2 (tensile) along x, y and z axes respectively. Determine the change in volume. Poisson's ratio is 0.25 and E = 2 × 10^5 N/mm^2.

Solution: **Given data:**

$$\sigma_x = 10 \text{ N/mm}^2 \text{ (Tensile)}$$

$$\sigma_y = -8 \text{ N/mm}^2 \text{ (Compressive)}$$

$$\sigma_z = 6 \text{ N/mm}^2 \text{ (Tensile)}$$

$$E = 2 \times 10^5 \text{ N/mm}^2$$

$$\mu = 0.25$$

For a member subjected to stresses along three mutually perpendicular directions, volumetric strain $\in_V = \left(\frac{\sigma_x + \sigma_y + \sigma_z}{E}\right)(1 - 2\mu)$

$$\in_V = \left(\frac{10 - 8 + 6}{2 \times 10^5}\right)(1 - 2 \times 0.25)$$

$$\in_V = \frac{\delta V}{V} = 2 \times 10^{-5}$$

$$\text{Volume of cube } V = l^3 = 100^3 \text{ mm}^3$$

$$\therefore \quad \text{Change in volume } \delta V = \in_V \times V$$

$$\delta V = 2 \times 10^{-5} \times 100^3 = 20 \text{ mm}^3.$$

Problem 21 A block of steel is subjected to forces as shown in Fig. 2.13. Determine the change in volume of the block. Take E = 200 GPa and M = 0.25.

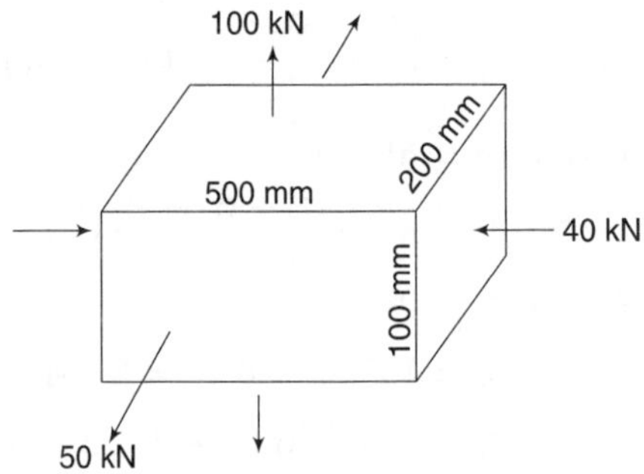

Fig. 2.13

Solution: **Given data:**

Forcc on 100 mm × 200 mm face P_x = – 40 kN = – 40 × 10^3N (Compressive)

Force on 500 mm × 200 mm face P_y = 100 kN = 100 × 10^3 N (Tensile)

Force on 500 mm × 100 mm face P_z = 50 kN = 50 × 10^3 N (Tensile)

$$\text{Stress } \sigma_x = \frac{P_x}{A_x} = \frac{-400 \times 10^3}{100 \times 200} = -2 \text{ N/mm}^2$$

$$\text{Stress } \sigma_y = \frac{P_y}{A_y} = \frac{100 \times 10^3}{500 \times 200} = 1 \text{ N/mm}^2$$

$$\text{Stress } \sigma_z = \frac{P_z}{A_z} = \frac{50 \times 10^3}{500 \times 100} = 1 \text{ N/mm}^2$$

$$\text{Volumetric strain } \epsilon_V = \frac{\delta V}{V} = (\sigma_x + \sigma_y + \sigma_z)(1 - 2\mu)$$

$$= (-2 + 1 + 1)(1 - 2 \times 0.3) = 0$$

∴ Change in volume $\delta V = \epsilon_V \times V = 0$

There is no change in volume of the block.

Problem 22 A test element is subjected to three mutually perpendicular unequal stresses.

(i) Find the change in volume of the element

(ii) If the algebraic sum of the these stresses is equal to zero, find the change in the volume of the element.

Solution:

Let stress in x-direction = σ_x

y-direction $= \sigma_y$ and

z-direction $= \sigma_z$

E = Modulus of elasticity

μ = Poisson's ratio

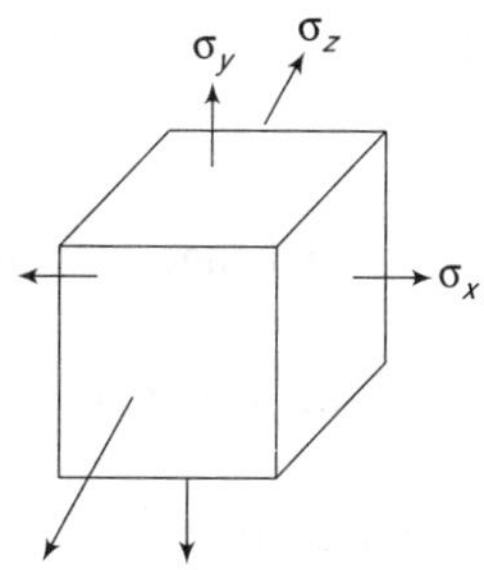

Fig. 2.14

Stress σ_x' will induce tensile strain $\frac{\sigma_x}{E}$ in x-direction and compressive strains $\frac{-\mu\sigma_x}{E}$ in y- and z-diresctions.

Similarly, σ_y will induce tensile strain $\frac{\sigma_y}{E}$ in y-direction and compressive strains $\frac{-\mu\sigma_y}{E}$ in x- and z-directions and σ_z will induce tensile strain $\frac{\sigma_z}{E}$ in z-direction and compressives strain $\frac{-\mu\sigma_z}{E}$ in x- and y-directions.

$\therefore$ Strains in x, y amd z directions are:

$$\in_x = \frac{\sigma_x}{E} - \frac{\mu\sigma_y}{E} - \frac{\mu\sigma_z}{E}$$

$$\in_y = \frac{\sigma_y}{E} - \frac{\mu\sigma_x}{E} - \frac{\mu\sigma_z}{E}$$

and

$$\in_z = \frac{\sigma_z}{E} - \frac{\mu\sigma_x}{E} - \frac{\mu\sigma_y}{E}$$

$$\text{Volume } V = lbt$$

$$\delta V = \delta l\, bt + \delta b\, lt + \delta t\, lb$$

Dividing by V

$$\text{Volumetric strain } \frac{\delta V}{V} = \frac{\delta lbt}{lbt} + \frac{\delta blt}{lbt} + \frac{\delta tlb}{lbt}$$

$$\frac{\delta V}{V} = \frac{\delta l}{l} + \frac{\delta b}{b} + \frac{\delta t}{t}$$

$$\frac{\delta V}{V} = e_x + e_y + e_z$$

$$\frac{\delta V}{V} = \left(\frac{\sigma_x}{E} - \frac{\mu\sigma_y}{E} - \frac{\mu\sigma_z}{E}\right) + \left(\frac{\sigma_y}{E} - \frac{\mu\sigma_x}{E} - \frac{\mu\sigma_z}{E}\right)$$

$$+ \left(\frac{\sigma_z}{E} - \frac{\mu\sigma_x}{E} - \frac{\mu\sigma_y}{E}\right)$$

$$= \frac{\sigma_x}{E}(1 - 2\mu) + \frac{\sigma_y}{E}(1 - 2\mu) + \frac{\sigma_z}{E}(1 - 2\mu)$$

$$\frac{\delta V}{V} = \left(\frac{\sigma_x + \sigma_y + \sigma_z}{E}\right)(1 - 2\mu)$$

$\therefore$ Change in volume $$\delta V = \left(\frac{\sigma_x + \sigma_y + \sigma_z}{E}\right)(1 - 2\mu)\cdot V$$

If the algebraic sum of $\sigma_x + \sigma_y + \sigma_z = 0$

then $\delta V = \frac{0}{E}(1 - 2\mu)\cdot V = 0$

Hence, change in volume = 0.

Problem 23 A steel bar of rectangular cross section 20 mm thick 50 mm wide is 300 mm long. A compresive force of 150 kN is applied axially. Find the change in dimensions and volume if the expansion along 50 mm × 300 mm sides is restricted. Take modulus of elesticity as 206 GPa and Poisson's, ratio as 0.3.

Also find the changes in dimensions and volume if half of expansion in 50 mm × 300 mm side is permitted.

Solution:

Given data: Load in x-direction $P_x = -150$ kN (Compressive)

$$E = 206 \text{ GPa} = 206 \times 10^3 \text{ MPa}$$

$$\mu = 0.3$$

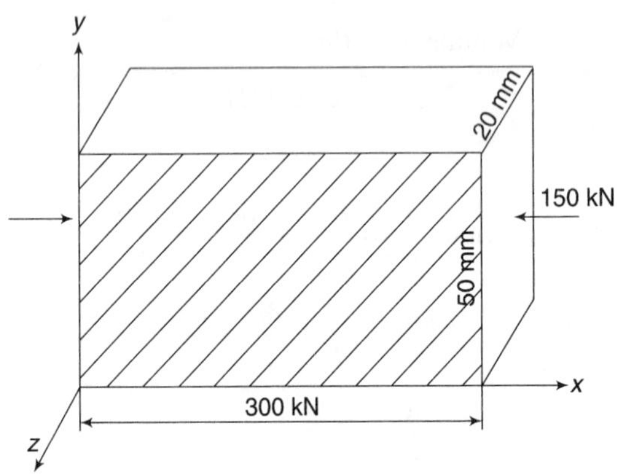

Fig. 2.15

Stress in x-direction $$\sigma_x = \frac{P_x}{A} = \frac{-150 \times 10^3}{20 \times 50} = -150 \text{ MPa}$$

The bar is restrained to expand in z direction. Hence, there will be compressive stress in z-direction = σ_z.

$$\in_x = \frac{\sigma_x}{E} - \frac{\mu\sigma_y}{E} - \frac{\mu\sigma_z}{E} \quad \text{Strain in } x\text{-direction}$$

Given $\sigma_y = 0$ (No expansion allowed)

$$\in_x = \frac{-150}{206 \times 10^3} + \frac{0.3(-\sigma_z)}{200 \times 10^3}$$

$$\in_x = \frac{0.3\sigma_z - 150}{206 \times 10^3} \quad ...(1)$$

Strain in y-direction

$$\in_y = \frac{\sigma_y}{E} - \frac{\mu\sigma_x}{E} - \frac{\mu\sigma_z}{E}$$

$$\in_y = 0 - \frac{0.3\ (-150)}{206 \times 10^3} - \frac{0.3(-\sigma_z)}{206 \times 10^3} = \frac{0.3(150 + \sigma_z)}{206 \times 10^3} \quad ...(2)$$

Strain in z-direction

$$\in_z = \frac{\sigma_z}{E} - \frac{\mu\sigma_x}{E} - \frac{\mu\sigma_y}{E}$$

$$\in_z = \frac{-\sigma_z}{206 \times 10^3} - \frac{0.3(-150)}{206 \times 10^3} - 0$$

$$= \frac{-\sigma_z + 50}{206 \times 10^3} \quad ...(3)$$

when the expansion in z-drection is fully restricted

Stran $\in_z = 0$

$$\frac{-\sigma_z + 50}{206 \times 10^3} = 0$$

∴ Stress in z-direction $\sigma_z = 50$ MPa (Compressive)

Substituting σ_z in Eqs. (1) and (2)

$$\text{Strain in } x \text{ direction } \in_x = \frac{0.3\ \sigma_z - 150}{206 \times 10^3}$$

$$\in_x = \frac{0.3 \times 50 - 150}{206 \times 10^3} = -6.5534 \times 10^4$$

$$\text{Strain } \in_y = \frac{0.3\ (150 + \sigma_z)}{206 \times 10^3} = \frac{0.3(150 + 50)}{206 \times 10^3}$$

$$= 2.9126 \times 10^4$$

$$\text{Change in length } \delta l = \in_x . l = -6.5534 \times 10^{-4} \times 300$$

$$\delta l = -0.0166 \text{ mm}$$

(– ve indicates decrease in length)

$$\text{Change in thickness } \delta t = \in_y \cdot t$$
$$= 2.9126 \times 10^{-4} \times 20 = 0.005825 \text{ mm}$$

(+ ve indicates increase in thickness)

$$\text{Change in width } \delta b = \in_z \cdot b = 0 \times 50 = 0$$
$$\text{Change in volume } \delta V = \in_V \cdot V$$
$$\text{But volumetric strain } \in_V = \in_x + \in_y + \in_z$$
$$\in_V = 06.5534 \times 10^{-4} + 2.9126 \times 10^{-4} + 0$$
$$= -3.6408 \times 10^{-4}$$
$$\text{Volume } V = lbt = 300 \times 50 \times 20$$
$$= 300{,}000 \text{ mm}^3$$
$$\therefore \quad \text{Change in volume} = -3.6408 \times 10^{-4} \times 300000$$
$$\delta V = -109.224 \text{ mm}^3$$

(–ve indicates decrease in volume)

half the expansion is permitted:

$$\text{Free strain } z \text{ direction due to } \sigma_x, = \frac{-\mu \cdot \sigma_x}{E}$$
$$= \frac{-0.3 \times (-150)}{206 \times 10^3} = \frac{50}{206 \times 10^3}$$

If half of this is permitted then strain in z direction $\in_z = \left(\dfrac{0.5}{206 \times 10^3}\right)$

$$\in_z = 1.2136 \times 10^{-4} \text{ (Tensile)}$$

From Eq. (3) $\in_z = \dfrac{-\sigma_z + 50}{206 \times 10^3} = 1.2136 \times 10^{-4}$

$$\therefore \quad \sigma_z = 25 \text{ MPa}$$

Subsituting in Eqs. (1), (2) and (3)

$$\therefore \quad \text{Strain in } x\text{-direction } \in_x = \frac{0.3\sigma_z - 150}{206 \times 10^3} = \frac{0.3 \times 25 - 150}{206 \times 10^3}$$
$$\in_x = -6.9175 \times 10^{-4}$$
$$\text{strain in } y\text{-direction } \in_y = \frac{0.3(\sigma_z + 150)}{206 \times 10^3}$$
$$\in_y = \frac{0.3(25 + 150)}{206 \times 10^3} = 2.5485 \times 10^{-4}$$

$$\text{Strain in } z \text{ direction } \in_z = \frac{-\sigma_z + 50}{206 \times 10^3}$$

$$\in_z = \frac{-25 + 50}{206 \times 10^3} = 1.2136 \times 10^{-4}$$

$$\text{Change in length } \delta l = \in_x \cdot l$$

$$= -6.9175 \times 10^{-4} \times 300 = -0.2075 \text{ mm}$$

−ve indicating decrease in length

$$\text{Change in width } \delta b = \in_z \cdot b$$

$$= 1.2136 \times 10^{-4} \times 50 = 0.006068 \text{ mm}$$

$$\text{Change in thickness } \delta t = \in_y \cdot t = 2.5484 \times 10^{-4} \times 20$$

$$= 0.005097 \text{ mm}$$

$$\text{Volumetric strain } \in_V = \in_x + \in_y + \in_z$$

$$= -6.9175 \times 10^{-4} + 2.5485 \times 10^{-4} + 1.2136 \times 10^{-4}$$

$$\in_V = -3.1554 \times 10^{-4} = \frac{\delta V}{V}$$

$$\therefore \quad \text{Change in volume } \delta V = \in_V \cdot V$$

$$\delta V = -3.1554 \times 10^{-4} \times 300000$$

$$= -94.662 \text{ mm}^3$$

(−ve indicates decrease in volume)

Problem 24 A steel bar of square cross section of side 10 mm carries an axial compressive load of 750 kN. The modulus of elasticity of the material is 200 GPa. Poisson's ratio = 0.3. Two lateral stresses act on the other two mutually perpendicular faces such that strains in these two lateral directions are zero. Find the modified value of modulus of elesticity under these conditions.

Solution:

Given data: Side of square a = 100 mm

$$\text{Load } P = 750 \text{ kN} = 750 \times 10^3 \text{ N}$$

$$\therefore \quad \text{Axial stress } \sigma_x = \frac{P}{A} = \frac{750 \times 103}{100 \times 100} = 75 \text{ MPa}$$

Young's modulus $E = 200 \text{ GPa} = 200 \times 10^3 \text{ MPa}$

Let $\sigma = \sigma_y = \sigma_z$ be the lateral compressive stresses act on other two mutually perpendicluar faces

$$\text{Lateral strain } \in_b = \in_t = \frac{\sigma}{E} - \frac{\mu . \sigma}{E} - \frac{\mu . \sigma x}{E}$$

$$= \frac{-\sigma}{200 \times 10^3} - \frac{0.3(-\sigma)}{200 \times 10^3} - \frac{0.3 \times (-75)}{200 \times 10^3}$$

It is given that the lateral strain is zero

$$\therefore \quad 0 = \frac{-\sigma}{200 \times 10^3} + \frac{0.3\sigma}{200 \times 10^3} + \frac{0.3 \times 75}{200 \times 10^3}$$

$$\therefore \quad \text{Streess } \sigma = 32.143 \text{ MPa}$$

$$\therefore \quad \text{Longitudinal strain } \in_x = \in_l = \frac{\sigma_x}{E} - \frac{\mu\sigma}{E} - \frac{\mu\sigma}{E}$$

$$= \frac{-75}{200 \times 10^3} - \frac{0.3\,(-32.143)}{200 \times 10^3} - \frac{0.3\,(-3\,2.143)}{200 \times 10^3}$$

$$\text{Strain} = 2.7857 \times 10^{-4}$$

$$\text{But modified longitudinal strain} = \frac{\sigma}{E_1} = \frac{-75}{E_1}$$

where E_1 = Modified Young's modulus

$$\text{Equating } 2.7857 \times 10^{-4} = \frac{75}{E_1}$$

$$\therefore \quad \text{Modified Young's modulus } E_1 = 269.23 \times 10^3 \text{ MPa.}$$

Problem 25 A perismatic bar is stretched in such a way that all the lateral strains are prevented. Find the value of the modified modulus of elasticity. What will be modified Poisson's ratio?

Solution:

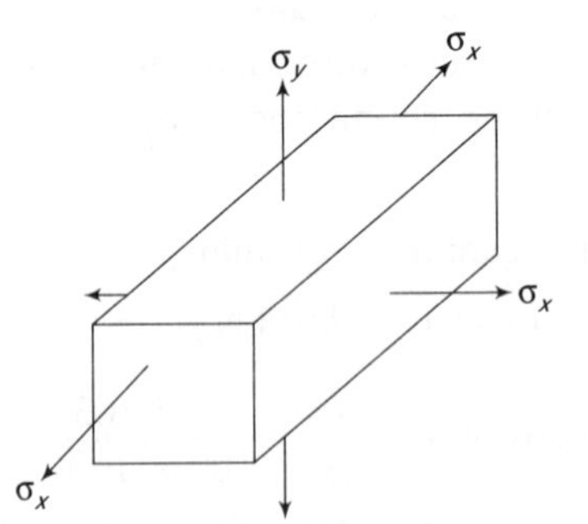

Fig. 2.16

Figure 2.16 shows a prismatic bar stretched in all the three directions.

Let σ_x = Stress in x-direction

σ_y = Stress in y-direction

σ_z = Stress in z-direction.
E = Original modulus of elasticity
μ = Poisson's ratio
E_1 = Modified modulus of elasiticity
μ_1 = Modified Poisson's ratio

It is given all the lateral strains are prevented.

$\therefore$ Lateral strain in y-direction $\in_y = 0$

Lateral strain in z-direction $\in_z = 0$

WKT strain in x-direction

$$\in_x = \frac{\sigma_x}{E} - \frac{\mu}{E}\sigma_y - \frac{\mu}{E}\sigma_z \quad ...(1)$$

Strain in y-direction $$\in_y = \frac{\sigma_y}{E} - \frac{\mu\sigma_x}{E} - \frac{\mu\sigma_z}{E} \quad ...(2)$$

Strain in z-direction $$\in_z = \frac{\sigma_z}{E} - \frac{\mu\sigma_x}{E} - \frac{\mu\sigma_y}{E} \quad ...(3)$$

From Eq. (2) $\frac{\sigma_y}{E} - \frac{\mu\sigma_x}{E} - \frac{\mu\sigma_z}{E} = 0$ (given $\in_y = 0$)

$$\frac{\in_y}{E} = \frac{\mu}{E}(\sigma_x + \sigma_z) \quad ...(4)$$

From Eq. (3)

$$\frac{\sigma_z}{E} - \frac{\mu\sigma_x}{E} - \frac{\mu\sigma_y}{E} = 0$$

$$\mu\frac{\sigma_y}{E} = \frac{\sigma_z}{E} - \frac{\mu\sigma_x}{E}$$

$$\frac{\sigma_y}{E} = \frac{1}{\mu E}[\sigma_z - \mu\sigma_x] \quad ...(5)$$

Equating Eqs. (4) and (5)

$$\frac{\mu}{E}(\sigma_x + \sigma_z) = \frac{1}{\mu E}(\sigma_z - \mu\sigma_x)$$

$$\mu(\sigma_x + \sigma_z) = \frac{1}{\mu}(\sigma_z - \mu\sigma_x)$$

$$\mu\sigma_x + \mu\sigma_z = \frac{\sigma_z}{\mu} - \sigma_x$$

$$\mu\sigma_x + \sigma_x = \frac{\sigma_z}{\mu} - \mu\sigma_z$$

$$\sigma_x (1 + \mu) = \frac{\sigma}{z}\left(\frac{1}{\mu} - \mu\right)$$

$$= \sigma_z \left(\frac{1 - \mu^2}{\mu}\right)$$

$$\sigma_x (1 + \mu) = \frac{\sigma_z (1 + \mu)(1 - \mu)}{\mu}$$

$$\therefore \qquad \sigma_z = \frac{\mu \sigma_x}{1 - \mu} \qquad \text{...(6)}$$

Substituting in Eq. (4) $\frac{\sigma_y}{E} = \frac{\mu}{E}\left(\frac{\sigma_x + \mu\sigma_x}{1 - \mu}\right)$

$$\sigma_y = \mu\sigma_x + \frac{\mu 2\sigma_x}{1-\mu}$$

$$= \mu\sigma_x \left(1 + \frac{\mu}{1 - \mu}\right)$$

$$= \mu\sigma_x \left[\frac{(1 - \mu) + \mu}{1 - \mu}\right]$$

$$\sigma_y = \frac{\mu\sigma_x}{1 - \mu} \qquad \text{...(7)}$$

Substituting for σ_y and σ_z from Eqs. (7) and (6) in Eq. (1)

$$\in_x = \frac{\sigma_x}{E} - \frac{\mu}{E}\left[\frac{\mu\sigma_x}{1 - \mu}\right] - \frac{\mu}{E}\left(\frac{\mu\sigma_x}{1 - \mu}\right)$$

$$= \frac{\sigma_x}{E}\left[1 - \frac{2\mu^2}{1 - \mu}\right]$$

$$= \frac{\sigma_x}{E}\left[\frac{1 - \mu - 2\mu^2}{1 - \mu}\right]$$

Factorising $(1 - \mu - 2\mu^2)$:

$$= 1 - 2\mu + \mu - 2\mu^2$$

$$= 1(1 - 2\mu) + \mu(1 - 2\mu)$$

$$= (1 - 2\mu)(1 + \mu)$$

$$\therefore \qquad \text{Strain in } x\text{-direction } \in_x = \frac{\delta l}{l} = \frac{\sigma_x}{E}\left[\frac{(1 + \mu)(1 - 2\mu)}{1-\mu}\right]$$

$$\text{Strain } \in_x = \frac{\delta l}{l} = \frac{\sigma_x}{E} \quad \left[\because E_1 = \frac{\sigma_x}{\in_x}\right]$$

$$E_1 = \text{Modified Young's modulus}$$

$$\frac{\sigma_x}{E_1} = \frac{\sigma_x}{E} \frac{(1 + \mu)(1 - 2\mu)}{(1 - \mu)}$$

$$\therefore \quad \text{Modified Young's modulus } E_1 = \frac{E(1 - \mu)}{(1 + \mu)(1 - 2\mu)}$$

$$\text{Modified Poisson's ratio } \mu_1 = \frac{\text{Lateral strain}}{\text{Longitudinal strain}}$$

Given lateral strain is prevented, hence = 0

$$= \mu_1 = \frac{0}{\in_x} = 0$$

Modified Poisson's ratio = μ_1 = 0.

Problem 26 A mild steel bar 50 mm square in section and 150 mm long is subjected to axial thrust of 200 kN. Half the lateral strain is prevented by the application of uniform external pressure of suitable intensity. If E = 200 GPa and Poisson's ratio is 0.3, calculate the change in length of the bar. Also find change in volume of the bar.

Solution:

Given data: Square cross section

$$\text{Side of square } a = 50 \text{ mm}$$

$$\text{Length } l = 150 \text{ mm}$$

$$\text{Load in axial direction } P_x = 200 \text{ kN} = 200 \times 10^3 \text{ N}$$

$$\text{Modulus of eleasticity } E = 200 \text{ GPa} = 200 \times 10^3 \text{ MPa}$$

$$\text{Poisson's ratio } \mu = 0.3$$

$$\text{Axial stress } \sigma_x = \frac{P_x}{A} = \frac{200 \times 10^3}{50 \times 50} = 80 \text{ MPa}$$

Lateral strain under the axial load

$$\in_b = \in_t = -\frac{\mu\,\sigma_x}{E} = \frac{-0.3 \times 80}{200 \times 10^3} = -1.2 \times 10^{-4}$$

Half of this lateral strain is prevented.

$$\text{Hence, actual lateral strain} = \frac{1}{2}(-1.2 \times 10^{-4}) = 6 \times 10^{-5}$$

With the application of 3 mutually perpendicular stresses

$$\text{Strain in } y \text{ direction } \epsilon_b = \frac{\sigma_y}{E} - \frac{\sigma_x}{\mu E} - \frac{\sigma_z}{\mu E}$$

$$6 \times 10^{-5} = \frac{\sigma_y}{200 \times 10^3} - \frac{0.3 \times 80}{200 \times 10^3} - \frac{0.3\sigma_y}{E}$$

$$6 \times 10^{-5} = \frac{0.7\sigma_y - 24}{200 \times 10^3}$$

$$\therefore \qquad \sigma_y = 51.428 \text{ MPa.}$$

$\therefore$ Strain in x-direction

$$\epsilon_x = \frac{\sigma_x}{E} - \frac{\mu\ \sigma_y}{E} - \frac{\mu\ \sigma_y}{E}$$

$$\epsilon_x = \frac{80}{200 \times 10^3} - \frac{0.3 \times 51.428}{200 \times 10^3} \times 2$$

$$\epsilon_x = \frac{\delta l}{l} = 2.4571 \times 10^{-4}$$

$\therefore$ Change in length $\delta l - \epsilon_x \cdot l = 2.457 \times 10^{-4} \times 150 = 0.0369$ mm

Change in volume:

$$\text{Volumetric strain } \frac{\delta V}{V} = \left(\frac{\sigma_x + \sigma_y + \sigma_z}{E}\right)(1 - 2\ \mu)$$

$$= \left(\frac{80 + 51.428 + 51.428}{200 \times 10^3}\right)(1 - 2 \times 0.3)$$

$$\frac{\delta V}{V} = 0.3657 \times 10^{-3}$$

$$\text{Volume } V = A \times l = (a \cdot a) \cdot l = 50 \times 50 \times 150 = 375000 \text{ mm}^3$$

$$\therefore \qquad \text{Change in volume } \delta V = \epsilon_V \cdot V = 0.3657 \times 10^3 \times 375000$$

$$\delta V = 137.1375 \text{ mm}^3.$$

Problem 27 A bar of 25 mm diameter carries a compressive load of 60 kN. Find the stresses in the bar and change in volume due to this compressive load in the bar if 50% of expansion in the diameter is restricted by suitable stress on the rod. Length of bar is 400 mm. E = 200 GPa. Poisson's ratio = 0.3.

Solution:

Given Data: Diameter d = 25 mm

Load = $P_x = -60$ kN = $-\ 60 \times 10^3$ N (Compressive)

Length l = 400 mm

Poisson's ratio μ = 0.3

$$E = 200 \text{ GPa} = 200 \times 10^3 \text{ MPa}$$

50% of lateral strain is prevented.

$$\text{Axial compressive stress } \sigma_x = \frac{P_x}{A} = \frac{-60 \times 10^3}{\frac{\pi}{4} \times 25^2} = -122.23^1 \text{ N}$$

Free lateral strain due to axial stress

$$\in_d = \frac{-\mu \sigma x}{E} = \frac{-0.3(-122.231)}{200 \times 10^3} = -1.83346 \times 10^{-4}$$

If 50% of expansion in the diameter is restricted.

Actual lateral strain = $0.5 \times 1.83346 \times 10^{-4} = 9.1673 \times 10^{-5}$

Under the application of suitable stresses

$$\text{Lateral strain } \in_d = \frac{\sigma_y}{E} - \frac{\mu\sigma_x}{E} - \frac{\mu\sigma_z}{E}$$

Since the strain in y and z directions is same

$$\sigma_y = \sigma_z$$

$$9.1673 \times 10^{-5} = \frac{\sigma_y}{200 \times 10^3} - \frac{0.3 \times (-122.231)}{200 \times 10^3} - \frac{0.3(\sigma_y)}{200 \times 10^3}$$

$$9.1673 \times 10^{-5} = \frac{\sigma_y + 36.6683 - 0.3\,\sigma_y}{200 \times 10^3}$$

$$\sigma_y = -26.192 \text{ MPa} = \sigma_z \quad \text{(Compressive)}$$

Stress in x-direction σ_x = 122.231 MPa

Stresses in y-direction $\sigma_y = \sigma_z$ = 26.192 MPa (All compressive)

For a member subjected to stresses in three mutually perpendicular directions,

$$\text{Volumetric strain } \in_V = \left(\frac{\sigma_x + \sigma_y + \sigma_z}{E}\right)(1 - 2\mu)$$

$$\in_V = \frac{\delta V}{V} = \left(\frac{-122.231 - 2 \times 26.192}{200 \times 10^3}\right)(1 - 2 \times 0.3)$$

$$= -3.4923 \times 10^{-4}$$

$\therefore$ Change in volume $\delta V = \in_V \times V$

$$\text{For round rod, volume } V = A \cdot l = \frac{\pi}{4} d^2 l = \frac{\pi}{4} \times 25^2 \times 400$$

$\therefore$

$$\delta V = -3.4923 \times 10^{-4} \times \left(\frac{\pi}{4} \times 25^2 \times 400\right)$$

$$\delta V = -68.571 \text{ mm}^3$$

(–ve indicates that reduction in volume)

THINGS TO REMEMBER

- Longitudinal strain = $\dfrac{\text{Change in length}}{\text{Origin length}}$
- Lateral strain = $\dfrac{\text{Change in lateral dimension (width, thickness or diameter)}}{\text{Origin lateral dimension}}$

$$= \frac{\delta b}{b} \text{ or } \frac{\delta t}{t} \text{ or } \frac{\delta d}{d}$$

- Poisson's ratio (μ) = $\dfrac{\text{Lateral strain}}{\text{Longitudinal strain}}$
- Shear stress due to pure shear = $\tau = \dfrac{P}{A}$
- Rigidity modulus or shear modulus (C, G or N)

$$G = \frac{\text{Shear stress}}{\text{Shear strain}} = \frac{\tau}{\varphi}$$

- Volumetric strain $\in_V = \dfrac{\delta V}{V} = \dfrac{\text{Change in volume}}{\text{Original volume}}$
- Volumetric strain of a rod (either rectangular or circular cross section) subjected to axial load,

$$\frac{\delta V}{V} = \frac{\sigma_x}{E}(1 - 2\mu)$$

or

$$\frac{\delta V}{V} = \frac{\delta l}{l}(1 - 2\mu)$$

- Poisson's ratio varies between 0.25 to 0.33 for metals and can have a maximum value of 0.5.
- For a block subjected to stresses in three mutually perpendicular directions, volumetric strain $\dfrac{\delta V}{V} = \left(\dfrac{\sigma_x + \sigma_y + \sigma_z}{E}\right)(1 - 2\mu)$
- Bulk modulus is the ratio of direct equal stress acting in three mutually perpendicular directions to the volumetric strain

$$K = \frac{\sigma}{\left(\dfrac{\delta v}{V}\right)}$$

- Relation between Young's modulus and rigidity modulus

$$E = 2G(1 + \mu) \text{ or } G = \frac{E}{2(1 + \mu)}.$$

- Relation between Young's modulus and bulk modulus

$$E = 3K\,(1 - 2\mu) \text{ or } K = \frac{E}{3K(1 - 2\mu)}$$

- Relation between the three modulii

$$E = \frac{9KG}{3K + G} \text{ or } \frac{9}{E} = \frac{3}{G} + \frac{1}{K}.$$

- If algebraic sum of three mutually perpendicular stresses acting on a body is zero, (i.e., $\sigma x + \sigma y + \sigma z = 0$) then the change in volume and volumetric strain are zero.

EXERCISES

1. Define Poisson's ratio, modulus of rigidity and bulk modulus.
2. What are elastic constants? Briefly explain.
3. Explain with an example, the difference between lateral strain and longitudinal strain, and define Poisson's ratio.
4. Find volumetric strain of a cylindrical bar subjected to an axial load.
5. Prove that volumetric strain is equal to the sum of the three principal strains. ($\in_V = \in_x + \in_y + \in_z$).
6. Define Poisson's ratio and prove that the maximum value of Poisson's ratio for any material is 0.5.
7. Explain volumetric strain and obtain the expression for volumetric strain for a prismatic bar subjected to an axial load.
8. Derive the relation between Young's modulus and rigidity modulus in terms of Poisson's ratio.
9. Derive the relation between Young's modulus, bulk modulus in terms of Poisson's ratio.
10. Derive the relation between the three elastic constants.
11. Derive expression for volumetric strain for a rectangular bar subjected to three mutually perpendicular tensile stresses.
12. For a block subjected to three mutually perpendicular unequal stresses, show that the change in volume is zero if the algebraic sum of these three stresses is zero.
13. Explain simple shear stress and shear strain.
14. State and explain shear modulus.
15. Determine the change in length, width and thickness of a steel bar 3 m long having rectangular cross section 50 mm wide and 30 mm thick subjected to an axial pull of 60 kN. Take Young's modulus as 206 GPa and Poisson's ratio as 0.3.

    (***Ans.*** δl = 0.5825 mm, δb = 0.002913 mm, δt = 0.0017476 mm)
16. A bar of rectangular cross section 80 mm wide 30 mm thick and 6 m long carries an axial tensile load of 120 kN. Modulus of elasticity is 2×10^5 MPa and Poisson's ratio is 0.28. Find the volumetric strain and change in volume.

17. A round rod of 60 mm diameter 4.5 m long carries a tensile load of 150 kN. Taking E = 200 GPa and Poisson's ratio = 0.3, find change in length, diameter, change in volume and volumetric strain.

(***Ans.*** δl = 1.19366 mm., δd = 0.004776 mm, δV = 1350 mm^3, $\epsilon_V = 1.06 \times 10^{-4}$)

18. Determine the Young's modulus, Poisson's ratio, bulk modulus and rigidity modulus for a bar 60 mm wide, 30 mm thick and 3 m long subjected to an axial load of 420 kN. The elongation in length is 3.6 mm and decrease in thickness is : 0.011 mm.

(***Ans.*** $E = 194.444 \times 10^3$ MPa, $\mu = 0.3056$, $K = 166.7 \times 10^3$ MPa, $G = 74.465 \times 10^3$ MPa)

19. A bar of circular cross section 100 mm diameter 4 m long carries a compressive load of 900 kN. The decrease in length is 2.30 mm and increase in diameter is = 0.0175 mm. Find the Poisson's ratio, and the values of three elastic constants.

(***Ans.*** $\mu = 0.3043$, $E = 199.29 \times 10^3$ MPa, $G = 76.4 \times 10^3$ MPa, $K = 169.72 \times 10^3$ MPa)

20. A metal rod of diameter 10 mm under axial load of 10 kN has shown a decrease in diameter of 0.002 mm. Find the modulus of elasticity and bulk modulus of the material.

(***Ans.*** $\mu = 0.3082$, $E = 196.24 \times 10^3$ MPa, $K = 170.57 \times 10^3$ MPa)

21. A block is subjected to forces as shown in Fig. 2.17. Determine the change in the volume of the block. Take E = 200 kN/mm^2 and Poisson's ratio = 0.3.

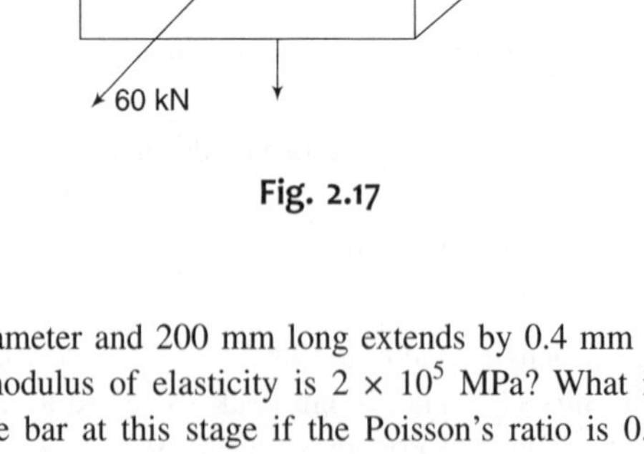

Fig. 2.17

(***Ans:*** δV = – 227.4 mm^3)

22. A bar of 20 mm diameter and 200 mm long extends by 0.4 mm upon a load P. What is the value of P if the modulus of elasticity is 2×10^5 MPa? What is the change in the cross-sectional area of the bar at this stage if the Poisson's ratio is 0.3. Also find the change in volume of the bar.

(***Ans.*** $P = 125.664 \times 10^3$ N, δA = 0.3769 mm^2, δV = 50.265 mm^3)

23. A block of steel is subjected to forces as shown in Fig. 2.18. Determine the change in volume of the block. E = 200 GPa and μ = 0.3.

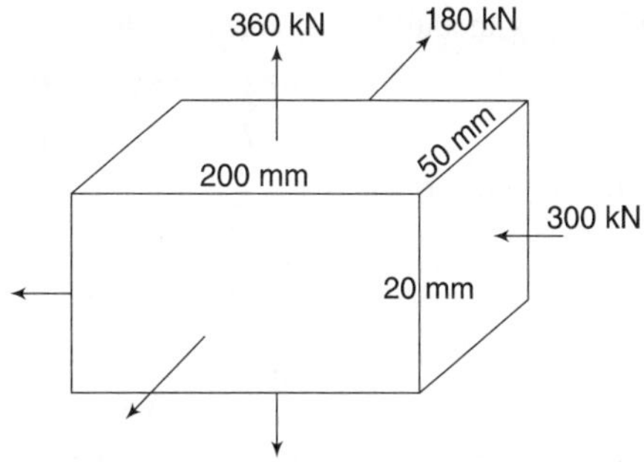

Fig. 2.18

(***Ans:*** δV = 100.8 mm^3)

24. A rectangular block 80 mm wide, 50 mm thick and 400 mm long is loaded width 300 kN tenside load along the axis of the member. A compressive force of 3 MN is applied on face 400 mm × 80 mm and another tensile force of 500 kN acts on the face 400 mm × 50 mm. If μ = 0.3 and E = 200 GPa, determine the change in length, width and thickness of bar. Also find change in volume of the bar. What load should be applied in place of 3 MN so that change in volume will be zero.

(***Ans:*** δl = 0.19125 mm, δb = 0.01225 mm, δt = – 0.03094 mm, δV = 20 mm^3, Py = 3.2 MN)

25. A bar of rectangular cross section carries an axial stress. The lateral strain on one of the faces is 50% of the strain that would be there without any constraint and there is no constraint on the other force. Show that the modified modulus of elasticity will be $[2E/(2 - \mu^2)]$.

Objective Type Questions

1. Poisson's ratio is the ratio of

(a) lateral strain to longitudinal strain
(b) longitudinal strain to lateral strain
(c) stress to strain
(d) stress to volumetric strain

2. Volumetric strain is the ratio of

(a) change in volume to length
(b) change in volume to volume
(c) change in volume to diameter
(d) change in volume in Young's modulus

3. For a member subjected to stress σ, the volumetric strain is given by

(a) $\frac{E}{\sigma}(1 - 2\mu)$
(b) $\frac{\sigma}{E}(1 - \mu)$
(c) $\frac{\sigma}{E}(1 - 2\mu)$
(d) $\frac{\sigma}{E}(1 + 2\mu)$

4. Maximum value of Poisson's ratio can be

(a) 0.3
(b) 0.28
(c) 1
(d) 0.5

5. The relation between Young's modulus and rigidity modulus is

(a) $E = 2G(1 + \mu)$
(b) $2G(1 - \mu)$
(c) $3G(1 - 2\mu)$
(d) $2G(1 + 2\mu)$

6. The relation between Young's modulus and bulk modulus is

(a) $E = 2K(1 + \mu)$ (b) $E = 3K(1 - 2\mu)$

(c) $E = 2K(1 - 2\mu)$ (d) $E = 3K(1 + 2\mu)$

7. The relation between three elastic modulii is

(a) $E = \dfrac{3KG}{9K + G}$ (b) $E = \dfrac{9KG}{K + 3G}$

(c) $E = \dfrac{9KG}{3K + G}$ (d) $E = \dfrac{9KG}{3K - G}$

8. The relation between three elastic modulii is

(a) $\dfrac{1}{E} = \dfrac{3}{G} + \dfrac{9}{K}$ (b) $\dfrac{1}{E} = \dfrac{9}{G} + \dfrac{3}{K}$

(c) $\dfrac{9}{E} = \dfrac{3}{K} + \dfrac{1}{G}$ (d) $\dfrac{9}{E} = \dfrac{3}{G} + \dfrac{1}{K}$

9. The ratio of modulus of elasticity and modulus of rigidity for a Poisson's ratio of 0.3 is

(a) 2.6 (b) 1.4

(c) 1.2 (d) $\dfrac{1}{2.6}$

10. The ratio of modulus of elasticity and bulk modulus for a Poisson's ratio of 0.3 is

(a) 2.6 (b) 1.2

(c) $\dfrac{1}{1.2}$ (d) $\dfrac{1}{2.6}$

11. The ratio of bulk modulus to rigidity modulus for a Poisson's ratio of 0.25 is

(a) $\dfrac{1}{2}$ (b) $\dfrac{3}{5}$

(c) $\dfrac{5}{3}$ (d) $\dfrac{3}{2}$

12. Which of the following is a true for Poisson's ratio μ.

(a) $-1 < \mu < 0$ (b) $0 < \mu < 0.3$

(c) $\dfrac{1}{2} < \mu < 1$ (d) $0 < \mu < \dfrac{1}{2}$

13. For a block subjected to a tensile stress of 90 MPa and a compressive stress of 60 MPa in perpendicular direction to it, the stress in 3rd perpendicular direction for the volumetric strain to be zero is

(a) 30 MPa (Compressive) (b) 30 MPa (Tensile)

(c) 150 MPa (Tensile) (d) 150 MPa (Compressive)

14. For a steel rod of cross-sectional area 200 mm^2 subjected to a tensile load of 4 kN. The Young's modulus is 2×10^5 N/mm^2 and Poisson's ratio is 0.3. The volumetric strain is

(a) 0.3×10^{-4} (b) 0.4×10^{-4}

(c) 0.4×10^{-5} (d) 0.5×10^{-4}

15. A certain material has Young's modulus of 200 GPa and rigidity modulus of 80 GPa. The poisson's ratio is

(a) 4 (b) 0.3
(c) 0.25 (d) 0.4

16. A cube of 10 mm side is subjected to a hydraulic pressure of 30 MPa on all sides. The reduction in volume if the bulk modulus is 150 GPa is

(a) 20 mm^3 (b) 2000 mm^3
(c) 2 mm^3 (d) 200 mm^3

17. The ratio of modulus of elasticity to modulus of rigidity is

(a) > 1 (b) < 1
(c) 0.3 (d) < 0.5

18. The ratio of modulus of rigidity to bulk modulus is

(a) > 1 (b) < 1
(c) 0.3 (d) $\frac{1}{2}$

19. For a certain material, the rigidity modulus is 80 GPa and Poisson's ratio is 0.25. The bulk modulus is

(a) 200 GPa (b) 100 GPa
(c) 133.33 GPa (d) 166.67 GPa

20. The Young's modulus for a material whose bulk modulus and rigidity modulus of 160 GPa and 80 GPa is

(a) $\frac{2400}{7}$ GPa (b) 288 GPa
(c) 200 GPa (d) $\frac{1440}{7}$

Key to Answers

1. (a)	2. (b)	3. (c)	4. (d)	5. (a)	6. (b)	7. (c)	8. (d)
9. (a)	10. (b)	11. (c)	12. (d)	13. (a)	14. (b)	15. (c)	16. (d)
17. (a)	18. (d)	19. (c)	20. (d)				

CHAPTER 3

Strain Energy and Impact Loading

3.1 INTRODUCTION

Whenever any external force is applied on a body, the body offers resistance and resisting force will be developed gradually in proportion to the external force. In other words, whenever a force acts on a body, the body is strained and energy is absorbed by the body. The energy which is absorbed and stored in a body due to straining of the body is called strain energy. Whenever external force is removed, this stored energy pushes back the material to original position.

The straining effect may be due to either gradually applied loads, suddenly applied loads or load with impact. The strain energy stored in the body is equal to the work done by the applied force on the body.

3.2 RESILIENCE, PROOF RESILIENCE AND UNIT RESILIENCE OR MODULUS OF RESILIENCE

3.2.1 Resilience

The total strain energy stored in a body is called resilience.

Figure 3.1 shows a diagram drawn for load P vs extension δl for a typical spacemen. As the load increases gradually from O to P, the deformation (extension or contraction) will increase from O to δl.

When deformation is zero, resistance $R = O$

When deformation is δl, resistance $R = P$

Work done by the resisting force = Average force × distance

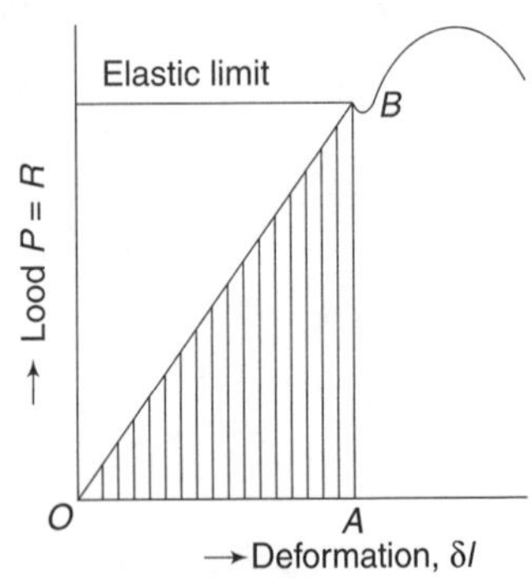

Fig. 3.1

$$\text{or} \qquad \text{Work done} = \left(\frac{O+P}{2}\right)\cdot \delta l = \frac{1}{2}P\cdot \delta l$$

$$\therefore \qquad \text{Work done} = \text{Area of } \Delta OAB = \frac{1}{2}P(\delta l)$$

$$\text{But} \qquad \text{Load } P = \sigma\cdot A$$

$$\therefore \qquad \text{Work done} = \text{Strain energy } U = \frac{1}{2}(\sigma\cdot A)\ \sigma l$$

$$\text{Strain } \in = \frac{\delta l}{l}$$

$$\therefore \qquad \delta l = \in \cdot l$$

$$\text{Substituting, strain energy } U \ \frac{1}{2}(\sigma A)\ (\in \cdot l)$$

$$\text{Volume } V = A\cdot l$$

$$\therefore \qquad U = \frac{1}{2}\sigma\in\cdot V \qquad ...(1)$$

$$\text{Young's modulus } E = \frac{\text{Stress}}{\text{Strain}} = \frac{\sigma}{E}$$

$$\therefore \qquad \text{Strain } \in = \frac{\sigma}{E}$$

$$U = \frac{1}{2}\sigma\left(\frac{\sigma}{E}\right)V$$

$$U = \frac{1}{2}\frac{\sigma^2}{E}V = \frac{1}{2}\frac{\left(\frac{P}{A}\right)^2\cdot(A\,l)}{E} = \frac{P^2 l}{2AE}$$

$$\text{Strain energy or resilience } U = \frac{\sigma^2 V}{2E} = \frac{P^2 l}{2AE} \qquad ...(2)$$

3.2.2 Proof Resilience

The maximum strain energy that can be stored by a body without undergoing any permanent deformation is called proof resilience. Hence, proof resilience is equal to the resilience in the body at elastic limit σ_y.

$$\text{Substituting } \sigma = \sigma_y \text{ in Eq. (2)}$$

$$\text{Proof resilience} = U_y = \frac{\sigma y^2}{2E}\cdot V \qquad ...(3)$$

3.2.3 Unit Resilience or Modulus of Resilience (U′)

Unit resilience or modulus of resilience is the strain energy or resilience per unit volume.

$$\text{Unit resilience} = \frac{\text{Resilience or total strain energy}}{\text{Volume of the body}}$$

$$U' = \frac{U}{V} = \frac{\frac{\sigma^2}{ZE}V}{V} = \frac{\sigma^2}{2E}$$

$$\therefore \qquad U' = \frac{\sigma^2}{2E} \qquad ...(4)$$

3.3 STRAIN ENERGY DUE TO VARIOUS TYPES OF LOADS

3.3.1 Strain Energy due to Gradually Applied Load

Consider a bar of length l and area of cross section A subjected to gradually applied load P. Since the load is gradually applied. It starts from zero and gradually increases to P. Hence, work done by load

$$= \text{Average force} \times \text{Distance}$$

$$U = \frac{O + P}{2} \times \delta l = \frac{P}{2}\,\delta l$$

But strain

$$\in = \frac{\delta l}{l}$$

or

$$\delta l = \in l$$

$$\therefore \qquad U = \frac{P}{2} \cdot \in l \qquad ...(5)$$

But from Eq. (1)

$$U = \frac{1}{2}\,\sigma \in V$$

$$\text{Volume } V = Al$$

$$U = \frac{\sigma}{2} \in (Al) \qquad ...(6)$$

Equating Eqs. (4) and (5)

$$\frac{P}{2} \in l = \frac{\sigma}{2} \in Al$$

$$\therefore \qquad \text{Stress } \sigma = \frac{P}{A} \qquad ...(7)$$

3.3.2 Strain Energy due to Suddenly Applied Load

As the load is applied suddenly, the load P is constant throughout. P is acting even when $\delta l = 0$ at the beginning and acts throughout the extension.

$$\text{Hence, work done} = \text{Load} \times \text{Distance}$$

$$= P \cdot \delta l \qquad ...(8)$$

Equating this to strain energy $U = \frac{\sigma^2}{2E} \cdot V$

$$\frac{\sigma^2}{2E} \cdot V = P\delta l$$

Substitute $\delta l = \frac{\sigma l}{E}$ and $V = Al$

$$\frac{\sigma^2}{2E} \cdot Al = P \cdot \frac{\sigma l}{E}$$

$$\therefore \qquad \text{Stress } \sigma = \frac{2P}{A} \qquad ...(9)$$

Hence, the stress developed due to suddenly applied load P is twice that due to the same load being applied gradually.

3.3.3 Strain Energy due to the Impact Load

Impact load is the load that falls from a height before it acts on the body.

Figure 3.2 shows a vertical rod of length l and area of cross section A

Let W = Weight of falling load

h = Height of fall

δl_i = Elongation under the impact

E = Modulus of elasticity

σ = Static stress $= \dfrac{W}{A}$

σ_i = Stress due to impact.

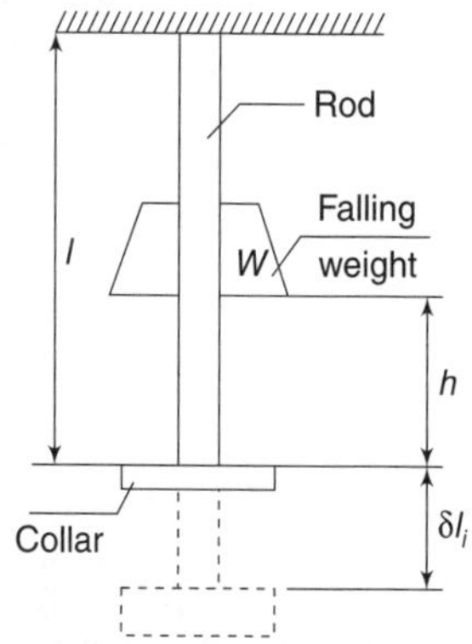

Fig. 3.2

Let the load W be dropped from a height h before it strikes the collar. Due to this impact the instantaneous elongation in the rod $= \delta l_i$

Total distance moved by load $H = h + \delta l_i$

Work done by the load = Potential energy

$$PE = \text{Load} \times \text{Distance moved} = W(h + \delta l_i)$$

Equate this to energy stored or resilience

$$U = \frac{\sigma_i^2}{2E} \cdot V$$

$$\therefore \quad \frac{\sigma_i^2}{2E} V = W(h + \delta l_i)$$

$$\sigma_i^2 = \frac{2WE}{V} = (h + \delta l_i)$$

But $$V = Al$$

$$\delta l_i = \frac{\sigma_i l}{E}$$

Substituting $$\sigma_i^2 = \frac{2WE}{Al}\left(h + \frac{\sigma_i l}{E}\right)$$

$$= 2\left(\frac{W}{A}\right)\frac{Eh}{l} + 2\,\frac{W}{A}\cdot\frac{E}{l}\cdot\left(\frac{\sigma_i l}{E}\right) = 2\,\frac{W}{A}\left(\frac{Eh}{l}\right) + 2\,\frac{W}{A}\,\sigma_i$$

$$\sigma i^2 - 2\left(\frac{W}{A}\right)\sigma_i = 2\left(\frac{W}{A}\right)\frac{Eh}{l}$$

Add $\left(\dfrac{W}{A}\right)^2$ on both sides

$$\sigma_i^2 - 2\left(\frac{W}{A}\right)\sigma_i + \left(\frac{W}{A}\right)^2 = \left(\frac{W}{A}\right)^2 + 2\left(\frac{W}{A}\right)\frac{Eh}{l}$$

LHS is of the form $a^2 - 2ab + b^2$ which is $= (a - b)^2$

$$\left(\sigma_i - \frac{W}{A}\right)^2 = \left(\frac{W}{A}\right)^2 \left(1 + \frac{2Eh}{l\left(\frac{W}{A}\right)}\right)$$

Take square root on both sides

$$\sigma_i - \frac{W}{A} = \frac{W}{A} \cdot \sqrt{1 + \frac{2hAE}{Wl}}$$

$$\therefore \qquad \sigma_i = \frac{W}{A} + \frac{W}{A}\sqrt{1 + \frac{2hAE}{Wl}}$$

$$\text{Impact stress } \sigma_i = \frac{W}{A}\left[1 + \sqrt{1 + \frac{2hAE}{Wl}}\right] \qquad ...(10)$$

$$\sigma_i = \sigma\left[1 + \sqrt{1 + \frac{2h}{\delta l}}\right] \qquad ...(11)$$

where $\sigma = \frac{W}{A}$ static stress and $\delta l = \frac{Wl}{AE}$ = static deformation.

It can be noticed that the impact stress is more than static stress produced by a load. If the load is applied suddenly without impact, then $h = 0$ in Eq. (10).

$$\sigma_i = \frac{W}{A}(1 + \sqrt{1 + 0}) = 2 \cdot \frac{W}{A} = 2\sigma$$

This is same as the suddenly applied load Eq. (9).

In many cases, elongation δl is very small compared to h, and if neglected, then

$$\text{Work done } PE = W \cdot h$$

$$\text{Equating this to resilience, } U = \frac{\sigma_i^2}{2E} \cdot V$$

$$\frac{\sigma_i^2}{2E} \cdot Al = Wh$$

$$\sigma_i^2 = \frac{2WhE}{Al}$$

$$\text{and impact stress} \qquad \sigma_i = \sqrt{\frac{2WhE}{Al}} \qquad ...(12)$$

If we neglect the value 1 which is small compared to $\frac{2hAE}{Wl}$ in the Eq. (10)

$$\sigma_i = \frac{W}{A}\left(1 + \sqrt{1 + \frac{2AEh}{Wl}}\right)$$

$$\sigma_i = \frac{W}{A} \cdot \sqrt{\frac{2hEA}{Wl}} = \sqrt{\frac{W^2}{A^2} \times \frac{2hAE}{Wl}}$$

$$\sigma_i = \sqrt{\frac{2hWE}{Al}} \text{ is same as Eq. (12)}$$

3.4 STRAIN ENERGY STORED IN A BODY DUE TO SIMPLE SHEAR STRESS

Figure 3.3 shows a rectangular block of length l width b and height h. It is fixed at bottom AB and a shear force P is gradually applied on the top face CD.

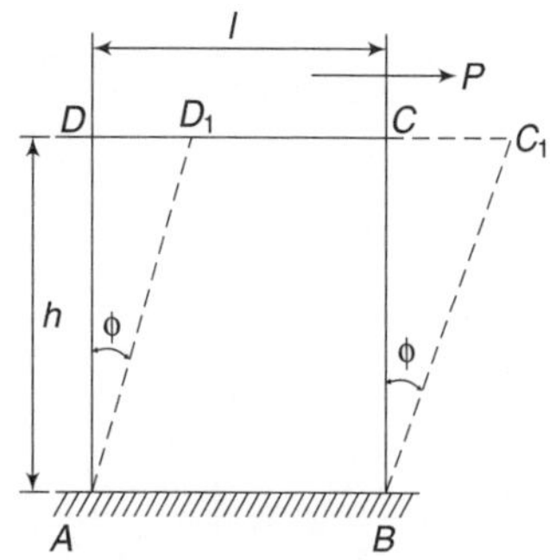

Fig. 3.3

Due to this shear force P.

The top face CD moves to new position $C_1\ D_1$

Shear stress produced on the surface

$$\tau = \frac{P}{\text{Area of top face}}$$

$$\tau = \frac{P}{l \times b}$$

$$\therefore \qquad P = \tau \cdot l \cdot b \qquad \text{...(13)}$$

From triangle BCC_1, $\tan \phi = \dfrac{CC_1}{BC} = \dfrac{CC_1}{h}$

For small angles $\tan \phi = \phi = \dfrac{CC_1}{h}$

$$\therefore \qquad CC_1 = h \cdot \phi.$$

Since force P is gradually applied,

$$\text{Work done} = \text{Average force} \times \text{Distance moved}$$

$$= \left(\frac{O + P}{2}\right) \cdot CC_1 = \frac{P}{2} \cdot h\phi$$

Substitute $P = lb.\tau$ from Eq. (13)

$$\text{Work done} = \frac{lb\tau\ h \cdot \phi}{2} = \frac{\tau}{2} \cdot \phi \cdot (lbh) = \frac{\tau}{2} \cdot \phi \cdot V$$

Since volume $V = lbh$

$$\text{WKT shear strain } \phi = \frac{\text{Shear stress}}{\text{Rigidity modulus}} = \frac{\tau}{G}$$

$$\therefore \qquad \text{Work done} = \frac{\tau}{2} \cdot \frac{\tau}{G} \cdot V = \frac{\tau^2}{2G} \cdot V$$

But Work done = Energy stored = Resilience

$$\text{Resilience in shear} = U_\tau = \frac{\tau^2}{2G}\cdot V \qquad \text{...(14)}$$

(It is similar to $U = \dfrac{\sigma^2}{2E}\cdot V$ for axial load).

3.5 STRAIN ENERGY IN TAPERING ROUND BAR

Figure 3.4 shows a taper round bar of length l tapering uniformly from bigger diameter d_2 to smaller diameter d_1.

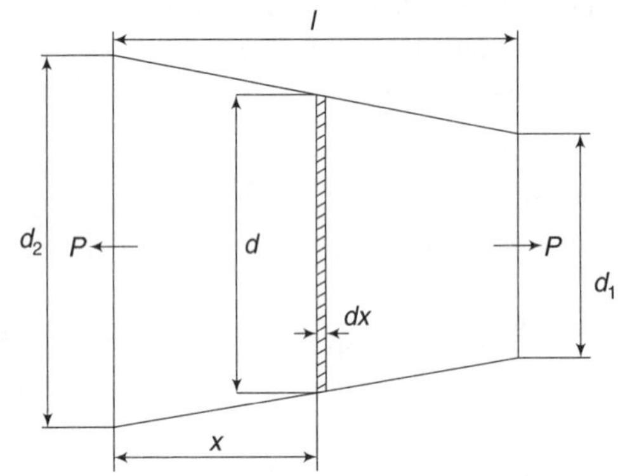

Fig. 3.4

Consider an elemental length dx of diameter d at a distance x from the bigger diameter.

$$P = \text{Load on the bar}$$

$$\text{From similar triangles } \frac{d_2 - d}{x} = \frac{d_2 - d_1}{l}$$

$$\therefore \quad d = d_2 - \left(\frac{d_2 - d_1}{l}\right) x = d_2 - Kx$$

where,
$$K = \frac{d_2 - d_1}{l}$$

$$\text{Strain Energy } U = \frac{\sigma^2}{2E}\cdot V = \frac{\sigma^2}{2E}\cdot Al = \frac{\sigma^2 A^2 l}{2AE}$$

$$U = \frac{P^2 l}{2AE}$$

For the elemental length dx, strain energy

$$U = \frac{P^2\cdot dx}{2Ax\cdot E} = \frac{P^2\cdot dx}{2\cdot\left(\frac{\pi}{4}\, d^2\right)E}$$

Substitute $d = d_2 - Kx$

$$U = \frac{2P^2 dx}{\pi(d_2 - Kx)^2 E}$$

∴ Strain energy for the entire bar

$$U = \int_o^l \frac{2P^2 dx}{\pi E(d^2 - Kx)^2} = \frac{2P^2}{\pi E}\int_o^l (d^2 - Kx)^{-2}\, dx$$

$$= \frac{2P^2}{\pi E}\left[\frac{(d^2 - Kx)^{-1}}{-1 \times -K}\right]_o^l = \frac{2P^2}{\pi EK}\left[\frac{1}{d_2 - Kx}\right]_o^l$$

But $$K = \frac{d_2 - d_1}{l}$$

∴ $$U = \frac{2P^2}{\pi E\left(\frac{d_2 - d_1}{l}\right)}\left[\frac{1}{d_2 - \left(\frac{d_2 - d_1}{l}\right)x}\right]_o^l$$

$$= \frac{2P^2 l}{\pi E(d_2 - d_1)}\left[\frac{1}{d_2\left(\frac{d_2 - d_1}{l}\right)l} - \frac{1}{d_2 - 0}\right]$$

$$= \frac{2P^2 l}{\pi E(d_2 - d_1)}\left[\frac{1}{d_1} - \frac{1}{d_2}\right] = \frac{2P^2\, l}{\pi E(d_2 - d_1)}\left[\frac{d_2 - d_1}{d_1 d_2}\right]$$

$$U = \frac{2P^2 l}{\pi E\, d_1 d_2} \qquad \ldots(15)$$

3.6 STAIN ENERGY FOR A TAPER RECTANGULAR PLATE OF UNIFORM THICKNESS

Figure 3.5 shows a taper plate of length l of uniform thickness t tapering uniformly from bigger width B to smaller width b.

It is subjected to a tansile load of P.

Consider an elemental length dx of width b_x at a distance x from bigger width.

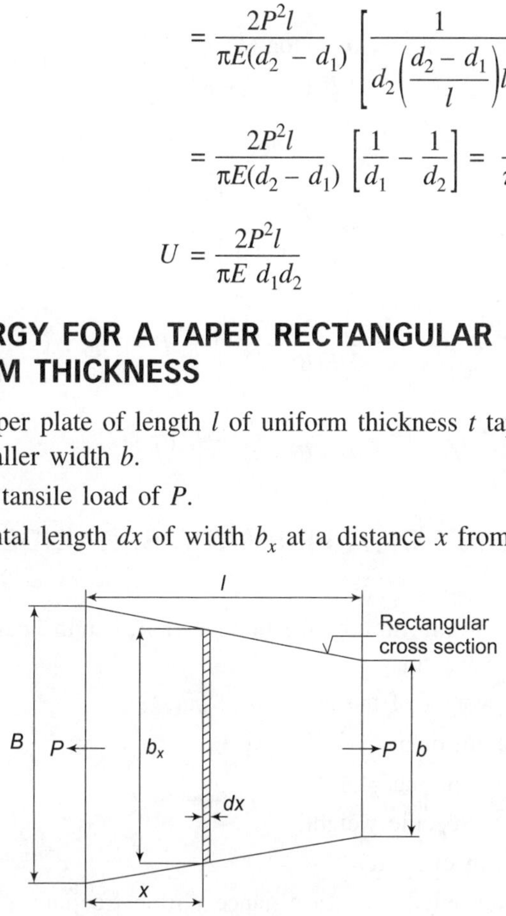

Fig. 3.5

From similar triangles $\frac{B - bx}{x} = \frac{B - b}{l}$

$$\therefore \quad b_x = B - \left(\frac{B-b}{l}\right)x = B - Kx$$

where,
$$K = \left(\frac{B-b}{l}\right)$$

$$\text{Strain energy } U = \frac{P^2 l}{2AE}$$

For the elemental length dx, strain energy

$$U = \frac{P^2 dx}{2AxE} = \frac{P^2 dx}{2(b_x \cdot t)E} = \frac{P^2 dx}{2(B - Kx)tE}$$

$\therefore$ Total strain energy for the entire plate

$$U = \int_o^l \frac{P^2 dx}{2tE(B - Kx)} = \frac{P^2}{2tE}\int_o^l \left(\frac{dx}{B - Kx}\right)$$

$$= \frac{P^2}{2tE}\left[\frac{\log_e(B - Kx)}{-K}\right]_o^l = \frac{P^2}{2tE}\left[-\log_e(B - Kx)\right]_o^l$$

$$= \frac{P^2}{2tE}\left[-\log_e(B - Kl) - (-\log_e(B - 0)\right]$$

$$= \frac{P^2}{2tE\left(\frac{B-b}{l}\right)}\left[-\log_e\left\{B - \left(\frac{B-b}{l}\right)l\right\} + \log_e B\right]$$

$$= \frac{P^2 l}{2tE(B - b)}\left[\log_e B - \log_e b\right]$$

$$U = \frac{P^2 l}{2tE(B - b)} \cdot \log_e\left(\frac{B}{b}\right) \qquad \ldots(16)$$

3.7 STRAIN ENERGY OF A PRISMATIC BAR UNDER ITS OWN WEIGHT

Figure 3.6 shows a bar of uniform cross-sectional area A, length l hanging freely under its own weight

Let w = Specific weight of the material of bar.

W = Total weight of bar

V = Volume of the bar = $A \cdot l$

W = Volume × Specific weight

E = Modulus of elasticity.

Consider an elemental length dx at a distance x from free end.

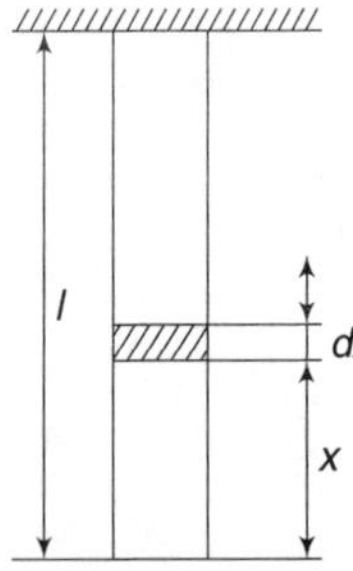

Fig. 3.6

Force applied on the element dx is the weight of the bar below it over a length x

$\therefore$ Weight of the bar of length x, $W_x = (A \cdot x)w$

$$Px = W_x = Axw$$

$$\therefore \qquad \text{Strain energy } U = \frac{P^2 \cdot l}{2AE} = \frac{(Axw)^2 \cdot dx}{2A \cdot E}$$

$$\text{Total strain energy } U = \int_o^l \frac{A^2x^2w^2dx}{2AE}$$

$$= \frac{Aw^2}{2E}\int_o^l x^2dx = \frac{Aw^2}{2E}\left(\frac{x^3}{3}\right)_o^l = \frac{Aw^2l^3}{6E} \qquad \ldots(17)$$

But $$Alw = W = \text{Total width}$$

$$U = \frac{Aw^2l^3}{6E} = \frac{A^2l^2w^2 \cdot l}{6AE} = \frac{W^2l}{6AE} \qquad \ldots(18)$$

Comparing with strain energy for a bar under load P

$$U = \frac{P^2l}{2AE},$$

the strain energy under self-weight of load P is one third of that under same external load.

Important Relations

- Strain energy or resilience $U = \dfrac{\sigma^2 V}{2E} = \dfrac{P^2l}{2AE}$
- Unit resilience $U' = \dfrac{\sigma^2}{2E}$
- Proof resilience $U_y = \dfrac{\sigma_y^{\,2}}{2E} \cdot V$
- Stress due to suddenly applied load

$$\sigma_i = \frac{2P}{A} = 2 \times \sigma$$

where, $\sigma = \dfrac{P}{A}$ is the static stress

- Impact stress $\sigma_i = \dfrac{W}{A}\left[1 + \sqrt{1 + \dfrac{2hAE}{Wl}}\right] = \sigma\left[1 + \sqrt{1 + \dfrac{2h}{\delta l}}\right]$
- Resilience in shear $= \dfrac{\tau^2}{2G} \cdot V$
- Strain energy of tapering round bar $= \dfrac{2P^2l}{\pi E d_1 d_2}$
- Strain energy of tapering rectangular plate $= \dfrac{P^2l}{2tE(B - b)} \log_e\left(\dfrac{B}{b}\right)$
- Strain energy of uniform bar under self-weight

$$U = \frac{W^2 l}{6AE} = \frac{Aw^2 l^3}{6E}$$

where, W = Total weight of bar = Alw.

SOLVED PROBLEMS

Problem 1 A circular rod of 60 mm diameter 6 m long carries an axial load of 120 kN. Young's modulus = 200 GPa. Determine the stress, strain and the strain energy absorbed by the rod. Also find unit resilience.

Solution: **Given Data:**

$$\text{Diameter } d = 60 \text{ mm}$$

$$\text{Length } l = 6 \text{ m} = 6000 \text{ mm}$$

$$\text{Load } P = 120 \text{ kN} = 120 \times 10^3 \text{ N}$$

$$\text{Young's modulus } E = 200 \text{ GPa} = 200 \times 10^3 \text{ MPa}$$

(i) $$\text{Stress } \sigma = \frac{P}{A} = \frac{120 \times 10^3}{\left(\frac{\pi}{4} \times 60^2\right)} = 42.441 \text{ MPa}$$

(ii) $$\text{Strain } \in = \frac{\sigma}{E} = \frac{42.441}{200 \times 10^3} = 2.122 \times 10^{-4}$$

$\therefore$ $$\text{Change in length } \delta l = \in \cdot l = 2.122 \times 10^{-4} \times 6000$$

$$\delta l = 1.2732 \text{ mm}$$

(iii) Strain energy $$U = \frac{P^2 l}{2EA} = \frac{(120 \times 10^3)^2 \times 6000}{2 \times 200 \times 10^3 \times \frac{\pi}{4} \times 60^2}$$

$$U = 76394.37 \text{ N-mm}$$

(iv) Unit resilience $$U = \frac{\sigma^2}{2E} = \frac{42.441^2}{2 \times 200 \times 10^3}$$

$$= 4.503 \times 10^{-3} \text{ N-mm/mm}^3.$$

Problem 2 A rectangular cross section bar whose width is twice the thickness and 3 m long has to resist an energy of 90 N-m. Modulus of elasticity is 206 kN/mm^2. Find the suitable cross section of the bar if the stress is limited to 60 MPa.

Solution: **Given Data:**

Rectangular cross section

$$\text{Width } b = \text{twice the thickness}$$

$$b = 2t$$

$$\text{Length } l = 3 \text{ m} = 300 \text{ mm}$$

$$\text{Modulus elasticity } E = 206 \text{ kN/mm}^2 = 206 \times 10^3 \text{ N/mm}^2$$

$$\text{Strain energy } U = 90 \text{ Nm} = 90 \times 10^3 \text{ N-mm.}$$

$$\text{Stress } \sigma = 60 \text{ MPa}$$

$$\text{Strain energy } U = \frac{\sigma^2}{2E} \cdot V$$

where,

$$V = \text{Volume} = A \cdot l$$

$$= (b \cdot t)\, l = (2t \times t)\, l = 2t^2 l$$

$$90 \times 10^3 = \frac{60^2}{2 \times 206 \times 10^3} \times 2t^2 \times 3000$$

$\therefore$

$$\text{Thickness } t = 41.43 \text{ mm}$$

$$\text{Width } b = 2t = 82.86 \text{ mm.}$$

Problem 3 A square cross section bar of side 50 mm and 2 m long is applied with a sudden load of 60 kN. Modulus of elasticity is 2×10^5 N/mm^2. Find the maximum stress induced, instantaneous elongation, strain energy stored by the rod and unit resilience.

Solution: **Given data:** Square cross section

$$\text{Side } a = 50 \text{ mm}$$

$$\text{Length } l = 2 \text{ m} = 2000 \text{ mm}$$

$$\text{Sudden load } P = 60 \text{ kN} = 60 \times 10^3 \text{ N}$$

$$\text{Modulus of elasticity } E = 2 \times 10^5 \text{ MPa}$$

$$\text{Volume} = V = l = (a \times a)l = (50 \times 50) \times 200 \text{ mm}^3$$

(i) For suddenly applied load, stress $\sigma = \dfrac{2P}{A}$

$$\sigma = 2 \times \frac{60000}{(50 \times 50)} = 48 \text{ MPa}$$

(ii) Instantaneous elongation in the bar

$$\delta l = \frac{\sigma}{E}\, l = \frac{48 \times 200}{20 \times 10^5} = 0.48 \text{ mm}$$

(iii) Strain energy stored in the bar $U = \dfrac{\sigma^2}{2E} \cdot v$

$$U = \frac{48^2}{2 \times 2 \times 10^5} \times (50 \times 50 \times 200) = 28800 \text{ N-mm}$$

(iv) Unit resilience

$$U' = \frac{\sigma^2}{2E} = \frac{48^2}{2 \times 2 \times 10^5} = 5.76 \times 10^{-3} \text{ Nmm/mm}^3.$$

Problem 4 A round bar of diameter 30 mm and 3 m long stretches by 1 mm under an unknown gradually applied axial load. Taking Young's modulus as 206 GPa, find the resilience in the bar.

Solution: **Given data:**

$$\text{Diameter } d = 30 \text{ mm}$$

$$\text{Length } l = 3 \text{ m} = 3000 \text{ mm}$$

$$\text{Increase in length } \delta l = 1 \text{ mm}$$

Gradually applied load.

$$\text{Young's modulus } E = 206 \text{ GPa} = 206 \times 10^3 \text{ MPa}$$

$$\text{Area of cross section } A = \frac{\pi}{4} d^2 = \frac{\pi}{4} \times 30^2$$

$$\text{We know that } \delta l = \frac{\sigma l}{E}$$

$$\therefore \quad \text{Stress } \sigma = \frac{\delta l \cdot E}{l} = \frac{1 \times 206 \times 10^3}{3000} = 68.67 \text{ MPa}$$

$$\text{Load } P = \sigma \cdot A = 68.67 \times \left(\frac{\pi}{4} \times 30^2\right)$$

$$P = 48,\ 537.61 \text{ N}$$

$$\text{Resilience } U = \frac{\sigma^2}{2E} \cdot V = \frac{\sigma^2}{2E} \cdot (Al)$$

$$= \frac{68.67^2}{2 \times 206 \times 10^3} \times \left(\frac{\pi}{4} \times 30^2 \times 3000\right)$$

$$U = 24\ 268.80 \text{ N-mm.}$$

Problem 5 A hollow round rod whose diameters ratio is 2 is 6 m long. An axial pull of 120 kN is suddenly applied to the rod and the maximum instantaneous stress is found to be 90 MPa. Find the diameters of the rod. Also find the maximum instantaneous elongation and strain energy induced in the rod. Take E = 200 GPa.

Solution: **Given data:**

$$\text{Hollow round rod, diameters ratio} = 2$$

$$\frac{\text{Outside diameter}}{\text{Inside diameter}} = 2 = \frac{d_o}{d_i}$$

$$d_o = 2\, d_i$$

$$\text{Area of cross section } A = \frac{\pi}{4} (d_o^2 - d_i^2)$$

$$A = \frac{\pi}{4} [(2d_i)^2 - d_i^2] = 3\, \frac{\pi}{4}\, d_i^2$$

$$\text{Length } l = 6 \text{ m} = 6000 \text{ mm}$$

$$\text{Suddenly applied load } P = 120 \text{ kN} = 120 \times 10^3 \text{ N}$$

$$\text{Maximum instantaneous stress } \sigma = 90 \text{ MPa}$$

We know that for suddenly applied load

$$\text{Stress } \sigma = \frac{2P}{A}$$

$$90 = \frac{2 \times 120 \times 10^3}{\left(\frac{3\pi}{4} d_i^2\right)}$$

$\therefore$ Inside diameter of rod $d_i = 33.64$ mm

Outside diameter of rod $d_o = 2d_i = 67.28$ mm

Instananeous elongation $\delta l = \dfrac{\sigma l}{E}$

$$\delta l = \frac{90 \times 6000}{2 \times 10^5} = 2.7 \text{ mm}$$

Strain energy stored by the rod $U = \dfrac{\sigma^2}{2E} \cdot V$

$$U = \frac{90^2}{2 \times 200 \times 10^3} \times A \cdot l$$

$$= \frac{90^2}{2 \times 200 \times 10^3} \left[\frac{\pi}{4}(67.28^2 - 33.64^2)\right] \times 6000$$

$$U = 323965.97 \text{ N-mm.}$$

Problem 6 A uniform rod of diameter 40 mm and length 1.8 m has elastic limit as 300 N/mm^2. Find the proof resilience of the bar. What suddenly applied load would produce the same energy as proof resilience. Also find the value of gradually applied load which will produce the same elongation as that is produced by suddenly applied load. Take E = 207 GPa.

Solution: **Given data:**

Diameter $d = 40$ mm

$\therefore$ Area $A = \dfrac{\pi}{4} d^2 = \dfrac{\pi}{4} \times 40^2 = 1256.64 \text{ mm}^2$

Length $l = 1.8$ m $= 1800$ mm.

Elastic limit $\sigma_y = 300$ N/mm^2

Modulus of elasticity $E = 207$ GPa $= 207 \times 10^3$ MPa

Volume $V = Al = 1256.\ 64 \times 1800 \text{ mm}^3$

$$\text{Proof resilience } U_y = \frac{\sigma_y^2}{2E} \cdot V = \frac{300^2}{2 \times 207 \times 10^3} \times (1256.64 \times 1800)$$

$$U_y = 491728.696 \text{ N-mm}$$

For suddenly applied load

$$\text{Stress } \sigma = \frac{2P}{A}$$

$$\therefore \quad \sigma = \sigma_y = \frac{2P}{A}$$

$$\therefore \qquad \text{Suddenly applied load } P = \frac{\sigma_y \cdot A}{2}$$

$$P_s = \frac{300 \times 1256.64}{2} = 188496 \text{ N}$$

$$\text{Gradually applied load } P_y = \sigma_y \cdot A = 300 \times 1256.64 = 376992 \text{ N.}$$

Problem 7 A stepped rod has a diameter of 50 mm over a length of 1200 mm and 20 mm over a length of 1800 mm. An axial load of 150 kN is gradually applied to the rod. The modulus of elasticity of the material is 200 kN/mm^2. Find the strain energy produced in each step and total strain energy in the rod. What is the percentage error in strain energy if calculated by taking a uniform bar of same volume and length.

Solution: **Given data:**

$$\text{Diameter of part (1)} = d_1 = 50 \text{ mm}$$

$$\text{Length of (1)} = l_1 = 1200 \text{ mm}$$

$$\text{Diameter of (2)} = d_2 = 20 \text{ mm}$$

$$\text{Length of (2) } l_2 = 1800 \text{ mm}$$

$$\text{Load } P = 150 \text{ kN} = 150 \times 10^3 \text{ N}$$

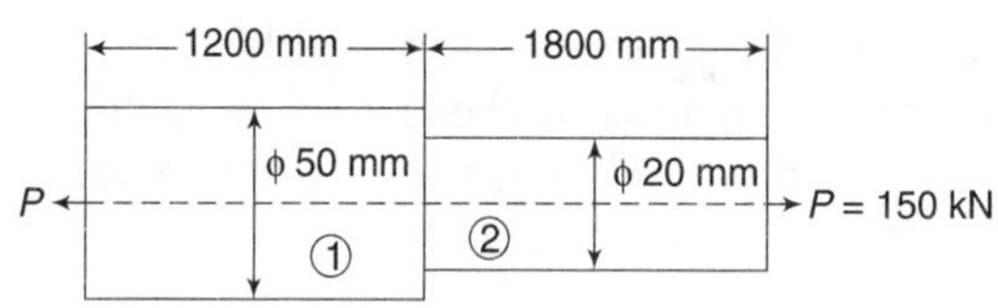

Fig. 3.7

$$\text{Modulus of elasticity } E = 200 \text{ kN/mm}^2 = 200 \times 10^3 \text{ N/mm}^2$$

$$\text{Area of (1) } A_1 = \frac{\pi}{4} d_1^2 = \frac{\pi}{4} \times 50^2 = 1963.495 \text{ mm}^2$$

$$\text{Volume of (1) } V_1 = A_1 \cdot l_1 = 1963.495 \times 1200 = 2.356 \times 10^6 \text{ mm}^3$$

$$\text{Area of (2) } A_2 = \frac{\pi}{4} d_2^2 = \frac{\pi}{4} \times 20^2 = 314.159 \text{ mm}^2$$

$$\text{Volume of (2) } V_2 = A_2 l_2 = 314.159 \times 1800 = 0.5655 \times 10^6 \text{ mm}^3$$

$$\text{Stress in (1) } \sigma_1 = \frac{P}{A_1} = \frac{150 \times 10^3}{1963.495} = 76.394 \text{ MPa}$$

$$\text{Stress in (2) } \sigma_2 = \frac{P}{A_2} = \frac{150 \times 10^3}{314.159} = 477.465 \text{ MPa}$$

$$\text{Strain energy in (1) } U_1 = \frac{\sigma_1^2}{2E} \cdot V_1 = \frac{76.394^2}{2 \times 200 \times 10^3} \times 2.356 \times 10^6$$

$$= 34.374 \times 10^3 \text{ Nmm}$$

$$\text{Strain energy in (2) } U_2 = \frac{\sigma_2^2}{2E} \cdot V_2$$

$$= \frac{477.465^2}{2 \times 200 \times 10^3} \times 0.5655 \times 10^6$$

$$= 322.296 \times 10^3 \text{ Nmm}$$

Total strain energy in the rod $U = U_1 + U_2 = 34.374 \times 10^3 + 322.296 \times 10^3$

$$U = 356.67 \times 10^3 \text{ N-mm}$$

(ii) Considering the stepped rod as uniform rod of same length and volume:

$$\text{Volume } V = V_1 + V_2 = 2.356 \times 10^6 + 0.5655 \times 10^6$$

$$V = 2.9215 \times 10^6 \text{ mm}^3$$

$$\text{Length } l = l_1 + l_2 = 1200 + 1800 = 3000 \text{ mm}$$

For uniform rod, volume $V = A \cdot l$

$$2.9215 \times 10^6 = A \times 3000.$$

$\therefore$ Uniform area $A = 973.833 \text{ mm}^2$

$\therefore$ Stress $\sigma = \dfrac{P}{A} = \dfrac{150 \times 10^3}{973.83} = 154.03$ MPa

$\therefore$ Strain energy for uniform rod

$$U_1 = \frac{\sigma^2}{2E} \cdot V = \frac{154\ .\ 03^2}{2 \times 200 \times 10^3} \times 2.9215 \times 10^6$$

$$U_1 = 173.284 \times 10^3 \text{ N-mm}$$

Percentage error in calculation

$$= \frac{\text{Strain energy of stepped rod} - \text{Strain energy of uniform rod}}{\text{Strain energy of stepped rod}} \times 100$$

$$= \frac{U - U_1}{U} \times 100$$

$$\%\text{error} = \frac{356.67 \times 10^3 - 173.284 \times 10^3}{356.67 \times 10^3} \times 100 = 51.416\%.$$

Problem 8 Figure 3.8 shows a stepped bar subjected to a tensile load of 60 kN. Find the strain energy produced in the bar. Take $E = 2 \times 10^5$ N/mm^2

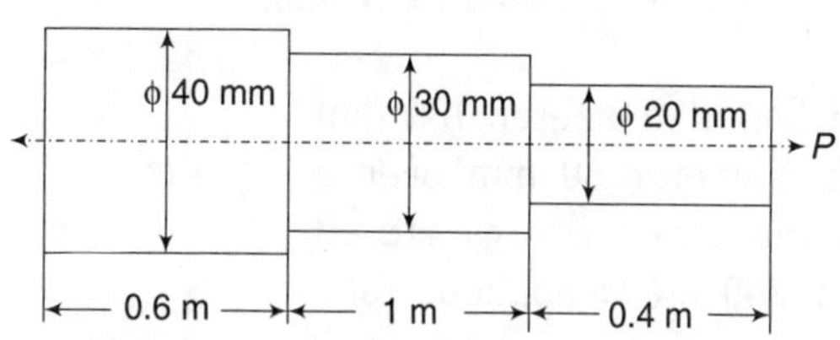

Fig. 3.8

Solution: **Given data:**

$$\text{Load } P = 60 \text{ kN} = 60 \times 10^3 \text{ N}$$

$$\text{Modulus of elasticity } E = 2 \times 10^5 \text{ N/mm}^2$$

$$\text{Diameter } d_1 = 40 \text{ mm}$$

$$d_2 = 30 \text{ mm}$$

$$d_3 = 20 \text{ mm}$$

$$\text{Length } l_1 = 0.6 \text{ m} = 600 \text{ mm}$$

$$l_2 = 1 \text{ m} = 1000 \text{ mm}$$

$$l_3 = 0.4 \text{ m} = 400 \text{ mm.}$$

$$\text{Stress } \sigma = \frac{P}{A}$$

$$\text{For part (1) } \sigma_1 = \frac{P}{A_1} = \frac{P}{\left(\frac{\pi}{4}d_1^2\right)} = \frac{60 \times 10^3}{\left(\frac{\pi}{4} \times 40^2\right)} = 47.75 \text{ MPa}$$

$$\text{For part (2) } \sigma_2 = \frac{P}{A_2} = \frac{60 \times 10^3}{\left(\frac{\pi}{4} \times 30^2\right)} = 84.88 \text{ MPa}$$

$$\text{For part (3) } \sigma_3 = \frac{P}{A^3} = \frac{60 \times 10^3}{\left(\frac{\pi}{4} \times 20^2\right)} = 190.99 \text{ MPa}$$

$$\text{Volume } V = A \cdot l$$

$$V_1 = A_1 l_1 = \left(\frac{\pi}{4} \times 40^2\right) \times 600 = 753.98 \times 10^3$$

$$V_2 = A_2 l_2 = \left(\frac{\pi}{4} \times 30^2\right)1000 = 706.86 \times 10^3 \text{ mm}^3$$

$$V_3 = A_3 l_3 = \left(\frac{\pi}{4} \times 20^2\right) \times 400 = 125.66 \times 10^3$$

$$\text{Total strain energy } U = U_1 + U_2 + U_3 = \frac{\sigma_1{}^2 V_1}{2E} + \frac{\sigma_2{}^2 V_2}{2E} + \frac{\sigma_3{}^2 V_3}{2E}$$

$$= \frac{1}{2 \times 200 \times 10^3} [47.75^2 \times 753.98 \times 10^3 + 84.88^2 \times 706.86 \times 10^3 + 190.99^2 \times 125.60 \times 10^3]$$

$$U = 28488.74 \text{ N-mm.}$$

Problem 9 A round bar of diameter 100 mm tapers uniformly to a diameter 60 mm over a length of 3 m. Find the strain energy stored when an axial load of 240 kN is applied. Take E = 200 GPa.

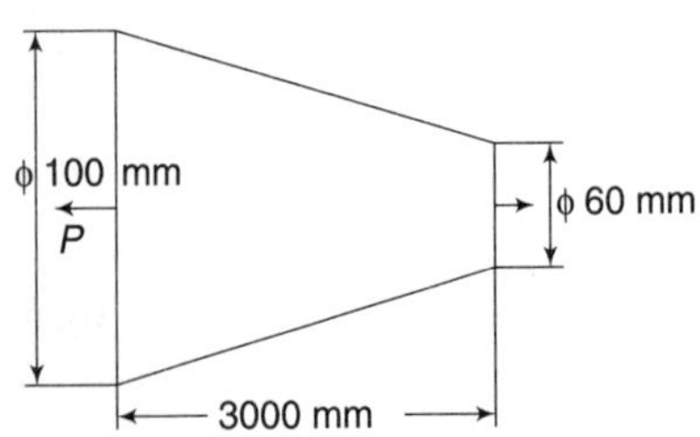

Fig. 3.9

Solution: **Given data:**

Taper round bar.

Bigger diameter $d_2 = 100$ mm

Smaller diameter $d_1 = 60$ mm

Length $l = 3$ m $= 3000$ mm

Load $P = 240$ kN $= 240 \times 10^3$ N

Modulus of elasticity $E = 200$ GPa $= 200 \times 10^3$ MPa.

For a taper round rod subjected to gradually applied load,

$$\text{Strain energy } U = \frac{2P^2 l}{\pi E d_1 d_2} = \frac{2 \times (240 \times 10^3)^2 \times 3000}{\pi \times 200 \times 10^3 \times 100 \times 60}$$

$$U = 91.673 \times 10^3 \text{ N-mm.}$$

Problem 10 Determine the total strain energy stored by a rod shown in Fig. 3.10. Take $E = 2 \times 10^5$ N/mm^2.

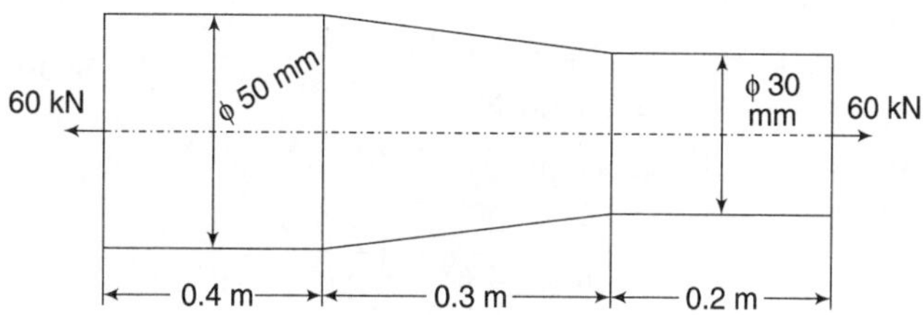

Fig. 3.10

Solution: **Given data:**

Load $P = 60$ kN $= 60000$ N

Modulus elasticity $E = 2 \times 10^5$ N/mm^2

Diameter $d_1 = 50$ mm

$d_2 = 30$ mm

Length $l_1 = 0.4$ m $= 400$ mm

$l_2 = 0.3$ m $= 300$ mm

$l_3 = 0.2$ m $= 200$ mm.

Total strain energy $U = U_1 + U_2 + U_3$

$$U = \frac{P^2 l_1}{2A_1 E} + \frac{2P^2 l_2}{\pi E d_1 d_2} + \frac{P^2 l_3}{2A_3 E}$$

where A_1 = Area of part (1)

$$= \frac{\pi}{4} d_1^2 = \frac{\pi}{4} 50^2 = 1963.495 \text{ mm}^2$$

A_3 = Area of part (3)

$$= \frac{\pi}{4} d_2^2 = \frac{\pi}{4} \times 30^2 = 706.86 \text{ mm}^2$$

$$U = \frac{60000^2}{2 \times 10^5}\left[\frac{400}{2 \times 1963.495} + \frac{2 \times 300}{\pi \times 50 \times 30} + \frac{200}{2 \times 706.86}\right]$$

$$U = 6671.26 \text{ Nmm}$$

Problem 11 A taper rectangular cross section plate of 18 mm thickness has width tapering uniformly from 120 mm to 60 mm over a length of 1.5 m. It is subjected to an axial load of 90 kN. Taking E = 200 GPa, find the resilience in the plate.

Solution: **Given data:**

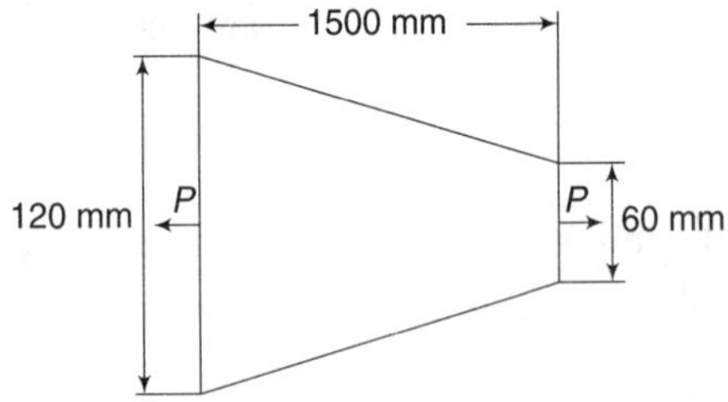

Fig. 3.11

Taper plate

Thickness of plate t = 18 mm

Bigger width B = 120 mm

Smaller width b = 60 mm

Length l = 1500 mm

$E = 200 \times 10^3$ MPa

P = 90000 N

Strain energy or resilience for taper plate

$$U = \frac{P^2 l}{2tE(B-b)} l_n\left(\frac{B}{b}\right)$$

$$= \frac{90000^2 \times 1500}{2 \times 18 \times 200 \times 10^3 \times (120-60)} . l_n\left(\frac{120}{60}\right)$$

$$U = 19494.76 \text{ N-mm}$$

Problem 12 For a plate of uniform thickness of 15 mm loaded as shown in Fig. 3.12, determine the energy stored taking $E = 2 \times 10^5$ MPa. Also find the total elongation in the plate.

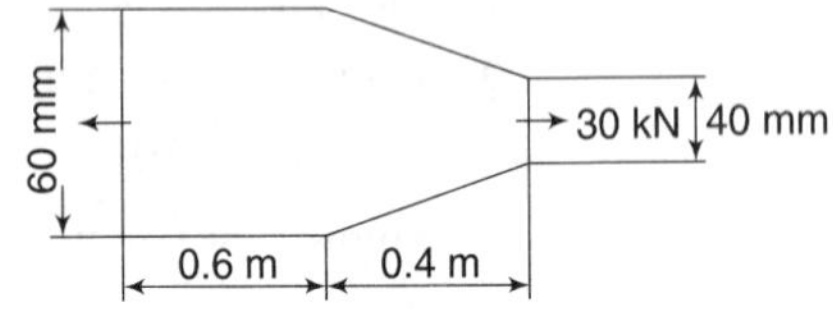

Fig. 3.12

Solution: **Given data:**

Plate of uniform thickness t = 15 mm

Modulus of elasticity $E = 2 \times 10^5$ N/mm^2

Load P = 30000 N

Length l_1 = 600 mm

l_2 = 400 mm

Width B = 60 mm

b = 40

$$\text{Total strain energy } U = U_1 + U_2 = \frac{P^2 l_1}{2AE_1} + \frac{P^2 l_2}{2tE(B-b)} l_n\left(\frac{B}{b}\right)$$

$$= \frac{30000^2 \times 600}{2 \times (60 \times 15)^2 \times 10^5} + \frac{30000^2 \times 400}{2 \times 15 \times 2 \times 10^5 (60-40)} l_n\left(\frac{60}{40}\right)$$

$$U = 1500 + 1216.4 = 2714.4 \text{ N-mm}$$

$$\text{Elongation } \delta l = \delta l_1 + \delta l_2 = \frac{Pl_1}{A_1 E} + \frac{Pl_2}{tE(B-b)}\ l_n\left(\frac{B}{b}\right)$$

$$= \frac{30000 \times 600}{60 \times 15 \times 2 \times 10^5} + \frac{30000 \times 400}{15 \times 2 \times 10^5\ (60-40)}\ l_n\left(\frac{60}{40}\right)$$

$$\delta l = 0.1 + 0.0811 = 0.1811 \text{ mm.}$$

Problem 13 A round rod of diameter 120 mm and length 3 m, hangs freely. It is made of steel having specific weight of 76.6 kN/m^3 and modulus of elasticity is 2×10^5 N/mm^2. Find the elongation under self weight and the energy stored by this rod.

Solution: **Given data:**

$$\text{Diameter } d = 120 \text{ mm}$$

$$\text{Length } l = 3 \text{ m} = 3000 \text{ mm}$$

$$\text{Specific weight } w = 76.6 \text{ kN/m}^3 = 76.6 \times 10^{-6} \text{ N/mm}^3$$

$$\text{Modulus of elasticity } E = 2 \times 10^5 \text{ N/mm}^2$$

$$\text{Elongation under self weight } \delta l = \frac{W \cdot l}{2AE} = \frac{wl^2}{2E}$$

$$\delta l = \frac{76.6 \times 10^{-6} \times 3000^2}{2 \times 2 \times 10^5} = 0.00172 \text{ mm}$$

$$\text{Strain energy to self weight } U = \frac{w^2 l^3 A}{6E}$$

$$U = \frac{(76.6 \times 10^{-6})^2 \times 3000^3 \times \left(\frac{\pi}{4} \times 120^2\right)}{6 \times 2 \times 10^5}$$

$$U = 1.4931 \text{ N-mm.}$$

Problem 14 Figure 3.13 shows two bars made of same material. Compare the strain energies stored in these two bars if

(i) load applied is same, and

(ii) maximum stress produced is same.

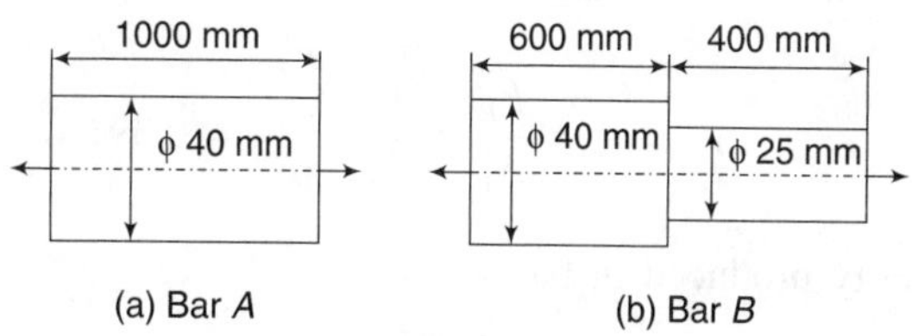

Fig. 3.13

Solution: **Given data**:

$$\text{For bar } A,\ d = 40 \text{ mm}$$
$$l = 1000 \text{ mm}$$
$$\text{For bar } B,\ d_1 = 40 \text{ mm},\ l_1 = 600 \text{ mm}$$
$$d_2 = 25 \text{ mm},\ l_2 = 400 \text{ mm}$$

(i) Applied load is same = P

Strain energy stored by bar A

$$U_A = \frac{P^2 l}{2AE} = \frac{P^2 \times 1000}{2 \times \left(\frac{\pi}{4} \times 40^2\right) \times E}$$

$$U_A = 0.3979 \frac{P^2}{E} \qquad \ldots(1)$$

Strain energy stored by bar B

$$U_B = U_1 + U_2 = \frac{P^2 l_1}{2A_1 E} + \frac{P^2 l_2}{2A_2 E}$$

$$= \frac{P^2 \times 600}{2 \times \left(\frac{\pi}{4} \times 40^2\right)E} + \frac{P^2 \times 400}{2 \times \left(\frac{\pi}{4} \times 25^2\right)E}$$

$$= 0.2387 \frac{P^2}{E} + 0.4074 \frac{P^2}{E}$$

$$U_B = 0.6461 \frac{P^2}{E} \qquad \ldots(2)$$

$$\text{Comparing, } \frac{U_A}{U_B} = \frac{0.3979}{0.6461} = 0.616.$$

(ii) When maximum stress produce is same.

$$\text{Let } \sigma = \text{Maximum stress produced}$$

For bar B, stress will be maximum in the smaller diameter portion (2), i.e., in 25 mm diameter protion.

$$\therefore \quad \text{Load on bar } B = A_i\, \sigma$$

$$P_B = \frac{\pi}{4} 25^2\, \sigma$$

$\therefore$ Strain in portion (1) where diameter = 40 mm

$$\sigma_1 = \frac{P_B}{A_1} = \frac{\frac{\pi}{4} 25^2\, \sigma}{\frac{\pi}{4} \times 40^2} = 0.391\, \sigma$$

$\therefore$ Strain in energy produced in bar A

$$U_A = \frac{\sigma^2}{2E} \cdot V = \frac{\sigma^2}{2E}\left(\frac{\pi}{4} \times 40^2 \times 1000\right)$$

$$U_A = 628.32 \times 10^3 \frac{\sigma^2}{E} \qquad \ldots(3)$$

Strain energy produced in bar B

$$U_B = U_1 + U_2 = \frac{{\sigma_1}^2}{2E} V_1 + \frac{{\sigma_2}^2 . V_2}{2E}$$

$$= \frac{(0.391\sigma)^2}{2E} \times \left(\frac{\pi}{4} \times 40^2 \times 600\right) + \frac{\sigma^2}{2E} \times \left(\frac{\pi}{4} \times 25^2 \times 400\right)$$

$$= 57.635 \times 10^3 \frac{\sigma^2}{E} + 98.175 \times 10^3 \frac{\sigma^2}{E}$$

$$U_B = 155.81 \times 10^3 \frac{\sigma^2}{E} \qquad \ldots(4)$$

$$\text{Comparing } \frac{U_A}{U_B} = \frac{628.32 \times 10^3}{155.81 \times 10^3} = 4.033.$$

Problem 15 A round bar of diameter 90 mm and length 900 mm carries a tensile load of 200 kN. The diameter at the middle for a length of 300 mm is reduced to 40 mm. Modulus of elasticity is 200 GPa. Calculate the strain energy stored in this bar. Compare it with the strain energy stored by a bar of uniform diameter of 90 mm over full length under same load and made of same material.

Solution: **Given data:**

$$\text{Load } P = 200 \text{ kN} = 200 \times 10^3 \text{ N}$$

$$\text{Modulus elasticity } E = 200 \text{ GPa} = 200 \times 10^3 \text{ MPa}$$

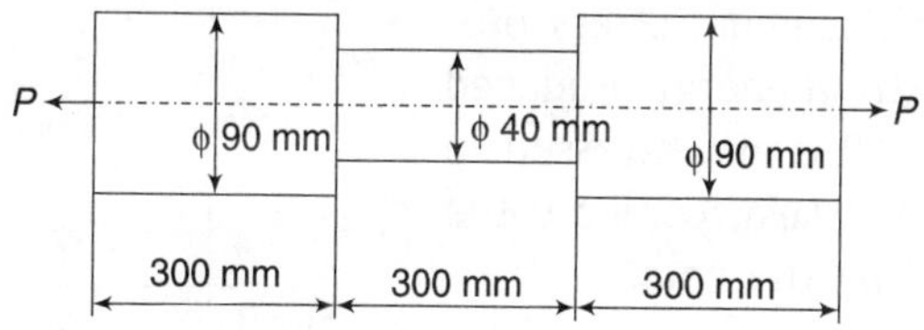

Fig. 3.14

(a) *Steep bar*

Let σ_1 = Stress induced in large diameter portion

σ_2 = Stress induced in smaller diameter portion

$$\sigma_1 = \frac{P}{A_1} = \frac{200 \times 10^3}{\left(\frac{\pi}{4} d_1^2\right)} = \frac{200 \times 10^3}{\left(\frac{\pi}{4} \times 90^2\right)} = 31.44 \text{ MPa}$$

$$\sigma_2 = \frac{P}{A_2} = \frac{200 \times 10^3}{\left(\frac{\pi}{4} d_2^2\right)} = \frac{200 \times 10^3}{\left(\frac{\pi}{4} \times 40^2\right)} = 159.15 \text{ MPa}$$

Length of 90 mm portion = l_1 = 300 + 300 = 600 mm

Length of 40 mm portion = l_2 = 300 mm

$$\text{Volume } V = Al$$

$$\text{Total strain energy } U = U_1 + U_2 = \frac{\sigma_1^2}{2E} V_1 + \frac{\sigma_2^2}{2E} \cdot V_2$$

$$= \frac{1}{2 \times 200 \times 10^3} \left[31.44^2 \times \frac{\pi}{4} \times 90^2 \times 600 + 159.15^2 \times \frac{\pi}{4} \times 40^2 \times 300\right]$$

$$U_A = 33304.35 \text{ N-mm}$$

(b) *Uniform diameter bar*

For a solid bar of same material and length and diameter 90 mm

$$\text{Strain energy } U_B = \frac{\sigma^2}{2E} V = \frac{31.44^2}{2 \times 200 \times 10^3} \times \left(\frac{\pi}{4} \times 90^2 \times 900\right)$$

$$U_B = 14148.89 \text{ N-mm}$$

$$\text{Comparing, } \frac{\text{Strain energy of stepped bar}}{\text{Strain energy of uniform bar}} = \frac{U_A}{U_B} = \frac{33304.35}{14148.89} = 2.3538$$

Energy storing capacity of bar A is 2.3538 times that of bar B.

Problem 16 Figure 3.15 shows two round bars of length 0.5 m. Bar A is applied with an axial force that produced a maximum stress of 120 MPa. Find the maximum stress produced in bar B if the strain energy produced is same in both. If bar B is also stressed to a maximum value of 120 MPa, compare the strain energies stored by the bars.

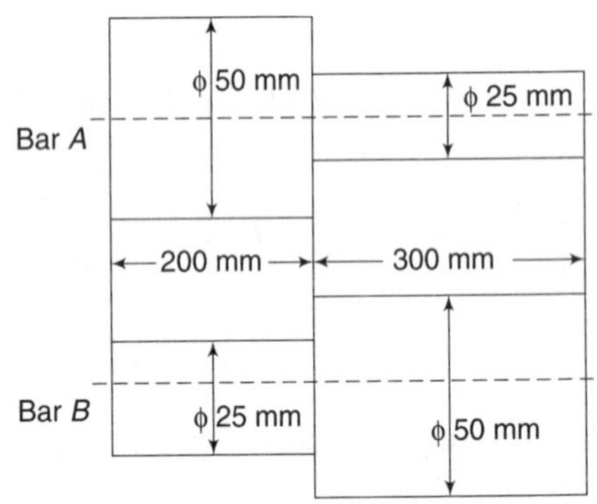

Fig. 3.15

Solution: **Given data:**

Maximum stress = 120 MPa

Part (a)

Consider bar A: The stress will be maximum in smaller diameter portion, i.e., 25 mm diameter portion.

$$\sigma_2 = 120 \text{ MPa}$$

$$\therefore \quad \text{Load } P = A_2 \sigma_2 = \frac{\pi}{4} \times 25^2 \times 120 = 58.904 \times 10^3 \text{ N}$$

$\therefore$ Stress in 50 mm diameter portion

$$\sigma_1 = \frac{P}{A_1} = \frac{58.904 \times 10^3}{\frac{\pi}{4} \times 50^2} = 30 \text{ MPa}$$

$$\text{Length } l_1 = 200 \text{ mm}$$

$$l_2 = 300 \text{ mm}$$

$$\text{Strain energy in bar } A \; U_A = \frac{\sigma_1^2}{2E} \cdot V_1 + \frac{\sigma_2^2}{2E} \cdot V_2$$

$$\text{Volume } v = A \cdot l$$

$$\therefore \quad U_A = \frac{30^2}{2E}\left(\frac{\pi}{4} \times 50^2 \times 200\right) + \frac{120^2}{2E}\left(\frac{\pi}{4} \times 25^2 \times 300\right)$$

$$= \frac{176.715 \times 10^6}{E} + \frac{1.06 \times 10^9}{E} = \frac{1.237 \times 10^9}{E} \text{ N-mm}$$

Bar B: Energy is same $U_B = U_A = \dfrac{1.237 \times 10^9}{E}$ N-mm

$$A_1 = \frac{\pi}{4} \times 25^2$$

$$l_1 = 208 \text{ m}$$

$$A_2 = \frac{\pi}{4} \times 50^2$$

$$l_2 = 300 \text{ mm}$$

Stress will be maximum in smaller diameter portion

$$\therefore \quad \sigma_{Max} = \sigma_1$$

For the load being same on the bar

$$P = \sigma_1 A_1 = \sigma_2 A_2$$

$$= \sigma_2 = \sigma_1 \frac{A_1}{A_2} = \frac{\sigma_1 \times \left(\frac{\pi}{4} \times 25^2\right)}{\left(\frac{\pi}{4} \times 50^2\right)} = \frac{\sigma_1}{4}$$

$$\sigma_2 = \frac{\sigma_1}{4}$$

Strain energy stored by bar *B*.

$$U_B = \frac{\sigma_1^2}{2E} V_1 + \frac{\sigma_2^2 V_2}{2E}$$

$$\frac{1.237 \times 10^9}{E} = \frac{\sigma_1^2 \times \left(\frac{\pi}{4} \times 25^2\right) \times 200}{2E} + \frac{\left(\frac{\sigma_1}{4}\right)^2 \times \left(\frac{\pi}{4} \times 50^2\right) \times 300}{2E}$$

$$\therefore \quad \sigma_{1B} = \sigma_{maxB} = 135.378 \text{ MPa.}$$

Part (b)

Bar B also is stressed to max. stress $\sigma_1 = 120$ MPa

$$\therefore \quad \sigma_2 = \frac{\sigma_1}{4} = 30 \text{ MPa}$$

$$\text{Strain energy stored by bar } B \; U_B = \frac{\sigma_1^{\,2} V_1}{2E} + \frac{\sigma_2^{\,2} V_2}{2E}$$

$$U_B = \frac{120^2 \times \left(\frac{\pi}{4} \times 25^2\right) \times 200}{2E} + \frac{30^2 \times \left(\frac{\pi}{4} \times 50^2\right) \times 300}{2E}$$

$$U_B = \frac{971.93 \times 10^3}{E}$$

$$\text{Comparing, } \frac{U_A}{U_B} = \frac{\left(\frac{1.237 \times 10^9}{E}\right)}{\left(\frac{971.93 \times 10^3}{E}\right)} = 1.2727$$

Bar A stores 1.2727 times the strain energy of bar B.

Problem 17 Figure 3.16 shows a bar of circular cross section of diameter 30 mm. Find the value of force P, the total elongation in the bar and the strain energies stored by each section.

Take E = 200 GPa.

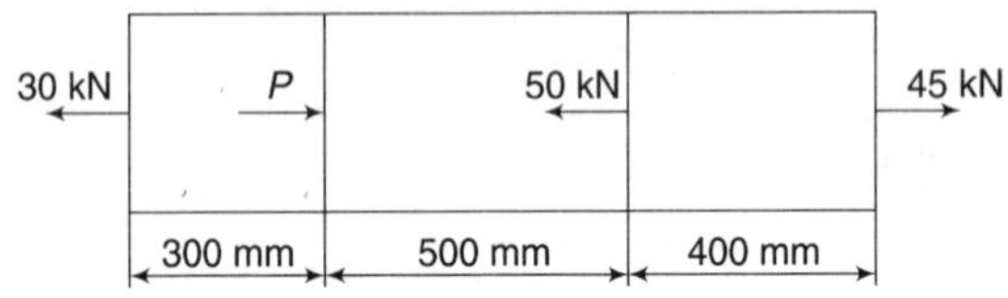

Fig. 3.16 (a)

Solution: **Given data:**

$$\text{Diameter } d = 300 \text{ mm } E = 200 \text{ GPa} = 200 \times 10^3 \text{ MPa}$$

$$\text{For equilibrium } \Sigma P = 0$$

$$30 + 50 = P + 45$$

$$\therefore \quad P = 35 \text{ kN}$$

$$P + 45 - 50 = 30 \text{ kN}$$

$$\text{Total elongation } \delta l = \delta l_1 + \delta l_2 + \delta l_3$$

$$\delta l = \frac{P_1 l_1}{A_1 E_1} + \frac{P_2 l_2}{A_2 E_2} + \frac{P_3 l_3}{A_3 E_3}$$

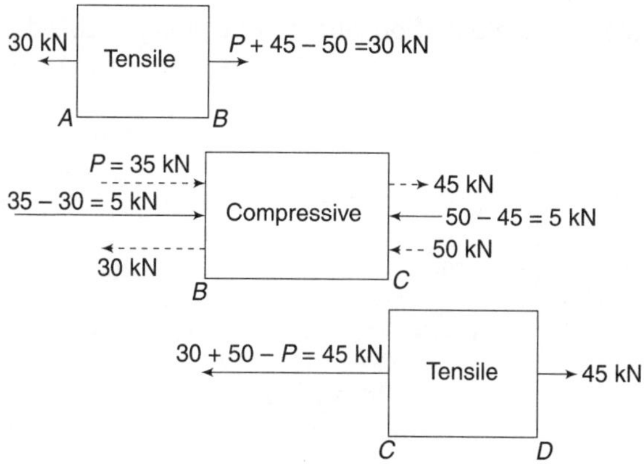

Fig. 3.16 (b) Free-body diagram

$$A = A_1 = A_2 = A_3 = \frac{\pi}{4} \times 30^2$$

$$E = 200 \times 10^3 \text{ MPa (same for all)}$$

$$P_1 = 30000 \text{ N}$$

$$P_2 = -5000 \text{ N (Compressive)}$$

$$P_3 = 45000 \text{ N}$$

$$\therefore \quad \delta l = \frac{1}{\frac{\pi}{4} \times 30^2 \times 200 \times 10^3} [30000 \times 300 - 5000 \times 500 + 45000 \times 200]$$

$$\delta l = 0.10964 \text{ mm}$$

$$\text{Strain energy } U = \frac{\sigma^2}{2E} V = \frac{P^2 l}{2AE}$$

$$U_1 = \frac{P_1^2 l_1}{2A_1 E} = \frac{30000^2 \times 300}{2 \times \left(\frac{\pi}{4} \times 30^2\right) \times 200 \times 10^3}$$

$$= 954.93 \text{ N mm}$$

$$U_2 = \frac{P_2^2 l_2}{2A_2 E} = \frac{50000^2 \times 500}{2 \times \left(\frac{\pi}{4} \times 30^2\right) \times 200 \times 10^3}$$

$$= 44.21 \text{ N mm}$$

$$U_3 = \frac{P_3^2 l_3}{2A_3 E} = \frac{45000^2 \times 200}{2 \times \left(\frac{\pi}{4} \times 30^2\right) \times 200 \times 10^3}$$

$$= 1432.39 \text{ N mm.}$$

Problem 18 Find the total elongation and strain energy stored in a bar of varying cross section loaded as shown in Fig. 3.17. Take $E = 2 \times 10^5$ N/mm^2.

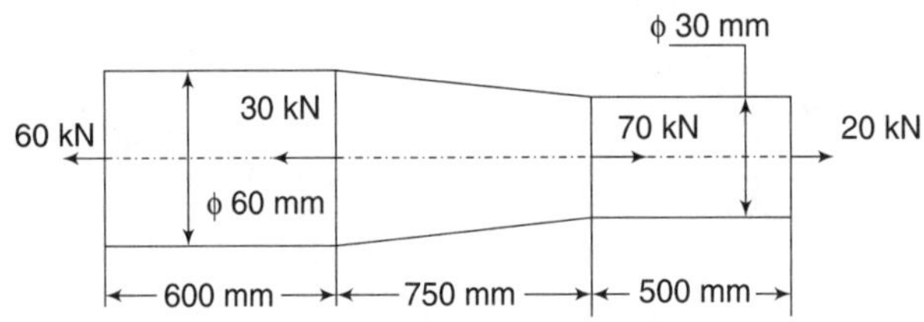

Fig. 3.17 (a)

Solution: **Given data:**

$$E = 2 \times 10^5 \text{ N/mm}^2$$

Free-body diagram: (Fig. 3.17 (b))

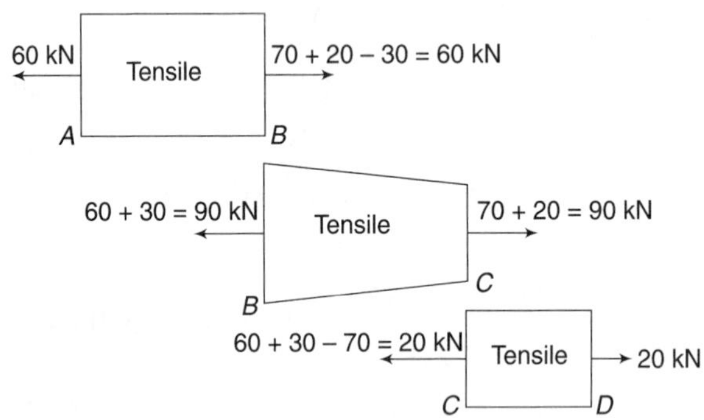

Fig. 3.17 (b)

$$\text{Total elongation } \delta l = \delta l_1 + \delta l_2 + \delta l_3 = \frac{P_1 l_1}{A_1 E_1} + \frac{4P_2 l_2}{\pi E_2 d_1\, d_2} + \frac{P_3 l_3}{A_3 E_3}$$

$$= \frac{60 \times 10^3 \times 600}{\left(\frac{\pi}{4} \times 60^2\right) \times 2 \times 10^5} + \frac{4 \times 90 \times 10^3 \times 750}{\pi \times 2 \times 10^5 \times 60 \times 30} + \frac{20 \times 10^3 \times 500}{\left(\frac{\pi}{4} \times 30^2\right) \times 2 \times 10^5}$$

$$= 0.06366 + 0.23873 + 0.07074$$

$$\delta l = 0.3731 \text{ mm}$$

$$\text{Total strain energy stored } U = U_1 + U_2 + U_3 = \frac{P_1^2 l_1}{2A_1 E} + \frac{2P_2^2 l_2}{\pi d_1 d_2 E} + \frac{P_3^2 l_3}{2A_3 E}$$

$$= \frac{1}{2 \times 10^5}\left[\frac{60000^2 \times 600}{2 \times \left(\frac{\pi}{4} \times 60^2\right)} + 2 \times \frac{90000^2 \times 750}{\pi \times 60 \times 30} + \frac{20000^2 \times 500}{2 \times \left(\frac{\pi}{4} \times 30^2\right)}\right]$$

$$U = 13\,360.17 \text{ N-mm.}$$

Problem 19 Find the total elongation and energy stored by a stepped bar dimensioned in mm as shown in Figure 3.18. Modulus of elasticity for the material is 206 GPa.

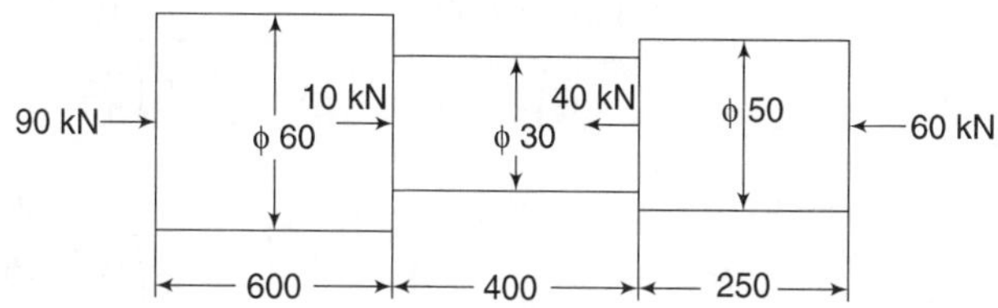

Fig. 3.18 (a)

Solution: **Given data:**

$$\begin{aligned} \text{Diameter } d_1 &= 60 \text{ mm} \\ d_2 &= 30 \text{ mm} \\ d_3 &= 50 \text{ mm} \\ \text{Length } l_1 &= 600 \text{ mm} \\ l_2 &= 400 \text{ mm} \\ l_3 &= 250 \text{ mm} \\ \text{Modulus of elasticity } E &= 206 \text{ GPa} = 206 \times 10^3 \text{ MPa} \end{aligned}$$

Free body diagram:

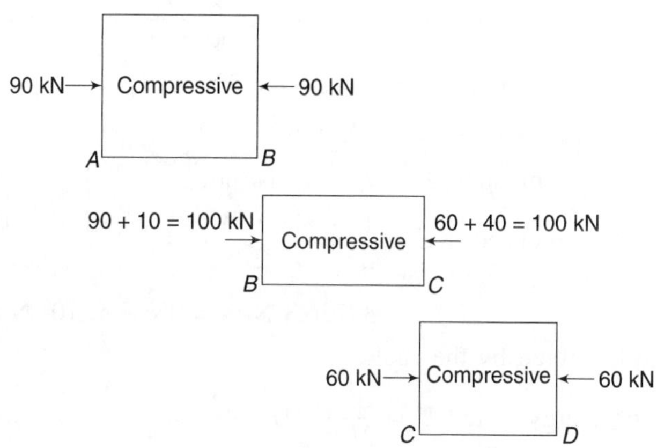

Fig. 3.18 (b)

$$\text{Total elongation } \delta l + \delta l_1 + \delta l_2 + \delta l_3 = \frac{P_1 l_1}{A_1 E} + \frac{P_2 l_2}{A_2 E} + \frac{P_3 l_3}{A_3 E}$$

$$= -\frac{90000 \times 600}{\left(\frac{\pi}{4} \times 60^2\right) 206 \times 10^3} - \frac{100000 \times 400}{\left(\frac{\pi}{4} \times 30^2\right) \times 206 \times 10^3} \frac{-60000 \times 250}{\left(\frac{\pi}{4} \times 50^2\right) \times 206 \times 10^3}$$

$$\delta l = -0.4045 \text{ mm}$$

(–ve indicates decrease in length)

$$\text{Total strain energy stored } U = U_1 + U_2 + U_3 = \frac{P_1^2 l_1}{2A_1E} + \frac{P_2^2 l_2}{2A_2E} + \frac{P_3^2 l_3}{2A_3E}$$

$$= \frac{90000^2 \times 600}{2 \times \left(\frac{\pi}{4} \times 60^2\right) \times 206 \times 10^3} + \frac{100000^2 \times 400}{2 \times \left(\frac{\pi}{4} \times 30^2\right) \times 206 \times 10^3}$$

$$+ \frac{60000^2 \times 250}{2 \times \left(\frac{\pi}{4} \times 50^2\right) \times 206 \times 10^3}$$

$$= 4172.02 + 13735.05 + 1112.54$$

$$U = 190\ 19.62 \text{ N-mm.}$$

Problem 20 A trolley weighing 120 kN is coming down a slope with a uniform speed of 9 kmph. It is connected to a steel rope and is suddenly stopped when the rope length is 300 m. The cross section area of the rope is 342 mm^2 and Young's modulus is 2 × 10^5 N/mm^2. Find the stress developed in the rope due to sudden braking. What will be the stress if the length of cable is 150 m?

Solution: **Given data:**

$$\text{Weight of trolley } W = 120 \text{ kN} = 120 \times 10^3 \text{ N}$$

$$\text{Speed of trolley } v = 9 \text{ kmph} = \frac{9 \times 1000}{60 \times 60} = 2.5 \text{ m/s}$$

$$\text{Length of rope } l = 300 \text{ mm} = 300 \times 10^3 \text{ mm}$$

$$\text{Area of cross section } A = 342 \text{ mm}^2$$

$$\text{Young's modulus } E = 2 \times 10^5 \text{ N/mm}^2$$

$$\text{Kinetic energy produced, KE} = \frac{1}{2} mv^2 = \frac{1}{2}\frac{W}{g}v^2 = \frac{1}{2} \times \frac{120 \times 10^3}{9.81} \times 2.5^2$$

$$= 38226.3 \text{ N-m} = 38.23 \times 10^6 \text{ N mm}$$

This energy is to be stored by the cable

$$\text{Resilience or energy stored } U = \frac{\sigma^2}{2E} \text{ (Vol)}$$

Equating resilience to kinetic energy

$$\frac{\sigma^2}{2E} \text{ (Vol)} = 38.23 \times 10^6$$

Hence, $\quad \text{Vol} = \text{Volume} = Al$

$$\frac{\sigma^2}{2 \times 2 \times 10^5} \times Al = 38.23 \times 10^6$$

$$\frac{\sigma^2}{2 \times 2 \times 10^5} \times (342 \times 300 \times 1000) = 38.23 \times 10^6$$

$$\therefore \quad \text{Stress indued } \sigma = 386.06 \text{ N/mm}^2$$

$$\text{If the length is } l = 150 \text{ m} = 150 \times 10^3 \text{ mm}$$

$$\frac{\sigma_2}{2 \times 2 \times 10^5} \times 342 \times 150 \times 10^3 = 38.23 \times 10^6$$

$$\text{Stress } \sigma = 545.98 \text{ N/mm}^2.$$

Problem 21 An elevator car carrying a load of 10 kN is descending by means of a steel rope at a speed of 1 m/s. The cross section area of the rope is 400 mm^2. The rope is suddenly brought to rest by braking after 30 seconds of travel. Calculate the stress induced in the rope due to sudden stoppage taking Young's modulus of rope as 200 GPa.

Solution: **Given data:**

$$\text{Weight or load } W = 10 \text{ kN} = 10 \times 10^3 \text{ N}$$

$$\text{Velocity } v = 1 \text{ m/s}$$

$$\text{Area of cross section } A = 400 \text{ mm}^2$$

$$\text{Time of stop } t = 30 \text{ secs}$$

$$\text{Young's modulus } E = 200 \text{ GPa} = 200 \times 10^3 \text{ MPa}$$

$$\text{Length of rope } l = \text{distance travelled in 30 sec(s)}$$

$$\text{WKT } S = Vt \text{ for uniform velocity}$$

$$S = l = 1 \times 30 = 30 \text{ m} = 30 \times 10^3 \text{ mm.}$$

$$\text{Kinetic energy (KE)} = \frac{1}{2} mv^2 = \frac{1}{2} \frac{W}{g} v^2$$

$$= \frac{1}{2} \times \frac{10000}{9.81} \times 1^2 = 509.684 \text{ Nm}$$

$$\text{KE} = 509.684 \times 10^3 \text{ N-mm}$$

$$\text{Equating this to resilience } U = \frac{\sigma^2}{2E} \cdot \text{Vol}$$

where, Vol = Volume = Area × Length = Al

$$\frac{\sigma^2}{2 \times 200 \times 10^3} \times (400 \times 30000) = 509.684 \times 10^3$$

$$\therefore \quad \text{Strain induced } \sigma = 130.34 \text{ MPa.}$$

Problem 22 A mass of 500 kg is being lowered by means of steel wire rope having cross sectional area 250 mm^2. The velocity of the falling mass is 0.5 m/sec. The mass is suddenly stopped as the sheave gets stuck up and the maximum instantaneous stress produced was known to be 75 MPa. Taking E = 200 kN/mm^2, find the length of rope at that instance.

Solution: **Given data:**

$$\text{Mass } m = 500 \text{ kg}$$

$$\text{Velocity } v = 0.5 \text{ m/s}$$

$$\text{Stress produced } \sigma = 75 \text{ MPa}$$

$$\text{Young's modulus } E = 200 \text{ kN/mm}^2 = 200 \times 10^3 \text{ N/mm}^2$$

$$\text{Area of cross section rope } A = 250 \text{ mm}^2$$

$$\text{Total kinetic energy KE} = \frac{1}{2} mv^2 = \frac{1}{2} \times 500 \times 6.5^2 = 62.\ 5 \text{ Nm}$$

$$= 62.5 \times 10^3 \text{ N-mm}$$

$$\text{Equating this to energy stored } U = \frac{\sigma^2}{2E} \times \text{Vol}$$

where $\text{Vol} = \text{Volume of rope} = A.l$

$$\frac{75^2}{2 \times 200 \times 10^3} \times 250 \times l = 62.5 \times 10^3$$

$\therefore$ Length of rope $l = 17976.89$ mm.

Problem 23 A weight of 3 kN falls through a height of 3 mm before it strikes the round head K of a round rod 30 mm diameter and 500 mm long as shown in Fig. 3.19. Modulus of elasticity is 210 GPa.

(i) Find the maximum stress induced

(ii) Compare it with static stress.

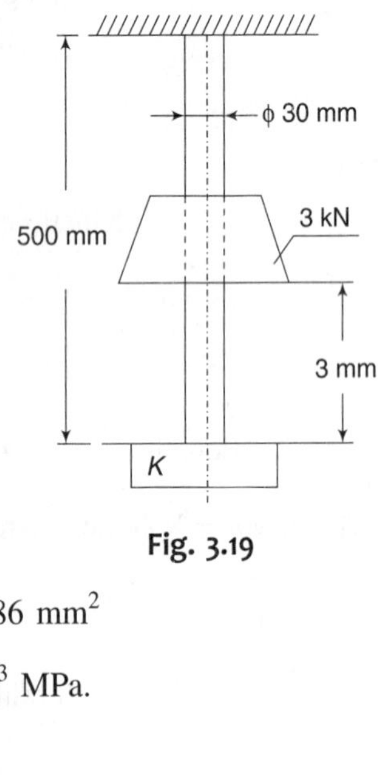

Fig. 3.19

Solution: **Given data:**

$$\text{Weight } W = 3 \text{ kN} = 3000 \text{ N}$$

$$\text{Height of fall } h = 3 \text{ mm}$$

$$\text{Diameter } d = 30 \text{ mm}$$

$$\text{Length } l = 500 \text{ mm}$$

$$\text{Area of cross section } A = \frac{\pi}{4} d^2 = \frac{\pi}{4} 30^2 = 706.86 \text{ mm}^2$$

$$\text{Modulus of elasticity } E = 210 \text{ GPa} = 210 \times 10^3 \text{ MPa.}$$

$$\text{Impact stress } \sigma_i = \frac{W}{A}\left[1 + \sqrt{1 + \frac{2hAE}{Wl}}\right]$$

$$\sigma_i = \frac{3000}{706.86}\left[1 + \sqrt{1 + \frac{2 \times 3 \times 706.86 \times 210 \times 10^3}{3000 \times 500}}\right]$$

$$= 107.75 \text{ MPa}$$

$$\text{Static stress } \sigma = \frac{W}{A} = \frac{3000}{706.86} = 4.244 \text{ MPa}$$

$$\text{Comparing } \frac{\sigma_i}{\sigma} = \frac{107.75}{4.244} = 25.39$$

The impact stress is 23.39 times the static stress.

Problem 24 A bar of rectangular cross section with sides ratio as 2 is 300 mm long. It is subjected to an axial impact by a load of 1.5 kN that falls on to it from a height of 10 mm. Determine the sectional dimensions of the bar if the stress induced is 120 MPa. Take E = 200 GPa.

Solution: **Given data:**

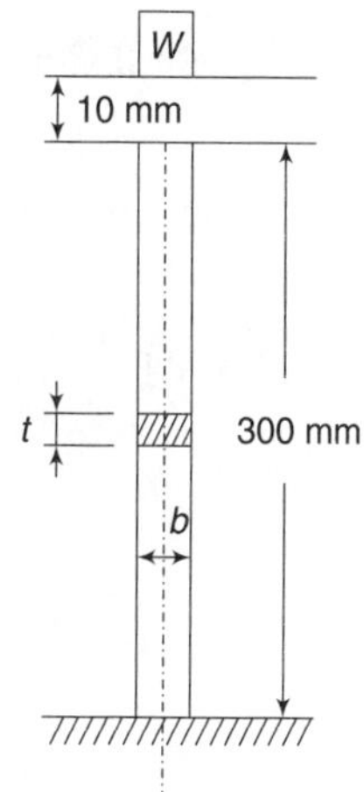

Fig. 3.20

$$\text{Sides ratio } = 2 = \frac{b}{t}$$

$$\therefore \quad b = 2t$$

$$\therefore \quad \text{Area } A = bt = 2t \times t = 2t^2$$

$$\text{Load } W = 1.5 \text{ kN} = -\ 1500 \text{ N}$$

$$\text{Height of fall } h = 10 \text{ mm}$$

$$\text{Length } l = 300 \text{ mm}$$

$$\text{Maximum stress } \sigma_i = 120 \text{ MPa}$$

$$\text{Modulus of elasticity } E = 200 \times 10^3 \text{ MPa}$$

$$\text{Impact stress } \sigma_i = \frac{W}{A}\left[1 + \sqrt{1 + \frac{2hAE}{Wl}}\right]$$

$$120 = \frac{1500}{A}\left[1 + \sqrt{1 + \frac{2 \times 10 \times A \times 200 \times 10^3}{1500 \times 300}}\right]$$

$$\frac{120A}{1500} = 1 + \sqrt{1 + 8.889A}$$

$$0.08A - 1 = \sqrt{1 + 8.8889A}$$

Squaring on both sides,

$$(0.08\ A - 1)^2 = 1 + 8.889A$$

$$(0.08A)^2 + 1 - 2 \times 1 \times 0.08A = 1 + 8.889A$$

$$0.0064A^2 = 8.889A + 0.16A$$

$$0.0064A^2 = 9.049A$$

$$\therefore \quad A = \frac{9.049}{0.0064} = 1413.91 \text{ mm}^2$$

But

$$A = bt = 2t^2 = 1413.91$$

$$t = 26.59 \text{ mm}$$

and

$$b = 2t = 2 \times 26.59 = 53.18 \text{ mm.}$$

Problem 25 A steel rod of length 1.5 m has to resist longitudinally an impact of 2.5 kN falling under gravity at a velocity of 0.9925 m/s. The maximum stress induced is not to exceed 150 MPa. Taking $E = 2 \times 10^5$ N/mm^2, find the diameter of the rod required.

Solution: **Given data:**

$$\text{Length } l = 1.5 \text{ m} = 1500 \text{ mm}$$

$$\text{Load } W = 2.5 \text{ kN} = 2500 \text{ N}$$

$$\text{Velocity } V = 0.9925 \text{ m/s}$$

$$\text{Maximum stress induced } \sigma_i = 150 \text{ MPa}$$

$$\text{Modulus of elasticity } E = 2 \times 10^5 \text{ MPa}$$

To find height of fall

WKT $\quad v^2 - u^2 = 2as$

$$\text{For falling under gravity, } u = 0,\ s = h,\ a = g$$

$\therefore \quad v^2 = 2gh$

$$\text{Height of fall } h = \frac{v^2}{2g} = \frac{0.9925^2}{2 \times 9.81} = 0.05021 \text{ m}$$

$$h = 50.21 \text{ mm}$$

$$\text{Impact stress } \sigma_i = \frac{W}{A}\left[1 + \sqrt{1 + \frac{2hAE}{Wl}}\right]$$

$$150 = \frac{2500}{A} = \left[1 + \sqrt{1 + \frac{2 \times 50.21 \times A \times 2 \times 10^5}{2500 \times 1500}}\right]$$

$$\frac{150A}{2500} = [1 + \sqrt{1 + 5.3557A}]$$

$$006A - 1 = \sqrt{1 + 5.3557A}$$

Squaring on both sides

$$(0.06A - 1)^2 = (1 + 5.3557A)$$

$$0.0036A^2 - 0.12A + 1 = 1 + 5.3557A$$

$$0.0036A^2 = 5.4757A$$

$$A = \frac{5.4757}{0.0036} = 1521.03$$

$$\frac{\pi}{4} d^2 = 1521.03$$

$$\text{Diameter } d = 44 \text{ mm.}$$

Problem 26 An unknown weight falls through a height of 15 mm on a collar rigidly attached to the lower end of a vertical bar 3 m long and 500 mm^2 area of cross section. If the maximum instantaneous elongation is known to be 2 mm under impact, find the maximum instantaneous stress induced and the value of the unknown weight. Take $E = 2 \times 10^5$ N/mm^2.

Solution: **Given data:**

$$\text{Height of fall } h = 15 \text{ mm}$$

$$\text{Length } l = 3 \text{ m} = 3000 \text{ mm}$$

$$\text{Area } A = 500 \text{ mm}^2$$

$$\text{Instantaneous elongation } \delta l_i = 2 \text{ mm.}$$

$$\text{Modulus of elasticity } E = 2 \times 10^5 \text{ N/mm}^2$$

$$\text{Elongation } \delta l_i = \frac{\sigma_i l}{E}$$

$$2 = \frac{\sigma_i \times 3000}{2 \times 10^5}$$

$$\therefore \text{ Maximum instantaneous stress } \sigma_i = 133.33 \text{ N/mm}^2$$

$$\text{Instantaneous stress } \sigma_i = \frac{W}{A}\left[1 + \sqrt{1 + \frac{2hAE}{Wl}}\right]$$

$$133.33 = \frac{W}{500}\left[1 + \sqrt{1 + \frac{2 \times 15 \times 500 \times 2 \times 10^5}{W \times 3000}}\right]$$

$$\frac{133.33 \times 500}{W} = 1 + \sqrt{1 + \frac{1 \times 10^6}{W}}$$

$$\frac{66666.67}{W} - 1 = \sqrt{1 + \frac{10^6}{W}}$$

Squaring on both sides

$$\left(\frac{66666.67}{W} - 1\right)^2 = 1 + \frac{10^6}{W}$$

$$\frac{44.44 \times 10^9}{W^2} - \frac{133.33 \times 10^3}{W} + 1 = 1 + \frac{10^6}{W}$$

$$\frac{44.44 \times 10^9}{W^2} = \frac{10^6 + 133.33 \times 10^3}{W}$$

$$\text{Weight } W = 3921.19 \text{ N.}$$

Problem 27 A round rod of 30 mm diameter 2.4 m long hangs freely from a bracket and has a collar rigidity fixed at the bottom. Modulus of elasticity is 200 kN/mm^2. Find the maximum instantaneous stress induced for the following cases.

(i) A weight of 30 kN falls through a height of 3 mm on the collar
(ii) A weight of 3 kN falls through a height of 30 mm on the collar.

Solution: **Given data:**

$$\text{Diameter } d = 30 \text{ mm}$$

$$\therefore \quad \text{Area } A = \frac{\pi}{4}d^2 = \frac{\pi}{4} \times 30^2 = 706.86 \text{ mm}^2$$

$$\text{Length } l = 2.4 \text{ m} = 2400 \text{ mm}$$

$$\text{Modulus of elasticity } E = 200 \text{ kN/mm}^2 = 200 \times 10^3 \text{ N/mm}^2$$

(i) Weight of 30 kN fall from height of 3 mm

$$W = 30{,}000 \text{ N}$$

$$h = 3 \text{ mm}$$

$$\text{Impact stress } \sigma_i = \frac{W}{A}\left[1 + \sqrt{1 + \frac{2hAE}{Wl}}\right]$$

$$= \frac{3000}{706.86}\left[1 + \sqrt{1 + \frac{2 \times 3 \times 706.8\ 6 \times 200 \times 10^3}{30000 \times 2400}}\right]$$

$$\sigma_i = 194.17 \text{ N/mm}^2.$$

(ii) Weight of 3 kN fall from height of 30 mm.

$$W = 3000 \text{ N}, \ h = 30 \text{ mm}$$

$$\text{Impact stress } \sigma_i = \frac{3000}{706.86}\left[1 + \sqrt{1 + \frac{2 \times 30 \times 706.86 \times 200 \times 10^3}{3000 \times 2400}}\right]$$

$$\sigma_i = 149.98 \text{ N/mm}^2.$$

Problem 28 Figure 3.21 shows a compound bar made of steel and copper. The diameter of the steel rod is 30 mm and the inside and outside diameters of the copper are 30 mm and 45 mm. The ends of rod and tube are rigidity fixed. A weight of 18 kN is falling through a height of 15 mm before it strikes the collar. Taking E_s = 200 GPa and E_c = 120 GPa, find the stresses induced in the rod and tube.

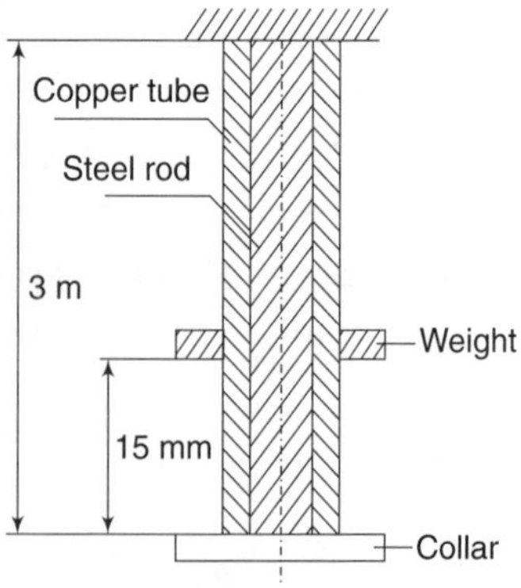

Fig. 3.21

Solution: **Given data:**

Diameter of the steel rod

$$d_s = 30 \text{ mm}$$

$$\therefore \quad \text{Area of steel } A_s = \frac{\pi}{4} d_s^2 = \frac{\pi}{4} \times 30^2 = 706.86 \text{ mm}^2$$

Inside diameter of copper tube d_i = 30 mm

Outside diameter of copper tube d_o = 45 mm

$\therefore$ Area of copper to be $A_c = \frac{\pi}{4}(d_o^2 - d_i^2) = \frac{\pi}{4}(45^2 - 30^2) = 833.57 \text{ mm}^2$

Weight W = 18 kN = 18000 N

Height of fall h = 15 mm

$E_s = 200 \text{ GPa} = 200 \times 10^3 \text{ MPa}$

$E_c = 120 \text{ GPa} = 120 \times 10^3 \text{ MPa}$

Length l = 3 m = 3000 mm

Instantaneous extension in the bar

$$\delta l_i = \frac{\sigma_i l}{E}$$

Maximum stress induced under impact $\sigma_i = E\left(\frac{\delta l_i}{l}\right)$

$$\text{For steel rod, stress } \sigma_{i(s)} = \frac{200 \times 10^3 \times \delta l_i}{3000} = 66.67\ \delta l_i$$

$$\text{For copper to be, stress } \sigma_{i(c)} = \frac{120 \times 10^3 \times \delta l_i}{3000} = 40\ \delta l_i$$

$$\text{Strain energy stored by the bar } U = U_s + U_c = \frac{\sigma {i_{(s)}}^2}{2Es} V_s + \frac{{\sigma_{i(c)}}^2}{2Ec} \cdot V_c$$

$$= \frac{(66.67\ \delta l_i)^2}{2 \times 200 \times 10^3} \times (706.86 \times 3000) + \frac{(40 \delta l_i)^2}{2 \times 120 \times 10^3} (833.57 \times 3000)$$

$$U = 40\ 233.4\ \delta l_i^2$$

Equating this to potential energy PE = $mg\ H$

$$W = mg \text{ and } H = h + \delta l_i$$

$$\text{PE} = W = (h + \delta l_i) = 18000\ (15 + \delta l_i)$$

$$40233.4\ \delta l_i^2 = 270000 + 18000\ \delta l_i$$

$$40233.4\ \delta l_i^2 = 18000\ \delta l_i - 270000 = 0$$

$$2.235\ \delta l_i^2 - \delta l_i - 15 = 0$$

$$\delta l_i = \frac{1 + \sqrt{1^2 - 4 \times 2.235(-15)}}{2 \times 2.235} = 2.824 \text{ mm}$$

$\therefore$ Stress in steel rod $\sigma_{i(s)} = 66.67\ \delta l_i = 66.67 \times 2.824 = 188.27$ MPa

Stress in copper tube $\sigma_{i(c)} = 40\ \delta l_i = 40 \times 2.824 = 112.96$ MPa

Problem 29 A steel rod 30 mm diameter and 600 mm long is struck by a weight of 120 N that falls freely under gravity. If the stress produced is not to exceed 300 MPa, find the maximum height from which the weight can be dropped. Take E = 200 GPa.

Solution: **Given data:**

$$\text{Weight } W = 120 \text{ N}$$

$$\text{Diameter } d = 30 \text{ mm}$$

$$\text{Area of cross section } A = \frac{\pi}{4} d^2 = \frac{\pi}{4} \times 30^2 = 706.86 \text{ mm}^2$$

$$\text{Length } l = 600 \text{ mm}$$

$$\text{Maximum stress} = \sigma_i = 300 \text{ MPa}$$

$$\text{Modulus of elasticity } E = 200 \text{ GPa} = 200 \times 10^3 \text{ MPa}$$

$$\text{Impact stress } \sigma = \frac{W}{A}\left[1 + \sqrt{1 + \frac{2hAE}{Wl}}\right]$$

$$300 = \frac{120}{706.86}\left[1 + \sqrt{1 + \frac{2h \times 706.86 \times 200 \times 10^3}{120 \times 600}}\right]$$

$$\frac{300 \times 706.86}{120} = 1 + \sqrt{1 + 3927h}$$

$$1767.15 - 1 = \sqrt{1 + 3927h}$$

$$1766.15 = \sqrt{1 + 3927h}$$

Squaring on both sides,

$$1766.15^2 = 1 + 3927h$$

$$\text{Maximum height of fall } h = 794.32 \text{ mm.}$$

Problem 30 A vertical 3 m long round rod is rigidly attached to a horizontal wall at the top. A weight fall from a height of 45 mm and the maximum stress induced was found to be 150 MPa. Find the stress, if the same load is applied gradually. Also find the stress if the load is dropped from a height of 90 mm. Take E = 200 GPa.

Solution: **Given data:**

$$\text{Length } l = 3 \text{ m} = 3000 \text{ mm}$$

$$\text{Height of fall } h = 45 \text{ mm}$$

$$\text{Maximum stress } \sigma_i = 150 \text{ MPa}$$

$$E = 200 \text{ GPa} = 200 \times 10^3 \text{ MPa}$$

$$\text{We know that } E = \frac{\sigma_i l}{\delta l_i}$$

$$\therefore \quad \text{Instantaneous elongation } \delta l_i = \frac{\sigma_i l}{E}$$

$$\delta l_i = \frac{150 \times 3000}{200 \times 10^3} = 2.25 \text{ mm}$$

$$\text{Potential energy induced PE} = mgh = W(H + \delta l_i) = W(45 + 2.25)$$

$$\text{PE} = 47.25 \text{ W} \quad \ldots(1)$$

$$\text{Resilience } U = \frac{\sigma_i^2}{2E} . V$$

$$V = \text{Volume} = Al$$

$$U = \frac{150^2}{2 \times 200 \times 10^3} \times A \times 3000$$

$$U = 168.75 \text{ A} \quad \ldots(2)$$

Equating resilience to potential energy induced,

$$47.25\ W = 168.75 \text{ A}$$

$$\therefore \quad \text{Static stress } \frac{W}{A} = \frac{168.75}{47.25} = 3.75 \text{ N/mm}^2$$

$$\text{Impact stress } \sigma_i = \frac{W}{A}\left[1 + \sqrt{1 + \frac{2hAE}{Wl}}\right] = \frac{W}{A}\left[1 + \sqrt{1 + \frac{2hE}{l\left(\frac{W}{A}\right)}}\right]$$

$$\sigma_i = 3.57\left[1 + \sqrt{1 + \frac{2 \times 90 \times 200 \times 10^3}{3000 \times 3.57}}\right]$$

$$\sigma_i = 120.58 \text{ N/mm}^2.$$

Problem 31 A railway wagon weighing 10 kN and moving with a velocity of 9 kmph is brought to rest by a buffer consisting of two round rods each 150 mm diameter and 300 mm long. Taking modulus of elasticity as 2×10^5 N/mm^2, find the instantaneous stress induced and instantaneous contraction in the buffer rods.

Solution: **Given data:**

$$\text{Weight } W = 10 \text{ kN} = 10000 \text{ N}$$

$$\text{Velocity } v = 9 \text{ kmph} = \frac{9 \times 1000}{60 \times 60} = 2.5 \text{ m/s}$$

$$\text{No. of rods} = 2$$

$$\text{Diameter of each rod } d = 150 \text{ mm}$$

$$\text{Length of each rod } l = 300 \text{ mm}$$

$$\text{Area of cross section of rod } A = \frac{\pi}{4} d^2 = \frac{\pi}{4} \times 150^2 = 17671.46 \text{ mm}^2$$

$$\text{Modulus of elasticity } E = 2 \times 10^5 \text{ N/mm}^2$$

$$\text{Total kinetic energy produced KE} = \frac{1}{2} mv^2$$

$$= \frac{1}{2}\frac{W}{g}v^2 = \frac{1}{2} \times \frac{10000}{9.81} \times 2.5^2$$

$$= 3185.5 \text{ Nm} = 3.1855 \times 10^6 \text{ N-mm}$$

Resilience or strain energy stored by 2 rods

$$U = 2\left(\frac{\sigma_i^2}{2E}V\right)$$

where V = Volume = Area × Length

$$= 2\left(\frac{\sigma_i^2}{2 \times 2 \times 10^5} \times 17671.46 \times 300\right)$$

Equate this to kinetic energy produced

$$= 2\left(\frac{\sigma_i^2}{2 \times 2 \times 10^5} \times 17671.46 \times 300\right)$$

$$= 3.1855 \times 10^6$$

∴ Maximum instantaneous stress $\sigma_i = 346.66 \text{ N/mm}^2$

Maximum instantaneous decrease in length

$$\delta l_i = \frac{\sigma_i l}{E} = \frac{346.66 \times 300}{2 \times 10^5} = 0.52 \text{ mm.}$$

Problem 32 Two elastic bars of same material and length, one of them circular cross section of diameter 90 mm and the other of square section of side 90 mm, absorb the same strain energy due to axial force. Compare the stresses induced.

Solution: **Given data:**

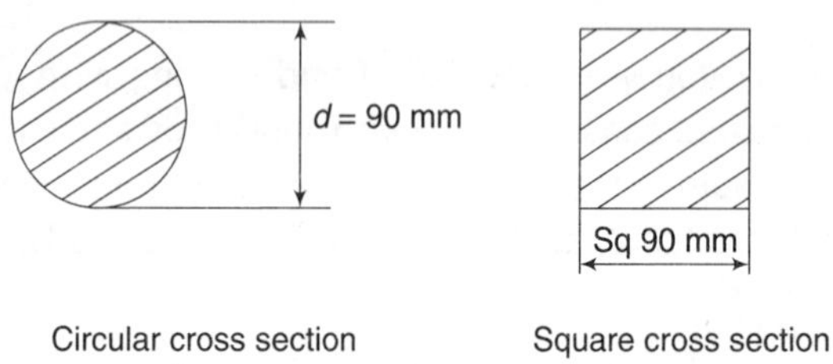

Fig. 3.22

$$\text{Area } A_c = \frac{\pi}{4}d^2 = \frac{\pi}{4}90^2 = 5361.73 \text{ mm}^2$$

$$A_s = a^2 = 90^2 = 8100 \text{ mm}^2$$

$$l = \text{Length}$$

$$\text{Volume } V = Al$$

$$E = \text{Modulus of elasticity}$$

$$\sigma_{ci} = \text{Impact stress in circular rod}$$

$$\sigma_{si} = \text{Impact stress in square rod.}$$

Strain energy is same for both

$$U_c = U_s$$

$$\frac{\sigma_{ci}^2}{2E} V_c = \frac{\sigma_{si}^2}{2E} \cdot V_s$$

$$\frac{\sigma_{ci}^2 \times 6361.73l}{2E} = \frac{\sigma_{si}^2 \times 8100l}{2E}$$

$$\therefore \qquad \frac{\sigma_{ci}}{\sigma_{si}} = \sqrt{\frac{8100}{6361.73}} = 1.128 = \sqrt{\frac{4}{\pi}}.$$

Stress in circular rod is 1.128 times that of square rod.

THINGS TO REMEMBER

- Strain energy or resilience or energy stored by a member
$$U = \frac{\sigma^2}{2E} \cdot V \text{ or } = \frac{P^2}{2EA} l$$
- Proof resilience $U_y = \frac{\sigma_y^2}{2E} \cdot V$
- Proof resilience is the area of triangle within elastic limit of a graph drawn for P vs δl.
- Unit resilience or modulus of resilience $U^1 = \frac{\sigma^2}{2E}$
- For gradually applied load, stress $\sigma = \frac{P}{A}$
- For suddenly applied load, stress $\sigma_i = \frac{2P}{A}$
- For axial impact, impact stress $\sigma_i = \frac{W}{A}\left[1 + \sqrt{1 + \frac{2hAE}{Wl}}\right]$
- Strain energy or resilience in shear $U_\tau = \frac{\tau^2}{2G} \cdot V$
- For tapper round rod, strain energy due to axial load $U = \frac{2P^2 l}{E\pi d_1 \, d_2}$
- For taper rectangular plate of uniform thickness, strain energy due to axial load
$$U = \frac{P^2 l}{2tE(B - b)} \log_e \left(\frac{B}{b}\right)$$
- Strain energy of a prismatic bar under its own weight
$$U = \frac{W^2 l}{6AE}$$
- Strain energy stored by member under gradually applied load P, extension δl,
$$U = \frac{1}{2} P.\delta l.$$

EXERCISES

1. Derive an expression for strain energy stored by a member subjected to axial load.
2. Define resilience, proof resilience and modulus of resilience.
3. Derive an expression for stress due suddenly applied axial load.
4. Derive an expression for stress due to impact load applied axially.
5. Derive an expression for strain energy due to shear load.
6. Derive an expression for strain energy due to gradually applied axial load in a round taper bar.
7. Derive an expression for strain energy due to gradually applied axial load in a taper rectangular cross section plate.
8. Derive an expression for strain energy due to self weight in a prismatic bar.
9. Compare the stress induced in two bars of same length and same material, subjected to same strain energy, one is round bar and the other is a square section having side equal to the diameter of bar. (***Ans.*** $U_c/U_s = 1.128$)
10. A steel rod of 40 mm diameter is 3 m long. Find the strain energy stored by it if a load of 42 kN is gradually applied on to it. Take $E = 2 \times 10^5$ MPa. (***Ans.*** U = 10528.1 Nmm)
11. A steel rod of 60 mm diameter is 2 m long. Find the maximum instantaneous stress induced when a load of 150 kN is suddenly applied to it. Also find the instantaneous elongation. Take $E = 2 \times 10^5$ MPa. (***Ans***. $\sigma = 10^6$ ./ Mpa, δl = 1.061 mm)
12. A uniform rod of 2 m long and 20 mm diameter has elastic limit of 240 MPa. Find the proof resilience of the bar. What suddenly applied load will produced same energy as proof resilience. Take E = 200 GPa

    (***Ans.*** U_y = 90,477.87 N-mm, P = 37,699.11 N)
13. A round rod has diameter 60 mm over a length of 1.2 m and diameter 30 mm for a length of 2 m. An axial load of 120 kN is applied to the rod. The modulus of elasticity is 200 GPa. Find the strain energy produced in the bar. What will be the diameter of the uniform rod of same total length that will produce the same strain energy. (U = 117,143.2 N mm, d = 35.385 mm)
14. A round bar of 120 mm diameter tapers uniformly to a diameter of 75 mm over a length of 1.2 m. Find the energy stored by it when it carries an axial load of 120 kN. E = 200 GPa

    (***Ans.*** U = 6111.55 N mm)
15. A rectangular cross section plate of thickness 20 mm tapers uniformly from a width of 90 mm to 40 mm over a length of 1.8 m. Taking young's modulus as 2×10^5 N/mm^2. Find the energy stored by it, if it is subjected to an axial load of 60 kN.
16. A round bar of diameter 60 mm and length 200 mm carries a gradually applied tensile load of 150 kN. For some reason, the diameter at the middle 200 mm length is reduced to 30 mm. E = 200 GPa Calculate the strain energy stored in the bar.

    (***Ans.*** Ʊ = 27,852.12 N mm)
17. Figure 3.23 shows a bar of square cross section of side 100 mm. Find the total elongation and strain energy stored by each section. Take modulus of elasticity as 2×10^5 N/mm^2.

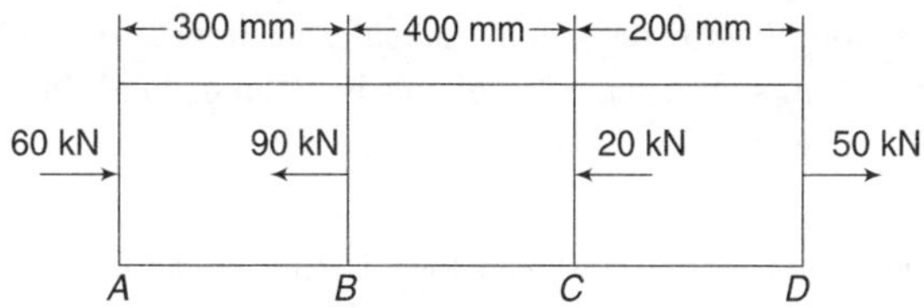

Fig. 3.23

(*Ans.* δl = 0.02 mm, U_1 = 270 N-mm, U_2 = 90 N-mm, U_3 = 125 N-mm)

18. Find the total elongation and total strain energy stored by a rod loaded as shown in Fig. 3.24. Take $E = 2 \times 10^5$ MPa.

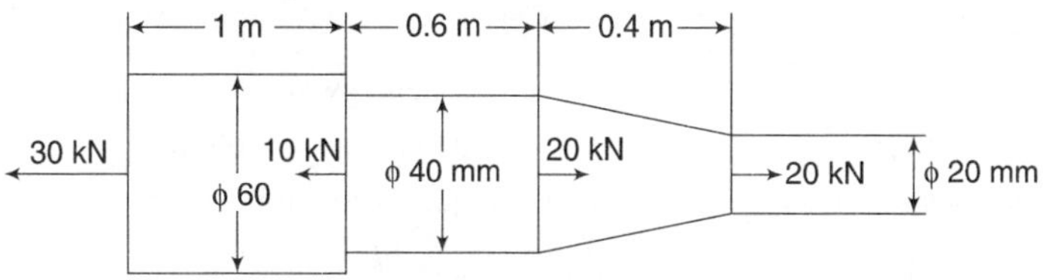

Fig. 3.24

(*Ans.* δl = 0.2122 mm, U = 33 42.25 N mm)

19. A trolley weighing 45 kN is moving with a velocity of 7.5 kmph. It is held by a wire rope and is suddenly stopped when the rope length is 0.5 km. The cross-sectional area of the rope is 300 mm^2 and E = 195 GPa. Find the stress developed in the rope due to sudden stop-page. (*Ans.* σ = 160.88 MPa)

20. A bar of 15 mm diameter gets stretched by 0.4 mm under a steady load of 12 kN. What stress would be produced in the same bar by a weight of 1.2 kN which falls through a distance of 50 mm on to a rigid collar attached at the end. Take Young's modulus as 120 GPa. (*Ans.* σ_i = 114.36 MPa)

21. An unknown weight falls through a height of 25 mm on a vertical bar 1 m long and 40 mm diameter. The instantaneous reduction in length in the rod is 1.2 mm. Taking E = 206 GPa, find max stress induced and magnitude of unknown weight.

(*Ans.* σ_i = 247.2 MPa, W = 7118.6 N)

22. Figure 3.25 shows a compound bar made of steel and brass. The diameter of brass rod is 50 mm and the inside and outside diameter of steel pipe are 30 mm and 60 mm. The ends of rod and pipe are rigidly fixed. A weight of 24 kN is freely falling through a height of 30 mm before it strikes the collar. Take $E_s = 2 \times 10^5$ N/mm^2 and $E_b = 75 \times 10^3$ N/mm^2. Find the instantaneous stresses produced in the members. The length of the compound bar is 2 m.

(*Ans.* σ_{is} = 331.96 MPa, σ_{iB} = 124.5 MPa)

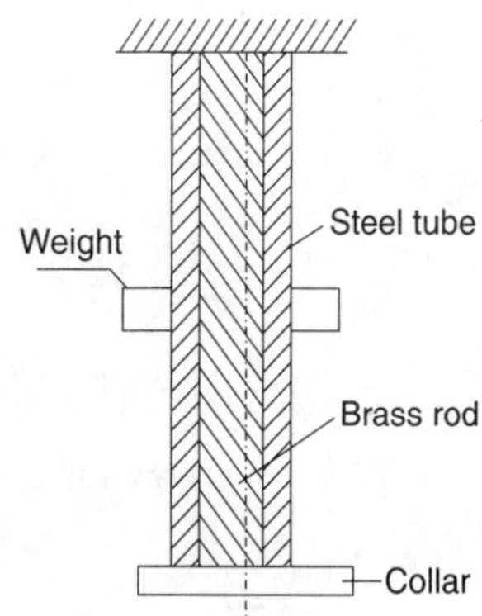

Fig. 3.25

23. A steel rod 2 m long has to resist longitudinally an impact of 3 kN falling under gravity at a velocity of 1.2 m/s. The diameter of rod is 50 mm. Find the maximum impact stress developed taking E as 2×10^5 N/mm^2. Compare it with static stress.

(***Ans.*** $\sigma_i = 151.3$, MPa, $\sigma_i/\sigma = 99.03$)

24. A bar of circular cross section 420 mm long is subjected to an axial impact by a load of 1 kN that falls on to it from a height of 30 mm. Determine the diameter of the bar if the instantaneous reduction in length is 0.32 mm. Taking $E = 2 \times 10^5$ N/mm^2

(***Ans:*** $d = 39.79$ mm)

Objective Type Questions

1. Resilience is

(a) internal stress (b) energy stored

(c) internal strain (d) change in volume

2. Equation for strain energy for a member subjected to axial load is

(a) $\dfrac{Pl}{A\delta l}$ (b) $\dfrac{Pl}{AE}$

(c) $\dfrac{P^2l}{2AE}$ (d) $\dfrac{2P^2l}{AE}$

3. Equation for strain energy for a member subjected to axial stress is

(a) $\dfrac{\sigma^2 V}{E}$ (b) $\dfrac{\sigma^2 l}{2E}$

(c) $\dfrac{2\sigma^2}{E}V$ (d) $\dfrac{\sigma^2}{2E}V$

4. Unit resilience is given by

(a) $\dfrac{\sigma^2}{2E}$ (b) $\dfrac{2\sigma^2}{E}$

(c) $\dfrac{2\sigma}{E}$ (d) $\dfrac{\sigma^2}{2E}V$

5. For a suddenly applied load, the stress produced compared to static stress is

(a) same (b) $\dfrac{1}{2}$

(c) twice (d) 4 times

6. The impact axial stress is given by $\sigma_i =$

(a) $\dfrac{W}{A}\left[1 + \sqrt{1 + \dfrac{2hAE}{Wl}}\right]$ (b) $\dfrac{W}{A}\left(\sqrt{1 + \dfrac{2hAE}{Wl}}\right)$

(c) $\dfrac{W}{A}[1 + \sqrt{1 + \dfrac{2hWl}{AE}}$ (d) $\dfrac{W}{A}\left(1 + \sqrt{\dfrac{2hAE}{Wl}}\right)$

7. Strain energy for a shear stress acting on a member is

(a) $\dfrac{\tau^2}{2G}\cdot V$ (b) $\dfrac{\tau^2}{2E}V$

(c) $\dfrac{2\tau^2}{V}E$ (d) $\dfrac{\tau^2}{2V}\cdot E$

8. Strain energy for a taper round bar under axial load is

(a) $\frac{4P^2}{\pi E d_1 d_2}$ (b) $\frac{2Pl}{\pi E d_1 d_2}$

(c) $\frac{2P^2 l}{\pi E d_1 d_2}$ (d) $\frac{4Pl}{\pi E d_1 d_2}$

9. Strain energy for a prismatic bar under self weight is

(a) $\frac{Wl^2}{2AE}$ (b) $\frac{W^2 l}{2AE}$

(c) $\frac{W^2 l}{3AE}$ (d) $\frac{W^2 l}{6AE}$

10. Strain energy of a member subjected to steady load P and a change in length of δl, is

(a) $P.\delta l$ (b) $\frac{P.\delta l}{2}$

(c) $2P.\delta l$ (d) $\frac{P.\delta l}{6}$

11. Two elastic bars of same material and length, one circular cross section and the other square section of side equal to the diameter, the ratio of strain energies stored by circular bar to square bar for same stress is

(a) $\frac{4}{\pi}$ (b) $\sqrt{\frac{4}{\pi}}$

(c) $\sqrt{\frac{\pi}{4}}$ (d) $\frac{\pi}{4}$

12. The strain energy is : ___________

(a) directly proportional to volume
(b) inversely proportional to volume
(c) not dependent on volume
(d) directly proportional to square times volume

Key to Answers

1. (b)	2. (c)	3. (d)	4. (a)	5. (c)	6. (a)	7. (a)	8. (c)
9. (d)	10. (b)	11. (d)	12. (a)				

CHAPTER 4

Shear Forces and Bending Moments

4.1 INTRODUCTION

In this chapter we will discuss shear forces and bending moments in beams, their relation with each other and with loads. To design any beam, one needs to estimate the shear forces and bending moments on that beam. Usually, one needs to know not only the maximum values of shear force and bending moment on a beam but also why they vary along the length of the beam. These shear force and bending moments help us to find the stresses, strains and deflections.

4.2 SHEAR FORCE AND BENDING MOMENT

Shear force at a section in a beam is the resultant force acting on any one of the parts normal to the axial of that member. It is the algebraic sum of all the forces (including reactions) acting normal to the axis of the beam either to the right side or left side of a section.

Bending moment at a section in a beam is the resultant moment acting on any one side of section. It is the algebraic sum of all the moments about that section (including moments due to reactions) acting on either to the right side or left side of a section.

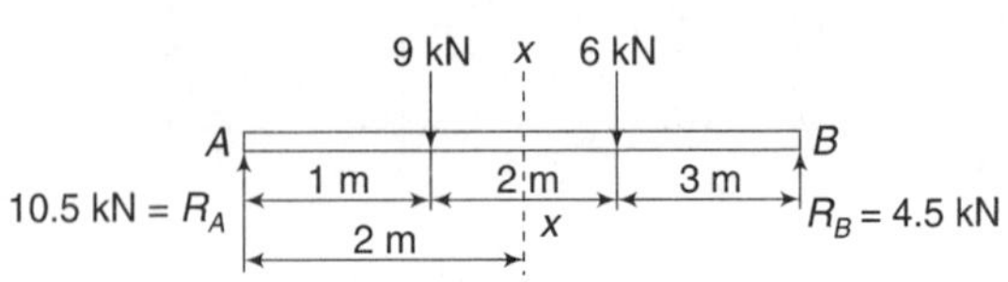

Fig. 4.1

To understand the concept of the shear force and bending moment, consider a simply supported beam as shown in Fig. 4.1.

R_A and R_B are reactions at A and B

$$R_A + R_B = 9 + 6 = 15 \text{ kN}$$

Taking moments about A,

$$M_A = 9 \times 1 + 6 \times 3 - R_B \times 6 = 0$$

$$\therefore \quad R_B = 4.5 \text{ kN}$$

$$R_A = 15 - 4.5 = 10.5 \text{ kN}$$

To find shear force and bending moment at a section X–X at a distance of 2 m from A, imagine that the beam is cut at XX and the two parts will be as shown in Fig. 4.2.

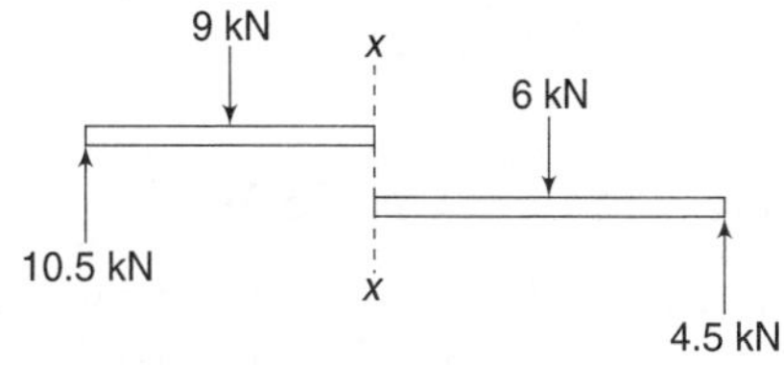

Fig. 4.2

To find shear force at XX, consider algebraic sum of forces acting on any one portion, say left side portion.

Shear force at $XX = SF_{XX} = 10.5 - 9 = 1.5$ kN

(or if right side portion is considered, then $SF_{XX} = 6 - 4.5 = 1.5$ kN)

For final bending moment at XX, consider algebraic sum of moments of all forces acting at either right side or left side of section XX.

For left side portion

$$M_x = 10.5 \times 2 - 9 \times (2 - 1) = 12 \text{ kN-m.}$$

(or for right side portion, $M_x = 4.5 \times 4 - 6 \times 1 = 12$ kN-m)

It is seen that the algebraic sum of forces on any one side of section XX is 1.5 kN and bending moment at section XX is 12 kN-m. This force is trying to cut or shear the beam and hence is called shear force. Similarly, the algebraic sum of moments at section XX either to the left or right is 12 kN-m and is trying to bend the beam at that section. It is called bending moment at section XX.

4.3 SIGN CONVENTION FOR SHEAR FORCE AND BENDING MOMENT

In the previous chapters, we regarded tensile loads as +ve and compressive loads as –ve. Similarly, for finding algebraic sum of either shear forces or bending moments, some sign convention is to be adopted.

The shear force that acts clockwise against the material is considered positive shear force and the shear force that acts counter- clockwise against the material is considered negative shear force as shown in Fig. 4.3.

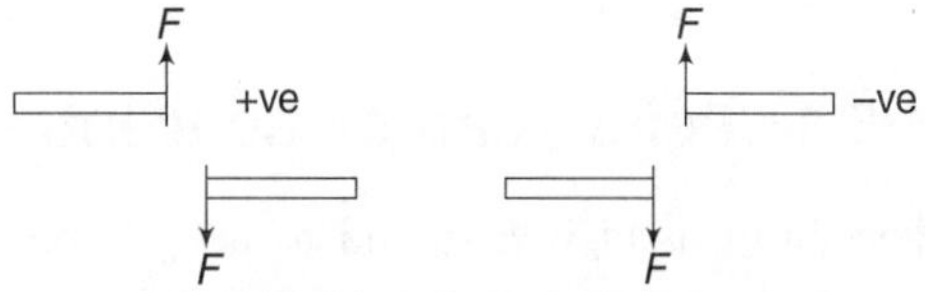

(a) +ve shear forces (b) –ve shear forces

Fig. 4.3 Sign convention for shear forces

In other words, left side upward and right side downward forces are considered positive and left side downward and right upward forces are considered negative.

Bending moments that cause sagging of beam are considered positive and bending moments that cause hogging are considered negative moments. In other words sagging moments are positive and hogging moments are negative as shown Fig. 4.4.

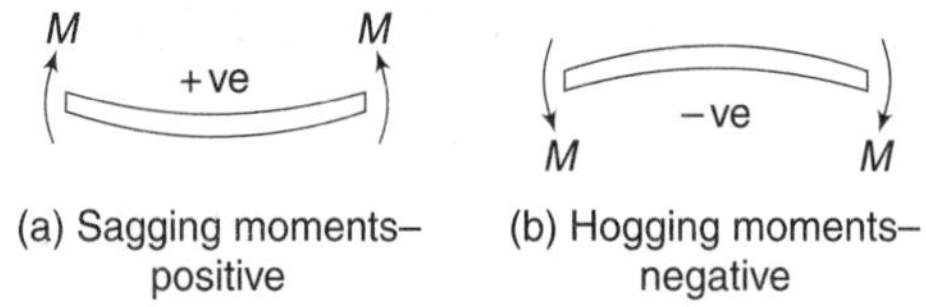

Fig. 4.4 Sign convention for bending moments

4.4 TYPES OF BEAMS

A beam is defined as any structural member, subjected to a system of external forces acting on it. Figure. 4.5 shows some typical beams.

(a) *Cantilever beam.* If a beam is fixed at one end and the other end is free, then it is called a cantilever beam.

(b) *Simply supported beam.* For a beam, if the ends are resting freely on supports, then it is called a simply supported beam.

(c) *Fixed beam.* If a beam is fixed at both ends then it is called a fixed beam.

(d) *Continuous beam.* A beam having more than two supports is called a continuous beam.

(e) *Overhanging beam.* A simply supported beam which over hangs beyond support on one side or both sides is called a overhanging beam.

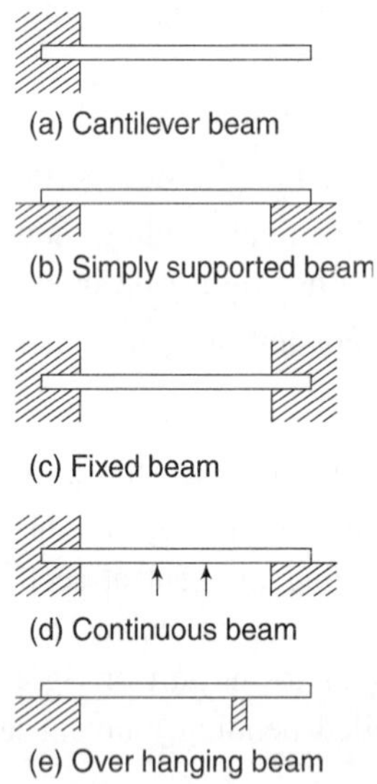

Fig. 4.5 Types of beams

4.5 TYPES OF LOADS

The loads are broadly classified into 4 types. They are:

- Point loads or concentrated loads
- Uniformly distributed load (UDL)
- Uniformly varying load (UVL)
- Moments.

4.5.1 Point Load or Concentrated Load

The entire load is assumed to act at one point and such load is known as concentrated load.

It is denoted by a single arrow as shown in Fig. 4.6.

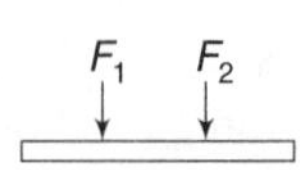

Fig. 4.6 Concentrated loads

4.5.2 Uniformly Distributed Load (UDL)

The load uniformly distributed over a length is called uniformly distributed load (UDL). For example, the weight of a slab or roof of a building is uniformly distributed over the entire length.

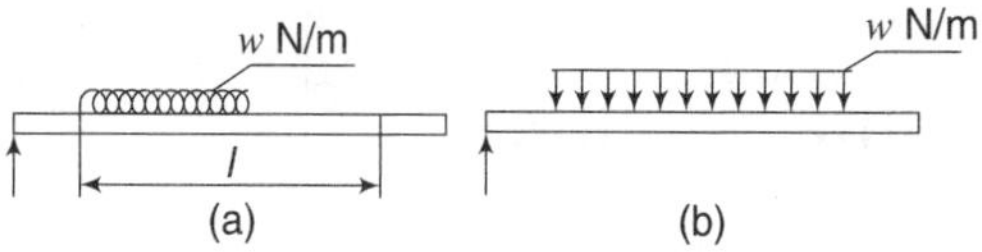

Fig. 4.7 UDL

This is represented in either way as shown in Fig. 4.7(a) or 4.7(b). It is denoted by w N/m length (w Newton per metre length).

For calculation purpose, the UDL is replaced by an equivalent point load of magnitude $W = w \cdot l$ acting in the middle of length l, i.e., at $l/2$ as shown in Fig. 4.8.

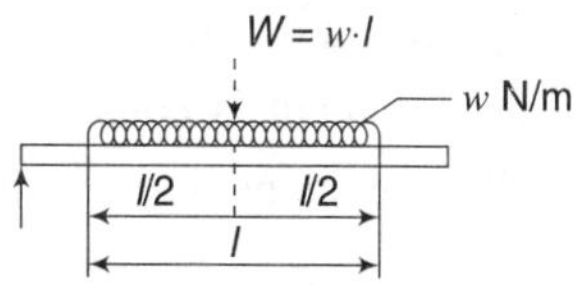

Fig. 4.8

4.5.3 Uniformly Varying Load (UVL)

Figure 4.9 shows a load that uniformly varies from 0 to w N/m over a length of l. In this the vertical line represents the load intensity at any point on the beam.

For calculation purpose, the UVL is replaced by an equivalent point load acting at the CG of the triangle, i.e., at $\frac{2}{3}\, l$ from tip or $\frac{1}{3}\, l$ from full intensity as shown in Fig. 4.9. The intensity of load is equal to the area of the UVL triangle $W = \frac{1}{2}\, w \cdot l$

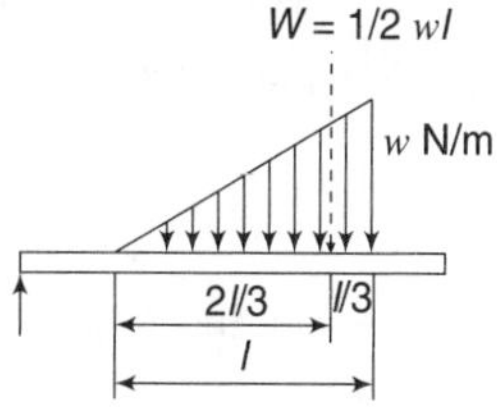

Fig. 4.9 UVL

4.5.4 Moments

A beam may be subjected to a moment (M) as shown in Fig. 4.10. The moment may be clockwise or counterclockwise.

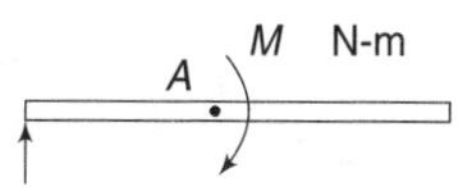

Fig. 4.10

4.5.5 Combined Effect

Figure 4.11(a) shows a load attached at point A acting at B at a distance of l from A. This has two effects, one is direct load F at A and a moment $M = F \cdot l$ as shown in Fig. 4.11(b).

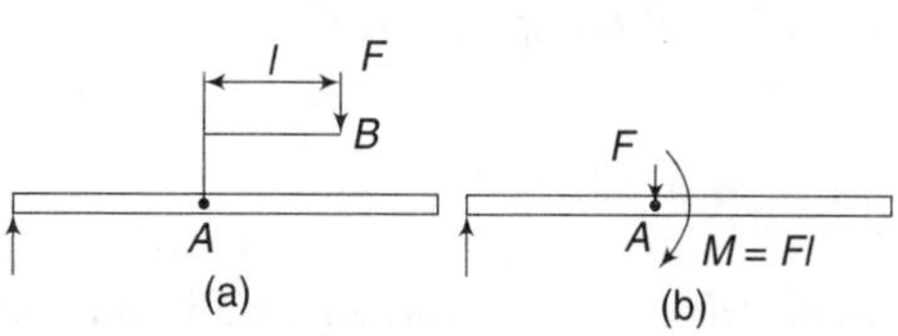

Fig. 4.11

4.6 RELATION BETWEEN LOAD INTENSITY, SHEAR FORCE AND BENDING MOMENT

Figure 4.12 shows a beam with general loading. Consider an elemental length of beam, dx, at a distance of x from left support.

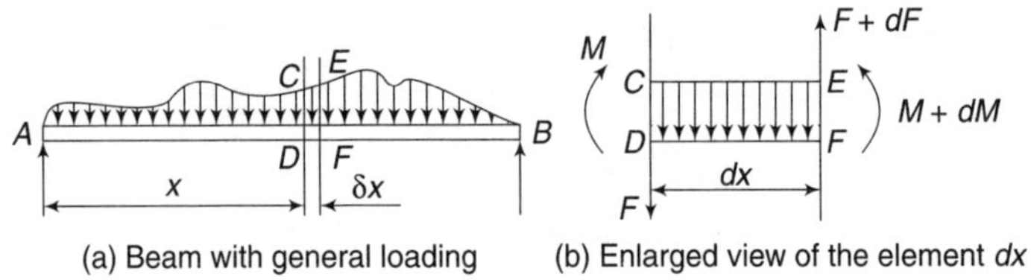

Fig. 4.12 (a) Beam with general loading, (b) Enlarged view of the elements dx

Figure 4.12 (b) shows the enlarged view of the element. The load intensity over this small element is considered to be constant (UDL) and equal to w/unit length.

The element is in equilibrium under the following forces:

- Force F acting downward at CD
- Force $F + dF$ acting upward at EF
- Force δF acting between CD and EF
- Moment M at CD
- Moment $M + dM$ at EF.

Consider vertical equilibrium of forces, $\Sigma F = 0$.

$$-F - (w \cdot dx) + (F + dF) = 0$$

(As per the sign convention of shear forces, left side downward and right side upward are –ve)

$$-wdx + dF = 0$$

$$\therefore \qquad w = \frac{dF}{dx}$$

Hence, it is seen that rate of change of force over unit length is equal to w.

Consider the equilibrium of moments, $\Sigma M = 0$ about face EF,

$$-M + F \cdot dx + (w \cdot dx)\frac{dx}{2} + (M + dM) = 0$$

$$-M + F \cdot dx + w\frac{(dx)^2}{2} + M + dM = 0$$

Since dx is very small, $(dx)^2$ will be further small and can be neglected.

$$Fdx + dM = 0$$

$$\therefore \qquad \frac{dM}{dx} = -F$$

Hence, the rate of change of moment is equal to the shearing force F.

4.7 STEP-BY-STEP PROCEDURE TO FIND SHEAR FORCES AND BENDING MOMENTS

The step-by-step method to find shear forces and bending moments is explained here.

Step 1: First find the total vertical force acting at the fixed end by taking the algebraic sum of forces acting on the beam for cantilever. For simply supported beam, find reactions.

Step 2: To find shear force at various points. Start from free end of the beam. Each point where there is a force or moment acting will be a salient point. For UDL and UVL, both ends will be salient points.

Step 3: At any point, if there is a point load, then shear force up to the point is calculated without considering that force and shear force at the point is calculated considering that force.

Step 4: At the fixed end, the shear force is first calculated without considering the vertical force at the fixed end and then, it will be zero when the vertical force is considered.

Step 5: Mark all the salient points and indicate the values of shear forces.

The shear force is constant between two point loads and is horizontal. The variation of shear force for UDL is triangular in shape and for UVL, is parabolic.

Step 6: Starting from free end, the bending moment at each salient point is calculated by taking algebraic sum of the moments acting on one side of beam.

Step 7: If there is any moment acting an the beam, then moment up to that point without considering the moment is calculated first and then the moment at that point is calculated by adding that moment.

The variation of bending moment between two point loads is triangular. For UDL, it is parabolic and for UVL it is cubic.

While calculating the shear forces and bending moments, it is important to follow some sign convention. As already discussed earlier in 4.3, we follow the following sign conventions.

Shear forces: Left side upward, right side downward is +ve

: Left side downward, right side upward is −ve

Bending moment: Sagging moments are positive and hogging moments are negative.

Step 8: After finding the shear forces and bending moments at various salient points, represent them graphically.

The nature of variation of shear force and bending moment for commonly applied loads is given in Table 4.1.

Table 4.1 Variation of SF and BM

Load	*SF*	*BM*
No load	Constant, horizontal	Linear (triangular)
Point load	Vertical line	Linear (triangular)
UDL	Linear (triangular)	Parabolic
UVL	Parabolic	Cubic
Moment	Constant	Vertical line

4.8 SHEAR FORCE AND BENDING MOMENT FOR CANTILEVER BEAMS

Shear force and bending moment diagrams for the following standard cases have been considered.

(a) Cantilever with point at free end
(b) Cantilever with point load at a distance a from the fixed end
(c) Cantilever with UDL over the entire length
(d) Cantilever with partial UDL towards the free end
(e) Cantilever with partial UDL towards the fixed end
(f) Cantilever with partial UDL in the middle
(g) Cantilever with UVL over the entire span
(h) Cantilever with UVL towards the free end
(i) Cantilever with UVL towards the fixed end
(j) Cantilever with UVL in the middle
(k) Cantilever with moment at the free end.

4.8.1 Cantilever with Point Load at the Free End

Consider a cantilever beam of length l carrying a point load w at its free end as shown in Fig. 4.13.

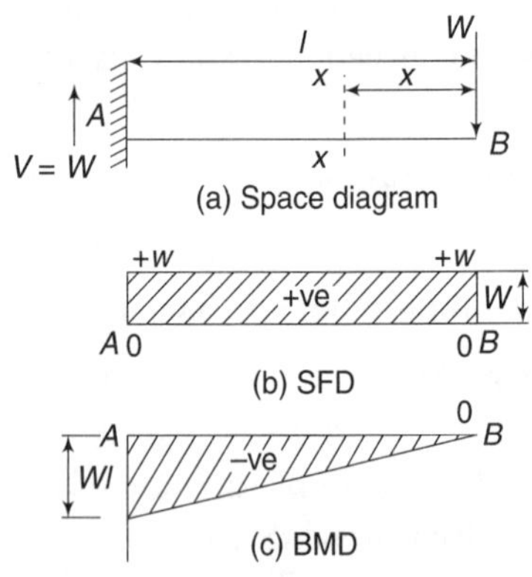

Fig. 4.13

(a) Considering the equilibrium of the vertical forces, $\Sigma v = 0$.

(b) **Shear forces.** Since there is a vertical force at B,

Consider $SF_{\text{up to } B} = 0$

$SF_{\text{at } B} = +W$

Again shear force upto A, $\neq lw$ SF up to $A = W$ (same as at B, since there is no other force between A and B)

Shear force at A, $SF_{\text{at } A} = W - V = W - W = 0$

Hence, $SF_{\text{at } x} = W$

(c) **Shear force diagram (SDF);** (Fig. 4.13b)

- From the above calculations, it is seen that $SF_{up\ to\ B} = 0$ and $SF_{at\ B} = +W$. At point B, draw a vertical line of W units upward above the horizontal line as W is +ve.
- Shear force up to A is also equal to $+W$. Hence, draw a horizontal line from B to A at a height of W. The shear force is same between B and A and there is no change in the SF diagram.
- Shear force at $A = 0$, hence, drop a vertical line downward from this horizontal line.

(d) **Bending moments:**

- Bending moment = Force × Perpendicular distance
- Sagging moments are considered as +ve and hogging moments are considered −ve
- Bending moment at B, $M_B = 0$
  (There are no forces to the right of B)
- Bending moment at A, $M_A = -W \times l = -Wl$ (Hogging)
- Bending moment at any other section $M_x = -W \cdot x$

Draw a horizontal line AB.

At B, BM = 0

At A, BM = $-Wl$. Hence, mark a vertical line downward of horizontal line AB. Equal to Wl units.

Since BM = $M \cdot x$

x is varying between 0 and l, it is linear. Hence, BMD is triangular in shape.

4.8.2 Cantilever with Point Load at a Distance (*a*) from Fixed End

Figure 4.14 shows a cantilever beam of a length l acted upon by a point load W at a distance of a from the fixed end.

(a) The vertical force at A, $V_A = W$

(b) **Shear forces (SFs):**

$SF_{at\ B} = 0$

$SF_{up\ to\ C} = 0$

$SF_{at\ C} = +W$

$SF_{up\ to\ A} = +W$

$SF_{at\ A} = +W - V = 0$

$SF_{at\ X} = +W$

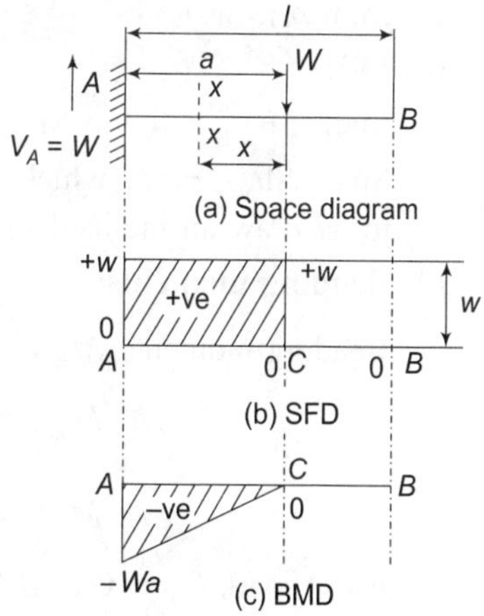

Fig. 4.14

(c) **SFD:** (↑+ve↓)

Draw a vertical line upwards at $B = W$ units

Draw a vertical line upwards at $A = W$ units

Between B and C, SF = 0

Between A and C, SF = W

At A and B, SF = 0

(d) **Bending moments:** ((↻+ve↺))

Bending moment $M_{\text{at } B} = 0$

$$M_{\text{at } C} = 0$$

$$M_{\text{at } A} = -\ W \cdot a \text{ (Hogging)}$$

$$M_{\text{at } x} = -\ W \cdot x \text{ (Hogging)}$$

(e) **BMD:**

Between A and C, x varies from $x = 0$ to a, hence, bending moment varies from O to $-Wa$ linearly. At B $M_B = 0$ at C, $M_C = 0$.

4.8.3 Cantilever with UDL over Full Length

Figure 4.15 shows a cantilever beam of length l carrying a UDL of w N/m over its full span

(a) Vertical force at $A = V = w.l$

(b) **Shear forces:** The UDL is treated point load acting

$SF_{\text{at } B} = 0$

$SF_{\text{at } x} = +\ w \cdot x$

$SF_{\text{at } A} = +\ w \cdot l$

as x is varying from $x = 0$ to l, the SF varies linearly from o to $w \cdot l$

(c) **SFD:**

Mark $SF_{\text{at } B} = 0$ and $SF_{\text{at } A} = +\ wl$ above line AB. Since $SF_{\text{at } x} = wx$ which is linear, varying from o to wl draw an inclined line joining o and wl.

(d) **Bending moments:**

Bending moments $M_{\text{at } B} = 0$

$$M_{\text{at } x} = -\ (w \cdot x)\frac{x}{2} \text{ (Hogging)}$$

$$M_{\text{at } A} = -\ (w \cdot l)\ \frac{l}{2} = -\ \frac{wl^2}{2}$$

(e) BMD mark $M_B = 0$ at B

$$M_{\text{at } x} = -\ \frac{wx^2}{2}$$

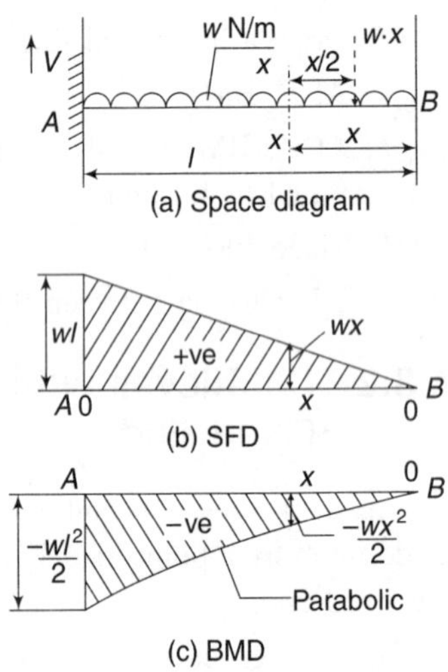

Fig. 4.15

$$M_{\text{at}A} = \frac{-wl^2}{2}$$

It is seen that $M_x = \dfrac{-wx^2}{2}$ is a parabolic equation. The variation of BM from B to A is parabolic in nature.

4.8.4 Cantilever with Partial UDL Towards the Free End

Figure 4.16 shows a cantilever of length l carrying an UDL of w N/m over a length of a from its free end.

The UDL is treated as point load acting at its middle.

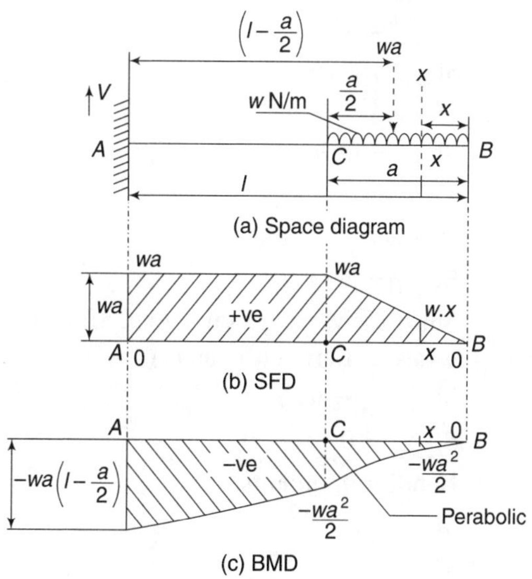

Fig. 4.16

(a) The vertical force at $A \cdot V = w \cdot a$

(b) **Shear forces:**

$SF_{\text{at } B} = 0$

$SF_{\text{at } X} = +(w \cdot x)$

$SF_{\text{at } C} = +(w \cdot a)$

$SF_{\text{upto } A} = +(w \cdot a)$

$SF_{\text{ at } A} = (w \cdot a) - v = 0$

(c) **SFD:**

Mark $SF_{\text{at } B} = 0$, $SF_x = wx$

$SF_c = wa$, $SF_{\text{upto } A} = wa$ and

$SF_{\text{at } A} = 0$

as x varies from 0 to a, SF linearly varies from o to $w \cdot a$ join horizontal line from A to C and inclined line from C to B.

(d) **Bending moments:**

$$M_{\text{at } B} = 0$$

$$M_{\text{at } x} = (-wx)\frac{x}{2} = \frac{-wx^2}{2} \text{ (Hogging)}$$

$$M_{\text{at } C} = (-w \cdot a)\frac{a}{2} = \frac{-wa^2}{2}$$

$$M_{\text{at } A} = (-w \cdot a)\left(l - \frac{a}{2}\right)$$

(e) **BMD:**

Mark $M_B = 0$, $M_C = \dfrac{-wa^2}{2}$ and $M_A = (-wa)\left(l - \dfrac{a}{2}\right)$

Since the variation g BM between B and is $M_x = -w\dfrac{x^2}{2}$, the shape will be parabolic.

Between C to A, it is an inclined line as shown.

4.8.5 Cantilever with Partial UDL Towards Fixed End

Figure 4.17 shows a cantilever of length l carrying an UDL of w N/m over a span of a from the fixed end.

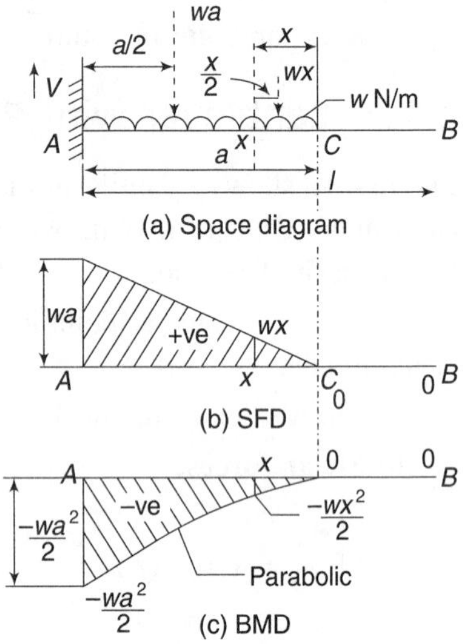

Fig. 4.17

(a) Vertical load at A, $V = w \cdot a$

(b) **Shear forces:**

$SF_{\text{at } B} = 0$

$SF_{\text{at } C} = 0$

$SF_{\text{at } X} = + w \cdot x$

$SF_{\text{up to } A} = w \cdot a$

$SF_{\text{at } A} = w \cdot a - V = 0$

(c) **SFD:**

at B and C, mark SF = 0

$SF_{\text{at } x} = wx$

Since x is varying from o to a, SF varies from o to wa. Between C and A.

$SF_{\text{up to } A} = w \cdot a$

$SF_{\text{at } A} = 0$

(d) **Bending moments:**

$M_{\text{at } B} = 0$

$M_{\text{at } C} = 0$

$$M_{\text{at } X} = -(w \cdot x)\frac{x}{2} = -\frac{wx^2}{2} \text{ (Hogging)}$$

$$M_{\text{at } A} = -(w \cdot a)\frac{a}{2} = \frac{-wa^2}{2} \text{ (Hogging)}$$

(e) **BMD:**

at B and C, mark $M = 0$

$$M_{\text{at } X} = -\frac{wx^2}{2}$$

$$M_{\text{at } A} = -\frac{wa^2}{2}$$

The variation between C to A is parabolic as

$$M_X = \frac{-wx^2}{2}.$$

4.8.6 Cantilever with Partial UDL at the Middle

Figure 4.18 shows a cantilever beam of length l, carrying an UDL over a span of a at a distance of b from fixed end.

(a) Vertical force at $A = V = w \cdot a$

(b) **Shear forces:**

$SF_{at\ B} = 0$

$SF_{at\ D} = 0$

$SF_{at\ X} = +(w \cdot x)$

$SF_{upto\ C} = +(w \cdot a)$

$SF_{upto\ A} = (w \cdot a)$

$SF_{at\ A} = (wa) - V = 0$

(c) **SFD:**

Draw horizontal line AB, mark C, X and D

At B and D, SF = 0

At C, SF = $w \cdot a$

At A. SF = 0 and $w \cdot a$

Between A and C, SF is constant, hence horizontal.

Between C and D, SF = $w \cdot a$

x varies for 0 to l which is linear.

Draw an inclined lined between C and D.

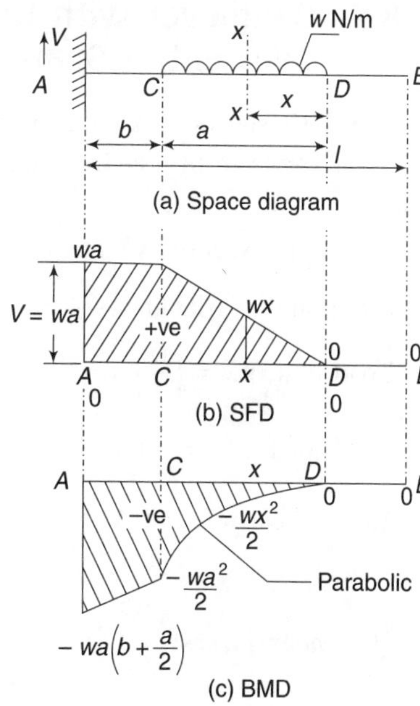

Fig. 4.18

(d) **Bending moments:**

$M_B = 0$

$M_D = 0$

$M_C = -(w \cdot a)\,\frac{a}{2} = \frac{-wa^2}{2}$ (Hogging)

$M_A = -(w \cdot a)\left(b + \frac{a}{2}\right)$

$M_x = -(w \cdot x)\,\frac{x}{2} = -\frac{wx^2}{2}$

(e) **BMD:**

Draw horizontal line AB. Mark C, D and X.

Mark $M = 0$ at B and D

Mark $-\frac{wa^2}{2}$ at C and $-wa\left(b + \frac{a}{2}\right)$ at A

$M_x = \frac{-wx^2}{2}$ is parabolic, draw a parabolic curve between C and D.

4.8.7 Cantilever with Uniformly Varying Load (UVL) Over Entire Span

Figure 4.19 shows a cantilever of length l carrying a UVL of varies from o at free end to w N/m at the fixed end.

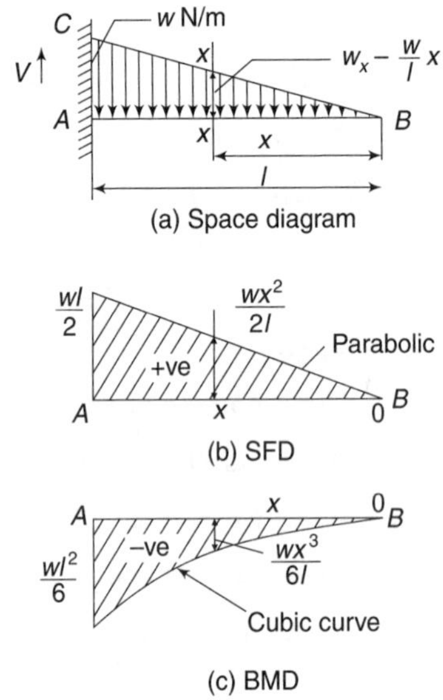

(a) Space diagram (b) SFD (c) BMD

Fig. 4.19

Consider a section XX at distance x from free end.

From similar triangles, $\frac{w_x}{x} = wl$

Load at x, $wx = \left(\frac{w}{l}\right) x.$

Total load at x, = Area of $\Delta = \frac{1}{2} x \left(\frac{w}{l}\right) x = \frac{wx^2}{2l}$

(a) Vertical force at A, V = Area of full Δ,

$$V = \frac{1}{2} w.l = \frac{wl}{2}$$

(b) **Shear forces:**

$$SF_{\text{at } B} = 0$$

$$SF_{\text{upto } A} = \text{UVL} = \frac{wl}{2}$$

$$SF_{\text{at } A} = \frac{wl}{2} - V = 0$$

$$SF_{\text{at } X} = \frac{wx^2}{2l}$$

(c) **SFD:** Draw horizontal line AB.

Mark C and B and $\frac{wl}{2}$ and A.

Since the variation of SF between AB is $\frac{wx^2}{2l}$ is parabolic, draw a parabolic curved between A and B

(d) Bending Moments:

$$M_{\text{at } B} = 0$$

$$M_{\text{at} A} = \text{UVL} \times \text{Distance to CG of } \Delta\, ABC$$

$$= -\left(\frac{w \cdot l}{2}\right) \times \frac{l}{3} = \frac{-wl^2}{6} \text{ (Hogging)}$$

Moment at any section XX, $M_x = -\left(\frac{wx^2}{2l}\right)\frac{x}{3} = \frac{-wx^3}{6l}$

(e) **BMD:** Draw a horizontal line AB make $M_B = 0$

Mark $M_A = \frac{-wl^2}{6}$

Since the variation of BM between A and B is $\frac{-wx^3}{6l}$, it is cubic. Draw a cubic curve between A and B.

4.8.8 Cantilever with Partial UVL Towards Free End

Figure 4.20 shows a cantilever of length l carrying an uniformly varying load varying from zero at free end to w N/m over a span of a from free end. Consider a section XX at a distance x from free end.

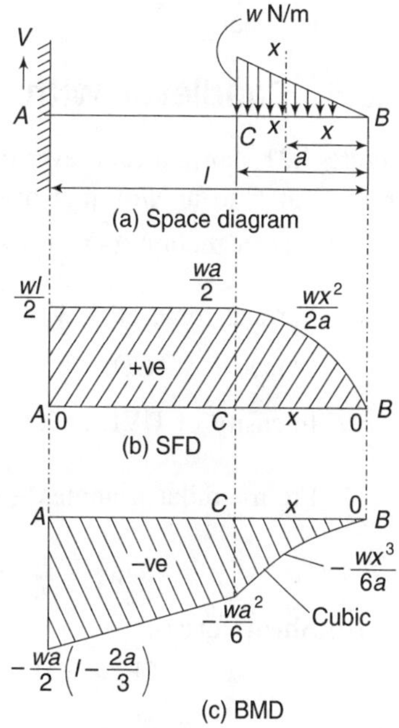

Fig. 4.20

(a) Vertical force at A, V = UVL

$$V = \frac{1}{2}\cdot w\cdot a = \frac{wa}{2}$$

Force at $x = w_x = \frac{wx}{l}$

(b) **Shear forces:**

$$SF_{\text{at } B} = 0$$

$$SF_{\text{at } C} = \text{UVL} = \frac{1}{2}\, w\cdot a = \frac{wa}{2}$$

$$SF_{\text{upto } A} = \frac{wa}{2}$$

$$SF_{\text{at} A} = \frac{wa}{2} - V = 0$$

$$SF_{\text{at } x} = \frac{w_x . x}{2} = \left(\frac{w}{l}.x\right)\frac{x}{2} = \frac{wx^2}{2l}$$

(c) **SFD:** Draw a horizontal line AB mark

$$SF_{\text{at } B} = 0$$

$$SF_{\text{at } C} = +\frac{wa}{2} \text{ and } SF_x = \frac{wx^2}{2l}$$

$$SF_{\text{upto} A} = \frac{wa}{2} \text{ and } SF_{\text{at} A} = 0$$

The variation of SF between B and C is $\frac{wx^2}{2l}$, is parabolic SF between A and C is constant.

(d) **Bending moments:**

$$M_{\text{at } B} = 0$$

$$M_{\text{at } C} = -\left(\frac{w\cdot a}{2}\right)\left(\frac{a}{3}\right) = \frac{-wa^2}{6} \text{ (Hogging)}$$

$$M_{\text{at } D} = -\left(\frac{w\cdot a}{2}\right)\left(l - \frac{2a}{3}\right)$$

$$M_{\text{at } x} = -\left(\frac{w\cdot x}{2a}\right).\frac{x}{3} = \frac{-wx^3}{6a}$$

(e) **BMD:**

Draw a horizontal line AB

$$M_{\text{at} B} = 0,$$

$$M_{\text{at}\,C} = \frac{wa^2}{6}$$

$$M_{\text{at}\,D} = \frac{-wa}{2}\left(1 - \frac{2a}{3}\right)$$

Variation of BM between B and C is cubic and variation of BM between A and C is linear.

4.8.9 Cantilever with Partial UVL Towards Fixed End

Figure 4.21 shows a cantilever of length l, carrying an uniformly varying load which varies from O at C to w N/m at A over a length of a towards fixed end.

Consider a section X-X at a distance x from C.

(a) Vertical force at A, $V = \frac{w \cdot a}{2}$

(Area of Δ ACD)

Intensity of UVL at XX.

From similar triangles $\frac{w_x}{x} = \frac{w}{a}$

$$\Rightarrow \quad w_x = \frac{w}{a} \cdot x$$

(b) **Shear forces:**

$$\text{SF}_{\text{at}\,B} = 0$$

$$\text{SF}_{\text{at}\,C} = 0$$

$$\text{SF}_{\text{up to}\,A} = +\frac{wa}{2}$$

$$\text{SF}_{\text{at}\,A} = \frac{w \cdot a}{2} - V = 0$$

$$\text{SF}_{\text{at}\,x} = \frac{w_x \cdot x}{2} = +\frac{w \cdot x^2}{2a}$$

(This is parabolic in nature.)

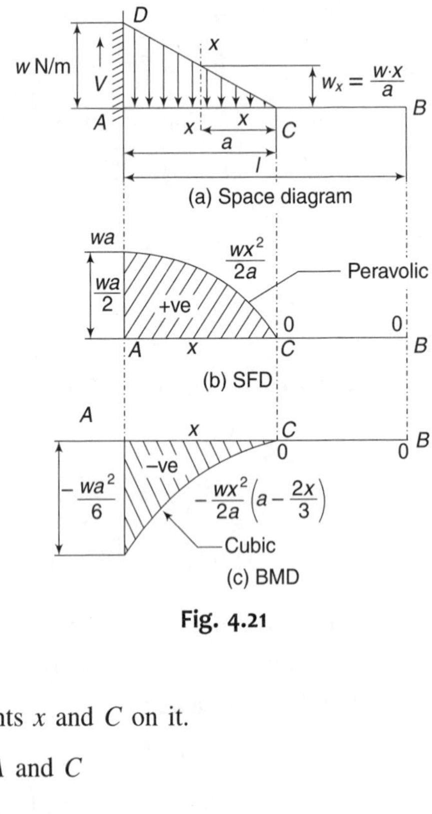

Fig. 4.21

(c) **SFD:** Draw a horizontal line AB. Mark points x and C on it.

At A, mark $\frac{wa}{2}$. Draw a parabola between A and C

(d) **Bending moments:**

$$M_{\text{at}\,B} = 0$$

$$M_{\text{at}\,C} = 0$$

$$M_{\text{at}\,A} = -\left(\frac{wa}{2}\right)\left(\frac{a}{3}\right) = \frac{-wa^2}{6} \qquad \text{(Hogging)}$$

$$M_{\text{at}\,x} = \frac{-wx^2}{2a}\left(a - \frac{2x}{3}\right)$$

The variation of BM between A and C is cubic in nature.

(e) **BMD:** Draw a horizontal line AB and Mark C and X on it.

At B and C, BM = 0

At A, BM = $\frac{-wa^2}{6}$. At x, make $M_X = \frac{-wx^2}{2a}\left(a - \frac{2x}{3}\right)$

Draw a cubic curve between A and C.

4.8.10 Cantilever with UVL at the Middle

Figure 4.22 shows a cantilever beam of length carrying an uniformly varying load that varies from o to w N/m over a span of a at a distance b from fixed end.

Consider section XX at a distance x from D.

Intensity at XX, $w_x = \frac{w}{a} \cdot x$

(a) Vertical force A, $V = \frac{1}{2}\, w.a$

(b) Shear forces:

$$SF_{\text{at } B} = 0$$
$$SF_{\text{at } D} = 0$$
$$SF_{\text{at } C} = +\frac{wa}{2}$$
$$SF_{\text{up to } A} = +\frac{wa}{2}$$
$$SF_{\text{at } A} = \frac{wa}{2} - V = 0$$
$$SF_{\text{at } x} = \frac{1}{2}\left(\frac{w}{a}.x\right) x = +\frac{wx^2}{2a}$$

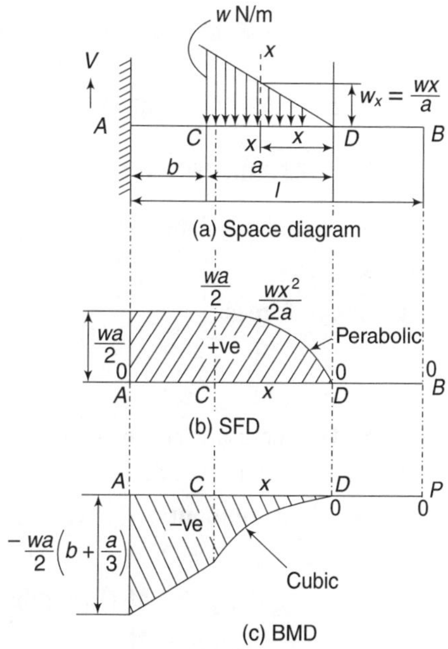

Fig. 4.22

The distribution SF between C to D is $\frac{wa}{2}$ to zero, parabolic.

(c) SFD: Draw a horizontal line AB. Make points C, D and x. Make SF at B and $D = 0$ at C equal to $\frac{wa}{2}$

$SF_{\text{upto } A}$ is $\frac{wa}{2}$ and at $A = 0$

SF between A and C is constant at $\frac{wa}{2}$ and between C and D is parabolic.

(d) Bending moment:

$$M_{\text{at } B} = 0$$
$$M_{\text{at } D} = 0$$
$$M_{\text{at } C} = \left[\frac{-1}{2}(w.a)\right]\frac{a}{3} = \frac{-wa^2}{6} \text{ (Hogging)}$$
$$M_{\text{at } A} = \left(\frac{-wa}{2}\right)\left(b + \frac{a}{3}\right)$$

$$M_{\text{at } x} = \left(\frac{-wx^2}{2a}\right)\left(a + b \frac{-2x}{3}\right)$$

This is cubic variation between C and D.

(e) **BMD:** Draw a horizontal line AB. Make C, D and X

Bending moment at B and $D = 0$

Bending moment at $C = \frac{-wa^2}{2}$

Bending moment at $A = \frac{-wa}{2}\left(b + \frac{a}{3}\right)$

The bending moment at $x = \frac{-wx^2}{2a}\left(a + b - \frac{x}{3}\right)$ is a cubic variation.

4.8.11 Cantilever with a Moment at the Free End

Figure 4.23 shows a cantilever of length l subjected to a moment M at its free end.

(a) $V = 0$

(b) **Shear forces:**

$$SF_{\text{at } B} = 0$$
$$SF_{\text{at } A} = 0$$

(c) **SFD:** There are no shear forces since it is subjected to pure bending

(d) **Bending moments:**

$$M_{\text{up to } B} = 0$$
$$M_{\text{at } B} = -M \text{ (Hogging)}$$
$$M_{\text{at } A} = -M$$

(e) **BMD:** The BMD is rectangle as shown in Fig. 4.23.

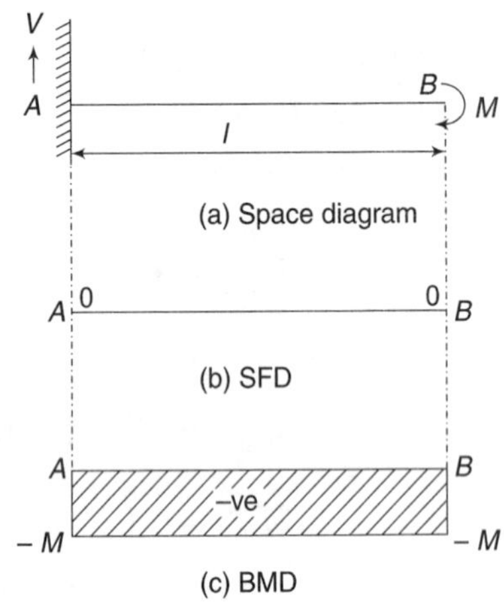

Fig. 4.23

SOLVED PROBLEMS ON CANTILEVERS

Problem 1 Draw the shear force and bending moment diagrams for a cantilever beam shown in Fig. 4.24. Indicate the values on it.

Solution:

(a) Vertical force at A

$$V = 30 + 20 + 10 = 60 \text{ kN}$$

(b) **Shear forces:**

$$SF_{\text{up to } D} = 0$$
$$SF_{\text{at } D} = +10 \text{ kN}$$
$$SF_{\text{up to }} C = +10 \text{ kN}$$

$SF_{\text{up to } C} = +10 + 20 = +30 \text{ kN}$

$SF_{\text{up to } B} = +30 \text{ kN}$

$SF_{\text{at } B} = 30 + 30 = +60 \text{ kN}$

$SF_{\text{up to } A} = 60 \text{ kN}$

$SF_{\text{at } A} = 60 - V = 0 \text{ kN}$

(c) The SFD as shown in Fig. 4.24 (b)

(d) **Bending moments:**

$M_{\text{at } D} = 0$

$M_{\text{at } C} = -10 \times 0.5 = -5 \text{ kNm}$ (Hogging)

$M_{\text{at } B} = -10 \times 0.5 - 20 \times 1 = -35 \text{ kN}$

$M_{\text{at } A} = -10 \times 3.5 - 20 \times 3 - 30 \times 2$

$= -155 \text{ kNm.}$

(e) Draw BMD as shown in Fig. 4.24 (c)

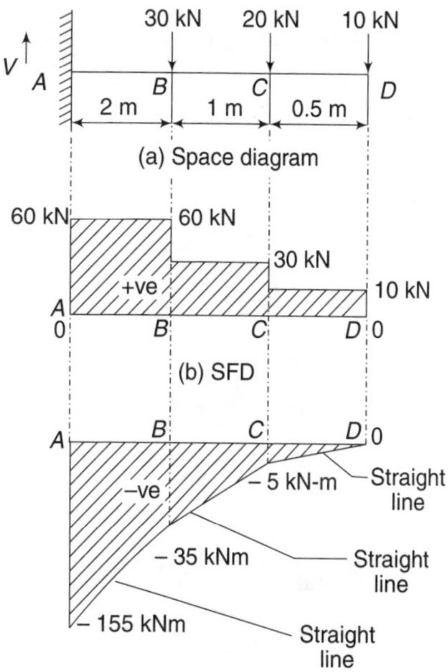

Fig. 4.24

Problem 2 Draw the shear force and bending moment diagram for a cantilever beam shown in Fig. 4.25. Indicate the values on it.

Solution:

(a) Vertical force at A

$$V = 20 + (10 \times 3) = 50 \text{ kN}$$

(b) **Shear forces:**

$SF_{\text{upto } B} = 0$

$SF_{\text{at } B} = +20 \text{ kN}$

$SF_{\text{at } A} = 20 + (10 \times 3) = 50 \text{ kN}$

$SF_{\text{up to } A} = 20 + (10 \times 3) = 50 \text{ kN}$

$SF_{\text{at } A} = 50 - V = 0$

(c) Draw SFD as shown at Fig. 4.25. It is triangle between A and B

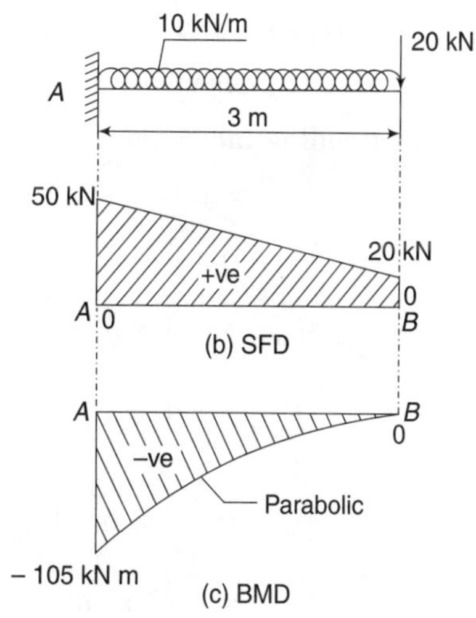

Fig. 4.25

(d) **Bending moments:**

$M_{\text{at } B} = 0$

$M_{\text{at } A} = -20 \times 3 - (10 \times 3)\left(\frac{3}{2}\right)$

$= -105 \text{ kNm}$

(e) Draw BMD as shown in Fig. 4.25. It is parabolic between A and B

Problem 3 Figure 4.26 shows a cantilever subjected to a system of loads. Draw SF and BM diagrams. Indicate the values on it.

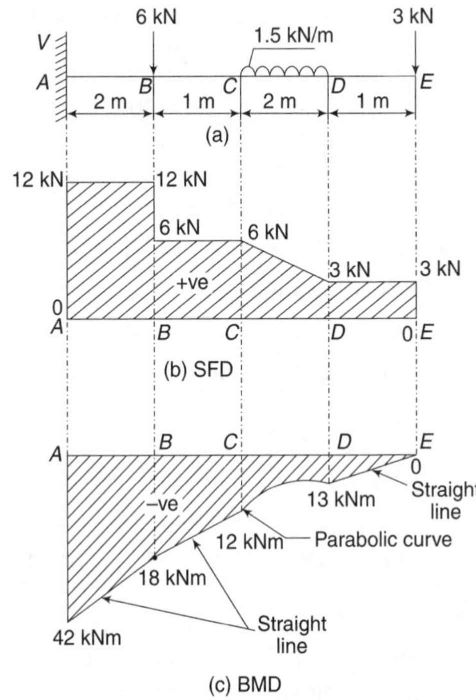

Fig. 4.26

Solution:

(a) $V = 6 + (1.5 \times 2) + 3 = 12$ kN

(b) **Shear forces:**

$$SF_{\text{up to } E} = 0$$

$$SF_{\text{at } E} = +3 \text{ kN}$$

$$SF_{\text{at } D} = +3 \text{ kN}$$

$$SF_{\text{at } C} = 3 + (1.5 \times 2) = 6 \text{ kN}$$

$$SF_{\text{up to } B} = 6 \text{ kN}$$

$$SF_{\text{at } B} = 6 + 6 = 12 \text{ kN}$$

$$SF_{\text{up to } A} = 12 \text{ kN}$$

$$SF_{\text{at } A} = 12 - V = 0$$

(c) Draw SFD as shown in Fig. 4.26 Between C and D for UDL

SFD is triangular

(d) **Bending moments:**

$$M_{\text{at } E} = 0$$

$$M_{\text{at } D} = -3 \times 1 = -3 \text{ kNm (Hogging)}$$

$$M_{\text{at } C} = -3 \times 3 - (1.5 \times 2)\frac{2}{2} = -12 \text{ kNm}$$

$$M_{\text{at } B} = -3 \times 4 - (1.5 \times 2)\left(1 + \frac{2}{2}\right) = -18 \text{ kNm}$$

$$M_{\text{at } A} = -3 \times 6 - (1.5 \times 2)\left(2 + 1 + \frac{2}{2}\right) - 6 \times 2$$

$$= -42 \text{ kNm}$$

(e) Draw BMD as shown in Fig. 4.26 For UDL, it is a parabolic curve.

Problem 4 Draw the shear force and bending moment diagrams for a cantilever loaded as shown in Fig. 4.27. Indicate the values on it.

Solution:

(a) $V = (2 \times 3) + 9 + 3 + 6 = 24$ kN

(b) **Shear forces:**

$$SF_{\text{up to } D} = 0$$
$$SF_{\text{at } D} = +6 \text{ kN } (+\text{ve})$$
$$SF_{\text{up to } C} = +6 \text{ kN}$$
$$SF_{\text{up to } C} = +6 + 3 = +9 \text{ kN}$$
$$SF_{\text{up to } B} = +9 \text{ kN}$$
$$SF_{\text{up to } B} = 9 + 9 = 18 \text{ kN}$$
$$SF_{\text{up to } A} = 18 + (2 \times 3) = 24 \text{ kN}$$
$$SF_{\text{at } A} = 24 - V = 0$$

(c) Draw SFD as shown. Between *A* and *B*, SFD is triangular (UDL).

(d) **Bending moments:**

$$M_D = 0$$
$$M_C = -6 \times 2 = -12 \text{ kNm}$$
$$M_B = -6 \times 3 - 3 \times 1 = -21 \text{ kNm}$$
$$M_A = -6 \times 6 - 3 \times 4 - 9 \times 3 - (2 \times 3)\frac{3}{2} = -84 \text{ kNm}$$

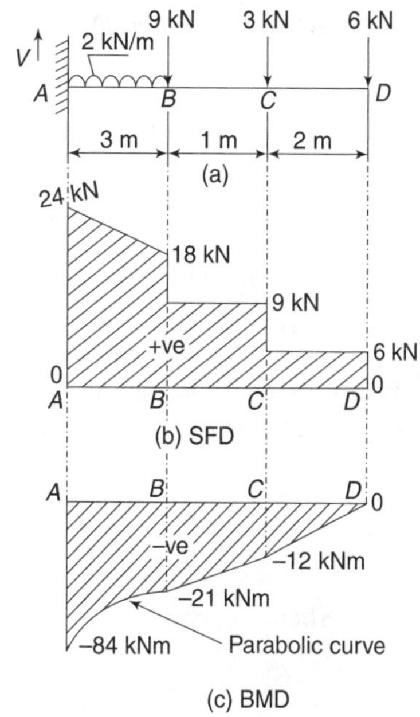

Fig. 4.27

(e) Draw BMD as shown in Fig. 4.27 (c). Between *A* and *B*, the BMD is parabolic for UDL.

Problem 5 Draw shear force and bending moment diagram for a cantilever loaded as shown in Fig. 4.28.

Solution:

(a) $V = 10 \times 5 + 9 + 6 = 65$ kN

(b) **Shear forces:**

$$SF_{\text{up to } C} = 0$$
$$SF_{\text{at } C} = +6 \text{ kN}$$
$$SF_{\text{up to } B} = +6 + (10 \times 2) = 26 \text{ kN}$$
$$SF_{\text{at } B} = 26 + 9 = 35 \text{ kN}$$
$$SF_{\text{up to } A} = 35 + 10 \times 3 = +65 \text{ kN}$$
$$SF_{\text{at } A} = 65 - V = 0$$

(c) Draw SFD as shown in Fig. 4.28 (b)

Between *A* and *B* and *B* and *C*, SFD is triangular

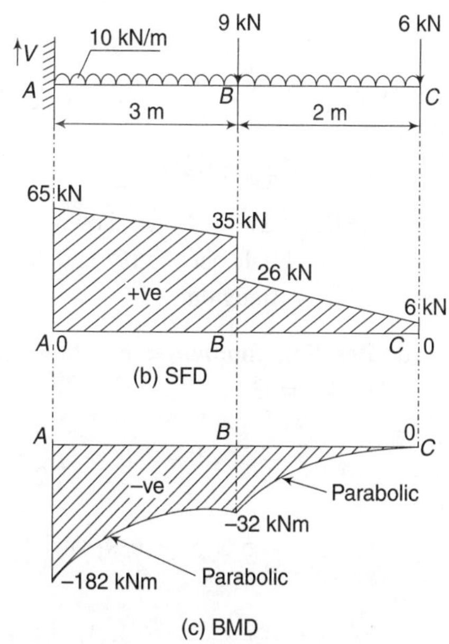

Fig. 4.28

(d) **Bending moments:**

$$M_{\text{at}\,C} = 0$$

$$M_{\text{at}\,B} = -\,6 \times 2 - (10 \times 2)\,\frac{2}{2} = -\,32 \text{ kNm}$$

$$M_{\text{at}\,A} = -\,6 \times 5 - 9 \times 3 - (10 \times 5)\,\frac{5}{2} = -\,182 \text{ kNm}$$

(e) Draw BMD as shown in Fig. 4.28 (c)

BMD between A and B, and B and C is parabolic for UDL.

Problem 6 Draw shear force on bending moment diagram for a cantilever shown in Fig. 4.29. Mark the corresponding values of SF and BM on the diagram.

Solution:

(a) $V = (12 \times 5\ ) + 9 + (6 \times 3) + 5$

$= 92$ kN

(b) **Shear forces:**

$$SF_{\text{upto}\,E} = 0$$

$$SF_{\text{at}\,E} = +\,5 \text{ kN}$$

$$SF_{\text{at}\,D} = 5 + 6 \times 3 = 23 \text{ kN}$$

$$SF_{\text{upto}\,C} = 23$$

$$SF_{\text{at}\,C} = 23 + 9 = 32 \text{ kN}$$

$$SF_{\text{at}\,B} = 32 \text{ kN}$$

$$SF_{\text{upto}\,A} = 32 + (12 \times 5) = 92 \text{ kN}$$

$$SF_{\text{at}\,A} = 92 - V = 0$$

(c) **SFD:** Draw SFD as shown in Fig. 4.29 (b) for UDL between A and B and D and E, it is triangular.

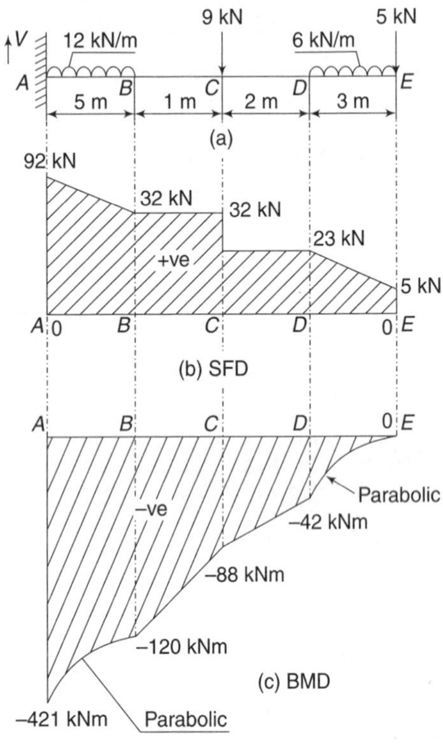

Fig. 4.29

(d) **Bending moments:** Bending moment at E, $M_E = 0$

$$M_{\text{at}\,D} = -\,5 \times 3 - (\,6 \times 3)\,\frac{3}{2} = -\,42 \text{ kNm}$$

$$M_{\text{at}\,C} = -\,5 \times 5 - (6 \times 3)\,\left(2 + \frac{3}{2}\right) = -\,88 \text{ kNm}$$

$$M_{\text{at}\,B} = -5 \times 6 - (6 \times 3)\,\left(1 + 2 + \frac{3}{2}\right) - 9 \times 1 = -\,120 \text{ kNm}$$

$$M_{\text{at } A} = -5 \times 11 - (6 \times 3)\left(5 + 1 + 2 + \frac{3}{2}\right) - 9 \times 5 - (12 \times 5)\frac{5}{2} = -421 \text{ kNm}$$

(e) Draw BMD as shown in Fig. 4.29 (c).

BM for UDL between *A* and *B* and *D* and *E* are parabolic.

Problem 7 A cantilever beam is loaded as shown in Fig. 4.30. Draw SF and BM diagrams. Indicate the values.

Solution:

(a) $V = 40 \times 2 + 30 \times 3 + 20 = 190$ kN

(b) **Shear forces:**

$SF_{\text{upto } E} = 0$

$SF_{\text{at } E} = 20$ kN

$SF_{\text{at } D} = 20$ kN

$SF_{\text{at } C} = 20 + (30 \times 3) = 110$ kN

$SF_{\text{at } B} = 110$ kN

$SF_{\text{upto } A} = 110 + 40 \times 2 = 190$ kN

$SF_{\text{at } A} = 190 - V = 0$

(c) **SFD:** Draw SFD as shown in Fig. 4.30 (b). For UDL between *AB* and *CD* it is triangular

(d) **Bending moments:**

$M_{\text{at } E} = 0$

$M_{\text{at } D} = -20 \times 2 = -40$ kNm

$$M_{\text{at } C} = -20 \times 5 - (30 \times 3)\frac{3}{2} = -235 \text{ kNm}$$

$$M_{\text{at } B} = -20 \times 8 - (30 \times 3)\left(3 + \frac{3}{2}\right) = -565 \text{ kNm}$$

$$M_{\text{at } A} = -20 \times 10 - (30 \times 3)\left(2 + 3 + \frac{3}{2}\right) - (40 \times 2)\frac{2}{2} = -856 \text{ kNm}$$

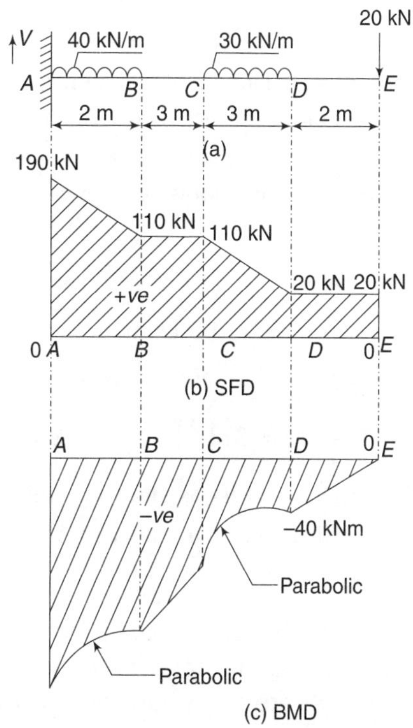

Fig. 4.30

(e) Draw BMD as shown in Fig. 4.30 (c). BMD for UDL at *AB* and *CD* are parabolic

Problem 8 A cantilever beam is loaded as shown in Fig. 4.31. Draw SF and BM diagrams indicating the values.

Solution:

(a) Vertical force at *A*, $V = 30 + 20 \times 2 + 12 \times 3 = 106$ kN

(b) **Shear forces:**

$SF_{\text{upto } C} = 0$

$SF_{\text{at } C} = 30$ kN

$SF_{\text{at } B} = 30 + 20 \times 2 = 70$ kN

$SF_{\text{upto } A} = 30 + 20 \times 2 + 12 \times 3 = 106$ kN

$SF_{\text{at } A} = 106 - V = 0$

(c) **SFD:** Draw SFD as shown in Fig. 4.31 (b) For UDL, between *AB* and *BC* is triangular.

(d) **Bending moments:**

$M_{\text{at } D} = 0$

$M_{\text{at } B} = -30 \times 2 - (20 \times 2)\dfrac{2}{2}$

$= -100$ kNm

$M_{\text{at } A} = -30 \times 5 - (20 \times 2)\left(3 + \dfrac{2}{2}\right)$

$-(12 \times 3)\left(\dfrac{3}{2}\right) = -364$ kNm

(e) Draw BMD as shown in Fig. 4.31 (c)

BMD for UDL between *AB* and *BC* is parabolic as shown.

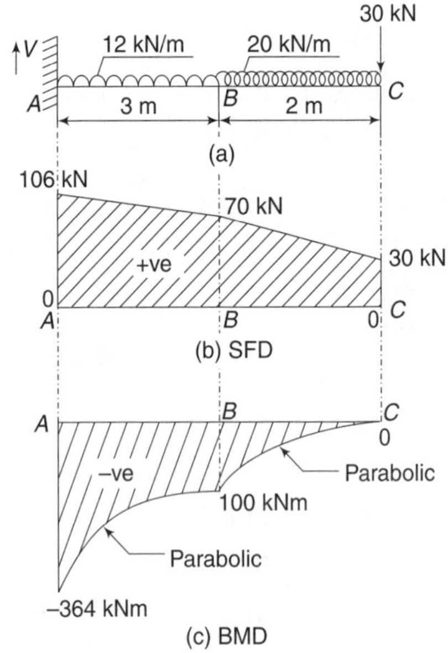

Fig. 4.31

Problem 9 Draw shear force and bending moment diagrams for a cantilever shown in Fig. 4.32. Indicate the values on the them.

Solution:

(a) Vertical force at *A*, $V = 20 + (10 \times 2) + 18$
$= 58$ kN

(b) **Shear forces:**

$SF_{\text{upto } D} = 0$

$SF_{\text{at } D} = 20$ kN

$SF_{\text{at } C} = 20 + (10 \times 2) = 40$ kN

$SF_{\text{upto } B} = 40$ kN

$SF_{\text{to } B} = 40 + 18 = 58$ kN

$SF_{\text{upto } A} = 58$ kN

$SF_{\text{at } A} = 58 - V = 0$

(c) **SFD:** Draw SFD as shown in Fig. 4.32 (b) SF for UDL at *CD* is triangular

(d) **Bending moments:**

$M_{\text{at } D} = 0$

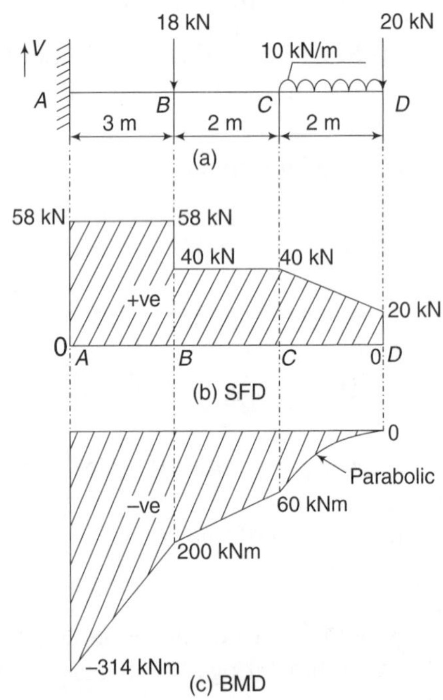

Fig. 4.32 (a)

$$M_{\text{at } C} = -20 \times 2 - (10 \times 2)\left(\frac{2}{2}\right) = -60 \text{ kNm}$$

$$\text{at } B,\ M_B = -20 \times 4 - (10 \times 2)\left(2 + \frac{2}{2}\right) = -200 \text{ kNm}$$

$$\text{at } A,\ M_A = -20 \times 7 - (10 \times 2)\left(3 + 2 + \frac{2}{2}\right) - 18 \times 3 = -314 \text{ kNm.}$$

(e) Draw BMD as shown in Fig. 4.32 (c). BM for UDL between *CD* is parabolic.

Problem 10 Draw the shear force and bending moment diagram for a cantilever beam loaded as shown in Fig. 4.32(a). Indicate the values on them.

Solution:

(a) Vertical force at A, $V = 6 + 20 \times 7 + 12 + 9 = 167$ kN

(b) **Shear forces:**

$SF_{\text{up to } D} = 0,$

$SF_{\text{at } D} = 6$ kN

$SF_{\text{up to } C} = 6 + (20 \times 5) = 106$ kN

$SF_{\text{at } C} = 106 + 12 = 118$ kN

$SF_{\text{up to } B} = 118 + (20 \times 2) = 158$ kN

$SF_{\text{at } B} = 158 + 9 = 167$ kN

$SF_{\text{up to } A} = 167$ kN

$SF_{\text{at } A} = 167 - V = 0$

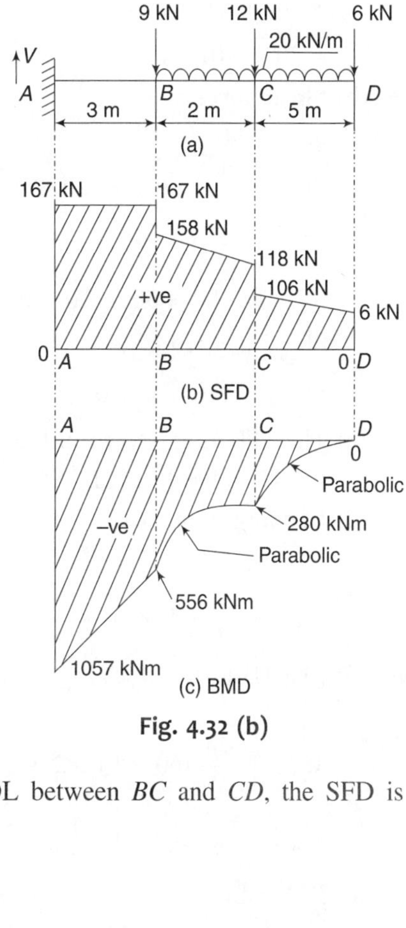

Fig. 4.32 (b)

(c) Draw SFD as shown in Fig. 4.32 (b) for UDL between *BC* and *CD*, the SFD is triangular.

(d) **Bending moments:**

$$M_{\text{at } D} = 0$$

$$M_{\text{at } C} = -6 \times 5 - (20 \times 5)\frac{5}{2} = -280 \text{ kNm}$$

$$M_{\text{at } B} = -6 \times 7 - (20 \times 7)\frac{7}{2} - 12 \times 2 = -556 \text{ kNm}$$

$$M_{\text{at } A} = -6 \times 10 - (20 \times 7)\left(3 + \frac{7}{2}\right) - 12 \times 5 - 9 \times 3 = -1057 \text{ kNm}$$

(e) Draw BM as shown in Fig. 4.32 (c). The variations of BMD for UDL between *BC* and *CD* are parabolic.

Problem 11 Draw shear force and bending moment diagrams for a cantilever beam loaded as shown in Fig. 4.33, indicating the values on them.

Solution:

(a) Vertical force at A, $V = \left(\frac{1}{2} \times 60 \times 6\right)$ $+ 30 = 210$ kN

(b) **Shear forces:**

$SF_{\text{at } D} = 0$

$\text{SF}_{\text{at } C} = \frac{1}{2} \times 60 \times 6 = 180 \text{ kN}$

$\text{SF}_{\text{upto } B} = 180 \text{ kN}$

$\text{SF}_{\text{at } B} = 180 + 30 = 210 \text{ kN}$

$\text{SF}_{\text{upto } A} = 210 \text{ kN}$

$\text{SF}_{\text{at } A} = 210 - V = 0$

(c) The SFD is as shown in Fig. 4.33 (b)

For UVL between CD, SF is parabolic.

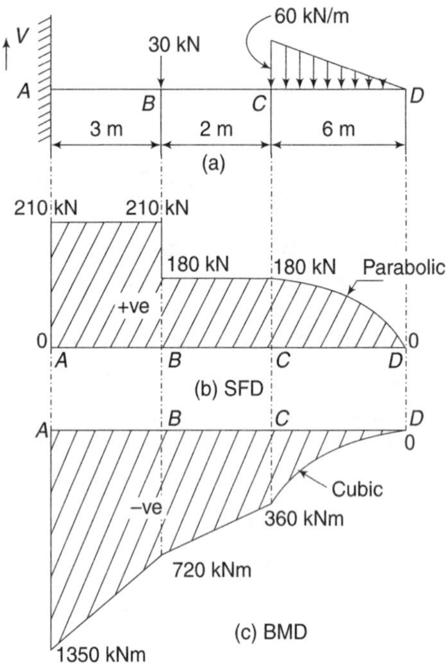

Fig. 4.33

(d) **Bending moments:**

$M_{\text{at } D} = 0$

$M_{\text{at } C} = \frac{-1}{2} (60 \times 6) \times \frac{6}{3} = -360 \text{ kNm}$

[For UVL, CG lies at $\frac{1}{3}$ of base $= \frac{1}{3} \times 6$ and intensity of UVL = Area of triangle $= \frac{1}{2} \times (60 \times 6)$

$M_{\text{at } B} = \frac{-1}{2} (60 \times 6) \left(2 + \frac{6}{3}\right) = -720 \text{ kNm}$

$M_{\text{at } A} = \frac{-1}{2} (60 \times 6) \left(3 + 2 + \frac{6}{3}\right) - 30 \times 3 = -1350 \text{ kNm}$

(e) **BMD:** Bending moments diagram is shown in Fig. 4.33 (c) for UVL, BMD is cubic in nature.

Problem 12 Draw shear force and bending moment diagrams for a cantilever beam loaded as shown in Fig. 4.34.

Solution:

(a) Vertical force at A, $V = 30 + \frac{1}{2} \times 60 \times 2 = 90 \text{ kN}$

(b) **Shear forces**

$\text{SF}_{\text{up to } D} = 0$

$\text{SF}_{\text{at } D} = 30 \text{ kN}$

$$SF_{at\ C} = 30 \text{ kN}$$

$$SF_{at\ B} = 30 + \left(\frac{1}{2} \times 60 \times 2\right) = 90 \text{ kN}$$

$$SF_{upto\ A} = 90 \text{ kN}$$

$$SF_{at\ A} = 90 - V = 0$$

(c) Draw SFD as shown in Fig. 4.34 (b). For UVL, SFD is parabolic for UVL and triangular for UDL.

(d) **Bending moments:**

Bending moment at D, $M_D = 0$

Bending moment at C, $M_C = -30 \times 6$

$$= -180 \text{ kN-m}$$

$$M_{at\ B} = -30 \times 8 - \left(\frac{1}{2} \times 60 \times 2\right)\left(\frac{2}{3}\right)$$

$$= -280 \text{ kNm}$$

$$M_{at\ A} = -30 \times 11 - \left(\frac{1}{2} \times 60 \times 2\right)\left(3 + \frac{2}{3}\right) = -550 \text{ kNm}$$

(e) **BDM:** The bending moment diagram is as shown in Fig. 4.34 (c)

For UVL, the BMD is a cubic curve.

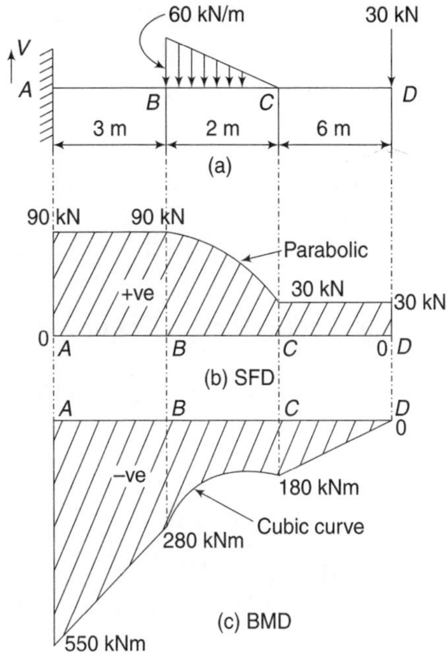

Fig. 4.34

Problem 13 Draw shear force and bending moment diagram for a cantilever loaded as shown in Fig. 4.35. Indicate values on them.

Solution:

(a) Vertical force at A,

$$V = 20 + \frac{1}{2}(30 \times 3) + (10 \times 2) = 85 \text{ kN}$$

(b) **Shear forces:**

$$SF_{upto\ E} = 0$$

$$SF_{at\ E} = 20 \text{ kN}$$

$$SF_{at\ D} = 20 \text{ kN}$$

$$SF_{at\ C} = 20 + \left(\frac{1}{2} \times 30 \times 3\right) = 65 \text{ kN}$$

$$SF_{at\ B} = 65 \text{ kN}$$

$$SF_{upto\ A} = 65 + 10 \times 2 = 85 \text{ kN}$$

$$SF_{at\ A} = 85 - V = 0$$

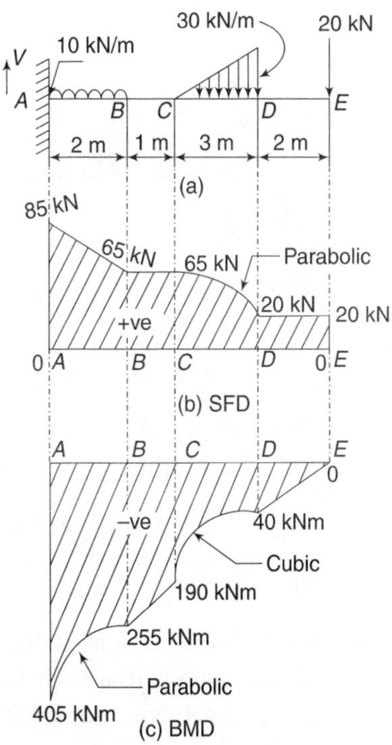

Fig. 4.35

(c) The SFD is as shown in Fig. 4.35 (b)

The variation of SF is parabolic for UVL and triangular for UDL

(d) **Bending moments:**

$M_{\text{at}E} = 0$

$M_{\text{at}D} = -20 \times 2 = -40$ kNm

$$M_{\text{at}C} = -20 \times 5 - \left(\frac{1}{2} \times 30 \times 3\right)\left(\frac{2}{3} \times 3\right) = -190 \text{ kNm}$$

(CG for triangle is 2/3 of 3 from C and $\frac{1}{3}$ of 3 from D)

$$M_{\text{at}B} = -20 \times 6 - \left(\frac{1}{2} \times 30 \times 3\right)\left(1 + \frac{2}{3} \times 3\right) = -255 \text{ kNm}$$

$$M_{\text{at}A} = -20 \times 8 - \left(\frac{1}{2} \times 30 \times 3\right)\left(2 + 1 + \frac{2}{3} \times 3\right) - (10 \times 2)\left(\frac{2}{2}\right)$$
$$= -405 \text{ kNm}$$

(e) **BMD:** The BMD is as shown in Fig. 4.35 (c). The BM variation is cubic for UVL between *CD* and parabolic for UDL between *AB*.

Problem 14 Draw shear force and bending moments diagrams for a cantilever shown in Fig. 4.35(a) Indicate the values on them.

Solution:

(a) Vertical force at A, $V = 30 + (15 \times 3) + 60 + \frac{1}{2} \times 5 \times 40 = 235$ kN

(b) **Shear forces:**

$SF_{\text{upto } D} = 0$

$SF_{\text{at } D} = 30$ kN

$SF_{\text{upto } C} = 30 + (15 \times 3) = 75$ kN

$SF_{\text{at } C} = 75 + 60 = 135$ kN

$SF_{\text{at } B} = 135$ kN

$$SF_{\text{upto } A} = 135 + \left(\frac{1}{2} \times 5 \times 40\right) = 235 \text{ kN}$$

$SF_{\text{at } A} = 235 - V = 0$

(c) **SFD:** The SFD is as shown in Fig. 4.35(b) SF variation is triangular for UDL between *CD* and parabolic for UVL between *AB*.

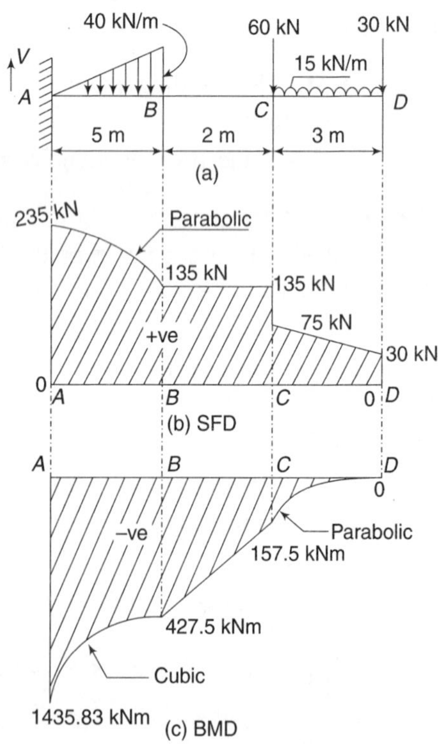

Fig. 4.35 (a)

(d) **Bending moments:**

$M_{\text{at}D} = 0$

$M_{\text{at}C} = -\ 30 \times 3 - (15 \times 3)\ \dfrac{3}{2} = -\ 157.5$ kNm

$M_{\text{at}B} = -\ 30 \times 5 - (15 \times 3)\ \left(2 + \dfrac{3}{2}\right) - 60 \times 2 = -\ 427.5$ kNm

$M_{\text{at}A} = -\ 30 \times 10 - (15 \times 3)\ \left(5 + 2 + \dfrac{3}{2}\right) - 60 \times 7$

$-\left(\dfrac{1}{2} \times 5 \times 40\right)\left(\dfrac{2}{3} \times 5\right) = -\ 1435.83$ kNm

(e) The BMD is as shown in Fig. 4.35 (c). The variation of BM is parabolic for UDL between *CD* and cubic for UVL between *AB*.

Problem 15 Draw bending moment and shear force diagram for a cantilever loaded as shown in Fig. 4.35(b). Indicate the values at salient points.

Solution: The trapezoidal shape load is split into two loads

- One UDL of 30 kN/m *CDEF* and
- One UVL of 90 – 30 = 60 kN/m *EFG* as shown in Fig. 4.35 (a).

UDL is assumed to be acting at the $cg = \dfrac{4}{2} = 2$ m for *C* or *D*

UVL is assumed to be acting at the *cg* of triangle, $= \dfrac{1}{3} \times 4$ from *C*

(a) Vertical force at *A*,

$V = 30 \times 4 + \dfrac{1}{2} \times 60 \times 4 + 30 = 270$ kN

(b) **Shear forces:**

$\text{SF}_{\text{at } D} = 0$

$\text{SF}_{\text{at } C}$ = UDL + UVL $= 30 \times 4 + \dfrac{1}{2} \times 60 \times 40 = 240$ kN

$\text{SF}_{\text{upto } B} = 240$ kN

$\text{SF}_{\text{at } B} = 240 + 30 = 270$ kN

$\text{SF}_{\text{upto } A} = 270$ kN

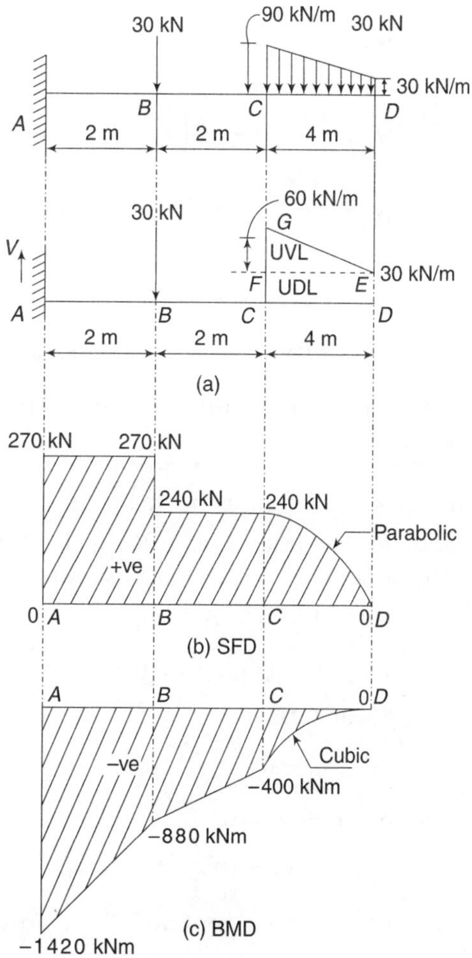

Fig. 4.35 (b)

$SF_{at\ A} = 270 - V = 0$

Consider a section *XX*, at a distance *x* from *D*.

Let w_x = vertical intensity of UVL at *x*

From similar triangles, $\frac{60}{4} = \frac{w_x}{x}$

$\therefore\ w_x = \frac{60}{4}.\ x = 15\ x$

$SF_{at x}$, $30.\ x + \left(\frac{1}{2} \times 15\ x \times x\right) = 30.\ x + 7.5x^2$

This is parabolic in nature between *C* and *D*.

(c) The shear force diagram is shown in Fig. 4.35(b)

(d) **Bending moments:**

$M_{at\ D} = 0$

$M_{at D}$ = BM due to UDL + BM due to UVL = $(-30 \times 4)\ \frac{4}{2} - \frac{1}{2}\ (60 \times 4) \times \left(\frac{4}{3}\right)$

$= -400$ kN-m

$M_{at B} = -(30 \times 4)\left(2 + \frac{4}{2}\right) - \left(\frac{1}{2} \times 60 \times 4\right)\left(2 + \frac{4}{3}\right) = -880$ kNm

$M_{at\ A} = (-30 \times 4)\left(2 + 2 + \frac{4}{2}\right) - \frac{1}{2} \times 60 \times 4\left(2 + 2 + \frac{4}{3}\right) - 30 \times 2 = -1420$ kNm

$M_{at x} = (-30 \times x)\ \frac{x}{2} - \left(\frac{1}{2} \times 15x \times x\right)\frac{x}{3} = -15\ x^2 - 2.5\ x^3$

The BMD variation is cubic between *CD*

(e) **BMD:** Draw BMD as shown in Fig. 4.35 (c)

Problem 16 Draw shear force and bending moment diagrams for a cantilever shown in Fig. 4.36. Indicate the values at all points.

Solution: The load between *BC* is divided into two, one UDL of 20 kN/M and the other UVL of 50 – 20 = 30 kN/m as shown in Fig. 4.36 (a)

(a) Vertical load at *A*,

$V = 10 + (6 \times 3) + (20 \times 6) + \frac{1}{2} \times 30 \times 6 = 238$ kN

(b) **Shear forces:**

$SF_{upto\ D} = 0$

$SF_{at\ D} = 10$ kN

$SF_{at\ C} = 10 + (6 \times 3) = 28$ kN

$SF_{at\ B} = 28 + (20 \times 6) + \left(\frac{1}{2} \times 30 \times 6\right) = 238$ kN

$SF_{upto\ A} = 238$ kN

$SF_{at\,A} = 238 - V = 0$

(c) SFD: The SFD is parabolic in Fig. 4.36 (b). SF is parabolic for UDL-UVL combination between *BC*.

(d) **Bending moments:**

$M_{at\,D} = 0$

$$M_{at\,C} = -10 \times 3 - (6 \times 3)\frac{3}{2}$$
$$= -57 \text{ kN-m}$$

$$M_{at\,B} = -10 \times 9 - (6 \times 3)\left(6 + \frac{3}{2}\right) - (20 \times 6)\frac{6}{2} - \left(\frac{1}{2} \times 30 \times 6\right)\left(\frac{2}{3} \times 6\right)$$
$$= -945 \text{ kNm}$$

$$M_{at\,A} = -10 \times 12 - (6 \times 3)\left(3 + 6 + \frac{3}{2}\right) - (20 \times 6)\left(3 + \frac{6}{2}\right) - \left(\frac{1}{2} \times 30 \times 6\right)\left(3 + \frac{2}{3} \times 6\right) = -1659 \text{ kNm}$$

(e) **BMD:** The BMD is as shown in Fig. 4.36 (c). The variation of BM is parabolic for UDL between *CD* and cubic for UDL-UVL combination at *BC*.

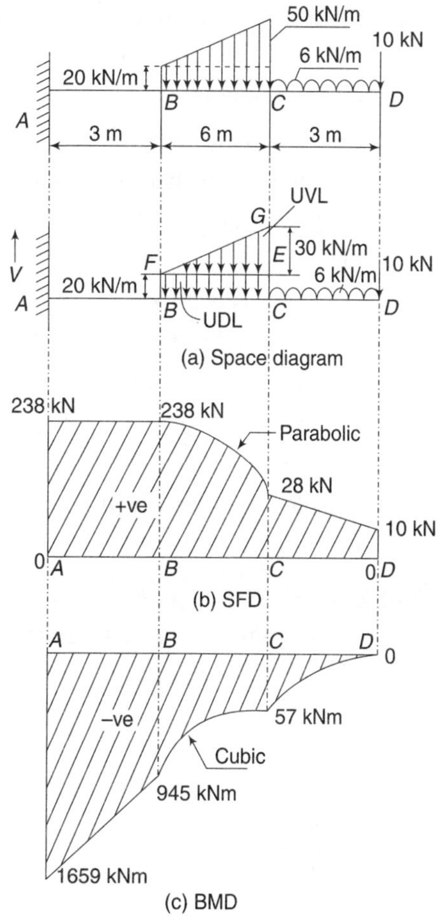

Fig. 4.36

Problem 17 Draw shear force and bending moment diagrams for a cantilever shown in Fig. 4.37. Indicate the values on them.

Solution:

(a) Vertical force $V = 30$ kN

(b) **Shear forces:**

$SF_{upto\,C} = 0$

$SF_{at\,C} = 30$ kN

$SF_{at\,B} = 30$ kN

$SF_{upto\,A} = 30$ kN

$SF_{at\,A} = 30 - V = 0$

(c) **SFD:** The shear force diagram is shown in Fig. 4.37 (b).

(d) **Bending moments:**

Bending moments at C, $M_C = 0$

For moment, if there is a moment,

Find moment upto B and moment at B separately

$$M_{\text{up to } B} = -30 \times 2$$
$$= -60 \text{ kNm}$$
$$M_{\text{at } B} = -60 - 40 = -100 \text{ kNm}$$
$$M_{\text{at } A} = -30 \times 5 - 40 = -190 \text{ kNm}$$

(e) Draw BMD as shown in Fig. 4.37 (c) for moment at B, there will be vertical line at B in BMD.

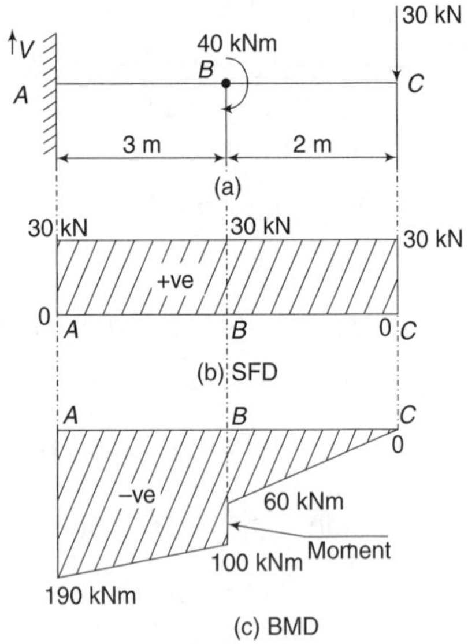

Fig. 4.37

Problem 18 Draw shear force and bending moment diagrams for a cantilever beam shown in Fig. 4.38.

Solution:

(a) Vertical Forces at A,

$$V = 20 + 12 \times 2 = 44 \text{ kN}$$

(b) **Shear forces:**

$$\text{SF}_{\text{at } D} = 0$$
$$\text{SF}_{\text{up to } C} = 0$$
$$\text{SF}_{\text{at } C} = 20 \text{ kN}$$
$$\text{SF}_{\text{at } B} = 20 \text{ kN}$$
$$\text{SF}_{\text{up to } A} = 20 + 12 \times 2$$
$$= 44 \text{ kN}$$
$$\text{SF}_{\text{at } A} = 44 - V = 0$$

(c) Figure 4.38 (b) shows the SFD

(d) **Bending noments:**

Moment up to D, $M_{\text{up to } D} = 0$

$$M_{\text{at } D} = -10 \text{ kNm (Hogging)}$$
$$M_{\text{at } C} = -10 \text{ kNm}$$
$$M_{\text{at } B} = -10 - 20 \times 1$$
$$= -30 \text{ kNm}$$
$$M_{\text{at } A} = -10 - (20 \times 3) - \left(12 \times 2 \times \frac{2}{2}\right)$$
$$= -94 \text{ kNm}$$

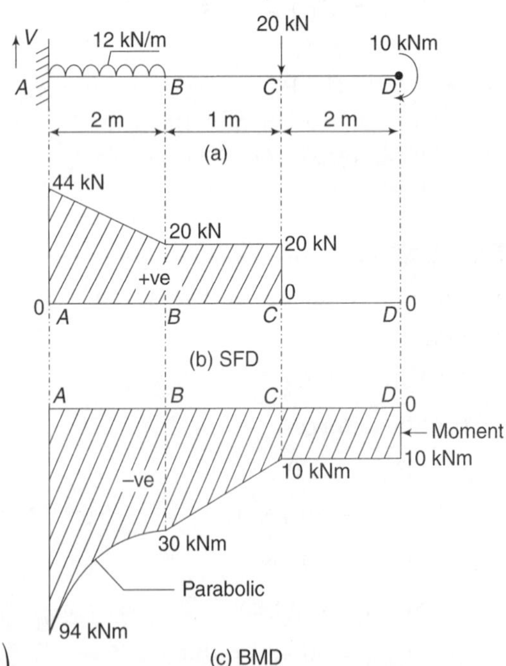

Fig. 4.38

(e) Figure 4.38 (c) show the BMD for moment at D, a vertical line of 10 kNm is drawn.

Problem 19 Draw shear force and bending moment diagram for a cantilever shown in Fig. 4.39. Indicate the values on them.

Solution:

(a) Vertical force at A, $V = 60 + 20 + (10 \times 3)$
$= 110$ kN

(b) **Shear forces:**

$SF_{\text{upto } D} = 0$

$SF_{\text{at } D} = 60$

$SF_{\text{up to } C} = 60$

$SF_{\text{at } C} = 60 + 20 = 80$ kN

$SF_{\text{at } B} = 80$ kN

$SF_{\text{up to } A} = 80 + 10 \times 3 = 110$ kN

$SF_{\text{at } A} = 110 - V = 0$

(c) Figure 4.39 (b) shown the SFD

(d) **Bending moments:**

$M_{\text{at } D} = 0$

$M_{\text{at } C} = -60 \times 1 = -60$ kNm

$M_{\text{upto } B} = -60 \times 3 - 20 \times 2$

$= -220$ kNm

$M_{\text{at } B} = -220 + 30 = -180$ kNm

($M = +30$ because of sagging)

at $A = -60 \times 6 - 20 \times 5 + 30 - (10 \times 3)\frac{3}{2}$

$M_{\text{at } A} = -475$ kNm

(e) Draw BMD as shown in Fig. 4.39 (c) for moment at B, draw a 30 kN upward line.

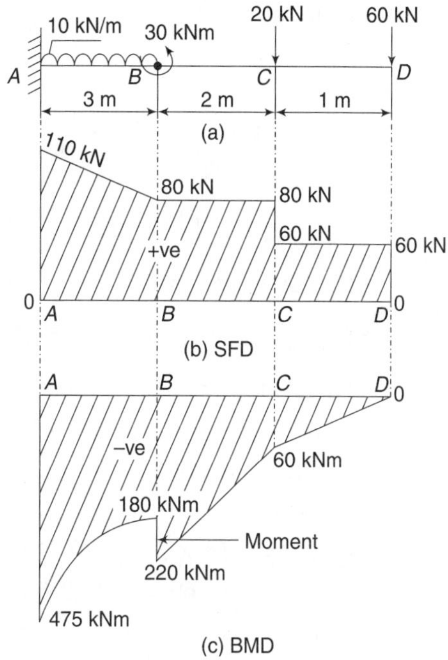

Fig. 4.39

Problem 20 Draw shear force and bending moment diagram for a cantilever shown in Fig. 4.40 Indicate the values on them.

Solution:

(a) Vertical force at A

$$V = 10 + \left(\frac{1}{2} \times 10 \times 3\right) + 5 \times 2$$

$= 35$ kN

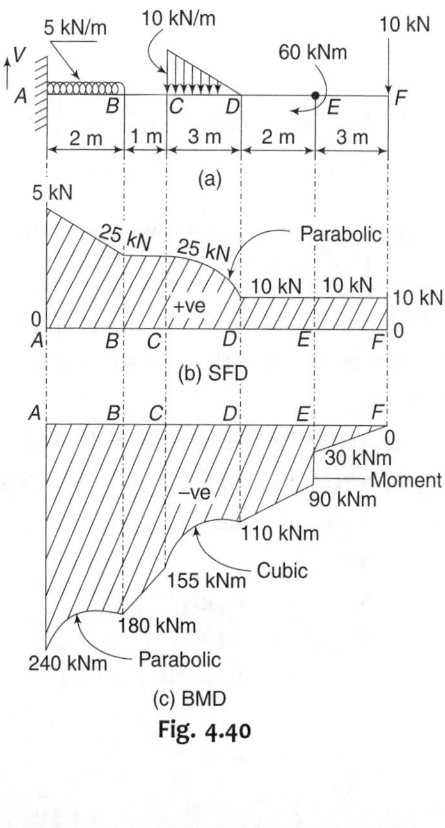

Fig. 4.40

(b) **Shear forces:**

$SF_{\text{upto } F = 0}$

$SF_{\text{at } F} = 10$ kN

$SF_{\text{at } E} = 10$ kN

$SF_{\text{at } D} = 10$ kN

$SF_{\text{at } C} = 10 + \left(\frac{1}{2} \times 10 \times 3\right) = 25$ kN

$SF_{\text{at } B} = 25$ kN

$SF_{\text{upto } A} = 25 + (5 \times 2) = 35$ kN

$SF_{\text{at } A} = 35 - V = 0$

(c) Figures 4.40 (b) shows the SFD. SFD is parabolic for UVL between *CD*.

(d) **Bending moments:**

$M_{\text{at } F} = 0$

$M_{\text{upto } E} = -10 \times 3 = -30$ kNm

$M_{\text{at } E} = -30 - 60 = -90$ kNm

(There is moment at *E*)

$M_{\text{at } D} = -10 \times 5 - 60 = -110$ kNm

$M_{\text{at } C} = -10 \times 8 - 60 - \left(\frac{1}{2} \times 10 \times 3\right)\left(\frac{1}{3} \times 3\right) = -155$ kNm

$M_{\text{at } B} = -10 \times 9 - 60 - \left(\frac{1}{2} \times 10 \times 3\right)\left(1 + \frac{1}{3} \times 3\right) = -180$ kNm

$M_{\text{at } A} = -10 \times 11 - 60 - \left(\frac{1}{2} \times 10 \times 3\right)\left(2 + 1 + \frac{1}{3} \times 3\right)$

$- (5 \times 2)\frac{2}{2} = -240$ kNm

(e) Figure 4.40 (c) shows the BMD. BMD is parabolic for UDL at *AB* and cubic for UVL at *CD*.

Problem 21 Draw shear force and bending moment diagrams for a cantilever beam loaded as shown in Fig. 4.41. Indicate the values on them.

Solution:

(a) Vertical force at *A*, $V = 40 + 20 + 10 = 70$ kN

(b) **Shear forces:** Starting from left side, i.e., from *D*, the sign convention is $\uparrow\downarrow$ is +ve and $\downarrow\uparrow$ is negative.

$SF_{\text{upto } D} = 0$

$SF_{\text{at } D} = -40$ kN

$SF_{\text{upto } C} = -40$

$SF_{\text{at } C} = -40 - 20 = -60$ kN

$SF_{\text{upto } B} = -60$ kN

$SF_{\text{at } B} = -60 - 10 = -70$ kN

$SF_{\text{upto } A} = -70$ kN

$SF_{\text{at } A} = -70 + V = 0$

(c) Draw SFD as shown in Fig. 4.41(b)

(d) **Bending moments:**

$M_{\text{at}D} = 0$

$M_{\text{at}C} = -40 \times 2 = -80$ kNm (Hogging)

$M_{\text{at}B} = -40 \times 5 - 20 \times 3 = -260$ kNm

$M_{\text{at}A} = -40 \times 6 - 20 \times 4 - 10 \times 1 = -330$ kNm

(e) Draw BMD as shown in Fig. 4.41(c)

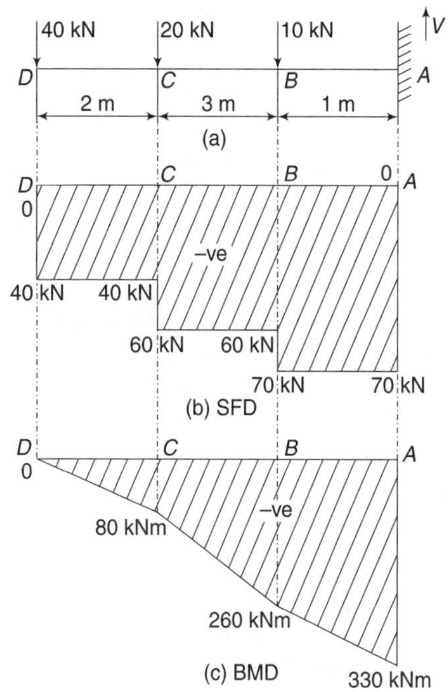

Fig. 4.41

Problem 22 Draw shear force and bending moment diagrams for a cantilever loaded as shown in Fig. 4.42. Indicate the values.

Solution:

(a) Vertical load at A, $V = 30 + \left(\frac{1}{2} \times 12 \times 3\right) + 20 \times 3 = 108$ kN

(b) **Shear forces:**

$SF_{\text{upto } E} = 0$

$SF_{\text{at } E} = -30$ kN

(As per sign convention ⌐↑ is – ve)

$SF_{\text{at } D} = -30$ kN

$SF_{\text{at } C} = -30 - \frac{1}{2} \times 12 \times 3 = -48$ kN

$SF_{\text{at } B} = -48$ kN

$SF_{\text{at } A} = -48 - 20 \times 3 = -108$ kN

$SF_{\text{at } A} = 0$

(c) Draw SFD as shown in Fig. 4.42 (b) For UVL, SFD is parabolic at CD and for UDL SFD is triangular at AB.

(d) **Bending moments:**

$M_{\text{at } E} = 0$

$M_{\text{at } D} = -30 \times 2 = -60$ kN-m

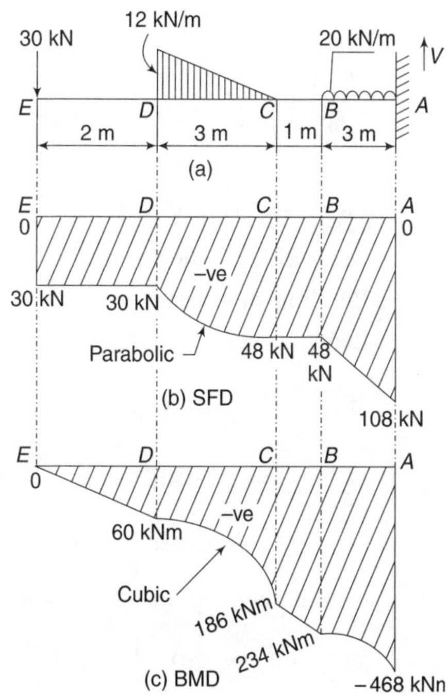

Fig. 4.42

$$M_{\text{at } C} = -30 \times 5 - \left(\frac{1}{2} \times 12 \times 3\right) \times \left(\frac{2}{3} \times 3\right) = -186 \text{ kNm}$$

(Distance from C to CG of UVL is $\frac{2}{3} \times 3$)

$$M_{\text{at } B} = -30 \times 6 - \left(\frac{1}{2} \times 12 \times 3\right)\left(1 + \frac{2}{3} \times 3\right) = -234 \text{ kNm}$$

$$M_{\text{at } A} = -30 \times 9 - \left(\frac{1}{2} \times 12 \times 3\right)\left(3 + 1 + \frac{2}{3} \times 3\right) - (20 \times 3) \times \frac{3}{2} = 468 \text{ kNm}$$

(e) Draw BMD as shown in Fig. 4.42 (c). For UVL at CD, BMD is cubic and for UDL at AB, BMD is parabolic

Problem 23 Draw shear force and bending moment diagram for a cantilever shown in Fig. 4.43. Indicate the values.

Solution:

(a) Vertical force at A, $V = 12 \times 2 + 10 = 34$ kN

(b) **Shear forces:**

$SF_{\text{at } E} = 0$

$SF_{\text{at } D} = -12 \times 2 = -24$ kN

$SF_{\text{at } C} = -24$ kN

$SF_{\text{up to } B} = -24$ kN

$SF_{\text{at } B} = -24 - 10 = -34$ kN

$SF_{\text{up to } A} = -34$ kN

$SF_{\text{at } A} = -34 + V = 0$

(c) Draw SFD as shown in Fig. 4.43 (b)

(d) **Bending moments:**

Bending moment at $M_E = 0$

$$M_{\text{at } D} - (12 \times 2) \times \frac{2}{2} = -24 \text{ kNm}$$

$$M_{\text{up to } C} = -(12 \times 2)\left(1 + \frac{2}{2}\right) = -48 \text{ kNm}$$

$$M_{\text{at } C} = -48 + 6 = -42 \text{ kNm}$$

(–48 is hogging and is –ve and 6 is sagging hence +ve)

$$M_{\text{at } B} = -(12 \times 2)\left(1 + 1 + \frac{2}{2}\right) + 6 = -66 \text{ kNm}$$

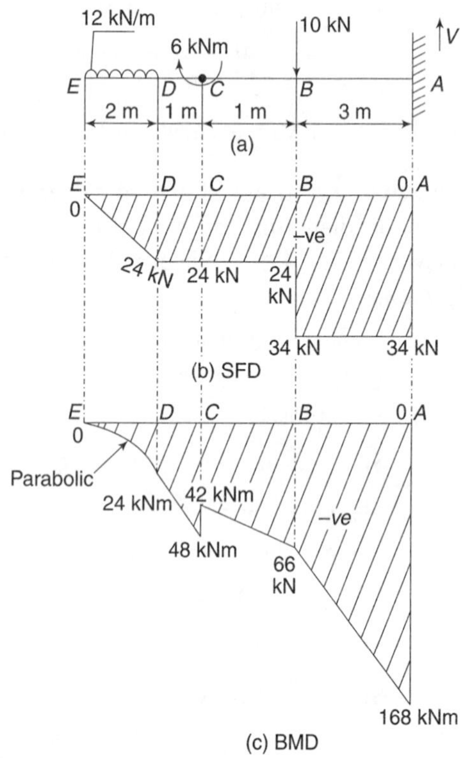

Fig. 4.43

$$M_{at\ A} = -\ (12 \times 2)\left(3 + 1 + 1 + \frac{2}{2}\right) + 6 - 10 \times 3 = -\ 168\ \text{kNm}$$

(e) Draw BMD as shown in Fig. 4.43 (c)

Draw a vertical line at B for a moment = 6 kNm.

Problem 24 Draw bending moment and shear force diagrams for a cantilever loaded as shown in Fig. 4.44. Indicate the values on them.

Solution:

(a) Vertical force at A

$$V = 10 + (20 \times 3) + \frac{1}{2} \times 12 \times 3$$
$$= 88\ \text{kN}$$

(b) **Shear forces:**

$SF_{up\ to\ F} = 0$

$SF_{at\ F} = -\ 10$ kN

$SF_{at\ E} = -\ 10 - (20 \times 3) = -\ 70$ kN

$SF_{at\ D} = -\ 70$ kN

$SF_{at\ C} = -\ 70$ kN

$$SF_{at\ B} = -\ 70\ \text{kN} - \left(\frac{1}{2} \times 12 \times 3\right)$$
$$= -\ 88\ \text{kN}$$

$SF_{up\ to\ A} = -\ 88$ kN

$SF_{at\ A} = -\ 88 + V = 0$

(c) The SFD is as shown in Fig. 4.44 (b). For UDL at EF, it is triangular and for UVL at BC, it is parabolic.

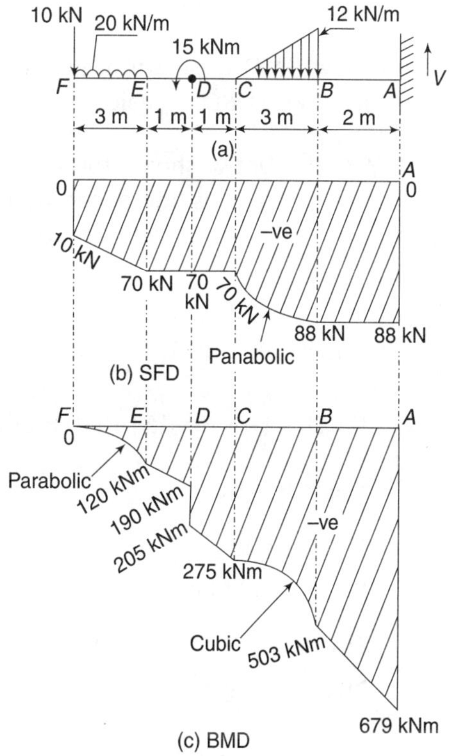

Fig. 4.44

(d) **Bending moments:**

$M_{at\ F} = 0$

$$M_{at\ E} = -\ 10 \times 3 - (20 \times 3)\ \frac{3}{2} = -\ 120\ \text{kNm}$$

$$M_{up\ to\ D} = -10 \times 4 - (20 \times 3)\left(1 + \frac{3}{2}\right) = -190\ \text{kNm}$$

$M_{at\ D} = -\ 190 - 15 = -\ 205$ kNm (Moment at D is hogging, hece –ve)

$$M_{at\ C} = -\ 10 \times 5 - (20 \times 3)\left(1 + 1 + \frac{3}{2}\right) - 15 = -\ 275\ \text{kNm}$$

$$M_{\text{at }B} = -10 \times 8 - (20 \times 3)\left(3 + 1 + 1 + \frac{3}{2}\right) - 15 - \left(\frac{1}{2} \times 12 \times 3\right)\left(\frac{1}{3} \times 3\right)$$
$$= -503 \text{ kNm}$$

$$M_{\text{at }A} = -10 \times 10 - (20 \times 3)\left(2 + 3 + 1 + 1 + \frac{3}{2}\right) - 15 - \left(\frac{1}{2} \times 12 \times 3\right)\left(2 + \frac{1}{3} \times 3\right)$$
$$= -679 \text{ kNm.}$$

(e) Draw BMD as shown in Fig. 4.44 (c). The BMD for UDL at *EF* is parabolic and for UVL at *BC* is cubic.

Problem 25 Draw shear force and bending moment diagrams for a cantilever loaded as shown in Fig. 4.45. Indicate the value on them.

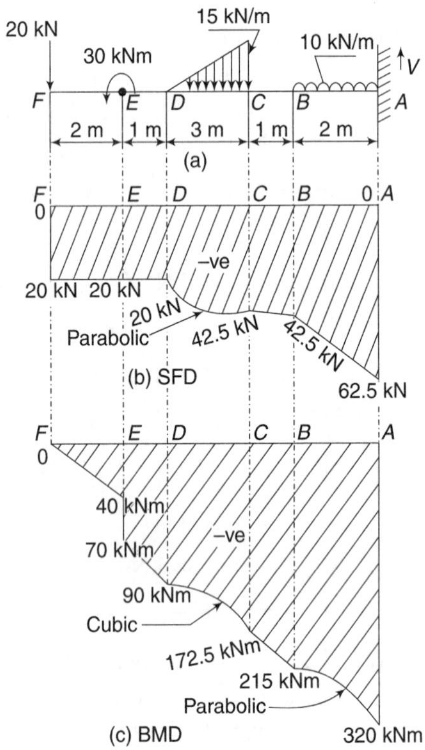

Fig. 4.45

Solution:

(a) Vertical force at *A*,

$$V = 20 + \left(\frac{1}{2} \times 3 \times 15\right) + (10 \times 2) = 62.5 \text{ kN}$$

(b) **Shear forces:**

$SF_{\text{up to }F} = 0$

$SF_{\text{at }F} = -20$ kN

$SF_{\text{at }E} = -20$ kN

$SF_{\text{at }D} = -20$ kN

$$SF_{\text{at }C} = -20 - \left(\frac{1}{2} \times 3 \times 15\right) = -42.5 \text{ kN}$$

$SF_{\text{at }B} = -42.5$ kN

$SF_{\text{up to }A} = -42.5 - (10 \times 2) = -62.5$ kN

$SF_{\text{at }A} = 0$

(c) The SFD is as shown in Fig. 4.45 (b) SFD is parabolic for UVL at *CD* and triangular for UDL at *AB*.

(d) **Bending moments:**

$M_{\text{at }F} = 0$

$M_{\text{up to }E} = -20 \times 2 = -40$ kNm

$M_{\text{at }E} = -40 - 30 = -70$ kNm

$M_D = -20 \times 3 - 30 = -90$ kNm

$$M_{\text{at } C} = -20 \times 6 - 30 - \left(\frac{1}{2} \times 15 \times 3\right)\left(\frac{1}{3} \times 3\right) = -172.5 \text{ kNm}$$

$$M_{\text{at } B} = -20 \times 7 - 30 - \left(\frac{1}{2} \times 15 \times 3\right)\left(1 + \frac{1}{3} \times 3\right) = -215 \text{ kNm}$$

$$M_{\text{at } A} = -20 \times 9 - 30 - \left(\frac{1}{2} \times 15 \times 3\right)\left(2 + 1 + \frac{1}{3} \times 3\right)$$

$$-(10 \times 2)\frac{2}{2} = -320 \text{ kNm}$$

(e) The BMD is as shown in Fig. 4.45 (c). For a moment at *E*, it is vertical at *E*, for UVL at *CD*, it is cubic and for UDL at *AB*, it is parabolic.

4.9 SHEAR FORCE AND BENDING MOMENTS FOR SIMPLY SUPPORTED BEAMS

Shear force and bending moment diagrams for the following standard cases for simply supported beams have been considered.

(a) Simply supported beam with point load

(b) Simply supported beam with UDL

(c) Simply supported beam with partial UDL

(d) Simply supported beam with UVL

(e) Simply supported beam with partial UVL

(f) Simply supported beam with moment.

4.9.1 Simply Supported Beam with Point Load

Figure 4.46 shows a simply supported beam of span *l* carrying a concentrated load *w* at a distance of *a* from left hand support and *b* from right hand support.

(a) To find reactions: $\Sigma V = 0$

$$R_A + R_C = W$$

Taking moments about *A*,

$$R_C \times l - W \cdot a = 0$$

$$\therefore \quad R_C = \frac{Wa}{l}$$

$$R_A = W \frac{-Wa}{l} = \frac{W(l-a)}{l} = \frac{W-b}{l}.$$

(b) **Shear forces:**

$$\text{SF}_{\text{upto } A} = 0$$

$$\text{SF}_{\text{at } A} = RA = \frac{Wb}{b}$$

$$\text{SF}_{\text{upto } B} = \frac{Wb}{l}$$

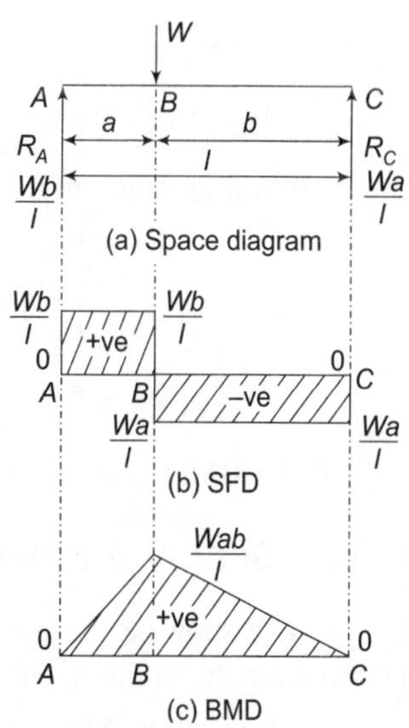

Fig. 4.46

$$SF_{\text{at } B} = \frac{Wb}{l} - W = \frac{-W(l-a)}{l} = \frac{-W.b}{l}$$

$$SF_{\text{up to } C} = \frac{-Wb}{l}$$

$$SF_{\text{at } C} = \frac{-Wb}{l}$$

(c) Draw SFD as shown in Fig. 4.25 (b)

(d) Bending moment at A, $M_A = 0$ (to the left)

$$M_{\text{at } B} = \left(\frac{Wb}{l}\right)\cdot a \text{ (sagging)}$$

$M_{\text{at } C} = 0$ (to the right)

(e) Draw BMD as shown in Fig. 4.25 (c)

Note: If the load acts in the middle, then $a = b = \dfrac{l}{2}$

(a) $R_A = R_B = \dfrac{Wa}{l} = \dfrac{Wb}{l} = \dfrac{W}{2}$

(b) $SF_{\text{up to } A} = 0$

$$SF_{\text{at } A} = \frac{Wl}{2}$$

$$SF_{\text{up to } B} = \frac{Wl}{2}$$

$$SF_{\text{up to } B} = +\frac{Wl}{2} - W = \frac{-Wl}{2}$$

$$SF_{\text{up to } C} = \frac{-Wl}{2}$$

$$SF_{\text{at } C} = \frac{Wl}{2}$$

(c) SFD is as shown in Fig. 4.25 (b)

(d) **Bending moments:**

$$M_A = 0$$

$$M_B = \left(\frac{W}{2}\right)\frac{l}{2} = +\frac{Wl}{4}$$

$$M_C = 0$$

(e) The BMD is as shown in Fig. 4.25 (c)

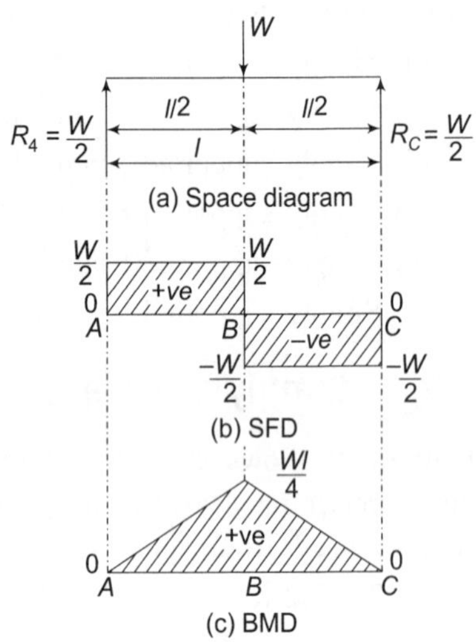

Fig. 4.47

4.9.2 Simply Supported Beam with UDL

Figure 4.48 shows a simply supported beam of length l carrying an UDL of intensity w/unit length over the entire span.

Consider section XX at a distance x from the left side support.

(a) **Reactions:**

$$\Sigma V = 0 \Rightarrow R_A + R_B = w \cdot l \qquad ...(1)$$

Taking moments about A,

$$R_B \cdot l = (w \cdot l)\, \frac{l}{2} = 0$$

$$R_B = \frac{wl}{2}$$

$$\therefore \quad R_A = wl - R_B = wl \frac{-wl}{2} = \frac{wl}{2}$$

$$R_A = R_B = \frac{wl}{2}$$

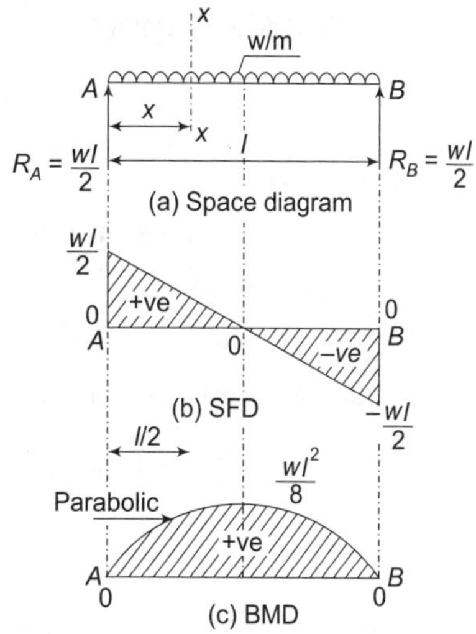

Fig. 4.48

(b) **Shear forces:**

Shear force $SF_{\text{upto } A} = 0$

$$SF_{\text{at } A} = R_A = \frac{wl}{2}$$

$$SF_{\text{upto } B} = \frac{wl}{2} - (wl) = \frac{-wl}{2}$$

$$SF_{\text{at } B} = \frac{-wl}{2} + R_B = \frac{-wl}{2} + \frac{wl}{2} = 0$$

$$SF_{\text{at } xn} = R_A - w.x$$

as x varies from 0 to l, SF varies from $\frac{wl}{2}$ to $\frac{-wl}{2}$

at $\quad x = \frac{l}{2}$, $SF_{\text{at}} \frac{l}{2} = R_A \frac{-w.l}{2} = \frac{wl}{2} \frac{-wl}{2} = 0$

(c) Draw SFD as shown in Fig. 4.48. It is a straight line joining $\frac{+wl}{2}$ at A to $\frac{-wl}{2}$ at B. at $x = \frac{l}{2}$, at the middle, SF = 0

(d) **Bending moments:**

$$M_{\text{at } A} = 0$$

$$M_{\text{at } B} = 0$$

at x, $\quad M_x = R_{A \cdot x} - (wx)\, \frac{x}{2} = \frac{wl}{2} x \frac{-wx^2}{2}$

This is parabolic in nature.

BM is maximum where SF is zero, changing its sign from +ve to −ve.

i.e., $\quad$ at $x = \frac{l}{2}$

$$\therefore \quad M_{\max} = M_{\text{at } x = \frac{l}{2}} = \frac{wl}{2}\frac{l}{2} - \frac{w}{2}\left(\frac{l}{2}\right)^2$$

$$= \frac{wl^2}{4} \frac{-wl^2}{8} = \frac{wl^2}{8}$$

(e) Draw BMD as shown in Fig. 4.48 (c)

4.9.3 Simply Supported Beam with Partial UDL

Figure 4.49 shows a simply supported beam of length l, carrying an UDL of w N/m over a length of a from left hand side support.

Consider a section XX at a distance x from A.

(a) **Reactions:** $\Sigma v = 0$

$$R_A + R_C = w{\cdot}a \qquad \ldots(1)$$

Taking moments about A,

$$R_C.l - (w.a)\,\frac{a}{2} = 0$$

$$\therefore \qquad R_C = \frac{wa^2}{2l}$$

$$\therefore \qquad R_A = wa - R_C$$

$$= wa \frac{-wa^2}{2l} = wa\left(1 - \frac{a}{2l}\right)$$

$$= \frac{wa(2l - a)}{2l}$$

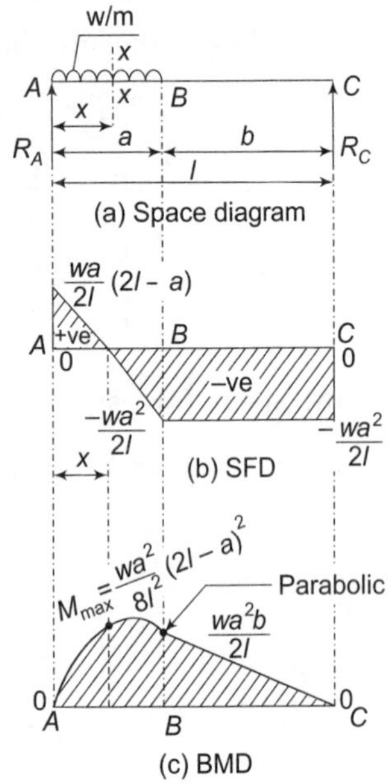

Fig. 4.49

(b) **Shear forces:**

$$SF_{\text{up to } A} = 0$$

$$SF_{\text{at } A} = R_A = \frac{wa}{2l}(2l - a)$$

$$SF_{\text{at } B} = R_A - wa = \frac{wa}{2l}(2l - a) - wa = wa\left(\frac{2l - a}{2l} - 1\right)$$

$$= wa\,\frac{(2l - a - 2l)}{2l} = \frac{-wa^2}{2}$$

$$SF_{\text{at } x} = R_A - wx = \frac{wa}{2l}(2l - a) - wx$$

$$SF_{\text{up to } C} = \frac{wa^2}{2l}$$

$$SF_{\text{at } C} = \frac{-wa^2}{2l} + R_C = 0$$

(c) Draw SFD as shown in Fig. 4.49 (b)

(d) **Bending moments:**

$$M_{\text{at } A} = 0$$

$$M_{\text{at } B} = R_A.a - (wa) = \frac{a}{2}$$

$$= \frac{wa}{2l}(2l - a)a \frac{-wa^2}{2} = \frac{wa^2}{2l}(2l - a - l)$$

$$= \frac{wa^2}{2l}(l - a) = \frac{wa^2 b}{2l}$$

$$M_{\text{at } C} = 0$$

at x,
$$M_{\text{at } X} = \frac{wa}{2l}(2l - a)\,x - \frac{wx^2}{2}$$

This is parabolic in nature

This will be max at SF = 0 or $\frac{dm}{dx} = 0$

i.e.,
$$\frac{wa}{2l}(2l - a) - wx = 0$$

$$x = \frac{a}{2l}(2l — a) \text{ at which BM is maximum}$$

$$M_{\max} = \frac{wa}{2l}(2l - a)\left[\frac{a}{2l}(2l - a)\right]\frac{-w}{2}\left[\frac{a}{2l}(2l - a)\right]^2$$

$$M_{\max} = \frac{wa^2}{4l^2}(2l - a)^2\,\frac{-wa^2}{8l^2}(2l - c)^2$$

$$M_{\max} = \frac{wa^2}{8l^2}(2l - a)^2 \text{ at } x = \frac{a}{2l}(2l - a)$$

(e) Draw BMD as shown in Fig. 4.49 (c)

4.9.4 Simply Supported Beam with UVL

Figure 4.50 shows a simply supported beam of span l carrying an UVL.

Consider section XX at a distance x from A

From similar triangles $\frac{w_x}{x} = \frac{w}{l}$

$\therefore$ Intensity at XX, $w_x = \frac{wx}{l}$

(a) **Reactions:** $\Sigma V = 0 \Rightarrow R_A + R_B = \frac{1}{2}(w \cdot l)$ ...(1)

Taking moments about A,

$$R_B\,l - \frac{1}{2}(wl)\left(\frac{2l}{3}\right) = 0$$

$$R_B = \frac{wl}{3}$$

Substituting in Eq. (1)

$$R_A = \frac{wl}{2} - R_B = \frac{wl}{2} - \frac{wl}{3} = \frac{wl}{6}$$

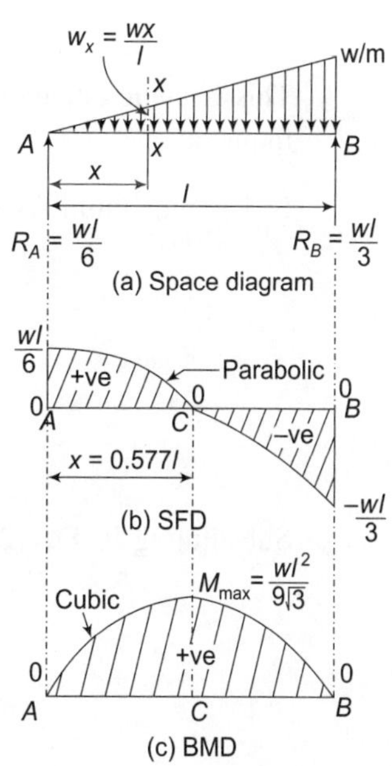

Fig. 4.50

(b) **Shear forces**

$$SF_{\text{upto } A} = 0$$

$$SF_{\text{at } A} = R_A = \frac{wl}{6}$$

$$SF_{\text{upto } B} = \frac{wl}{6} - \frac{1}{2}wl = \frac{-wl}{3}$$

$$SF_{\text{at } B} = \frac{-wl}{3} + R_B = \frac{-wl}{3} + \frac{wl}{3} = 0$$

$$SF_{\text{at } x} = R_A - \left(\frac{1}{2}wx\cdot x\right) = \frac{wl}{6} \frac{-wx}{2l}\cdot x = \frac{wl}{6} \frac{-wx^2}{2l}$$

This is parabolic in nature

$$\left(\text{at } x = 0, \text{ at } A, SF_A = \frac{wl}{6} \text{ at } x = l, \text{ at } B, SF_B = \frac{wl}{6} \frac{-wl^2}{2l} = \frac{-wl}{3}\right)$$

(c) The SFD is drawn as shown in Fig. 4.50 (b)

(d) **Bending moments:**

$$M_{\text{at } A} = 0$$

$$M_{\text{at } B} = 0$$

$$M_{\text{at } x} = R_A\cdot x - \left(\frac{1}{2}wx\cdot x\right)\left(\frac{1}{3}x\right) = \frac{wl}{6}\cdot x \frac{-1}{2}\left(\frac{wx}{l}\right)\cdot x\cdot \frac{x}{3}$$

$$M_x = \frac{wl}{6}x - \frac{wx^3}{6l} \qquad \ldots(2)$$

This is cubic equation. Hence, the BMD is cubic curve. To find maximum bending moment:

BM is maximum where SF = 0 changing its sign from +ve to –ve. i.e., at C.

$$SF_x = \frac{wl}{6} - \frac{wx^2}{2l} = 0$$

$$\frac{wl}{6} = \frac{wx^2}{2l} \text{ or } x^2 = \frac{l^2}{3}$$

$$x = \frac{l}{\sqrt{3}} = 0.577\,l$$

Substituting in Eq. (2)

$$M_{\max} = M \text{ at } x = \frac{l}{\sqrt{3}} = \frac{wl}{6}\left(\frac{l}{\sqrt{3}}\right) \frac{-w}{6l}\left(\frac{l}{\sqrt{3}}\right)^3$$

$$= \frac{wl^2}{6\sqrt{3}} - \frac{w}{6l}\frac{l^3}{3\sqrt{3}} = \frac{wl^2}{6\sqrt{3}}\left(1-\frac{1}{3}\right)$$

$$M_{\max} = \frac{wl^2}{9\sqrt{3}} = 0.06415\,wl^2$$

(e) The BMD is cubic in nature with maximum value of $M_{max} = \frac{wl^2}{9\sqrt{3}}$ at a distance of $\frac{l}{\sqrt{3}}$ from left end A, as shown in Fig. 4.50 (c).

4.9.5 Simply Supported Beam with Partial UVL

Figure 4.51 shows a simply supported beam of length l. Carrying an UVL over a span of a at the left side.

Consider section XX at a distance x from A.

From similar triangles $w_x = wl \cdot x$

(a) **Reactions:**

$$\Sigma V = 0 \Rightarrow R_A + R_C = \frac{1}{2} w \cdot a \quad ...(1)$$

Taking moments about A,

$$R_C . l - \left(\frac{1}{2} w \cdot a\right) \frac{2xa}{3} = 0$$

$$R_C = \frac{wa^2}{3l}$$

Substituting in Eq. (1)

$$R_A + \frac{wa^2}{3l} = \frac{1}{2} wa$$

$$R_A = \frac{1}{2} wa - \frac{wa^2}{3l} = wa\left(\frac{1}{2} - \frac{a}{3l}\right)$$

$$R_A = wa\left(\frac{3l - 2a}{6l}\right) = \frac{wa}{6l}(3l - 2a)$$

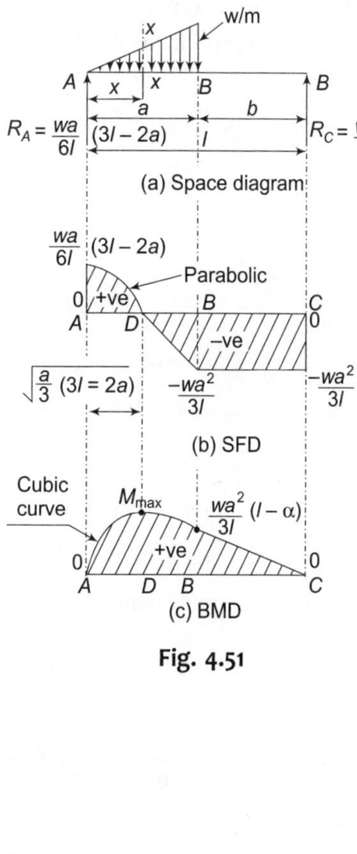

Fig. 4.51

(b) **Shear forces:**

$$SF_{upto\ A} = 0$$

$$SF_{at\ A} = R_A = \frac{wa}{6l}(3l - 2a)$$

$$SF_{at\ B} = R_A - UVL = \frac{wa}{6l}(3l - 2a) - \frac{1}{2} w.a$$

$$= \frac{wa}{2}\left[\left(\frac{3l - 2a}{3l}\right) - 1\right] = \frac{wa}{2}\left[\frac{3l - 2a - 3l}{3l}\right] = -\frac{wa^2}{3l}$$

$$SF_{up\ to\ C} = \frac{-wa^2}{3l}$$

$$SF_{at\ C} = \frac{-wa^2}{3l} + R_C = \frac{-wa^2}{3l} + \frac{wa^2}{3l} = 0$$

$$SF_{at\ x} = \frac{wa}{6l}(3l - 2a) - \left(\frac{1}{2} wx \cdot x\right)$$

$$= \frac{wa}{6l}(3l - 2a) - \left(\frac{1}{2} \cdot \frac{w \cdot x}{l} \cdot x\right)$$

$$= \frac{wa}{6l}(3l - 2a) \frac{wx^2}{2l}.$$

This is parabolic in nature between A and B

(c) The SFD is as shown in Fig. 4.51 (b)

(d) **Bending moments:**

Bending moment at A, $M_{\text{at } A} = 0$

$$M_{\text{at } B} = R_A \times a - \left(\frac{1}{2} \times w \cdot a\right) \frac{1}{3} \times a = \frac{wa}{6l}(3l - 2a) \cdot a \frac{-wa^2}{6}$$

$$= \frac{wa^2}{6l}(3l - 2a - l) = \frac{wa^2}{6l}(2l - 2a)$$

$$= \frac{wa^2}{6l} 2(l - a) = \frac{wa^2}{3l}(l - a)$$

$$M_{\text{at } C} = 0$$

$$M_{\text{at } x} = R_a \cdot x - \left(\frac{w_x \cdot x}{2}\right) \cdot \frac{1}{3} x = \frac{wa}{6l}(3l - 2a) \cdot x - \frac{w}{l} \cdot x \cdot \frac{x}{2} \cdot \frac{x}{3}$$

$$= \frac{wax}{6l}(3l - 2a) - \frac{wx^3}{6l}$$

This is cubic distribution.

To find $M_{\max}$ which is between A and B.

For maximum bending moment SF = 0 or $\frac{dM}{dx} = 0$

$$\text{SF}_x = \frac{wa}{6l}(3l - 2a) \frac{-wx^2}{2l} = 0$$

$$\frac{wa}{6l}(3l - 2a) = \frac{wx^2}{2l}$$

$$x^2 = \frac{a}{3}(3l - 2a)$$

$$x = \sqrt{\frac{a}{3}(3l - 2a)}$$

(e) Draw BMD as shown in Fig. 4.51 (e).

4.9.6 Simply Supported Beam with Moment

Figure 4.52 shows a simply supported beam of length l, carrying a moment M at a distance a from support A.

(a) **Reactions:**

$$R_A + R_C = 0 \quad \ldots(1)$$

Taking moments about A,

$$-M + R_C \cdot l = 0$$

$$\therefore \quad R_C = \frac{M}{l}$$

Substituting in Eq. (1), $\quad R_A = -R_C$

$$R_A = -\frac{M}{l} \text{ (downward)}$$

(b) **Shear forces:**

$$SF_{\text{upto } A} = 0$$

$$SF_{\text{at } A} = R_A = -\frac{M}{l}$$

$$SF_{\text{at } B} = -\frac{M}{l}$$

$$SF_{\text{upto } C} = -\frac{M}{l}$$

$$SF_{\text{at } C} = -\frac{M}{l} + R_C = -\frac{M}{l} + \frac{M}{l} = 0$$

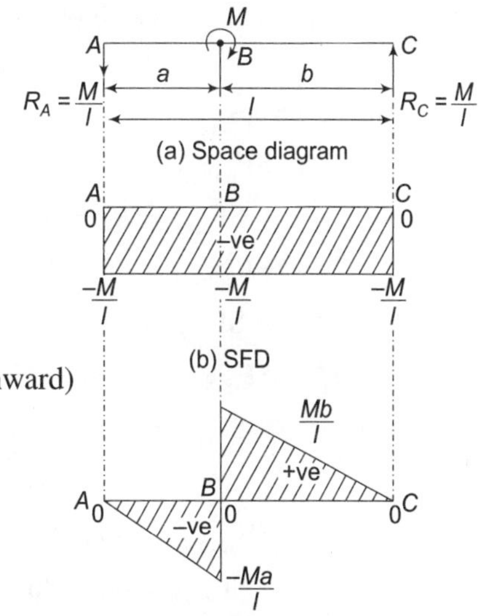

Fig. 4.52

(c) Draw SFD as shown in Fig. 4.52 (b)

(d) **Bending moments:**

$$M_{\text{at } A} = 0$$

$$M_{\text{upto } B} = -R_A \cdot a = \frac{-Ma}{l}$$

$$M_{\text{at } B} = \frac{-Ma}{l} + M = \frac{M}{l}(-a + l) = \frac{Mb}{l}$$

$$M_{\text{at } C} = 0$$

(e) Draw BMD as shown in Fig. 4.52 (c)

PROBLEMS ON SIMPLY SUPPORTED BEAMS

Problem 26 Draw shear force and bending moment diagrams for a simply supported beam loaded as shown in Fig. 4.53. Indicate the values on them.

Solution:

(a) **Reactions:** $\quad \Sigma V = 0$

$$R_A + R_D = 30 + 50 = 80 \text{ kN} \quad \ldots(1)$$

Taking moments about A

$$-30 \times 2 - 50 \times 5 + R_D \times 7 = 0$$

$$R_D = 44.286 \text{ kN}$$

$$\therefore \quad R_A = 80 - R_D = 35.714 \text{ kN}$$

(b) **Shear forces:**

$$SF_{\text{upto } A} = 0$$
$$SF_{\text{at } A} = R_A = 35.714 \text{ kN}$$
$$SF_{\text{upto } B} = 35.714 \text{ kN}$$
$$SF_{\text{at } B} = 35.714 - 30 = 5.714 \text{ kN}$$
$$SF_{\text{upto } C} = 5.714 \text{ kN}$$
$$SF_{\text{at } C} = 5.714 - 50 = -44.286 \text{ kN}$$
$$SF_{\text{upto } D} = -44.286 \text{ kN}$$
$$SF_{\text{at } D} = -44.286 + R_D = 0$$

(c) The SFD is as shown in Fig. 4.53 (b)

(d) **Bending moments:**

$$M_{\text{at } A} = 0$$
$$M_{\text{at } B} = R_A \times 2 = 35.714 \times 2 = 71.428 \text{ kNm}$$
$$M_{\text{at } C} = R_A \times 5 - 30 \times 3 = 35.714 \times 5 - 30 \times 3 = 88.57 \text{ kNm}$$
$$M_{\text{at } D} = 0$$

Note: For a simply supported beam, bending moment at the first and last points are always equal to zero

(e) The BMD is as shown in Fig. 4.53 (c)

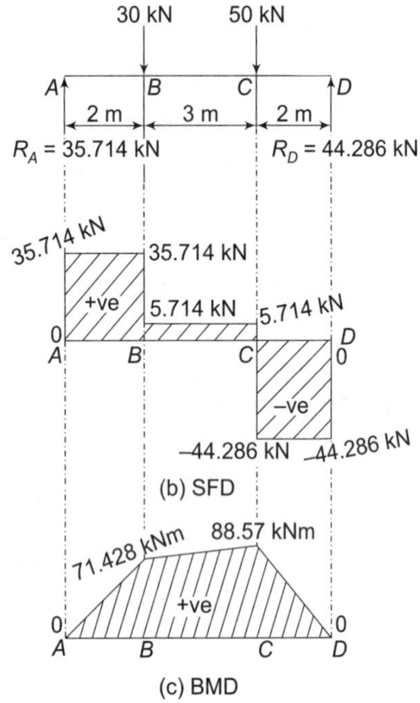

Fig. 4.53

Problem 27 Draw shear and bending moment diagrams for a simply supported beam as shown in Fig. 4.54. Indicate the values at salient points.

Solution:

(a) **Reactions:** $\Sigma V = 0 \Rightarrow R_A + R_D = 20 + 15 \times 4 = 80$ kN

Taking moments about A, $-20 \times 3 - (15 \times 4)\left(3 + 5 + \frac{4}{2}\right) + R_D \times 12 = 0$

$$R_D = 55 \text{ kN}$$
$$R_A = 80 - R_D = 80 - 55 = 25 \text{ kN.}$$

(b) **Shear forces:**

$$SF_{\text{upto } A} = 0$$
$$SF_{\text{at } A} = R_A = 25 \text{ kN}$$
$$SF_{\text{upto } B} = 25 \text{ kN}$$
$$SF_{\text{at } B} = 25 - 20 \text{ kN}$$

$$SF_{at\ C} = 5 \text{ kN}$$

$$SF_{upto\ D} = 5 - (15 \times 4)$$

$$= -55 \text{ kN}$$

$$SF_{at\ D} = -55 + R_D = 0$$

(c) Figure 4.54 (b) shows SFD. For UDL, SFD is triangular

(d) **Bending moments:**

$$M_{at\ A} = 0$$

$$M_{at\ B} = R_A \times 2 = 25 \times 3 = 75 \text{ kNm}$$

$$M_{at\ C} = R_A \times 8 - 20 \times 5$$

$$= 25 \times 8 - 20 \times 5 = 100 \text{ kNm}$$

$$M_{at\ D} = 0$$

To find maximum bending moment:

BM is maximum at SF = 0 and changes its sign (at X).

From similar triangules of SF diagrams between CD

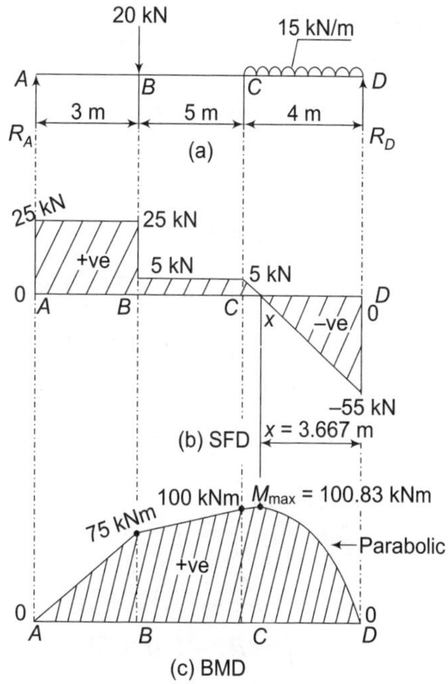

Fig. 4.54

$$\frac{5}{4-x} = \frac{55}{x}$$

$$5x = 55\ (4 - x)$$

$$5x = 220 - 55\ x$$

$$\therefore \quad 60x = 220$$

$$x = \frac{220}{60} = 3.667 \text{ m from } D$$

BM to the right of XX.

$$M_{max} = R_D.x - (w \cdot x)\frac{x}{2}$$

$$M_{max} = 55 \times 3.667 - (15 \times 3.667) \times \frac{3.667}{2} = 100.83 \text{ kNm}$$

(e) Draw BMD as shown in Fig. 4.54 (c)

For UDL, BMD is parabolic.

Problem 28 Draw shear force and bending moment diagrams for a simply supported beam shown in Fig. 4.55. Indicate the values at salient points.

Solution:

(a) **Reactions:**

$\Sigma V = 0 \Rightarrow R_A - (10 \times 5) - 12 - 9 + R_D = 0$

$R_A + R_D = 71$ kN

Taking moments about A,

$- (10 \times 5)\dfrac{5}{2} - 12 \times 5 - 9 \times 6 + R_D \times 8 = 0$

$R_D = 29.875$ kN

$\therefore R_A = 41.125$ kN

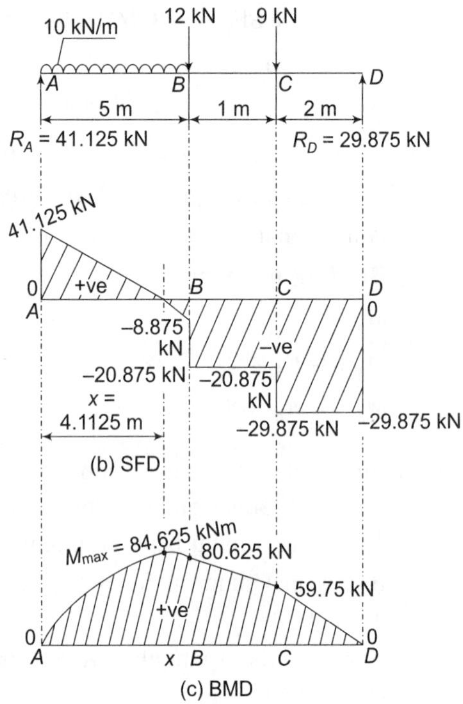

Fig. 4.55

(b) **Shear forces:**

$SF_{\text{upto } A} = 0$

$SF_{\text{at } A} = R_A = 41.125$ kN

$SF_{\text{upto } B} = R_A - (10 \times 5)$

$= 41.125 - 50 = -8.875$ kN

$SF_{\text{at } B} = -8.875 - 12 = -20.875$ kN

$SF_{\text{upto } C} = -20.875$ kN

$SF_{\text{at } C} = -20.875 - 9 = -29.875$ kN

$SF_{\text{upto } D} = -29.875$ kN

$SF_{\text{at } D} = -29.875 + R_D = 0$

(c) Draw SFD as shown in Fig. 4.55 (b). For UDL, SFD is triangular at *AB*

(d) **Bending moments:**

$M_{\text{at } A} = 0$

$M_{\text{at } B} = R_A \times 5 - (10 \times 5)\dfrac{5}{2} = 41.125 \times 5 - (10 \times 5)\dfrac{5}{2} = 80.625$ kNm

$M_{\text{at } C} = 41.125 \times 6 - (10 \times 5)\left(1 + \dfrac{5}{2}\right) - 12 \times 1 = 59.75$ kNm

$M_{\text{at } D} = 0$

To find maximum bending moment.

BM is maximum at SF = 0 and change its sign (at *X*) from similar triangles at *AB* in SFD

$$\frac{41.125}{x} = \frac{8.875}{5 - x}$$

$$41.125\,(5 - x) = 8.875.x$$

$$x = \frac{41.125 \times 5}{50} = 4.1125 \text{ m from } A$$

$$M_x = 41.125 \cdot x\,(10 \cdot x)\,\frac{x}{2} = 41.125 \times 4.1125 - 10 \times \frac{4.1125^2}{2}$$

$$M_x = 84.563 \text{ kNm}$$

(e) Draw BMD as shown in Fig. 4.55 (c)

BMD for UDL is parabolic with maximum at x.

Problem 29 Draw shear force bending moment diagram for a simply supported beam as shown in Fig. 4.56. Indicate the values at salient points.

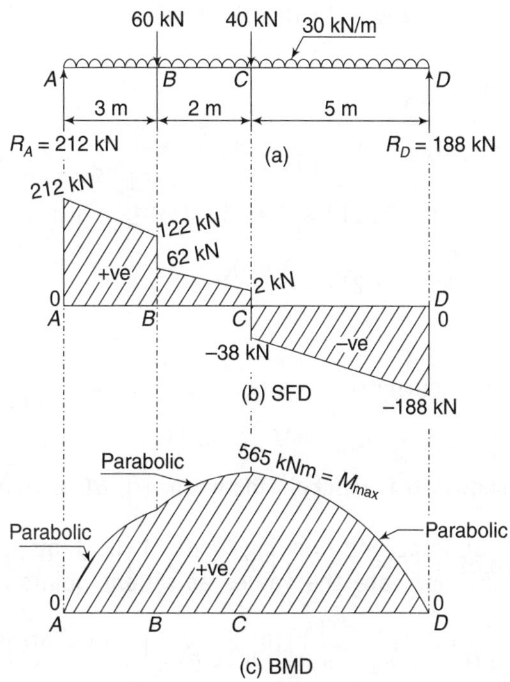

Fig. 4.56

Solution:

(a) **Reactions:**

$$\Sigma V = 0 \Rightarrow R_A + R_D = 60 + 40 + (30 \times 10) = 400 \text{ kN}$$

Taking moments about A, $-\,60 \times 3 - 40 \times 5 - (30 \times 10)\,\dfrac{10}{2} + R_D \times 10 = 0$

$$\therefore \quad R_D = 188 \text{ kN}$$

$$R_A = 400 - R_D = 400 - 188 = 212 \text{ kN}$$

(b) **Shear forces:**

Shear force $SF_{\text{up to } A} = 0$

$$SF_{\text{at } A} = R_A = 212 \text{ kN}$$
$$SF_{\text{up to } B} = 212 - (30 \times 3) = 122 \text{ kN}$$
$$SF_{\text{at } B} = 122 - 60 = 62 \text{ kN}$$
$$SF_{\text{up to } C} = 62 - (30 \times 2) = 2 \text{ kN}$$
$$SF_{\text{at } C} = 2 - 40 = -38 \text{ kN}$$
$$SF_{\text{up to } D} = -38 - (30 \times 5) = -188 \text{ kN}$$
$$SF_{\text{at } D} = -188 + R_D = 0$$

(c) Draw SFD as shown in Fig. 4.56 (b). For UDL between *AB*, *BC* and *CD*, it is triangular

(d) **Bending moment:** $M_{\text{at } A} = 0$

$$M_{\text{at } B} = R_A \times 3 - (30 \times 3)\frac{3}{2}$$
$$= 212 \times 3 - (30 \times 3)\frac{3}{2} = 501 \text{ kNm}$$
$$M_{\text{at } C} = 212 \times 5 - (30 \times 5)\frac{5}{2} - 60 \times 2 = 565 \text{ kNm}$$
$$M_{\text{at } D} = 0$$

Maximum bending moment occurs where SF = 0 and charge its sign i.e., at *C*

$$M_{max} = M_D = 565 \text{ kNm}$$

(e) Draw BMD as shown. It is parabolic between *AB*, *BC* and *CD* repeatedly.

Problem 30 Draw shear force and bending moment diagrams for a simply supported beam shown in Fig. 4.57. Indicate the values at salient points on them.

Solution:

(a) **Reactions:** $\Sigma V = 0 \Rightarrow = R_A + R_E = (40 \times 6) + 30 + 20 \times 4 = 350 \text{ kN}$ ...(1)

Taking moments about *A*, $-(40 \times 6)\frac{6}{3} - 30 \times 7 - (20 \times 4)\left(6 + 1 + 2 + \frac{4}{2}\right) + R_E \times 13 = 0$

$$\therefore \quad R_E = 139.23 \text{ kN}$$
$$R_A = 350 - R_E = 350 - 139.23 = 120.77 \text{ kN}$$

(b) **Shear forces:**

, $\quad SF_{\text{up to } A} = 0$

at *A*, $\quad SF_{\text{at } A} = R_A = 210.77 \text{ kN}$

$$SF_{\text{at }B} = 210.77 - (40 \times 6) = -29.23 \text{ kN}$$

$$SF_{\text{up to }C} = -29.23 \text{ kN}$$

$$SF_{\text{at }C} = -29.23 - 30 = -59.23 \text{ kN}$$

$$SF_{\text{at }D} = -59.23 \text{ kN}$$

$$SF_{\text{at }E} = -59.23 - 20 \times 4 = -139.23 \text{ kN}$$

$$SF_{\text{at }E} = -139.23 + R_E = 0$$

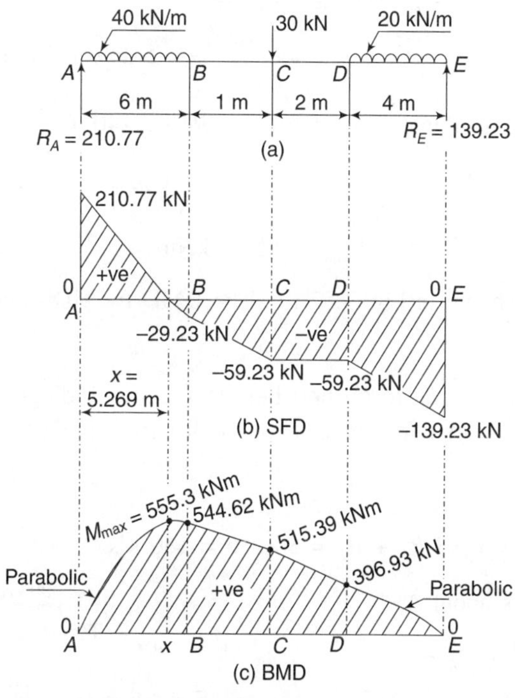

Fig. 4.57

(c) Draw SFD as shown in Fig. 4.57 (b). For UDL at *AB* and *DF*, the SFD is triangular

(d) **Bending moments:**

Bending moment $M_{\text{at }A} = 0$

$$M_{\text{at }B} = 210.77 \times 6 - (40 \times 6)\,\frac{6}{2} = 544.62 \text{ kNm}$$

$$M_{\text{at }C} = 210.77 \times 7 - (40 \times 6)\left(1 + \frac{6}{2}\right) = 515.39 \text{ kNm}$$

$$M_{\text{at }D} = 210.77 \times 9 - (40 \times 6)\left(2 + 1 + \frac{6}{2}\right) - 30 \times 2 = 396.93 \text{ kNm}$$

$$M_{\text{at }E} = 0$$

To find maximum bending moment:

Bending moment is maximum at SF = 0, changing its sign.

From similar triangular at *AB* in SFD

$$\frac{210.77}{x} = \frac{29.23}{6 - x}$$

$$210.77\,(6 - x) = 29.23x$$

$$1264.62 - 210.77x = 29.33x$$

$$x = \frac{1264.62}{240} = 5.269 \text{ m from } A$$

Maximum bending moment, $M_{max} = M_x = R_A.x - (40 \cdot x)\,\frac{x}{2}$

$$M_{max} = 210.77 \times 5.269 - (40 \times 5.269)\,\frac{5.269}{2}$$

$$M_{max} = 555.3 \text{ kNm}$$

(e) Draw BMD as shown in Fig. 4.57 (c). The BMD for UDL between *AB* and *DE* are parabolic.

Problem 31 Draw the shear force and bending moment diagrams for a simply supported beam of span 6 m carrying UDL of 30 kN/m over the entire span. Indicate the values on them.

Solution:

(a) **Reactions:** $\Sigma V = 0 \Rightarrow R_A + R_B = 30 \times 6 = 180$ kN

Taking moments about *A*,

$$-(30 \times 6) \times \frac{6}{2} + R_B = 0$$

$$R_B = 90 \text{ kN}$$

$$\therefore \quad R_A = 180 - R_B$$

$$= 90 \text{ kN}$$

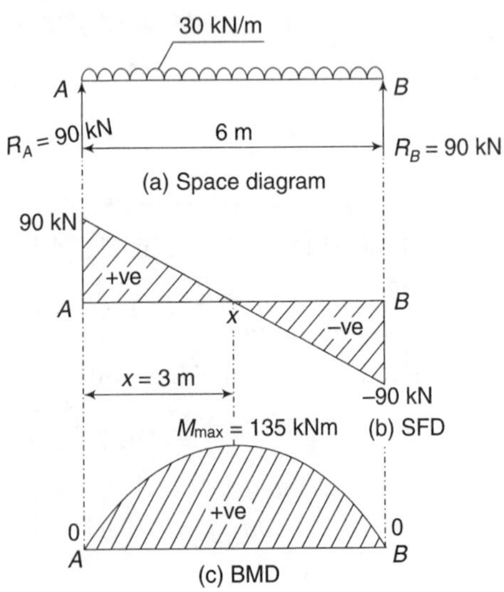

Fig. 4.58

(b) **Shear forces:**

$$SF_{\text{upto } A} = 0$$

$$SF_{\text{at } A} = R_A = 90 \text{ kN}$$

$$SF_{\text{upto } B} = R_A - (30 \times 6)$$

$$= 90 - 180$$

$$= -90 \text{ kN}$$

$$SF_{\text{at } B} = -90 + R_B = 0$$

(c) Draw SFD as shown in Fig. 4.58 (b) It is triangular for UDL.

(d) **Bending moments**

$$M_{\text{at } A} = 0$$

$$M_{\text{at } B} = 0$$

$$M_{\text{at } x} = 90\,x - (30\,x)\,\frac{x}{2}$$

This is maximum at the middle where $x = \frac{l}{2} = \frac{6}{2} = 3$ m where SFD = 0

$$M_{max} = 90 \times 3 - (30 \times 3) \times \frac{3}{2} = 135 \text{ kNm.}$$

(e) Draw BMD as shown in Fig. 4.58 (c). BMD is parabolic for UDL over full length between AB,

Problem 32 Draw shear forces and bending moment diagrams for a simply supported beam shown in Fig. 4.59. Indicate the values on them

Solution:

(a) **Reactions:**

$$\Sigma V = 0 \Rightarrow R_A + R_D$$

$$= \left(\frac{1}{2} \times 30 \times 5\right) + 12 = 87 \text{ kN} \quad ...(1)$$

Taking moments about A,

$$-\left(\frac{1}{2} \times 30 \times 5\right)\left(\frac{2}{3} \times 5\right) - 12 \times 8 + R_D \times 10 = 0$$

$$R_D = 34.6 \text{ kN}$$

$$\therefore \quad R_A = 87 - R_D = 52.4 \text{ kN}$$

(b) **Shear forces:**

Shear force upto A, $SF_{\text{upto } A} = 0$

$$SF_{\text{at } A} = R_A = 52.4 \text{ kN}$$

$$SF_{\text{at } B} = 52.4 - \frac{1}{2} \times 30 \times 5$$

$$= -22.6 \text{ kN}$$

$$SF_{\text{at } B} = -22.6 \text{ kN}$$

$$SF_{\text{at } C} = -22.6 - 12 = -34.6 \text{ kN}$$

$$SF_{\text{upto } D} = -34.6 \text{ kN}$$

$$SF_{\text{at } D} = -34.6 + R_D = 0$$

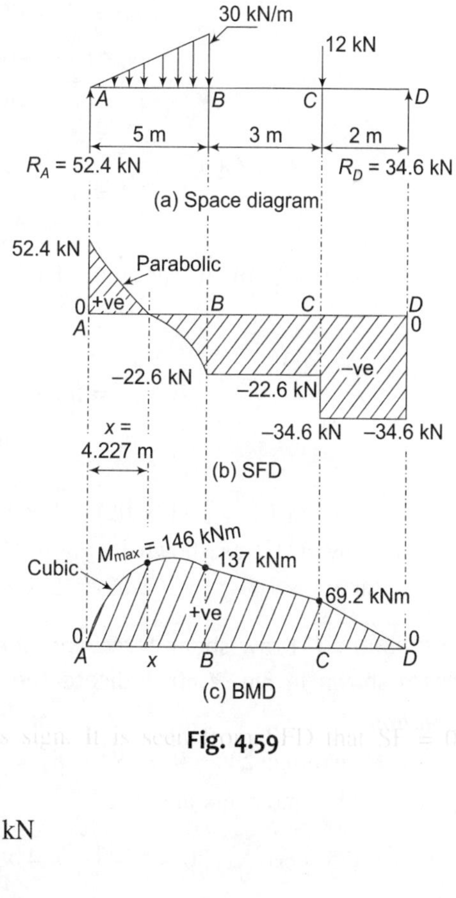

Fig. 4.59

(c) Draw SFD as shown in Fig. 4.59 (b) SFD is parabolic for UVL between A and B.

(d) **Bending moments:**

Bending moment at A, $M_A = 0$

$$M_{\text{at}B} = R_A \cdot 5 - \left(\frac{1}{2} \times 30 \times 5\right)\left(\frac{1}{3} \times 5\right) = 52.4 \times 5 - \left(\frac{1}{2} \times 30 \times 5\right)\frac{5}{3} = 137 \text{ kNm}$$

$$M_{\text{at}C} = 52.4 \times 8 - \left(\frac{1}{2} \times 30 \times 5\right)\left(3 + \frac{1}{3} \times 5\right) = 69.2 \text{ kNm}$$

$$M_{\text{at}D} = 0$$

Maximum bending moment

BM is maximum at SF = 0, changing its sign

$$\text{SF}_x = 52.4 - \left(\frac{1}{2} \times w_x.x\right)$$

For UVL, $$\frac{w}{l} = \frac{w_x}{x}$$

$\therefore$ $$w_x = \frac{w}{l} \cdot x = \frac{30}{5} \times x = 6\,x$$

$$\text{SF}_x = 52.4 - \frac{1}{2} \times 6\,x.x = 0$$

$$52.4 = 3x^2$$

$\therefore$ $$x = 4.179 \text{ m from } A$$

$\therefore$ $$M_x = 52.4\,x - \left(\frac{1}{2} \cdot w_x \cdot x\right)\frac{1}{3}.x$$

$$= 52.4 \times 4.179 - \left(\frac{1}{2} \times 6x.x\right)\frac{x}{3}$$

$$= 52.4 \times 4.179 \frac{-1}{2} \times 6 \times \frac{(4.179)^3}{3} = 146 \text{ kNm}$$

(e) Draw BMD as shown in Fig. 4.59 (c). The BMD for UVL between AB is cubic curve.

Problem 33 Draw shear force and bending moment diagram for a simply supported beam shown in Fig. 4.60. Indicate the values on them.

Solution:

(a) **Reactions:** $\Sigma V = 0 \Rightarrow R_A + R_F = 25 \times 3 + 30 + \frac{1}{2} \times 12 \times 4 = 129$ kN ...(1)

Taking moments about A,

$$-(25 \times 3)\frac{3}{2} - 30 \times 5 - \left(\frac{1}{2} \times 4 \times 12\right)\left(3 + 2 + 2 + \frac{2}{3} \times 4\right) + R_F \times 12 = 0$$

$$R_F = 41.21 \text{ kN}$$
$$R_A = 129 - R_F = 87.79 \text{ kN}$$

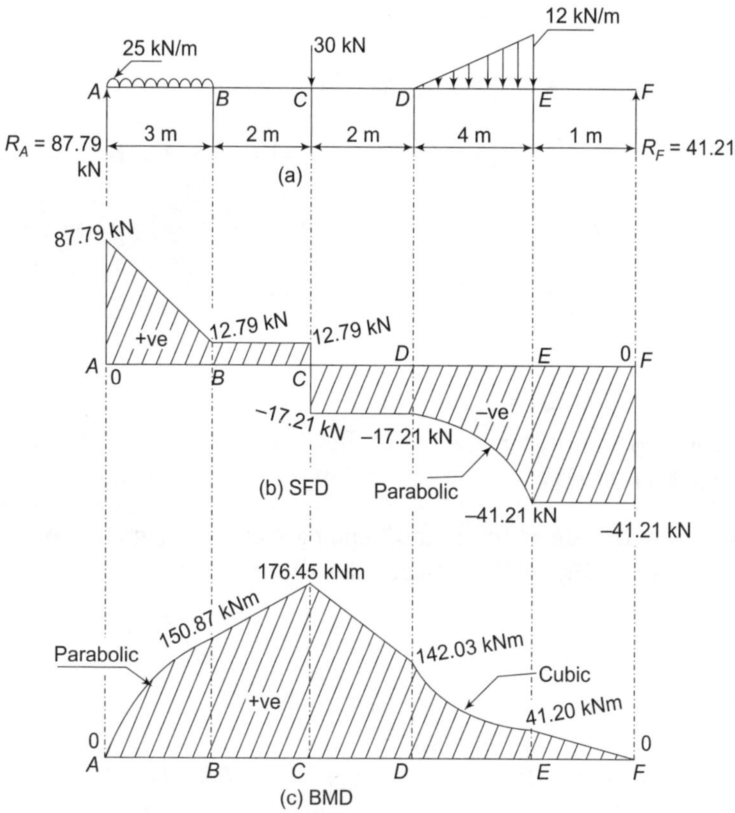

Fig. 4.60

(b) **Shear forces:**

$$SF_{\text{up to } A} = 0$$
$$SF_{\text{at } A} = R_A = 87.79 \text{ kN}$$
$$SF_{\text{at } B} = 87.79 - 25 \times 3 = 12.79 \text{ kN}$$
$$SF_{\text{up to } C} = 12.79 \text{ kN}$$
$$SF_{\text{at } C} = 12.79 - 30 = -17.21 \text{ kN}$$
$$SF_{\text{at } D} = -17.21 \text{ kN}$$
$$SF_{\text{at } E} = -17.21 - \frac{1}{2} \times 12 \times 4 = -41.21 \text{ kN}$$
$$SF_{\text{up to } F} = -41.21 \text{ kN}$$
$$SF_{\text{at } F} = -41.21 + R_F = 0$$

(c) Draw SFD as shown in Fig. 4.60 (b). For UDL at *AB*, SFD is triangular and for UVL at *DE*, it is parabolic.

(d) **Bending moments:**

$$M_{\text{at}A} = 0$$

$$M_{\text{at}B} = 87.79 \times 3 - (25 \times 3)\,\frac{3}{2} = 150.87 \text{ kNm}$$

$$M_{\text{at}C} = 87.79 \times 5 - (25 \times 3)\left(2 + \frac{3}{2}\right) = 176.45 \text{ kNm}$$

$$M_{\text{at}D} = 87.79 \times 7 - (25 \times 3)\left(2 + 2 + \frac{3}{2}\right) - 30 \times 2 = 142.03 \text{ kNm}$$

$$M_{\text{at}E} = 87.79 \times 11 - (25 \times 3)\left(4 + 2 + 2 + \frac{3}{2}\right) - 30 \times 6 - \left(\frac{1}{2} \times 12 \times 4\right)\left(\frac{1}{3} \times 4\right) = 41.2 \text{ kNm}$$

$$M_{\text{at}F} = 0$$

BM is maximum at SF = 0, i.e., at *C*.

(e) Draw BMD as shown in Fig. 4.60 (c). BMD is parabolic for UVL at *AB* and cubic curve for UVL at *DE*.

Problem 34 Draw the shear force and bending moment diagrams for a simply supported beam shown in Fig. 4.61. Indicate the values on it.

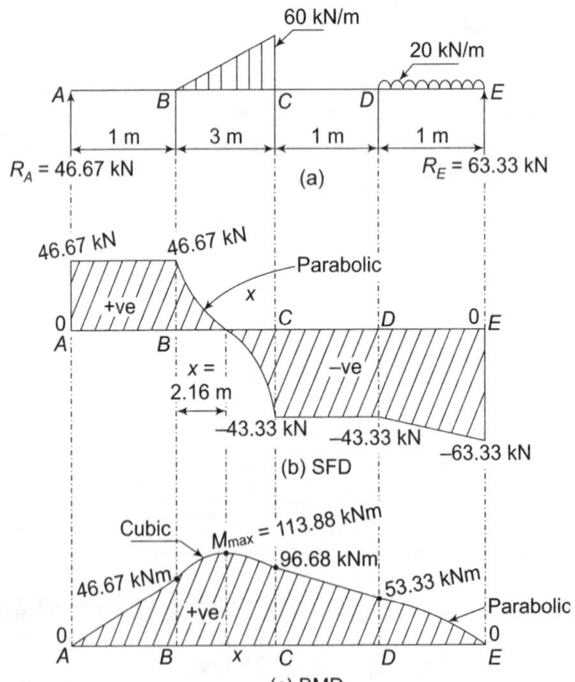

Fig. 4.61

Solution:

(a) **Reactions:** $\Sigma V = 0 \Rightarrow R_A + R_E = \left(\frac{1}{2} \times 60 \times 3\right) + 20 \times 1 = 110$ kN ...(1)

Taking moments about A,

$$-\left(\frac{1}{2} \times 60 \times 3\right)\left(1 + \frac{2}{3} \times 3\right) - (20 \times 1)\left(1 + 3 + 1 + \frac{1}{2}\right) + R_E \times 6 = 0$$

$\therefore$ $R_E = 63.33$ kN

$R_A = 110 - R_E = 46.67$ kN

(b) **Shear forces:**

$$\text{SF}_{\text{up to } A} = 0$$

$$\text{SF}_{\text{at } A} = R_A = 46.67 \text{ kN}$$

$$\text{SF}_{\text{at } B} = 46.67 \text{ kN}$$

$$\text{SF}_{\text{at } C} = 46.67 - \left(\frac{1}{2} \times 60 \times 3\right) = -43.33 \text{ kN}$$

$$\text{SF}_{\text{at } D} = -43.33 \text{ kN}$$

$$\text{SF}_{\text{at } E} = -43.33 - (20 \times 1) = -63.33 \text{ kN}$$

$$\text{SF}_{\text{at } E} = -63.33 + R_E = 0$$

(c) The SFD is as shown in Fig. 4.61 (b). SFD for UVL at BC is parabolic and for UVL at DE is triangular.

(d) **Bending moments:**

$$M_{\text{at} A} = 0$$

$$M_{\text{at} B} = 46.67 \times 1 = 46.67 \text{ kNm}$$

$$M_{\text{at} C} = 46.67 \times 4 - \left(\frac{1}{2} \times 60 \times 3\right)\left(\frac{1}{3} \times 3\right) = 96.68 \text{ kNm}$$

$$M_{\text{at} D} = 46.67 \times 5 - \left(\frac{1}{2} \times 60 \times 3\right)\left(1 + \frac{1}{3} \times 3\right) = 53.35 \text{ kNm}$$

$$M_{\text{at} E} = 0$$

To find maximum bending moment:

BM is maximum at SF = 0

SF at $x = R_A - \frac{1}{2}(w_x \cdot x) = 0$

From similar triangles at UVL,

$$\frac{w_x}{x} = \frac{w}{l}$$

$$\therefore \quad w_x = \frac{w}{l}.x = 60.\frac{x}{3} = 20\,x$$

$$\therefore \text{ SF at } \quad x = 46.67 - \frac{1}{2} \times 20\, x.x = 0$$

$$46.67 = 10\, x^2$$

$$x = 2.16 \text{ m from } B$$

$$M_{max} = M_x = 46.67\,(1 + 2.16) - \left(\frac{1}{2} \times 20 \times 2.16^2\right)\left(\frac{1}{3} \times 2.16\right)$$

$$= 113.88 \text{ kNm}$$

(e) Draw BMD as shown in Fig. 4.61 (c). For UVL at *BC*, BMD is cubic curve and for UDL at *DE*, it is parabolic.

Problem 35 Draw the shear force and bending moment diagram for a simply supported beam as shown in Fig. 4.62.

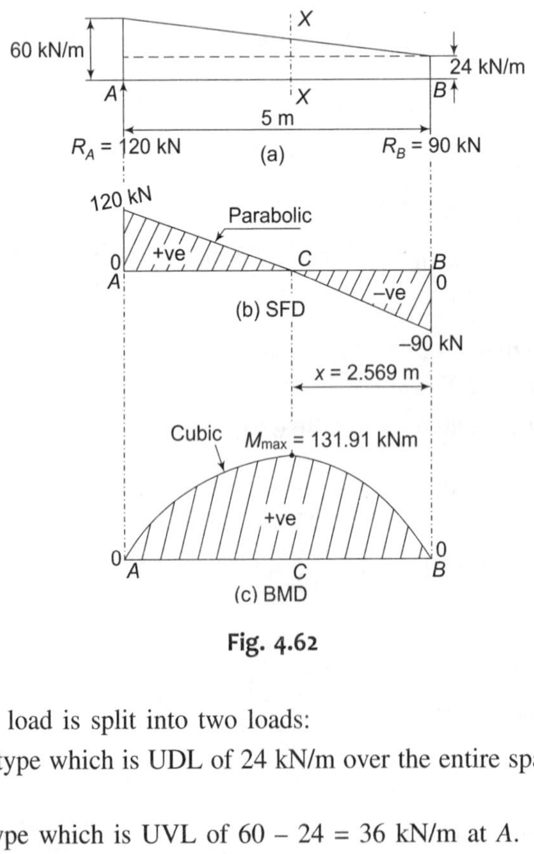

Fig. 4.62

Solution: The given load is split into two loads:

(i) a rectangular type which is UDL of 24 kN/m over the entire span as shown in dotted line, and

(ii) a triangular type which is UVL of 60 – 24 = 36 kN/m at *A*.

(a) **Reactions:** $\Sigma V = 0 \Rightarrow R_A + R_B = (24 \times 5) + \left(\frac{1}{2} \times 36 \times 5\right)$

$$R_A + R_B = 210 \text{ kN} \qquad \ldots(1)$$

Taking moments about *A*,

$$-(24 \times 5)\frac{5}{2} - \left(\frac{1}{2} \times 36 \times 5\right)\left(\frac{1}{3} \times 5\right) + R_B \times 5 = 0$$

$$R_B = 90 \text{ kN}$$

$$\therefore \quad R_A = 210 - R_B = 210 - 90 = 120 \text{ kN}$$

(b) **Shear forces:**

$$SF_{\text{up to } A} = 0$$

$$SF_{\text{at } A} = R_A = 120 \text{ kN}$$

$$SF_{\text{up to } B} = 120 - (24 \times 5) - \left(\frac{1}{2} \times 36 \times 5\right) = -90 \text{ kN}$$

$$SF_{\text{at } B} = -90 + R_B = 0$$

at x, at a distance x from right side support B

$$\text{UVL} = w_x = \frac{w \cdot x}{l} = \frac{36.x}{5} = 7.2x$$

SF at $\quad x = -90 + (24.\ x) + \left(\frac{1}{2} \times 7.2\ x \cdot x\right)$

This is parabolic in nature.

(c) Draw SFD as shown in Fig. 4.62 (b).

(d) **Bending moments:**

$$M_{\text{at}A} = 0$$

$$M_{\text{at}B} = 0$$

$$M_{\text{at}x} = 90 \cdot x - (24 \cdot x)\frac{x}{2} - \left(\frac{1}{2}\ w_x \cdot x\right)\left(\frac{1}{3} \cdot x\right)$$

$$= 90x - 12x^2 - \frac{1}{2}(7.2x \cdot x)\frac{1}{3}x \text{ (considering to the right side)}$$

$$M_x = 90x - 12x^2 - 1.2x^3$$

This is cubic variation between A, and B.

To find maximum bending moment, Equate $SF_{\text{at}x} = 0$

$$SF_{\text{at}x} = -90 + 24x + \left(\frac{1}{2} \cdot 72\ x^2\right) = 0$$

$$3.6x^2 + 24x - 90 = 0$$

$$x = \frac{-24 \pm \sqrt{24^2 - 4 \times 3.6\,(-90)}}{2 \times 3.6} = 2.676 \text{ m from } B$$

$$\therefore \quad M_{max} = M_x = 90 \times 2.676 - 12 \times 2.676^2 - 1.2 \times 2.676^3$$

$$= 131.91 \text{ kNm.}$$

(e) Draw BMD as shown in Fig. 4.62 (c). It is cubic curve for the given load.

Problem 36 Draw shear force and bending moment diagram for a simply supported beam loaded as shown in Fig. 4.63. Indicate the values on them.

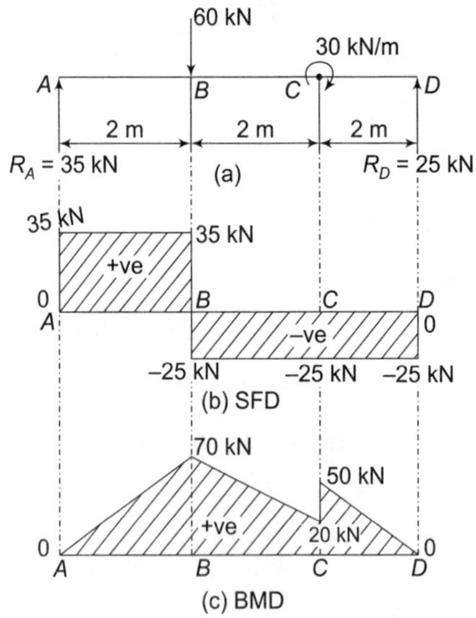

Fig. 4.63

Solution:

(a) **Reactions:** $\Sigma V = 0 \Rightarrow R_A + R_D = 60 \text{ kN}$ ...(1)

Taking moments about A, $-60 \times 2 - 30 + R_D \times 6 = 0$

$$\therefore \quad R_D = \frac{150}{6} = 25 \text{ kN}$$

$$R_A = 60 - 25 = 35 \text{ kN}$$

(b) **Shear forces:**

$$SF_{\text{up to } A} = 0$$
$$SF_{\text{at } A} = R_A = 35 \text{ kN}$$
$$SF_{\text{up to } B} = 35 \text{ kN}$$
$$SF_{\text{at } B} = 35 - 60 = -25 \text{ kN}$$
$$SF_{\text{at } C} = -25 \text{ kN}$$
$$SF_{\text{up to } D} = -25 \text{ kN}$$
$$SF_{\text{at } D} = -25 + R_D = 0$$

(c) Draw SFD as shown in Fig. 4.63 (b)

(d) **Bending moments:**

$$M_{\text{at} A} = 0$$
$$M_{\text{at} B} = 35 \times 2 = 70 \text{ kNm}$$
$$M_{\text{up to } C} = 35 \times 4 - 60 \times 2 = 20 \text{ kNm}$$
$$M_{\text{at} C} = 20 + M = 20 + 30 = 50 \text{ kNm}$$
$$M_{\text{at} D} = 0$$

(e) Draw BMD as shown in Fig. 4.63 (c). Draw a vertical line at C for the moment M = 30 kNm.

Problem 37 Draw shear force and bending moment diagrams for a beam as shown in Fig. 4.64. Indicate the values at salient points on the diagrams.

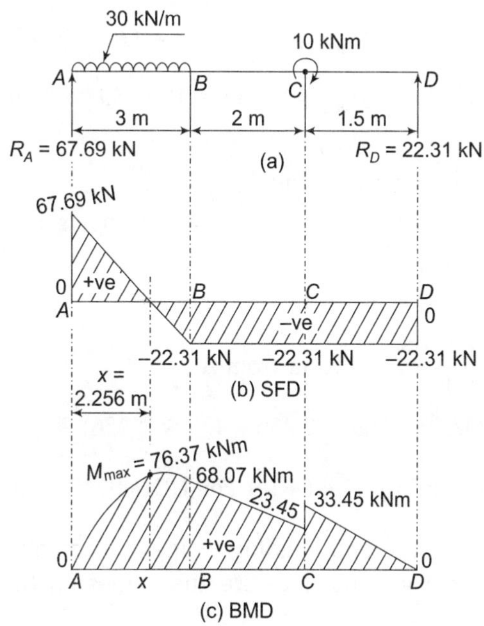

Fig. 4.64

Solution:

(a) **Reactions:** $R_A + R_D = 30 \times 3 = 90$ kN

Taking moment about A, $-(30 \times 3)\dfrac{3}{2} - 10 + R_D \times 6.5 = 0$

$$\therefore \quad R_D = \frac{145}{6.5} \; 22.31 \text{ kN}$$

$$\therefore \quad R_A = 90 - R_D = 67.69 \text{ kN}$$

(b) **Shear forces:**

$$SF_{\text{up to } A} = 0$$

$$SF_{\text{at } A} = 67.69 \text{ kN}$$

$$SF_{\text{at } B} = 67.69 - (30 \times 3) = -22.31 \text{ kN}$$

$$SF_{\text{at } C} = -22.31 \text{ kN}$$

$$SF_{\text{up to } D} = -22.31 \text{ kN}$$

$$SF_{\text{at } D} = -22.31 + R_D = 0$$

$$SF_{\text{at } x} = 67.69 - 30x$$

(c) Draw SFD as shown in Fig. 4.64 (b). For UDL at *AB*, SFD is triangular.

(d) **Bending moments:**

$$M_{\text{at}A} = 0$$

$$M_{\text{at}B} = 67.69 \times 3 - (30 \times 3)\,\frac{3}{2} = 68.07 \text{ kNm}$$

upto *C* (left side) $$M_{\text{at}C} = 69.69 \times 5 - (30 \times 3)\left(2 + \frac{3}{2}\right) = 23.45 \text{ kNm}$$

$$M_{\text{at}C} = 23.45 + 10 = 33.45 \text{ kNm}$$

$$M_{\text{at}D} = 0$$

To find maximum bending moment:

Bending moment is maximum, where SF = 0 changing its sign.

This is between *A* and *B*

$$\text{SF}_x = 67.69 - 30x = 0$$

$$x = \frac{67.69}{30} = 2.256 \text{ m from } A$$

$$\therefore \quad M_{\max} = M_x = 67.69 \times 2.256 - (30 \times 2.256)\,\frac{2.256}{2} = 76.37 \text{ kNm}$$

(e) Draw BMD as shown in Fig. 4.64 (c) For UDL BMD is parabolic.

Problem 38 Draw shear force and bending moment diagrams for a simply supported beam loaded as shown in Fig. 4.65. Indicate the values on them.

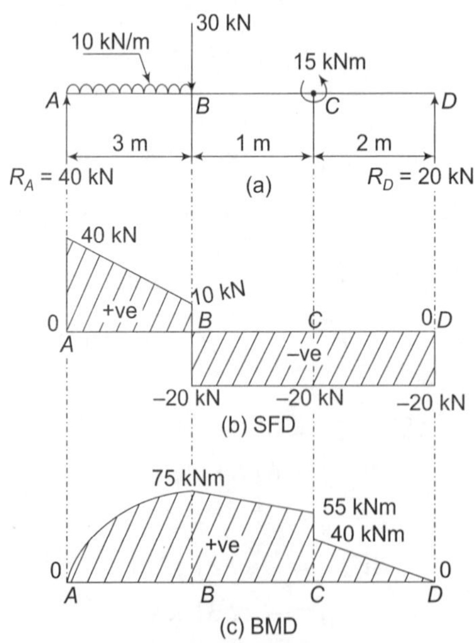

Fig. 4.65

Solution:

(a) **Reactions:** $\Sigma V = 0 \Rightarrow R_A + R_D = (10 \times 3) + 30 = 60$ kN

Taking moments about A, $-(10 \times 3)\frac{3}{2} - 30 \times 3 + 15 + R_D \times 6 = 0$

$$R_D = \frac{120}{6} = 20 \text{ kN}$$

$\therefore$ $$R_A = 60 - R_D = 40 \text{ kN}$$

(b) **Shear forces:**

$$\text{SF}_{\text{up to } A} = 0$$
$$\text{SF}_{\text{at } A} = 40 \text{ kN}$$
$$\text{SF}_{\text{up to } B} = 40 - 10 \times 3 = 10 \text{ kN}$$
$$\text{SF}_{\text{at } B} = 10 - 30 = -20 \text{ kN}$$
$$\text{SF}_{\text{at } C} = -20 \text{ kN}$$
$$\text{SF}_{\text{up to } D} = -20 \text{ kN}$$
$$\text{SF}_{\text{at } D} = -20 + R_D = 0$$

(c) The SFD is as shown in Fig. 4.65 (b) For UDL at AB, it is triangular

(d) **Bending moments:**

$$M_{\text{at}A} = 0$$
$$M_{\text{at}B} = 40 \times 3 - (10 \times 3)\frac{3}{2} = 75 \text{ kNm}$$

Just left of C, $$M_{\text{up to } C} = 40 \times 4 - (10 \times 3)\left(1 + \frac{3}{2}\right) - 30 \times 1 = 55 \text{ kNm}$$

$$M_{\text{at}C} = 55 - 15 = 40 \text{ kNm}$$
$$M_{\text{at}D} = 0$$

(e) Draw BMD as shown in Fig. 4.65 (c).

For UDL at AB, it is parabolic. For moment at C, it is a vertical line.

Problem 39 Draw shear force and bending moment diagrams for a simply supported beam loaded as shown in Fig. 4.66. Indicate the values on them.

Solution:

(a) **Reactions:** $\Sigma V = 0 \Rightarrow R_A + R_D = \frac{1}{2} \times 3 \times 30 = 45$ kN

Taking moments about A, $-\left(\frac{1}{2} \times 30 \times 3\right) \times \frac{2}{3} \times 3 - 20 + R_D \times 5 = 0$

$\therefore$ $$R_D = 22 \text{ kN}$$

$\therefore$ $$R_A = 45 - R_D = 23 \text{ kN}$$

(b) **Shear forces:**

$$\text{SF}_{\text{up to } A} = 0$$

$$SF_{\text{at } A} = 23 \text{ kN}$$

$$SF_{\text{at } B} = 23 - \left(\frac{1}{2} \times 30 \times 3\right) = -22 \text{ kN}$$

$$SF_{\text{at } C} = -22 \text{ kN}$$

$$SF_{\text{up to } D} = -22 \text{ kN}$$

$$SF_{\text{at } D} = -22 + R_D = 0$$

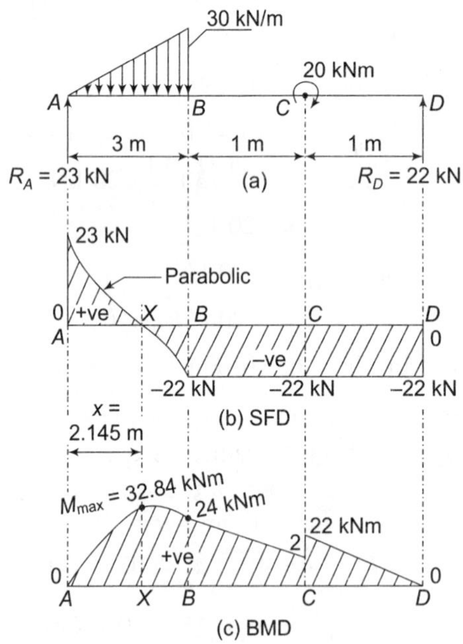

Fig. 4.66

(c) Draw SFD as shown in Fig. 4.66 (b). SFD is parabolic for UVL at *AB*.

(d) **Bending moments:**

$$M_{\text{at}A} = 0$$

$$M_{\text{at}B} = 23 \times 3 - \left(\frac{1}{2} \times 30 \times 3\right)\left(\frac{1}{3} \times 3\right) = 24 \text{ kNm}$$

Just to left of *C*,
$$M_{\text{up to } C} = 23 \times 4 - \left(\frac{1}{2} \times 30 \times 3\right)\left(\frac{1}{3} \times 3 + 1\right) = 2 \text{ kNm}$$

$$M_{\text{at}C} = 2 + 20 = 22 \text{ kNm}$$

$$M_{\text{at}D} = 0$$

Maximum bending moment:

Bending moment is maximum at SF = 0, changing its sign

$$SF_{\text{at } x} = 23 - \left(\frac{1}{2} \times w_x \cdot x\right) = 0$$

For UVL, from similar triangular

$$\frac{w_x}{x} = \frac{w}{l}$$

$$\therefore \quad w_x = \frac{w}{l}\cdot x = \left(\frac{30}{3}\right)\cdot x = 10x$$

$$\therefore \quad SF_x = 23 - \frac{1}{2}\cdot 10x\cdot x = 0$$

$$5x^2 = 23$$

$$x = \sqrt{\frac{23}{5}} = 2.145 \text{ m from } A$$

$$\therefore \quad M_{max} = M_x = 23\cdot x - \left(\frac{1}{2}\cdot w_x\cdot x\right)\left(\frac{1}{3}\cdot x\right)$$

$$= 23 \times 2\cdot 145 - \left(\frac{1}{2}\ 10\ x\cdot x\right)\frac{x}{3}$$

$$= 23 \times 2.145 - \frac{1}{2} \times 10 \times \frac{2.145^3}{3} = 32.84 \text{ kNm.}$$

(e) Draw BMD as shown in Fig. 4.66 (c). BMD for UVL it AB cubic curve. For moment at C, draw a vertical line.

Problem 40 Draw shear force and bending moment diagram for a simply supported beam loaded as shown in Fig. 4.67. Indicate the values on them.

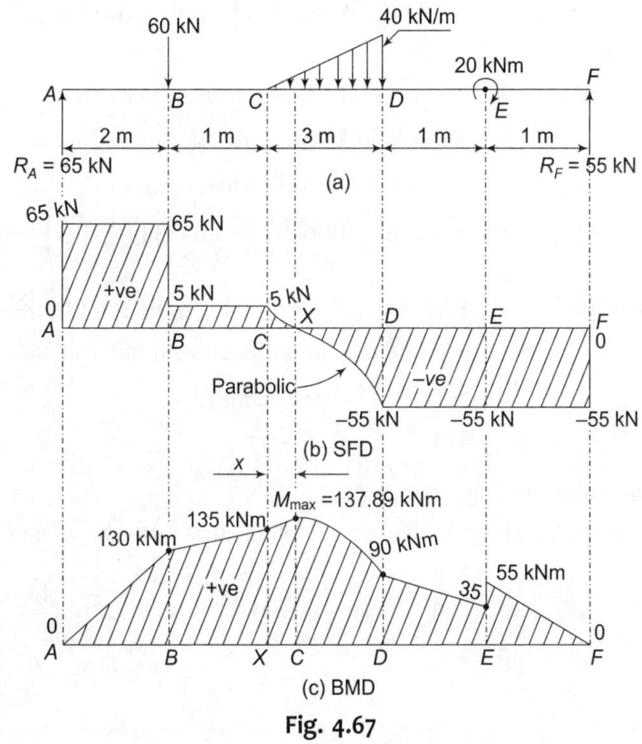

Fig. 4.67

Solution:

(a) **Reactions:** $\Sigma V = 0 \Rightarrow R_A + R_F = 60 + \left(\frac{1}{2} \times 40 \times 3\right) = 120$ kN ...(1)

Taking moment about A, $-60 \times 2 - \left(\frac{1}{2} \times 40 \times 3\right)\left(2 + 1 + \frac{2}{3} \times 3\right)$

$$- 20 + R_F \times 8 = 0$$

$$\therefore \quad R_F = \frac{440}{8} = 55 \text{ kN}$$

$$\therefore \quad R_A = 120 - R_F = 65 \text{ kN}$$

(b) **Shear forces:**

$$SF_{\text{up to } A} = 0$$

$$SF_{\text{at } A} = 65 \text{ kN}$$

$$SF_{\text{up to } B} = 65 \text{ kN}$$

$$SF_{\text{at } B} = 65 - 60 = 5 \text{ kN}$$

$$SF_{\text{at } C} = 5 \text{ kN}$$

$$SF_{\text{at } D} = 5 - \left(\frac{1}{2} \times 40 \times 3\right) = -55 \text{ kN}$$

$$SF_{\text{at } E} = -55 \text{ kN}$$

$$SF_{\text{up to } F} = -55 \text{ kN}$$

$$SF_{\text{at } F} = -55 + R_F = 0$$

(c) Draw SFD as shown in Fig. 4.67 (b). For UVL at CD, SFD is parabolic

(d) **Bending moments:**

$$M_{\text{at}A} = 0$$

$$M_{\text{at}B} = 65 \times 2 = 130 \text{ kNm}$$

$$M_{\text{at}C} = 65 \times 3 - 60 \times 1 = 135 \text{ kNm}$$

$$M_{\text{at}D} = 65 \times 6 - 60 \times 4 - \left(\frac{1}{2} \times 40 \times 3\right)\left(\frac{1}{3} \times 3\right) = 90 \text{ kNm}$$

at just left of E, $M_{\text{up to } E} = 65 \times 7 - 60 \times 5 - \left(\frac{1}{2} \times 40 \times 3\right)\left(\frac{1}{3} \times 3 + 1\right)$

$$= 35 \text{ kNm}$$

$$M_{\text{at}E} = 35 + 20 = 55 \text{ kNm}$$

$$M_{\text{at}F} = 0$$

To find maximum bending moment.

BM is maximum at SF = 0, changing its sign. It is between C and D at a distance x from C.

$$SF_x = 65 - 60 - \frac{1}{2}.(w_x.x) = 0$$

For UVL $\quad \dfrac{w_x}{x} = \dfrac{w}{l}$

$$\therefore \quad w_x = \frac{w}{l} \cdot x = \frac{40}{3} x$$

$$\therefore \quad SF_x = 65 - 60 - \frac{1}{2} \times \frac{40}{3} x.x = 0$$

$$20 \frac{x^2}{3} = 65 - 60$$

$$x^2 = \frac{3 \times 5}{20} = \frac{3}{4}$$

$$x = 0.866 \text{ m from } C$$

$$\therefore \quad M_{max} = M_x = 65(2 + 1 + x) - 60(1 + x) - \left(\frac{1}{2} \times w_x \cdot x\right) \frac{1}{3} x$$

$$M_{max} = 65\ (2 + 1 + 0.866) - 60\ (1 + 0.866) - \frac{1}{2} \times \frac{40}{3} \times \frac{0.866^3}{3}$$

$$= 137.89 \text{ kNm}$$

(e) Draw BMD as shown in Fig. 4.67 (c). For UVL at *CD* BMD is cubic curve. For moment at *E*, draw a vertical line M = 20 kNm.

Problem 41 Draw shear force and bending moment diagrams for a simply supported beam shown in Fig. 4.68. Indicate the values on them.

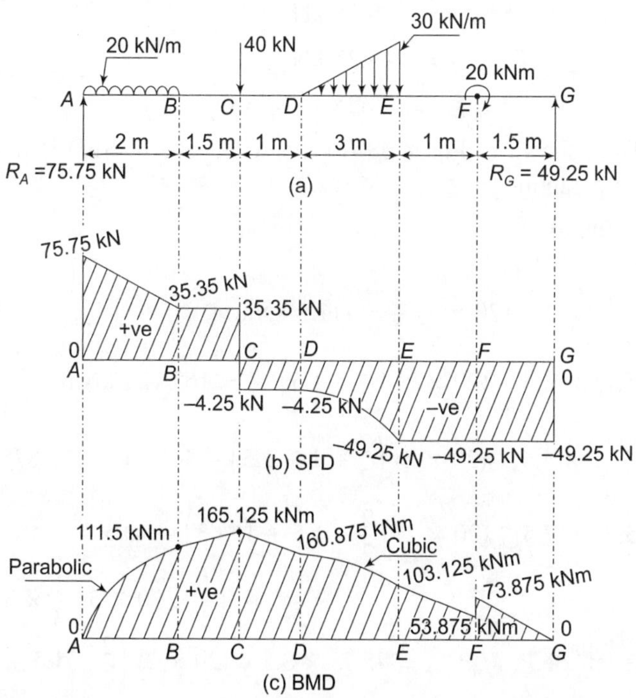

Fig. 4.68

Solution:

(a) **Reactions:** $\Sigma V = 0 \Rightarrow R_A + R_G = (20 \times 2) + 40 + \frac{1}{2} \times 3 \times 30 = 125$ kN ...(1)

Taking moments about A,

$$-(20 \times 2)\frac{2}{2} - 40 \times 3.5 - \left(\frac{1}{2} \times 30 \times 3\right)\left(2 + 1.5 + 1 + \frac{2}{3} \times 3\right) - 20 + R_G \times 10 = 0$$

$$\therefore \qquad R_G = 49.25 \text{ kN}$$

$$R_A = 125 - 49.25 = 75.75 \text{ kN}$$

(b) **Shear forces:**

$$SF_{\text{up to } A} = 0$$

$$SF_{\text{at } A} = 75.75 \text{ kN}$$

$$SF_{\text{at } B} = 75.75 - 20 \times 2 = 35.75 \text{ kN}$$

$$SF_{\text{up to } C} = 35.75 \text{ kN}$$

$$SF_{\text{at } C} = 35.75 - 40 = -4.25 \text{ kN}$$

$$SF_{\text{at } D} = -4.25 \text{ kN}$$

$$SF_{\text{at } E} = -4.25 - \left(\frac{1}{2} \times 30 \times 3\right) = -49.25 \text{ kN}$$

$$SF_{\text{at } F} = -49.25 \text{ kN}$$

$$SF_{\text{up to } G} = -49.25 \text{ kN}$$

$$SF_{\text{at } G} = -49.25 + R_G = 0$$

(c) Draw SFD as shown in Fig. 4.68 (b). For UDL at AB, SFD is triangular for UVL at DE, it is parabolic.

(d) **Bending moments:**

$M_{\text{at}A} = 0$

$M_{\text{at}B} = 75.75 \times 2 - (20 \times 2)\frac{2}{2} = 111.5$ kNm

$M_{\text{at}C} = 75.75 \times 3.5 - (20 \times 2)\left(\frac{2}{2} + 1.5\right) - = 165.125$ kNm

$M_{\text{at}D} = 75.75 \times 4.5 - (20 \times 2)\left(\frac{2}{2} + 1 + 1.5\right) - 40 \times 1 = 160.875$ kNm

$$M_{\text{at}E} = 75.75 \times 7.5 - (20 \times 2)\left(\frac{2}{2} + 1.5 + 1 + 3\right) - 40 \times 4 - \left(\frac{1}{2} \times 30 \times 3\right)\frac{1}{3} \times 3 = 103.125 \text{ kNm}$$

Just to the left of F, $M_{\text{up to } F} = 75.75 \times 8.5 - (20 \times 2)\left(\frac{2}{2} + 1.5 + 1 + 3 + 1\right) - 40 \times 5 - \left(\frac{1}{2} \times 30 \times 3\right)\left(\frac{1}{3} \times 3 + 1\right) = 53.875$ kNm

$M_{at\,F} = 53.875 + 20 = 73.875$ kNm

$M_{at\,G} = 0$

To find maximum bending moment: BM is maximum at SF = 0, changing its sign, i.e., at C

(e) Draw BMD as shown in Fig. 4.68 (c).

For UDL at AB, BMD is parabolic and for UVL at DE, it is cubic.

Problem 42 Draw shear forces and bending moment diagrams for a simply supported beam loaded as shown in Fig. 4.69. Indicate the values at salient points on them.

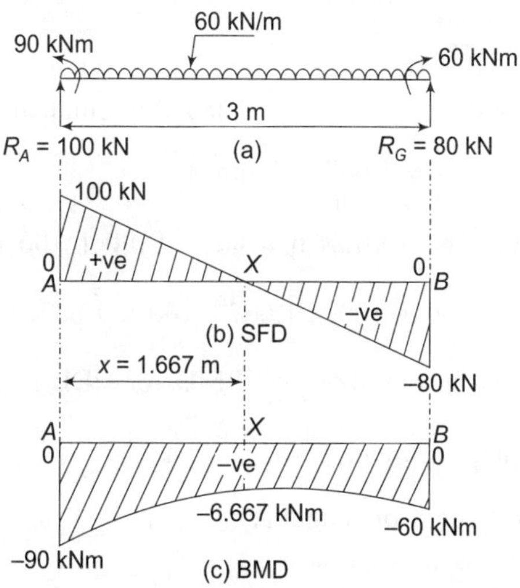

Fig. 4.69

Solution:

(a) **Reactions:** $\Sigma V = 0 \Rightarrow R_A + R_B = 60 \times 3 = 180$ kN ...(1)

Taking moments about A, $90 - (60 \times 3)\dfrac{3}{2} - 60 + R_B \times 3 = 0$

$$R_B = 80 \text{ kN}$$

$$\therefore \quad R_A = 180 - 80 = 100 \text{ kN}$$

(b) **Shear forces:**

$$SF_{\text{up to } A} = 0$$

$$SF_{\text{at } A} = R_A = 100 \text{ kN}$$

$$SF_{\text{up to } B} = 100 - 60 \times 3 = -80 \text{ kN}$$

$$SF_{\text{at } B} = -80 + R_B = 0$$

(c) Draw SFD as shown in Fig. 4.69 (b). For UDL, SFD is triangular

(d) **Bending moments:**

To the left of A, $M_{\text{up to } A} = 0$

$$M_{\text{at}A} = -90 \text{ kNm}$$

Just to left of B, $M_{\text{up to } B} = -90 + 100 \times 3 - (60 \times 3)\dfrac{3}{2} = -60 \text{ kNm}$

$$M_{\text{at}B} = -60 + 60 = 0$$

To find maximum bending moment:

BM is maximum at SF = 0 at x at a distance x from A

From similar triangles of SFD

$$\frac{100}{x} = \frac{80}{3 - x}$$

$$100\,(3 - x) = 80\,x$$

$$x = \frac{300}{180} = 1.667 \; M \text{ from } A$$

$\therefore$ Maximum bending moment, $M_{\max} = M_x = -90 + 100\,x - (60 \cdot x)\dfrac{x}{2}$

$$M_{\max} = -90 + 100 \times 1.667 - (60 \times 1.667) \times \frac{1.667}{2} = -6.667 \text{ kNm}$$

(e) Draw BMD as shown in Fig. 4.69 (c). BMD for UDL is parabolic.

4.10 PURE BENDING

Consider simply supported beam of length l, carrying two identical loads W, each at a distance of a from each of two supports as shown in Fig. 4.70.

(a) **Reactions:** $R_A = R_D = W$ (Symmetry)

(b) **Shear forces:**

$$SF_{\text{up to } A} = 0$$

$$SF_{\text{at } A} = W$$

$$SF_{\text{up to } B} = W$$

$$SF_{\text{at } B} = W - W = 0$$

$$SF_{\text{up to } C} = 0$$

$$SF_{\text{at } C} = -W$$

$$SF_{\text{up to } D} = -W$$

$$SF_{\text{at } D} = -W + W = 0$$

(c) Draw SFD as shown in Fig. 4.70 (b)

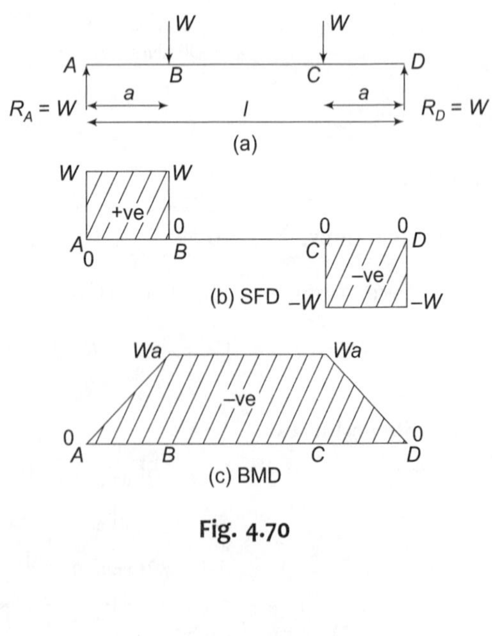

Fig. 4.70

(d) **Bending moments:**

$$M_{\text{at}A} = 0$$
$$M_{\text{at}B} = W \cdot a$$
$$M_{\text{at}C} = W(l - a) - W(l - 2a) = Wa$$
$$M_{\text{at}D} = 0$$

(e) Draw BMD as shown in Fig. 4.70 (c)

It is seen that SF between B and C is zero and BM between B and C is constant at $(W \cdot a)$. The region between B and overhang is subjected to only bending moment and the shear force is zero and is called case of "PURE BENDING"

4.11 OVERHANGING BEAMS

A simply supported beam which extends beyond the support on one side or both sides is called overhanging beam. Some portion of the beam overhangs beyond the support. Figure 4.71 shows some examples of overhanging beams.

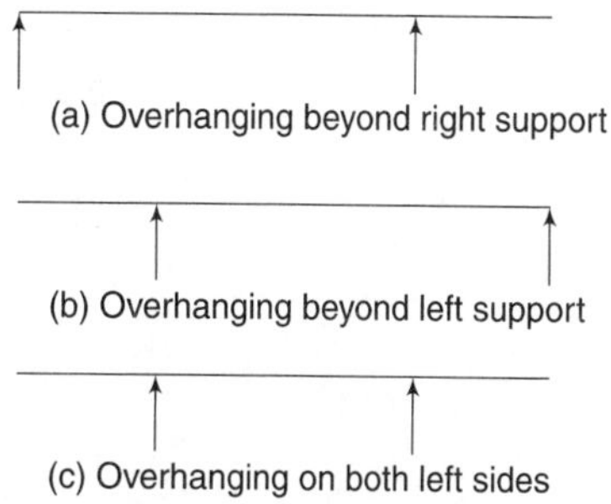

Fig. 4.71 Overhanging beams

4.12 POINT OF CONTRAFLEXURE

For some simply supported overhanging beams, the bending moment changes its sign from positive to negative.

The point on the beam where the bending moment is zero and changing its sign from positive to negative is called point of contraflexure.

Some examples of point of contraflexure are shown in Figs. 4.72 and 4.73.

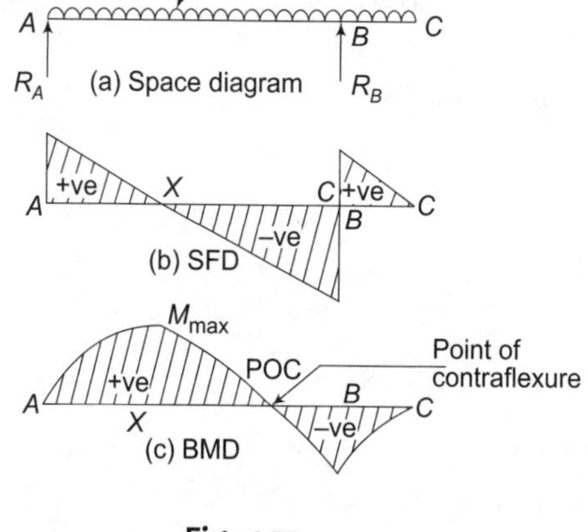

Fig. 4.72

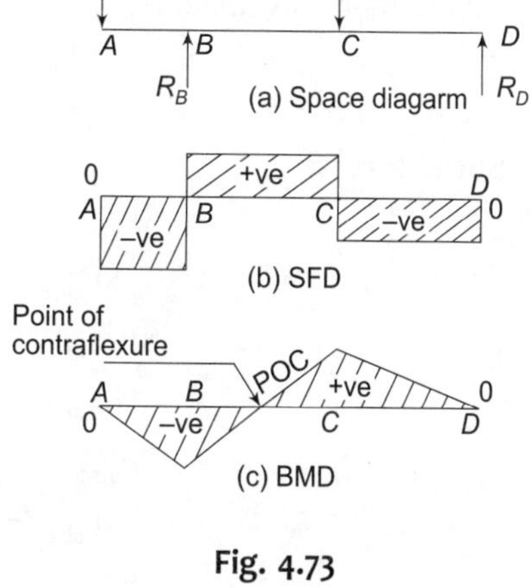

Fig. 4.73

PROBLEMS ON OVERHANGING BEAMS

Problem 43 Draw the shear force and bending moment diagrams for a beam loaded as shown in Fig. 4.74. Mark the values of SF and BM at various points. Locate the points of contraflexure (PoC) if any.

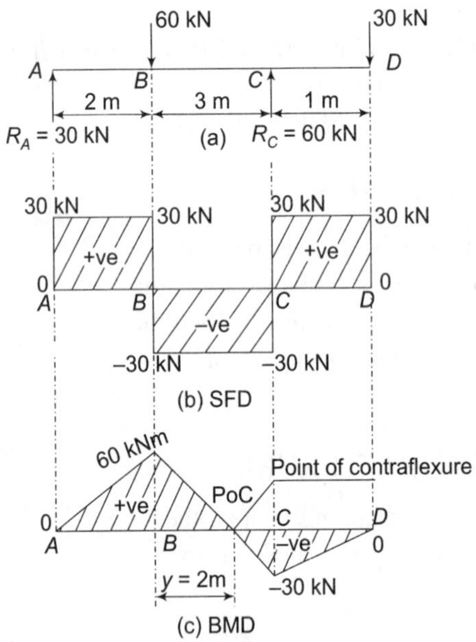

Fig. 4.74

(a) **Reactions:** $\Sigma V = 0 \Rightarrow R_A + R_C = 60 + 30 = 90$ kN ...(1)

Taking moments about A, $-60 \times 2 + R_C \times 5 - 30 \times 6 = 0$

$$\therefore \quad R_C = 60 \text{ kN}$$

$$R_A = 90 - R_C = 30 \text{ kN}$$

(b) **Shear forces:**

$$SF_{\text{up to } A} = 0$$

$$SF_{\text{at } A} = 30 \text{ kN}$$

$$SF_{\text{up to } B} = 30 \text{ kN}$$

$$SF_{\text{at } B} = 30 - 60 = -30 \text{ kN}$$

$$SF_{\text{up to } C} = -30 \text{ kN}$$

$$SF_{\text{at } C} = -30 + 60 = 30 \text{ kN}$$

$$SF_{\text{up to } D} = 30 \text{ kN}$$

$$SF_{\text{at } D} = 30 - 30 = 0 \text{ kN}$$

(c) Draw SFD as shown in Fig. 4.74 (b)

(d) **Bending moments:**

$$M_{\text{at}A} = 0$$
$$M_{\text{at}B} = R_A \times 2 = 30 \times 2 = 60 \text{ kNm}$$
$$M_{\text{at}C} = 30 \times 5 - 60 \times 3 = -30 \text{ kNm}$$
$$M_{\text{at}D} = 0$$

(e) Draw BMD as shown n Fig. 4.74 (c)

(f) To locate point contraflexure (PoC)

At PoC, $\text{BM} = 0$

$$M_{\text{at PoC}} = 30 \times (2 + y) - 60\,(y) = 0$$
$$60 + 30\,y - 60\,y = 0$$
$$y = \frac{60}{30} = 2 \text{ m from } B$$

Point of contraflexure is at 2 m from *B*.

Problem 44 Draw shear force and bending moment diagrams for a beam shown in Fig. 4.75. Locate point of contraflexure if any. Mark the values of SF and BM at various points.

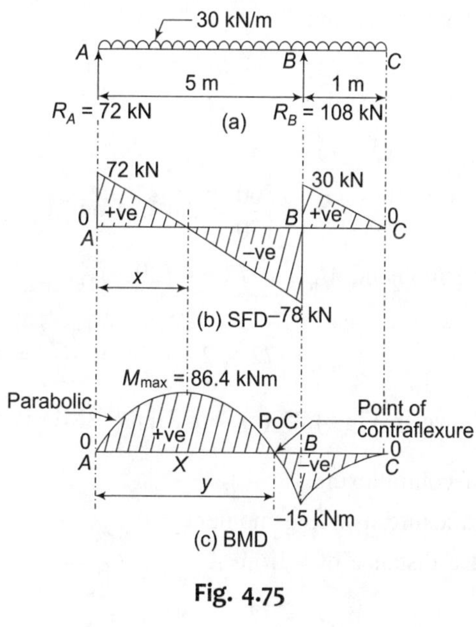

Fig. 4.75

(a) **Reactions:** $\Sigma V = 0 \Rightarrow R_A + R_B = 30 \times 6 = 180$ kN ...(1)

Taking moments about *A*, $-(30 \times 6)\,\dfrac{6}{2} + R_B \times 5 = 0$

$$\therefore \quad R_{\text{at}B} = \frac{540}{5} = 108 \text{ kN}$$

$$\therefore \quad R_{\text{at}A} = 180 - R_B = 72 \text{ kN.}$$

(b) **Shear forces**

$$\text{SF}_{\text{up to } A} = 0$$

$$\text{SF}_{\text{at } A} = R_A = 72 \text{ kN}$$

$$\text{SF}_{\text{up to } B} = 72 - 30 \times 5 = -78 \text{ kN}$$

$$\text{SF}_{\text{at } B} = -78 + 108 = 30 \text{ kN}$$

$$\text{SF}_{\text{at } B} = -78 + R_B = -78 + 108 = 30 \text{ kN}$$

$$\text{SF}_{\text{at } C} = 30 - (30 \times 1) = 0$$

(c) Draw SFD as shown in Fig. 4.75 (b). For UDL, SFD is triangular.

(d) **Bending moments:**

$$M_{\text{at}A} = 0$$

$$M_{\text{at}B} = 72 \times 5 - (30 \times 5)\frac{5}{2} = -15 \text{ kNm}$$

$$M_{\text{at}C} = 0$$

To find maximum bending moment.

BM is maximum where SF = 0, changing its sign, say at x from A. From similar triangles of SFD between A and B,

$$\frac{72}{x} = \frac{78}{5 - x}$$

$$72\,(5 - x) = 78x$$

$$\therefore \quad x = \frac{360}{150} = 2.4 \text{ m from } A$$

Maximum bending moment, $M_{\max} = 72.x - (30.x)\dfrac{x}{2}$

$$M_{\max} = 72 \times 2.4 - 30 \times \frac{2.4^2}{2} = 86.4 \text{ kNm.}$$

(e) Draw BMD as shown in Fig. 4.75 (c). For UDL, BMD is parabolic

(f) To locate point of contraflexure

At point of contraflexure, bending moment = 0

Let the PoC is at a distance of y from A.

$$M_{\text{PoC}} = 72y - (30 \times y)\frac{y}{2} = 0$$

$$72y = 15y^2$$

$$\therefore \quad y = \frac{72}{15} = 4.8 \text{ m from } A$$

Point of contraflexure is at a distance of 4.8 m from A.

Problem 45 Draw shear force and bending moment diagrams for a beam shown in Fig. 4.76. Locate the point of contraflexure if any. Marks the values.

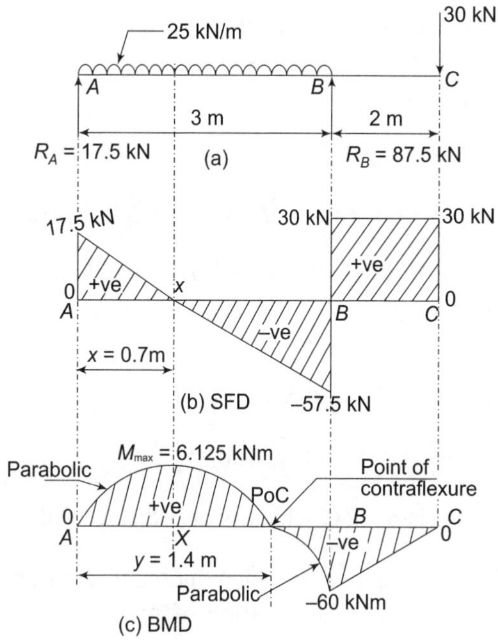

Fig. 4.76

Solution:

(a) **Reactions:** $\Sigma V = 0 \Rightarrow R_A + R_B = (25 \times 3) + 30 = 105$ kN ...(1)

Taking moments about A, $-(25 \times 3) \times \frac{3}{2} + R_B \times 3 - 30 \times 5 = 0$

$$\therefore \quad R_B = 87.5 \text{ kN}$$

$$\therefore \quad R_A = 105 - R_B = 17.5 \text{ kN}$$

(b) **Shear forces:**

$$SF_{\text{up to } A} = 0$$
$$SF_{\text{at } A} = R_A = 17.5 \text{ kN}$$
$$SF_{\text{up to } B} = 17.5 - 25 \times 3 = -57.5 \text{ kN}$$
$$SF_{\text{at } B} = -57.5 + R_B = -57.5 + 87.5 = 30 \text{ kN}$$
$$SF_{\text{up to } C} = 30 \text{ kN}$$
$$SF_{\text{at } C} = 30 - 30 = 0$$

(c) Draw SFD as shown in Fig. 4.76 (b). UDL, SFD is triangular.

(d) **Bending moments:**

$$M_{\text{at} A} = 0$$

$$M_{\text{at}B} = R_A \times 3 - (25 \times 3)\,\frac{3}{2} = 17.5 \times 3 - (25 \times 3)\,\frac{3}{2} = -\,60 \text{ kNm}$$

$$M_{\text{at}C} = 0$$

To find maximum bending moment

BM is maximum where SF = 0, changing its sign,

$$\text{SF}_{\text{at}x} = 17.5 - 25 \times x = 0$$

$$\therefore \quad x = 0.7 \text{ m from } A$$

$$\therefore \text{ Maximum} \quad \text{BM} = M_{\text{max}} = M_x = 17.5 \times x - (25 \times x)\,\frac{x}{2}$$

$$M_{\text{max}} = 17.5 \times 0.7 - 25 \times 0.7 \times \frac{0.7}{2} = 6.125 \text{ kNm}$$

(e) Draw BMD as shown in Fig. 4.76 (c). BMD is parabolic for UDL between *AB*.

(f) To locate point of contraflexure: It is seen from BMD that the BM is zero at PoC changing its sign.

$$M_{\text{PoC}} = 17.5\,y - (25 \times y)\,\frac{y}{2} = 0$$

$$\therefore \quad y = \frac{17.5}{12.5} = 1.4 \text{ m from } A$$

Point of contraflexure is at PoC which is at a distance of 1.4 m from *A*.

Problem 46 Draw shear force and bending moment diagrams for a beam loaded as shown in Fig. 4.77. Determine the location of maximum bending moment and its magnitude. Also locate the point of contraflexure if any.

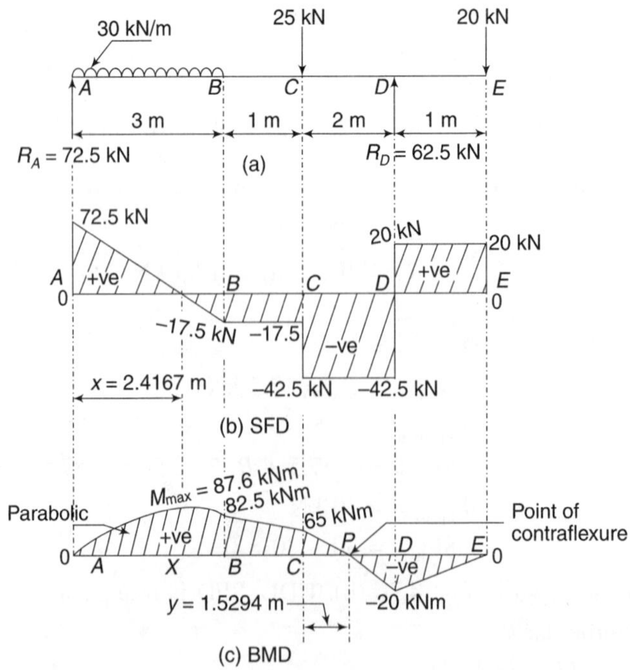

Fig. 4.77

Solution:

(a) **Reactions:** $\Sigma V = 0 \Rightarrow R_A - (30 \times 3) - 25 + R_D - 20 = 0$

$$R_A + R_D = 135 \text{ kN} \quad ...(1)$$

Taking moment about A, $-(30 \times 3)\frac{3}{2} - 25 \times 4 + R_D \times 6 - 20 \times 7 = 0$

$$\therefore \quad R_D = 62.5 \text{ kN}$$

$$R_A = 135 - R_D = 72.5 \text{ kN}$$

(b) **Shear forces:**

$$SF_{\text{up to } A} = 0$$
$$SF_{\text{at } A} = R_A = 72.5 \text{ kN}$$
$$SF_{\text{at } B} = 72.5 - (30 \times 3) = -17.5 \text{ kN}$$
$$SF_{\text{up to } C} = -17.5 \text{ kN}$$
$$SF_{\text{at } C} = -17.5 - 25 = -42.5 \text{ kN}$$
$$SF_{\text{up to } D} = -42.5 \text{ kN}$$
$$SF_{\text{at } D} = -42.5 + 62.5 = +20 \text{ kN}$$
$$SF_{\text{up to } E} = 20 \text{ kN}$$
$$SF_{\text{at } E} = 20 - 20 = 0$$

(c) Draw SFD as shown in Fig. 4.77 (b). For UDL at AB SFD is triangular

(d) **Bending moments:**

$$M_{\text{at}A} = 0$$
$$M_{\text{at}B} = 72.5 \times 3 - (30 \times 3)\frac{3}{2} = 82.5 \text{ kNm}$$
$$M_{\text{at}C} = 72.5 \times 4 - (30 \times 3)\left(\frac{3}{2} + 1\right) = 65 \text{ kNm}$$
$$M_{\text{at}D} = 72.5 \times 6 - (30 \times 3)\left(\frac{3}{2} + 2 + 1\right) - 25 \times 2 = -20 \text{ kN}$$
$$M_{\text{at}E} = 0$$

To find maximum bending moment:

BM is maximum at x where SF = 0, changing its sign (between A and B)

$$SF_x = 72.5 - (30.x) = 0$$

$$\therefore \quad x = \frac{72.5}{30} = 2.4167 \text{ m from } A.$$

$$\therefore \quad M_{\max} = M_x = 72.5.x - (30.x)\frac{x}{2} = 72.5 \times 2.4167 - 30 \times \frac{2.4167^2}{2}$$

$$M_{\max} = 87.6 \text{ kNm}$$

(e) Draw BMD as shown in Fig. 4.77 (c). For UDL at AB, BMD is parabolic.

(f) To locate point of contraflexure:

At point of contraflexure, BM is zero, changing its sign. Let this distance = y from C.

From similar triangles of BMD between CD

$$\frac{65}{y} = \frac{20}{2-y}$$

$$65\,(2-y) = 20y$$

$$y = \frac{65 \times 2}{85} = 1.5294 \text{ m from } C$$

Point of contraflexure at 1.5294 from C.

Problem 47 Draw shear force and bending moment diagram for a beam loaded as shown in the Fig. 4.78. Locate point of contraflexure and calculate maximum bending moment.

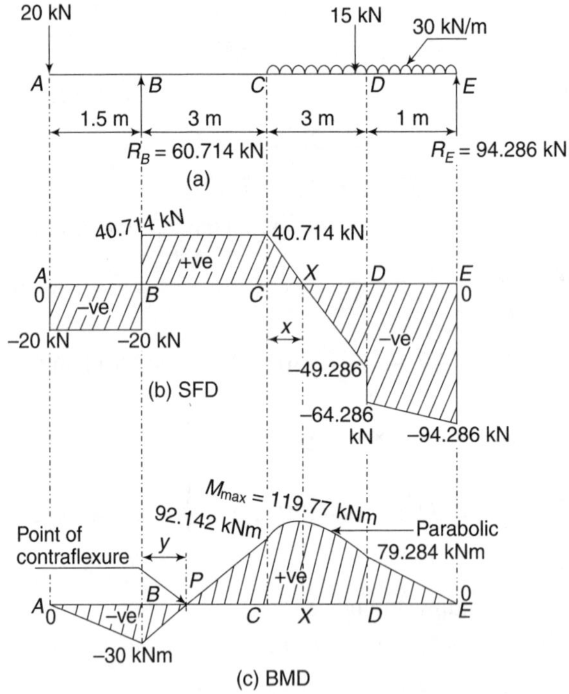

Fig. 4.78

Solution:

(a) **Reactions:** $\Sigma V = 0 \Rightarrow R_B + R_E = 20 + 15 + (30 \times 4) = 155$ kN ...(1)

Taking moments about B, $-\,20 \times 1.5 + 15 \times 6 + (30 \times 4)\left(3 + \frac{4}{2}\right) - R_E \times 7 = 0$

$$\therefore \qquad R_E = 94.286 \text{ kN}$$

$$R_B = 155 - R_E = 60.714 \text{ kN}$$

(b) **Shear forces:**

$$SF_{\text{up to } A} = 0$$

$$SF_{\text{at } A} = -20 \text{ kN}$$

$$SF_{\text{up to } B} = -20 \text{ kN}$$

$$SF_{\text{at } B} = -20 + 60.714 = 40.714 \text{ kN}$$

$$SF_{\text{at } C} = 40.714 \text{ kN}$$

$$SF_{\text{up to } D} = 40.714 - (30 \times 3) = -49.286 \text{ kN}$$

$$SF_{\text{at } D} = -49.286 - 15 = -64.286 \text{ kN}$$

$$SF_{\text{up to } E} = -64.286 - (30 \times 1) = -94.286 \text{ kN}$$

$$SF_{\text{at } E} = -94.286 + R_E = 0$$

(c) Draw SFD as shown in Fig. 4.78 (b). For UDL at *CD* and *DE* SFD is triangular.

(d) **Bending moments:**

$$M_{\text{at}A} = 0$$

$$M_{\text{at}B} = -20 \times 1.5 = -30 \text{ kNm}$$

$$M_{\text{at}C} = -20 \times 4.5 + 60.714 \times 3 = 92.142 \text{ kNm}$$

$$M_{\text{at}D} = -20 \times 7.5 + 60.714 \times 6 - (30 \times 3)\frac{3}{2} = 79.284 \text{ kNm}$$

$$M_{\text{at}E} = 0$$

To find maximum bending moment

BM is maximum between *C* and where SF = 0, changing its sign.

Let $CX = x$. Consider SF between *CD*. For similar triangles

$$\frac{40.714}{x} = \frac{49.286}{3 - x}$$

$$40.714\,(3 - x) = 49.286\,x$$

$$x = \frac{40.714 \times 3}{90} = 1.357 \text{ m}$$

$$\therefore \text{ Max. BM} = M_{\max} = M_x = -20\,(1.5 + 3 + x) + R_B\,(3 + x) - (30 \times x)\frac{x}{2}$$

$$M_{\max} = -20\,(1.5 + 3 + 1.357) + 60.714\,(3 + 1.357) - (30 \times 1.357)\frac{1.357}{2}$$

$$= 119.77 \text{ kNm}$$

(e) Draw BMD as shown in Fig. 4.78 (c). For UDL, BMD is parabolic

(f) To locate point of contraflexure:

At point of contraflexure, BM = 0, between B and C.

Moment at PoC, $$M_P = -20 \times (1.5 + y) + 60.714 \times y = 0$$

$$y = 0.737 \text{ m from } B.$$

(since the BMD is triangular between B and in this problem, distance y can be found using similar triangles

$$\frac{30}{y} = \frac{92.142}{3 - y} \Rightarrow y = 0.737 \text{ m from } B)$$

PoC is at a distance of 0.737 m from B.

Problem 48 Draw the shear force and bending moment diagrams for the beam shown in Fig. 4.79. Locate point of contraflexure if any. Indicate the values.

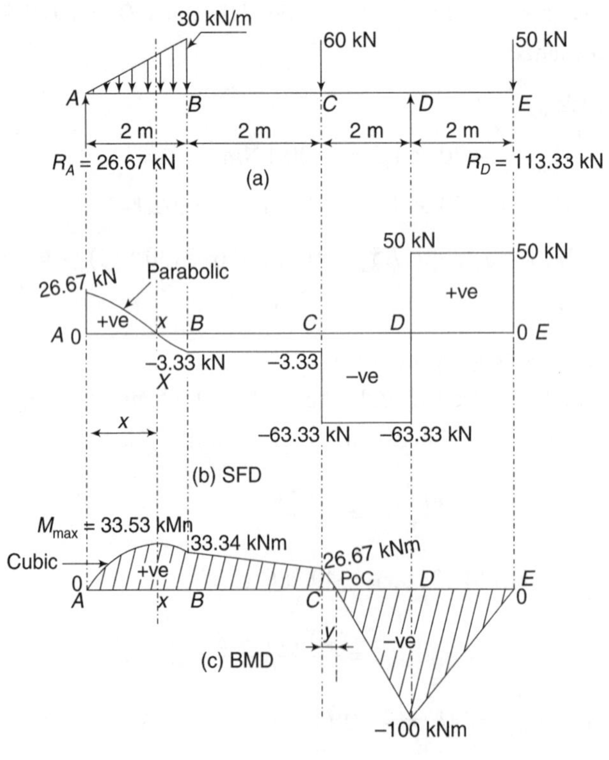

Fig. 4.79

Solution:

(a) **Reactions:** $\Sigma V = 0 \Rightarrow R_A + R_D = \left(\frac{1}{2} \times 2 \times 30\right) + 60 + 50 = 140$ kN

Taking moments about A, $-\left(\frac{1}{2} \times 30 \times 2\right)\left(\frac{2}{3} \times 2\right) - 60 \times 4 + R_D \times 6 - 50 \times 8 = 0$

$$\therefore \quad R_D = 113.33 \text{ kN}$$
$$R_A = 26.67 \text{ kN}$$

(b) **Shear forces:**

$$\text{SF}_{\text{up to } A} = 0$$
$$\text{SF}_{\text{at } A} = 26.67 \text{ kN}$$
$$\text{SF}_{\text{at } B} = 26.67 - \left(\frac{1}{2} \times 2 \times 30\right) = -3.33 \text{ kN}$$
$$\text{SF}_{\text{up to } C} = -3.33 \text{ kN}$$
$$\text{SF}_{\text{at } C} = -3.33 - 60 = -63.33 \text{ kN}$$
$$\text{SF}_{\text{up to } D} = -63.33 \text{ kN}$$
$$\text{SF}_{\text{at } D} = -63.33 + 113.33 = 50 \text{ kN}$$
$$\text{SF}_{\text{up to } E} = 50 \text{ kN}$$
$$\text{SF}_{\text{at } E} = 50 - 50 = 0$$

(c) Draw SFD as shown in Fig. 4.79 (b). For UVL at AB, SFD is parabolic

(d) **Bending moments:**

$$M_{\text{at}A} = 0$$
$$M_{\text{at}B} = 26.67 \times 2 - \left(\frac{1}{2} \times 2 \times 30\right)\frac{1}{3} \times 2 = 33.34 \text{ kNm}$$
$$M_{\text{at}C} = 26.67 \times 4 - \left(\frac{1}{2} \times 2 \times 30\right)\left(\frac{1}{3} \times 2 + 2\right) = 26.67 \text{ kNm}$$
$$M_{\text{at}D} = (26.76 \times 6) - \left(\frac{1}{2} \times 2 \times 30\right)\left(\frac{1}{3} \times 2 + 2 + 2\right) - 60 \times 2 = -100 \text{ kNm}$$
$$M_{\text{at}E} = 0$$

To find maximum bending moment at X, where SF = 0, Changing its sign

To find w_x at XX, from similar triangles of UVL

$$\frac{w_x}{x} = \frac{w}{l}$$

$$\therefore \quad w_x = \frac{w}{l} \cdot x = \frac{30}{2} \cdot x = 15x$$

$$\text{SF}_{\text{at } x} = 26.67 - \left(\frac{1}{2} \cdot w_x \cdot x\right) = 0$$

$$= 26.67 - \left(\frac{1}{2} \times 15x\ .x\right) = 0$$

$\therefore \quad x = 1.886$ m from A

$\therefore$ Bending moment at X,

$$M_x = M_{max} = 26.67 \times 1.886 - \left(\frac{1}{2} \times 15 \times 1.886 \times 1.886\right) \times \frac{2}{3} \times 1.886$$

$$= 33.53 \text{ kNm.}$$

(e) Draw BMD as shown in Fig. 4.79 (c). For UVL at AB, BMD is cubic curve

(f) To locate point of contraflexure:

It is seen that BM is zero at PoC, changing its sign.

Let distance to this from $C = y$

From similar triangles of BMD between C and D,

$$\frac{26.67}{y} = \frac{100}{2 - y}$$

$$26.67\ (2 - y) = 100\ y$$

$$y = \frac{26.67 \times 2}{126.67} = 0.421 \text{ m from } C$$

Point of contraflexure is at 0.421 m from C.

Problem 49 For a beam shown in Fig. 4.80, obtain shear force and bending moment diagrams. Locate point of contraflexure if any. Indicate the values.

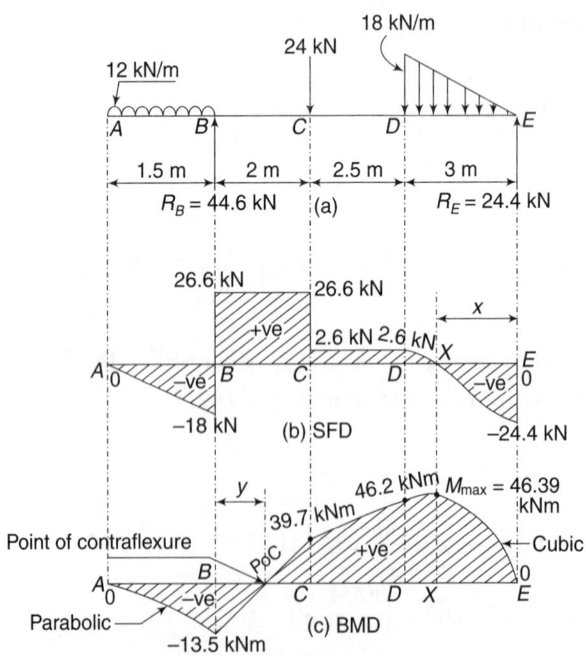

Fig. 4.80

Solution:

(a) **Reactions:** $\Sigma V = 0 \Rightarrow R_B + R_E = (12 \times 1.5) + 24 + \left(\frac{1}{2} \times 18 \times 3\right) = 69$ kN ...(1)

Taking moments about B,

$$-(12 \times 1.5)\frac{1.5}{2} + 24 \times 2 + \left(\frac{1}{2} \times 18 \times 3\right)\left(2 + 2.5 + \frac{1}{3} \times 3\right) - R_E \times 7.5 = 0$$

$$R_E = 24.4 \text{ kN}$$

$$R_B = 69 - R_E = 44.6 \text{ kN}$$

(b) **Shear forces:**

$$SF_A = 0$$

$$SF_{\text{up to } B} = -(12 \times 1.5) = -18 \text{ kN}$$

$$SF_{\text{at } B} = -18 + 44.6 = 26.6 \text{ kN}$$

$$SF_{\text{up to } C} = 26.6 \text{ kN}$$

$$SF_{\text{at } C} = 26.6 - 24 = 2.6 \text{ kN}$$

$$SF_{\text{at } D} = 2.6 \text{ kN}$$

$$SF_{\text{up to } E} = 2.6 - \left(\frac{1}{2} \times 18 \times 3\right) = -24.4 \text{ kN}$$

$$SF_{\text{at } E} = -24.4 + R_E = 0$$

(c) Draw SFD as shown in Fig. 4.80 (b). For UDL at AB, SFD is triangular and for UVL at DE, SFD is parabolic.

(d) **Bending moments:**

$$M_{\text{at}A} = 0$$

$$M_{\text{at}B} = -(12 \times 1.5) \times \frac{1.5}{2} = -13.5 \text{ kNm}$$

$$M_{\text{at}C} = -(12 \times 1.5) \times \left(2 + \frac{1.5}{2}\right) + 44.6 \times 2 = 39.7 \text{ kNm}$$

$$M_{\text{at}D} = -(12 \times 1.5)\left(2.5 + 2 + \frac{1.5}{2}\right) + 44.6 \times 4.5 - 24 \times 2.5 = 46.2 \text{ kNm}$$

$$M_{\text{at}E} = 0$$

To find maximum bending moment.

BM is maximum at SF = 0 and changing its sign.

Let the distance to X be x from E.

$SF_{\text{at } x} = -24.4 + \left(\frac{1}{2} \times w_x.x\right)$ (Considering right side of X)

From similar triangles of UVL,

$$\frac{w}{l} = \frac{w_x}{x}$$

$$\therefore \quad w_x = \frac{w.x}{l} = \frac{18 \times x}{3} = 6\,x$$

$$\therefore -24.4 + \left(\frac{1}{2} \times 6x \cdot x\right) = 0$$

$$\therefore \quad x = 2.852 \text{ m from } E$$

Considering right side of 'X'

Maximum BM, $M_{max} = M_x = 24.4 \times x - \left(\frac{1}{2}\, w_x \cdot x\right) \cdot \frac{1}{3}\, x$

$M_{max} = 24.4 \times 2.852 - \frac{1}{2} \cdot (6 \times 2.852)\ 2.852 \times \frac{1}{3} \times 2.852 = 46.39$ kNm.

(e) Draw BMD as shown in Fig. 4.80 (c). For UDL at *AB*, BMD is parabolic. For UVL at *DE*, it is cubic BMD is parabolic. For UVL at *DE*, it is cubic.

(f) To locate point of contraflexure

Point of contraflexure is at BM = 0, changing its sign. From BMD, it is seen that BM = 0 at PoC between *B* and, at a distance *y* from *B*.

From similar triangles of BMD between *B* and *C*

$$\frac{13.5}{y} = \frac{39.7}{2 - y}$$

$$13.5\ (2 - y) = 39.7\ y$$

$$y = 0.5075 \text{ m from } B$$

Point of contraflexure is at 0.5057 m from *B*.

Problem 50 Draw shear force bending moment diagrams for a beam loaded as shown in Fig. 4.81. Locate point of contraflexure if any. Indicate the values.

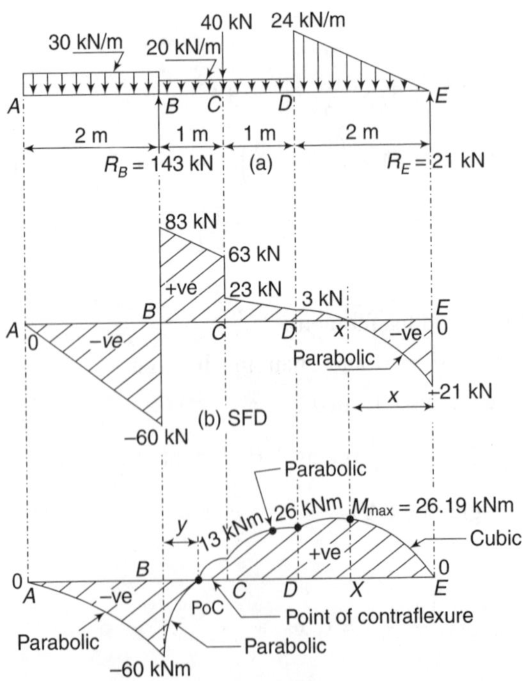

Fig. 4.81

Solution:

(a) **Reactions:**

$$\Sigma V = 0 \Rightarrow R_B + R_E = (30 \times 2) + (20 \times 2) + 40 + \left(\frac{1}{2} \times 24 \times 2\right) = 164 \text{ kN}$$

Taking moment about A

$$-(30 \times 2)\frac{2}{2} + (20 \times 2)\frac{2}{2} + 40 \times 1 + \left(\frac{1}{2} \times 24 \times 2\right)\left(2 + \frac{1}{3} \times 2\right) - R_E \times 4 = 0$$

$$\therefore \qquad R_E = 21 \text{ kN}$$

$$R_B = 164 - R_E = 143 \text{ kN}$$

(b) **Shear forces:**

$$\text{SF}_{\text{at } A} = 0$$

$$\text{SF}_{\text{up to } B} = -(30 \times 2) = -60 \text{ kN}$$

$$\text{SF}_{\text{at } B} = -60 + 143 = 83 \text{ kN}$$

$$\text{SF}_{\text{up to } C} = 83 - (20 \times 1) = 63 \text{ kN}$$

$$\text{SF}_{\text{at } C} = 63 - 40 = 23 \text{ kN}$$

$$\text{SF}_{\text{at } D} = 23 - (20 \times 1) = 3 \text{ kN}$$

$$\text{SF}_{\text{up to } E} = 3 - \left(\frac{1}{2} \times 24 \times 2\right) = -21 \text{ kN}$$

$$\text{SF}_{\text{at } E} = -21 + R_E = 0$$

(c) Draw SFD as shown in Fig. 4.81 (b). For UDL between AB, BC and CD, it is triangular. For UVL at DE it is parabolic.

(d) **Bending moments:**

$$M_{\text{at}A} = 0$$

$$M_{\text{at}B} = -(30 \times 2)\frac{2}{2} = -60 \text{ kNm}$$

$$M_{\text{at}C} = -(30 \times 2)\left(\frac{2}{2} + 1\right) + 143 \times 1 - (20 \times 1)\frac{1}{2} = 13 \text{ kNm}$$

$$M_{\text{at}D} = -(30 \times 2)\left(\frac{2}{2} + 2\right) + 143 \times 2 - (20 \times 2)\frac{2}{2} - 40 \times 1 = 26 \text{ kNm}$$

$$M_{\text{at}E} = 0$$

To find maximum bending moment:

BM is maximum at SF = 0, changing its sign. It is seen from SFD that SF = 0 between DE at a distance x from E.

Consider right side of section XX.

$$\text{SF}_x = -21 + \frac{1}{2}(w_x \cdot x) = 0$$

From similar triangles of UVL, $\dfrac{24}{2} = \dfrac{w_x}{x}$

$\therefore \qquad w_x = 12x$

$$-21 + \frac{1}{2} \times 12x \cdot x = 0$$

$\therefore \qquad x = 1.8708$ m from E.

Bending moment at X, $M_x = M_{max} = 21.x - \left(\frac{1}{2}.w_x.x\right)\frac{x}{3}$

(Considering to the right side of X)

$$M_{max} = 21 \times 1.8708 - \left(\frac{1}{2} \times 12x \cdot x\right)\frac{x}{3}$$

$$M_{max} = 21 \times 1.8708 - \frac{1}{2} \times 12 \times \frac{1.8708^3}{3} = 26.19 \text{ kNm}$$

(e) Draw BMD as shown in Fig. 4.81 (c). For UDL at AB, BC and CD BMD is parabolic. For UVL at DE, BMD is cubic.

(f) To locate point of contraflexure

Point of contraflexure is where BM = 0, changing its sign.

It is seen from BMD that PoC is between B and, at a distance of y from B.

$$\text{At PoC, } BM_{PoC} = -(30 \times 2)\left(y + \frac{2}{2}\right) + 143 \times y - (20.y)\frac{y}{2} = 0$$

$$-60y - 60 + 143y - 10y^2 = 0$$

$$10y^2 - 83y + 60 = 0$$

$$y = \frac{83 \pm \sqrt{83^2 - 4 \times 10 \times 60}}{2 \times 10} = 0.8 \text{ or } 7.5$$

Taking the possible value $y = 0.8$ m from B

Point of contraflexure is at a distance of 0.8 m from B.

Problem 51 Draw shear force and bending moment diagram for a beam loaded as shown in Fig. 4.82. Locate the point of contraflexure if any. Indicate the values.

Solution:

(a) **Reactions:** $\Sigma V = 0 \Rightarrow R_B + R_D = 15 + (20 \times 2.5) + \left(\frac{1}{2} \times 30 \times 2.5\right) = 102.5$ kN

Taking moments about B, $-15 \times 2 + (20 \times 2.5)\frac{2.5}{2} + \left(\frac{1}{2} \times 30 \times 2.5\right) \times \left(2.5 + \frac{1}{3} \times 2.5\right) - R_D \times 5 = 0$

$$\therefore \qquad R_D = 31.5 \text{ kN}$$

$$R_B = 102.5 - R_D = 71 \text{ kN}$$

(b) **Shear forces:**

$$SF_{\text{up to } A} = 0$$

$$SF_{\text{at } A} = -15 \text{ kN}$$

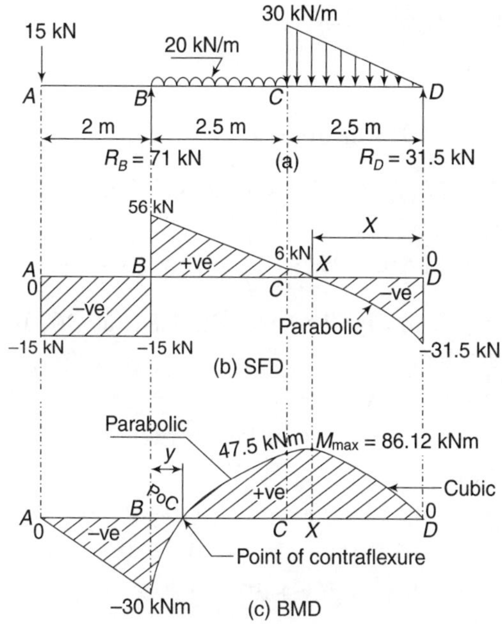

Fig. 4.82

$$SF_{\text{up to } B} = -15 \text{ kN}$$
$$SF_{\text{at } B} = -15 + 71 = 56 \text{ kN}$$
$$SF_{\text{at } C} = -15 + 71 - (20 \times 2.5) = 6 \text{ kN}$$
$$SF_{\text{up to } D} = -15 + 71 - (20 \times 2.5) - \left(\frac{1}{2} \times 30 \times 2.5\right) = -31.5 \text{ kN}$$
$$SF_{\text{at } D} = -31.5 + R_D = 0$$

(c) Draw SFD as shown in Figure 4.82 (b). For UDL at *BC* it is triangular and for UVL at *CD* it is parabolic.

(d) **Bending moments:**

$$M_{\text{at}A} = 0$$
$$M_{\text{at}B} = -15 \times 2 = -30 \text{ kNm}$$
$$M_{\text{at}C} = -15 \times 4.5 + 71 \times 2.5 - (20 \times 2.5)\frac{2.5}{2} = 47.5 \text{ kNm}$$
$$M_{\text{at}D} = 0$$

To find maximum bending moment

BM is maximum at SF = 0. It is seen that SF = 0 between *CD* at a distance x from *D*. Consider to the right of section.

$$SF_x = -31.5 + \left(\frac{1}{2} w_x . x\right) = 0$$

For UVL at *CD*, from similar triangles, $\dfrac{w_x}{x} = \dfrac{30}{2.5}$

$$w_x = \frac{30.\ x}{2.5} = 12x$$

$$SF_x = -\ 31.5 + \frac{1}{2} \times 12\ x \cdot x = 0$$

$$x = 2.2913 \text{ m from } D$$

$$M_x = M_{max} = 31.2 \cdot x - \left(\frac{1}{2}\ w_x \cdot x\right) \frac{1}{3} \times x$$

$$= 31.5 \times 2.2913 - \frac{1}{2} \times 12 \times \frac{2.2913^3}{3} = 48.12 \text{ kNm.}$$

(e) Draw BMD as shown in Fig. 4.82 (c). For UDL at *BC*, BM is parabolic and for UVL at *DC*, it is cubic.

(f) To locate point of contraflexure:

It is seen from BMD that BM = 0 at PoC between *B* and *C* changing its sign, at a distance of *y* from *B*.

$$M_{PoC} = -\ 15\ (2 + y) + 71y - (20 \times y)\ \frac{y}{2} = 0$$

$$-\ 30 - 15y + 71y - 10y^2 = 0$$

$$-\ 10y^2 - 56y + 30 = 0$$

Solving y = 5 or 0.6 m.

Taking the possible value, y = 0.6 m from *B*.

Problem 52 Find the value of *W* for both reactions to be equal. Draw shear force and bending moment diagrams for a beam loaded as shown in Fig. 4.83. Locate the position of point of contraflexure on the beam. Indicate the values.

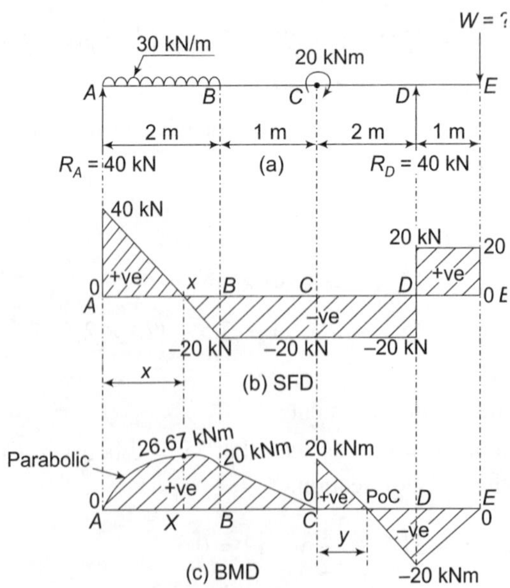

Fig. 4.83

Solution:

(a) **Reactions:** $\Sigma V = 0 \Rightarrow R_A + R_D = (30 \times 2) + W$ ...(1)

Taking moments about A, $-\left(30 \times 2 \times \frac{2}{2}\right) - 20 + R_D \times 5 - W \times 6 = 0$

$$5R_D = 80 + 6W \quad ...(2)$$

Given both reactions are to be equal, $R_A = R_B$

$$\therefore \quad R_A + R_D = 2R_D = 60 + W \quad ...(3)$$

Substitute for R_D from Eq. (2)

$$2\left(\frac{80 + 6W}{5}\right) = 60 + W$$

$$32 + 2.4\,W = 60 + W$$

$$\therefore \quad W = 20 \text{ kN}$$

Substituting Eq. (3), $2R_D = 60 + W$

$$2R_D = 60 + 20 = 80$$

$$R_D = R_A = 40 \text{ kN}$$

(b) **Shear forces:**

$$SF_{\text{up to } A} = 0$$

$$SF_{\text{at } A} = R_A = 40 \text{ kN}$$

$$SF_{\text{at } B} = 40 - (30 \times 2) = -20 \text{ kN}$$

$$SF_{\text{at } C} = -20 \text{ kN}$$

$$SF_{\text{up to } D} = -20 \text{ kN}$$

$$SF_{\text{at } D} = -20 + 40 = 20 \text{ kN}$$

$$SF_{\text{up to } E} = 20 \text{ kN}$$

$$SF_{\text{at } E} = 20 - 20 = 0$$

(c) Draw SFD as shown in Fig. 4.83 (b). For UDL, SFD between AB is triangular.

(d) **Bending moments:**

$$M_{\text{at}A} = 0$$

$$M_{\text{at}B} = R_A \times 2 - (30 \times 2)\frac{2}{2} = 40 \times 2 - (30 \times 2)\frac{2}{2} = 20 \text{ kNm}$$

$$M_{\text{up to } C} = 40 \times 3 - (30 \times 2)\left(1 + \frac{2}{2}\right) = 0$$

$$M_{\text{at}C} = 0 + 20 = 20 \text{ kNm}$$

$$M_{\text{at}D} = 40 \times 5 - (30 \times 2)\left(2 + 1 + \frac{2}{2}\right) + 20 = -20 \text{ kNm}$$

$$M_{\text{at}E} = 0$$

To find maximum bending moment:

BM is maximum where SF = 0, changing its sign.

Let it be at a distance x from A

From similar triangles of SFD between A and B,

$$\frac{40}{x} = \frac{20}{2-x} \quad \Rightarrow \quad 40\,(2-x) = 20\,x$$

$$x = \frac{80}{60} = 1.333 \text{ m}$$

∴ Maximum BM, $$M_{max} = M_x = 40\cdot(1.333) - (30 \times 1.333)\,\frac{1.333}{2}$$

$$M_{max} = 26.67 \text{ kNm.}$$

(e) Draw BMD as shown in Fig. 4.83 (c). For UDL between AB it is parabolic for moment at C, it is a vertical line.

(f) **To locate point of contraflexure:**

Point of contraflexure is one where BM = 0, changing its sign from +ve to –ve. Let distance from C to PoC = y in BMD.

From similar triangles of BMD between CD

$$\frac{20}{y} = \frac{20}{2-y}$$

∴ $$20\,(2-y) = 20\,y$$

∴ $$y = \frac{40}{40} = 1 \text{ m from } C.$$

Point of contraflexure is at 1 m from C, between CD.

Problem 53 Draw shear force and bending moment diagrams for a beam loaded as shown in Fig. 4.84. Locate the point of contraflexure.

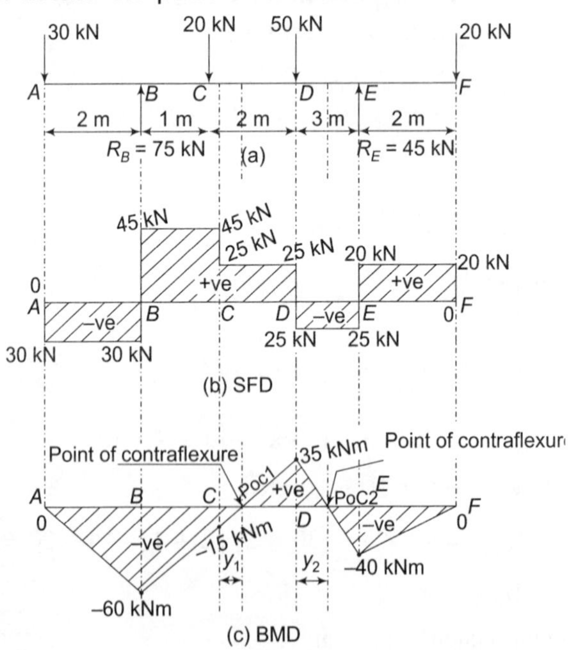

Fig. 4.84

Solution:

(a) **Reactions:** $\Sigma V = 0 \Rightarrow R_B + R_E = 30 + 20 + 50 + 20 = 120$ kN ...(1)

Taking moments about B,

$$-30 \times 2 + 20 \times 1 + 50 \times 3 - R_E \times 6 + 20 \times 8 = 0$$

$$R_E = 45 \text{ kN}$$

$$\therefore \quad R_B = 120 - R_E = 75 \text{ kN}$$

(b) **Shear forces:**

$$\begin{aligned}
SF_{\text{up to } A} &= 0 \\
SF_{\text{at } A} &= -30 \text{ kN} \\
SF_{\text{up to } B} &= -30 \text{ kN} \\
SF_{\text{at } B} &= -30 + 75 = 45 \text{ kN} \\
SF_{\text{up to } C} &= 45 \text{ kN} \\
SF_{\text{at } C} &= 45 - 20 = 25 \text{ kN} \\
SF_{\text{up to } D} &= 25 \text{ kN} \\
SF_{\text{at } D} &= 25 - 50 = -25 \text{ kN} \\
SF_{\text{up to } E} &= -25 \\
SF_{\text{at } E} &= -25 + 45 = 20 \text{ kN} \\
SF_{\text{up to } F} &= 20 \text{ kN} \\
SF_{\text{at } F} &= 20 - 20 = 0
\end{aligned}$$

(c) Draw SFD as shown in Fig. 4.84 (b)

(d) **Bending moments:**

$$\begin{aligned}
M_{\text{at} A} &= 0 \\
M_{\text{at} B} &= -30 \times 2 = -60 \text{ kNm} \\
M_{\text{at} C} &= -30 \times 3 + 75 \times 1 = -15 \text{ kNm} \\
M_{\text{at} D} &= -30 \times 5 + 75 \times 3 - 20 \times 2 = 35 \text{ kNm} \\
M_{\text{at} E} &= -30 \times 8 + 75 \times 6 - 20 \times 5 - 50 \times 3 = -40 \text{ kNm} \\
M_{\text{at} F} &= 0
\end{aligned}$$

(e) Draw BMD as shown in Fig. 4.84 (c)

(f) To locate point of contraflexure:

It is seen from BMD that there are two places where BM = 0, changing its sign, one between C and D and the other between D and E.

(i) Let y_1 be the distance from C to PoC 1.

From similar triangles of BMD between CD,

$$\frac{15}{y_1} = \frac{35}{2 - y_1}$$

$$15\,(2 - y_1) = 35\,y_1$$

$$30 = 35\ y_1 + 15\ y_1$$

$$y_1 = \frac{30}{50} = 0.6 \text{ m from } C$$

PoC1 is at 0.6 m from *C*.

(ii) Let y_2 be the distance from *D* to PoC 2.

From similar triangles of BMD between *DE*,

$$\frac{35}{y_2} = \frac{40}{3 - y_2}$$

$$35\ (3 - y_2) = 40\ y_2$$

$$105 - 35\ y_2 = 40\ y_2$$

$$y_2 = \frac{105}{75} = 1.4 \text{ m from } D$$

PoC 2 is at 1.4 m from *D*.

Problem 54 Draw shear force and bending moment diagrams for a beam shown in Fig. 4.85. Locate the point of contraflexure. Indicate the values at salient points on the diagrams.

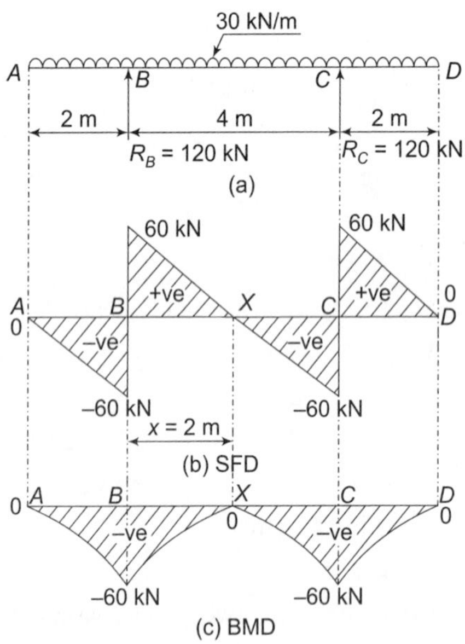

Fig. 4.85

Solution:

(a) **Reactions:** $\Sigma V = 0,\ R_B + R_C = 30 \times 8 = 240$ kN

Taking moments about *B*, $-(30 \times 2) \times \frac{2}{2} + (30 \times 4)\ \frac{4}{2}$

$$- R_C \times 4 + 30 \times 2\left(4 + \frac{2}{2}\right) = 0$$

$$\therefore \quad R_C = 120 \text{ kN}$$

$$R_B = 240 - R_C = 120 \text{ kN}$$

(b) **Shear forces:**

$$\text{SF}_{\text{at } A} = 0$$

$$\text{SF}_{\text{up to } B} = -30 \times 2 = -60 \text{ kN}$$

$$\text{SF}_{\text{at } B} = -60 + 120 = 60 \text{ kN}$$

$$\text{SF}_{\text{up to } C} = 60 - (30 \times 4) = -60 \text{ kN}$$

$$\text{SF}_{\text{at } C} = -60 + 120 = 60 \text{ kN}$$

$$\text{SF}_{\text{up to } D} = 60 - 30 \times 2 = 0$$

(c) Draw SFD as shown in Fig. 4.85 (b). For UDL between *AB*, *BC* and *CD*, SFD is triangular

(d) **Bending moments:**

$$M_{\text{at}A} = 0$$

$$M_{\text{at}B} = -(30 \times 2)\frac{2}{2} = -60 \text{ kNm}$$

$$M_{\text{at}C} = -(30 \times 6) \times \frac{6}{2} + 120 \times 4 = -60 \text{ kNm}$$

$$M_{\text{at}D} = 0$$

To find maximum bending moment.

BM is maximum where SF = 0, changing its sign.

From SFD, it is seen that, SF = 0 between *B* and *C*.

From similar triangles of SFD between *BC*,

$$\frac{60}{x} = \frac{60}{4 - x}$$

$$\therefore \quad x = \frac{4}{2} = 2 \text{ m, it is in the middle of } BC.$$

Maximum bending moment $= M_{\text{max}} = M_x$

$$M_{\text{max}} = -(30 \times 4)\frac{4}{2} + 12 \times 2 = 0$$

(e) Draw BMD as shown in Fig. 4.85 (c). BMD is parabolic for UDL, between *AB*, *BC* and *CD*.

Since the BMD is just touching *O* at *X*, there is no point of contraflexure.

Problem 55 Draw shear force and bending moment diagrams for a beam loaded as shown in Fig. 4.86. Locate the point of contraflexure. Indicate the values at the salient points.

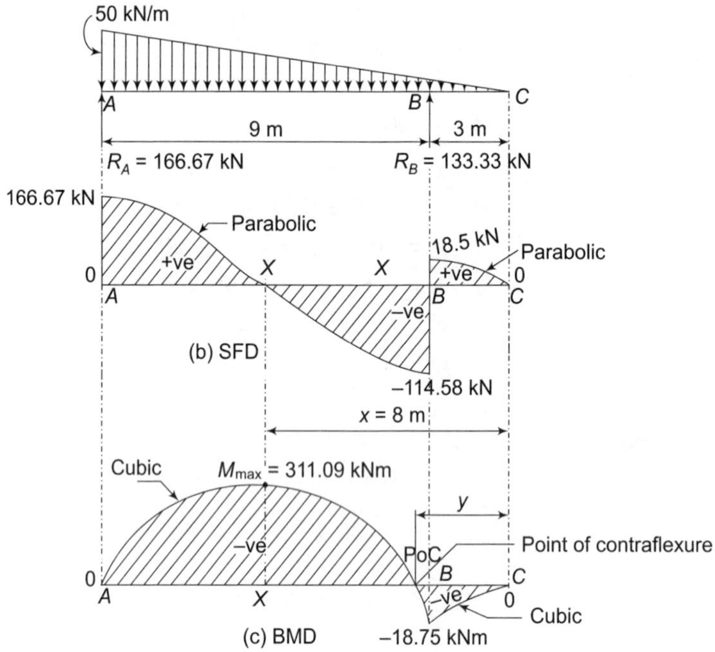

Fig. 4.86

Solution:

(a) **Reactions:** $\Sigma V = 0.\ R_A + R_B = \left(\frac{1}{2} \times 50 \times 12\right) = 300$ kN

Taking moments about A

$$R_B \times 9 - \left(\frac{1}{2} \times 50 \times 12\right)\left(\frac{1}{3} \times 12\right) = 0$$

$$R_B = 133.33 \text{ kN}$$

$$R_A = 300 - R_B = 166.67 \text{ kN}$$

(b) **Shear forces:**

$$\text{SF}_{\text{up to } A} = 0$$

$$\text{SF}_{\text{at } A} = 166.67 \text{ kN}$$

$$\text{SF}_{\text{up to } B} = 166.67 - (\text{Full UVL} - \text{UVL beyond } B)$$

$$= 166.67 - \left(\frac{1}{2} \times 50 \times 12 - \frac{1}{2} \times w_B.3\right)$$

From similar triangles of UVL

$$\frac{w}{l} = \frac{w_B}{3}$$

$$\frac{50}{12} = \frac{w_B}{3}$$

$$\therefore \quad w_B = 12.5 \text{ kN}$$

$$\text{SF}_{\text{up to } B} = 166.67 - \left(\frac{1}{2} \times 50 \times 12 - \frac{1}{2} \times 12.5 \times 3\right)$$

$$= -114.58 \text{ kN}$$

$$\text{SF}_{\text{at } B} = -114.58 + 133.33 = 18.75 \text{ kN}$$

$$\text{SF}_{\text{at } C} = 18.75 - \left(\frac{1}{2} \times w_B \times 3\right)$$

$$= 18.75 - \left(\frac{1}{2} \times 12.5 \times 3\right) = 0$$

(c) Draw SFD as shown in Fig. 4.86 (b). For UVL, SFD is parabolic.

(d) **Bending moments:**

$$M_{\text{at } A} = 0$$

$$M_{\text{at } B} = -\left(\frac{1}{2} w_B \times 3 \times \frac{1}{2} \times 3\right) \text{ (Considering right side)}$$

$$= -\left(\frac{1}{2} \times 12.5 \times 3 \times \frac{1}{3} \times 3\right) = -18.75 \text{ kNm}$$

$$M_{\text{at } C} = 0$$

To find maximum bending moment:

BM is maximum where SF = 0, changing its sign

Let the distance of X from C be x

Consider forces to the right of section for a convenience.

Shear force intensity at x, $\frac{w_x}{x} = \frac{w}{l}$

$$\therefore \quad w_x = \frac{w}{l} \cdot x = \frac{50}{12} \cdot x$$

$$\therefore \quad \text{SF}_x = -133.33 + \left(\frac{1}{2} w_x \cdot x\right) = 0$$

$$= -133.33 + \frac{1}{2} \times \frac{50}{12} x \cdot x = 0$$

$$\therefore \quad x = 8 \text{ m form } C \text{ or } 12 - 8 = 4 \text{ m from } A$$

Maximum bending moment, $M_{\max} = M_x = 133.33 \times (x - 3) - \left(\frac{1}{2} \cdot w_x x\right) \frac{x}{3}$

$$M_{\max} = 133.33 \times (8 - 3) - \frac{1}{2} \times \frac{50}{12} \times 8 \times 8 \times \frac{8}{3}$$

$$= 311.09 \text{ kNm}$$

(e) Draw BMD as shown in Fig. 4.86 (c). For the UVL, BMD is cubic curve.

(f) **To locate PoC:** Point of contraflexure is a point where BM = 0, changing its sign.

It is seen from BMD that, BM is zero at a distance y from C.

Load intersity of PoC, $w_y = \frac{50}{12} y$

Consider bending moment to the right of PoC,

$$133.33 (y - 3) - \left[\frac{1}{2} \times w_y (y)\right] \frac{1}{3} y = 0$$

$$133.33 (y - 3) - \left[\frac{1}{2} \times \frac{50}{12} (y).(y)\right] \frac{1}{3} y = 0$$

$$133.33 (y - 3) = \frac{50}{72} (y)^3$$

$$\frac{50}{72} y^3 - 133.33 y + 400 = 0$$

Solving y = 3.165 m from C

Point of contraflexure is at a distance of 3.165 m from C.

Problem 56 A horizontal beam 10 m long carries a UDL of 360 kN. The beam is supported at *A*, the left end and at *B* at a distance *x* from right end *C*. Determine the value of *x* if the point of contraflexure is at the mid point of the beam. For this, draw shear force and bending moment diagrams. Indicate the values on BMD and SFD.

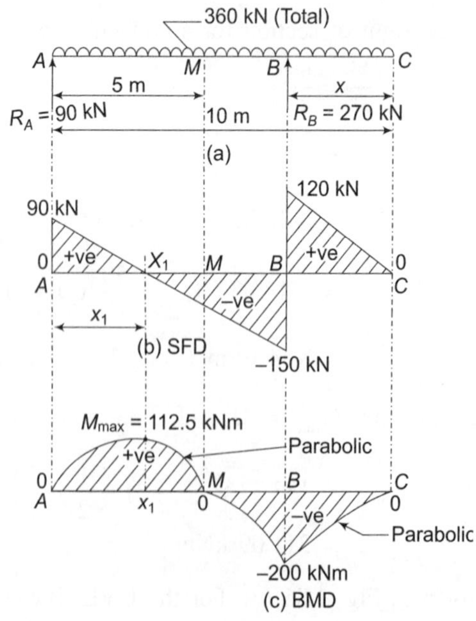

Fig. 4.87

Solution:

(a) **Reactions:** $\Sigma V = 0,\ R_A + R_B = 360$ kN ...(1)

Taking moment about A,

$$-360 \times 5 + R_B (10 - x) = 0$$

$$R_B = \frac{1800}{10 - x}$$

Substituting in Eq. (1) $R_A + \dfrac{1800}{10 - x} = 360$

$$\therefore \quad R_A = 360 - \frac{1800}{10 - x} = \frac{360\,(10 - x) - 1800}{10 - x} = \frac{1800 - 360x}{10 - x}$$

Total load = 360 over a span of 10 m.

$$\therefore \quad \text{UDL } w = \frac{360}{10} = 36 \text{ kN/m.}$$

For the point of contraflexure to be at the middle

i.e., at M at a distance $\dfrac{10}{2} = 5$ m from A

Bending moment at $M = 0$

$$M_M = R_A\,.5 - (36 \times 5) \times \frac{5}{2} = 0$$

$$\left(\frac{1800 - 360x}{10 - x}\right)5 = 36 \times 5 \times \frac{5}{2}$$

$$5(1800 - 360x) = 450\,(10 - x)$$

$$9000 - 1800x = 4500 - 450x$$

$$(1800 - 450)x = 9000 - 4500$$

$$x = \frac{4500}{1350} = 3.333 \text{ m from } C$$

The supported is to be at 3.333 m from right end C

For the point of contraflexure to be at middle of beam

$$R_A = \frac{1800 - 360 \times 3.333}{10 - x} = 90 \text{ kN}$$

$$R_B = 360 - 90 = 270 \text{ kN}$$

(b) **Shear forces:**

$$SF_{\text{up to } A} = 0$$

$$SF_{\text{at } A} = 90 \text{ kN}$$

$$SF_{\text{up to } B} = 90 - 36 \times (10 - x) = 90 - 36\,(10 - 3.333) = -150 \text{ kN}$$

$$SF_{\text{at } B} = -150 + 270 = 120 \text{ kN}$$

$$SF_{\text{at } C} = 120 - 36 \times 3.333 = 0$$

(c) Draw SFD as shown in Fig. 4.87 (b). SF is triangular for UDL between *AB* and *BC*

(d) **Bending moments:**

$$M_{\text{at }A} = 0$$

$$M_{\text{at }B} = 90 \times (10 - 3.333) - 36 \times (10 - 3.333)\left(\frac{10 - 3.333}{2}\right)$$

$$= -\ 200 \text{ kNm}$$

$$M_{\text{at }C} = 0$$

To find maximum bending moment

BM is maximum where SF = 0, changing its sign.

at section *X*, *X*, $\text{SF}_{\text{at }x1} = 90 - 36 \cdot x_1 = 0$

$$x_1 = \frac{90}{36} = 2.5 \text{ m from } A$$

Maximum bending moment, $M_{max} = M_{x1} = 90\ x_1 - 36 \cdot x_1 \dfrac{x_1}{2}$

$$= 90 \times 2.5 - 36 \times \frac{2.5 \times 2.5}{2} = 112.5 \text{ kNm.}$$

(e) Draw BMD as shown in Fig. 4.87 (c). BMD is parabolic for UDL between *AB* and *BC*.

(f) Point of contraflexure is at the middle of the beam as discussed above.

Problem 57 Draw shear force and bending moment diagrams for a beam shown in Fig. 4.88. Mark the values of important ordinates.

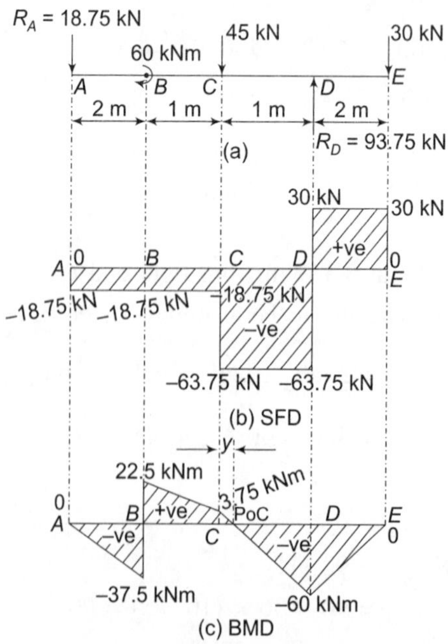

Fig. 4.88

Solution:

(a) To find reactions $\Sigma V = 0 \Rightarrow R_A + R_D = 45 + 30 = 75$ kN ...(1)

Taking moments about A, $-60 - 45 \times 3 + R_D \times 4 - 30 \times 6 = 0$

$$R_D = 93.75 \text{ kN}$$

$$\therefore \quad R_A = 75 - 93.75 = -18.75$$

Since R_A is –ve, it is acting downward

(b) **Shear forces:**

$$SF_{\text{up to } A} = 0$$
$$SF_{\text{at } A} = -18.75 \text{ kN}$$
$$SF_{\text{at } B} = -18.75 \text{ kN}$$
$$SF_{\text{up to } C} = -18.75 \text{ kN}$$
$$SF_{\text{at } C} = -18.75 - 45 = -63.75 \text{ kN}$$
$$SF_{\text{up to } D} = -63.75 \text{ kN}$$
$$SF_{\text{at } D} = -63.75 + 93.75 = 30 \text{ kN}$$
$$SF_{\text{up to } E} = 30 \text{ kN}$$
$$SF_{\text{at } E} = 30 - 30 = 0 \text{ kN}$$

(c) Draw SFD as shown in Fig. 4.88 (b)

(d) **Bending moments:**

$$M_{\text{at } A} = 0$$

Just to left of B, $$M_{\text{up to } B} = -18.75 \times 2 = -37.5 \text{ kNm}$$
$$M_{\text{at } B} = -37.5 + 60 = 22.5 \text{ kNm}$$
$$M_{\text{at } C} = -18.75 \times 3 + 60 = +3.75 \text{ kNm}$$
$$M_{\text{at } D} = -18.75 \times 4 + 60 - 45 \times 1 = -60 \text{ kNm}$$
$$M_{\text{at } E} = 0$$

(e) Draw BMD as shown in Fig. 4.88 (c).

(f) To locate point of contraflexure

Point of contraflexure is where BM = 0, changing its sign

It is seen from BMD that BM = 0 at PoC between C and D at a distance y from C.

From similar triangles of BMD between C and

$$\frac{3.75}{y} = \frac{60}{1 - y}$$

$$3.75\,(1 - y) = 60\,y$$

$$y = 0.0588 \text{ m from } C$$

Point of contraflexure is at 0.0588 m from C.

Problem 58 Draw shear force and bending moment diagram for a beam as shown in Fig. 4.89. Locate point of inflexion if any. Mark values at all points

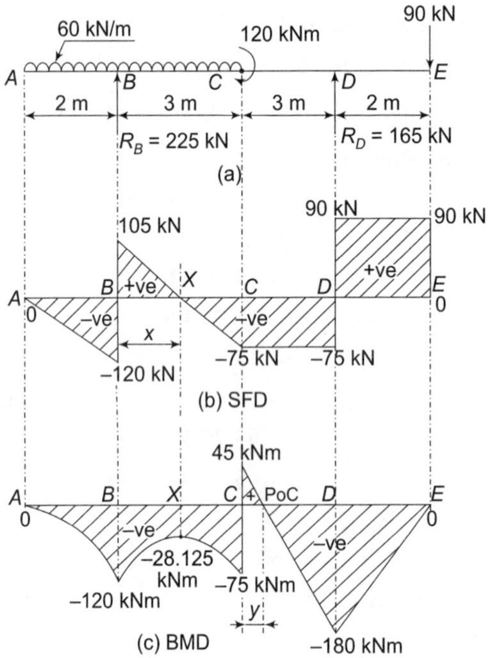

Fig. 4.89

Solution:

(a) **Reactions:** $\Sigma V = 0 \Rightarrow R_B + R_D = (60 \times 5) + 90 = 390$ N ...(1)

Taking moments about B,

$$-(60 \times 5) \times (2.5 - 2) - 120 + R_D \times 6 - 90 \times 8 = 0$$

$$R_D = 165 \text{ kN}$$

$$\therefore \quad R_B = 390 - R_D = 225 \text{ kN}$$

(b) **Shear forces:**

$$SF_{at\ A} = 0$$

$$SF_{up\ to\ B} = -60 \times 2 = -120 \text{ kN}$$

$$SF_{at\ B} = -120 + 225 = -105 \text{ kN}$$

$$SF_{at\ C} = 105 - (60 \times 3) = -75 \text{ kN}$$

$$SF_{at\ D} = -75 \text{ kN}$$

$$SF_{at\ D} = -75 + 165 = 90 \text{ kN}$$

$$SF_{up\ to\ E} = 90 \text{ kN}$$

$$SF_{at\ E} = 90 - 90 = 0$$

(c) Draw SFD as shown in Fig. 4.89 (b). For UDL between AB and BC, SF is triangular.

(d) **Bending moments:**

$$M_{\text{at } A} = 0$$

$$M_{\text{at } B} = -(60 \times 2) \times \frac{2}{2} = -120 \text{ kNm}$$

Just upto C, $$M_{\text{up to } C} = -(60 \times 5)\frac{5}{2} + 225 \times 3 = -75 \text{ kNm}$$

$$M_{\text{at } C} = -75 + 120 = 45 \text{ kNm}$$

$$M_{\text{at } D} = -(60 \times 5)\left(3 + \frac{5}{2}\right) + 225 \times 6 + 120 = -180 \text{ kNm}$$

$$M_{\text{at } E} = 0$$

To find maximum bending moment:

BM is maximum where SF = 0, changing the sign.

It is seen that SFD that between B and C, at x, SF = 0 at a distance x from B.

From similar triangles of SFD between B and C,

$$\frac{105}{x} = \frac{75}{3 - x}$$

$$75x = 105\,(3 - x)$$

$$x = \frac{105 \times 3}{75 + 105} = 1.75 \text{ m from } B$$

Maximum bending moment, $M_{max} = M_x$

$$= -60 \times (2 + 1.75) \times \frac{(2 + 1.75)}{2} + 225 \times 1.75$$

$$= -28.125 \text{ kNm.}$$

(e) Draw BMD as shown in Fig. 4.89 (c). For UDL between AB and BC, BMD is parabolic. For moment at C, it is a vertical line.

(f) To locate the point of inflexion (point of contraflexure):

The point of inflexion is where BM = 0, changing its sign.

Bending moment is zero and changing its sign at PoC, between C and D at a distance y from C.

From similar triangles of BMD between C and D,

$$\frac{45}{y} = \frac{180}{3 - y}$$

$$180\,y = 45\,(3 - y)$$

$$y = \frac{45 \times 3}{(180 + 45)} = 0.6 \text{ m from } C.$$

The point of inflexion is at PoC between C and D at a distance of 0.6 m from C.

Problem 59 Draw shear force and bending moment diagrams for a beam shown in Fig. 4.90. Locate the point of inflexion if any. Indicate the values on diagrams.

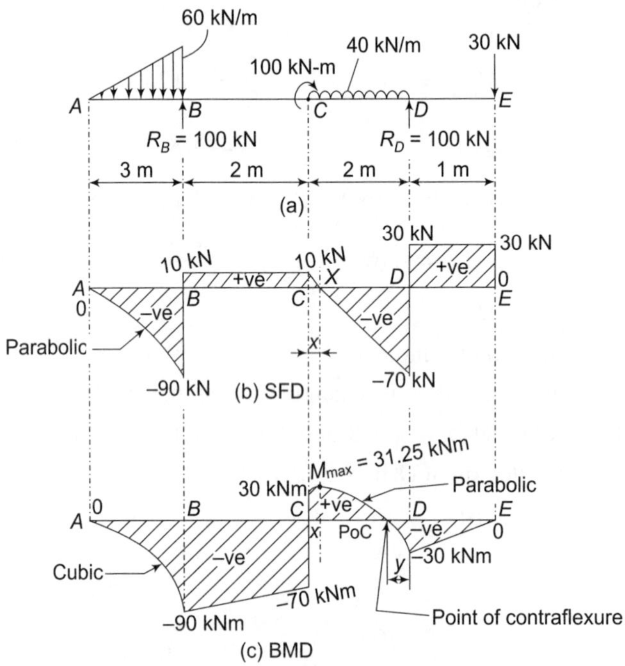

Fig. 4.90

Solution:

(a) **Reactions:** $\Sigma V = 0 \Rightarrow R_B + R_D = \left(\frac{1}{2} \times 60 \times 3\right) + 40 \times 2 + 30 = 200$ kN ...(1)

Taking moments about B,

$$-\left(\frac{1}{2} \times 60 \times 3\right)\frac{1}{3} \times 3 + 100 + (40 \times 2)\left(2 + \frac{2}{2}\right) - R_D \times 4 + 30 \times 5 = 0$$

$\therefore$ $R_D = 100$ kN

$\therefore$ $R_B = 200 - R_D = 100$ kN.

(b) **Shear forces:**

$$\text{SF}_{\text{at } A} = 0$$

$$\text{SF}_{\text{up to } B} = -\frac{1}{2} \times 60 \times 3 = -90 \text{ kN}$$

$$\text{SF}_{\text{at } B} = -90 + 100 = 10 \text{ kN}$$

$$\text{SF}_{\text{at } C} = 10 \text{ kN}$$

$$\text{SF}_{\text{up to } D} = 10 - 40 \times 2 = -70 \text{ kN}$$

$$SF_{at\ D} = -70 + 100 = 30 \text{ kN}$$

$$SF_{upto\ E} = 30 \text{ kN}$$

$$SF_{at\ E} = 30 - 30 = 0$$

(c) Draw SFD as shown in Fig. 4.90 (b). For UVL at *AB*, SFD is parabolic. For UDL at *CD*, SFD is triangular.

(d) **Bending moments:**

$$M_{at\ A} = 0$$

$$M_{at\ B} = -\left(\frac{1}{2} \times 60 \times 3\right) \times \frac{1}{3} \times 3 = -90 \text{ kNm}$$

Just to the left of *C*, $M_{up\ to\ c} = -\left(\frac{1}{2} \times 60 \times 3\right)\left(\frac{3}{3} + 2\right) + 100 \times 2 = -70$ kNm

$$M_{at\ C} = -70 + 100 = +30 \text{ kNm}$$

$$M_{at\ D} = -\left(\frac{1}{2} \times 60 \times 3\right)\left(\frac{3}{3} + 2 + 2\right) + 100 \times 4 + 100 - (40 \times 2)\frac{2}{2}$$

$$= -30 \text{ kNm}$$

$$M_{at\ E} = 0$$

To find maximum bending moment

BM is maximum at SF = 0, changing its sign.

It is seen from SFD, SF = 0 between *C* and *D*

Let distance from *C* to *X* be *x*

From similar triangles of SFD at *CD*

$$\frac{10}{x} = \frac{70}{2 - x}$$

$$10(2 - x) = 70\,x$$

$$x = \frac{10 \times 2}{(10 + 70)} = 0.25 \text{ m form } C.$$

Maximum bending moment, $M_{max} = M_x$

$$= -\left(\frac{1}{2} \times 60 \times 3\right)\left(\frac{3}{3} + 2 + 0.25\right) + 100\,(2 + 0.25) + 100 - (40 \times 0.25)\,\frac{0.25}{2}$$

= 31.25 kNm.

(e) Draw BMD as shown in Fig. 4.90 (c). For UVL at *AB* BMD is cubic curve and for UDL at *CD*, BMD is parabolic. For moment at *C*, it is a vertical line.

(f) To locate the point of inflexion or point of contraflexure.

Point of inflexion is where BM = 0 changing its sign. It is seen from BMD that BM = 0 and changing its sign between *C* and at PoC at a distance *y* from *D*.

Considering to the right of PoC, $M_{PoC} = 0$

$$-30 \times (1 + y) + 100 \times y - (40 \times y) \times \frac{y}{2} = 0$$

$$20y^2 + 30 + 30y - 100\,y = 0$$

$$20y^2 - 70y + 30 = 0$$

Solving $y = 3$ or 0.5.

Adopting the feasible value, $y = 0.5$ m from D.

The point of inflexion is at a distance of 1.5 m from right side end E or at 0.5 from right side support D.

Problem 60 Draw shear force and bending moment diagrams for a loaded as shown in Fig. 4.91. Locate the point of maximum bending moment and point of contraflexure. Indicate the values on diagrams.

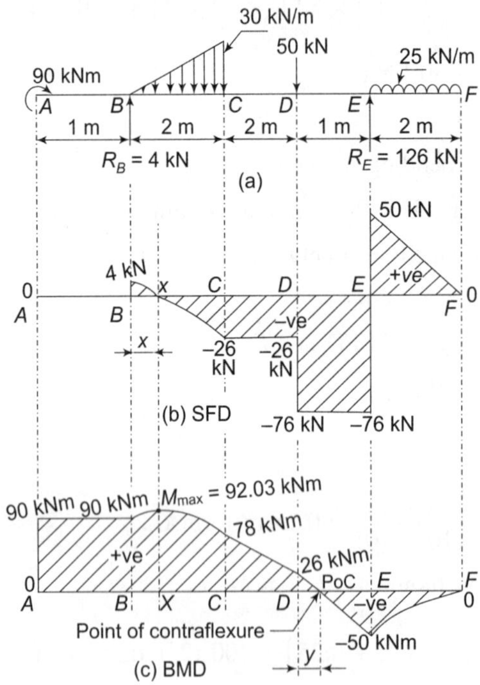

Fig. 4.91

Solution:

(a) **Reactions:** $\Sigma V = 0 \Rightarrow R_B + R_E = \frac{1}{2} \times 30 \times 2 + 50 + 25 \times 2 = 130$ kN ...(1)

Taking moments about B,

$$90 + \left(\frac{1}{2} \times 30 \times 2\right)\left(\frac{2}{3} \times 2\right) + 50 \times 4 - R_E \times 5 + (25 \times 2)\left(2 + 2 + 1 + \frac{2}{2}\right) = 0$$

$\therefore \qquad R_E = 126 \text{ kN}$

$\therefore \qquad R_B = 130 - 126 = 4 \text{ kN.}$

(b) **Shear forces:**

$$SF_{at\ A} = 0$$
$$SF_{up\ to\ B} = 0$$
$$SF_{at\ B} = 4 \text{ kN}$$
$$SF_{at\ C} = 4 - \left(\frac{1}{2} \times 30 \times 2\right) = -26 \text{ kN}$$
$$SF_{up\ to\ D} = -26 \text{ kN}$$
$$SF_{at\ D} = -26 - 50 = -76 \text{ kN}$$
$$SF_{up\ to\ E} = -76 \text{ kN}$$
$$SF_{at\ E} = -76 + 126 = 50 \text{ kN}$$
$$SF_{at\ F} = 50 - (25 \times 2) = 0$$

(c) Draw SFD as shown in Fig. 4.91 (b). For UVL at *BC*, SFD is parabolic. For UDL at *EF*, SFD is triangular.

(d) **Bending moments:**

Just to left of *A*, $M_{up\ to\ A} = 0$

$$M_{at\ A} = 90 \text{ kNm}$$
$$M_{at\ B} = 90 \text{ kNm}$$
$$M_{at\ C} = 90 + 4 \times 2 - \left(\frac{1}{2} \times 30 \times 2\right)\frac{1}{3} \times 2 = 78 \text{ kNm}$$
$$M_{at\ D} = 90 + 4 \times 4 - \left(\frac{1}{2} \times 30 \times 2\right)\left(\frac{1}{3} \times 2 + 2\right) = 26 \text{ kNm}$$
$$M_{at\ E} = 90 + 4 \times 5 - \left(\frac{1}{2} \times 30 \times 2\right)\left(\frac{1}{3} \times 2 + 2 + 1\right) - 50 \times 1 = -50 \text{ kNm}$$
$$M_{at\ F} = 0$$

To find maximum bending moment:

Bending moment is maximum, where SF = 0, changing its sign, i.e., at point *X*, at a distance *x* from *B*.

Intensity of UVL at *X*, $\qquad \frac{30}{2} = \frac{w_x}{x}$

$\therefore \qquad w_x = 15x$

$$SF_{at\ x} = 4 - \frac{1}{2}(15x.x) = 0$$
$$7.5\,x^2 = 4$$
$$x = \sqrt{\frac{4}{7.5}} = 0.73 \text{ m form } B.$$

$\therefore$ Maximum bending moment, $M_{max} = M_x = 90 + (4 \cdot x) - \left(\frac{1}{2} \times w_x \cdot x\right)\frac{1}{3}.x$

$$= 90 + 40 \times 0.73 - \frac{1}{2} \times 15 \times \frac{0.73^3}{3} = 92.03 \text{ kNm}$$

(e) Draw BM as shown in Fig. 4.91 (c). For UVL at *BC,* BMD is cubic. For UDL at *EF*, BMD is parabolic. For moment at *A*, it is a vertical line.

(f) To locate the point of contraflexure.

Point of contraflexure is where BM = 0, changing its sign. It is seen that PoC is between *D* and *E* at a distance *y* from *D*.

From similar triangles of BMD between *DE*

$$\frac{26}{y} = \frac{50}{(1 - y)}$$

$$26\ (1 - y) = 50\ y$$

$$y = \frac{26 \times 1}{(26 + 50)} = 0.342 \text{ m from } D.$$

Point of contraflexure is at a distance of 0.342 m to the right of *D*.

Problem 61 Draw shear force and bending moment diagrams for a beam loaded as shown in Fig. 4.92. Locate the point of contraflexure and find maximum bending moment. Indicate the values.

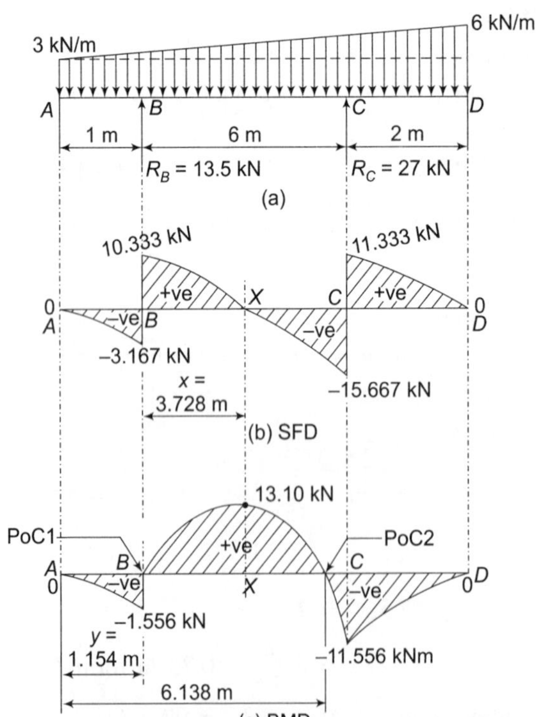

Fig. 4.92

Solution:

(a) **Reactions:** Consider the trapezoidal load as a UDL of 3 kN/m on entire length and a UVL that varies from O at A to 6 – 3 = 3 kN/m at D.

$$\Sigma V = 0 \Rightarrow R_B + R_C = (3 \times 9) + \frac{1}{2} \times 3 \times 9 = 40.5 \text{ kN}$$

Taking moment about B,

$$-3 \times 9 \times \left(\frac{9}{2} - 1\right) - \frac{1}{2} \times 9 \times 3 \left(\frac{2}{3} \times 9 - 1\right) + R_C \times 6 = 0$$

$$R_C = 27 \text{ kN}$$

$$\therefore \quad R_B = 40.5 - 27 = 13.5 \text{ kN}$$

(b) **Shear forces:**

$$\text{SF}_{\text{at } A} = 0$$

$$\text{SF}_{\text{up to } B} = -(3 \times 1) - \left(\frac{1}{2} \times w_B\, .1\right)$$

From similar triangles for UVL, $\frac{w}{l} = \frac{w_x}{x}$

$$\therefore \quad w_x = \frac{3}{9} \times x = \frac{x}{3}$$

$$x = 1\ m,\ w_B = \frac{1}{3}$$

$$\text{SF}_{\text{up to } B} = -(3 \times 1) \times \left(\frac{1}{2} \times \frac{1}{3} \times 1\right) = -3.167 \text{ kN}$$

$$\text{SF}_{\text{at } B} = -3.167 + 13.5 = +10.\ 333 \text{ kN}$$

$$\text{SF}_{\text{up to } C} = -(3 \times 7) - \left(\frac{1}{2}.w_c.7\right) + 13.5$$

$$w_c = \frac{x}{3} = \frac{7}{3} \text{ since distance } x = 7 \text{ m}$$

$$\text{SF}_{\text{upto } C} = -(3 \times 7) - \left(\frac{1}{2} \times \frac{7}{3} \times 7\right) + 13.5 = -15.667 \text{ kN}$$

$$\text{SF}_{\text{at } C} = -15.667 + 27 = 11.333 \text{ kN}$$

$$\text{SF}_{\text{at } D} = -(3 \times 9) - \left(\frac{1}{2} \times 3 \times 9\right) + 13.5 + 27 = 0$$

(c) Draw SFD as shown in Fig. 4.92 (b).

$\text{SF}_{\text{at } X}$ between B and C, $\text{SF}_x = -(3.\ x) - \left(\frac{1}{2}.w_x.x\right) + 13.5$

$$= -3x - \frac{1}{2}.\frac{x^2}{3} + 13.5$$

SF is parabolic in nature.

(d) **Bending moments:**

$$M_{\text{at } A} = 0$$

$$M_{\text{at } B} = -(3 \times 1)\frac{1}{2}\left(\frac{1}{2} \times \frac{1}{3} \times 1\right) \cdot \frac{1}{3} \times 1 = 1.556 \text{ kNm}$$

$$M_{\text{at } C} = -(3 \times 7)\frac{7}{2} - \left(\frac{1}{2} \times \frac{7}{3} \times 7\right)\left(\frac{1}{3} \times 7\right) + 13.5 \times 6 = -11.556 \text{ kNm}$$

$$M_{\text{at } D} = 0$$

Maximum bending moment.

BM is maximum at SF = 0, changing its sign. It is seen from SFD, that at X, between B and C, $SF_x = 0$

$$SF_x = -3x - \frac{x^2}{6} + 13.5 = 0$$

$$x^2 + 18x - 81 = 0$$

$$x = 3.728 \text{ m from } A$$

$$M_{ax} \text{ BM} = M_{max} = M_x = -(3.x) \cdot \frac{x}{2} - \left(\frac{1}{2} \times w_x.x\right)\frac{1}{3} \cdot x + 13.5\,(x - 1)$$

But $\quad w_x = \frac{x}{3}$ and $x = 3.728$ m

$$M_{max} = -3 \times \frac{3.728^2}{2} - \left(\frac{1}{2} \times \frac{3.728}{3} \times 3.728\right) \cdot \frac{3.728}{3} + 13.5\,(3.728 - 1)$$

$$= +13.\ 10 \text{ kN-m}$$

(e) Draw BMD as shown in Fig. 4.92 (e). BMD is cubic in nature as seen above.

(f) To locate point of contraflexure: Point of contraflexure is where BM = 0. It is seen from BMD that BM is zero between B and C at two planes.

$$M_y = -(3.y)\left(\cdot\frac{y}{2}\right) - \frac{1}{2}(w_y \cdot y)\frac{y}{3} + 13.5\,(y - 1) = 0$$

but $\quad w_y = \frac{y}{3} = -1.5\,y^2 - \frac{y^3}{18} + 13.5\,(y - 1) = 0$

$$\frac{y^3}{18} + 1.5\,y^2 - 13.5\,y + 13.5 = 0$$

$$y = 6.138 \text{ and } y = 1.154 \text{ m from } A.$$

There are two points of contraflexure at 1.154 m and 6.138 m from left end A.

4.13 LOAD DIAGRAM FROM GIVEN SHEAR FORCE DIAGRAM

It is sometimes necessary to prepare load diagram from a shear force diagram and find the loads and reactions. Further, bending moments are calculated and bending moment diagram will be drawn.

The following points may be kept in mind while analysing the given SFD.

(i) For a vertical line on SFD, there will be a vertical force or a vertical reaction.

(ii) If SFD is horizontal between two points, then there is no change in load between any two points.

(iii) If SFD is triangular between any two points, then it represents a UDL between those two points. If F_1 and F_2 are the shear forces over a span of l, then UDL $= \dfrac{F_1 - F_2}{a}$ N/m.

(iv) If SFD is parabolic between any two points then it represents a UVL between those two points.

Consider the following example:

For the SFD shown in Fig. 4.93, the following point may be noted.

- Since all the SFD is only on one side of horizontal line, it should be a cantilever.

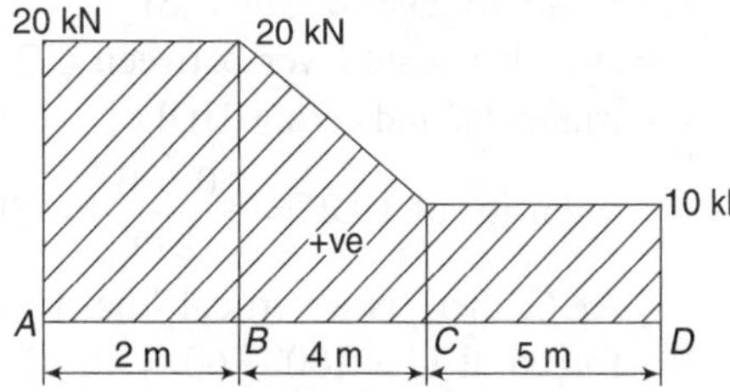

Fig. 4.93

SOLVED PROBLEMS

Problem 62 For a given force diagram shown in Fig. 4.94 prepare the load diagram and draw bending moment diagram indicating values.

Solution:

(a) **Load diagram:**

- Since all the shear forces are on one side of horizontal line, it is a cantilever.
- Each vertical line indicates a vertical force.

At D, there is a vertical line, hence the vertical force at D = 30 kN.

At C, there is another vertical line indicating another force = 50 – 30 = 20 kN.

At B, vertical force = 90 – 50 = 40 kN.

A is fixed end, the V = 90 kN = is the sum of all forces. Draw the load diagram as shown in Fig. 4.94 (b)

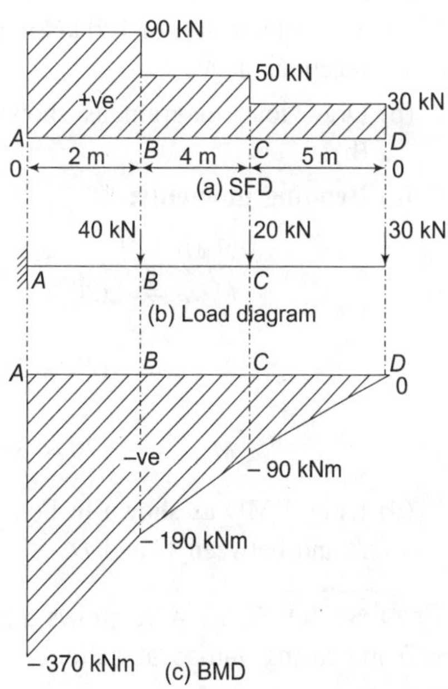

Fig. 4.94

(b) **Bending moment:**

$$M_{\text{at } D} = 0$$

$$M_{\text{at } C} = -\ 30 \times 3 = -190 \text{ kNm}$$

at B, $M_B = -30 \times 5 - 20 \times 2 = -90$ kNm

at A, $M_A = -30 \times 7 - 20 \times 4 - 40 \times 2 = -370$ kNm

(c) Draw BMD as shown in Fig. 4.94 (c)

Problem 63 For a given force diagram in Fig. 4.95, identify the load diagram and draw bending moment diagram, indicating the values.

Solution:

(a) **Load diagram:** Since SFD is only on one side, it is cantilever between CD, the SFD is triangular indicating UDL.

Intensity of UDL $= \dfrac{60 - 0}{3 \text{ m}} = 20$ kN/m

At C, there is a vertical line. Hence, vertical forced at $C = 100 - 60 = 40$ kN.

Between B and C, the SFD is horizontal, hence no force.

Between A and B, the SFD is triangular, hence UDL.

Intensity of UDL $= \dfrac{120 - 100}{4 \text{ m}} = 5$ kN/m.

Vertical line at A of 120 kN indicates a hidden reactions at A.

(b) Draw load diagram as shown in Fig. 4.95 (b)

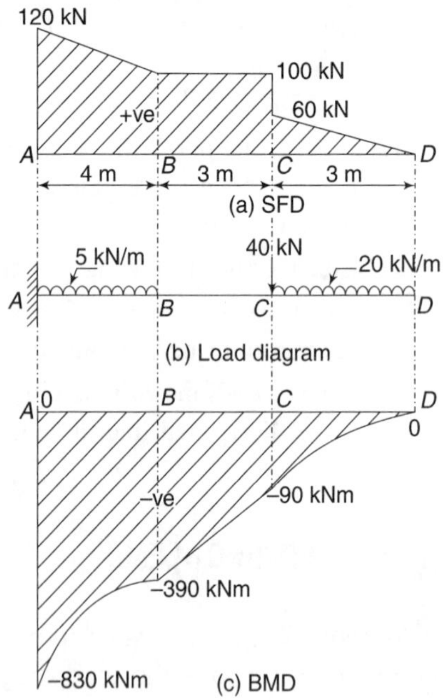

Fig. 4.95

(c) **Bending moments:**

$$M_{\text{at } D} = 0$$

$$M_{\text{at } C} = -(20 \times 3)\frac{3}{2} = -90 \text{ kNm}$$

$$M_{\text{at } B} = -(20 \times 3)\left(\frac{3}{2} + 3\right) - 40 \times 3 = -390 \text{ kNm}$$

$$M_{\text{at } A} = -(20 \times 3)\left(\frac{3}{2} + 3 + 4\right) - 40 \times 7 - (5 \times 4)\frac{4}{2} = -830 \text{ kNm}$$

(d) Draw BMD as shown in Fig. 4.95 (a). For UDL, BMD is parabolic between A and B and between C and D.

Problem 64 Figure 4.96 shows a SFD for a cantilever. Prepare load diagram and draw BMD indicating values at salient points.

Solution:

(a) **Load diagram:** Between $C - D$, the SFD is triangular, indicating UDL

Intensity of UDL $= \dfrac{60 - 0}{2 \text{ m}} = 30$ kN/m

At C, the vertical line indicates a concentrated force = 80 – 60 = 20 kN

Between B and C, the SFD is triangular and hence UDL

$$\text{Intensity of UDL} = \frac{140 - 80}{2 \text{ m}} = 30 \text{ kN/m}$$

At B, the vertical line corresponds to a vertical force = 150 – 140 = 10 kN

Between A and B, the SFD is triangle and hence UDL

$$\text{Intensity of UDL} = \frac{210 - 150}{2 \text{ m}} = 30 \text{ kN/m}$$

Hence, the entire UDL is of same intensity, 30 kN/m.

(b) Draw load diagram as shown in Fig. 4.96 (b)

(c) **Bending moment:**

$$M_{\text{at}\,D} = 0$$

$$M_{\text{at}\,C} = -(30 \times 2)\frac{2}{2} = -60 \text{ kNm}$$

$$M_{\text{at}\,B} = -(30 \times 4)\frac{4}{2} - 20 \times 2 = -280 \text{ kNm}$$

$$M_{\text{at}\,A} = -(30 \times 6) \times \frac{6}{2} - 20 \times 4 - 10 \times 2 = -640 \text{ kNm}$$

(d) Draw BMD as shown in Fig. 4.96 (c). For UVL, BMD is parabolic.

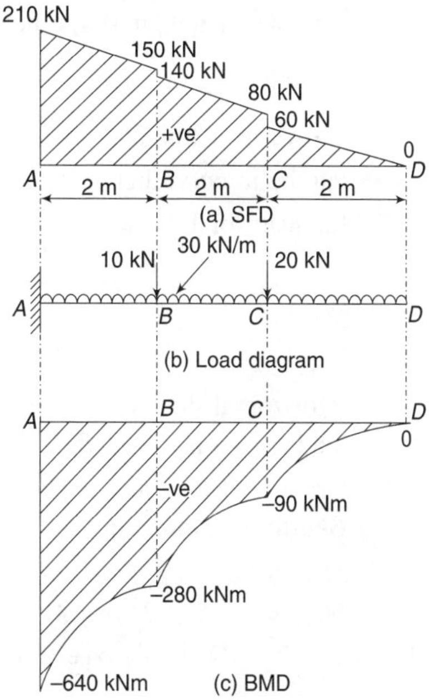

Fig. 4.96

Problem 65 For Fig. 4.97, draw load and bending moment diagrams. Indicate values at salient points.

Solution:

(a) **Load diagram:**

Since the SF is only on one side of horizontal line, the beam is a cantilever.

- The vertical line at E indicates a downward force of 25 kN at E.
- The horizontal line between D and E indicates no change in SF and hence there is no force between D and E.

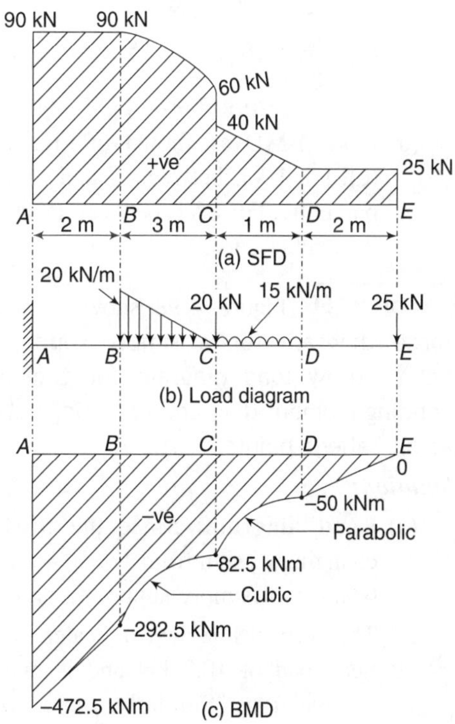

Fig. 4.97

- The triangular shape between C and D indicates a UDL.

  $$\text{Intensity of UDL} = \frac{40 - 25}{1 \text{ m}} = 15 \text{ kN/m}$$

- Vertical line at C is a point load at $C = 60 - 40 = 20$ kN

- Parabolic curve between B and C indicates a UVL

  Intensity of UVL = w

  $$\text{Force} = \frac{1}{2}(w.l) = 90 - 60 = 30$$

  $$w = \frac{30 \times 2}{3} = 20 \text{ kN/m}$$

- Horizontal line between A and B shows that the SF is constant and there is no other load between A and B.

(b) Draw load diagram as shown in Fig. 4.97 (b).

(c) **Bending moments:**

$M_{\text{at}E} = 0$

$M_{\text{at}D} = -25 \times 2 = -50$ kN-m

$M_{\text{at}C} = -25 \times 3 - (15 \times 1)\frac{1}{2} = -82.5$ kNm

$$M_{\text{at}B} = -25 \times 6 - (15 \times 1)\left(\frac{1}{2} + 3\right) - \left(\frac{1}{2} \times 20 \times 3\right)\frac{1}{3} \times 3 - 20 \times 3$$

$$= -292.5 \text{ kNm}$$

$$M_{\text{at}A} = -25 \times 8 - (15 \times 1)\left(\frac{1}{2} + 3 + 2\right) - \left(\frac{1}{2} \times 20 \times 3\right)\left(\frac{1}{3} \times 3 + 2\right) - 20 \times 5 = -472.5 \text{ kNm.}$$

(d) Draw BMD as shown in Fig. 4.97 (c). For UVL between C and D, it is parabolic. For UVL between B and C, it is cubic curve.

Problem 66 Figure 4.98 shows a shear force diagram for a simply supported beam. Draw load diagram for this and bending moment diagram indicating values at all salient points.

Solution:

(a) **Load diagram:** Since the SFD is changing its sign from +ve to –ve, the beam is of simply supported type.

The vertical line at A indicates a vertical load of 10.2 kN and is one of the reactions. Similarly, the vertical line at D indicates another reaction of magnitude 7.8 kN.

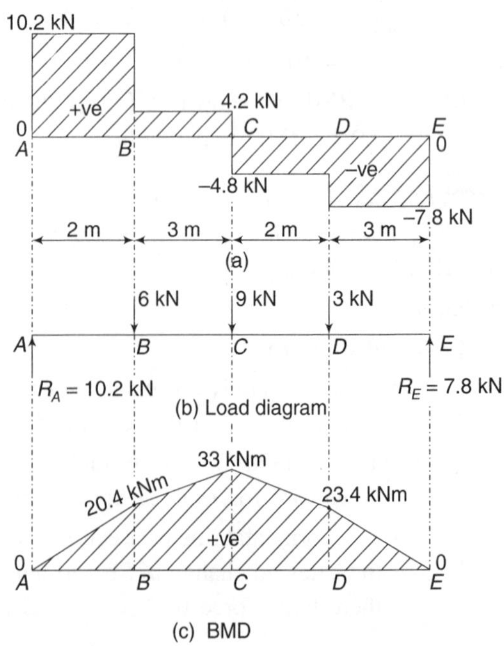

Fig. 4.98

The vertical lines at B, C and D indicate downward loads of

$$10.2 - 4.2 = 6 \text{ kN at } B,$$
$$4.2 - (-4.8) = 9 \text{ kN at } C,$$
$$-4.8 - (-7.8) = 3 \text{ kN at } D,$$

The horizontal lines between AB, BC, CD and DE indicate there are no other loads.

(b) Draw the load diagram as shown in Fig. 4.98 (b)

(c) **Bending moments:**

$$M_{\text{at}A} = 0$$
$$M_{\text{at}B} = 10.2 \times 2 = 20.4 \text{ kNm}$$
$$M_{\text{at}C} = 10.2 \times 5 - 6 \times 3 = 33 \text{ kN-m}$$
$$M_{\text{at}D} = 10.2 \times 7 - 6 \times 5 - 9 \times = 23.4 \text{ kNm}$$
$$M_{\text{at}E} = 0$$

(d) Draw BMD as shown in Fig. 4.98 (c).

Problem 67 From the shear force diagram shown in Fig. 4.99, draw load diagram and bending moment diagrams indicating values at salient points.

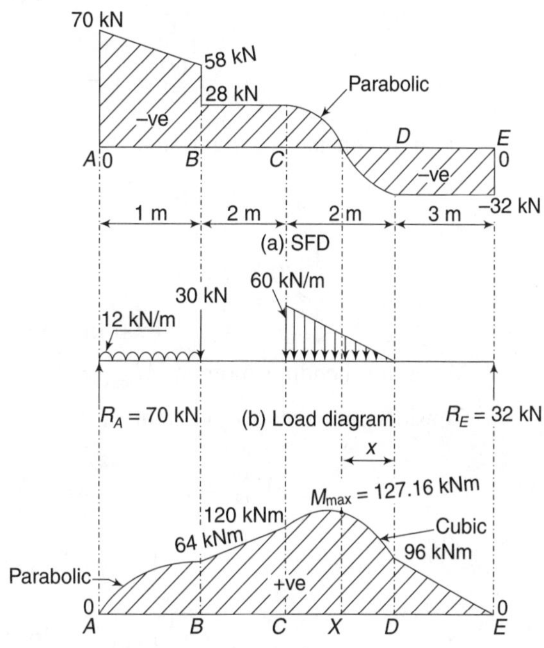

Fig. 4.99

Solution:

(a) **Load diagram:**

It can be seen from given SFD that there are two vertical lines as reactions at A and E.

$R_A = 70$ kN and $R_E = 32$ kN

Between A and B, the SFD is inclined line indicating UDL.

$$\text{Intensity of UDL} = \frac{70 - 58}{1 \text{ m}}$$
$$= 12 \text{ kN/m}.$$

Between B and C, SFD is horizontal, hence there is no other load.

Between C and D, SFD is parabolic, hence there will be UVL.

$$\text{UVL} = \frac{1}{2} w \cdot l = 28 - (-32) = 60 \text{ kN}$$

$$\text{UVL intensity } w = \frac{60 \times 2}{2} = 60 \text{ kN/m}$$

(b) Draw the load diagram as shown in Fig. 4.99 (b)

(c) **bending moments:**

$M_{\text{at } A} = 0$

$M_{\text{at } B} = 70 \times 1 - (12 \times 1)\, 0.5 = 64$ kNm

$M_{\text{at } C} = 70 \times 3 - (12 \times 1)\left(\frac{1}{2} + 2\right) - 30 \times 2 = 120$ kN-m

$$M_{\text{at } D} = 70 \times 5 - (12 \times 1)\left(\frac{1}{2} + 2 + 2\right) - 30 \times 4 - \left(\frac{1}{2} \times 60 \times 2\right)\frac{2}{3} \times 2$$

$= 96$ kN-m

$M_{\text{at } E} = 0$

To find maximum bending moment,

BM is maximum, where SF = 0, changing its sign.

Consider x, at x from C

Load intensity, $\quad \dfrac{w}{l} = \dfrac{w_x}{x}$

$$w_x = \frac{60}{2} \times x = 30\,x$$

SF at x, considering right side,

$$\frac{1}{2}(w_x \cdot x) - 32 = 0$$

$$\frac{1}{2} \cdot 30\, x \cdot x = 32$$

$$x = 1.4606 \text{ m from } D.$$

Maximum bending moment $M_{\max} = M_x$

Consider moments to the right of X,

$$M_{\max} = -\frac{1}{2} \cdot (w_x \cdot x) \times \frac{1}{3} . x + 32\,(x + 3)$$

$$= -\frac{1}{2} \times 30 \cdot \frac{x^3}{3} + 32\,(x + 3)$$

$$= -\frac{1}{2}\, 30 \times \frac{1.4606^3}{3} + 32\,(1.4606 + 3) = 127.16 \text{ kNm}$$

(d) Draw BM D as shown in Fig. 4.99 (c).

For UDL between AB, BMD is parabolic

For UVL between CD, it is cubic.

Problem 68 For the shear force diagram shown in Fig. 4.100 draw load diagram and bending moment diagram indicating at salient points.

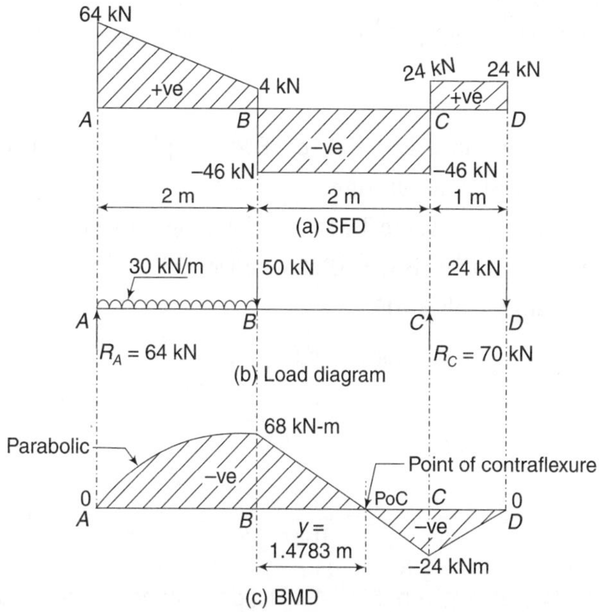

Fig. 4.100

Solution:

(a) **Load diagram:** Since SFD is crossing the horizontal line twice, it is overhanging beam.

Reactions at A, $\qquad R_A = 64$ kN

Reactions at C, $\qquad R_C = 24 + 46 = 70$ kN

SFD is triangular between A and B, hence, it is UDL

$$\text{UDL} = \frac{64-4}{2} = 30 \text{ kN/m}$$

Vertical downward line at B is a point load = 4 + 46 = 50 kN

Horizontal line between B and C indicates, no change in loading.

Vertical upward line at C is a vertical upward load

Reaction at C = 46 + 24 = 70 kN

Horizontal line between C and D shows no change in load between C and D.

Vertical line at D is a downward force = 24 kN.

(b) Draw load diagram as shown in Fig. 4.100 (b)

(c) **Bending moments:**

$$M_{\text{at } A} = 0$$

$$M_{\text{at } B} = 64 \times 2 - \left(\frac{1}{2} \times 60 \times 2\right)\frac{1}{2} \times 2 = 68 \text{ kNm}$$

$$M_{\text{at } C} = 64 \times 4 - \left(\frac{1}{2} \times 60 \times 2\right)\left(\frac{1}{2} \times 2 + 2\right) - 50 \times 2 = -24 \text{ kNm}$$

$$M_{\text{at } D} = 0$$

(d) Draw BMD as shown in Fig. 4.100 (c). BMD is parabolic between *AB* for UDL.

(e) To locate the point of contraflexure

Point of contraflexure is where BM = 0, changing its sign.

It is seen from BMD that, BM = 0 at *y* between *B* and *C*

From similar triangles of BMD

$$\frac{68}{y} = \frac{24}{2 - y}$$

$$68\,(2 - y) = 24\,y$$

$$\therefore \quad y = \frac{68 \times 2}{(68 + 24)} = 1.4783 \text{ m from } B.$$

Problem 69 From the SFD given Fig. 4.101, draw load diagram and draw bending moment diagram indicating values at the salient points.

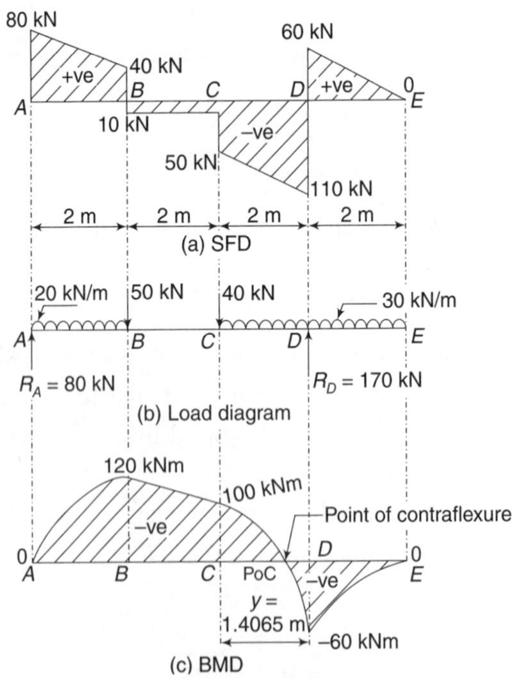

Fig. 4.101

Solution:

(a) **Load diagrams:** Since SFD is zero, changing its sign at two places, it is a simply supported beam with one side overhanging. It can be seen by observation that there will be one support at *A* and the other a *D*.

Reaction *A* $R_A = 80$ kN

Reaction *D* $R_D = 60 + 110 = 170$ kN

Between *A* and *B*, the SFD is triangular indicating UDL.

Intensity of UDL, $w = \dfrac{80 - 40}{2 \text{ m}} = 20$ kN/m

Sudden vertical line at *B* indicates a downward force = 40 + 10 = 50 kN

Horizontal line between *B* and *C* shows SF is constant and no other force between *B* and *C*.

Vertical line at *C* indicates a downward force = 50 – 10 = 40 kN

Inclined line between *C* and is a UDL

Intensity $w = \dfrac{110 - 50}{2 \text{ m}} = \dfrac{60}{2} = 30$ kN/m

Upward vertical line at *D* is the reaction at *D* = 110 + 60 = 170 kN

Inclined line between *D* and *E* is UDL.

Intensity $w = \dfrac{60 - 0}{2 \text{ m}} = 30$ kN/m

It is observed that the UDL is same between *CD* and *DE*.

(b) Draw the load diagram as shown in Fig. 4.101 (b).

(c) **Bending moments:**

$M_{\text{at } A} = 0$

$M_{\text{at } B} = 80 \times 2 - (20 \times 2)\dfrac{2}{2} = 120$ kNm

$M_{\text{at } C} = 80 \times 4 - (20 \times 2)\left(2 + \dfrac{2}{2}\right) - 50 \times 2 = 100$ kN-m

$M_{\text{at } D} = 80 \times 6 - (20 \times 2)\left(4 + \dfrac{2}{2}\right) - 50 \times 4 - 40 \times 2 - (30 \times 2)\dfrac{2}{2} = -60$ kN

$M_{\text{at } E} = 0$

To find maximum bending moment.

BM is maximum where SF = 0, changing its sign at *B*.

$M_{\max} = 120$ kNm.

(d) Draw BMD as shown in Fig. 4.101 (c). For UDL, BM is parabolic between *AB*, *CD* and *DE*.

(e) To locate the point of contraflexure

Point of contraflexure is where BM = 0, changing its sign,

It is seen from BMD that BM is changing sign between C and D at a distance y from C.

$$\text{BM at } y,\ M_y = 80\ (4 + y) - (20 \times 2)\left(y + 2 + \frac{2}{2}\right)$$

$$- 50 \times (y + 2) - 40 \times y - (30 \times y)\frac{y}{2} = 0$$

$$320 + 80y - 40y - 120 - 50y - 100 - 40\ y - 15y^2 = 0$$

$$15y^2 + 50\ y - 100 = 0$$

$$y = 1.4065 \text{ m}$$

The point of contraflexure is at a distance of 1.4065 m from C.

Problem 70 For the SFD shown in Fig. 4.102, draw load diagram and bending moment diagram. Indicate values at various points.

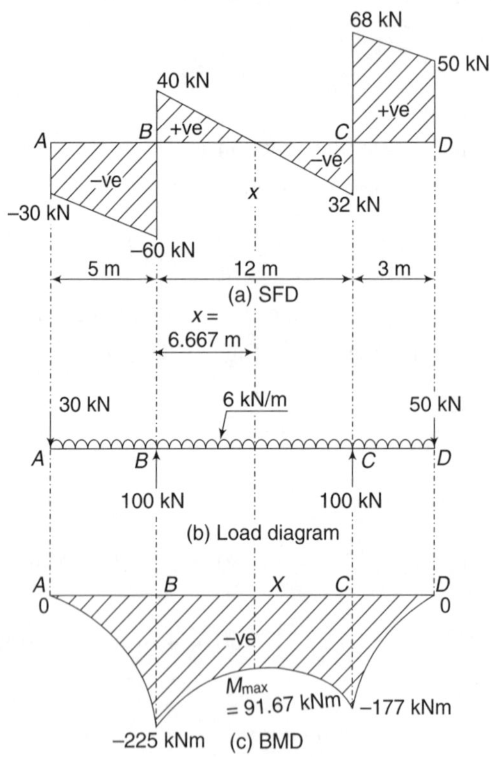

Fig. 4.102

Solution:

(a) **Load diagram:** Since the SFD is changing its direction thrice, it is overhanging on both sides. It can also be observed that there will be reactions at B and C.

- At A, the vertical downward line indicates a downward force of magnitude 30 kN
- The sloping line between A and B indicates a UDL.

  Intensity UDL $w = \dfrac{60 - 30}{5m} = 6$ kN/m

- The vertical line upward at B is the vertical reaction $R_B = 60 + 40 = 100$ kN
- The sloping line between B and C is again UDL $= \dfrac{40 - (-32)}{12} = 6$ kN/m
- Again the vertical upward line at C is the reaction at C

$$R_C = 32 + 68 = 100 \text{ kN}$$

- Sloping line between C and D is UDL $= \dfrac{68 - 50}{3} = 6$ kN/m

  Thus, all the UDLs are same = 6 kN/m.

  That is, the UDL from A to D is 6 kN/m (same)

- The vertical downward line at D is a downward point load = 50 kN.

(b) Draw load diagram as shown in Fig. 4.102 (b)

(c) **Bending moments:**

$$M_{\text{at}A} = 0$$

$$M_{\text{at}B} = -30 \times 5 - (6 \times 5)\frac{5}{2} = -225 \text{ kNm}$$

$$M_{\text{at}C} = -30 \times 17 - (6 \times 17)\left(\frac{17}{2}\right) + 100 \times 12 = 177 \text{ kNm}$$

$$M_{\text{at}D} = 0$$

To find maximum bending moment: BM is maximum where SF = 0, changing its sign. Say at x, at a distance x from B. From similar triangles of SFD, between B and C.

$$\frac{40}{x} = \frac{32}{12 - x}$$

$$40(12 - x) = 32x$$

$$x = \frac{40 \times 12}{40 + 32} = 6.667 \text{ m from } B$$

$$M_{max} \text{ BM} = M_{max} = M_x = -30(x + 5) + 100x - 6(x + 5)\frac{(x + 5)}{2}$$

$$M_{max} = -30x - 150 + 100x - 3(x^2 + 10x + 25) = -91.67 \text{ kNm}$$

(d) Draw BMD as shown in Fig. 4.102 (c) There is no point of contraflexure.

Problem 71 : For a given SFD in Fig. 4.103, draw load diagram and bending moment diagram.

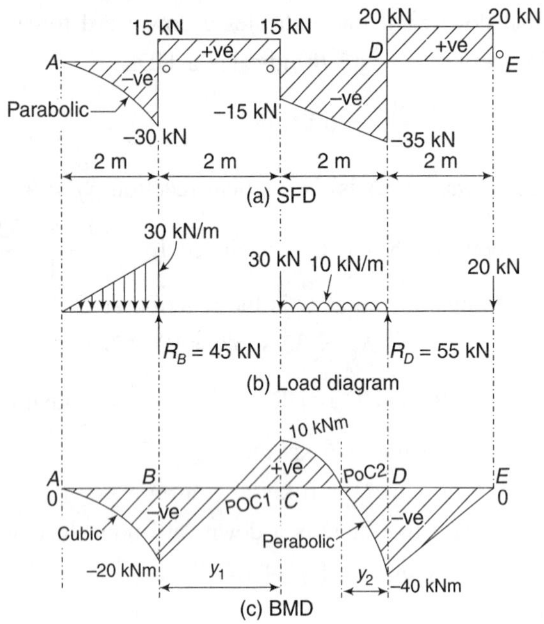

Fig. 4.103

Solution:

(a) **Load diagram:** Since the SFD is reversing at 3 places, it is an overhanging beam on both sides.

It can be seen that there are supports at B and D.

- Between A and B, the parabolic distribution indicates UVL.

$$\text{Load of UVL} = \frac{1}{2} \times w.l$$

$$30 = \frac{1}{2} \times w \times 2$$

$\therefore$ Intensity of UVL = 30 kN/m

- Vertical line at B is the reaction $R_B = 30 + 15 = 45$ kN
- Between B and C, there is no load
- Downward line at C is a downward load = 15 + 15 = 30 kN
- Sloping line between C and D is a UDL, intensity of UDL = $\dfrac{35 - 15}{2 \text{ m}}$ = 10 kN/m
- Vertical line at D is an upward reaction = 35 + 20 = 55 kN

$$R_D = 55 \text{ kN}$$

- Horizontal line between DE is an indication of no change in loading
- Vertical downward line at E is a downward force = 20 kN.

(b) Draw the load diagram as shown in Fig. 4.103 (b)

(c) **Bending moments:**

$M_{\text{at } A} = 0$

$M_{\text{at } B} = -\left(\frac{1}{2} \times 30 \times 2\right) \frac{1}{3} \times 2 = -20 \text{ kNm}$

$M_{\text{at } C} = -\left(\frac{1}{2} \times 30 \times 2\right) \left(\frac{1}{3} \times 2 + 2\right) + 45 \times 2 = +10 \text{ kNm}$

$M_{\text{at } D} = -\left(\frac{1}{2} \times 30 \times 2\right) \left(\frac{1}{3} \times 2 + 4\right) + 45 \times 4 - 30 \times 2 - (10 \times 2) \times \frac{2}{2}$

$= -40 \text{ kNm}$

$M_{\text{at } E} = 0$

(d) Draw BMD as shown in Fig. 4.103 (c), BMD is cubic for UVL at *AB* and parabolic at *CD* for UDL.

(e) To locate the point of contraflexure: It is seen from BMD that there are two points of contraflexure, one between *B* and *C* and the other between *C* and *D*.

(i) Let y_1 be the distance from *B* to PoC1. From similar triangles of BMD between *BC*

$$\frac{20}{y_1} = \frac{10}{2 - y_1}$$

$$20\,(2 - y_1) = 10\,y_1$$

$$y_1 = \frac{20 \times 2}{20 + 10} = 1.333 \text{ m from } B.$$

(ii) Let y_2 be the distance of PoC2 from *D*.

$$\text{BM at PoC2} = M_{\text{PoC 2}} = -20\,(2 + y_2) + 55 \times y_2 - (10 \times y_2)\,\frac{y_2}{2} = 0$$

(Considering to the right side)

$$= -40 - 20\,y_2 + 55y_2 - 5y_2^2 = 0$$

$$5y^2 - 35y^2 + 40 = 0$$

$$y_2 = 1.438 \text{ or } 5.562$$

Considering the possible value, $y_2 = 1.438$ m to left of *D*.

THINGS TO REMEMBER

- For point load, the shear force is a vertical line
- For UDL, the shear force is triangular (inclined line)
- For UVL, the shear force is parabolic
- For point load, the bending moment is triangular
- For UDL, the bending moment is parabolic
- For UVL, the bending moment is cubic.

- Sign conversation is:
  - For shear force, left side upward and right side downward is taken as positive and left side downward and right side upward is taken as negative
  - For bending moment, sagging moments are considered positive and hogging moment are taken as negative.
- Rate of change of bending moment is equal to shear force $\frac{dM}{dX} = -F.$
- Rate of change of force over unit length is load intensity $\frac{dF}{dX} = w.$
- Shear force at a section in a beam is the force which tries to shear off the beam at that section. It is the algebraic sum of the forces including reactions if any, acting perpendicular to the axis of the beam taken either to the left side or right side of the section.
- Bending moment at a section in a beam is the moment which bends the beam. It is the algebraic sum of moments about that section taken for all the forces including reactions if any, acting on either left side or right side of the beam
- A graphical representation of shear force against position of a section is called shear force diagram.
- A graphical representation of bending moment against position of a section is called bending moment diagram.
- Bending moment is maximum at a section where shear force is equal to zero and changing its sign.
- The section where bending moment is zero changing its sign is called point of contraflexure or point of inflexion.

Shear force and bending moment for same standard cases.

Type of beam and load	*Shear force*	*Bending moment*
1. Cantilever with point load at force end.	$F_{max} = W$ at any section	$M_{max} = -Wl$ at fixed end
2. Cantilever with UDL of w over full span	$F_{max} = wl$ at fixed end	$M_{max} = \frac{-wl^2}{2}$ at fixed end
3. Cantilever with UVL varying from zero at free end to w at fixe end	$F_{max} = \frac{wl}{2}$ at fixed end	$M_{max} = -\frac{wl^2}{6}$ at fixed end
4. Simply supported beam with central load W	$F_{max} = \frac{W}{2}$ at the supports	$M_{max} = \frac{Wl}{4}$ at the middle under the load
5. Simply supported beam with point load W at a distance of a from left end and b from right end	$F_{max} = \frac{Wb}{l}$ or $\frac{Wa}{l}$ at left side or at right side whichever is higher	$M_{max} = \frac{Wab}{l}$ under the load

Contd.

Contd...

6 Simply supported beam with UDL over entire span	$F_{max} = \frac{wl}{2}$ at the supports	$M_{max} = \frac{wl^2}{8}$ at the mid span
7. Simply supported beam with UVL over entire span	$F_{max} = \frac{wl}{3}$ at the end of max intensity side	$M_{max} = \frac{wl^2}{9\sqrt{3}}$ at a distance of 0.577 l from zero end
8. Simply supported beam with a distance a from left end and b from right end	$F_{max} = \frac{M}{l}$ at any section	$M_{max} = \frac{Ma}{l}$ or $\frac{Mb}{l}$ whichever is maximum at the moment

EXERCISES

1. Draw shear force and bending moment diagrams for a cantilever of length l carrying a concentrated load of W at its free end.
2. Draw shear force and bending moment diagrams for a cantilever of l carrying a concentrated load of W at a distance of a from the fixed end.
3. Draw shear force and bending moment diagrams for a cantilever of length l carrying a UDL of w N/m over the entire span.
4. Draw shear force and bending moment diagrams for a cantilever of length l carrying a UDL of w over a length a from the fixed end.
5. Draw shear force and bending moment diagrams for a cantilever of length l carrying a UDL w N/m over a length of a at the free end.
6. Draw shear force and bending moment diagrams for a cantilever of length l carrying a uniformly varying load varies from zero at free end to w N/m at the fixed end.
7. Draw shear force and bending moment diagrams for a cantilever of length l carrying a UVL varying from zero at fixed end to w at free end.
8. Draw shear force and bending moment diagrams for a simply supported beam of length l carrying a central load W.
9. Draw shear force and bending moment diagrams for a simply supported beam of length l carrying a concentrated load W at a distance a from left end.
10. Draw shear force and bending moment diagrams for a simply supported beam carrying a UDL of w N/m over full length.
11. Draw shear force and bending moment diagrams for a simply supported beam subjected to a moment M at a distance a from left end.
12. Draw shear force and bending moment diagrams for a simply supported beam carrying a UVL that varies from o to w N/m over entire span.
13. Draw shear force and bending moment diagrams for a beam loaded as shown in Fig. 4.104.

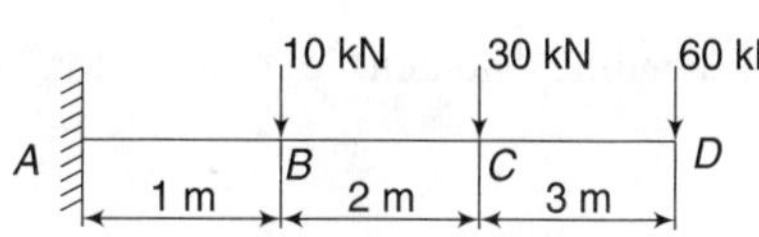

Fig. 4.104

14. Draw shear force and bending moment diagrams for a beam loaded as shown in Fig. 4.105.

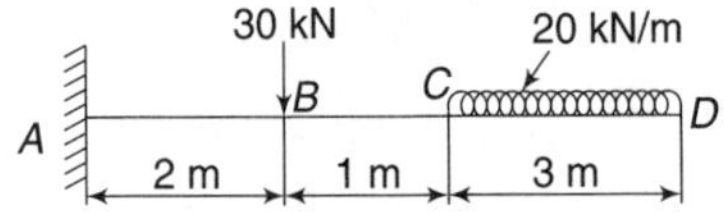

Fig. 4.105

15. Draw shear force and bending moment diagrams for a beam loaded as shown in Fig. 4.106

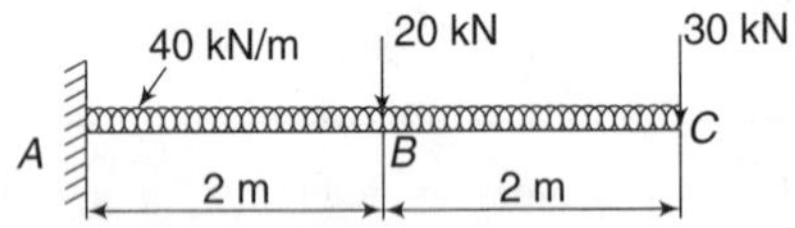

Fig. 4.106

16. Draw shear force and bending moment diagrams for a beam loaded as shown in Fig. 4.107.

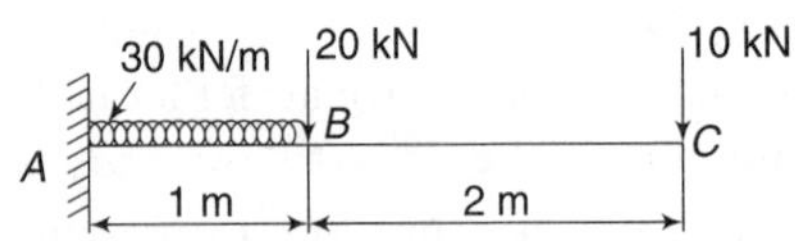

Fig. 4.107

17. Draw shear force and bending moment diagrams for a beam loaded as shown in Fig. 4.108.

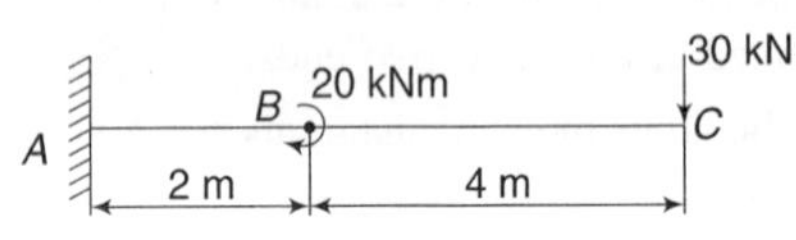

Fig. 4.108

18. Draw shear force and bending moment diagrams for a beam loaded as shown in Fig. 4.109.

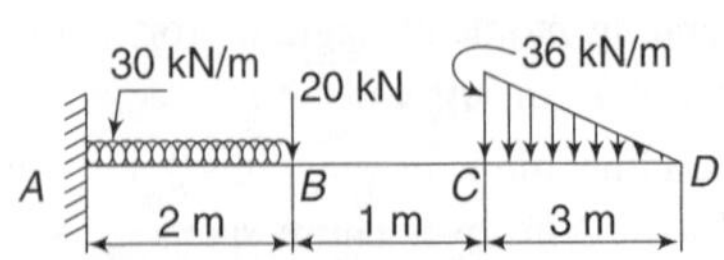

Fig. 4.109

19. Draw shear force and bending moment diagrams for a beam loaded as shown in Fig. 4.110.

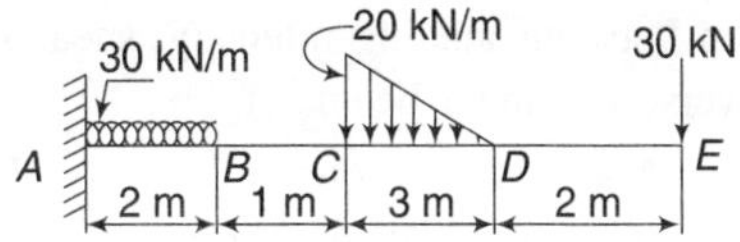

Fig. 4.110

20. Draw shear force and bending moment diagrams for a beam loaded as shown in Fig. 4.111.

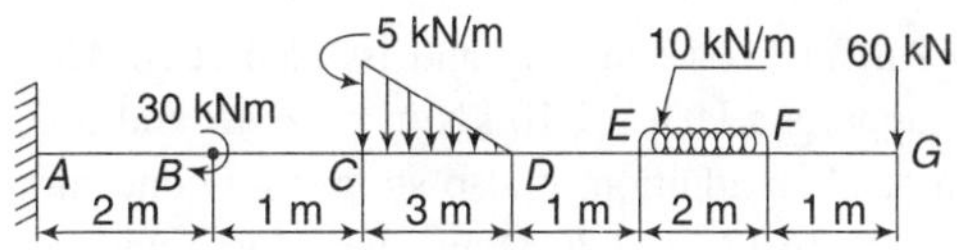

Fig. 4.111

21. Draw shear force and bending moment diagrams for a beam loaded as shown in Fig. 4.112.

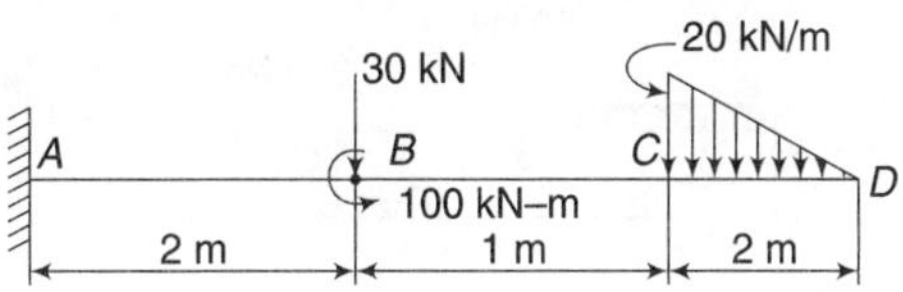

Fig. 4.112

22. Draw shear force and bending moment diagrams for a beam loaded as shown in Fig. 4.113.

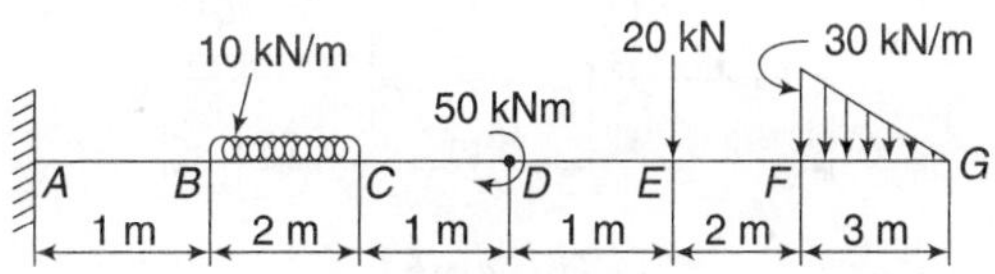

Fig. 4.113

23. Draw bending moment and shear force diagrams for a beam loaded as shown in Fig. 4.114.

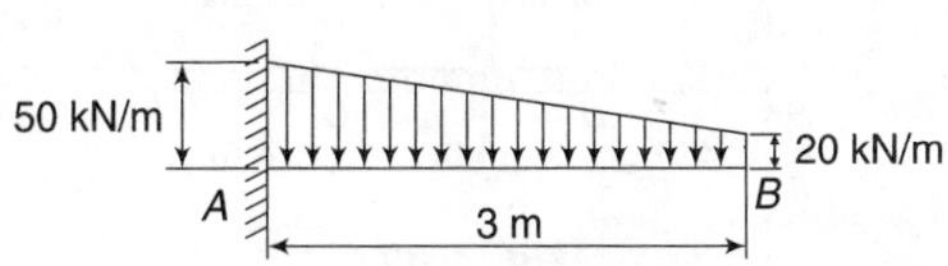

Fig. 4.114

24. Establish the relationship between loading intensity, shear force and bending moment at a cross section of a transversely straight beam.

25. A beam *ABCD* is fixed at *A* and is 8 m long. It carries a UDL of 10 kNm over a span of 4 m at *AB*, There is a clockwise couple of 60 kN m at the free and *D*. It also carries a point load of 30 kN at *C*. Draw shear force and bending moment diagrams.

26. A cantilever beam *ABCDE* is fixed at *A* and carries a UDL of 30 kN/m on *AC* which is 3 m. It carries a point load of 10 kN at *B* at a distance of 1.5 m from *A* and a UVL that varies from *O* at free end *E* to 40 kN/m at *E* over a span of 2 m. The distance between *C* and *D* is 1 m. Draw shear force and bending moment diagrams.

27. A cantilever beam *ABCDE* is 12 m long and is fixed at *A*. *AB* = *DE* = 2 m and *BC* = *CD* = 4 m. The beam supports a UDL of 10 kN/m over *AB* and a UVL on *CD* that varies from *O* at *D* to 30 kN/m at *C*. In addition, it also supports a concentrated load of 10 kN at *E* and a clockwise moment of 16 kNm at *B*. Draw shear force and bending moment diagrams.

28. A beam *ABCD* is simply supported at *A* and *D* at a distance of 6 m. It carries a concentrated load of 30 kN at *B* at a distance of 1.5 m from *A* and another one of 40 kN at *C* at a distance of 2 m from *D*. Draw shear force and bending moment diagrams.

29. Draw shear force and bending moment diagrams for a beam shown in Fig. 4.115.

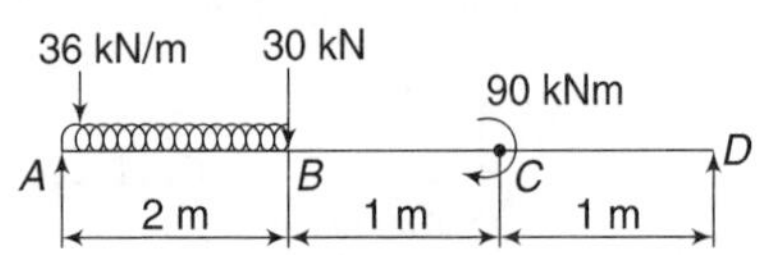

Fig. 4.115

30. Draw shear force and bending moment diagrams for a beam shown in Fig. 4.116.

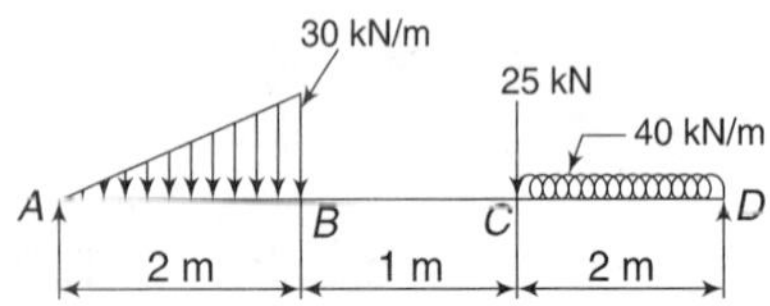

Fig. 4.116

31. Draw shear force and bending moment diagrams for a simply supported beam with loads as shown in Fig. 4.117.

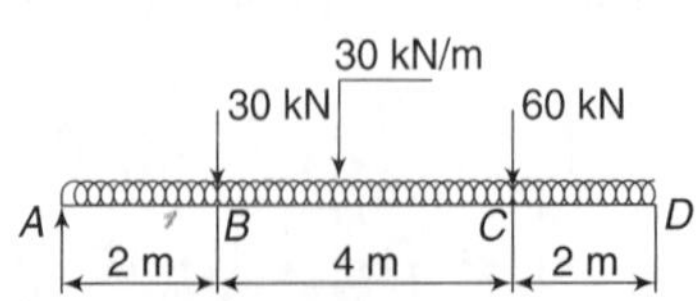

Fig. 4.117

32. Draw shear force and bending moment diagrams for a simply supported beam shown in Fig. 4.118.

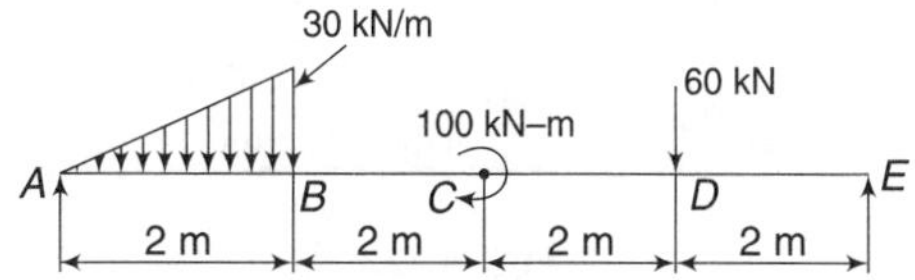

Fig. 4.118

33. Draw shear force and bending moment diagrams for a simply supported beam shown in Fig. 4.119. Indicate values at salient points.

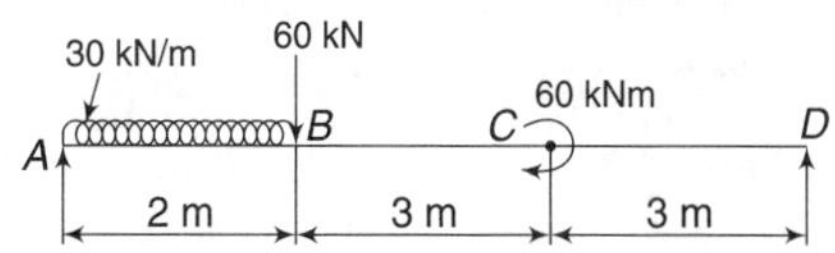

Fig. 4.119

34. A beam *ABCD*, is simply supported at *B* and *C*, 4.5 m apart and overhanging parts *AB* and *CD* are 1.5 m and 2 m long respectively. It carries a UDL of 10 kN/m between *A* and *C*. There is a clockwise couple of 60 kN-m at *D*. Draw shear force and bending moment diagrams. Find the position of points of contraflexure.

35. Draw shear force and bending moment diagrams for a beam as shown in Fig. 4.120. Locate the point of contraflexure if any.

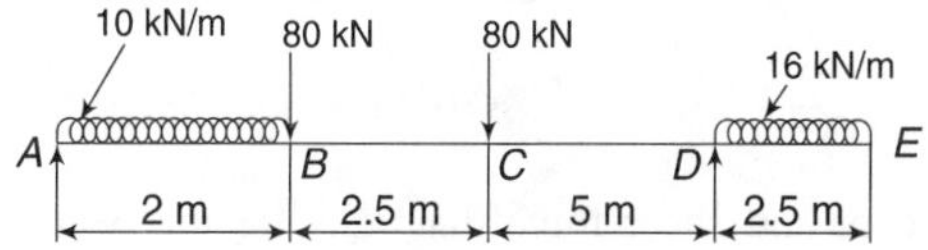

Fig. 4.120

36. Construct SFD and BMD for the beam loaded as shown in Fig. 4.121. Determine the location and magnitude of maximum bending moment and point of contraflexure.

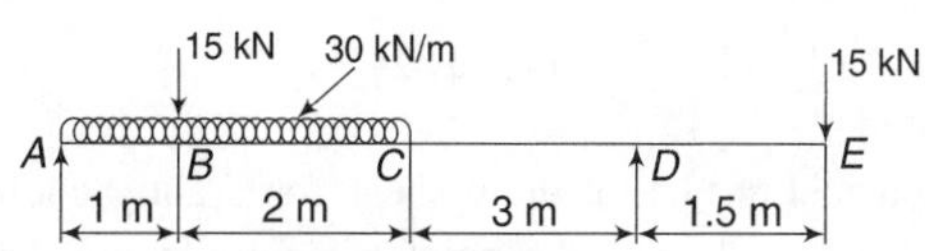

Fig. 4.121

37. A beam *ABCDE* is 12 m long simply supported at point *B* and *D*. *AB* = *DE* = 2 m and are overhanging *BC* = *CD* = 4 m. The beam supports a UDL of 10 kN/m over *AB* and 20 kN/m over *CD*. In addition, it also supports concentrated load of 10 kN at *E* and a clockwise moment of 16 kNm at point *C*. Sketch BMD and SFD.

38. Draw SFD and BMD for a beam loaded as shown in Fig. 4.122.

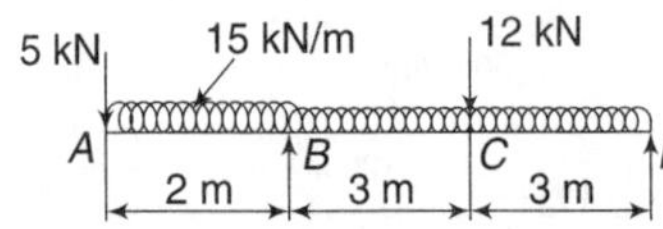

Fig. 4.122

39. Draw SFD and BMD for a beam loaded as shown in Fig. 4.123. Indicate the values at all salient points. Locate point of contraflexure. Further calculate the maximum bending moment.

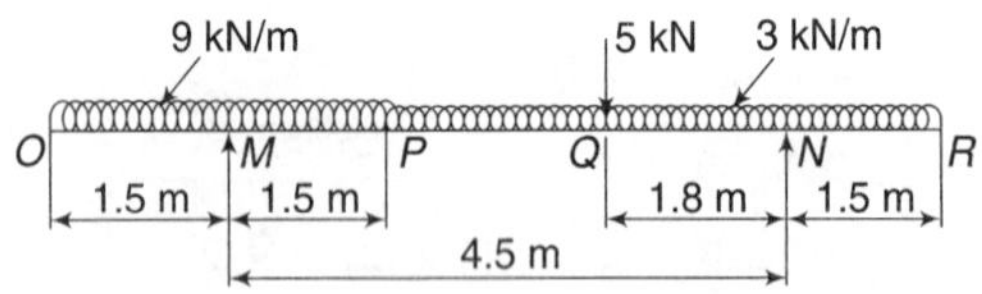

Fig. 4.123

40. Draw SFD and BMD for a beam loaded as shown ion Fig. 4.124. Locate the point of contra- flexure. Mark the values at salient points.

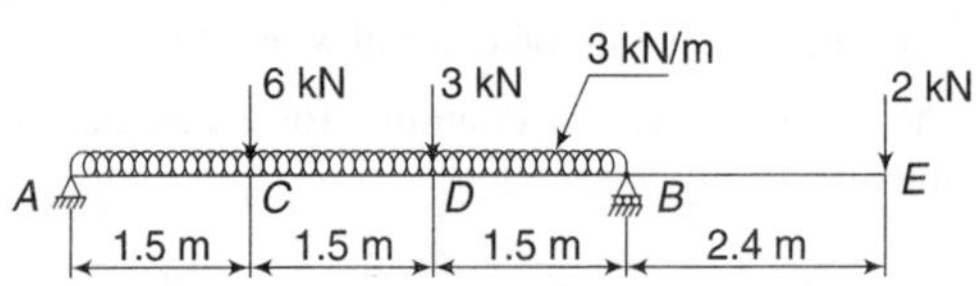

Fig. 4.124

41. Draw bending moment and shear force diagrams for the beam shown in Fig. 4.125. Mark the salient values on the diagram. Locate point of inflexion.

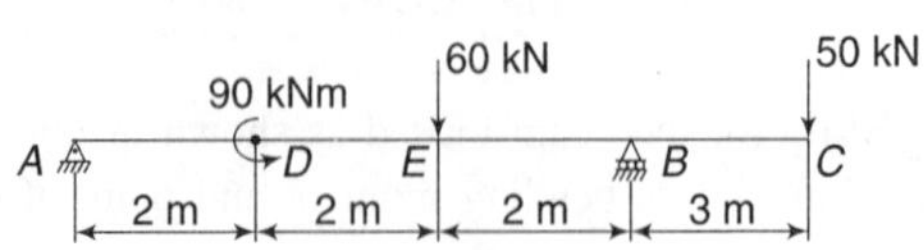

Fig. 4.125

42. For the beam shown in Fig. 4.126, draw shear force and bending moment diagram. Mark the values at salient points. Locate the point of contraflexure if any.

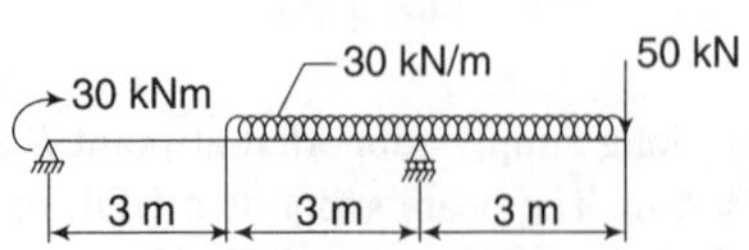

Fig. 4.126

43. A beam *AB*, 20 m long supported on two intermediate props 12 m apart carries a UDL of 10 kN/m together with concentrated loads of 32 kN at the left end *A* and 60 kN at the right end *B*. The props are so located that the reactions are same at each support. Determine the position of props and draw SFD and BMD. Locate the point of contraflexure.

44. For a beam shown in Fig. 4.127, obtain shear force and bending moment at salient points. Hence, draw SFD and BMD. Mark maximum value of bending moment and locate point of contraflexure if any.

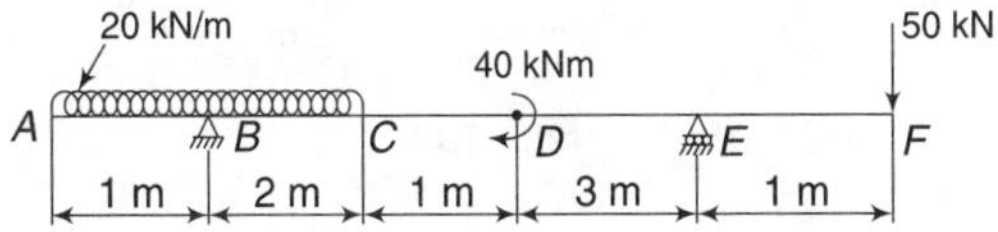

Fig. 4.127

45. For a beam shown in Fig. 4.128, obtain SFD and BMD indicate the values at salient points.

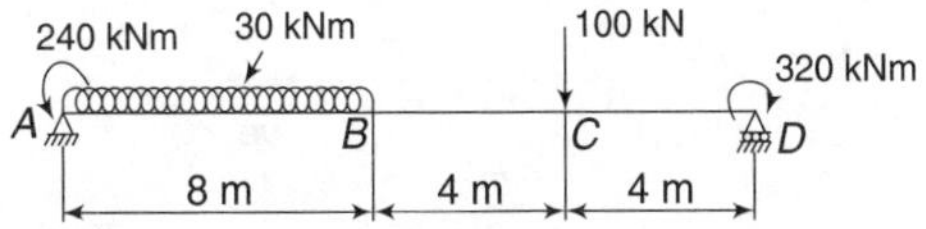

Fig. 4.128

46. For the beam shown in Fig. 4.129, find the load *P* to have equal reactions at *A* and *C*. Draw SF and BM diagrams indicating their values at significant points. Locate the point of contraflexure.

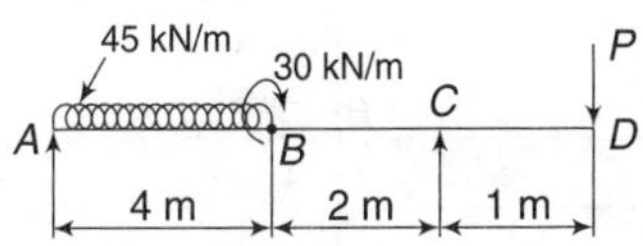

Fig. 4.129

47. For a beam loaded as shown in Fig. 4.130, find the overhang value *a* if the maximum positive and negative bending moments are same. Draw SFD and BMD and indicate the values. Locate the point of contraflexure.

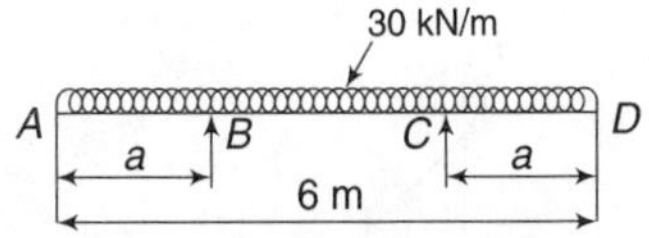

Fig. 4.130

48. A beam 8.5 m long rests on supports 5 m apart, the right hand end overhanging by 2 m. The beam carries a UDL of 50 kN/m between the supports. The beam also carries a point load of 60 kN at the extreme right end 40 kN at the extreme left end. Construct SF and BMDs and indicate all important valves on them. Locate point of inflexion of the beam.

49. Draw load diagram and BMD for a shear force diagram shown below (Fig. 4.131).

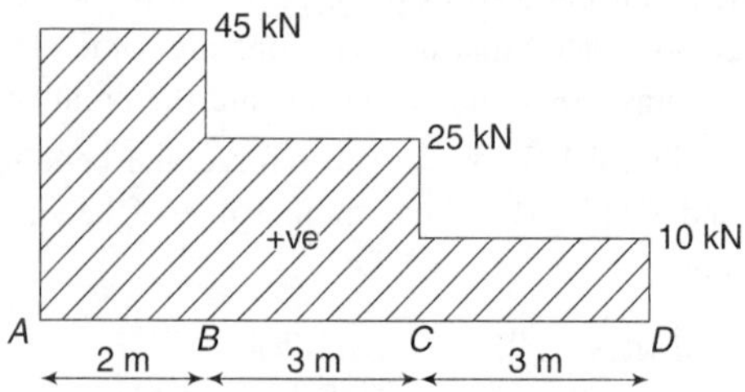

Fig. 4.131

50. Draw load diagram and BMD for a shear force diagram shown in Fig. 4.132.

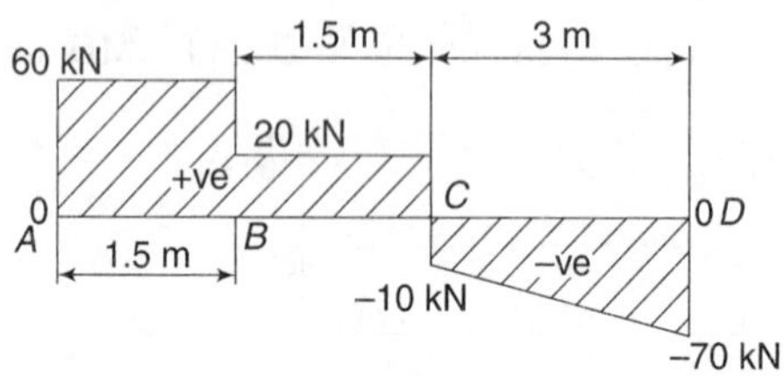

Fig. 4.132

51. Draw load diagram and BMD for a SFD given in Fig. 4.133.

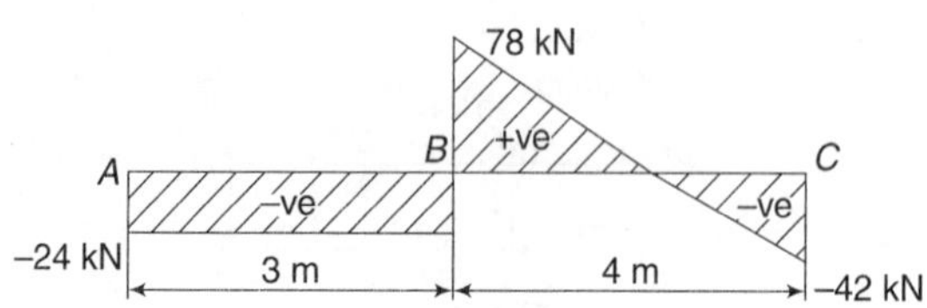

Fig. 4.133

52. Draw load and bending moment diagram for a SFD shown in Fig. 4.134.

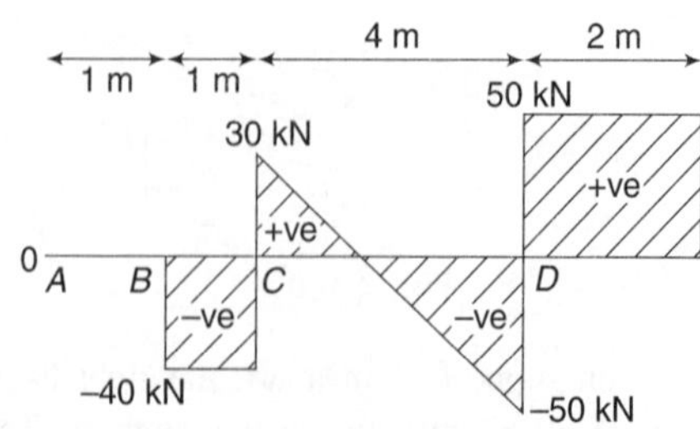

Fig. 4.134

Objective Type Questions

1. The relationship between load intensity and shear force is

(a) $F = \frac{dw}{dx}$ (b) $F = \frac{dx}{dw}$

(c) $w = \frac{dF}{dx}$ (d) $w = \frac{dx}{dF}$

2. The relationship between shear force and bending moment is

(a) $-F = \frac{dM}{dx}$ (b) $-F = \frac{dx}{dM}$

(c) $-M = \frac{dF}{dx}$ (d) $M = \frac{dx}{dF}$

3. The maximum shear force for a cantilever of length l subjected to a concerted load W at free end is

(a) Wl (b) $\frac{Wl}{4}$

(c) W (d) $\frac{Wl}{2}$

4. The maximum bending moment for a cantilever of length l subjected to a concentrated load W at free end is

(a) W (b) Wl

(c) $\frac{Wl}{4}$ (d) $\frac{W}{l}$

5. The maximum shear force for a cantilever of length l subjected to a UDL of w N/m over entire span is

(a) $\frac{wl^2}{2}$ (b) $\frac{wl}{2}$

(c) w^2l (d) wl

6. The maximum bending moment for a cantilever of l subjected to a UDL of w N/m over entire span is

(a) wl (b) $\frac{wl}{2}$

(c) $\frac{w^2l}{2}$ (d) $\frac{wl^2}{2}$

7. The maximum shear force for a cantilever of span l subjected to a UVL, that varies from w at fixed end to zero at free end is

(a) $\frac{wl}{2}$ (b) $\frac{wl^2}{4}$

(c) $\frac{wl}{3}$ (d) $\frac{wl^2}{6}$

8. The maximum bending moment for a cantilever of span l subjected to a UVL that varies from w at fixed end to zero at all free end is

(a) $\frac{wl^2}{6}$ (b) $\frac{wl^2}{2}$

(c) $\frac{wl}{4}$ (d) $\frac{wl}{3}$

9. For a simply supported beam of span l carrying a central load W, the maximum shear force is

(a) $\frac{Wl}{4}$ (b) $\frac{W}{2}$

(c) $\frac{Wl}{2}$ (d) W

10. For a simply supported beam of span l carrying a central load W, the maximum bending moment is

(a) $\frac{Wl}{4}$ (b) $\frac{Wl^2}{2}$

(c) $\frac{Wl}{2}$ (d) W

11. For a simply supported beam of span l carrying a UDL of w over the entire span, the maximum shear force is

(a) $\frac{wl}{4}$ (b) $\frac{w}{2}$

(c) $\frac{wl}{2}$ (d) $\frac{wl^2}{4}$

12. For a simply supported beam of span l carrying a UDL of w over the entire span, the maximum bending moment is

(a) $\frac{wl}{2}$ (b) $\frac{wl^2}{8}$

(c) $\frac{wl^2}{4}$ (d) $\frac{wl}{4}$

13. Point of contraflexure is the point on the beam where
    (a) shear force is zero changing its sign
    (b) shear force is maximum
    (c) bending moment is zero changing its sign
    (d) bending moment is maximum

14. On a loaded beam, bending moment is maximum where

(a) $SF = 0$ (b) SF is maximum

(c) SF is negative (d) At point of contraflexure

15. For a cantilever subjected to loads, the point of contraflexure will be

(a) at fixed end (b) at free end

(c) does not exist (d) at middle

16. For a cantilever of length 3 m, carrying a concentrated load of 30 kN at its free end, the maximum shear force is

(a) 90 kN (b) 30 kN
(c) 10 kN (d) 0

17. For a cantilever of length 3 m, carrying a concentrated load of 30 kN at its free end, the maximum bending moment is

(a) 30 kNm (b) 10 kNm
(c) 90 kNm (d) 270 kNm

18. For a cantilever of length 3 m, carrying a UDL of 30 kN/m over full length, the maximum shear force is

(a) 45 kN (b) 90 kN
(c) 135 kN (d) 30 kN

19. For a cantilever of length 3 m, carrying a UDL of 30 kN/m over full length, the maximum bending moment is

(a) 90 kNm (b) 10 kNm
(c) 30 kNm (d) 135 kNm

20. For a cantilever of length 3 m, carrying a concentrated load of 30 kN at the middle, the minimum shear force is

(a) zero (b) 45 kN
(c) 90 kN (d) 30 kN

21. For a cantilever of length 3 m, carrying a concentrated load of 30 kN at the middle, the minimum bending moment is

(a) 90 kNm (b) 45 kNm
(c) 30 kNm (d) zero

22. For a simply supported beam of span 3 m, carrying a central load of 30 kN, the maximum shear force is

(a) 30 kN (b) 15 kN
(c) 90 kN (d) 45 kN

23. For a simply supported beam of span 3 m, carrying a central load of 30 kN, the maximum bending moment is

(a) 30 kNm (b) 45 kNm
(c) 22.5 kNm (d) 90 kNm

24. For a simply supported beam of span 3 m carrying a UDL of 30 kN/m over the entire span, maximum shear force is

(a) 90 kN (b) 45 kN
(c) 30 kN (d) 33.75 kN

25. For a simply supported beam of span 3 m carrying a UDL of 30 kN/m over the entire span, maximum bending moment is

(a) 90 kNm (b) 45 kNm
(c) 30 kNm (d) 33.75 kNm

26. For a simply supported beam of span l carrying moment M at the middle, the maximum shear force is

(a) Ml (b) $\frac{M}{l}$

(c) $\frac{l}{M}$ (d) zero

27. For a simply supported beam of span l carrying a moment M at the middle, the maximum bending moment is

(a) Ml (b) $\frac{M}{l}$

(c) $\frac{M}{2}$ (d) zero

28. For a simply supported beam of span l carrying a UVL of zero at the left end to w at right end, the maximum bending moment from left support occurs at

(a) $\frac{l}{2}$ (b) $\frac{l}{3}$

(c) $\frac{2l}{3}$ (d) $\frac{l}{\sqrt{3}}$

29. For a cantilever of span 3 m carrying a UVL zero at free end 30 kN/m at fixed end, the maximum shear force is

(a) 30 kN (b) 22.5 kN

(c) 45 kN (d) 15 kN

30. For a cantilever of span 3 m carrying a UVL zero at free end to 30 kN/m at fixed end, the maximum bending moment is

(a) 45 kNm (b) 90 kNm

(c) 135 kNm (d) 270 kNm

Key to Answers

1. (c)	2. (a)	3. (c)	4. (b)	5. (b)	6. (d)	7. (a)	8. (a)
9. (b)	10. (a)	11. (c)	12. (b)	13. (c)	14. (a)	15. (c)	16. (b)
17. (c)	18. (b)	19. (d)	20. (a)	21. (d)	22. (b)	23. (c)	24. (b)
25. (d)	26. (b)	27. (c)	28. (d)	29. (c)	30. (a)		

CHAPTER 5

Bending and Shear Stress in Beams

5.1 INTRODUCTION

Generally, beams are straight and prismatic members. In most cases the loads are perpendicular to axis of the beam. Such perpendicular load is called transverse load. These transverse loads cause bending and shear stresses in beams. Due to the shear force and bending moments, the beams undergo deformation.

It is convenient to imagine a beam to be composed of an infinite number of their longitudinal fibres. Each longitudinal fibre is independent of every other fibre.

In this chapter the stresses due to bending moments, called bending stresses, and shear stresses, due to direct shear forces will be discussed.

5.2 PURE BENDING

If a beam is subjected to only bending moment and if there is no shear force, then the beam is said to be in pure bending. If a beam is bent with the application of constant bending moment without being subjected to any shear, then it is said to be in the state of pure bending or simple bending. Figure 5.1 shows a beam simply supported at *AB* and two equal forces '*W*' are applied at *C* and *D* at equal distance '*a*' from each support *A* and *B*.

It can be seen that the bending moment is constant between *C* and *D* and there is no shear force in this length. The length *CD* is in pure bending because the bending moment is constant and there is no shear force.

However, the bending moment between *AC* and *BD* is not constant and there is shear force present. This is said to be non-uniform bending.

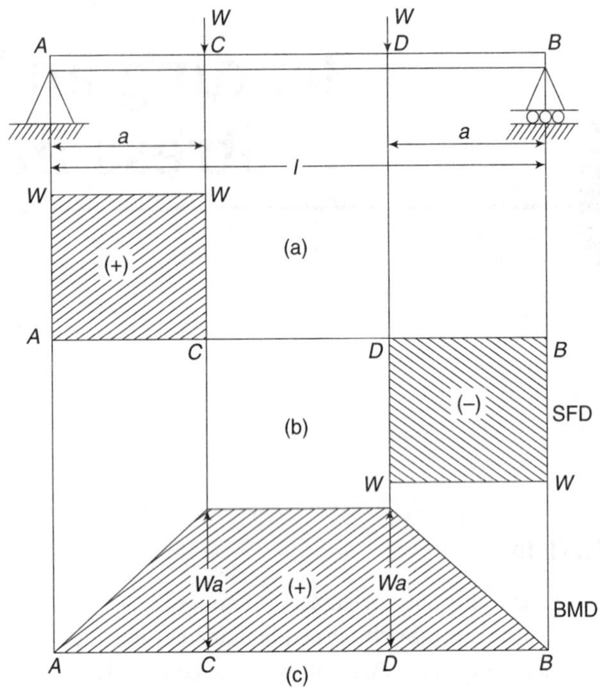

Fig. 5.1 Pure bending

5.3 SIMPLE BENDING

Bending and shear usually associate together. For simplicity, in this chapter, we calculate the stresses at a section induced due to pure bending, Hence, it is called simple bending. It was seen that moments applied may sag or hog the beam as shown in Fig. 5.2 (a) and (b).

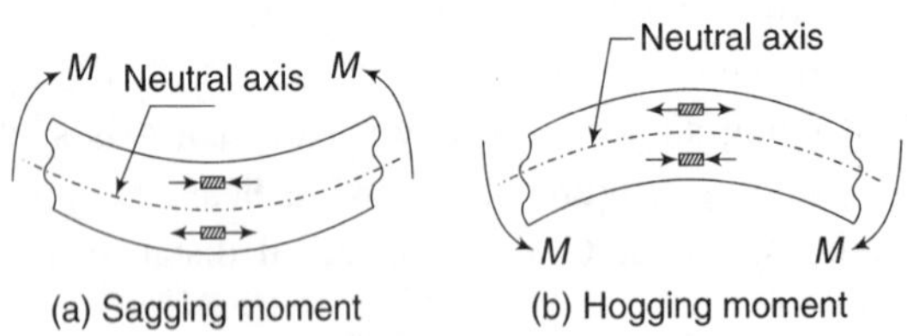

Fig. 5.2

It can be seen from Fig. 5.2 (a) that sagging moments cause bending of the beam in such a way that all the outer fibres (bottom fibres) are stretched and inner fibres (top fibres) are compressed. Accordingly the bottom fibres experience tensile stress and the top fibres experience compressive stress. In case of hogging moments shown in Fig. 5.2 (b), the situation is opposite. All top fibres experience tensile stress and bottom fibres experience compressive stress.

Hence, any bending will induce tensile stresses on one side and compressive stress on the other side. In between these two opposite nature of stresses, there is a layer where the stress is neither tensile nor compressive and the stress is zero. Such layer is called neutral layer and the axis is called neutral axis.

Hence, neutral axis is one, in a beam subjected to bending, the stress along this axis is zero. One side of neutral axis the beam experiences tenside stress and the other side, compressive stress.

Usually, for straight beams, neutral axis is same as centroidal axis.

5.4 ASSUMPTIONS IN SIMPLE THEORY OF BENDING

The following assumptions are made in simple bending.

1. The material is homogeneous and isotropic.
2. The modulus of elasticity is same in tension and compression.
3. The beam is stressed within elastic limit and the Hooke's law is obeyed.
4. The beam is initially straight and every layer is free to elongate or contract.
5. The radius of curvature is large compared to the depth of the beam.
6. Plane sections before bending remain plane even after bending.

The assumption "plane sections before bending remain plane section even after bending" is known as Bernauli's assumption.

According to Bernauli, plane section XX shown in Fig. 5.3 remains plane even after bending as X_1X_1. This is valid only for the case of pure bending. If there is shear force also associated with bending then, due distortion caused by shear force X_1X_1 will not be plane after bending. Deformation due to shear forces will be smaller and can be ignored in beams having depth less than $\frac{1}{10}$ of its span. If the depth of the beam is more than $\frac{1}{10}$ of span, the deformation due to shear forces will be considerably large and hence the simple bending theory will not be valid.

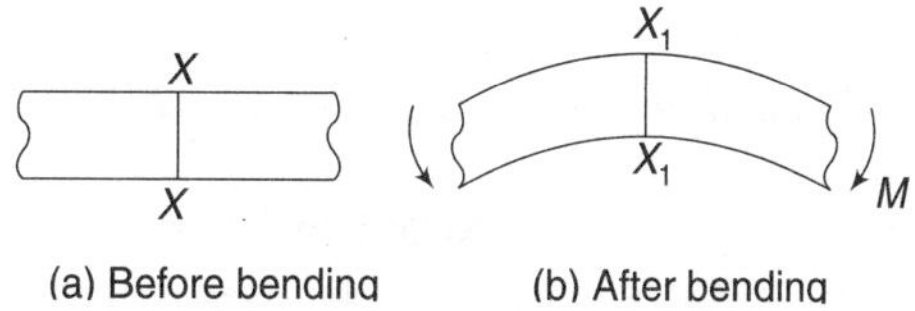

Fig. 5.3

5.5 RELATION BETWEEN BENDING STRESS AND RADIUS OF CURVATURE OF A BEAM SUBJECTED TO SIMPLE BENDING

Figure 5.4 (a) shows a beam before bending. Consider a portion of beam at section AB CD. EF is the neutral axis. Consider any element GH at a distance y from the neutral axis.

Figure 5.4 (b) shows the beam after subjected to bending moment M. The beam bends into a circular arc of radius R to neutral axis.

where, R = Radius of curvature at neutral axis

θ = Angle of sector formed between A_1 C_1 and B_1 D_1 subtends the centre of curvature.

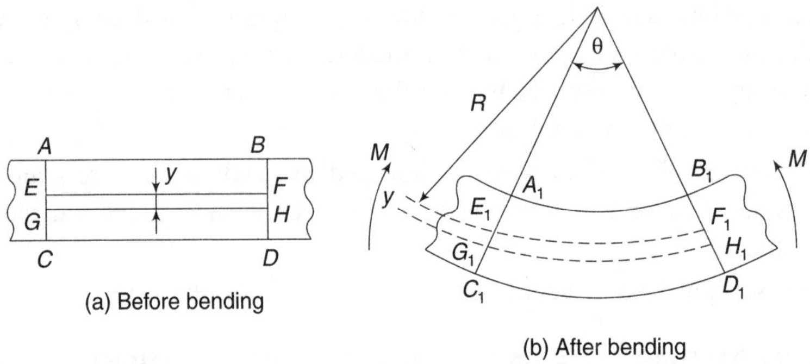

Fig. 5.4

Since EF is a neutral axis, there is no change in its length, $EF = E_1F_1$

From Fig. 5.4 (b) sector length $E_1F_1 = R\theta$ ...(1)

Length GH will change to G_1H_1 after bending.

$$\text{Radius to layer } G_1H_1 = R + y$$

$$\therefore \quad \text{Length of sector } G_1H_1 = (R + y)\theta \qquad \text{...(2)}$$

$$\text{Strain in layer } GH = \frac{G_1H_1 - GH}{GH}$$

From Fig. 5.4 (a) $GH = EF = R\theta$

$$\therefore \quad \text{Strain in layer } GH = \frac{(R + y)\,\theta - R\theta}{R\theta} = \frac{y}{R} \qquad \text{...(3)}$$

We know that modulus of elasticity, $E = \dfrac{\text{Stress}}{\text{Strain}}$

$$\therefore \quad \text{Strain} = \frac{\text{Stress}}{E} = \frac{\sigma}{E} \qquad \text{...(4)}$$

Equating Eqs. (3) and (4)

$$\text{Strain in } GH = \frac{\sigma}{E} = \frac{y}{R}$$

or

$$\frac{\sigma}{y} = \frac{E}{R} \qquad \text{...(5)}$$

$$\text{Stress } \sigma = \frac{E \cdot y}{R} \qquad \text{...(6)}$$

Since for a given problem, E and R are constant, the bending stress in a beam is proportional to its distance from neutral axis and various linearly across the depth.

$$\sigma \propto y$$

Since at neutral axis $y = 0$, the stress at neutral axis is zero.

Consider a beam of any cross section as shown in Fig. 5.5. Consider an elemental area δa at a distance y from neutral axis.

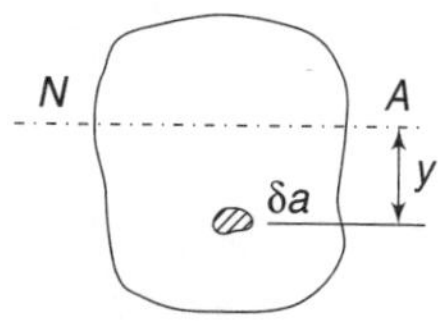

Fig. 5.5

Let σ = Stress on this area.

Force of the elemental area

$$\delta F = \sigma \delta a$$

Total force on the entire section $F = \int \delta F = \Sigma \sigma \delta a$

From Eq. (6) stress $\sigma = \frac{E}{R} \cdot y$

Substituting, $$F = \Sigma \frac{E}{R} \cdot y\ \delta a.$$

Since there is no axial force on the beam, $F = 0$

$$\therefore \quad \Sigma \frac{E}{R} \cdot y \cdot \delta a = 0$$

or $$\frac{E}{R} \cdot \Sigma y\ \delta a = 0$$

Since $$\frac{E}{R} \neq 0,\ \Sigma y \delta a = 0 \text{ or } \frac{\Sigma\ y\ \delta a}{A} = 0$$

where A = Total area of cross section.

From the first moments theorem about neutral axis, $\frac{\Sigma y\ \delta a}{A}$ is equal to zero.

Here there is no distance between CG axis and neutral axis. Thus, neutral axis coincides with centroidal axis of the cross section.

5.6 RELATION BETWEEN BENDING MOMENT AND RADIUS OF CURVATURE OF THE BEAM SUBJECTED TO SIMPLE BENDING

Figure 5.6 shows a beam subjected to bending moment 'M'. Let R be the radius of curvature

E = Modulus of elasticity

I = Moment of intertia of the cross section.

NA = Neutral axis

Consider an elemental area δa at a distance of y from neutral axis of the beam.

From Eq. (6) stress $$\sigma = \frac{E}{R} \cdot y$$

Force on the element $$\delta F = \sigma \cdot \delta a = \frac{E}{R}\, y \cdot \delta a$$

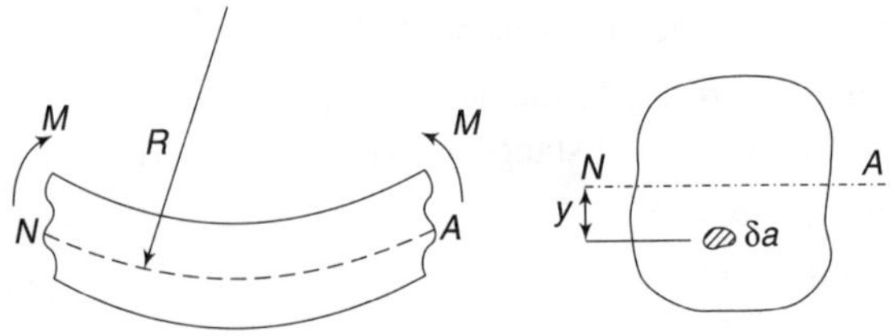

Fig. 5.6

Moment of this resisting force about neutral axis

$$\delta M = (\delta F)y = \frac{E}{R}\cdot y^2\,\delta a$$

Total moment

$$M = \int \delta M = \int \frac{E}{R}\, y^2\,\delta a$$

$$M = \frac{E}{R}\cdot \int \delta a\cdot y^2 = \frac{E}{R}\Sigma\,\delta a\cdot y^2$$

From the definition of moment of inertia, moment of moments of an area is defined as moment of inertia (second moments theorem).

i.e.,

$$\Sigma\,\delta a\cdot y^2 = \Sigma(\delta a\cdot y)y = I$$

∴

$$M = \frac{E}{R}\, I$$

or

$$\frac{M}{I} = \frac{E}{R} \qquad \ldots(7)$$

Equating Eqs. (5) and (7)

$$\frac{M}{I} = \frac{\sigma}{y} = \frac{E}{R} \qquad \ldots(8)$$

This is known as equation of bending.

where M = Bending moment

I = Moment of inertia of given cross section about its centroidal axis

σ = Bending stress

y = Distance of layer from neutral axis (or centroidal axis)

E = Modulus of elasticity of the material

R = Radius of curvature to neutral axis.

5.7 NATURE OF BENDING STRESS AND STRESS DISTRIBUTION ACROSS THE CROSS SECTION

The bending equation is $\frac{M}{I} = \frac{\sigma}{y}$

∴ Bending stress

$$\sigma = \frac{M}{I}\cdot y$$

For the given section of beam, bending moment (M) and moment of inertia (I) are constant. Hence, the stress is proportional to y.

$$\sigma \propto y$$

where y is the distance from neutral axis to the layer. This distribution is linear in nature.

- will be zero at neutral axis ($y = 0$)
- will be maximum tensile at $y = y_{max}$ on tensile side
- will be maximum compressive at $y = y_{max}$ on compressive side.

Hence, it varies linearly from maximum tensile value (+ve) from one side of neutral axis, becomes zero at neutral axis and will be maximum compressive (–ve) on the opposite side of neutral axis.

Figure 5.7 shows a typical distribution of bending stress across the cross section.

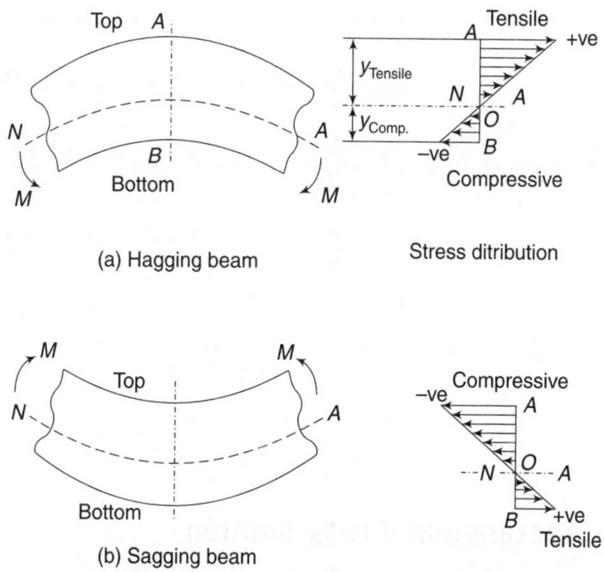

Fig. 5.7 Stress distribution across section

5.8 MOMENT CARRYING CAPACITY AND SECTION MODULUS

From bending equation $\dfrac{M}{I} = \dfrac{\sigma}{y}$

$\therefore$ Bending stress $\qquad \sigma = \dfrac{M \cdot y}{I}$

$$\sigma = \sigma_{allowable}$$

then $\qquad M = \dfrac{\sigma \cdot I}{y}$

$$M_{allowable} = \sigma_{all} \cdot \frac{I}{y_{max}}$$

as y increases, M decreases

$$M = \sigma \cdot Z$$

where $Z = \frac{I}{y}$ is known as section modulus.

$y = y_{max}$ is the distance from neutral axis to the outer fibre. For non-symmetric sections, the CG axis will not be in the middle and $y_{\text{tensile side}}$ will not be equal to $y_{\text{Comp side}}$.

In such cases, $Z = \frac{I}{y}$ for both tensile side and compressive side are to be calculated and the smaller (Z) has to be used to find the moment capacity.

5.9 SECTION MODULUS FOR VARIOUS CROSS SECTIONS

5.9.1 Rectangular Section

Figure 5.8 shows a rectangule of width b and depth d. NA is neutral axis

$$y_{max} = \frac{d}{2} \text{ since the centre of gravity lies in the middle.}$$

$$\text{Moment of inertia } I = \frac{bd^3}{12}$$

$$y = \frac{d}{2}$$

$$\therefore \quad \text{Section modulus } Z = \frac{I}{y} = \frac{\left(\frac{bd^3}{12}\right)}{\left(\frac{d}{2}\right)} = \frac{bd^2}{6}$$

$$Z = \frac{bd^2}{6}.$$

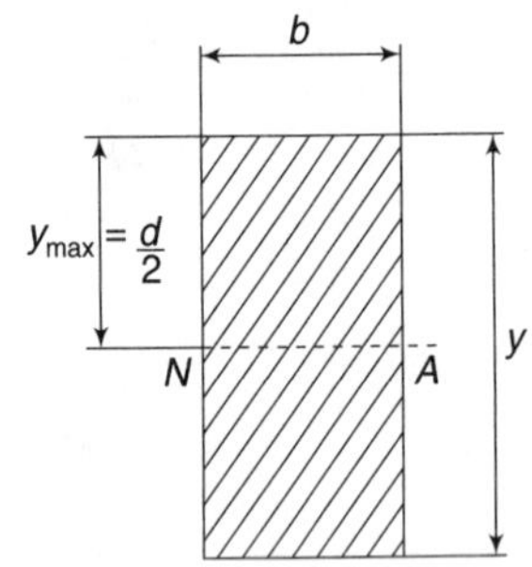

Fig. 5.8

5.9.2 Hollow Rectangular Cross Section

Figure 5.9 shows a hollow rectangular cross section of

Outer width = B

Inner width = b

Outer depth = D

Inner depth = d

NA is neutral axis.

Due to symmetry, CG lies in the middle.

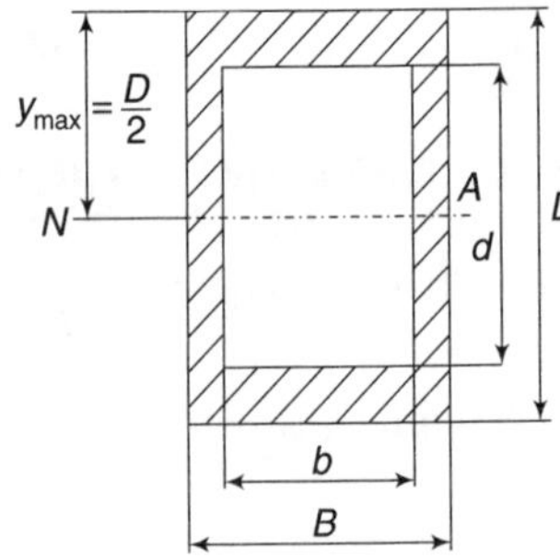

Fig. 5.9

$$y_{max} = \frac{D}{2}$$

$$I = I_o - I_i = \frac{BD^2 - bd^3}{12}$$

$$\text{Section modulus } Z = \frac{I}{y} = \frac{\left(\dfrac{BD^3 - bd^3}{12}\right)}{\left(\dfrac{D}{2}\right)}$$

$$Z = \frac{BD^3 - bd^3}{6D}$$

5.9.3 Circular Cross Section

Figure 5.10 shows a circular cross section of diameter = d

Due to symmetry,

CG lies in the middle

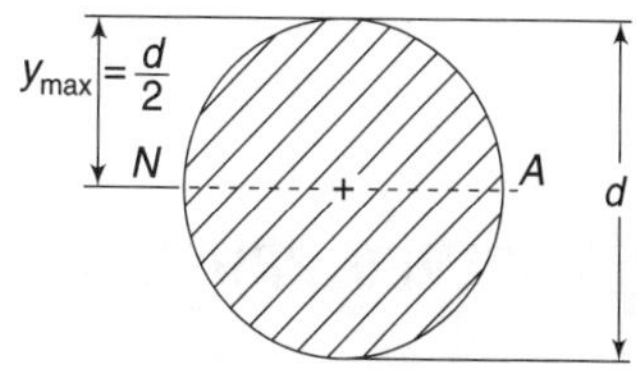

Fig. 5.10

$$y_{\max} = \frac{d}{2}$$

$$\text{Moment of inertia } I = \frac{\pi d^4}{64}$$

$$\therefore \quad \text{Section modulus } Z = \frac{I}{y} = \frac{\left(\dfrac{\pi d^4}{64}\right)}{\left(\dfrac{d}{2}\right)} = \frac{\pi}{32} d^3$$

$$Z = \frac{\pi d^3}{32}.$$

5.9.4 Hollow Circular Section

Figure 5.11 shows a hollow circular cross section of outer diameter D and inner diameter d. NA is neutral axis. Due to symmetry, CG lies in the middle. $y_{\max} = \frac{D}{2}$.

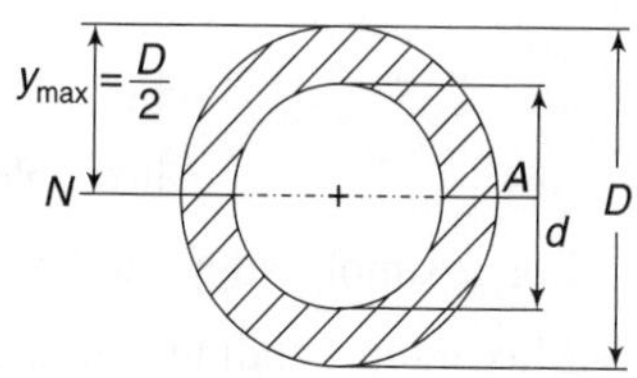

Fig. 5.11

$$I = I_o - I_i = \frac{\pi D^4 - \pi d^4}{64}$$

$$\therefore \text{ Section modulus } Z = \frac{I}{y}$$

$$Z = \frac{\left(\dfrac{\pi D^4 - \pi d^4}{64}\right)}{\left(\dfrac{D}{2}\right)} = \frac{\left(\dfrac{\pi(D^4 - d^4)}{64}\right)}{\left(\dfrac{D}{2}\right)}$$

$$Z = \frac{\pi(D^4 - d^4)}{32D}.$$

5.9.5 Triangular Section

Figure 5.12 shows a triangular cross section of base = b and depth = h

NA is neutral axis CG for triangle is a distance of $\frac{2h}{3}$ from vertex.

$$\therefore \qquad y_{max} = \frac{2h}{3}$$

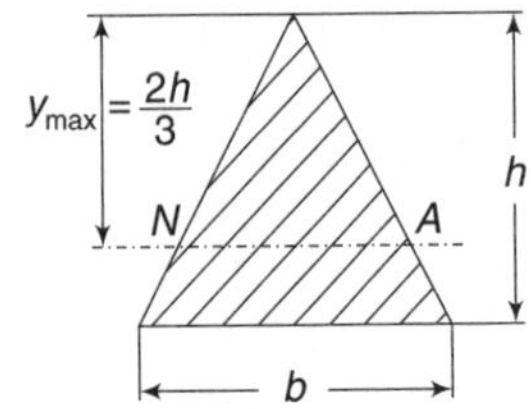

Fig. 5.12

$$\text{Moment of inertia about } NA,\ I = \frac{bh^3}{36}$$

$$\text{Section modulus } Z = \frac{I}{y}$$

$$Z = \frac{\left(\frac{bh^3}{36}\right)}{\left(\frac{2h}{3}\right)} = \frac{bh^2}{24}$$

$$Z = \frac{bh^2}{24}.$$

SOLVED PROBLEMS

Problem 1 A simply supported beam (100 mm × 200 mm) carries a central load of *W*. The permissible bending stress is 15 N/mm^2. Determine the safe load, if the span is 3 m.

Solution:

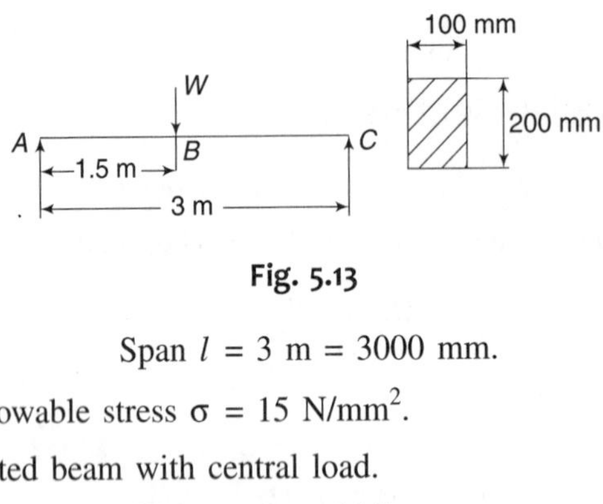

Fig. 5.13

$$\text{Span } l = 3 \text{ m} = 3000 \text{ mm}.$$

$$\text{Allowable stress } \sigma = 15 \text{ N/mm}^2.$$

For a simply supported beam with central load.

Maximum bending moment will be at the middle.

$$M = \frac{Wl}{4} = \frac{W \times 3000}{4} = 750\ W$$

For rectangular cross section, moment of inertia

$$I = \frac{bd^3}{12} = \frac{100 \times 200^3}{12} = 66.67 \times 10^6 \text{ mm}^4$$

$$y = \frac{d}{2} = \frac{200}{2} = 100 \text{ mm}$$

$$\text{From bending equation } \frac{M}{I} = \frac{\sigma}{y}$$

$$\frac{750\ W}{66.67 \times 10^6} = \frac{15}{100}$$

$\therefore$ Load $W = 13,\ 333.33$ N.

Problem 2 A steel wire of diameter 12 mm is bent into a circular are of radius 5 m. Find the maximum stress induced and the bending moment required to bend the wire. Take $E = 2 \times 10^5$ MPa.

***Solution:* Given data**

Diameter of the wire $d = 12$ mm

Radius of curvature $R = 5$ m $= 5000$ mm

Modulus of elasticity $E = 2 \times 10^5\ \text{N/mm}^2$

Consider $\frac{\sigma}{y} = \frac{E}{R}$

For circular cross section $y = \frac{d}{2} = \frac{12}{2} = 6$ mm

Bending stress $\sigma = \frac{E}{R} \cdot y = \frac{2 \times 10^5}{5000} \times 6 = 240$ MPa.

Again consider $\frac{M}{I} = \frac{E}{R}$

$\therefore$ Moment required $M = \frac{E}{R} \cdot I$

For circular cross section $I = \frac{\pi d^4}{64} = \frac{\pi \times 12^4}{64} = 1017.88\ \text{mm}^4$

$\therefore$ $M = \frac{2 \times 10^5}{5000} \times 1017.88 = 40{,}715$ N-mm.

Problem 3 A beam of rectangular cross section 100 mm wide 300 mm deep is simply supported at the ends over a span of 3 m. It carries a point load of 10 kN at a distance of 1 m from the left support. Find the maximum stress induced taking modulus of elasticity as 200 GPa.

***Solution:* Given data:**

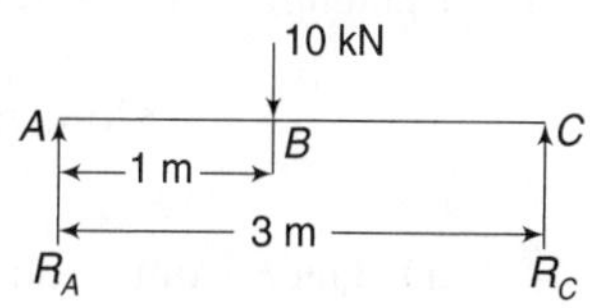

Fig. 5.14

Span $l = 3$ M $= 300$ mm

Load $W = 10$ kN

$E = 200$ GPa $= 200 \times 10^3$ MPa

Reactions: $R_A + R_C = 10$ kN

Taking moments about A

$$-10 \times 1 + R_C \times 3 = 0$$

$\therefore$ $R_C = \frac{10}{3} = 3.333$ kN

$$R_A = 10 - R_e = 6.667 \text{ kN}$$

Bending moments $\quad M_A = 0$

$$M_B = R_A \times 1000 = 6667 \times 1000 = 6.667 \times 10^6 \text{ Nmm}$$

$$M_C = 0$$

Maximum bending moment $M = 6.667 \times 10^6$ Nmm. From bending equation, $\frac{M}{I} = \frac{\sigma}{y}$

where, I = Moment of inertia $= \frac{bd^3}{12} = \frac{100 \times 300^3}{12} = 225 \times 10^6 \text{ mm}^4$

y = Distance from neutral axis to outer force $= \frac{d}{2} = \frac{300}{2} = 150$ mm

$$\therefore \quad \text{Bending stress } \sigma = \frac{M.y}{I} = \frac{6.667 \times 10^6 \times 150}{225 \times 10^6}$$

$$\sigma = 445 \text{ N/mm}^2.$$

Problem 4 A timber beam of rectangular cross section carries a UDl of 3 kN/m over a span of 3.6 m. It is simply supported at the ends. The depth of section is twice its width. Taking allowable bending stress for timber as 30 MPa. Find its cross sectional dimensions.

If it is to be replaced by a round beam, find its diameter.

Solution: **Given data:**

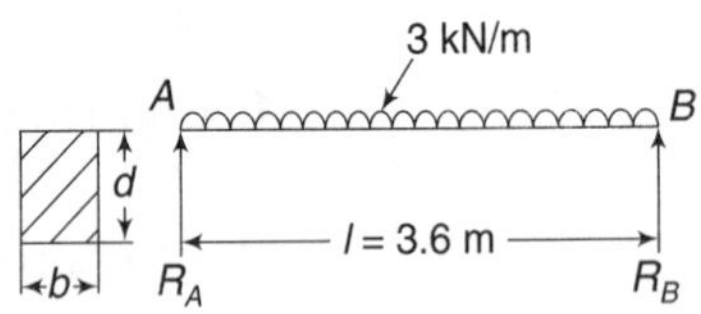

Fig. 5.15

$$\text{Depth } d = 2 \times \text{width} = 2b$$

$$\text{Allowable stress } \sigma = 30 \text{ MPa.}$$

$$\text{Span } l = 3.6 \text{ m} = 3600 \text{ mm}$$

For simply supported beam with UDL, maximum bending moment

$$M = \frac{wl^2}{8} = \frac{3000 \times 3600^2}{8} = 4.86 \times 10^9 \text{ Nmm}$$

(a) Rectangular cross section.

$$\text{Moment of inertia } I = \frac{bd^3}{12} = \frac{b(2b)^3}{12} = \frac{8b^4}{12}$$

Distance from neutral axis to outer layer $y = \frac{d}{2}$

Consider equation of bending $\frac{M}{I} = \frac{\sigma}{y}$

$$\frac{4.86 \times 10^9}{\left(\frac{8b^4}{12}\right)} = \frac{30}{\left(\frac{d}{2}\right)}$$

$$\therefore \quad \text{Width } b = 495.29 \text{ mm}$$

$$\text{Depth } d = 2b = 2 \times 495.29 = 990.58 \text{ mm.}$$

(b) Round cross section: Moment of inertia $I = \frac{\pi d^4}{64}$ $y = \frac{d}{2}$

Substituting in bending equation.

$$\frac{4.86 \times 10^9}{\left(\frac{\pi d^4}{64}\right)} = \frac{30}{\left(\frac{d}{2}\right)}$$

$\therefore$ Diameter of beam $= d = 1181.69$ mm.

Problem 5 A cantilever beam of rectangular cross section has a moment of inertia as 45 000 mm^4. Calculate the radius to which the beam may be bent if the stress is limited to 90 MPa. The depth of beam is 30 mm and modulus of elasticity = 200 kN/mm^2. Also find the bending moment to be applied to bend the beam.

Solution: **Given data:**

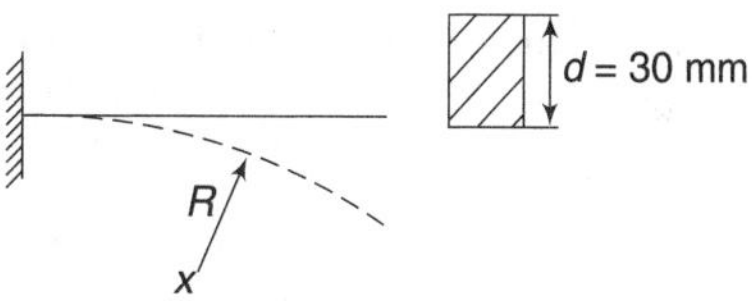

Fig. 5.16

$$\text{Depth } d = 30 \text{ mm}$$

$$\text{Moment of inertia } I = 45000 \text{ mm}^4$$

$$\text{Bending stress } \sigma = 90 \text{ MPa}$$

$$\text{Modulus of elasticity } E = 200 \text{ kN/mm}^2 = 200 \times 10^3 \text{ N/mm}^2$$

Distance from neutral axis to farthest layer

$$y = \frac{d}{2} = \frac{30}{2} = 15 \text{ mm}$$

$$\text{Bending equation } \frac{M}{I} = \frac{\sigma}{y} = \frac{E}{R}$$

$$\text{Consider } \frac{\sigma}{y} = \frac{E}{R}$$

$$\frac{90}{15} = \frac{200 \times 10^3}{R}$$

$\therefore$ Radius of curvature $R = 33{,}333.33$ mm

$$\text{Again consider } \frac{M}{I} = \frac{\sigma}{y}$$

$\therefore$ Bending moment $M = \frac{\sigma I}{y} = \frac{90 \times 45000}{15}$

$$M = 270 \times 10^3 \text{ N mm.}$$

Problem 6 Calculate the maximum bending moment that the section shown in Fig. 5.17, can withstand limiting the stress to 60 MPa?

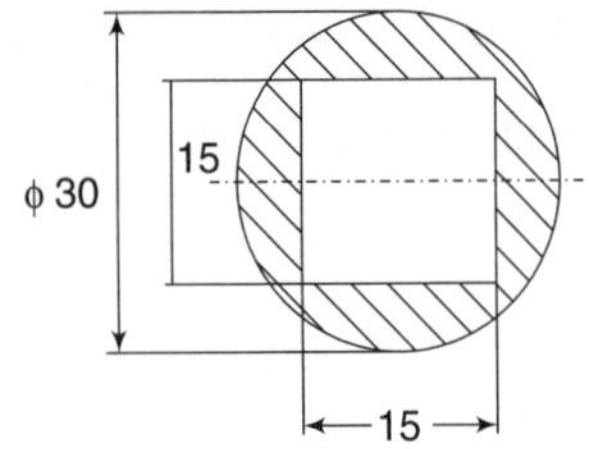

Fig. 5.17

Solution: **Given data:**

$$\text{Allowable stress } \sigma = 60 \text{ MPa}$$

$$\text{Diameter } d = 30 \text{ mm}$$

$$\text{Side of square } b = 15 \text{ mm}$$

Moment of inertia of the given cross section

$$I = I_C - I_{sq}$$

$$I = \frac{\pi}{64} d^4 - \frac{b^4}{12}$$

$$= \frac{\pi}{64} \times 30^4 - \frac{15^4}{12} = 35{,}542.03 \text{ mm}^4$$

y = Distance from neutral axis to the outer fibre

$$= \frac{d}{2} = \frac{30}{2} = 15 \text{ mm}$$

$$\text{From bending equation } \frac{M}{I} = \frac{\sigma}{y}$$

$$\therefore \quad \text{Bending moment } M = \frac{\sigma \cdot I}{y} = \frac{60 \times 35.542.03}{15} = 142{,}168.13 \text{ N-mm}$$

$$M = 142.168 \times 10^3 \text{ N-mm.}$$

Problem 7 A simply supported beam of span 6 m has a rectangular cross section of width 90 mm and depth 150 mm. It is made of steel having allowable stress as 120 MPa. Determine

(a) the maximum central load it can withstand, and

(b) the intensity of UDL that can be put on it.

Solution: **Given data:**

$$\text{Width } b = 90 \text{ mm}$$

$$\text{Depth } d = 150 \text{ mm}$$

$$\text{Allowable stress } \sigma = 120 \text{ MPa}$$

$$\text{Moment of inertia } I = \frac{bd^3}{12} = \frac{90 \times 150^3}{12} = 25.3125 \times 10^6$$

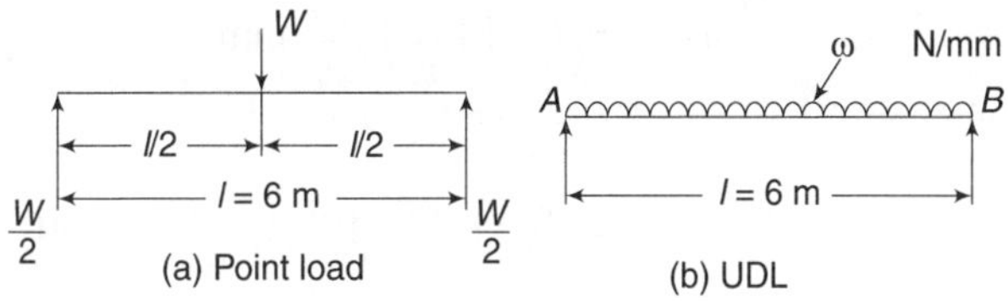

Fig. 5.18

Distance from neutral axis to outer fibre

$$y = \frac{d}{2} = \frac{150}{2} = 75 \text{ mm}$$

(a) Maximum central load. For simply supported beam with central load,

Maximum bending moment, $M = \dfrac{Wl}{4}$

$$M = \frac{W \times 6000}{4} = 1500\ W$$

From bending equation.

$$\frac{M}{I} = \frac{\sigma}{y}$$

$$\frac{1500\ W}{25.3125 \times 10^6} = \frac{120}{75}$$

Central load $W = 33{,}750$ N

(b) **UDL**: For simply supported beam with UDL, maximum bending moment

$$M = \frac{wl^2}{8} = w \times 6000^2/8 = 4.5 \times 10^6\ w$$

Substituting in bending equation

$$\frac{4.5 \times 10^6\ w}{25.3125 \times 10^6} = \frac{120}{75}$$

$\therefore$ UDL = w = 9 N/mm = 9 kN/m.

Problem 8 A circular tube of external diameter 120 mm and wall thickness 10 mm is fixed at one end and carries a concentrated load at a distance of 1.8 M from the fixed end. Find the safe value of this load limiting the stress to 120 MPa. If the modulus of elasticity is 206 GPa, find the radius to which the beam will be bent.

Solution: **Given data:**

Outside diameter D = 120 mm

Wall thickness t = 10 mm

$\therefore$ Inside diameter $d = D - 2t = 120 - 2 \times 10 = 100$ mm

Stress σ = 120 MPa

$$\text{Span } l = 1.8 \text{ M} = 1800 \text{ mm}$$

$$\text{Moment of inertia } I = \frac{\pi}{64}(D^4 - d^4)$$

$$I = \frac{\pi}{64}(120^4 - 100^4) = 5.27 \times 10^6 \text{ mm}^4$$

Distance from neutral axis to outer fibre

$$y = \frac{D}{2} = \frac{120}{2} = 60 \text{ mm}$$

For cantilever with point load, maximum bending moment

$$M = W \cdot l = W \times 1800$$

(a) The bending equations $\frac{M}{I} = \frac{\sigma}{y}$

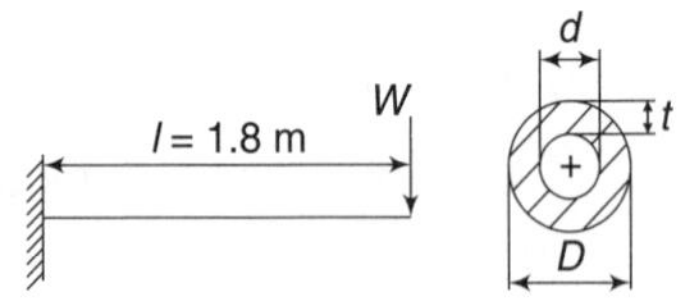

Fig. 5.19

$$\frac{W \times 1800}{5.27 \times 10^6} = \frac{120}{60}$$

$$\therefore \quad W = 5855.58 \text{ N}$$

(b) Consider $\frac{\sigma}{y} = \frac{E}{R}$

$$\therefore \quad \text{Radius } R = \frac{E.y}{\sigma}$$

$$R = \frac{206 \times 10^6 \times 60}{120} = 103000 \text{ mm.}$$

Problem 9 A cross iron test beam 25 mm × 25 mm cross section and 1 m long supported at its ends, fails when a central load of 800 N is applied on it. What UDL will break a cantilever of the same material 50 mm wide, 100 mm and 2 m long?

Solution:

(a) Simply supported beam with central load (Fig. 5.20 (a))

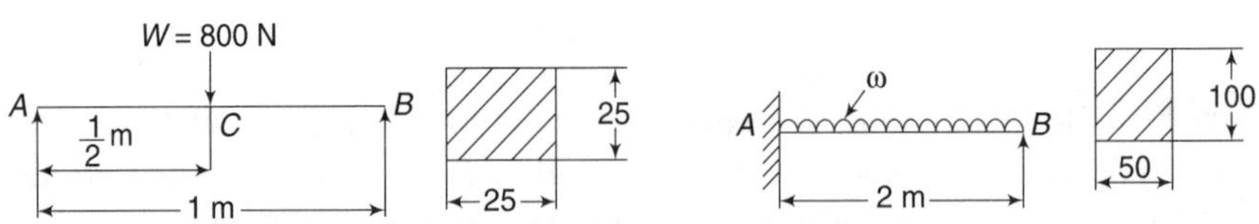

(a) Simply supported beam with central load

(b) Cantilever with UDL

Fig. 5.20

$$\text{Maximum bending moment } M = \frac{Wl}{4}$$

$$M = \frac{800 \times 1000}{4} = 20 \times 10^3 \text{ N/mm}$$

$$\text{Moment of inertia } I = \frac{b^4}{12} \text{ for square section}$$

$$I = \frac{25^4}{12} = 32552.08 \text{ mm}^4$$

$$y = \frac{b}{2} = \frac{25}{2} = 12.5 \text{ mm}$$

$$\text{Bending equation is } \frac{M}{I} = \frac{\sigma}{y}$$

$$\therefore \quad \text{Bending stress } \sigma = \frac{My}{I}$$

$$\sigma = \frac{20 \times 10^3 \times 12.5}{32552.08} = 7.68 \text{ MPa}$$

(b) Cantilever with UDL (Fig. 5.20 (b))

$$\text{Maximum bending moment, } M = \frac{wl^2}{2} = \frac{w \times 2000^2}{2} = 2 \times 10^6 \, w$$

$$\text{Moment of inertia, } I = \frac{bd^3}{12} = \frac{50 \times 100^3}{12} = 4.1667 \times 10^6 \text{ mm}^4$$

$$\text{Distance, } y = \frac{d}{2} = \frac{100}{2} = 50 \text{ mm}$$

Since the material is same, bending stress σ = 7.68 MPa

Substituting in bending equation,

$$\frac{2 \times 10^6 \, w}{4.1667 \times 10^6} = \frac{7.68}{50}$$

$$\therefore \quad \text{UDL, } w = 0.32 \text{ N/mm} = 0.32 \text{ kN/m.}$$

Problem 10 In a overhanging beam of steel shown in Fig. 5.21, the allowable bending stress is 90 N/mm^2. Determine the minimum size of square section required for the beam.

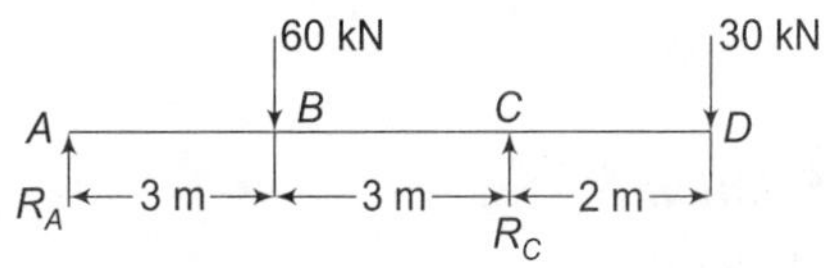

Fig. 5.21

Solution: **Given data:**

$$\text{Allowable stress } \sigma = 9 \text{ N/mm}^2$$

Reactions:

$$R_A + R_C = 60 + 30 = 90 \text{ kN}$$

Taking moments about A, $-\,60 \times 3 + R_C \times 6 - 30 \times 8 = 0$

$$\therefore \quad R_C = 70 \text{ kN}$$

$$R_A = 90 - R_C = 20 \text{ kN}$$

Bending moments:

$$\text{At } A,\ M_A = 0$$

$$\text{At } B,\ M_B = 20 \times 3 = 60 \text{ kNm}$$

$$M_C = 20 \times 6 - 60 \times 3 = -60 \text{ kNm}$$

$$M_D = 0$$

Maximum bending moment of the beam $M = 60$ kNm

$$M = 60 \times 10^6 \text{ Nmm.}$$

For square section beam of side b, moment of inertia $I = \frac{b^4}{12}$. Distance from neutral axis to outer fibre $y = \frac{b}{2}$

Bending equation is $\frac{M}{I} = \frac{\sigma}{y}$

$$\frac{60 \times 10^6}{\left(\frac{b^4}{12}\right)} = \frac{90}{\left(\frac{b}{2}\right)}$$

$\therefore$ Side of square section $b = 158.74$ mm.

Problem 11 A simply supported rectangular beam is loaded as shown in Fig. 5.22. Determine the maximum flexural stress and flexural stress at a cross section located 2 m from the left support. Sketch the cross stress distribution across this section.

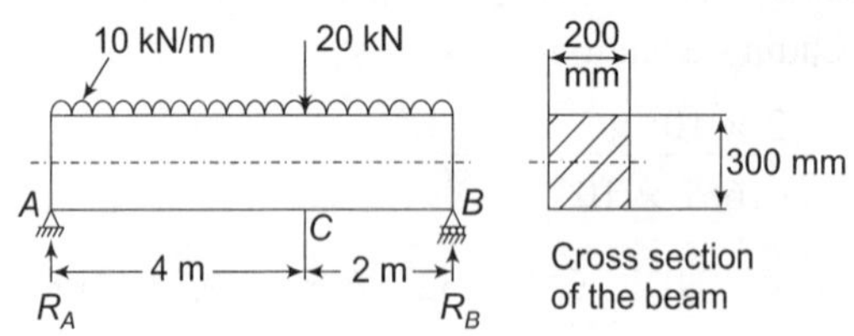

Fig. 5.22

Solution:

Reaction:

$$R_A + R_B = 20 + (10 \times 6) = 80 \text{ kN}$$

Taking moments about A, $-(10 \times 6)\frac{6}{2} - 20 \times 4 + R_B \times 6 = 0$

$$R_B = 43.33 \text{ kN}$$

$$R_A = 80 - 43.33 \text{ kN} = 36.67 \text{ kN}$$

Moment of inertia $I = \frac{bd^3}{12} = \frac{200 \times 300^3}{12} = 450 \times 10^6 \text{ mm}^4$

$$y = \frac{d}{2} = \frac{300}{2} = 150 \text{ mm}$$

(a) To find maximum flexural stress.

Stress is maximum where bending moment is maximum. Maximum bending moment will be maximum where shear force is equal to zero

$$SF = R_A - (w \cdot x) = 0$$

$$36.67 - (10 \cdot x) = 0$$

$$\therefore \quad x = 3.667 \text{ m from } A$$

$$\text{Maximum bending moment} = M_{max} = R_A \cdot x - (wx)\frac{x}{2}$$

$$= 36.67 \times 3.667 - (10 \times 3.667)\frac{3.667}{2}$$

$$M_{max} = M = 67.234 \text{ kN-m} = 67.234 \times 10^6 \text{ Nmm}$$

$$\text{Bending equation, } \frac{M}{I} = \frac{\sigma}{y}$$

$$\therefore \quad \text{Maximum stress } \sigma_{max} = \frac{M}{I} \cdot y$$

$$\sigma_{max} = \frac{67.234 \times 10^6}{450 \times 10^6} \times 150 = 22.41 \text{ MPa.}$$

(b) To find flexural stress at 2 M from A

$$\text{Bending moment } M = R_A \cdot 2 - (w \times 2)\frac{2}{2}$$

$$= 36.67 \times 2 - (10 \times 2)\frac{2}{2} = 53.34 \text{ kNm}$$

$$M = 53.34 \times 10^6 \text{ Nmm}$$

$$\therefore \quad \text{Bending stress } \sigma = \frac{M \cdot y}{I} = \frac{53.34 \times 10^6}{450 \times 10^6} \times 150$$

$$\sigma = 17.78 \text{ MPa.}$$

(c) Stress distribution across the section

It is seen from Fig. 5.23 that all the top fibres are subjected to compressive stress and bottom fibres to compression. Stress at neutral axis is zero. The stress distribution is shown in Fig. 5.23.

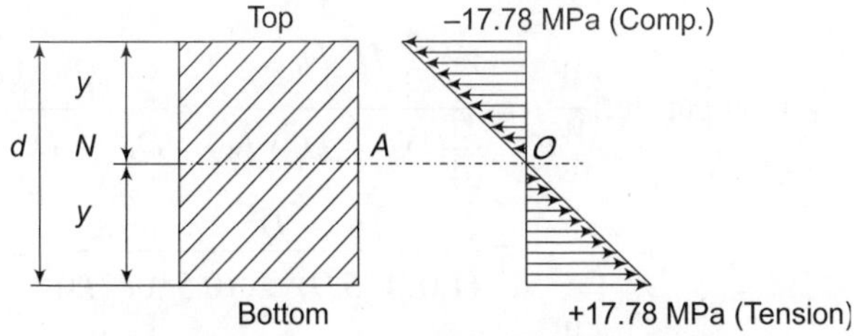

(a) Cross section (b) Stress distribution

Fig. 5.23

Problem 12 Compare the weights of two equally strong beams of circular cross section made of same material, one is of solid section and the other of hollow section, with diameters ratio as 0.5.

Solution: Given both beams of same material and area are equally strong

$$M_S = M_H$$

We know that $\frac{M}{I} = \frac{\sigma}{y}$

$$\therefore \quad M = \sigma \frac{I}{y} = \sigma Z$$

$$\therefore \quad (\sigma Z)_S = (\sigma Z)_H$$

Since material is same, σ is same

$$\therefore \quad Z_S = Z_H$$

$$Z_S = \text{Section of modulus of solid} = \frac{\pi D^3}{32}$$

$$Z_H = \text{Section of modulus of hollow} = \frac{\pi}{32}\frac{(D_o^4 - d_i^4)}{D_o}$$

$$\text{Given diameter ratio} = 0.5 = \frac{D_i}{D_o}$$

$$\therefore \quad D_i = 0.5\ D_o$$

$$\therefore \quad Z_H = \frac{\pi}{32}\frac{(D_o^4 - (0.5\ D_o)^4)}{D_o} = \frac{\pi}{32} D_o^3\ (0.9375)$$

$$\text{Equating,} \quad \frac{\pi}{32} D^3 = \frac{\pi}{32} D_o^3\ (0.9375)$$

$$\therefore \quad D_o = 1.02175\ D$$

$$\therefore \quad D_i = 0.5\ D_o = 0.51087\ D$$

$$\text{Weight} = \text{Vol} \times \text{Specific weight}$$

$$W = Al\ w$$

$$\therefore \quad \text{Weight of solid } W_S = A_S\ lw = \frac{\pi}{4} D^2 lw$$

$$\text{Weight of hollow } W_H = A_H\ lw = \frac{\pi}{4}(D_o^2 - D_i^2)\ lw$$

$$\text{Comparing,} \quad \frac{W_S}{W_H} = \frac{\frac{\pi}{4} D^2\ lw}{\frac{\pi}{4}(D_o^2 - D_i^2)\ lw} = \frac{D^2}{D_o^2 - D_i^2}$$

$$= \frac{D^2}{(1.02175\,D)^2 - (0.51087D)^2} = \frac{D^2}{0.7898\,D^2}$$

$$\frac{W_S}{W_H} = 1.277.$$

∴ Weight of solid shaft is 1.277 times the weight of the equally strong hollow shaft.

Problem 13 Compare the section modulus of two beams of same weight, length and material, one being solid of diameter d and the other hollow of outer diameter d_o and inner diameter d_i.

Solution: Material, length and weights are same

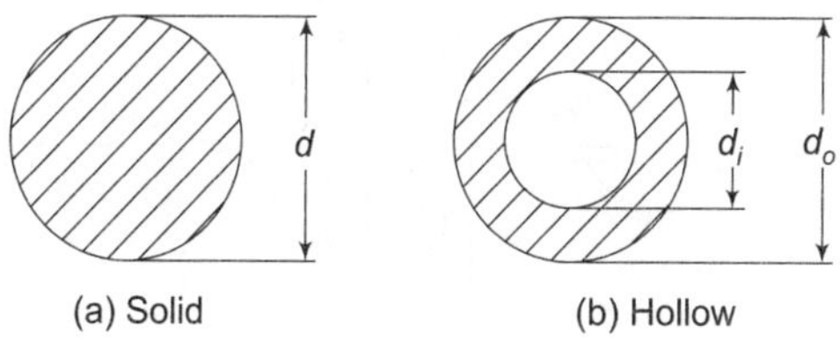

Fig. 5.24

$$\text{Weight } W = \text{Volume} \times \text{Specific weight}$$

$$= (Al) \times w$$

$$\text{For solid section, } W_S = A_S\, lw = \frac{\pi}{4} d^2 \cdot lw$$

$$\text{For hollow section } W_H = A_H\, lw = \frac{\pi}{4}(d_o^2 - d_i^2)\, lw$$

$$\text{For weights being same, } W_S = W_H$$

$$\frac{\pi}{4} d^2 lw = \frac{\pi}{4}(d_o^2 - d_i^2)\, lw$$

$$\therefore \quad d = \sqrt{d_o^2 - d_i^2}$$

$$\text{Section modulus } Z = \frac{I}{C}$$

$$\text{For solid section } Z_S = \frac{\pi d^3}{32}$$

$$\text{For hollow section } Z_H = \frac{\pi}{32} \frac{(d_o^4 - d_i^4)}{d_o}$$

$$\text{Comparing } \frac{Z_S}{Z_H} = \frac{\left[\frac{\pi}{32} d^3\right]}{\left[\frac{\frac{\pi}{32}(d_o^4 - d_i^4)}{d_o}\right]} = \frac{(\sqrt{d_o^2\ d_i^2})^3}{\left[\frac{d_o^4 - d_i^4}{d_o}\right]}$$

$$\frac{Z_S}{Z_H} = \frac{(d_o^2 - d_i^2)\sqrt{d_o^2 - d_i^2}}{(d_o^2 - d_i^2)(d_o^2 + d_i^2)} d_o = \frac{d_o\sqrt{d_o^2 - d_i^2}}{(d_o^2 + d_i^2)}$$

$$= \frac{d_o\sqrt{d_o^2\left[1-\left(\frac{d_i}{d_o}\right)^2\right]}}{d_o^2\left[1+\left(\frac{d_i}{d_o}\right)^2\right]} = \frac{d_o^2\sqrt{1-\left(\frac{d_i}{d_o}\right)^2}}{d_o^2\left[1+\left(\frac{d_i}{d_o}\right)^2\right]}$$

$$\frac{Z_S}{Z_H} = \frac{\sqrt{1-\left(\frac{d_i}{d_o}\right)^2}}{\left[1+\left(\frac{d_i}{d_o}\right)^2\right]} = \frac{\sqrt{1-k^2}}{(1+k^2)}$$

where, $k = \dfrac{d_i}{d_o}$

Problem 14 For a given stress, compare the moments of resistance of a beam of square section when placed with its two sides horizontal and with its one diagonal horizontal.

Solution:

(a) Square section when placed with two of its sides horizontal Fig. 5.25 (a)

$$I = \frac{a^4}{12}$$

$$y = \frac{a}{2}$$

$\therefore$ Moment of resistance $M = \sigma Z$

$$M_a = \sigma\left(\frac{I}{C}\right) = \sigma\frac{\left(\frac{a^4}{12}\right)}{\left(\frac{a}{2}\right)} = \frac{\sigma a^3}{6}$$

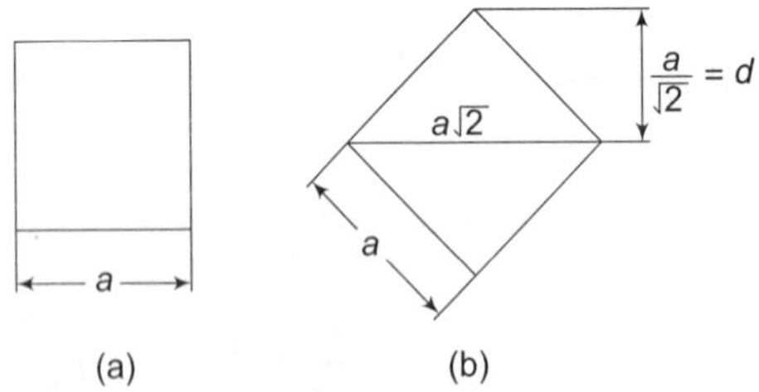

Fig. 5.25

(b) For square section placed with one diagonal horizontal, Fig. 5.25 (b)

Consider the square as two triangles, each of base $b = a/\sqrt{2}$ and height, $d = \dfrac{a}{\sqrt{2}}$

$$I = 2 \times I = 2\left(\frac{bd^3}{12}\right)$$

$$I = 2 \times \frac{a\sqrt{2}\left(\frac{a}{\sqrt{2}}\right)^3}{12} = \frac{a^4}{12}$$

Distance from neutral axis to outer fibre:

$$y = \frac{a}{\sqrt{2}}$$

$$\therefore \quad \text{Moment of resistance } M_b = \sigma \cdot \frac{\left(\frac{a^4}{12}\right)}{\left(\frac{a}{\sqrt{2}}\right)} = \frac{\sigma a^3}{6\sqrt{2}}$$

$$\text{Comparing } \frac{M_a}{M_b} = \frac{\left(\frac{\sigma a^3}{6}\right)}{\left(\frac{\sigma a^3}{6\sqrt{2}}\right)} = \sqrt{2} = 1.4142.$$

Problem 15 Find the dimensions of the strongest rectangular section that can be obtained from a circular log of diameter d to resist moment in bending?

Solution: **Given data:**

$$\text{Diameter} = d$$

Dimension of rectangular cross section = $b \times h$

Moment of inertia of rectangular cross section

$$I = \frac{bh^3}{12}$$

Distance to farthest layer from neutral axis $y = \frac{h}{2}$

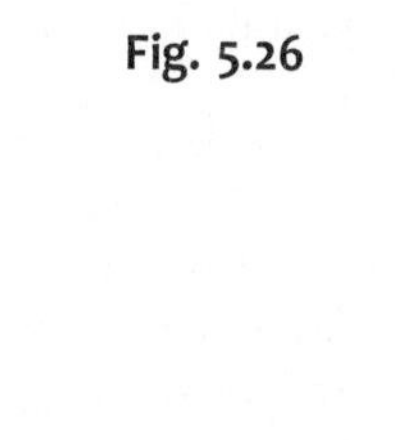

Fig. 5.26

$$\therefore \quad \text{Section modulus } Z = \frac{I}{C} = \frac{\left(\frac{bh^3}{12}\right)}{\left(\frac{h}{2}\right)} = \frac{bh^2}{6}$$

From the right angle triangle,

$$d^2 = b^2 + h^2$$

$$\therefore \quad h^2 = d^2 - b^2$$

$$\text{Substituting in the } Z \text{ equation } Z = \frac{b(d^2 - b^2)}{6} = \frac{bd^2}{6} - \frac{b^3}{6}$$

For the beam to be strongest, its section modulus is to be maximum.

$$Z \text{ is maximum when } \frac{dZ}{ab} = 0$$

$$\frac{dZ}{db} = \frac{d^2}{6} - \frac{3b^2}{6} = 0$$

$$d^2 = 3b^2$$

$$\therefore \qquad b = \frac{d}{\sqrt{3}}$$

But
$$h^2 = d^2 - b^2 = d^2 - \left(\frac{d}{\sqrt{3}}\right)^2$$
$$= d^2 - \frac{d^2}{3} = \frac{2d^2}{3}$$

$$\therefore \qquad h = d\frac{\sqrt{2}}{3}$$

The dimension of strongest rectangular,

$$b = \frac{d}{\sqrt{3}} \text{ and } h = d\sqrt{\frac{2}{3}}.$$

Problem 16 Show that the moment of resistance of a square beam with one of its diagonals horizontal, will increase by making it flat at the top and bottom corners. Also find the distance to the flat from neutral axis.

Solution: Consider a square of diagonal = d with one of this diagonal horizontal as shown in Fig. 5.27.

Consider an element strip of thickness δx and width b, at a distance of x from neutral axis (NA).

Moment of inertia of this strip about axis NA

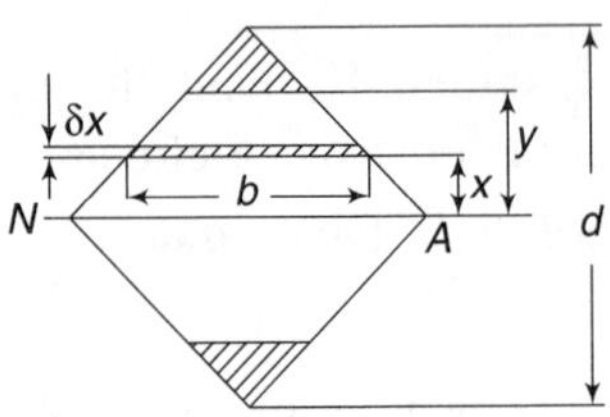

Fig. 5.27

$$I = I_{cg} + Ak^2$$

$$I = \frac{b(\delta x)^3}{12} + (b \cdot \delta x)\, x^2$$

(Parallel axis theorem of moment of inertia)

Since δx is very smaller, $(\delta x)^3$ will be further small and can be neglected.

$$\therefore \qquad I = b \cdot \delta x \cdot x^2 \qquad \ldots(1)$$

(a) When corners are not flattened, considering the square as two triangles, moment of inertia of the square about axis NA

= 2 × Moment of inertia of each triangle about base

$$I_a = 2 \times \frac{d \times \left(\frac{d}{2}\right)^3}{12} = \frac{d^4}{48}$$

Distance to farthest layer from NA = $y = \frac{d}{2}$

$$\therefore \qquad \text{Section modulus } Z_a = \frac{I}{y} = \frac{\left(\frac{d^4}{48}\right)}{\left(\frac{d}{2}\right)} = \frac{d^3}{24} \qquad \ldots(2)$$

(b) When corners are flattened, y

$$\text{Moment of inertia } I_b = 2\int_o^y b.x^2 dx$$

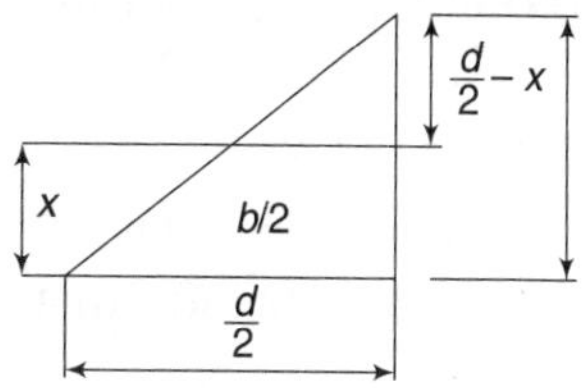

Fig. 5.27 (a)

From similar triangles,

$$\frac{\left(\frac{d}{2}\right)}{\left(\frac{d}{2}\right)} = \frac{\left(\frac{d}{2} - x\right)}{\left(\frac{b}{2}\right)}$$

$$\therefore \quad \frac{b}{2} = \frac{d}{2} - x$$

$$b = d - 2x$$

$$\therefore \quad I_b = 2\int_o^y (d - 2x)\, x^2 \cdot dx = 2\left[\int_o^y dx^2\, dx - \int_O^y 2x^3\, dx\right]$$

$$I_b = 2\left[\frac{dx^3}{3} - \frac{2 \cdot x^4}{4}\right]_o^y = \frac{2dy^3}{3} - y^4$$

Distance $y = y$

$$\therefore \quad \text{Section modulus } Z_b = \frac{I}{y} = \frac{\left(\frac{2by^3}{3} - y^4\right)}{y} = \frac{2dy^2}{3} - y^3$$

For the section modulus to be maximum $\frac{dZ}{dy} = 0$

$$\frac{dZ}{dy} = \frac{4dy}{6} - 3y^2 = 0$$

$$\therefore \quad y = \frac{4d}{9}$$

$$\therefore \quad Z_b = \frac{2.d\left(\frac{4d}{9}\right)^2}{3} - \left(\frac{4d}{9}\right)^3 = \frac{32d^3}{243} - \frac{64d^3}{729} = \frac{96d^3 - 64d^3}{729}$$

$$Z_b = \frac{32d^3}{729} \quad \ldots(3)$$

Moment of resistance for full square section $M_a = \sigma.Z_a$

Moment of resistance for flattened section $M_b = \sigma\, Z_b$

$$\text{Comparing, } \frac{M_b}{M_a} = \frac{\sigma\, Z_b}{\sigma\, Z_a} = \frac{\left(\frac{32}{729}d^3\right)}{\left(\frac{d^3}{24}\right)} = 1.0535$$

Hence, moment of inertia of flattened section is greater than that of full square section.

Distance to the flat from NA = $y = \frac{4d}{9}$.

Problem 17 A cantilever beam of span 9 m has a depth of 300 m and is symmetric in cross section. The moment of inertia is 6×10^6 mm^4. Taking the allowable bending stresses 90 MPa,

Find (i) Safe point load that can be applied at the free end

(ii) Safe UDL that can be applied over entire span.

Solution:

$$\text{Span } l = 9 \text{ m} = 9000 \text{ mm}$$

$$\text{Depth } d = 300 \text{ mm}$$

Symmetric cross section

$$\therefore \quad y = \frac{d}{2} = \frac{300}{2} = 150 \text{ mm}$$

$$\text{Moment of inertia } I = 6 \times 10^6 \text{ mm}^4$$

$$\text{Allowable stress } \sigma = 90 \text{ MPa}$$

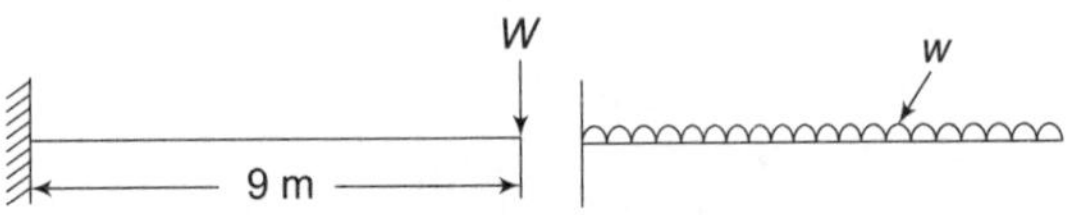

(a) Point load at free end (b) UDL over entire spam

Fig. 5.28

(a) Point load at free end, maximum bending moment $M = Wl = W \times 9000$

$$\text{Bending equation } \frac{M}{I} = \frac{\sigma}{y}$$

$$\frac{W \times 9000}{6 \times 10^6} = \frac{90}{150}$$

$$\therefore \quad \text{Point load at the end } W = 400 \text{ N}$$

(b) UDL over the entire span, maximum bending moment $M = \frac{wl^2}{2} = \frac{w \times 9000^2}{2}$

Substituting in bending equation

$$\frac{w \times 9000^2}{2 \times 6 \times 10^6} = \frac{90}{150}$$

$$\therefore \quad \text{Intensity of UDL } w = 0.08889 \text{ N/m} = 0.08889 \text{ kN/m.}$$

Problem 18 A *I*-section beam of each flange 200 mm wide × 10 mm thick and the web of 380 mm × 8 mm thick is made of steel having permissible stress of 150 N/mm^2. Find its moment of resistance. Compare it with equivalent square section of same area.

Solution:

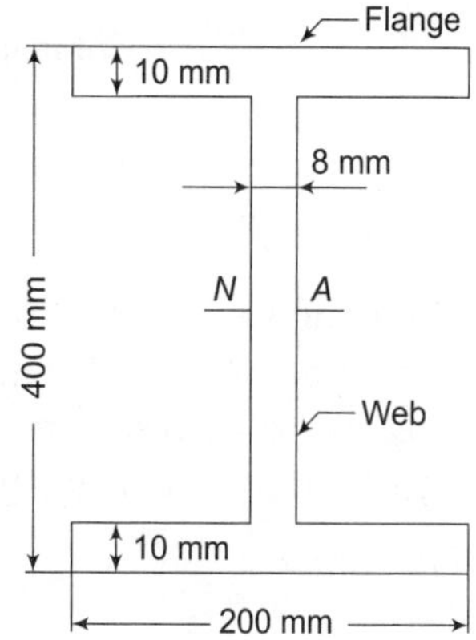

Fig. 5.29

Width of each flange $B = 200$ mm

Thickness of each flange $t_F = 10$ mm

Height of web $= d = 380$ mm

Thickness of web $= t_W = 8$ mm

Permissible stress $\sigma = 150$ MPa.

(a) For symmetric I-section by considering

Total rectangle of $B = 200$ mm,

and depth $D = 380 + 10 + 10$

$= 400$ mm

Less unwanted rectangle of $b = 200 - 8 = 192$ mm and $d = 380$ mm

$$I = I_1 - I_2 = \frac{BD^3 - bd^3}{12} = \frac{200 \times 400^3 - 192 \times 380^3}{12}$$

$$I = 188.715 \times 10^6 \text{ mm}^4$$

$$y = y_{\max} = \frac{D}{2} = \frac{400}{2} = 200 \text{ mm}$$

$\therefore$ Moment of resistance $M = \sigma \cdot \dfrac{I}{y}$

$$M_I = 150 \times \frac{188.715 \times 10^6}{200} = 141.536 \times 10^6 \text{ N-mm}$$

(b) Square section of same area

$$A = b^2 = A_1 + A_2 + A_3$$

$$= (200 \times 10) + (380 \times 8) + (200 \times 10)$$

$$b^2 = 7040$$

Side by square $b = 83.905$ mm

$\therefore$ Section modulus $Z = \dfrac{I}{C} = \dfrac{\left(\dfrac{b^4}{12}\right)}{\left(\dfrac{b}{2}\right)} = \dfrac{b^3}{6}$

$$Z = \frac{83.905^3}{6} = 98.448 \times 10^3 \text{ mm}^3$$

$\therefore$ Moment of resistance of square section

$$M = \sigma \cdot Z = 150 \times 98.448 \times 10^3$$

$$M_{sq} = 14.767 \times 10^6 \text{ N mm}$$

$$\text{Comparing } \frac{M_I}{M_{sq}} = \frac{141.536 \times 10^6}{14.767 \times 10^6} = 9.585$$

The moment of resistance of I section is 9.585 times. The moment of resistance of square section of equal area.

Problem 19 A hollow rectangular cross section beam of outside width 120 mm and outside depth of 180 mm has a wall thickness of 20 mm and is simply supported at ends over a span of 3 m. Find the UDL that this beam can be carried over the entire span limiting the bending stress to 90 MPa.

Solution:

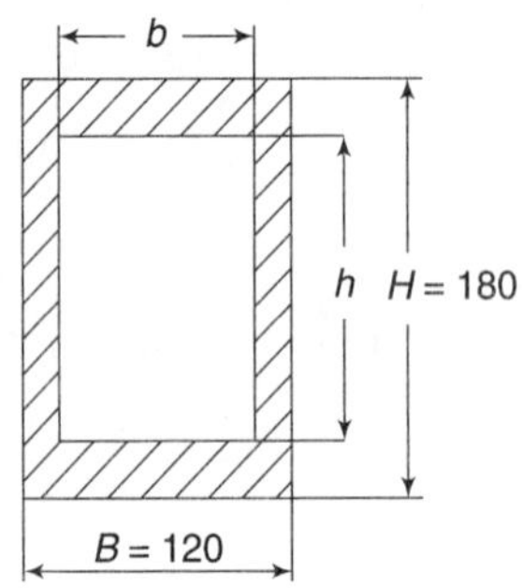

Fig. 5.30

Outer width $B = 120$ mm

Outer height $H = 180$ mm

Wall thickness $t = 20$ mm

$\therefore$ Inner width $b = B - 2t = 80$ mm

Inner height $h = H - 2t = 140$ mm

Span $l = 3$ m $= 3000$ mm

Bending stress $\sigma = 90$ MPa

UDL $= w$ N/mm

For simply supported beam with full UDL, maximum bending moment

$$M = \frac{wl^2}{8} = \frac{w \times 3000^2}{8}$$

For hollow rectangular cross section, moment of inertia

$$I = \frac{BH^3 - bh^3}{12}$$

$$= \frac{120 \times 180^3 - 80 \times 140^3}{12} = 40.027 \times 10^6 \text{ mm}^4$$

Distance from neutral axis to outer fibre

$$y = \frac{H}{2} = 90 \text{ mm}$$

$\therefore$ From bending equation $\frac{M}{I} = \frac{\sigma}{y}$

$$\frac{w \times 3000^2}{8 \times 40.027 \times 10^6} = \frac{90}{90}$$

$\therefore$ Intensity of UDL $w = 35.579$ N/mm $= 35.579$ kN/m.

Problem 20 A symmetric I section of has the following dimensions.

Width of each flange = 160

Thickness of flange = 15

Overall depth = 300

Thickness of web = 10 mm.

The beam is simply supported over a span of 6 m and carries a concentrated load of 300 kN at the middle, find the bending stress induced. Sketch the stress distribution across the depth.

Solution:

$$\text{Total width } B = 160$$

$$\text{Total depth } H = 300$$

$$\text{Thickness of each } F \text{ lang } t = 15 \text{ mm}$$

$$\therefore \quad \text{Depth of inner portion } h = H - 2t = 300 - 2 \times 15 = 270 \text{ mm}$$

$$\text{Thickness of web } t_1 = 10 \text{ mm}$$

$$\text{Width of inner portion } b = B - t_1 = 160 - 10 = 150 \text{ mm}$$

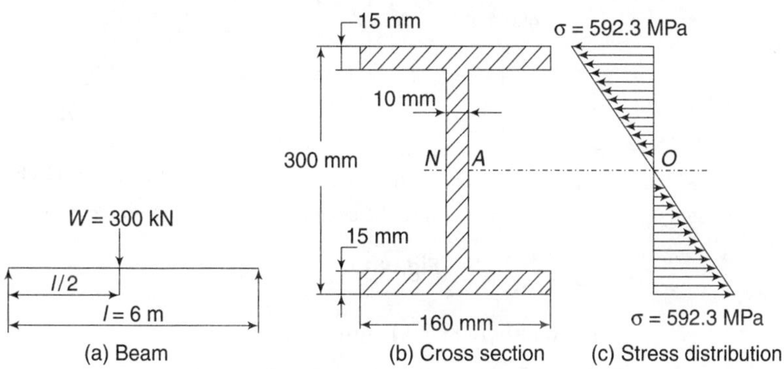

Fig. 5.31

Moment of inertia of the I section about neutral axis

$$I = \frac{BH^3 - bh^3}{12} = \frac{160 \times 300^3 - 150 \times 270^3}{12}$$

$$= 113.9625 \times 10^6 \text{ mm}^4$$

Distance from neutral axis to farthest layer

$$y = \frac{H}{2} = \frac{300}{2} = 150 \text{ mm}$$

For simply supported beam with central load maximum bending

$$M = \frac{Wl}{4} = \frac{300 \times 10^3 \times 6000}{4} = 450 \times 10^6 \text{ N-mm}$$

$$\text{Bending equation} = \frac{M}{I} = \frac{\sigma}{y}$$

$$\frac{450 \times 10^6}{113.9625 \times 10^6} = \frac{\sigma}{150}$$

$\therefore$ Maximum stress $\sigma = 592.3 \text{ N/mm}^2$

This is compressive at the top fibre and tensile at the bottom fibres and zero at neutral axis. The stress distribution is shown in Fig. 5.31(c).

Problem 21 Calculate and plot the bending stress distribution for a I-section of flang 300 mm × 12 mm and web 376 mm × 10 mm, span 4 m, simply supported at ends and carrying UDL of 10 kN/m.

Solution:

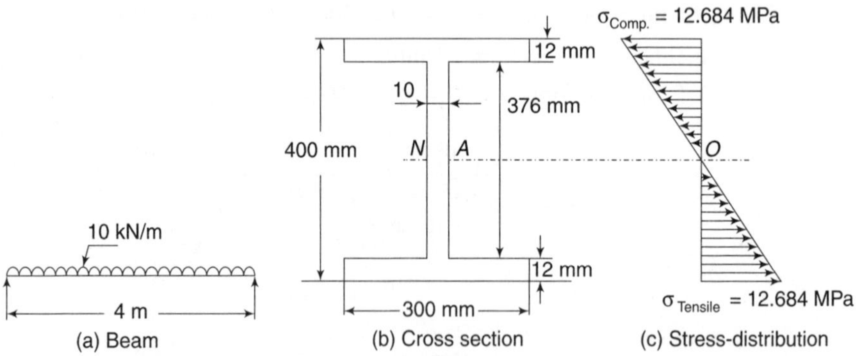

Fig. 5.32

$$\text{Total width } B = 300 \text{ mm}$$

$$\text{Total height } H = 376 + 12 + 12 = 400 \text{ mm}$$

$$\text{Inner width } b = 300 - 10 = 290 \text{ mm}$$

$$\text{Inner height } h = 376 \text{ mm}$$

$$\text{Moment of inertia of I-section} = \frac{BH^3 - bh^3}{12} = \frac{300 \times 400^3 - 290 \times 376^3}{12}$$

$$I = 315.363 \times 10^6 \text{ mm}^4$$

Maximum distance to outer fibre from neutral axis

$$y = \frac{H}{2} = \frac{400}{2} = 200 \text{ mm}$$

For simply supported beam with UDL, maximum bending moment

$$M = \frac{wl^2}{8} = \frac{10000 \times 4^2}{8} = 20000 \text{ Nm}$$

$$= 2000 \times 10^3 \text{ N mm}$$

The bending equations $\frac{M}{I} = \frac{\sigma}{y}$

$$\frac{20000 \times 10^3}{315.363 \times 10^6} = \frac{\sigma}{200}$$

$\therefore$ Maximum bending stress σ = 12.684 N/mm^2

This is compressive at top fibre, tensile at bottom fibre and zero at neutral axis as shown in Fig. 5.32(c).

Problem 22 The compressive flange of a cast iron girder is 300 mm wide and 50 mm deep. The tension flange is 100 mm wide and 50 mm deep. The wed is 250 mm × 30 mm. Find the maximum uniformly distributed load the beam can carry when simply supported over a span of 3 m, given the maximum permissible stresses are 95 MPa in compression and 150 MPa in tension. Neglect the weight of the beam.

Solution: **Given data:**

$$\sigma_{\text{Comp}} = 95 \text{ MPa}$$

$$\sigma_{\text{Tensile}} = 950 \text{ MPa}$$

$$\text{Span } l = 3\ M = 3000 \text{ mm.}$$

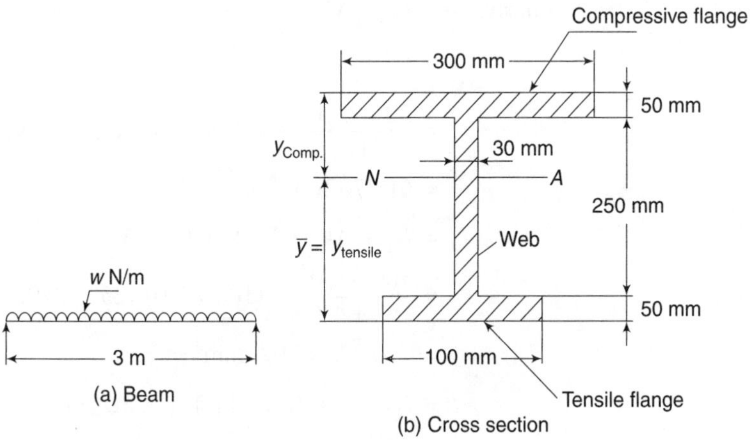

Fig. 5.33

For simply supported beam with UDL, maximum bending moment

$$M = \frac{wl^2}{8} = \frac{w \times 3000^2}{8} = 1.125 \times 10^6\ w \text{ N-mm.}$$

Consider the unsymmetric I-section as 3 rectangles.

$$\text{Centre of gravity } \bar{y} = \frac{A_1Y_1 + A_2Y_2 + A_3Y_3}{A_1 + A_2 + A_3}$$

$$\bar{y} = \frac{(300 \times 50)\left(50 + 250 + \frac{50}{2}\right) + (250 \times 30)\left(50 + \frac{250}{2}\right) + (100 \times 50)\left(\frac{50}{2}\right)}{(300 \times 50 + 250 \times 30 + 100 \times 50)}$$

$$\bar{y} = 229.545 \text{ mm} = y_{\text{Ten}}$$

$$y_{\text{comp}} = H - y_{\text{Ten}} = (50 + 250 + 50) - 229.545$$

$$= 120.455 \text{ mm}$$

$$\text{Moment of inertia } I = I_1 + I_2 + I_3$$

$$\text{From parallel axis theorem } I_1 = I_{cg} + Ak^2$$

$$I_1 = \frac{300 \times 50^3}{12} + (300 \times 50)\,(y_1 - \bar{y})^2$$

$$= \frac{300 \times 50^3}{12} + (300 \times 50)\,(325 - 229.545)^2$$

$$= 139.8 \times 10^6 \text{ mm}$$

$$\text{Similarly, } I_2 = I_{cg} + Ak^2$$

$$= I_{cg} + A\,(y_2 - \bar{y})^2$$

$$= \frac{30 \times 250^3}{12} + (30 \times 250)\,(175 - 229.545)^2$$

$$= 61.376 \times 10^6 \text{ mm}^4$$

$$I_3 = I_{cg} + Ak^2 = I_{cg} + A\,(y_3 - \bar{y})^2$$

$$= \frac{100 \times 50^3}{12} + (100 \times 50)\,(25 - 229.545)^2$$

$$= 210.235 \times 10^6 \text{ mm}^4$$

$$\therefore \quad I = I_1 + I_2 + I_3 = 411.411 \times 10^6 \text{ mm}^4$$

$$\text{Bending equation is } \frac{M}{I} = \frac{\sigma}{y}$$

$$\text{Consider compressive side, } \frac{M}{I} = \frac{\sigma_{\text{Comp.}}}{y_{\text{Comp.}}}$$

$$\frac{1.125 \times 10^6\, w}{411.411 \times 10^6} = \frac{95}{120.455}$$

$$\therefore \quad \text{Intensity of UDL } w = 288.42 \text{ N/mm} = 288.42 \text{ kN/m}$$

$$\text{Considering tensile side } \frac{M}{I} = \frac{\sigma_{\text{Ten.}}}{y_{\text{Ten.}}}$$

$$\frac{1.25 \times 10^6\, w}{411.411 \times 10^6} = \frac{150}{229.545}$$

$\therefore$ Intensity of UDL $w = 238.97$ N/mm $= 238.97$ kN/m

$\therefore$ Permissible UDL $= 238.97$ kN/m, (smaller of the above two values.)

Problem 23 Calculate the bending stress values for an unsymmetric I-section with top flange 200 mm × 30 mm, bottom flange 150 mm × 24 mm, web thickness of 40 mm and overall height of section of 140 mm when it carries a concentrated central load of 90 kN of a simply supported beam of span 4 m. Sketch the stress distribution across the section.

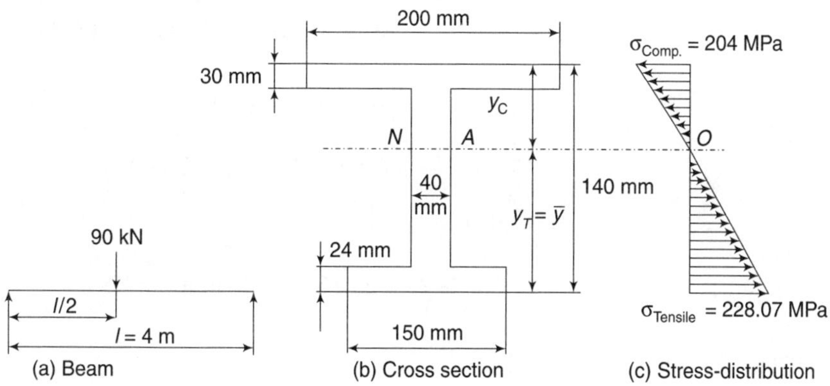

Fig. 5.34

Solution: For simply supported beam with central load, maximum

$$\text{Bending moment } M = \frac{Wl}{4} = \frac{90000 \times 4000}{4} = 90 \times 10^6 \text{ N-mm}$$

$$\text{Height of web} = 140 - 30 - 24 = 86 \text{ mm}$$

Considering the I-section as 3 rectangle, position of centre of gravity

$$\bar{y} = \frac{A_1y_1 + A_2y_2 + A_3y_3}{A_1 + A_2 + A_3}$$

$$= \frac{(200 \times 30)\left(24 + 86 + \frac{30}{2}\right) + (40 \times 86)\left(24 + \frac{86}{2}\right) + (150 \times 24) \times \frac{24}{2}}{200 \times 30 + 40 \times 86 + 150 \times 24}$$

$$\bar{y} = 73.90 \text{ mm} = y_T$$

$\therefore$ $y_C = 140 - y_T = 140 - 73.90 = 66.10$

The top fibres will experience compressive stress and the bottom fibres tensile.

Moment of inertia of the cross section about neutral axis

$$I = I_1 + I_2 + I_3$$

where, $I_1 = I_{cg} + Ak^2$ (Parallel axis theorem)

$$= \frac{bh^3}{12} + (bh)\,(y_1 - \bar{y})^2$$

$$= \frac{200 \times 30^3}{12} + (200 \times 30)\,(115 - 73.90)^2$$

$$= 10.585 \times 10^6 \text{ mm}^4$$

Similarly, $I_2 = I_{cg} + Ak^2 = \frac{bh^3}{12} + (bh)\,(y_2 - \bar{y})^2$

$$= \frac{40 \times 86^3}{12} + (40 \times 86)\,(47 - 73.90)^2 = 4.61 \times 10^6 \text{ mm}^4$$

$$I_3 = I_{cg} + Ak^2 = \frac{bh^3}{12} + bh\,(y_3 - \bar{y})^2$$

$$= \frac{150 \times 24^3}{12} + (150 \times 24)\,(12 - 73.90)^2$$

$$= 13.967 \times 10^6 \text{ mm}^4$$

$\therefore$ $I = I_1 + I_2 + I_3 = 29.162 \times 10^6 \text{ mm}^4$

The bending equation, $\frac{M}{I} = \frac{\sigma}{y}$

$\therefore$ Bending stress $\sigma = \frac{M}{I} \cdot y$

$\therefore$ Maximum compressive stress at the top

$$\sigma_C = \frac{M}{I} \cdot y_C$$

$$\sigma_C = \frac{90 \times 10^6}{29.162 \times 10^6} \times 66.1 = 204 \text{ MPa}$$

Maximum tensile stress at the bottom

$$\sigma_T = \frac{M}{I}\, y_T$$

$$\sigma_T = \frac{90 \times 10^6}{29.162 \times 10^6} \times 73.9 = 228.07 \text{ MPa}$$

Stress at neutral axis = 0.

Figure 5.34(c) shows the stress distribution across the section.

Problem 24 A simply supported beam of I-section shown in Fig. 5.35 carries a UDL of 4 kN/m run over the entire span of beam of length 10 m. Determine the maximum stresses produced and sketch the stress distribution across section.

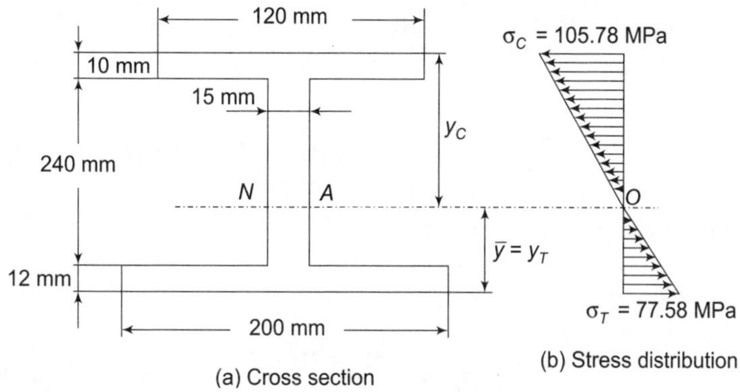

Fig. 5.35

Solution:

$$\text{Span } l = 10 \text{ m} = 10000 \text{ mm}$$

$$\text{UDL } w = 40 \text{ kN/m} = 400 \text{ N/mm}$$

For simply supported beam with UDL, maximum bending

$$\text{Moment } M = \frac{wl^2}{8} = \frac{40 \times 10000^2}{8} = 500 \times 10^6 \text{ Nmm}$$

Considering I section as three rectangles, centre of gravity

$$\bar{y} = \frac{A_1 y_1 + A_2 y_2 + A_3 y_3}{A_1 + A_2 + A_3}$$

$$= \frac{(120 \times 10)\left(12 + 240 + \frac{10}{2}\right) + (240 \times 15)\left(12 + \frac{240}{2}\right) + (200 \times 12)\left(\frac{12}{2}\right)}{120 \times 10 + 240 \times 15 + 200 \times 12}$$

$$\bar{y} = 110.833 \text{ mm} = y_T$$

$$\therefore \quad y_C = \text{Total height} - \bar{y} = (10 + 240 + 12) - 110.833 = 151.167 \text{ mm}$$

Moment of inertia of I section about material axis

$$I = I_1 + I_2 + I_3$$

$$I_1 = I_{cg} + Ak^2 = \frac{bh^3}{12} + (bh)\,(y_1 - \bar{y})^2$$

$$= \frac{120 \times 10^3}{12} + (120 \times 10)\,(257 - 110.833)^2 = 25.648 \times 10^6 \text{ mm}^4$$

Similarly, $$I_2 = \frac{15 \times 240^3}{12} + (15 \times 240)\,(132 - 110.833)^2 = 18.893 \times 10^6 \text{ mm}^4$$

$$I_3 = \frac{200 \times 12^3}{12} + (200 \times 12)\,(6 - 110.833)^2 = 26.910 \times 10^6 \text{ mm}^4$$

$$\therefore \quad I = I_1 + I_2 + I_3 = 71.451 \times 10^6 \text{ mm}^4$$

Maximum compressive stress at top fibre

$$\sigma_C = \frac{M}{I} \cdot y_C$$

$$\sigma_C = \frac{50 \times 10^6}{71.451 \times 10^6} \times 151.167 = 105\ .78 \text{ MPa}$$

Maximum tensile stress at bottom fibre

$$\sigma_T = \frac{M}{I} y_T$$

$$\sigma_T = \frac{50 \times 10^6}{71.451 \times 10^6} \times 110.833 = 77.58 \text{ MPa}$$

Stress at neutral axis = 0.

The stress distribution is plotted as shown in Fig. 5.35(b).

Problem 25 A cantilever has a length of 3 m. Its cross section is of T-section with flange 100 mm × 20 mm and web 200 mm × 12 mm and with flange in tension. What is the intensity of UDL that can be applied if maximum tensile and compressive stresses are limited to 30 MPa and 45 MPa respectively.

Solution: For cantilever in bending, all top fibres will be in tension and bottom fibres in compression. For cantilever with UDl, maximum bending moment

$$M = \frac{wl^2}{2} = \frac{w \times 3000^2}{2}$$

$$M = 4.5 \times 10^6\ w \text{ N-mm}$$

Considering T-section as two rectangles.

$$\text{Centre of gravity } \bar{y} = \frac{A_1 y_1 + A_2 y_2}{A_1 + A_2}$$

$$= \frac{(100 \times 20)\left(200 + \frac{20}{2}\right) + (200 \times 12)\left(\frac{200}{2}\right)}{100 \times 20 + 200 \times 12}$$

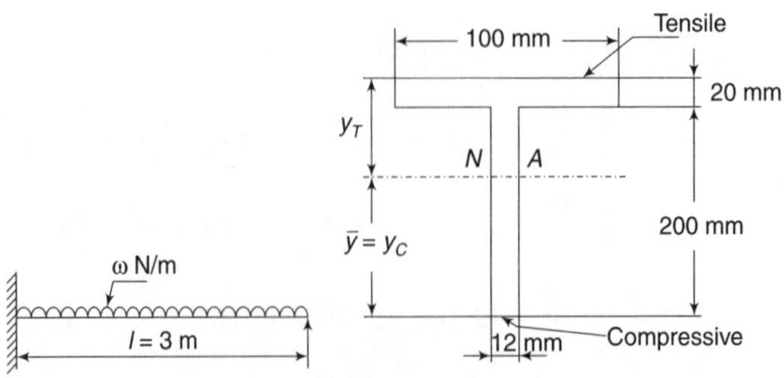

Fig. 5.36

$$\bar{y} = y_C = 150 \text{ mm}$$

$$y_T = 220 - \bar{y} = 220 - 150 = 70 \text{ mm}$$

Moment of inertia $= I = I_1 + I_2$

Applying parallel axis theorem.

$$I_1 = I_{cg} + Ak^2 = \frac{bh^3}{12} + bh\,(y_1 - \bar{y})^2$$

$$= \frac{100 \times 20^3}{12} + (100 \times 20)\,(210 - 150)^2 = 7.267 \times 10^6 \text{ mm}^4$$

$$I_2 = \frac{12 \times 200^3}{12} + (12 \times 200)\,(100 - 150)^2 = 14 \times 10^6 \text{ mm}^4$$

$\therefore$ $$I = I_1 + I_2 = 21.267 \times 10^6 \text{ mm}^4$$

Bending equation $\dfrac{M}{I} = \dfrac{\sigma}{y}$

Considering top fibre in tension.

$$\frac{M}{I} = \frac{\sigma_T}{y_T}$$

$$\frac{4.5 \times 10^6\, w}{21.267 \times 10^6} = \frac{30}{70}$$

$\therefore$ Intensity of UDL $= w = 2.025$ N/mm

Considering bottom fibres in compression

$$\frac{M}{I} = \frac{\sigma_C}{y_C}$$

$$\frac{4.5 \times 10^6\, w}{21.267 \times 10^6} = \frac{45}{150}$$

$\therefore$ Intensity of UDL $w = 1.4178$ N/mm

Recommended intensity of UDL $w = 1.4178$ N/mm $= 1.4178$ kN/m.

Problem 26 A T-section is formed by cutting the bottom flange of an I-section. The flange is 120 mm × 24 mm and the web is 150 mm × 20 mm. Draw the bending stress distribution if the bending moment at a section of the beam is 10 kN-m sagging.

Solution: Sagging moment will induce compressive stress at top and tensile stress at bottom fibres

$$\text{Moment } M = 10 \text{ kN-m} = 10 \times 10^6 \text{ N-mm}$$

$$\text{Centre of gravity } \bar{y} = \frac{A_1 y_1 + A_2 y_2}{A_1 + A_2}$$

$$\bar{y} = \frac{(120 \times 24)(150 + 24/2) + (150 \times 20)(150/2)}{120 \times 24 + 150 \times 20}$$

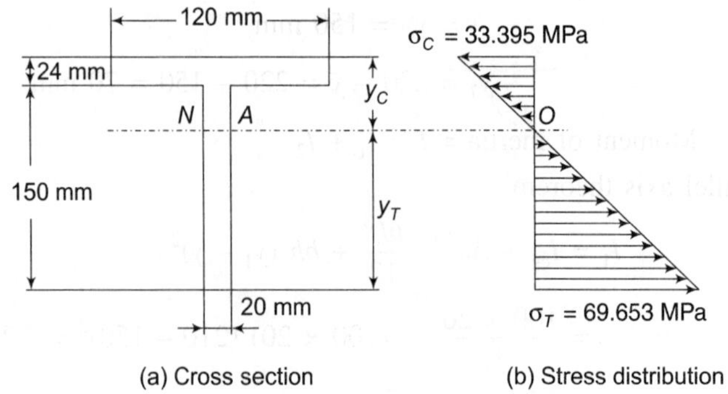

Fig. 5.37

$$\bar{y} = 117.612 \text{ mm} = y_T$$

$$y_C = 174 - y_T = 174 - 117.612 = 56.388 \text{ mm}$$

$$\text{Moment of inertia } I = I_1 + I_2$$

$$I_1 = \frac{bh^3}{12} + (bh)\,(y_1 - \bar{y})^2$$

$$= \frac{120 \times 24^3}{12} + (120 \times 24)\,(162 - 117.612)^2$$

$$= 5.813 \times 10^6 \text{ mm}^4$$

$$I_2 = \frac{bh^3}{12} + (bh)\,(y_2 - \bar{y})^2$$

$$= \frac{20 \times 150^3}{12} + (20 \times 150)(75 - 117.612)^2$$

$$= 11.072 \times 10^6 \text{ mm}^4$$

$$\therefore \quad I = I_1 + I_2 = 16.885 \times 10^6 \text{ mm}^4$$

Compressive stress at top fibre

$$\sigma_C = \frac{M}{I} \cdot y_C$$

$$\sigma_C = \frac{10 \times 10^6}{16.875 \times 10^6} \times 56.388 = 33.395 \text{ MPa}$$

Tensile stress at bottom fibre

$$\sigma_T = \frac{M}{I} \cdot y_T$$

$$\sigma_T = \frac{10 \times 10^6}{16.885 \times 10^6} \times 117.612 = 69.653 \text{ MPa}$$

Stress at neutral axis = 0

The stress distribution is plotted as shown in Fig. 5.37 (b).

Problem 27 The T-section as shown in Fig. 5.38 is used as a simply supported beam over a span of 4 metres. It carries a uniformly distributed load of 8 kN/m over the entire span. Calculate the maximum tensile and compressive stresses occurring in the section. Sketch the stress distribution.

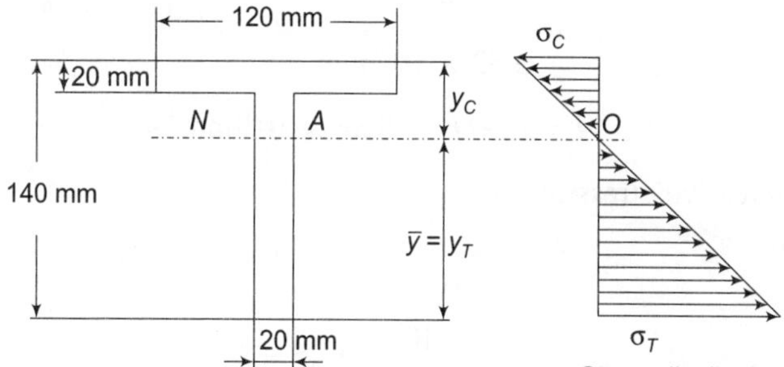

Fig. 5.38

Solution:

$$\text{Span } l = 4 \text{ m} = 4000 \text{ mm}$$

$$\text{UDL } w = 8 \text{ kN/m} = 8 \text{ N/mm}$$

For simply supported beam with UDL, maximum bending

$$\text{Moment, } M = \frac{wl^2}{8} = \frac{8 \times 4000^2}{8} = 16 \times 10^6 \text{ N mm}$$

For simply supported beam with downward loads, top fibre will be in compression and bottom fibres will be in tension.

$$\text{Centre of gravity, } \bar{y} = \frac{A_1 y_1 + A_2 y_2}{A_1 + A_2}$$

$$\bar{y} = \frac{(120 \times 20)\left(140 - \frac{20}{2}\right) + (120 \times 20) \times \left(\frac{120}{2}\right)}{120 \times 20 + 120 \times 20}$$

$$= 95 \text{ mm} = y_\tau$$

$$\therefore \quad y_C = 140 - \bar{y} = 140 - 95 = 45 \text{ mm}$$

$$\text{Moment of inertia, } I = I_1 + I_2$$

$$I_1 = \frac{bh^3}{12} + bh\,(y_1 - \bar{y})^2$$

$$I_1 = \frac{120 \times 20^3}{12} + (120 \times 20)\,(130 - 95)^2$$

$$= 3.02 \times 10^6 \text{ mm}^4$$

Similarly,

$$I_2 = \frac{bh^3}{12} + bh\,(y_2 - \bar{y})^2$$

$$= \frac{20 \times 120^3}{12} + (20 \times 120)\,(60 - 95)^2$$

$$= 5.82 \times 10^6 \text{ mm}^4$$

$\therefore$ $$I = I_1 + I_2 = 8.84 \times 10^6 \text{ mm}^4$$

Maximum compressive stress at the top

$$\sigma_C = \frac{M}{I} \cdot y_C$$

$$\sigma_C = \frac{16 \times 10^6}{8.84 \times 10^6} \times 45 = 81.45 \text{ N/mm}^2$$

Maximum tensile stress at the bottom

$$\sigma_T = \frac{M}{I} \cdot y_T$$

$$\sigma_T = \frac{16 \times 10^6}{8.84 \times 10^6} \times 95 = 171.95 \text{ N/mm}^2$$

The stress distribution is shown in Fig. 5.38.

Problem 28 A simply supported beam of span 6 m has a cross section as shown in Fig. 5.39. It carries two point loads each of 30 kN at a distance of 2 m from each support. Calculate the bending stress corresponding to maximum value of bending moment. Draw neat diagram of variation of bending stress across the section.

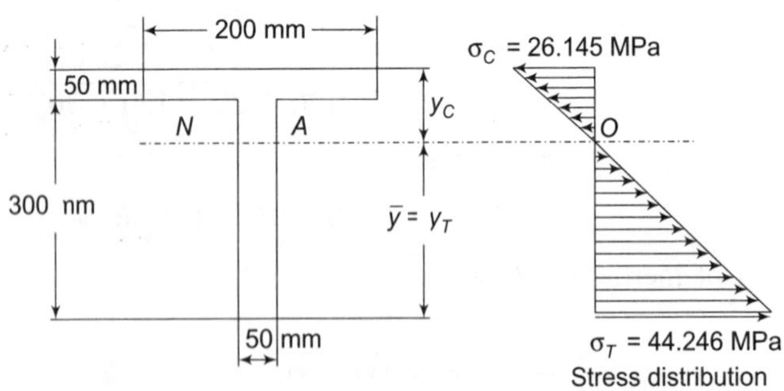

Fig. 5.39

Solution: Reaction $R_A + R_D = 30 + 30 = 60$ kN

Taking moments about A, $-30 \times 2 - 30 \times 4 + R_D \times 6 = 0$

$\therefore$ $$R_D = 30 \text{ kN and } R_A = 60 - R_D = 30 \text{ kN}$$

Bending moments

$$M_A = 0$$

$$M_B = 30 \times 2 = 60 \text{ kNm}$$

$$M_C = 30 \times 2 = 60 \text{ kNm}$$

$$M_D = 0$$

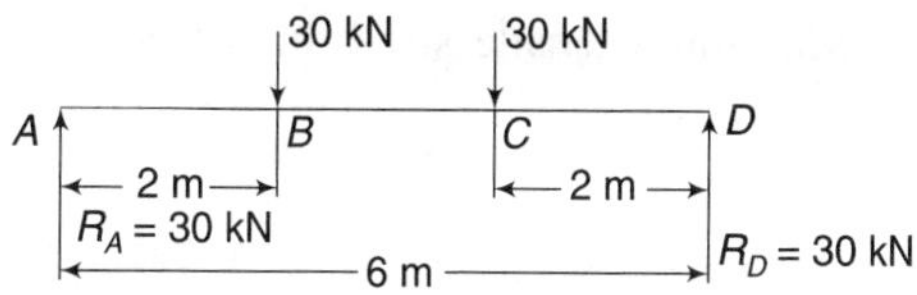

Fig. 5.39 (a)

∴ Maximum bending moment $M = 60$ kNm $= 60 \times 10^6$ Nmm The top of beam is in compression and bottom in tension.

Centre of gravity:

$$\bar{y} = \frac{A_1 y_1 + A_2 y_2}{A_1 + A_2}$$

$$= \frac{(200 \times 50)(300 + 50/2) + (300 \times 50)(300/2)}{(200 \times 50 + 300 \times 50)}$$

$$\bar{y} = 220 \text{ mm} = y_T$$

$$\therefore \quad y_C = 350 - \bar{y} = 130 \text{ mm}$$

Moment of inertia: $I = I_1 + I_2$

$$I_1 = I_{cg} + Ak^2 = \frac{bh^3}{12} + bh\,(y_1 - \bar{y})^2$$

$$= \frac{200 \times 50^3}{12} + (200 \times 50)\,(325 - 220)^2 = 112.333 \times 10^6$$

$$I_2 = \frac{bh^3}{12} + Ak^2 = \frac{bh^3}{12} + bh\,(y_2 - \bar{y})^2$$

$$= \frac{50 \times 300^3}{12} + (50 \times 300)\,(150 - 220)^2 = 186 \times 10^6$$

$$\therefore \quad I = I_1 + I_2 = 298.333 \times 10^6 \text{ mm}^4$$

Bending equation $\dfrac{M}{I} = \dfrac{\sigma}{y}$

Bending stress $\sigma = \dfrac{M}{I} \cdot y$

Maximum compressive top due to bending,

$$\sigma_C = \frac{M}{I} \cdot y_C$$

$$\sigma_C = \frac{60 \times 10^6}{298.333 \times 10^6} \times 130 = 26.145 \text{ N/mm}^2$$

Maximum tensile stress at bottom due to bending,

$$\sigma_T \cdot = \frac{M}{I} y_T$$

$$\sigma_T = \frac{60 \times 10^6}{298.333 \times 10^6} \times 220 = 44.246 \text{ N/mm}^2$$

The stress distribution across section is plotted as shown in Fig. 5.39 (a).

Problem 29 A beam of T-section has a length of 2.5 metres and is subjected to a point load as shown in Fig. 5.40. Calculate the compressive bending stress and plot the stress distribution across the cross section of the beam. The maximum tensile stress is limited to 300 MPa. Calculate the value of load *W*.

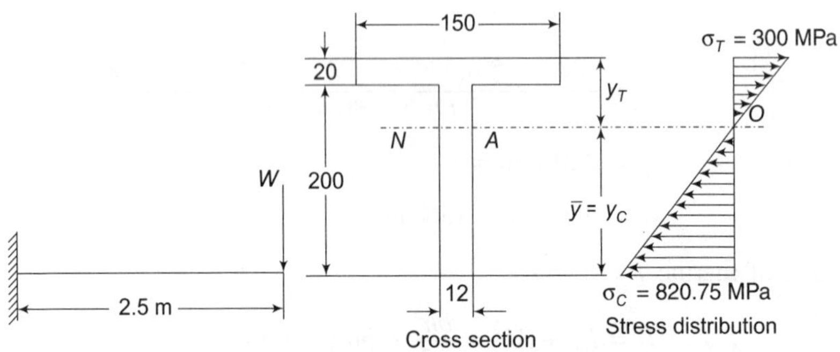

Fig. 5.40

Solution: For cantilever with end load, maximum bending moment

$$M = Wl = W \times 2500 \text{ N-mm.}$$

The cantilever bends down due to the load. Inducing tensile stress at top and compressive stress at bottom.

$$\text{Centre of gravity } \bar{y} = \frac{A_1 y_1 + A_2 y_2}{A_1 y_2}$$

$$\bar{y} = \frac{(150 \times 20)(200 + 20/2) + (200 \times 12)(200/2)}{(150 \times 20 + 200 \times 12)}$$

$$= 161.111 \text{ mm} = y_C$$

$$y_T = 220 - \bar{y} = 220 - 161.111 = 58.889 \text{ mm}$$

$$\text{Moment of inertia } I = I_1 + I_2$$

$$I_1 = \frac{150 \times 20^3}{12} + 150 \times 20\ (210 - 161.111)^2$$

$$= 7.270 \times 10^6 \text{ mm}^4$$

$$I_2 = \frac{200 \times 12^3}{12} + 200 \times 12\ (100 - 161.111)^2$$

$$= 8.992 \times 10^6 \text{ mm}^4$$

$$I = I_1 + I_2 = 16.262 \times 10^6 \text{ mm}^4$$

Given tensile stress $\sigma_T = 300$ MPa

Equation of bending is $\dfrac{M}{I} = \dfrac{\sigma_T}{y_T}$

$$\frac{2500\ W}{16.262 \times 10^6} = \frac{300}{58.889}$$

$\therefore$ Load $W = 33{,}137.6$ N

Compressive stress $\sigma_C = \dfrac{M}{I} y_C = \dfrac{2500 \times 33137.6}{16.262 \times 10^6} \times 161.111$

$$\sigma_C = 820.75 \text{ N/mm}^2$$

The stress distribution is as shown in Fig. 5.40.

Problem 30 A symmetric I-section has flange 120 mm × 10 mm and a overall depth of 200 m. The thickness of web is 10 mm. Find its moment of resistance if allowable stress is 120 N/mm². Compare it with same area square section, rectangular section with depth as twice its width and circular cross section.

Solution: Stress $\sigma = 120$ N/mm²

(a) I-section: $\bar{y} = \dfrac{200}{2} = 100 \text{ mm} = y$

Moment of inertia $I = \dfrac{BH^3 - bh^3}{12} = \dfrac{120 \times 200^3 - 110 \times 180^3}{12}$

$$= 26.54 \times 10^6 \text{ mm}^4$$

Moment of resistance $M_I = \dfrac{\sigma \cdot I}{y} = \dfrac{120 \times 26.54 \times 10^6}{100}$

$$= 31.848 \times 10^6 \text{ N-mm}$$

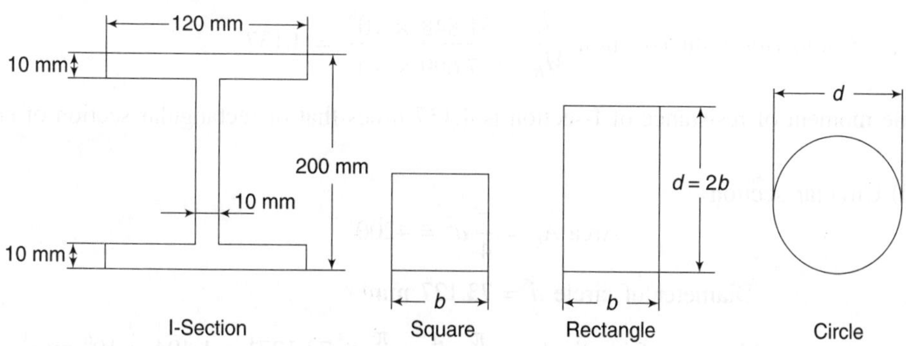

Fig. 5.41

(b) Square section of equal area:

$$\text{Area of I section } A_I = (120 \times 10) \times 2 + 180 \times 10 = 4200 \text{ mm}^2$$

$$\text{Area of square } A_S = b^2 = 4200$$

$$\therefore \quad \text{Side of square } b = 64.807 \text{ mm}$$

$$\therefore \quad \text{Moment of inertia } I_S = \frac{b^4}{12} = 1.47 \times 10^6 \text{ mm}^4$$

$$y = \frac{b}{2} = 32.\ 4035 \text{ mm}$$

$$\therefore \quad \text{Moment of resistance } M_S = \frac{\sigma \cdot I}{y} = \frac{120 \times 1.47 \times 10^6}{32.4035} = 5.444 \times 10^6 \text{ Nmm}$$

Comparing the moment of resistance with I section.

$$\frac{M_I}{M_S} = \frac{31.848 \times 10^6}{5.44 \times 10^6} = 5.85$$

The moment of resistance of I section is 5.85 times that of square section of equal area.

(c) Rectangular cross section of equal area

$$\text{Area of rectangle } A_R = b.d$$

$$\text{Given depth } d = 2b,\ A_R = 2b^2 = 4200$$

$$\therefore \quad b = 45.826 \text{ mm}$$

$$d = 91.652 \text{ mm}$$

$$\text{Moment of inertia } I_R = \frac{bd^3}{12} = \frac{45.826 \times 91.652^3}{12} = 2.94 \times 10^6 \text{ mm}^4$$

$$y = \frac{d}{2} = \frac{91.652}{2} = 45.826 \text{ mm}$$

$$\text{Moment of resistance } M_R = \frac{\sigma I}{y} = \frac{120 \times 2.94 \times 10^6}{45.826} = 7.699 \times 10^6 \text{ Nmm}$$

$$\text{Comparing with I-section } \frac{M_I}{M_R} = \frac{31.848 \times 10^6}{7.699 \times 10^6} = 4.137$$

The moment of resistance of I-section is 4.137 times that of rectangular section of equal area.

(d) Circular section

$$\text{Area } A_C = \frac{\pi}{4} d^2 = 4200$$

$$\therefore \quad \text{Diameter of circle } d = 73.127 \text{ mm}$$

$$\text{Moment of inertia } I_C = \frac{\pi}{64} d^4 = \frac{\pi}{64} \times 73.127^4 = 1.404 \times 10^6 \text{ mm}^4$$

$$y = \frac{d}{2} = \frac{73.127}{2} = 36.0635 \text{ mm}$$

$$\text{Moment of resistance } M_C = \frac{\sigma I}{y} = \frac{120 \times 1.404 \times 10^6}{36.0635} = 4.672 \times 10^6 \text{ N mm}$$

Comparing with moment of resistance of I-section

$$\frac{M_I}{M_C} = \frac{31.848 \times 10^6}{4.672 \times 10^6} = 6.817$$

The moment of resistance of I-section is 6.817 times that of circular section of same area.

Problem 31 A symmetric I-section has flange of size 200 × 20 mm and overall height of 400 mm. The web thickness is 20 mm. It is welded with a plate of 300 mm × 16 mm on the top side of I section. Find the moment of resistance of the section, if the permissible stress is 120 N/mm². How much uniformly distributed load it can safely carry if a 4 metres length beam is simply supported at the ends.

Solution: For simply supported beam with UDL, maximum bending moment

$$M = \frac{wl^2}{8} = \frac{w \times 4000^2}{8} = 2 \times 10^6 \; w \text{ N-mm}$$

$$\text{Centre of gravity } \bar{y} = \frac{A_1y_1 + A_2y_2 + A_3y_3 + A_4y_4}{A_1 + A_2 + A_3 + A_4}$$

$$\bar{y} = \frac{(300\times16)\left(400+\frac{16}{2}\right)+(200\times20)\left(400-\frac{20}{2}\right)+(360\times20)\left(20+\frac{360}{2}\right)+(200\times20)\;\frac{20}{2}}{(300 \times 16) + (200 \times 20) + (360 \times 20) + (200 \times 20)}$$

$$\bar{y} = 249.92 \text{ mm} = y_T$$

It is seen that for simply supported beam with downward roads, the top fibres are in compression and bottom ones in tension.

$$y_C = (400 + 16) - \bar{y} = 166.08 \text{ mm}$$

$$\text{Moment of inertia } I = I_1 + I_2 + I_3 + I_4$$

$$I_1 = \frac{300 \times 16^3}{12} + (300 \times 16)\,(408 - 249.92)^2$$

$$= 120.05 \times 10^6 \text{ mm}^4$$

$$I_2 = \frac{200 \times 20^3}{12} + (200 \times 20)\,(390 - 249.92)^2$$

$$= 78.62 \times 10^6 \text{ mm}^4$$

$$I_3 = \frac{20 \times 360^3}{12} + (20 \times 360)\,(200 - 249.9)^2$$

$$= 95.70 \times 10^6 \text{ mm}^4$$

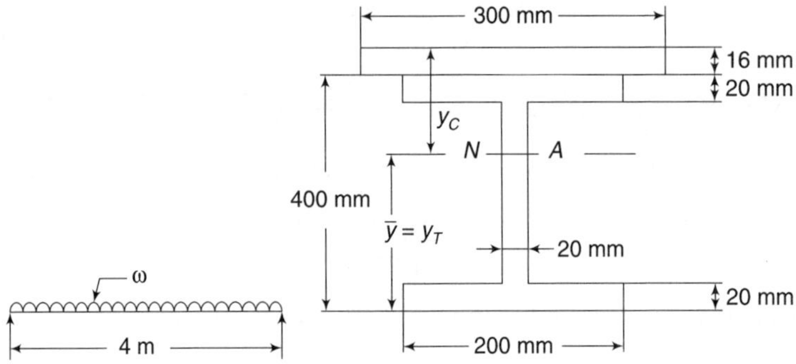

Fig. 5.42

$$I_4 = \frac{200 \times 20^3}{12} + (200 \times 20)\,(10 - 249.92)^2$$

$$= 230.38 \times 10^6 \text{ mm}^4$$

$$\therefore \qquad I = I_1 + I_2 + I_3 + I_4 = 524.75 \times 10^6 \text{ mm}^4$$

Bending equation, $\frac{M}{I} = \frac{\sigma}{y}$

$\therefore$ Moment of resistance, $M = \sigma \frac{I}{y}$

Hence, moment of resistance is inversely proportional to y

It is seen that $y_T > y_C$, $M = \sigma \frac{I}{y_T}$

$$M = 120 \times \frac{524.75 \times 10^6}{249.92} = 251.96 \times 10^6 \text{ N-mm}$$

Equating, $M = 2 \times 10^6\, w = 251.96 \times 10^6$

$\therefore$ UDL, $w = 125.98$ N/mm $= 125.98$ kN/m.

Problem 32 Figure 5.43 shows a T-section beam. Find the width of flange b such the ratio of maximum extreme fibre stress in bending is 2.5. Take tensile stress at the top.

Solution:

Ratio of stresses = 2.5

It is seen that $y_C > y_T$, σ_C will be greater than σ_T as stress $\sigma \propto y$

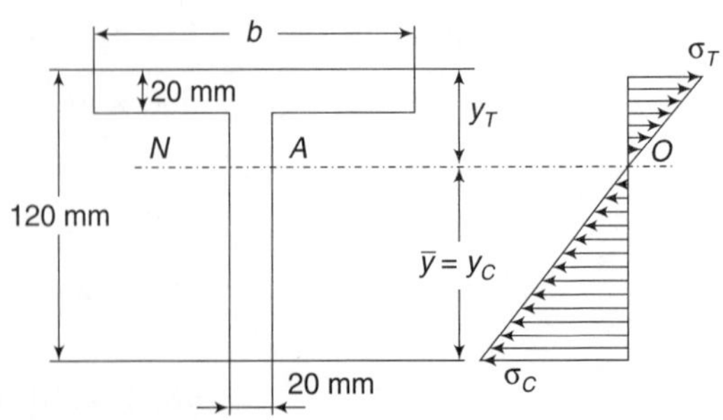

Fig. 5.43

$$\therefore \qquad \frac{\sigma_C}{\sigma_T} = 2.5$$

or $$\sigma_C = 2.5\ \sigma_T$$

$$\bar{y} = \frac{A_1 y_1 + A_2 y_2}{A_1 + A_2}$$

$$= \frac{(b \times 20)\left(120 - \frac{20}{2}\right) + (100 \times 20)\frac{100}{2}}{b \times 20 + 100 \times 20}$$

$$\bar{y} = \frac{2200b + 100000}{20b + 2000} = \frac{110b + 5000}{b + 100} = y_C$$

$$y_T = 120 - \bar{y} = 120 - \left(\frac{110b + 5000}{b + 100}\right)$$

$$y_T = \frac{10b + 7000}{b + 100}$$

From bending equation $\frac{M}{I} = \frac{\sigma}{y}$

Since for a given situation M and I are constant, $\frac{\sigma}{y}$ is constant

$$\therefore \qquad \frac{\sigma_T}{y_T} = \frac{\sigma_C}{y_C} \text{ or } \frac{\sigma_C}{\sigma_T} = \frac{y_C}{y_T} = 2.5$$

$$\frac{\left(\frac{110b + 5000}{b + 100}\right)}{\left(\frac{10b + 7000}{b + 100}\right)} = 2.5$$

$$110b + 5000 = 2.5\ (10b + 7000)$$

$$110b - 25b = 17500 - 5000$$

$$\therefore \qquad \text{Width } b = 147.059 \text{ mm.}$$

Problem 33 A T-section beam with flange 100 mm wide and overall height of 200 mm has a web thickness as half of flange thickness. The ratio of extreme stresses is 2. Find the thickness of flange and web.

Solution: The stress on one side of neutral axis tensile and the other side is compressive

$$\text{Stress ratio } \frac{\sigma_2}{\sigma_1} = 2$$

From similar triangle of stress distribution,

$$\frac{\sigma_2}{y_2} = \frac{\sigma_1}{y_1} \text{ or } \frac{\sigma_2}{\sigma_1} = \frac{y_2}{y_1} = 2$$

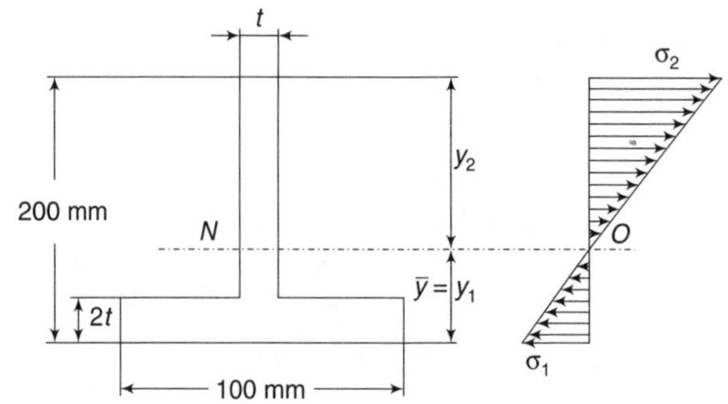

Fig. 5.44

$$y_2 = 2y_1$$

But from Fig. 5.44 $y_1 + y_2 = 200$

$$y_2 = 200 - y_1$$

$$200 - y_1 = 2y$$

$$\therefore \quad y_1 = \frac{200}{3} = 66.67 \text{ mm}$$

$$y_2 = 2y_1 = 133.33 \text{ mm}$$

Centre of gravity $\bar{y} = y_1 = \dfrac{A_1 y_1 + A_2 y_2}{A_1 + A_2}$

$$66.67 = \frac{(100 \times 2t)\left(\frac{2t}{2}\right) + (200 - 2t)t \times \left(2t + \frac{200 - 2t}{2}\right)}{100 \times 2t + (200 - 2t)\, t}$$

$$66.67 = \frac{2t\,[100t + (100 - t) \cdot (100 + t)]}{2t\,[100 + 100 - t]}$$

$$66.67\,(200 - t) = (100\,t + 10000 - t^2)$$

$$13334 - 66.67t - 100t - 10000 + t^2 = 0$$

$$t^2 - 166.67t + 3334 = 0$$

$$t = 143.42 \text{ or } 23.246$$

Taking the feasible value $t = 23.246$ mm.

Problem 34 A channel in the form of a simply supported beam, supported at the ends over span of 4 metres. The cross-sectional dimensions are as shown in Fig. 5.45. Find the central load that can be applied limiting the bending stress to 90 MPa. Find the maximum tensile and compressive stresses and draw the stress distribution across the section.

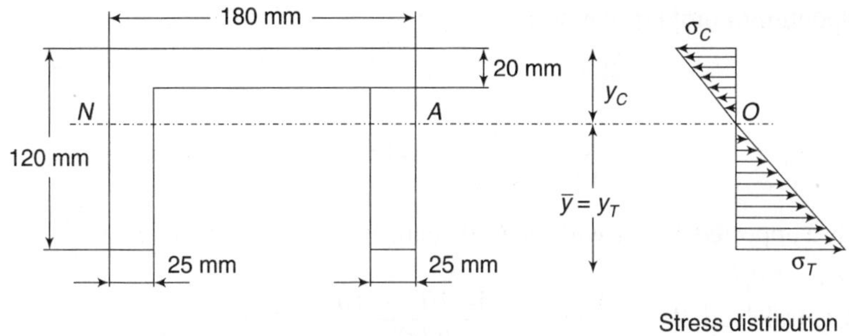

Fig. 5.45

Solution: For simply supported beam with central load, top fibers will be compressive and bottom fibers will be tenside. It is seen that $y_C < y_T$, the tensile stress will be more compared to compressive stress.

Hence, take the given allowable stress as σ_T

$$\sigma_{max} = \sigma_T = 90 \text{ MPa}$$

$$\bar{y} = \frac{A_1 y_1 + A_2 y_2 + A_3 y_3}{A_1 + A_2 + A_3}$$

$$\bar{y} = \frac{(180 \times 20) \times \left(120 - \frac{20}{2}\right) + 2 \times (25 \times 100)\left(\frac{100}{2}\right)}{180 \times 20 + 2 \times (25 \times 100)}$$

$$= 75.116 \text{ mm}$$

$$\bar{y} = y_T = 75.116 \text{ mm}$$

$$y_C = 120 - \bar{y} = 44.884 \text{ mm}$$

Moment of inertia about material axis: $= I = I_1 + 2I_2$

$$I_1 = \frac{bh^3}{12} + bh\,(y_1 - \bar{y})^2$$

$$= \frac{180 \times 20^3}{12} + (180 \times 20)\,(110 - 75.116)^2$$

$$= 4.501 \times 10^6 \text{ mm}^4$$

$$I_2 = \frac{25 \times 100^3}{12} + (25 \times 100)\,(50 - 75.116)^2 = 3.66 \times 10^6 \text{ mm}^4$$

$$I = I_1 + 2I_2 = 11.821 \times 10^6 \text{ mm}^4$$

$\therefore$ Moment of resistance $M = \frac{\sigma}{y} \cdot I$

$$M = \frac{\sigma_T}{y_T} I = \frac{90 \times 11.821 \times 10^6}{75.116} = 14.163 \times 10^6 \text{ N mm}$$

∴ Corresponding compressive stress

$$\sigma_C = \frac{M}{I} \cdot y_C$$

$$= \frac{14.163 \times 10^6}{11.821 \times 10^6} \times 44.884 = 53.78 \text{ N/mm}^2$$

For simply supported beam with central load, bending moment $M = \frac{Wl}{4}$

$$\therefore \quad \text{Central load } W = \frac{4M}{l} = \frac{4 \times 14.163 \times 10^6}{4000} = 14163 \text{ N.}$$

The stress distribution is as shown in Fig. 5.45.

Problem 35 Figure 5.46 shows a channel of length 5 metres as cantilever carrying UDL. If the allowable stresses are 72 MPa in tension and 108 MPa in compression, find the safe value of UDL for the following conditions.

(i) The flange in tension

(ii) The flange in compression

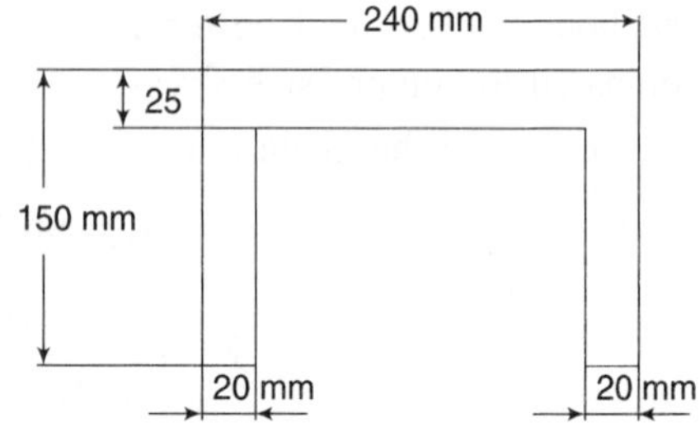

Fig. 5.46

Solution:

$$\text{Length } l = 5 \text{ m} = 5000 \text{ mm}$$

$$\text{UDL} = w \text{ N/mm}$$

$$\sigma_T = 72 \text{ MPa}$$

$$\sigma_C = 108 \text{ MPa}$$

For cantilever with UDL, maximum bending moment $M = \frac{wl^2}{2}$

$$M = \frac{w \times 5000^2}{2} = 12.5 \times 10^6 \, w \text{ N-mm.}$$

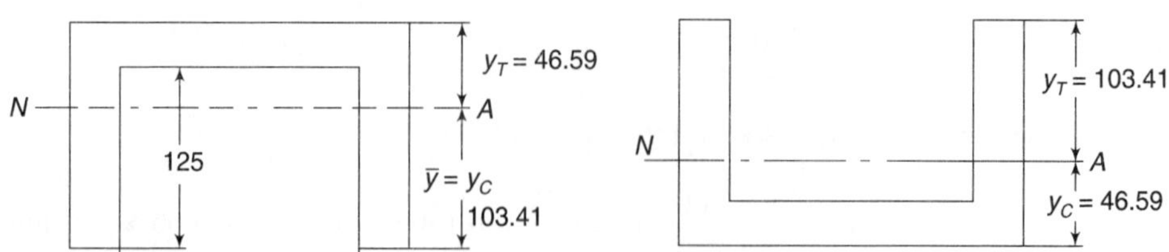

(a) Flange in tension (b) Flange in compression

The top fibres are in tension and bottom fibres in compression.

$$\text{Consider Fig. 5.46(a) } \bar{y} = \frac{A_1 Y_1 - A_2 Y_2}{A_1 + A_2}$$

where, A_1 is area of full rectangle 240×150

A_2 is area of inside rectangle $= 200 \times 125$

$$\bar{y} = \frac{(240 \times 150)\frac{150}{2} - (200 \times 125)\frac{125}{2}}{240 \times 150 - 200 \times 125} = 103.41 \text{ mm}$$

(a) For flange in tension, $\bar{y} = y_C = 103.41$ mm

$$y_T = 150 - \bar{y} = 46.59 \text{ mm}$$

$$\text{Moment of inertia, } I = I_1 + 2I_2$$

$$I_1 = \frac{bh^3}{12} + bh(\bar{y} - y_1)^2$$

$$= \frac{240 \times 25^3}{12} + 240 \times 25\,(103.41 - 137.5)^2$$

$$= 7.285 \times 10^6 \text{ mm}^4$$

$$I_2 = \frac{20 \times 125^3}{12} + 20 \times 125\,(103.41 - 62.5)^2$$

$$= 7.439 \times 10^6 \text{ mm}^4$$

$$\therefore \quad I = I_1 + 2I_2 = 22.163 \times 10^6 \text{ mm}^4$$

Moment of inertia based on tenside stress

$$M = \frac{\sigma_T}{y_T} \times I$$

$$M = \frac{72}{46.59} \times 22.163 \times 10^6 = 34.251 \times 10^6 \text{ N mm}$$

Moment of resistance based on compressive stress

$$M = \frac{\sigma_C}{y_C} \cdot I$$

$$M = \frac{108}{103.41} \times 22.163 \times 10^6 = 23.147 \times 10^6 \text{ Nmm}$$

$\therefore$ Maximum carrying capacity $M = 23.147 \times 10^6$ N m (least value)

$$\text{Equating to } M = 12.5 \times 10^6\, w = 23.147 \times 10^6$$

$$\text{Intensity of UDL } w = 1.85 \text{ N/mm} = 1.85 \text{ kN/m}$$

(b) For flange in compression: $y_C = 46.59$ mm

$$y_T = 1063.41 \text{ mm}$$

$\therefore$ Moment of resistance based on tensile stress

$$M = \frac{\sigma_T}{y_T} \times I = \frac{72}{103.41} \times 23.163 \times 10^6$$

$$= 15.431 \times 10^6 \text{ N mm}$$

Moment of resistance based on compressive stress

$$M = \frac{\sigma_C}{y_C} \times I = \frac{108}{46.59} \times 22.163 \times 10^6$$

$$= 51.376 \times 10^6 \text{ N mm}$$

Maximum carrying capacity $M = 15.431 \times 10^6$ N mm

$$M = 12.5 \times 10^6\, w = 15.431 \times 10^6$$

$\therefore$ Intensity of UDL, $w = 1.234$ N/mm $= 1.234$ kN/m.

Problem 36 Figure 5.47 shows a channel used as a simply supported beam, supported over a span of 5 m. It carries a central load of 60 kN. Find the maximum tensile and compressive stresses developed and sketch the stress distribution across the section

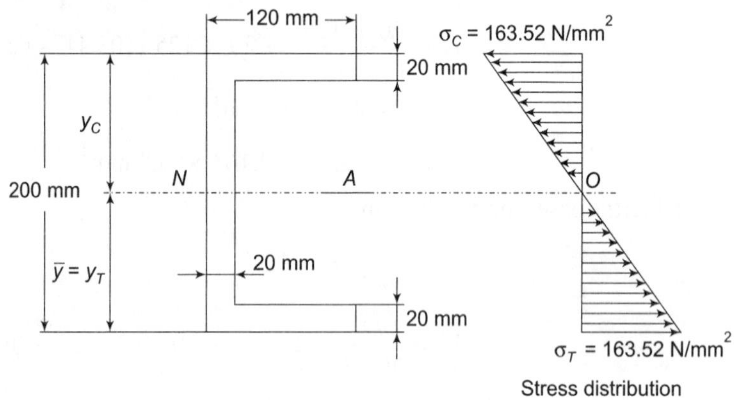

Fig. 5.47

Solution: Due to symmetry $\bar{y} = \frac{200}{2} = 100 \text{ mm} = y_T = y_C$

Length $l = 5000$ mm

Central load $W = 60 \text{ kN} = 60 \times 10^3$ N

For simply supported beam with central load, maximum bending moment

$$M = \frac{Wl}{4} = \frac{60 \times 10^3 \times 5000}{4} = 75 \times 10^6 \text{ N mm}$$

Moment inertia of the channel about neutral axis.

$$I = \frac{Bh^3 - bh^3}{2}$$

where, $B = 120$, $H = 200$, $b = 120 - 20 = 100$, $h = 200 - 2 \times 20 = 160$

$$I = \frac{120 \times 200^3 - 100 \times 160^3}{12} = 45.867 \times 10^6 \text{ mm}^4$$

$$\text{Bending stress } \sigma = \frac{M}{I} y = \frac{75 \times 10^6}{45.867 \times 10^6} \times 100 = 163.52 \text{ N/mm}^2$$

This is tensile at the bottom and compressive at the top $\sigma_T = \sigma_C = 163.52$ N/mm^2

Stress at neutral axis = 0.

The stress distribution is as shown in Fig. 5.47.

Problem 37 A hollow rectangular cross-sectional beam as shown in Fig. 5.48 is used in EOT crane. The allowable stresses in tension and compression are 30 MPa and 75 MPa respectively. Calculate the moment of resistance about horizontal neutral axis for both sagging and hogging bending moments.

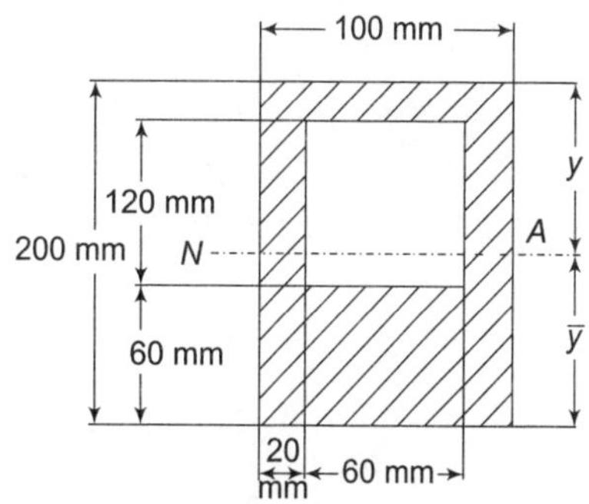

Fig. 5.48

Solution:

Allowable tensile stress $\sigma_T = 30$ N/mm^2

Allowable compressive stress $\sigma_C = 75$ N/mm^2

For sagging moments, top fibres will be in compression and bottom ones in tension.

For hogging moments, top fibres will be in tension and bottom ones in compression.

$$\text{Centre of gravity } \bar{y} = \frac{A_1 y_1 - A_2 y_2}{A_1 - A_2}$$

$$\bar{y} = \frac{(200 \times 100)\dfrac{200}{2} - (60 \times 120)\left(60 + \dfrac{120}{2}\right)}{200 \times 100 - 60 \times 120} = 88.75 \text{ mm}$$

$$\therefore \quad y = 200 - \bar{y} = 111.25 \text{ mm.}$$

Moment of inertia of the cross section about neutral axis,

$$I = I_1 - I_2$$

$$I_1 = \frac{BH^3}{12} + BH\,(y_1 - \bar{y})^2$$

$$= \frac{100 \times 200^3}{12} + 100 \times 200\,(100 - 88.75)^2$$

$$= 69.198 \times 10^6 \text{ mm}^4$$

$$I_2 = \frac{bh^3}{12} + bh\,(y_2 - \bar{y})^2$$

$$= \frac{60 \times 120^3}{12} + 60 \times 120\,(120 - 88.75)^2$$

$$= 15.671 \times 10^6 \text{ mm}^4$$

$$\therefore \qquad I = I_1 - I_2 = 53.527 \times 10^6 \text{ mm}^4$$

(a) Considering sagging moment

$$y_T = \bar{y} = 88.75 \text{ mm}$$

$$y_C = y = 111.25 \text{ mm}$$

Moment of resistance in tension of side

$$M = \frac{\sigma_T}{y_T} \cdot I$$

$$M = \frac{30}{88.75} \times 53.527 \times 10^6 = 18.09 \times 10^6 \text{ N-mm}$$

Moment of resistance in compression side

$$M = \frac{\sigma_C}{y_C} \cdot I$$

$$M = \frac{75}{111.25} \times 53.527 \times 10^6 = 36.086 \times 10^6 \text{ N-mm}$$

$\therefore$ Moment of resistance capacity for sagging moment $M = 18.09 \times 10^6$ N-mm (smaller value)

(b) For hogging moments

$$y_T = y = 111.25 \text{ mm}$$

$$y_C = y = 88.75 \text{ mm}$$

$\therefore$ Moment of resistance on tension side

$$M = \frac{\sigma_T}{y_T} \cdot I$$

$$M = \frac{30}{111.25} \times 53.526 \times 10^6 = 14.43 \times 10^6 \text{ N-mm}$$

Moment of resistance on compression side

$$M = \frac{\sigma_C}{y_C} I$$

$$M = \frac{75}{88.75} \times 53.526 \times 10^6 = 45.233 \times 10^6 \text{ N-mm}$$

$\therefore$ Moment of resistance capacity for hogging moment

$$M = 14.43 \times 10^6 \text{ N-mm (smaller value)}$$

Problem 38 The cross section of a cantilever beam is shown in Fig. 5.49. The beam carries a point load of 3 kN at its free end and the self weight is 4.6 kN/m. Calculate the maximum stresses induced at the critical section. Sketch the stress distribution at that section. Also find stresses at 1 metre from fixed end.

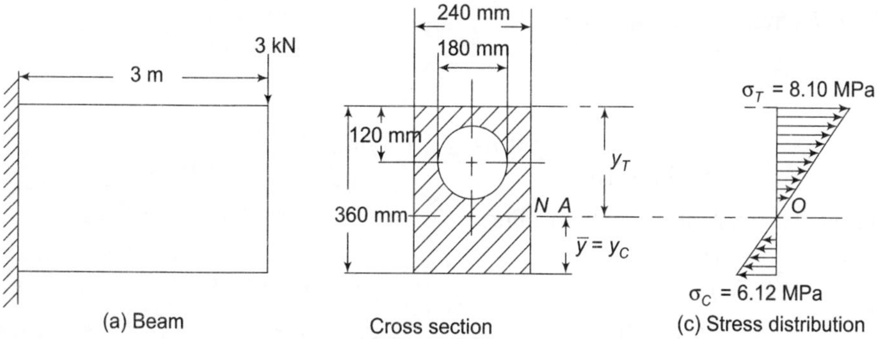

Fig. 5.49 (a)

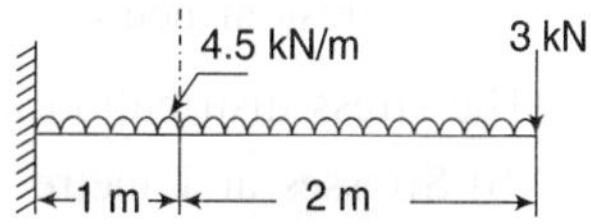

Fig. 5.49 (b)

Solution: The self weight is always considered UDL.

$$w = 4.6 \text{ kN/m} = 4.6 \text{ N/mm}$$

Maximum bending moment occurs at fixed end.

$$M = Wl + \frac{wl^2}{2} = 3000 \times 3000 + \frac{4.6 \times 3000^2}{2} = 29.7 \times 10^6 \text{ N mm}$$

For cantilever top fibres are in tension and bottom in compression.

$$\text{Centre of gravity } \bar{y} = \frac{A_1 y_1 - A_2 y_2}{A_1 - A_2}$$

$$\bar{y} = \frac{(240 \times 360) \times \frac{360}{2} - \left(\frac{\pi}{4} \times 180^2\right)(360 - 120)}{240 \times 360 - \frac{\pi}{4} \times 180^2} = 154.95 \text{ mm}$$

$$\bar{y} = y_C = 154.95 \text{ mm}$$

$$y_T = 360 - \bar{y} = 205.05 \text{ mm}$$

Moment of inertia of the section about neutral axis

$$I = I_1 - I_2$$

$$I_1 = \frac{BH^3}{12} + BH\,(y_1 - \bar{y})^2 = \frac{240 \times 360^3}{12} + 240 \times 360\,(180 - 154.95)^2$$

$$= 987.34 \times 10^6 \text{ mm}^4$$

$$I_2 = \frac{\pi d^4}{64} + \frac{\pi}{4} d^2\,(y_2 - \bar{y})^2 = \frac{\pi \times 180^4}{64} + \frac{\pi}{4} \times 180^2\,(240 - 154.95)^2$$

$$= 235.6 \times 10^6 \text{ mm}^4$$

$$\therefore \qquad I = I_1 - I_2 = 751.74 \times 10^6 \text{ mm}^4$$

∴ Tensile stress at the top fibre

$$\sigma_T = \frac{M}{I} \cdot y_T$$

$$\sigma_T = \frac{29.7 \times 10^6}{751.74 \times 10^6} \times 205.05 = 8.10 \text{ N/mm}^2$$

Compressive stress at the bottom fibre

$$\sigma_C = \frac{M}{I} y_C$$

$$\sigma_C = \frac{29.7 \times 10^6}{751.74 \times 10^6} \times 154.95 = 6.12 \text{ N/mm}^2$$

Stress at neutral axis = 0

The stress distribution is shown in Fig. 5.49.

(b) Stresses at 1 metre

Bending moment at 1 m distance from fixed end

$$M = w \times 2000 + \frac{w \cdot 2000^2}{2} = 3000 \times 2000 + \frac{4.6 \times 2000^2}{2} = 15.2 \times 10^6 \text{ N mm}$$

Tensile stress at top fibre

$$\sigma_T = \frac{15.2 \times 10^6}{751.74 \times 10^6} \times 205.05 = 4.17 \text{ N/mm}^2$$

Compressive stress at the bottom fibre

$$\sigma_C = \frac{15.2 \times 10^6}{751.74 \times 10^6} \times 154.95 = 3.13 \text{ N/mm}^2.$$

Problem 39 Figure 5.50 shows a steel beam simply supported over a span of 4 m. Calculate the maximum stress in the section when it carries uniformly distributed load of 600 N/m over the entire span a concentrated load of 500 N at a distance of 3 m from the left support.

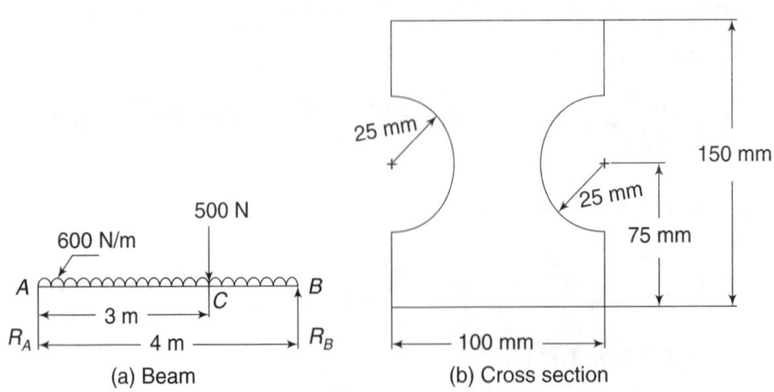

Fig. 5.50

Solution: To find reactions at A and B, consider $\Sigma V = 0$

$$R_A + R_B = (600 \times 4) + 500 = 2900 \text{ N}$$

Taking moments about A, $(600 \times 4) \dfrac{4}{2} + 500 \times 3 - R_B \times 4 = 0$

$$\therefore \quad R_B = 1575 \text{ N and } R_A = 2900 - 1575 = 1325 \text{ N}$$

Maximum bending moment occurs where SF = 0, changing its sign consider section XX at a distance x from A between A and

$$\text{SF}_{\text{at } x} = R_A - wx = 1325 - 600\,x = 0$$

$$\therefore \quad x = 2.2083 \text{ m from } A$$

$\therefore$ Maximum bending moment $= M_{\text{at } x} = R_A \cdot x - \dfrac{wx^2}{2}$

$$M = 1325 \times 2.2083 - 600 \times \frac{2.2083^2}{2} = 1463.02 \text{ Nm}$$

$$M = 1.463 \times 10^6 \text{ N-mm}$$

Moment of inertia $I = I_1 - I_2$

where $I = MI$ of rectangle and $I_2 = MI$ of 2 semicircles.

$$I = \frac{BH^3}{12} - \frac{\pi d^4}{64}$$

(The section is symmetric about XX axis)

$$I = \frac{100 \times 150^3}{12} - \frac{\pi}{64} \times 50^4 = 27.8182 \times 10^6 \text{ mm}^4$$

$$y = \frac{H}{2} = \frac{150}{2} = 75 \text{ mm}$$

$$\therefore \quad \text{Bending stress } \sigma = \frac{M}{I} \cdot y = \frac{1.463 \times 10^6}{27.8182 \times 10^6} \times 75$$

$$\sigma = 3.944 \text{ N/mm}^2$$

This is tensile at the bottom and compressive at the top.

Problem 40 For a beam shown in Fig. 5.51, find the maximum moment of resistance that the cross section can withstand about XX and YY axis taking the safe bending stress as 90 MPa.

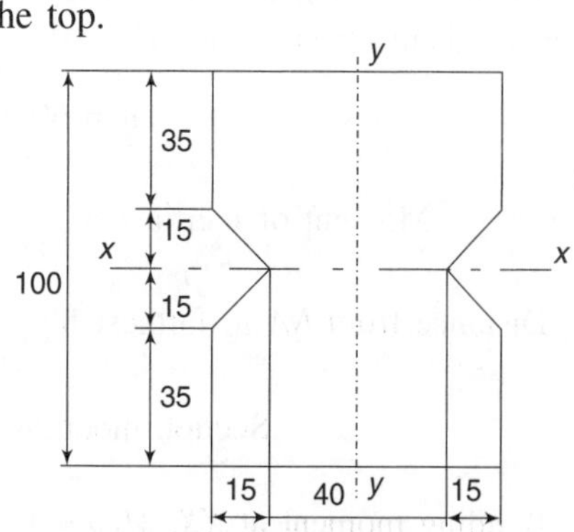

Fig. 5.51

Solution: Bending stress σ = 90 MPa

(a) Consider XX: Distance to outer member from NA

$$y = \frac{100}{2} = 50 \qquad \text{(Symmetric)}$$

Moment of inertia $I_{XX} = I_{Rectangle} - 4 \times I_{Triangles} = I_1 - 4I_2$

$$I_{XX} = \frac{70 \times 100^3}{12} - 4\left[\frac{15 \times 15^3}{12}\right] = 5.816 \times 10^6 \text{ mm}^4$$

$\therefore$ Moment of resistance about XX, $M_{XX} = \dfrac{\sigma . I_{XX}}{Y_{XX}}$

$$M_{XX} = \frac{90 \times 5.816 \times 10^6}{50} = 10.696 \times 10^6 \text{ N-mm}$$

(b) Consider YY:

$$Y_{YY} = \frac{70}{2} = 35 \text{ mm}$$

$$I_{YY} = I_{Rectangle} - 4 \times I_{Triangle}$$

$$I_{Triangle} = I_{cg} + AK^2$$

$$I_{YY} = \frac{100 \times 70^3}{2} - 4\left[\frac{15 \times 15^3}{36} + \frac{1}{2} \times 15 \times 15\left(20 + \frac{2}{3} \times 15\right)^2\right]$$

$$= 2.448 \times 10^6 \text{ mm}^4$$

$\therefore$Moment of resistance about $YY = M_{YY} = \dfrac{\sigma \cdot I_{YY}}{y_{YY}}$

$$M_{YY} = 90 \times \frac{2.448 \times 10^6}{35} = 6.295 \times 10^6 \text{ N-mm.}$$

Problem 41 A vertical pole of circular cross section has diameter varying from 180 mm at the bottom to 60 mm at the top and is 9 m long and is fixed at the bottom end. A horizontal load of 600 N acts at the top. Find the magnitude and position of maximum stress.

Solution: Consider any section XX at a distance x from the top. Let the diameter at section be = d

$$d = 60 + \frac{x}{9}(180 - 60) = 60 + \frac{120x}{9}$$

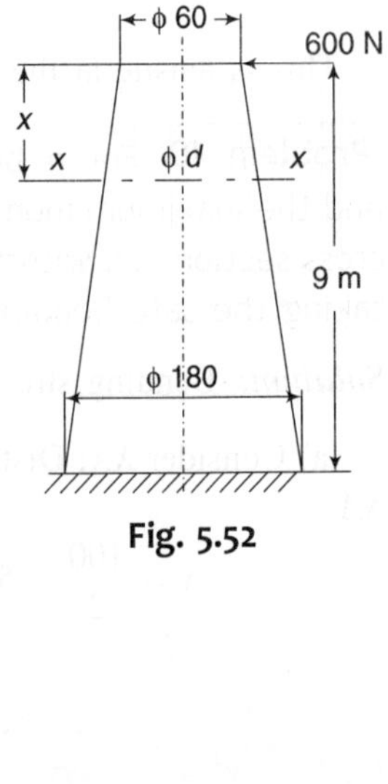

Fig. 5.52

Moment of inertia $I = \dfrac{\pi d^4}{64}$

Distance from NA to farthest fibre $y = \dfrac{d}{2}$

Section modulus $Z = \dfrac{I}{y} = \left(\dfrac{\pi d^4}{64}\right)\left(\dfrac{d}{2}\right) = \dfrac{\pi d^3}{32}$

Bending moment at XX, $M_{XX} = F \cdot x = 600 \cdot x$ Nm

$$M_{XX} = (600.x) \times 1000 = 60000\, x \text{ N-mm}$$

$$\text{Bending stress } \sigma = \frac{M \cdot y}{I} = \frac{M}{Z} = \frac{60000\ x}{\frac{\pi}{32} d^3}$$

$$\text{But } d = 60 + \frac{120}{9}x$$

$$\sigma = \frac{60000\ x}{\frac{\pi}{32}\left[60 + \frac{120}{9}x\right]^3} = \frac{32}{\pi} \times 60000 \frac{x}{\left(60 + \frac{120}{9}x\right)^3} \qquad ...(1)$$

For stress to be maximum, $\frac{d\sigma}{dx} = 0$.

$$\frac{d}{dx}\left[\frac{32}{\pi} \times 60000.x\left(60 + \frac{120}{9}x\right)^{-3}\right] = 0$$

$$x(-3)\left(60 + \frac{120x}{9}\right)^{-4} \times \frac{120}{9} + \left(60 + \frac{120x}{9}\right)^{-3} \times 1 = 0$$

$$\left(60 + \frac{120x}{9}\right)^{-3}\left[1 - 40x\left(60 + \frac{120x}{9}\right)^{-1}\right] = 0$$

Since $60 + \frac{120x}{9} \neq 0$, $1 - 40x\left(60 + \frac{120x}{9}\right)^{-1} = 0$

$$1 = 40x\left(60 + \frac{120x}{9}\right)^{-1}$$

$$1 = \frac{40x}{\left(60 + \frac{120x}{9}\right)}$$

$$60 + \frac{120x}{9} = 40x.$$

$\therefore$ $x = 2.25$ m

The maximum stress occurs at a distance of 2.25 m from the top. Substituting in (1)

$$\sigma_{max} = \frac{60000 \times 2.25}{\frac{\pi}{32}\left(60 + \frac{120}{9} \times 2.25\right)^3} = 1.186 \text{ N/mm}^2.$$

Problem 42 A vertical pole is 8 m long above the ground is of square section 200 mm × 200 mm at the ground and 100 mm × 100 mm at the top. A horizontal pull of 200 N is applied at the top in the direction of a diagonal of the section. Find the maximum stress due to bending in the pole.

Solution: Figure 5.53 shows the front and top views of the pole. Consider a section *XX* at a distance *x* from the top.

Let the sides of square at section *XX* be = *a*.

$\therefore$ $$a = 100 + \frac{x}{8}(200 - 100)$$

$$= 100 + 12.5x$$

$$\text{Diagonal, } d = \sqrt{a^2 + a^2} = \sqrt{2}\,.a$$

Distance from NA to extreme fibre

$$y = \frac{d}{2} = \frac{a.\sqrt{2}}{2} = \frac{a}{\sqrt{2}}$$

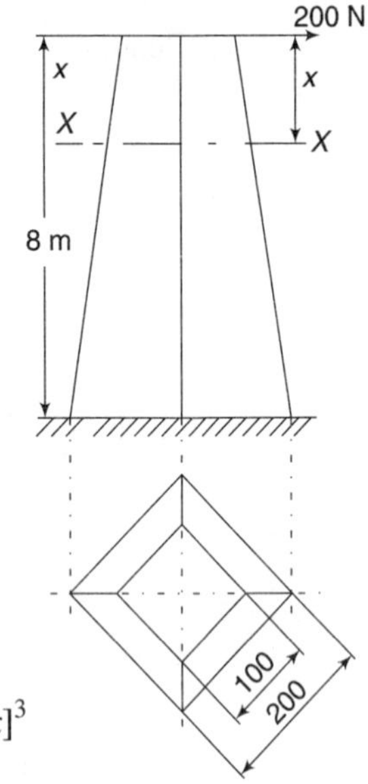

Fig. 5.53

Moment of inertia of a square about diagonal

$$I = \frac{a^4}{12}$$

$$\therefore \quad \text{Section modulus } z = \frac{I}{y} = \frac{\left(\frac{a^4}{12}\right)}{\left(\frac{a}{\sqrt{2}}\right)}$$

$$z = \frac{\sqrt{2}a^3}{12} = \frac{\sqrt{2}}{12}\,[100 + 12.5x]^3$$

$$\text{Bending moment } M = F \cdot x = 200x \text{ Nm}$$

$$= 200.x \times 1000 = 200000\;x \text{ N-mm}$$

$$\therefore \quad \text{Bending stress } \sigma = \frac{M \cdot y}{I} = \frac{M}{z} = \frac{200 \times 10^3 x}{\left[\frac{\sqrt{2}}{12}\,(100 + 12.5x)^3\right]}$$

$$\sigma = \frac{12 \times 200 \times 10^3}{\sqrt{2}}.x.(100 + 12.5x)^{-3}$$

$$\text{For stress } \sigma \text{ to be maximum, } \frac{d\sigma}{dx} = 0$$

$$\text{Differentiating, } x\,[(-3) \times (100 + 12.5x)^{-4} \times 12.5] + (100 + 12.5x)^{-3} = 0$$

$$= (100 + 12.5x)^{-3}\,[1 - 37.5\,x(100 + 12.5x)^{-1}] = 0$$

Since $100 + 12.5\,x \neq 0$,

$$1 - 37.5x\,(100 + 12.5x)^{-1} = 0$$

$$1 = \frac{37.5x}{100 + 12.5x}$$

$$\therefore \quad x = 4 \text{ m}$$

$$\therefore \quad \text{Maximum stress } \sigma = \frac{12 \times 200 \times 10^3}{\sqrt{2}} \times 4\,(100 + 12.5 \times 4)^{-3}$$

$$= 2.01 \text{ N/mm}^2.$$

Problem 43 Two steel plates of same width and length are simply supported at the ends and carry a central load. Find ratio of maximum stresses in the plates it their thickness are t_1 and t_2.

Solution: Figure 5.54 shows the arrangement of plates. The plates are resting one above another, they bend together and have their own neutral axis. The radius of curvature being very large compared to the thickness of the plates, we can assume that the radius of curvature is same for both plate. $R_1 = R_2$.

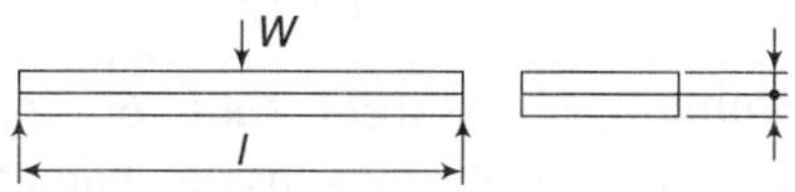

Fig. 5.54

$$\text{We know that } \frac{M}{I} = \frac{E}{R}$$

or
$$R = \frac{EI}{M}$$

$$\therefore \quad \frac{EI_1}{M_1} = \frac{EI_2}{M_2}$$

where, I_1 = Moment of inertia of plate (1) and I_2 is moment of inertia of plate (2)

M_1 = Maximum bending moment for plate (1)

M_2 = Maximum bending moment for plate (2)

$$\therefore \quad \frac{M_1}{M_2} = \frac{I_1}{I_2}$$

For rectangular cross section $I = \dfrac{bt^3}{12}$

$$\frac{M_1}{M_2} = \frac{\left(\dfrac{b_1 t_1^3}{12}\right)}{\left(\dfrac{b_2 t_2^3}{12}\right)} = \frac{t_1^3}{t_2^3} \qquad (\because b_1 = b_2) \ldots(1)$$

$$\text{Also } \frac{M}{I} = \frac{\sigma}{y} \text{ where } I = \frac{bt^3}{12} \text{ and } y = \frac{t}{2}$$

$$\therefore \quad \text{Moment } M = \frac{\sigma \cdot I}{y} = \frac{\sigma \cdot bt^2}{6}$$

$$\therefore \quad \frac{M_1}{M_2} = \frac{\left(\dfrac{\sigma_1 bt_1^2}{6}\right)}{\left(\dfrac{\sigma_2 bt_2^2}{6}\right)} = \frac{\sigma_1 t_1^2}{\sigma_2 t_2^2} \qquad \ldots(2)$$

Equating Eqs. (1) and (2)

$$\frac{M_1}{M_2} = \frac{t_1^3}{t_2^3} = \frac{\sigma_1 t_1^2}{\sigma_2 t_2^2}$$

$$\therefore \quad \frac{\sigma_1}{\sigma_2} = \frac{t_1}{t_2}$$

Hence, the ratio of maximum bending stresses in the plates is equal to the ratio of thickness of the two plates.

Problem 44 A steel tube 60 mm outside diameter and 40 mm inside diameter is 3 m long and is simply supported at the ends. The central load that it can safely carry is 1.5 kN. Four of such tubes are placed as shown in Fig. 5.55 which are firmly fixed together to form a single beam. Find safe uniformly distributed load that this beam can carry.

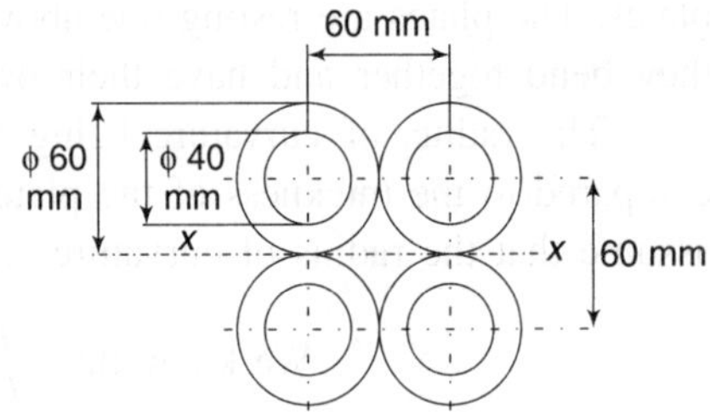

Fig. 5.55

Solution: **Single tube:**

$$\text{Moment of inertia } I = \frac{\pi}{64}(d_o^4 - d_i^4)$$

$$I_1 = \frac{\pi}{64}(60^4 - 40^4) = 510.51 \times 10^3 \text{ mm}^4$$

$$y = \frac{d_o}{2} = \frac{60}{2} = 30 \text{ mm}$$

For simply supported beam with central load,

$$\text{Maximum bending moment } M = \frac{Wl}{4} = \frac{1500 \times 3000}{4} = 1.125 \times 10^6 \text{ N mm}$$

$$\text{Bending stress } \sigma = \frac{M \cdot y}{I_1} = \frac{1.125 \times 10^6 \times 30}{510.51 \times 10^3} = 66.11 \text{N/mm}^2$$

Four tubes: Moment of inertia of four tubes about XX axis

$$I = 4\, I_1 = 4\,[I_{cg} + Ak^2]$$

$$I_{cg} = \frac{\pi}{64}(d_o^4 - d_i^4) = 510.51 \times 10^3 \text{ mm}^4$$

$$A = \frac{\pi}{4}(d_o^2 - d_i^2) = \frac{\pi}{4}(60^2 - 40^2) = 1570.8 \text{ mm}^2$$

k = Distance from XX to CG of each tube = 30 mm

$$\therefore \quad I = 4[510.51 \times 10^3 + 1570.8 \times 30^2] = 7.697 \times 10^6 \text{ mm}^4$$

y = Distance from XX (Neutral axis) to outer fibre = 60 mm

Stress σ = 66.11 N/mm^2 (Same as calculated above.)

For simply supported beam with UDL, bending maximum

$$M = \frac{wl^2}{8} = \frac{w \times 3000^2}{8} = 1.125 \times 10^6\ w$$

Substituting in the bending equation $\frac{M}{I} = \frac{\sigma}{y}$

$$\frac{1.125 \times 10^6\ w}{7.697 \times 10^6} = \frac{66.11}{60}$$

$\therefore$ UDL intensity $w = 7.5385$ N/mm $= 7.5385$ kN/m.

5.9.6 Flitched Beams

Flitched beams are beams of composite section consisting of a wooden beam strengthened by steel plates. Figure 5.56 shows the arrangement of flitched beams.

The wood and steel are so connected that there is no slipping between them and hence the components act together as one beam.

Since there is no slip between the steel and wooden members, the strains in both are same.

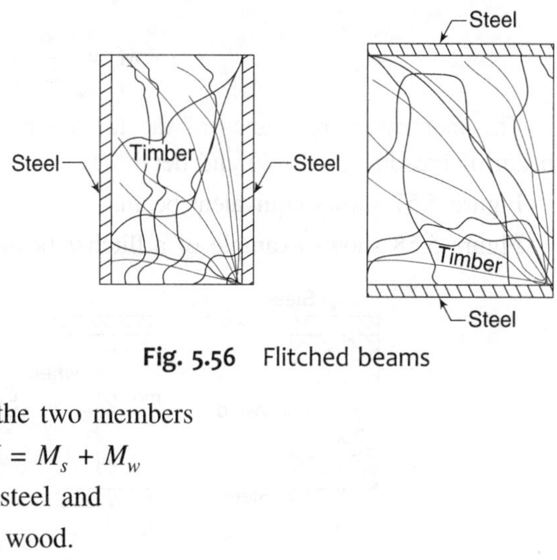

Fig. 5.56 Flitched beams

Strain $\in = \frac{\sigma_s}{E_s} = \frac{\sigma_w}{E_w}$

where, σ is the stress and E is modulus of elasticity.

$$\therefore \quad \sigma_s = \frac{E_s}{E_w} \cdot \sigma_w$$

$$\sigma_s = m \cdot \sigma_w \quad \text{where } m = \frac{E_s}{E_w}$$

m is called ratio of the moduli or modular ratio between steel and wood.

Consider the moment of resistance of the two members

Moment of resistance of the section $M = M_s + M_w$

where M_s = Moment of resistance of steel and

M_w = Moment of resistance of wood.

We know that $\frac{M}{I} = \frac{\sigma}{y}$ or $M = \frac{\sigma . I}{y} = \frac{\sigma . bd^2}{6}$

$$\therefore \quad M_s = 2 \times \frac{\sigma_s \cdot td^2}{6} \quad \text{and} \quad M_w = \sigma_w \cdot \frac{bd^2}{6}$$

$$\therefore \quad M = 2 \times \frac{\sigma_s td^2}{6} + \sigma_w \cdot \frac{bd^2}{6}$$

Since $\sigma_S = m \cdot \sigma_w$

$$M = 2\ m\ \sigma_w \frac{td^2}{6} + \sigma_w \frac{bd^2}{6}$$

$$M = \frac{\sigma_w\, d^2}{6} [b + 2mt]$$

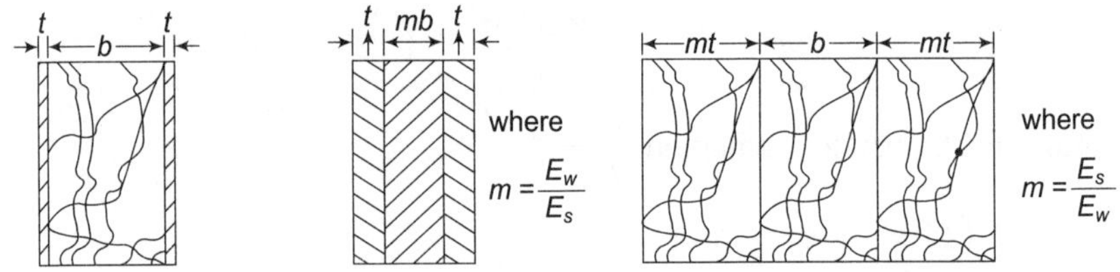

(a) Flitched beam (b) Equivalent steel section (c) Equivalent timber section

Fig. 5.57

Hence, the moment of resistance of a the flitcher beam is same as that of a wooden member of breadth $b + m(2t)$ and depth d. This rectangular section beam of width $[b + m(2t)]$ and depth d is called equivalent wooden section.

$$\text{Similarly, substituting } \sigma_w = \frac{E_w}{E_s} \cdot \sigma_s = m\ \sigma_S$$

$$M = \sigma_w \frac{bd^3}{6} + 2.\sigma_S \frac{td^3}{6} = m\ .\ \sigma_S \frac{bd^3}{6} + 2\ \sigma_S \frac{td^3}{6}$$

$$M = \frac{\sigma_S d^3}{6}(mb + 2t)$$

The moment of resistance of the flitcher beam is same as that of an equivalent steel beam of breadth $(2t + mb)$ and depth d.

Figure 5.57 shows equivalent beams.

Figure 5.58 shows example of a flitcher beam with their equivalent beams.

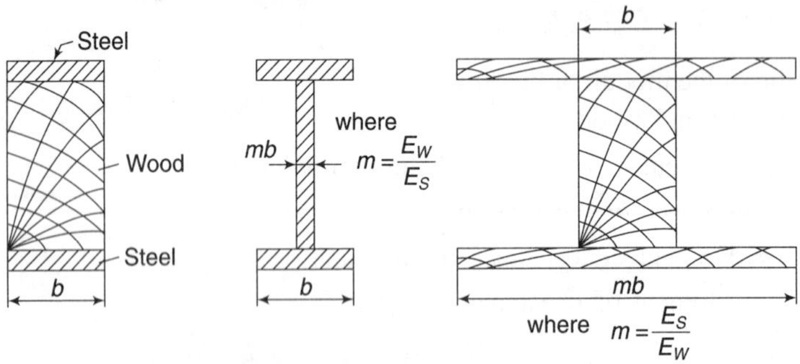

(a) Flitched beam (b) Equivalent steel beam (c) Equivalent timber beam

Fig. 5.58

SOLVED PROBLEMS

Problem 45 A fticher beam consists of a timber beam of 120 mm wide and 240 mm deep with two steel plates of 10 mm thick and 240 mm deep one each on either side of the timber beam. If the maximum stress in the steel is 120 MPa, find the corresponding stress in the timber beam. Also find out the moment of resistance of the flitcher beam. Take $E_{steel} = 2 \times 10^5$ N/mm^2 and $E_{timber} = 1 \times 10^4$ N/mm^2.

Solution: $E_{steel} = 2 \times 10^5$ N/mm^2 $E_{timber} = 1 \times 10^4$ N/mm^2

Thickness of steel plate $t = 10$ mm

Width of timber $b = 120$ mm

Depth of beam $d = 240$ mm

Stress in steel $\sigma_S = 120$ MPa

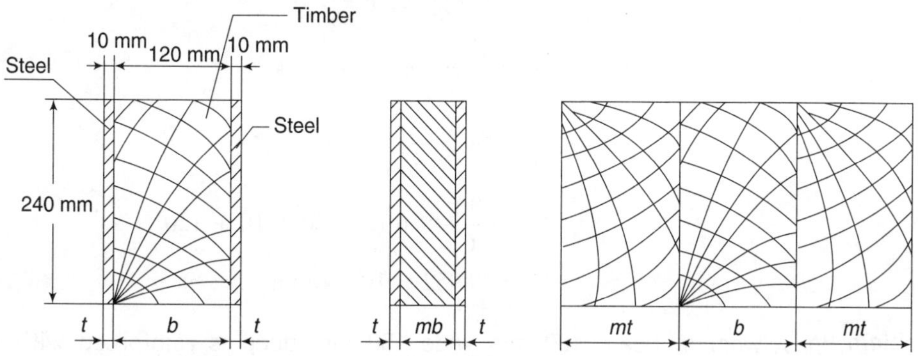

(a) Flitched beam (b) Equivalent steel beam (c) Equivalent timber beam

Fig. 5.59

Consider equivalent steel beam,

For the strains to be equal, $\dfrac{\sigma_S}{E_S} = \dfrac{\sigma_t}{E_t}$

$$\sigma_S = \frac{E_S}{E_T} \cdot \sigma_t = \frac{2 \times 10^5}{1 \times 10^4}\ \sigma_t = 20\ \sigma_t$$

$$\text{Modular ratio } m = \frac{E_t}{E_S} = \frac{1 \times 10^4}{2 \times 10^5} = \frac{1}{20} = 0.05 = \frac{\sigma_t}{\sigma_S}$$

$\therefore$ Stress in timber $\sigma_t = \dfrac{\sigma_S}{20} = \dfrac{120}{20} = 6$ N/mm^2

$$\text{Moment of resistance } M = M_S + M_t = \frac{2td^2}{6} \cdot \sigma_S + \frac{bd^2}{6} \cdot \sigma_t$$

$$= \frac{2td^2}{6}\ \sigma_S + \frac{(mb)d^2\sigma_s}{6} \quad (\because\ \sigma_t = \sigma_s)$$

$$M = \frac{d^2}{6}\ \sigma_S\ (2t + mb)$$

$$= \frac{240^2}{6} \times 120\ [2 \times 10 + 0.05 \times 120]$$

$$M = 29.952 \times 10^6 \text{ N-mm}$$

Note: This problem can also be solved by considering equivalent timber beam:

$$\text{Modular ratio } m = \frac{E_S}{E_t} = \frac{\sigma_S}{\sigma_t} = \frac{2 \times 10^5}{1 \times 10^4} = 20$$

$$\therefore \quad \sigma_S = m\,\sigma_t = 20\,\sigma_t$$

$$\text{Moment of resistance } M = M_S + M_t = \frac{2td^2}{6}\sigma_S + \frac{bd^2}{6}\sigma_t$$

$$= 2\frac{mtd^2}{6}\sigma_t + \frac{bd^2}{6}\sigma_t$$

$$M = \frac{d^2}{6}\sigma_t\,(2mt + b)$$

$$\sigma_t = \frac{\sigma_S}{m} = \frac{120}{20} = 6 \text{ N/mm}^2$$

$$\therefore \quad M = \frac{240^2}{6} \times 6\,(2 \times 20 \times 10 + 120)$$

$$M = 29.952 \times 10^6 \text{ N-mm.}$$

Problem 46 A wooden beam 100 mm wide 150 mm deep is reinforced with two steel plates each 100 mm wide and 10 mm thick at top and bottom. Find moment of resistance if the allowable stress in wood is 7.2 N/mm². What is the corresponding stress in steel plates. Take $E_s = 20\,E_w$.

Solution:

$$\text{Modular ratio } m = \frac{E_S}{E_W} = 20$$

$$\sigma_w = \text{Stress in wood} = 7.2 \text{ N/mm}^2$$

$$\sigma_w = \text{Stress in steel}$$

$$\text{For wooden beam, } y = \frac{d}{2} = \frac{150}{2} = 75 \text{ mm}$$

$$\text{For steel beam, } y_{\max} = \frac{D}{2} = \frac{170}{2} = 85 \text{ mm}$$

$$y_{\min} = \frac{d}{2} = \frac{150}{2}\ 75 \text{ mm}$$

At the common layer, the strain will be equal.

$$\text{At } y = 75, \text{ at } AB, \ \sigma_S = m\cdot\sigma_w = 20 \times 7.2 = 144 \text{ N/mm}^2$$

At the top layer for the steel plate, $y = 85$ mm

$$\text{Max. stress in steel } \sigma_{s\max} = \frac{85}{75} \times 144 = 163.2 \text{ N/mm}^2$$

$$\text{Moment of resistance } M = M_w + M_s$$

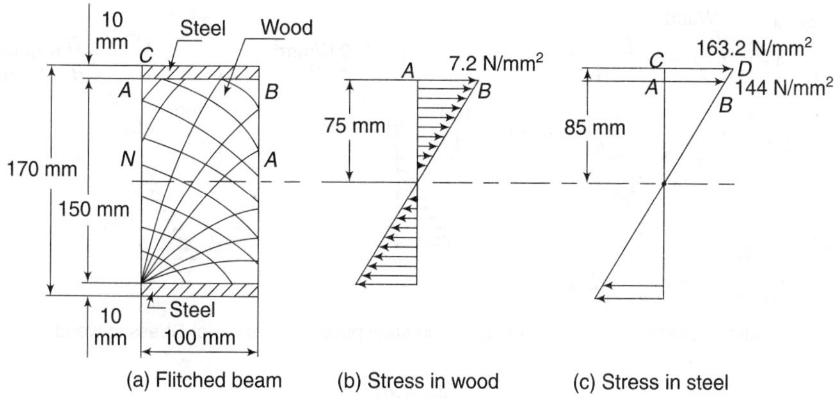

Fig. 5.60

$$M = \frac{1}{6} bd^2 \sigma_w + \left(\frac{1}{6} b D^2 \cdot \sigma_{s\max} - \frac{1}{6} bd^2 . \sigma_S\right)$$

$$= \frac{1}{6} \times 100 \times 150^2 \times 7.2 + \left(\frac{1}{6} \times 100 \times 170^2 \times 163.2 - \frac{1}{6} \times 100 \times 150^2 \times 144\right)$$

$$M = 27.308 \times 10^6 \text{ N-mm.}$$

Problem 47 A flitcher beam consists of a wooden piece 150 mm wide and 240 mm deep and two steel plates of 12 mm thick and 200 mm deep on either side that are placed symmetrically. If the stress in steel and wood are not to exceed 150 N/mm^2 and 6 N/mm^2, find the moment of resistance of the flitcher beam. Take $\frac{E_s}{E_w} = 20$.

Solution: Modular ratio $m = \frac{E_s}{E_w} = 20$

$$\sigma_{w\ \max} = 6 \text{ N/mm}^2 \text{ and } \sigma_{s\ \max} = 150 \text{ N/mm}^2$$

Consider stress in wood $\sigma_w = 6$ N/mm^2 at CD (Fig. 5.60)

$\therefore$ Stress in wood at AB, $\sigma_{w\ AB} = 6 \times \frac{100}{120} = 5$ N/mm^2

Corresponding stress in steel plate $\sigma_{s\ AB} = m\ \sigma_{w\ AB} = 20 \times 5 = 100$ N/mm^2

Since this is less than $\sigma_s = 150$ MPa, it is safe.

Consider steel at $AB = 150$ N/mm^2

$$\therefore \quad \sigma_{w \text{ at } AB} = \frac{\sigma_{sAB}}{m} = \frac{150}{20} = 7.5 \text{ N/mm}^2$$

Since this is $> \sigma_w = 6$ MPa, not safe.

Hence, take $\sigma_S = 100$ N/mm^2 and $\sigma_w = 5$ N/mm^2 at AB, so that

$$\sigma_{w \text{ at CD}} = \sigma_{w(\max)} = 6 \text{ N/mm}^2$$

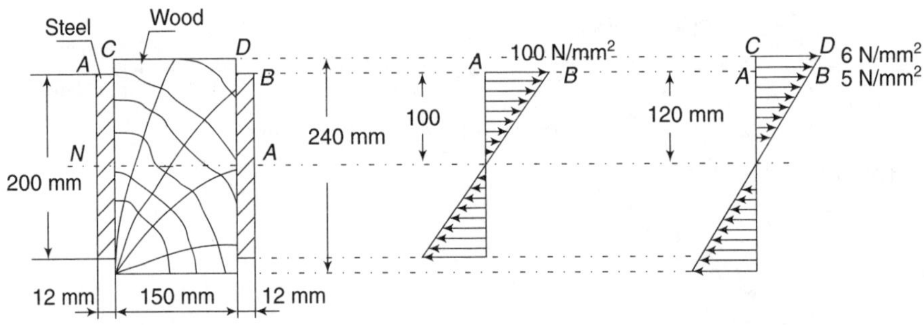

(a) Flitcher beam (b) Stress in steel plate (c) Stress in wood

Fig. 5.61

$$\text{Moment of resistance } M = M_w + M_s = \frac{1}{6} bD^2 \sigma_w + 2 \times \frac{1}{6} td^2 \sigma_{S.}$$

$$= \frac{1}{6} \times 150 \times 240^2 \times 6 + 2 \times \frac{1}{6} \times 12 \times 200^2 \times 100$$

$$M = 24.64 \times 10^6 \text{ N-mm.}$$

Problem 48 Find the moment of resistance of a flitcher beam shown in Fig. 5.62, if the stresses in steel and wood are limited to 120 MPa and 6 MPa respectively. Take $E_W = 1 \times 10^4$ MPa and $E_S = 20 \times 10^4$ MPa.

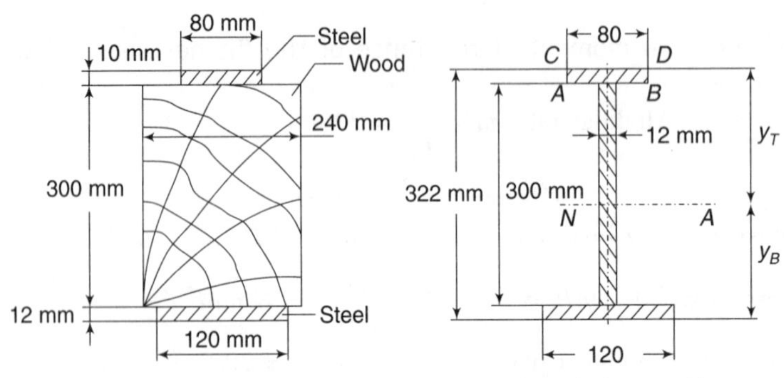

(a) Flitcher beam (b) Equivalent steel beam

Fig. 5.62

Solution: Modular ratio $m = \dfrac{E_S}{E_W} = \dfrac{200 \times 10^4}{1 \times 10^4} = 20$

The wooden beam can be replaced by an equivalent steel plate of thickness $= \dfrac{b}{m} = \dfrac{240}{20} = 12$ mm in the middle as shown in Fig. 5.62 (b).

$$\text{Centre of gravity } \bar{y} = y_B = \frac{A_1y_1 + A_2y_2 + A_3y_3}{A_1 + A_2 + A_3}$$

$$= \frac{80 \times 10 \times 317 + 12 \times 300 \times 162 + 120 \times 12 \times 6}{80 \times 10 + 12 \times 300 + 120 \times 12}$$

$$= 144.77 \text{ mm}$$

$$\therefore \quad y_T = 322 - y_B = 177.23 \text{ mm}$$

Consider $\sigma_w = 6$ MPa at the top of wooden beam.

Then stress in steel at the bottom = $\sigma_{s\,AB} = m \times \sigma_w = 20 \times 6 = 120$ MPa

∴ Stress is steel at the top of plate

$$\sigma_{sCD} = \sigma_{sAB} \times \frac{y_T}{(y_T - 10)}$$

$$\sigma_{sCD} = 120 \times \frac{177.23}{167.23} = 127.17 \text{ MPa}$$

This is greater than allowable stress in steel of 120 MPa hence not safe.

Consider $\sigma_s = 120$ MPa at *CD*.

$$\sigma_{sAB} = 120 \times \frac{167.23}{177.23} = 113.23 \text{ MPa}$$

$$\therefore \quad \sigma_{wAB} = \frac{\sigma_S}{m} = \frac{113.23}{20} = 566 \text{ MPa.}$$

This is safe.

$$\text{Moment of inertia } I = I_1 + I_2 + I_3$$

$$I_1 = I_{cg} + Ak^2 \text{ for top steel plate} = \frac{bd^3}{12} + (bd)(y_1 - \bar{y})^2$$

$$I_1 = \frac{80 \times 10^3}{12} + (80 \times 10)(317 - 144.77)^2$$

$$= 23.81 \times 10^6 \text{ mm}^4$$

Similarly,

$$I_2 = \frac{12 \times 300^3}{12} + (12 \times 300)(162 - 144.77)^2$$

$$= 28.07 \times 10^6 \text{ mm}^4$$

$$I_3 = \frac{120 \times 12^3}{12} + (120 \times 12)(6 - 144.77)^2$$

$$= 27.75 \times 10^6 \text{ mm}^4$$

$$\therefore \quad I = I_1 + I_2 + I_3 = 79.63 \times 10^6 \text{ mm}^4$$

$\therefore$ Moment of resistance $M = \sigma \cdot \dfrac{I}{y_T}$

$$M = \frac{120 \times 79.63 \times 10^6}{177.23} = 53.915 \times 10^6 \text{ N-mm.}$$

Problem 49 Figure 5.64 shows a flitcher beam made of a timber beam of 100 mm wide and depth 200 mm and on the top of it a steel plate of width 100 mm and thickness of 5 mm. The beam is simply supported over a span of 3 m and carry two point loads of 1 kN and 3 kN at distance 1 m and 2 m from left support. Find the stresses induced in steel and timber due to bending. Take E_T = 10 GPa and E_S = 200 GPa.

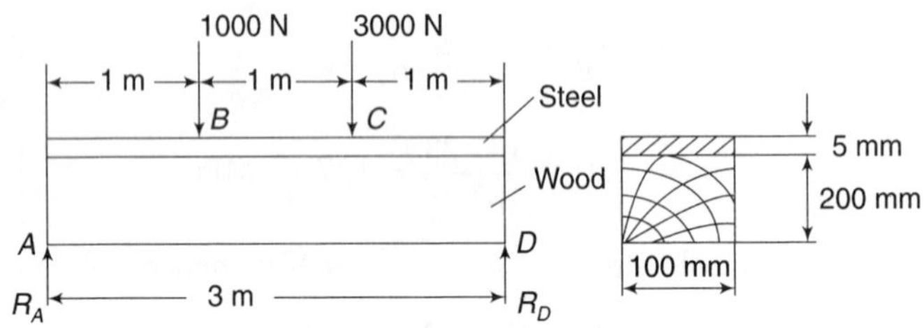

Fig. 5.63

Solution:

$$E_S = 200 \times 10^3 \text{ MPa}$$

$$E_T = 10 \times 10^3 \text{ MPa}$$

$$\text{Modular ratio } m = \frac{E_S}{E_T} = 20$$

Figure 5.64 shows an equivalent steel beam.

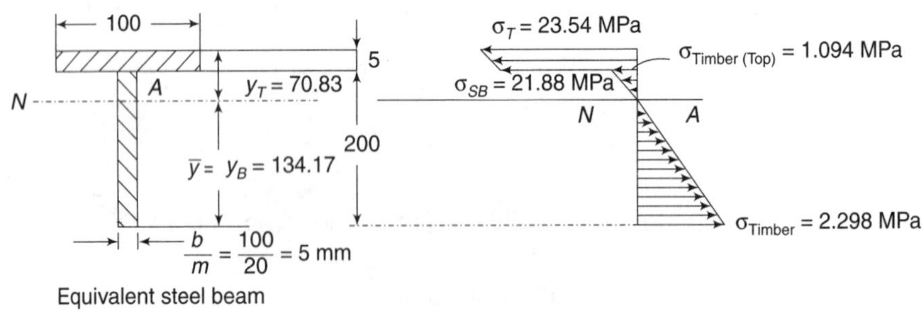

Fig. 5.64

$$\text{Centre of gravity } \bar{y} = y_B = \frac{100 \times 5 \times 202.5 + (5 \times 200)\,100}{100 \times 5 + 5 \times 200}$$

$$= 134.17 \text{ mm}$$

$$\therefore \qquad y_T = 205 - y_B = 70.83 \text{ mm}$$

Moment of inertia about NA. $I = I_1 + I_2$

$$I = \frac{100 \times 5^3}{12} + 100 \times 5\ (205 - 134.17)^2 + \frac{5 \times 200^3}{12} + 5 \times 200\ (100 - 134.17)^2$$

$$I = 7.01 \times 10^6 \text{ mm}^4$$

Maximum bending moment:

$$R_A + R_D = 1000 + 3000 = 4000 \text{ N}$$

Taking moments about A, $- 1000 \times 1000 - 3000 \times 2000 + R_D \times 3000 = 0$

$$\therefore \quad R_D = 2333.33 \text{ N}$$

$$R_A = 4000 - R_D = 1666.67 \text{ N}$$

Bending moment at A, $= 0$

at $B = 1666{,}67 \times 1000 = 1.67 \times 10^6$ N mm

at $C = 2333.33 \times 1000 = 2.33 \times 10^6$ N mm

at $D = 0$

$\therefore$ Maximum bending moment $M = 2.33 \times 10^6$ N mm

Maximum bending stress in steel at the top (compressive)

$$\sigma_T = \frac{M \cdot y_T}{I} = \frac{2.33 \times 10^6 \times 70.83}{7.01 \times 10^6} = 23.54 \text{ MPa}$$

Maximum bending stress in equivalent steel at bottom (tension)

$$\sigma_B = \frac{M \cdot y_B}{I} = \frac{2.33 \times 10^6 \times 134.17}{7.01 \times 10^6} = 44.596 \text{ MPa}$$

$$\text{Modular ratio } m = 20 = \frac{\sigma_S}{\sigma_{\text{Timber}}}$$

$$\therefore \quad \sigma_{\text{Timber}} = \frac{44.59}{\text{m}} = \frac{44.596}{20} = 2.298 \text{ MPa (tension)}$$

Maximum stress in the bottom of steel plate.

$$y = y_T - 5 = 70.83 - 5 = 65.83 \text{ mm}$$

$$\sigma_{SB} = \frac{2.33 \times 10^6}{7.01 \times 10^6} \times 65.83 = 21.88 \text{ MPa}$$

Maximum compressive stress at the top of timber

$$\sigma_{\text{Timber (Top)}} = \frac{\sigma_{SB}}{m} = \frac{21.88}{20} = 1.094 \text{ MPa}$$

The stress distribution is as shown in Fig. 5.64(b).

Problem 50 Two timber beams 200 mm wide, 300 mm deep are reinforced with a rectangular steel plate of 12 mm wide and 250 mm deep in between them. They are bolted firmly with bottom edges of the plate and timber beams are at the same level. Determine the moment of resistance of the composite beam if the stress in the timber beam is to be limited to 6N/mm^2. Take modular ratio = 20.

Solution: Modular ratio $m = 20$

Allowable stress in timber $= \sigma_T = 6$ MPa

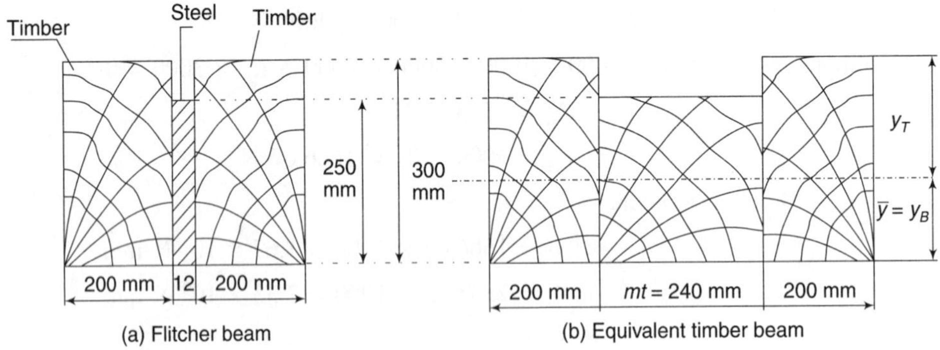

Fig. 5.65

Figure 5.65 shows the flitcher beam and equivalent timber beam.

$$\text{Equivalent width of steel plate} = mt = 20 \times 12 = 240 \text{ mm}$$

$$\text{Centre of gravity } \bar{y} = y_B = \frac{A_1 y_1 + A_2 y_2 + A_3 y_3}{A_1 + A_2 + A_3}$$

$$\bar{y} = y_B = \frac{200 \times 300 \times 150 + 240 \times 250 \times 125 + 200 \times 300 \times 150}{200 \times 300 + 240 \times 250 + 200 \times 300}$$

$$y_B = 141.67 \text{ mm}$$

$$y_T = 300 - y_B = 158.33 \text{ mm}$$

Moment of inertia of equivalent timber beam about NA, $I = 2I_1 + I_2$

$$I = 2\left[\frac{200 \times 300^3}{12} + 200 \times 300 \times (150 - 141.67)^2\right] + \frac{240 \times 250^3}{12} + 240 \times 250(125 - 141.67)^2$$

$$I = 1.2375 \times 10^9 \text{ mm}^4$$

It is seen that maximum stress in steel will occur at the bottom and in timber at the top where y is maximum.

$$\text{For steel } y = y_B = 141.67 \text{ mm}$$

$\therefore$ Maximum stress in timber at bottom

$$\sigma_T = 6\,\frac{y_B}{y_T} = 6 \times \frac{141.67}{158.33} = 5.37 \text{ MPa}$$

$$\therefore \qquad \sigma_S = m\,\sigma_T = 20 \times 5.37 = 107.37 \text{ MPa}$$

$$\text{Moment of resistance } M = \frac{I\sigma}{y_B} = \frac{1.2375 \times 10^9 \times 5.37}{141.67}$$

$$M = 46.907 \times 10^6 \text{ N-mm.}$$

5.10 SHEAR STRESSES IN BEAMS

As seen earlier, the transverse section of a beam is subjected to bending moment and shear force. The relation between bending moment (M) and shear force (F) is $\frac{dM}{dx} = F$.

Let us consider a beam as shown in Fig. 5.66. Consider an element length of beam dx between two section AA and BB. Let the moment and shear force at AA be M and F and the moment and shear force at BB are $M + dM$ and $F + dF$, respectively.

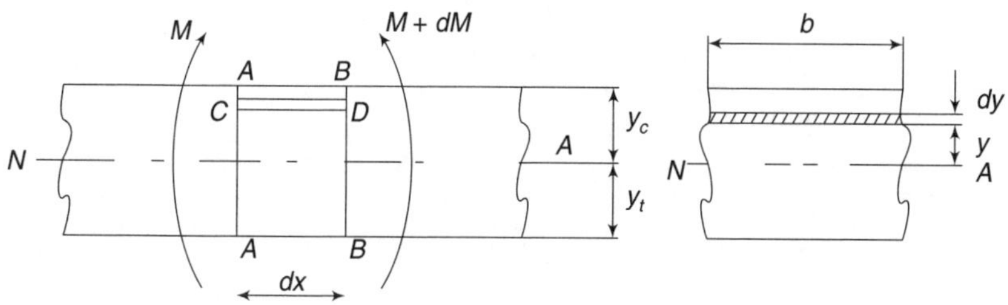

Fig. 5.66

Consider any element CD at a distance y from neutral axis whose thickness is dy. Area of the element = $b.dy$.

$$b = \text{width of beam.}$$

Bending stress at the left side of element = $\frac{M}{I} \cdot y$

$\therefore$ Force on the element on the left side = Stress × Area = $\frac{My}{I}(b \cdot dy)$

Similarly, bending stress at the right side of element = $\frac{(M + dM)y}{I}$

Force on the element on the right side = $\frac{(M + dM)y}{I} \cdot (b \cdot dy)$

$\therefore$ Unbalanced force towards right side of element

$$= \frac{(M + dM)y}{I}(bdy) - \frac{M}{I}y\,(bdy) = \frac{dM}{I}\,yb\,dy$$

Total unbalanced force above section $CD = \int_{y}^{y_C} \frac{dM}{I}.yb\,dy$

This is resisted by a horizontal force = $\tau\,(b \cdot dx)$

where τ = Shear stress.

Equating these two forces, $\tau\, b\, dx = \int_{y}^{y_C} \frac{dM}{I} . y b dy$

$$\tau\, bdx = \frac{dM}{I} \int_{y}^{y_C} y b dy$$

but $b \cdot dy = da$ area of the element

$$\tau\, bdx = \frac{dM}{I} \int_{y}^{y_C} y \cdot da$$

$$\text{Shear stress } \tau = \frac{dM}{dx} \frac{1}{Ib} \int_{y}^{y_C} y da$$

But $\int_{y}^{y_C} y\, da = a\bar{y}$ (First moment theorem)

$\therefore$ $$\tau = \frac{dM}{dx} \frac{a\bar{y}}{Ib}$$

$$\text{Substituting } \frac{dM}{dx} = F$$

$$\boxed{\tau = \frac{F}{Ib} a\bar{y}} \qquad \ldots(1)$$

From the above equation, it may be seen that the shear stress at the extreme fibres is equal to zero.

5.11 SHEAR STRESS VARIATION ACROSS A RECTANGULAR SECTION

Figure 5.67 shows a rectangle of width b and depth d subjected of a shearing force F.

Consider section AA at a distance y from neutral axis.

From Eq. (1), shear stress $\tau = \frac{F}{bI} a \cdot y$

For the shaded area above the section AA, area of section $a = b\left(\frac{d}{2} - y\right)$

$\bar{y}$ = Distance to CG of the shaded area from NA

$$\bar{y} = y + \frac{1}{2}\left[\frac{d}{2} - y\right] = \frac{1}{2}\left(\frac{d}{2} + y\right)$$

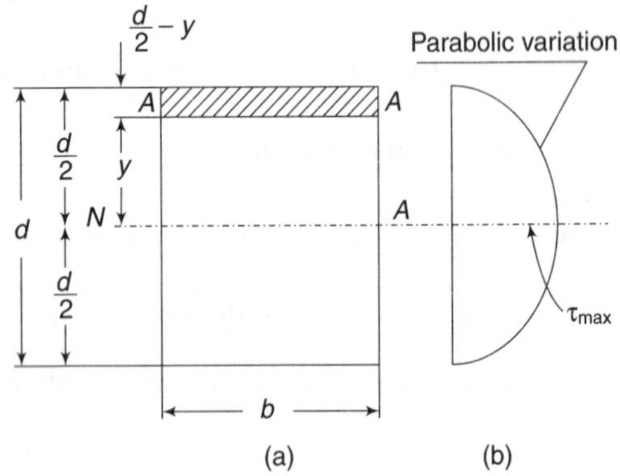

Fig. 5.67

$$I = \text{Moment of inertia of total section} = \frac{bd^3}{12}$$

$$\text{Substituting, } \tau = \frac{F}{b \cdot \left(\frac{bd^3}{12}\right)} \left[b\left(\frac{d}{2} - y\right)\right] \frac{1}{2}\left(\frac{d}{2} + y\right)$$

$$\tau = \frac{6F}{bd^3}\left(\frac{d^2}{4} - y^2\right)$$

This is parabolic variation. Applying boundary conditions,

$$\text{At } y = \frac{d}{2}, \ \tau = \frac{6F}{bd^3}\left(\frac{d^2}{4} - \left(\frac{d}{2}\right)^2\right) = 0$$

$$\text{At } NA, \ y = 0, \ \tau = \frac{6F}{bd^3}\left(\frac{d^2}{4} - 0\right) = \frac{6F}{bd^3}\frac{d^2}{4} = \frac{3}{2}\frac{F}{bd}$$

This is maximum at $y = 0$.

$$\therefore \quad \text{Maximum stress } \tau_{max} = \frac{3}{2} \cdot \frac{F}{A} = \frac{3}{2}\tau_{avg}$$

$$\text{Since } bd = \text{Area } A \text{ and } \frac{F}{A} = \tau_{avg}$$

$$\therefore \quad \text{Maximum shear stress } \tau_{max} = 1.5\ \tau_{avg}$$

At $y = 0$ at centroid.

Thus, for a rectangular section, the maximum shear stress is 1.5 times the average shear stress and occurs at the neutral axis. It is seen that the shear stress varies according to y^2, the variation is parabolic. The shear stress variation across rectangular section is shown in Fig. 5.67 (b).

5.12 SHEAR STRESS VARIATION ACROSS CIRCULAR SECTION

Figure 5.68 shows a solid circular section of diameter d

$$\text{Radius } r = \frac{d}{2}$$

Consider a section CD at a distance y from neutral axis.

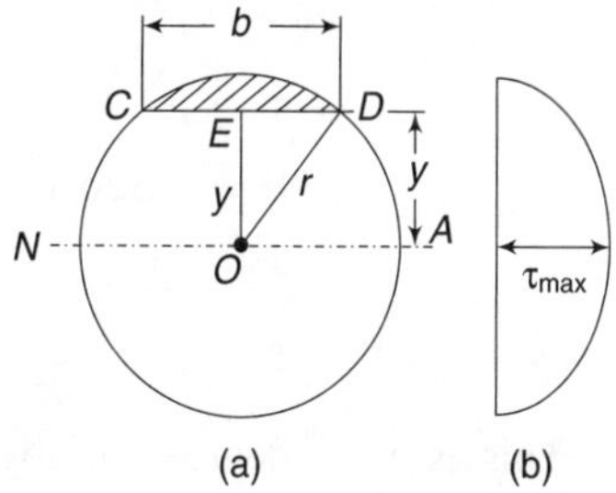

Fig. 5.68

From triangle OED, $OD^2 = OE^2 + ED^2$

$$\therefore \quad ED = \sqrt{OD^2 - OE^2} = \sqrt{r^2 - y^2}$$

$$\therefore \quad b = CD = CE + ED = 2ED$$

$$= 2\sqrt{r^2 - y^2}$$

Area of shaded portion about $CD = b \cdot dy$

$$a = 2\sqrt{r^2 - y^2} \cdot dy$$

Moment of the area above *CD* about neutral axis

$$a\bar{y} = \int_y^r 2y\sqrt{r^2 - y^2} \cdot dy \qquad ...(1)$$

Put $$2\sqrt{r^2 - y^2} = b \qquad ...(2)$$

Squaring on both sides, $4(r^2 - y^2) = b^2$

Differentiating, $-8y \cdot dy = 2b \cdot db$

$$\therefore \quad ydy = -\frac{bdb}{4} \qquad ...(3)$$

Substituting (2) and (3) in (1) $a\bar{y} = \int_r^y 2y\sqrt{r^2 - y^2} \cdot dy = \int_b^o -\frac{b.db}{4}.b = \frac{1}{4}\int_b^o -b^2 db$

when y varies from r to y, b varies from b to 0

$$a\bar{y} = \frac{1}{4}\left[\frac{-b^3}{3}\right]_b^o = \frac{b^3}{12}$$

Shear stress at level $CD = \tau = \frac{F}{Ib} \times a\bar{y}$

where, $$I = \frac{\pi d^4}{64} = \frac{\pi r^4}{4}$$

$$b = 2\sqrt{r^2 - y^2}$$

$$a\bar{y} = \frac{b^3}{12}$$

$$\tau = \frac{F}{\left(\frac{\pi r^4}{4}\right) \cdot b}\left(\frac{b^3}{12}\right) = \frac{4}{12}\frac{F}{\pi r^4} \cdot b^2$$

Substitute for $t = 2\sqrt{r^2 - y^2}$

$$\tau = \frac{4}{12}\frac{F}{\pi r^4} \cdot \left[2\sqrt{r^2 - y^2}\right]^2 = \frac{4}{3}\frac{F}{\pi r^4}(r^2 - y^2)$$

This is a 2nd degree variation in *y*, hence, the shear stress distribution across the cross section is parabolic.

At the extreme layers (top and bottom) $y = r$, shear stress $\tau = 0$

and at neutral axis, $y = 0$, $\tau = \tau_{max}$

$$\therefore \quad \tau_{max} = \frac{4}{3}\frac{F}{\pi r^4}(r^2) = \frac{4}{3}\frac{F}{\pi r^2} = \frac{4}{3}\frac{F}{A}$$

$$\tau_{max} = \frac{4}{3}\tau_{avg}$$

$$\left(\text{Since } \pi r^2 = A \text{ and } \frac{F}{A} = \tau_{avg}\right)$$

Hence, the maximum shear stress in a circular cross section is $\frac{4}{3}$ times the average shear stress and occurs at the middle at the centroidal axis. The shear stress variation is shown in Fig. 5.68(b).

5.13 SHEAR STRESS VARIATION ACROSS TRIANGULAR SECTION

Consider an isosceles triangular section shown in Fig. 5.69 (a)

b = Width of triangle and h = Height of triangle.

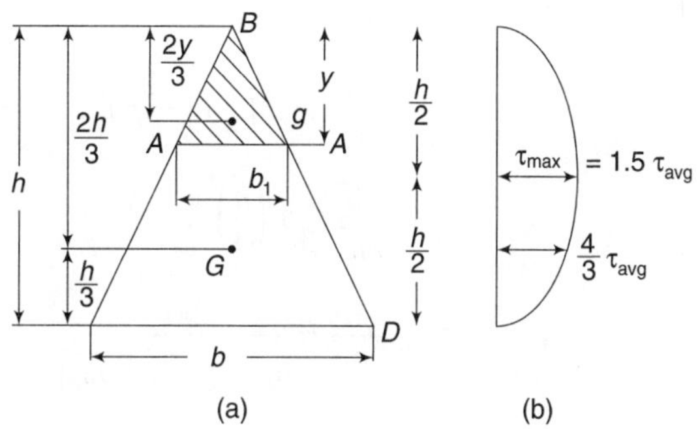

Fig. 5.69

G is the position of centroid of the triangle which lies at a distance of $\frac{2h}{3}$ from the top.

Consider section AA at the top of triangle, at a distance y from the top as shown in the Fig. 5.69 (a). The centroid of this section is g at a distance $\frac{2y}{3}$ from the top. Let b_1 be the width of this section at AA.

$$\text{From similar triangles, } \frac{b_1}{y} = \frac{b}{h}$$

$$\therefore \qquad \text{Width } b_1 = \frac{b}{h}\cdot y$$

Moment of inertia of the triangle about centroid

$$I = \frac{bh^3}{36}$$

$$\text{Shear stress at } AA \text{ is } \tau = \frac{F}{b_1 I}\cdot a\bar{y}$$

where, $$a\bar{y} = \left(\frac{1}{2} b_1 y\right) \times \left(\frac{2h}{3} - \frac{2y}{3}\right)$$

$\therefore$ $$\tau = \frac{F}{b_1\left(\frac{bh^3}{36}\right)} \left(\frac{1}{2} b_1 y\right)\left(\frac{2h}{3} - \frac{2y}{3}\right) = \frac{36F}{bh^3} \cdot \frac{1}{3} y\,(h - y)$$

$$\tau = \frac{12F}{bh^3} y\,(h - y)$$

This is parabolic variation.

at $y = 0$ at B, shear stress $\tau = 0$

at $y = h$ at CD, shear stress $\tau = 0$

At the centroid G, $y = \frac{2h}{3}$

$$\tau = \frac{12F}{bh^3}\left(\frac{2h}{3}\right)\left(h - \frac{2h}{3}\right)$$

$$\tau = \frac{12F}{bh^3} \frac{2}{3} h \left(\frac{h}{3}\right)$$

$$\tau = \frac{24F}{9bh} = \frac{24F}{9 \times 2 \times \left(\frac{1}{2} bh\right)} = \frac{24F}{18\left(\frac{1}{2} bh\right)}$$

$$\tau = \frac{4}{3}\frac{F}{A} = \frac{4}{3}\tau_{avg}$$

Since $\frac{1}{2} bh = A$ = Area of triangle.

Hence, for a triangle, shear stress at centroid

$$\tau = \frac{4}{3} \cdot \tau_{avg}$$

To find maximum shear stress, $\frac{d\tau}{dy} = 0$

$$\tau = \frac{12F}{bh^3} y\,(h - y)$$

$$\frac{d\tau}{dy} = \frac{12F}{bh^3}(h - y) + \frac{12F}{bh^3} y(-1)$$

$$= \frac{12F}{bh^3}(h - y - y) = \frac{12F}{bh^3}(h - 2y) = 0$$

Since F is not equal to zero.

$$h - 2y = 0 \text{ or } y = \frac{h}{2}$$

The shear stress will be maximum at $y = \frac{h}{2}$, i.e., at the middle of the section.

$$\therefore \qquad \tau_{max} = \frac{12F}{bh^3} \cdot \frac{h}{2}\left(h - \frac{h}{2}\right) = \frac{12F}{bh^3} \cdot \frac{h^2}{4} = \frac{12F}{4bh}$$

$$\tau_{max} = \frac{12F}{4 \times 2\left(\frac{1}{2}\, bh\right)} = \frac{12}{8}\,\frac{F}{A} = 1.5\ \tau_{avg}$$

Hence, the maximum shear stress in a triangle is 1.5 times the average shear stress and occurs at half the depth. The shear stress variation across the section is shown in Fig. 5.69 (b)

5.14 SHEAR STRESS VARIATION ACROSS SQUARE SECTION PLACED WITH ONE OF ITS DIAGONALS HORIZONTAL

Consider a square section of side a with one of its diagonals horizontal as shown in Fig. 5.70 (a).

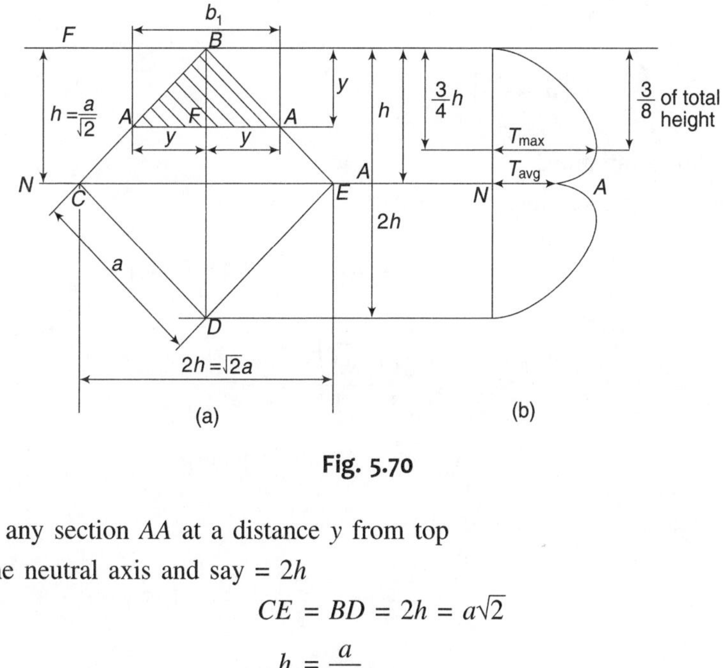

Fig. 5.70

Consider any section AA at a distance y from top

CE is the neutral axis and say $= 2h$

$$CE = BD = 2h = a\sqrt{2}$$

$$\therefore \qquad h = \frac{a}{\sqrt{2}}$$

Moment of inertia of the square about neutral axis CE

$$I = 2\, I_{\text{Triangle}} \text{ about base.}$$

$$= 2 \times \frac{bh^3}{12} = \frac{2\cdot(a\sqrt{2})}{12}\left(\frac{a}{\sqrt{2}}\right)^3 = \frac{a^4}{12}$$

Width of element at $AA = 2y = b_1$

The centroid for the shaded area is at $\frac{2}{3}h$ from the top.

WKT Shear stress $\tau = \frac{F}{bI} \cdot a\bar{y}$

where $b = b_1$

$$I = \frac{a^4}{12}$$

$$a = \text{Area of triangle} = \frac{1}{2} b_1 y$$

$\bar{y}$ = distance from centroid axis to CG axis of shaded area.

$$y = \left(h - \frac{2y}{3}\right)$$

$$\therefore \quad \tau = \frac{F}{b_1\left(\frac{a^4}{12}\right)} \times \frac{1}{2} b_1 y \left(h - \frac{2y}{3}\right)$$

$$\tau = \frac{6F}{a^4} \cdot y\left(h - \frac{2y}{3}\right)$$

This is parabolic variation.

Applying bounding conditions, at $y = 0$ at B, $\tau = 0$

$$\text{at } y = h \text{ at } NA, \ \tau \ \frac{6F}{a^4} h\left(h - \frac{2h}{3}\right) = \frac{6F}{a^4}\frac{h^2}{3}$$

$$\tau = \frac{6F}{a^4}\frac{\left(\frac{a}{\sqrt{2}}\right)^2}{3} = \frac{F}{a^2} = \frac{F}{A} = \tau_{avg}$$

(Since $A = a^2$ = Area of square)

For maximum shear stress $\frac{d\tau}{dy} = 0$

$$\frac{d\tau}{dy} = \frac{6F}{a^4}\left(h - \frac{2y}{3}\right) + \frac{6F}{a^4} y\left(\frac{-2}{3}\right) = 0$$

$$= \frac{6F}{a^4}\left(h - \frac{4y}{3}\right) = 0$$

Since F and a^4 are not equal to zero,

$$h - \frac{4y}{3} = 0$$

$$\therefore \quad y = \frac{3h}{4}$$

The shear stress will be maximum at $y = \dfrac{3h}{4}$ substituting,

$$\tau_{\max} = \frac{6F}{a^4}\left(\frac{3h}{4}\right)\left(h - \frac{2}{3} \times \frac{3h}{4}\right)$$

$$= \frac{18Fh}{4\,a^4} \times \left(\frac{12h - 6h}{12}\right) = \frac{9}{4}\frac{Fh^2}{a^4}$$

Substitute $h = \dfrac{a}{\sqrt{2}}$,

$$\tau_{\max} = \frac{9}{4}\,\frac{F.\left(\frac{a}{\sqrt{2}}\right)^2}{a^4} = \frac{9}{4}\,\frac{Fa^2}{a^4 \times 2}$$

$$\tau_{\max} = \frac{9}{8}\frac{F}{a^2} = \frac{9}{8}\frac{F}{A} = \frac{9}{8}\,\tau_{avg}$$

$$\tau_{\max} = \frac{9}{8}\,\tau_{avg}$$

The maximum shear stress for a square section with one diagonal horizontal is $\dfrac{9}{8}$ times average shear stress and occurs at a distance of $\dfrac{3h}{4}$ from the top where h is semi-diagonal or $\dfrac{3}{8}$ of total height. The variation of shear stress is same for the lower half of the square.

The shear stress variation is shown in Fig. 5.70(b).

5.15 SHEAR STRESS VARIATION IN HEXAGONAL CROSS SECTION WITH ONE OF ITS DIAGONALS HORIZONTAL

Figure 5.71 shows a hexagonal section of a beam of side a and is placed such that one of its diagonals is horizontal. The beam section is subjected to a shear force F. The total height is $\sqrt{3}\ a$. Total width = $2a$.

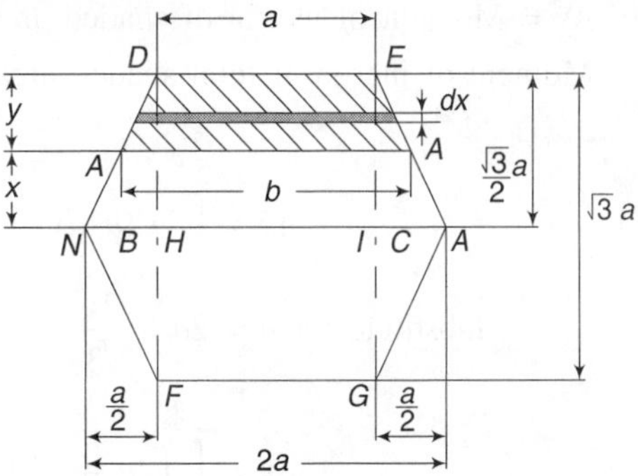

Fig. 5.71

I = Moment of inertia of the hexagonal section about neutral axis NA.

= Moment of inertia of rectangle about NA + Moment of inertia of 4 triangles about base

= I_{DEFG} about NA + 4 × I_{DNH} about base

$$I = \frac{Bh^3}{12} + 4\cdot\frac{b_1 h_1{}^3}{12}$$

where, B = Base of rectangle = a, H = Height of rectangle = $\sqrt{3}a$

b_1 = Base of each triangle = $\frac{a}{2}$ and h_1 = height of each triangle = $\frac{\sqrt{3}}{2}a$

$$I = \frac{a \times (\sqrt{3}a)^3}{12} + \frac{4 \times \frac{a}{2}\left(\frac{\sqrt{3}}{2}a\right)^3}{12}$$

$$= \frac{3\sqrt{3}\,a^4}{12} + \frac{\frac{4a}{2} \times \left(3\frac{\sqrt{3}}{8}\right)a^3}{12}$$

$$I = \frac{3\sqrt{3}}{12}a^4\left[1 + \frac{1}{4}\right] = \frac{5}{16}\sqrt{3}a^4$$

Consider a section AA at a distance y from top.

Let the width of section = $b = NA - 2NB$

$NA = 2a$ and $NB = CA$

Let the distance from NA to section $AA = x$

From triangle ANB, $\tan 60 = \frac{AB}{NB}$

$$\therefore \quad NB = \frac{AB}{\tan 60} = \frac{x}{\sqrt{3}}$$

$$\therefore \quad b = 2a - \frac{2x}{\sqrt{3}}$$

Shear stress $\tau = \frac{F}{Ib}.A\bar{y}$

$A\bar{y}$ = Moment of inertia of shaded area about $NA = (b \cdot dx)\,x$

Moment of inertia of total shaded area.

$$A\bar{y} = \int_x^{\frac{\sqrt{3}}{2}a} a\,(b \cdot dx)\,x$$

Substitute for $b = 2a - \frac{2x}{\sqrt{3}}$

$$A\bar{y} = \int_x^{\frac{\sqrt{3}}{2}a}\left(2a - \frac{2x}{\sqrt{3}}\right)x \cdot dx = \frac{2}{\sqrt{3}}\int_x^{\frac{\sqrt{3}a}{2}}(\sqrt{3}a - x)x\,dx$$

$$= \frac{2}{\sqrt{3}}\int_x^{\frac{\sqrt{3}a}{2}}(\sqrt{3}a\,x - x^2)\,dx = \frac{2}{\sqrt{3}}\left[\sqrt{3}\,a \cdot \frac{x^2}{2} - \frac{x^3}{3}\right]_x^{\frac{\sqrt{3}a}{2}}$$

$$= \frac{2}{\sqrt{3}}\left[\frac{\sqrt{3}a}{2}\left(\frac{\sqrt{3}a}{2}\right)^2 - \frac{1}{3}\left(\frac{\sqrt{3}a}{2}\right)^3 - \sqrt{3}a\,\frac{x^2}{2} + \frac{x^3}{3}\right]$$

$$= \frac{2}{\sqrt{3}}\left[\frac{3\sqrt{3}}{8}a^3 - \frac{3\sqrt{3}}{3\times 8}a^3\,\frac{\sqrt{3}}{2}ax^2 + \frac{x^3}{3}\right]$$

$$A\bar{y} = \frac{2}{\sqrt{3}}\left[\frac{a^3\sqrt{3}}{4} - \frac{\sqrt{3}}{2}ax^2 + \frac{x^3}{3}\right]$$

$$\therefore \quad \tau = \frac{F}{Ib}A\bar{y} = \frac{F}{\left(\frac{5}{16}\sqrt{3}\,a^4\right)\left(2a - \frac{2}{\sqrt{3}}x\right)} \times \frac{2}{\sqrt{3}}\left[a^3\frac{\sqrt{3}}{4} - \frac{\sqrt{3}}{2}ax^2 + \frac{x^3}{3}\right]$$

$$\tau = \frac{16F}{5\sqrt{3}\cdot a^4\,(\sqrt{3}a - x)} \times \left[\frac{\sqrt{3}}{4}a^3 - \frac{\sqrt{3}}{2}ax^2 + \frac{x^3}{3}\right]$$

For shear stress at neutral NA, $x = 0$

$$\tau_{NA} = \frac{16F}{5\sqrt{3}\,a^4\cdot(\sqrt{3}a)}\left[\frac{\sqrt{3}}{4}a^3\right] = \frac{4F}{5\sqrt{3}\,a^2}$$

Area of hexagon = Area of rectangle $DEFG$ + 4 × Area of triangle DNH

$$A = (FG \times EG) + 4 \times \left(\frac{1}{2}NH \times DH\right)$$

$$A = a \times \sqrt{3}\,a + 4 \times \frac{1}{2} \times \frac{a}{2} \times \frac{\sqrt{3}}{2}a$$

$$A = \sqrt{3}\,a^2 + \frac{\sqrt{3}}{2}a^2 = \frac{3}{2}\sqrt{3}\,a^2$$

$$\text{Average shear stress } \tau_{avg} = \frac{F}{A} = \frac{F}{\frac{3\sqrt{3}}{2}a^2}$$

But

$$\tau_{NA} = \frac{4F}{5\sqrt{3}\,a^2} = \frac{4F}{5 \times \left(\frac{3\sqrt{3}}{2}a^2\right) \times \frac{2}{3}} = \frac{12}{10}\,\frac{F}{\left(\frac{3\sqrt{3}}{2}a^2\right)}$$

$$\tau_{NA} = \frac{12}{10}\tau_{avg} = 1.2\,\tau_{avg}$$

The shear stress at the neutral axis for a hexagonal cross section is 1.2 times the average shear stress.

5.16 SHEAR STRESS DISTRIBUTION IN I-SECTION

Figure 5.72 shows an I-section

B = Width of flange

b = Width of web

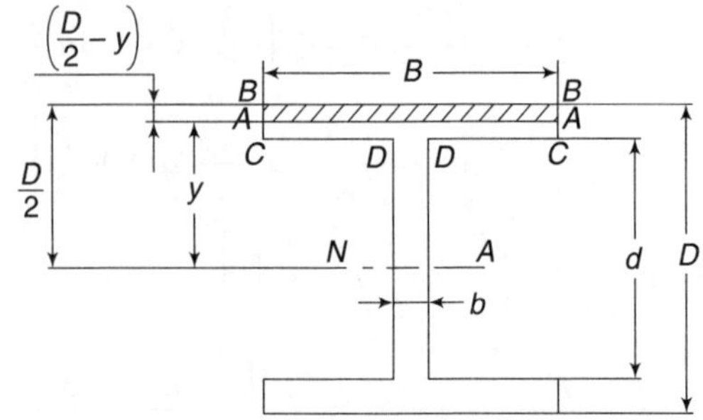

Fig. 5.72

D = Total height of I-section.

d = Height of web.

(a) Shear stress distribution in the flange.

Consider section AA at a distance y from neutral axis as shown in Fig. 5.72.

Width of F lange = B

Thickness of shaded portion $= \dfrac{D}{2} - y$

$\therefore$ Area of shaded portion of flange $= B\left(\dfrac{D}{2} - y\right)$

Distance to centroid of shaded portion from neutral axis

$$= y + \frac{1}{2}\left(\frac{D}{2} - y\right)$$

$$\bar{y} = y + \frac{D}{4} - \frac{y}{2} = \frac{1}{2}\left(\frac{D}{2} + y\right)$$

$$\text{Shear stress } \tau = \frac{F}{Ib} \cdot a\bar{y}$$

where, I = Moment of inertia of I-section

b = Width = B

$a = \text{Area} = B\left(\dfrac{D}{2} - y\right)$

$\bar{y} = \dfrac{1}{2}\left(\dfrac{D}{2} + y\right)$

$$\therefore \quad \tau = \frac{F}{IB} \cdot B \cdot \left(\frac{D}{2} - y\right) \times \frac{1}{2}\left(\frac{D}{2} + y\right) = \frac{F}{IB} \cdot \frac{B}{2}\left(\frac{D^2}{4} - y^2\right)$$

$$\tau = \frac{F}{2I}\left(\frac{D^2}{4} - y^2\right)$$

This is parabolic distribution.

At the top at BB, $y = \dfrac{D}{2}$, $\tau = 0$

At the bottom of flange at CC, $y = \dfrac{d}{2}$

$$\tau_{CC} = \frac{F}{2I}\left(\frac{D^2}{4} - \frac{d^2}{4}\right) = \frac{F}{8I}(D^2 - d^2)$$

(b) Shear stress distribution in the web

Refer Fig. 5.73 width of section at any distance y from the neutral axis $NA = b$.

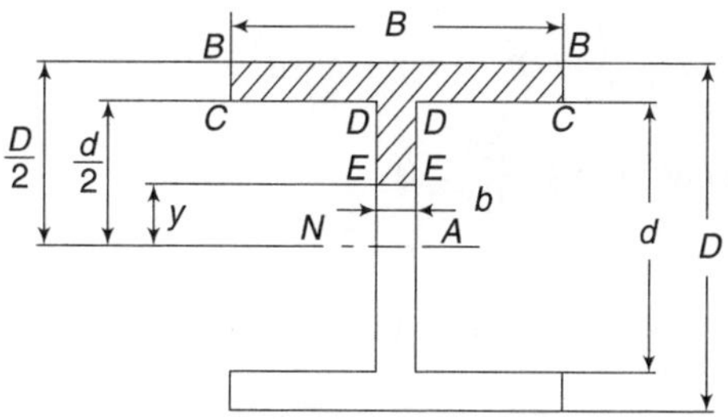

Fig. 5.73

Moment of inertia of shaded protion $a\bar{y} = a_1\bar{y}_1 + a_2\bar{y}_2$

where, a_1 = Area of flange $= B\cdot\left(\dfrac{D}{2} - \dfrac{d}{2}\right)$

$\bar{y}_1$ = Distance from NA to centroid of flange

$$\frac{D}{2} - \frac{1}{2}\left(\frac{D}{2} - \frac{d}{2}\right) = \frac{1}{2}\left(\frac{D}{2} + \frac{d}{2}\right)$$

a_2 = Area of shaded protion of web $= b\left(\dfrac{d}{2} - y\right)$

$\bar{y}_2$ = Distance from NA to centroid of shaded portion of web

$$= \frac{d}{2} - \frac{1}{2}\left(\frac{d}{2} - y\right) = \frac{1}{2}\left(\frac{d}{2} + y\right)$$

$$\therefore \quad a\bar{y} = B\left(\frac{D}{2} - \frac{d}{2}\right)\frac{1}{2}\left(\frac{D}{2} + \frac{d}{2}\right) + b\left(\frac{d}{2} - y\right)\frac{1}{2}\left(\frac{d}{2} + y\right)$$

$$a\bar{y} = \frac{B}{8}(D^2 - d^2) + \frac{b}{2}\left(\frac{d^2}{4} - y^2\right)$$

$$\text{Shear stress } \tau = \frac{F}{Ib}\cdot a\bar{y} = \frac{F}{Ib}\left[\frac{B}{8}(D^2 - d^2) + \frac{b}{2}\left(\frac{d^2}{4} - y^2\right)\right]$$

$$= \frac{F}{Ib}\frac{B}{8}(D^2 - d^2) + \frac{F}{Ib}\cdot\frac{b}{2}\left(\frac{d^2}{4} - y^2\right)$$

$$\tau = \frac{B}{b}\frac{F}{8I}(D^2 - d^2) + \frac{F}{2I}\left(\frac{d^2}{4} - y^2\right) \qquad \ldots(1)$$

This has two parts. First part is a constant value and second part is parabolic varying with y.

At the neck of flange and web at DD, $y = \frac{d}{2}$

$$\therefore \qquad \tau_{DD} = \frac{B}{b}\frac{F}{8I}(D^2 - d^2)$$

At the neutral axis at NA, $y = 0$,

From Eq. (1), $\tau_{NA} = \frac{B}{b}\frac{F}{8I}(D^2 - d^2) + \frac{F}{8I}d^2$

For maximum shear stress, $\frac{d\tau}{dy} = 0$

or $\qquad y = 0$

Hence, the shear stress is maximum at the neutral axis

$$\tau_{max} = \tau_{NA} = \frac{B}{b}\frac{F}{8I}(D^2 - d^2) + \frac{F}{8I}d^2$$

Figure 5.74 shows the shear stress distribution across the section. It may be observed that the shear stress at the neck of flange and web suddenly increases from $\frac{F}{8I}(D^2 - d^2)$ to $\frac{F}{8I}\frac{B}{b}(D^2 - d^2)$ and in the flange and web, the shear stress distribution is parabolic. The shear stress distribution is symmetric about neutral axis NA.

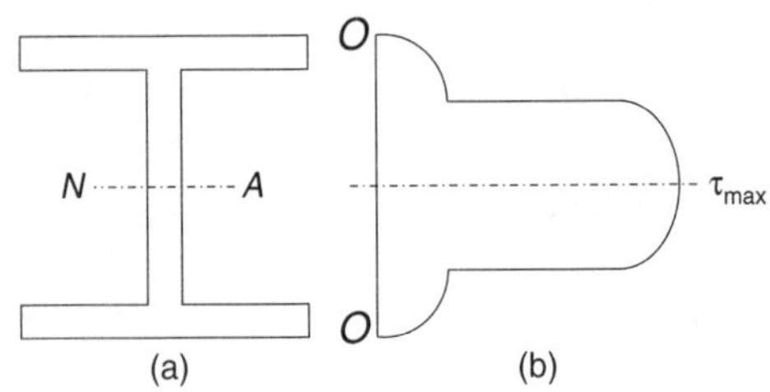

Fig. 5.74

SOLVED PROBLEMS

Problem 51 A rectangular beam 90 mm wide and 200 mm deep is subjected to a shear force of 600 kN. Find the maximum shear stress induced in the beam. If the maximum shear stress is not to exceed 40 N/mm^2. Find the depth of the beam.

Solution: **Given data:**

Rectangular cross section

$$\text{Width } b = 90 \text{ mm}$$

$$\text{Depth } d = 200 \text{ mm}$$

$$\text{Load } F = 600 \text{ kN} = 600 \times 10^3 \text{ N}$$

$$\text{Maximum shear stress } \tau_{max} = 40 \text{ N/mm}^2$$

(a) For rectangular cross section maximum shear stress

$$\tau_{max} = 1.5\ \tau_{avg} = 1.5\ \frac{F}{A} = 1.5\ \frac{F}{bd}$$

$$\tau_{max} = 1.5 \times \frac{600 \times 10^3}{90 \times 200} = 50 \text{ MPa}$$

(b) For $\tau_{max} = 40$ MPa, $\tau_{max} = 1.5\ \frac{F}{bd}$

$$40 = 1.5 \times \frac{600 \times 10^3}{90 \times d}$$

$\therefore$ Required depth $d = 250$ mm.

Problem 52 A 6 m long timber beam is simply supported at the ends and carries a UDL of 30 kN/m over the entire span. Find the maximum shear stress induced in the beam. The beam is of circular cross section of diameter 45 mm. If the shear stress is to be limited to 33 N/mm², find the suitable diameter.

Solution: **Given data:**

$$\text{UDL} = 30 \text{ kN/m} = 30 \times 10^3 \text{ N/m}$$

$$\text{Span} = 6 \text{ m}$$

$$\therefore \quad \text{Total load} = F = 30 \times 10^3 \times 6 = 180 \times 10^3 \text{ N}$$

$$\text{Diameter } d = 45 \text{ mm}$$

$$\tau_{max} = 63 \text{ N/mm}^2$$

(a) For circular cross section, maximum shear stress

$$\tau_{max} = \frac{4}{3}\ \tau_{avg} = \frac{4}{3}\frac{F}{A} = \frac{4}{3}\ \frac{F}{\left(\frac{\pi}{4}\ d^2\right)}$$

$$\tau_{max} = \frac{4}{3} \cdot \frac{180 \times 10^3}{\frac{\pi}{4} \times 45^2} = 150.90 \text{ N/mm}^2$$

(b) To limit the shear stress $\tau_{max} = 33$ N/mm²,

$$\tau_{max} = \frac{4}{3}\ \frac{F}{\frac{\pi}{4}\ d^2}$$

$$33 = \frac{4}{3}\ \frac{180 \times 10^3}{\frac{\pi}{4} \times d^2}$$

Required diameter $d = 96.29$ mm.

Problem 53 A timber beam 120 mm wide and 180 mm deep in the form of cantilever supports a UDL over a length of 5 m. If the safe shear stress is 3 N/mm². Find the intensity of UDL that can be supported by the beam.

Solution: **Given data:**

$$\text{Width } b = 120 \text{ mm, depth } d = 180 \text{ mm}$$

$$\text{Length } l = 5 \text{ m}$$

$$\text{Safe shear stress } \tau_{max} = 3 \text{ N/mm}^2$$

$$\text{Let UDL} = w \text{ N/m}$$

$$\therefore \quad \text{Total force } F = w \cdot l = (w \times 5) \text{ N}$$

For rectangular cross section beam, maximum shear stress

$$\tau_{max} = \frac{3}{2}\tau_{avg} = \frac{3}{2}\frac{F}{A} = \frac{3}{2}\frac{F}{b.d}$$

$$3 = \frac{3}{2} \cdot \frac{(5\,w)}{120 \times 180}$$

$$\therefore \quad \text{Intensity of UDL } w = 8640 \text{ N/m} = 8.64 \text{ kN/m.}$$

Problem 54 A simply supported beam 3 m long has rectangular cross section having depth as twice the width carries a concentrated load of 60 kN in the middle. Find the width and depth of cross section taking allowable shear stress as 72 N/mm^2.

Solution: **Given data:**

$$\text{Depth } d = 2 \times \text{Width} = 2b$$

$$\text{Load } F = 60 \times 10^3 \text{ N}$$

$$\text{Maximum shear stress } \tau_{max} = 72 \text{ N/mm}^2$$

$$\text{Area of cross section} = bd = 2b^2$$

For rectangular cross, maximum shear stress $\tau_{max} = \frac{3}{2}\tau_{avg}$

$$72 = \frac{3}{2} \times \frac{60000}{2b^2}$$

$$\therefore \quad \text{Width of section } b = 25 \text{ mm and depth of section } d = 2b = 50 \text{ mm.}$$

Problem 55 A beam of rectangular cross section of width 60 mm and depth 100 mm is simply supported at ends over a span of 9 m and carries a central load of 240 kN. For some reason it is to be replaced by a beam of circular cross section. Find the suitable diameter of the beam made of same material.

Solution: **Given data:**

$$\text{Width } b = 60 \text{ mm, Depth } d = 100 \text{ mm}$$

$$\text{Load } F = 240 \text{ kN} = 240 \times 10^3 \text{ N}$$

For rectangular cross section, maximum shear stress $\tau_{max} = \frac{3}{2}\frac{F}{A}$

$$\tau_{max} = \frac{3}{2} \times \frac{240 \times 10^3}{60 \times 100} = 60 \text{ N/mm}^2$$

For circular cross section, maximum shear stress $\tau_{max} = \frac{4}{3}\frac{F}{A}$

$$60 = \frac{4}{3} \times \frac{240 \times 10^3}{\frac{\pi}{4}D^2}$$

$$\therefore \quad \text{Required diameter of beam } D = 82.41 \text{ mm.}$$

Problem 56 Find the ratio of transverse shear loads that can be carried by a rectangular cross sectional beam and a circular cross-sectional beam made of same material and same cross-sectional area.

Solution: **Given data:**

$$\text{Area of cross section} = A$$

$$\text{Load} = F$$

$$\text{Maximum shear stress} = \tau_{max}$$

For rectangular cross section. Maximum shear stress $\tau_{max} = \dfrac{3}{2}\dfrac{F}{A}$

$$\therefore \quad \text{Load } F_1 = F = \frac{2}{3}\tau_{max}\cdot A$$

For circular cross section, maximum shear stress $\tau_{max} = \dfrac{4}{3}\dfrac{F}{A}$

$$\therefore \quad \text{Load } F_2 = F = \frac{3}{4}\tau_{max}\cdot A$$

$$\text{Comparing } \frac{F_1}{F_2} = \frac{\left(\frac{2}{3}\tau_{max}\cdot A\right)}{\left(\frac{3}{4}\tau_{max}\cdot A\right)} = \frac{8}{9}.$$

The load capacity of the rectangular cross section beam is $\dfrac{8}{9}$ times that of circular cross section.

Problem 57 A beam of I-section consists of 80 mm × 15 mm flanges and web of 280 mm depth × 15 mm thickness. It is subjected to a shear force of 60 kN. Sketch the shear stress distribution along the depth of the section indicating value.

Solution:

$$\text{Width of each flange } B = 180 \text{ mm}$$

$$\text{Thickness of each flange } t = 15 \text{ mm}$$

$$\text{Thickness of web } b = 15 \text{ mm}$$

$$\text{Height of web } d = 280 \text{ mm}$$

$$\text{Total height } D = d + 2t = 280 + 2 \times 15 = 310 \text{ mm}$$

Due to symmetry, the centroid lies in the middle

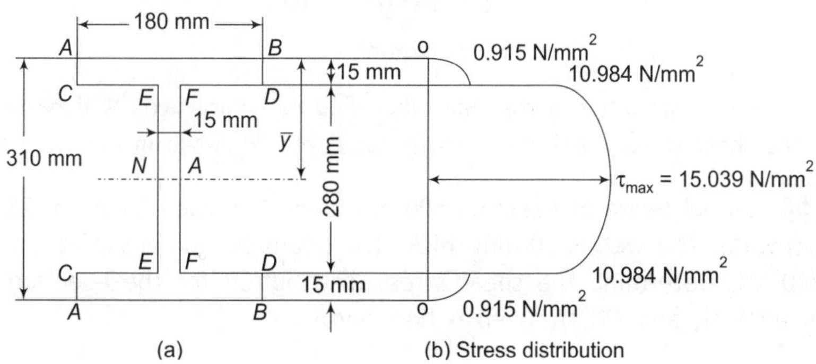

Fig. 5.75

$$\therefore \quad \bar{y} = \frac{D}{2} = \frac{310}{2} = 155 \text{ mm.}$$

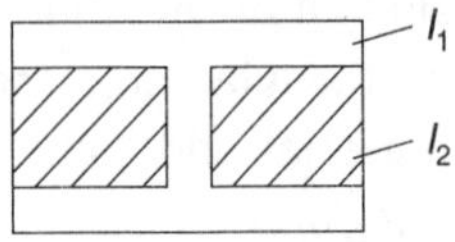

$$\text{Load } F = 60000 \text{ N}$$

Moment of inertia of I section about NA

$$I = I_1 - I_2$$

$$I_1 = MI \text{ of total rectangle } 180 \times 310$$

$$I_2 = MI \text{ of middle rectangle excluding web}$$

$$I_1 = \frac{BD^3}{12} = \frac{180 \times 310^3}{12} = 446.865 \times 10^6 \text{ mm}^4$$

$$I_2 = \frac{(B - b)\, d^3}{12} = (180 - 15) \times \frac{280^3}{12} = 301.84 \times 10^6 \text{ mm}^4$$

$$\therefore \quad I = I_1 - I_2 = 446.865 \times 10^6 - 301.84 \times 10^6$$

$$I = 145.025 \times 10^6 \text{ mm}^4$$

Shear stress at the top of flange, $\tau_{AB} = 0$

At the bottom of flange $\tau_{CD} = \frac{F}{8I}(D^2 - d^2) = \frac{60000}{8 \times 145.025 \times 10^6}(310^2 - 280^2)$

$$\tau_{CD} = 0.915 \text{ N/mm}^2$$

Shear stress at the neck of flange and web

$$\tau_{EF} = \frac{B}{b}\frac{F}{8I}(D^2 - d^2)$$

$$\tau_{EF} = \frac{B}{b}\tau_{CD} = \frac{180}{15} \times 0.915 = 10.984 \text{ N/mm}^2$$

Shear stress is maximum at neutral axis and is equal to τ_{max}

$$= \tau_{NA} = \frac{B}{b}\frac{F}{8I}(D^2 - d^2) + \frac{F}{8I}d^2$$

$$= 10.984 + \frac{60000}{8 \times 145.025 \times 10^6} \times 280^2$$

$$\tau_{NA} = \tau_{max} = 15.039 \text{ N/mm}^2$$

The shear stress distribution across the other side of neutral axis will be same due to symmetry. The shear stress distribution across the depth is shown in Fig. 5.75 (b).

Problem 58 A steel beam of I-section 500 mm deep has each flange of 25 mm thick and 160 mm wide. The web is 20 mm thick. The beam section is subjected to a shear force of 360 kN. Determine the shear stress distribution for the I-section where (i) the web is vertical, and (ii) the web is horizontal.

Solution: Total depth $D = 500$ mm

Thickness of flange $t = 25$ mm

Thickness of web $b = 20$ mm

Width of flange $B = 160$ mm

Depth of web $d = D - 2t = 500 - 2 \times 25 = 450$ mm.

Load $F = 360 \times 10^3$ N

(i) When the web is vertical

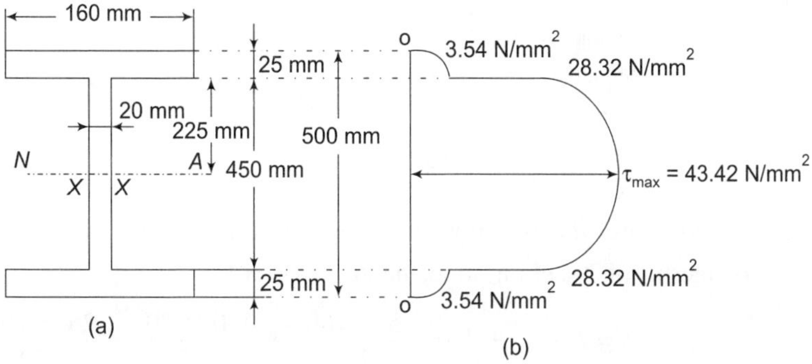

Fig. 5.76

Figure 5.76 shows the I-section with the web vertical moment of inertia of I-Section

$$I = I_1 - I_2$$

$$I = \frac{160 \times 500^3}{12} - \frac{(160 - 20)\ (450)^3}{12} = 603.54 \times 10^6 \text{ mm}^4$$

Shear stress at the top of flange = 0

Shear stress at the flange of the neck

$$\tau = \frac{F}{8I}(D^2 - d^2) = \frac{360 \times 10^3}{8 \times 603.54 \times 10^6}(500^2 - 450^2)$$

$$\tau = 3.54 \text{ N/mm}^2$$

Shear in the web at the neck,

$$\tau = \frac{B}{b}\cdot\frac{F}{8I}(D^2 - d^2) = \frac{160}{20} \times 3.54 = 28.32 \text{ N/mm}^2$$

Maximum shear stress occurs at neutral axis

$$\tau_{max} = \tau_{NA} = \frac{B}{b}\frac{F}{8I}(D^2 - d^2) + \frac{Fd^2}{8I}$$

$$\therefore \quad 28.32 + \frac{360 \times 10^3 \times 450^2}{8 \times 603.54 \times 10^6} = 43.42 \text{ N/mm}^2$$

The shear stress distribution is shown in Fig. 5.76 (b).

(ii) When the web is horizontal

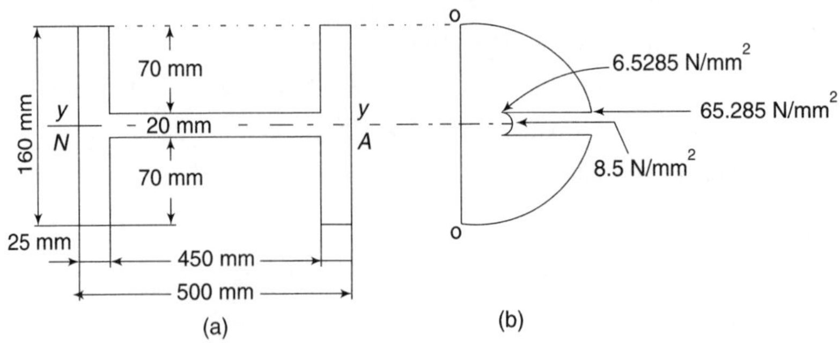

Fig. 5.77

Figure 5.77 (a) shows the I-section with web in horizontal position.

Moment of inertia of I-section about neutral axis *NA*

$$I = I_1 + I_2 + I_3 = \frac{25 \times 160^3}{12} + \frac{450 \times 20^3}{12} + \frac{25 \times 160^3}{12}$$

$$= 17.37 \times 10^6 \text{ mm}^4$$

(a) Shear stress at the top = 0

(b) Shear stress in the flange just above the web.

$$\tau = \frac{F}{Ib} \cdot a\bar{y}$$

$$b = 25 + 25 = 50 \text{ for (2 flange)}$$

For 2 flange portions above the web.

$$a\bar{y} = 2 \times (25 \times 70) \left(\frac{70}{2} + \frac{20}{2}\right) = 157500 \text{ mm}^2.$$

$$\therefore \qquad \tau = \frac{360 \times 10^3}{17.37 \times 10^6 \times 50} \times 157500 = 65.285 \text{ N/mm}^2$$

(c) Shear stress in the web at 10 mm from *NA*,

$$\tau = \frac{b}{B} \text{ (shear stress in the flange above web)}$$

$$= \frac{50}{500} (65.285) = 6.5285 \text{ N/mm}^2$$

(d) Shear stress at neutral axis

Consider area above neutral axis

$$a\bar{y} = 2 \times (25 \times 80) \times \frac{80}{2} + (450 \times 10) \frac{10}{2}$$

$$a\bar{y} = 205000 \text{ mm}^3$$

$$b \text{ at } NA = 500 \text{ mm}$$

$$\tau_{NA} = \frac{F}{Ib} \cdot a\bar{y} = \frac{360 \times 10^3}{17.37 \times 10^6 \times 500} \times 205000 = 8.50 \text{ N/mm}^2$$

The shear stress distribution is as shown in Fig. 5.77(b)

Problem 59 **Figure 5.78 shows the cross section of an unsymmetric I-section and is subjected to a shear force of 60 kN. Determine and plot the shear variation across the depth of section.**

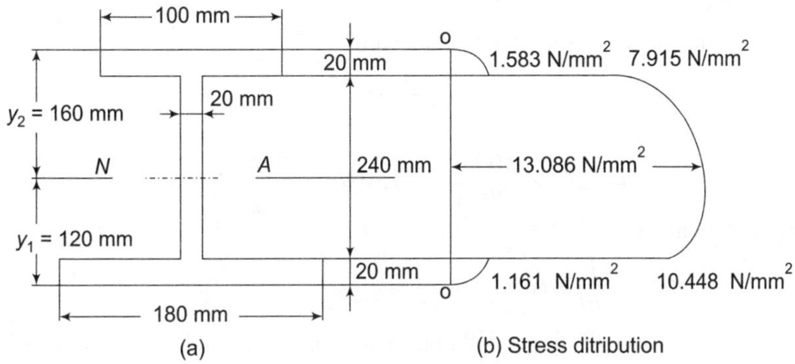

Fig. 5.78

Solution: Distance to neutral axis (Centroidal axis) from bottom,

$$y_1 = \frac{A_1 y_1 + A_2 y_2 + A_3 y_3}{A_1 + A_2 + A_3}$$

$$y_1 = \frac{100 \times 20 \times (20 + 240 + 10) + 20 \times 240\,(20 + 120) + 180 \times 20)(10)}{100 \times 20 + 20 \times 240 + 180 \times 20}$$

$$= 120 \text{ mm}$$

$\therefore$ Distance to NA from top = 280 – 120 = 160 mm

Moment of inertia of I-section about NA

$$I = I_1 + I_2 + I_3$$

where, $I_1 = I_{cg} + Ak^2$

$$= \frac{bd^3}{12} + (bd)\,(y_1 - \bar{y}_1)^2$$

$$= \frac{100 \times 20^3}{12} + 100 \times 20\,(270 - 120)^2 = 45.067 \times 10^6 \text{ mm}^4$$

$$I_2 = I_{cg} + Ak^2$$

$$= \frac{20 \times 240^3}{12} + (20 \times 240)\,(140 - 120)^2 = 24.96 \times 10^6 \text{ mm}^4$$

$$I_3 = I_{cg} + Ak^2$$

$$= \frac{180 \times 20^3}{12} + (180 \times 20)\,(10 - 120)^2 = 43.68 \times 10^6 \text{ mm}^4$$

$$\therefore \quad I = I_1 + I_2 + I_3 = 113.707 \times 10^6 \text{ mm}^4$$

Shear stress at top of top flange = 0

Shear stress at the bottom of top flange

$$\tau = \frac{F}{BI} \cdot a\bar{y} = \frac{60 \times 10^3}{100 \times 113.707 \times 10^6} \times (100 \times 20)\,(160 - 10)$$

$$\tau_{\text{bottom of } TF} = 1.583 \text{ N/mm}^2$$

Shear stress at the neck of top flange and web

$$\tau_{\text{top neck}} = \frac{F}{BI}\, a \cdot \bar{y} = \left(\frac{F}{BI} \cdot a\bar{y}\right) \cdot \frac{B}{b}$$

$$= 1.58 \times \frac{100}{20} = 7.915 \text{ N/mm}^2$$

Shear stress at neutral axis (*NA*)

$$\tau_{NA} = \frac{F}{bI} \cdot (a\bar{y})$$

$$a\bar{y} = a\bar{y} \text{ for top flange} + a\bar{y} \text{ for portion of web above } NA$$

$$= 100 \times 20 \times (160 - 10) + 20\,(160 - 20) \times \left(\frac{160 - 20}{2}\right)$$

$$= 496000 \text{ mm}^4$$

$$\therefore \quad \tau_{NA} = \frac{60000}{20 \times 113.707 \times 10^6} \times 496000 = 13.086 \text{ N/mm}^2$$

Shear stress at the neck at bottom flange

$$\tau_{\text{bottom neck}} = \frac{F}{bI} \cdot a\bar{y}$$

$$= \frac{60000}{20 \times 113.707 \times 10^6} \times (180 \times 20)\,(120 \times 10)$$

$$= 10.448 \text{ N/mm}^2$$

Shear stress at the top of bottom flange

$$\tau_{\text{top of } B\cdot F} = \frac{F}{B \cdot I}\, a\bar{y} = \frac{60000}{180 \times 113.707 \times 10^6} \times (180 \times 20)\,(120 - 10)$$

or

$$10.448 \times \frac{b}{B} = 10.448 \times \frac{20}{180} = 1.161 \text{ N/mm}^2$$

Shear stress at the bottom of flange = 0

The shear stress distribution is shown in Fig. 5.78 (b)

Problem 60 Figure 5.79 shows the cross section of a beam subjected to a sharing force of 360 kN at a section. Calculate the maximum intensity of shear stress in the section. Also calculate the shear force carried by the web and flanges. Sketch the shear stress distribution across the section.

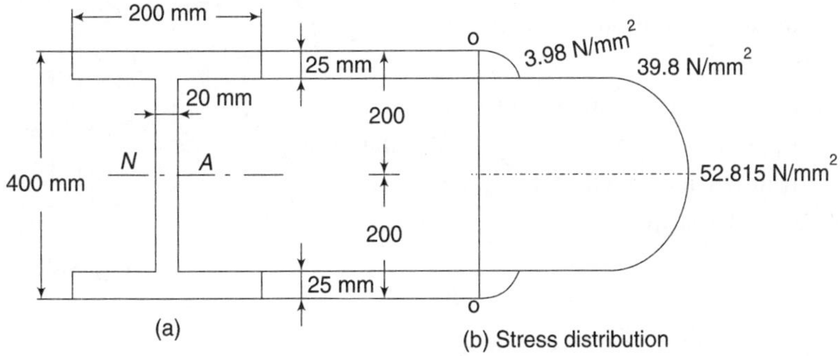

Fig. 5.79

$$\text{Force } F = 360 \times 10^3 \text{ N}$$

Due to symmetry GG lies at the middle

$$\bar{y} = 200$$

Moment of inertia about neutral axis

$$I = I_o - I_i$$

I_o = *MI* of outside rectangle

I_i = *MI* of inside rectangle

$$= \frac{BH^3 - bh^3}{12} = \frac{200 \times 400^3 - (200 - 20)(400 - 40)^3}{12}$$

$$I = 423.54 \times 10^6 \text{ mm}^4$$

Shear stress at the top of top flange = 0

Shear stress at the junction of flange $\tau = \frac{F}{I \cdot B} \cdot a\bar{y}$

$$\tau = \frac{360 \times 10^3}{423.54 \times 10^6 \times 200} \times (200 \times 25)(200 - 12.5) = 3.98 \text{ N/mm}^2$$

Shear stress at the junction of web

$$= \frac{F}{Ib} a\bar{y} = \left(\frac{F}{Ib} \cdot a\bar{y}\right) \cdot \frac{B}{b}$$

$$= 3.98 \times \frac{200}{20} = 39.8 \text{ N/mm}^2$$

Maximum shear stress at the neutral axis

$$\tau = \frac{F}{8I}(D^2 - d^2)\cdot\frac{B}{b} + \frac{F}{8I}d^2 = 39.8 + \frac{360 \times 10^3}{8 \times 423.54 \times 10^6} \times 350^2$$

$$\tau_{NA} = 52.815 \text{ N/mm}^2$$

The shear stress distribution across the depth is given in Fig. 5.79(b).

To find forces carried by flange and web:

Consider a small strip of thickness dy in the flange at a distance y from NA as shown in Fig. 5.80.

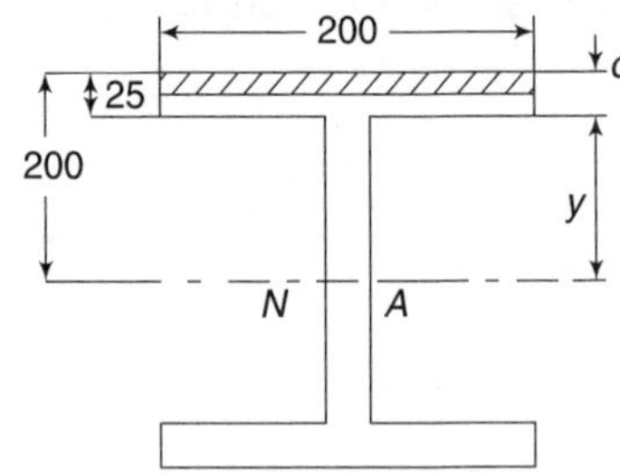

Fig. 5.80

Shear stress in the strip

$$\tau = \frac{F}{B{\cdot}I}\cdot a\bar{y}$$

where a = Area = 200 (200 – y)

$$\bar{y} = y + \frac{(200 - y)}{2} = \frac{200 + y}{2}$$

$\therefore$

$$\tau = \frac{360 \times 10^3}{423.54 \times 10^6 \times 200}\left[200(200 - y)\left(\frac{200 + y}{2}\right)\right]$$

$$= 4.2499 \times 10^{-4}(200^2 - y^2)$$

$\therefore$ Shear force carried by the element strip = $\tau\cdot$Area.

$$F = 4.2499 \times 10^{-4}(200^2 - y^2) \times (200 \times dy)$$

For shear force carried by flange, integrate this w.r.t y for y = 175 to 200.

$$F_{\text{lange}} = \int_{175}^{200} 4.2499 \times 10^{-4}(40000 - y^2) \times 200\, dy$$

$$= 0.085 \int_{175}^{200} (40000\, dy - y^2\, dy)$$

$$= 0.085\left[40000y - \frac{y^3}{3}\right]_{175}^{200} = 453333.33 - 443151.04$$

$$F_F = 10182.288 \text{ N}$$

$\therefore$ Shear force carried by two flanges = 2 × 10182.29 N = 20,364.58 N

$\therefore$ Shear force carried by web = F – 20364.58

$$F_w = 339.635 \times 10^3 \text{ N.}$$

Problem 61 The cross section of a beam is box section as shown in Fig. 5.81. The beam is subjected to a transverse load of 50 kN. Sketch the shear stress distribution across the depth of section.

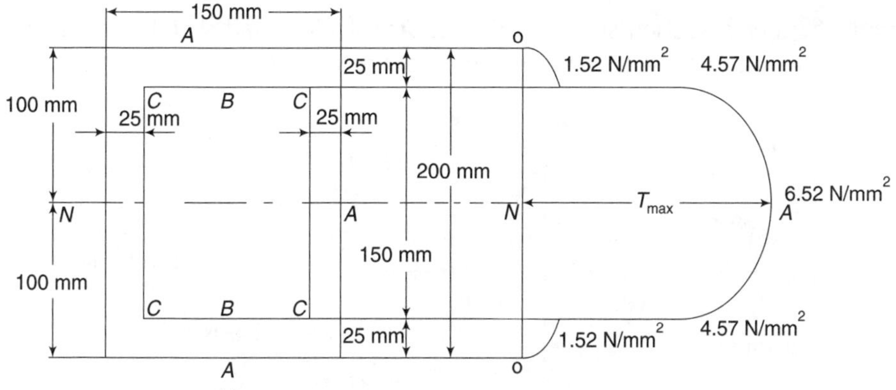

(a) (b) Stress distribution

Fig. 5.81

Solution: Load F = 50 kN = 50000 N

The centre of gravity lies in the middle at *NA*

The moment of inertia of the section about *NA*,

$$I = I_o - I_i$$

$$= \frac{BD^3 - b_i d^3}{12} = \frac{150 \times 200^3 - 100 \times 150^3}{12} = 71.875 \times 10^6 \text{ mm}^4$$

Shear stress at the top of flange

$$\tau_A = 0$$

Shear stress at the bottom of flange

$$\tau_B = \frac{F}{8I}(D^2 - d^2)$$

$$= \frac{50000}{8 \times 71.875 \times 10^6}(200^2 - 150^2) = 1.52 \text{ N/mm}^2$$

Shear stress at the web at the neck at C

$$\tau_C = \frac{F}{8I}(D^2 - d^2)\cdot\frac{B}{b}$$

$$\tau_C = 1.52 \times \frac{150}{50} = 4.57\text{N/mm}^2$$

Shear stress at the neutral axis $\tau_{NA} = \frac{F}{8I}(D^2 - d^2)\frac{B}{b} + \frac{F}{8I}d^2$

$$= 4.57 + \frac{50000}{8 \times 71.875 \times 10^6} \times 150^2 = 6.52 \text{ N/mm}^2$$

The stress distribution is shown in Fig. 5.81 (b).

Problem 62 For a section shown in Fig. 5.82, calculate and sketch the shear stress distribution across the depth. The member carries a shear force of 120 kN.

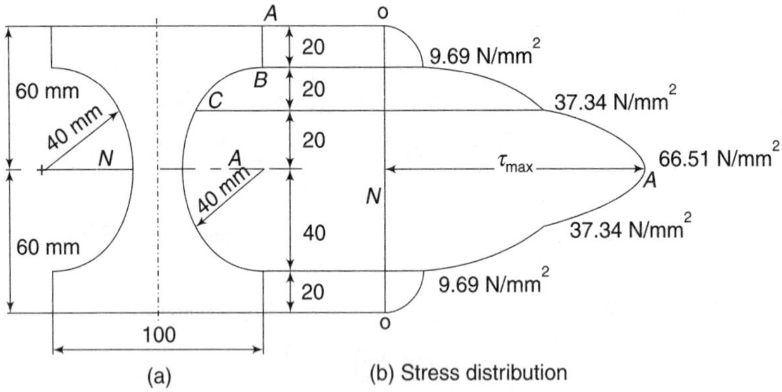

Fig. 5.82

Solution: Force $F = 120 \times 10^3$ N

Due to symmetry the neutral axis lies in the middle, at a height of $\frac{120}{2} = 60$ mm

Moment of inertia $I = I_{\text{Rectangle}} - 2 \times I_{\text{Semicircle}}$

$$= \frac{100 \times 120^3}{12} - 2 \times \frac{\pi}{128} \times 80^4 = 12.389 \times 10^6 \text{ mm}^4$$

Shear stress at $A = \tau_A = 0$

Shear stress at $B = \tau_B = \frac{F}{BI} \cdot a\bar{y}$

For the flange at A, $a\bar{y} = (100 \times 20)\left(40 + \frac{20}{2}\right) = 100000 \text{ mm}^3$

$$\therefore \qquad \tau_B = \frac{120 \times 10^3}{100 \times 12.389 \times 10^6} \times 100000 = 9.69 \text{ N/mm}^2$$

Shear stress at C, $\tau_C = \frac{F}{BI} a\bar{y}$

Width at C, $b = B - 2\sqrt{R^2 - x^2}$

$$= 100 - 2\sqrt{40^2 - 20^2} = 30.718 \text{ mm}$$

$$a\bar{y} = a_1\bar{y}_1 - a_2\bar{y}_2$$

$$a_1\bar{y}_1 = (100 \times 40) \times \left(20 + \frac{40}{2}\right) = 160000 \text{ mm}^3$$

(Considering a fall rectangle of 100 × 40 from top)

$$a_2\bar{y}_2 = \int_{20}^{40} 2b_1 dy \cdot y$$

(Consider a segment of circle)

$$= \int_{20}^{40} \sqrt{40^2 - y^2}\; y \cdot dy$$

$$= \int_{20}^{40} (40^2 - y^2)\; y\; dy$$

Put $[40^2 - y^2] = x$

$$-2ydy = dx$$

$$2ydy = -dx$$

$$= \int -\sqrt{x} \cdot dx = -\frac{x^{3/2}}{3/2} = \frac{-2}{3}(40^2 - y^2)^{3/2}$$

$$a_2\overline{y}_2 = \left[\frac{-2}{3}(40^2 - y^2)^{3/2}\right]_{20}^{40} = \frac{-2}{3}\left[0 - (40^2 - 20^2)^{3/2}\right]$$

$$= 41569.22 \text{ mm}^3$$

$$\therefore \quad a\overline{y} = 160000 - 41569.22 = 118430.78 \text{ mm}^3$$

$$\therefore \quad \tau_C = \frac{120 \times 103}{12.389 \times 10^6 \times 30.718} \times 118430.78 = 37.34 \text{ N/mm}^2$$

At D at neutral axis $\tau_{NA} = \dfrac{F}{Ib}\, a\overline{y}$

$$b = 20 \text{ mm}$$

$$a\overline{y} = \left(100 \times 60 \times \frac{60}{2}\right) - \int_{o}^{40} \sqrt{40^2 - y^2}\; 2y\; dy$$

$$= 180000 - \frac{2}{3}(40^2)^{3/2} = 137333.33 \text{ mm}^3$$

$$\therefore \quad \tau_{NA} = \frac{120000}{12.389 \times 10^6 \times 20} \times 137333.33 = 66.51 \text{ N/mm}^2$$

The stress distribution is as shown in Fig. 5.82(b).

Problem 63 Figure. 5.83 shows the section of a bar carrying a shear force of 210 kN. Find the maximum shear stress induced.

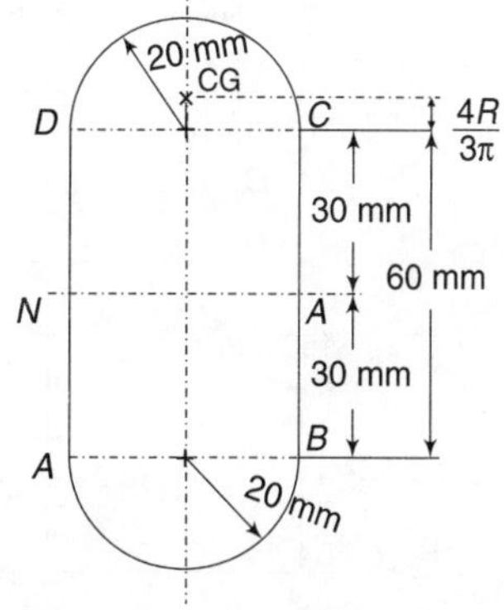

Fig. 5.83

Solution: Load $F = 210 \times 10^3$ N

The CG for the semicircle portion is at a height of $\dfrac{4R}{3\pi}$ from its base (CD)

The moment of inertia of semicircle about the base

$$I_{CD} = \frac{\pi d^4}{128} = \frac{\pi R^4}{8}$$

But $\quad I_{CD} = I_{cg} + Ak^2$

$\therefore$ Moment of inertia of the semicircle about CG, $I_{cg} = I_{CD} - Ak^2$

$$I_{cg} = \frac{\pi R^4}{8} - \frac{\pi}{2} R^2 \left(\frac{4R}{3\pi}\right)^2 = 0.1098\ R^4 = 0.1098 \times 20^4 = 17561.11\ \text{mm}^4$$

$\therefore$ Moment of inertia of whole section about *NA*,

$$I = I_{\text{Rectangle}} + 2\ I_{\text{Semi-circle}} = \frac{40 \times 60^3}{12} + 2\ [I_{cg} + Ak^2]$$

$$= \frac{40 \times 60^3}{12} + 2\left[17561.11 + \frac{\pi}{2} \times 20^2 \left(30 + \frac{4R}{3\pi}\right)^2\right]$$

$$= 720000 + 2\left[17561.11 + \frac{\pi}{2} \times 20^2\left(30 + \frac{4 \times 20}{3 \times \pi}\right)^2\right] = 2.617 \times 10^6\ \text{mm}^4$$

We know that maximum shear stress occurs at neutral axis

$$\tau_{max} = \tau_{NA} = \frac{F}{Ib} \cdot a\bar{y}$$

$$a\bar{y} = a_1\bar{y}_1 + a_2\bar{y}_2 = 40 \times 30 \times \frac{30}{2} + \frac{\pi}{2} \times 20^2 \times \left(30 + \frac{4R}{3\pi}\right)$$

$$= 18000 + \frac{\pi}{2}\ 20^2 \left(30 + \frac{4 \times 20}{3\pi}\right) = 42182.89\ \text{mm}^3$$

$$\therefore \quad \tau_{max} = \frac{210 \times 10^3}{2.617 \times 10^6 \times 40} \times 42182.89 = 84.62\ \text{N/mm}^2.$$

Problem 64 Figure 5.84 shows the cross section of a beam subjected to a shear force of 90 kN. Draw the shear stress distribution across the depth marking values at salient points.

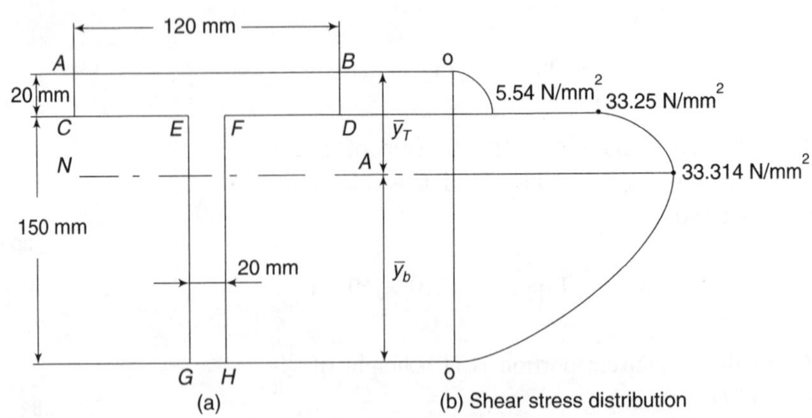

Fig. 5.84

Solution: Load $F = 90000$ N

Position of CG from bottom GH $\bar{y}_b = \dfrac{A_1 y_1 + A_2 y_2}{A_1 + A_2}$

$$\bar{y}_b = \frac{120 \times 20 \times \left(150 + \dfrac{20}{2}\right) + 150 \times 200\ (75)}{120 \times 20 + 150 \times 20}$$

$$= 112.78 \text{ mm}$$

$\therefore$ From top $\bar{y}_T = (150 + 20) - \bar{y}_b = 57.22$ mm

Moment of inertia of the section about neutral axis NA

$$I = I_1 + I_2$$

$$= \left\{\frac{120 \times 20^3}{12} + (120 \times 20)\ (57.22 - 10)^2\right\} + \left\{\frac{20 \times 150^3}{12} + 150 \times 20\ (112.78 - 75)^2\right\}$$

$$= 5431348.16 + 9906985.2 = 15.338 \times 10^6 \text{ mm}^4$$

Shear stress at the top of flange at AB

$$\tau_{AB} = 0$$

Shear stress at the bottom of flange at CD,

$$\tau_{CD} = \frac{F}{IB} \cdot a\bar{y}$$

$$\tau_{CD} = \frac{90000}{15.338 \times 10^6 \times 120}\ (120 \times 20)\ (57.22 - 10) = 5.54 \text{ N/mm}^2$$

Shear stress at the neck of web at EF_1 $\tau_{EF} = \dfrac{F}{Ib} . a_y$

$$\tau_{EF} = \frac{90000}{15.338 \times 10^6 \times 20}\ (120 \times 20)\ (57.22 - 10) = 33.25 \text{ N/mm}^2$$

Maximum shear stress at NA, $\tau_{NA} = \dfrac{F}{Ib}\, a\bar{y}$

$a\bar{y} = a\bar{y}$ of flange + $a\bar{y}$ of web above NA

$$= (120 \times 20)\ (57.22 - 10) + 20 \times (57.22 - 20) \left(\frac{57.22 - 20}{2}\right)$$

$$= 127181.28 \text{ mm}^3$$

$$\tau_{NA} = \frac{90000}{15.338 \times 10^6 \times 20} \times 127181.28 = 37.314 \text{ N/mm}^2$$

Shear stress at GH at the bottom

$$\tau_{GH} = 0$$

The shear stress distribution is shown in Fig. 5.84(b).

Problem 65 A simply supported beam is subjected to a uniformly distributed load of 24 kN/m over the full span. The length of the beam is 2 m. The beam is of T-section with flange of 200 mm wide and 25 mm thick and web of 300 mm and thickness of 25 mm. Find the shear stress at various levels and draw the stress distribution across the depth.

Solution: The T-section is shown in Fig. 5.85 (a). For a simply supported beam with UDL, each reaction $= \frac{wl}{2} = \frac{24000 \times 2}{2} = 24000$ N.

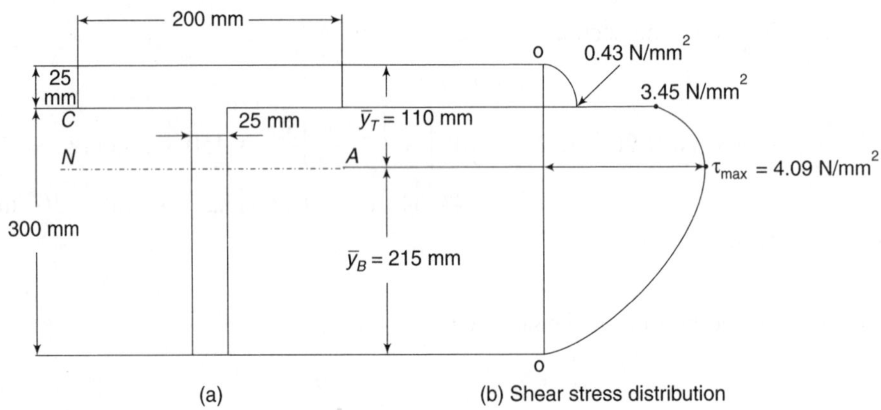

Fig. 5.85

This is the maximum shear force on the beam.

$$\therefore \quad F = 24000 \text{ N}$$

$$\text{Centre of gravity } \bar{y}_B = \frac{A_1 y_1 + A_2 y_2}{A_1 + A_2}$$

$$\bar{y}_B = \frac{200 \times 25 \times \left(300 + \frac{25}{2}\right) + 300 \times 25 \times \left(\frac{300}{2}\right)}{200 \times 25 + 300 \times 25} = 215 \text{ mm}$$

$$\therefore \quad \bar{y}_T = (300 + 25) - \bar{y}_B = 110 \text{ mm}$$

Moment of inertia of the section about neutral axis

$$I = I_1 + I_2 = (I_{cg} + Ak^2)_1 + (I_{cg} + Ak^2)_2$$

$$= \left\{\frac{200 \times 25^3}{12} + 200 \times 25\ (312.5 - 215)^2\right\} + \left\{\frac{25 \times 300^3}{12} + 25 \times 300\ (150 - 215)^2\right\}$$

$$I = 135.729 \times 10^6 \text{ mm}^4$$

Shear stress at the top of flange = 0

Shear stress at the bottom of flange $= \frac{F}{IB} \cdot a\bar{y}$

For flange B = 200 mm

$$a\bar{y} = 200 \times 25 \times (110 - 12.5) = 487500 \text{ mm}^4$$

$$\therefore \quad \text{Shear stress} = \frac{24000}{135.729 \times 10^6 \times 200} \times 487500 = 0.43 \text{ N/mm}^2$$

Shear stress at the neck of web.

$$\tau = \frac{F}{Ib} a\bar{y} = \left(\frac{F}{IB} \cdot a\bar{y}\right) \times \frac{B}{b} = 0.43 \times \frac{200}{25} = 3.45 \text{ N/mm}^2$$

Maximum shear stress occurs at neutral axis.

$$\tau_{max} = \frac{F}{Ib} a\bar{y}$$

$$a\bar{y} = \text{Moment of area above neutral axis}$$

$$= a_1\bar{y}_1 \text{ for flange} + a_2\bar{y}_2 \text{ for the web above } NA$$

$$= 200 \times 25 (110 - 12.5) + 25 \times (110 - 25) \times \left(\frac{110 - 25}{2}\right)$$

$$= 577812.5 \text{ mm}^3$$

$$\therefore \quad \tau_{max} = \frac{24000}{135.729 \times 10^6 \times 25} \times 577812.5 = 4.09 \text{ N/mm}^2$$

The shear stress distribution across the depth is shown in Fig. 5.85(b).

Problem 66 Figure 5.86 shows the section of a beam and is subjected to shear force of 45 kN. Calculate and sketch the shear stress distribution across the depth.

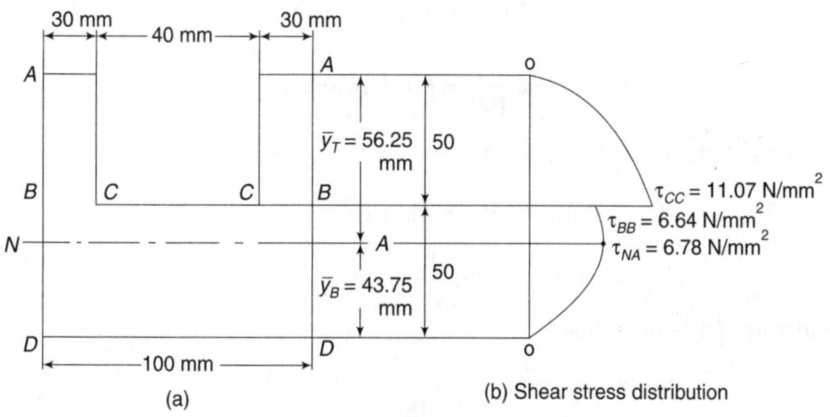

Fig. 5.86

Solution: Load F = 45000 N

Centre of gravity of the channel

Consider full square – Inner rectangle (40 × 50)

$$\bar{y}_B = \frac{A_1y_1 - A_2y_2}{A_1 - A_2} = \frac{100 \times 100 \times 50 - 40 \times 50 \times 75}{100 \times 100 - 40 \times 50}$$

$$\bar{y}_B = 43.75 \text{ mm}$$

$$\therefore \quad \bar{y}_T = 100 - \bar{y}_B = 56.25 \text{ mm}$$

Moment of inertia of the channel about neutral axis

$$I = I_1 - I_2 = (I_{cg} + Ak^2)_1 - (I_{cg} + Ak^2)_2$$

$$= \left\{\frac{100 \times 100^3}{12} + 100 \times 100\ (50 - 43.75^2)\right\} - \left\{\frac{40 \times 50^3}{12} + 40 \times 50\ (75 - 43.75)^2\right\}$$

$$I = 6.354 \times 10^6 \text{ mm}^4$$

Shear stress at the top $\tau_{AA} = 0$

Shear stress at the neck of CC $\tau_{CC} = \frac{F}{I.b}\, a\bar{y}$

$$b = 30 + 30 = 60 \text{ mm}$$

$$a_y = 2 \times (30 \times 50) \times \left(56.25 - \frac{50}{2}\right) = 93750 \text{ mm}^3$$

$$\therefore \quad \tau_{CC} = \frac{45000}{6.354 \times 10^6\ (2 \times 30)} \times 93750 = 11.07 \text{ N/mm}^2$$

Shear stress at the beginning of flange BB,

$$\tau_{BB} = \frac{F}{Ib} \cdot a\bar{y} = \left(\frac{F}{Ib}\, ay\right)\frac{b}{B}$$

$$\tau_{BB} = 11.07 \times \frac{60}{100} = 6064 \text{ N/mm}^2$$

Shear stress at the neutral axis $\tau_{NA} = \frac{F}{Ib} \cdot a\bar{y}$

$$B = 100 \text{ mm of neutral axis}$$

$$a\bar{y} = (100 \times 43.75)\,\frac{43.75}{2}$$

(Considering the bottom side)

$$\therefore \quad \tau_{NA} = \frac{45000}{6.354 \times 10^6 \times 100} \times (100 \times 43.75)\,\frac{43.75}{2} = 6.78 \text{ N/mm}^2$$

Shear stress at the bottom of cross section

$$\tau_{DD} = 0$$

The shear stress distribution across the depth is shown in Fig. 5.86 (b)

Problem 67 Figure 5.87 shows the cross section of a beam dimensioned in mm. Find and sketch the shear stress distribution across the depth of the beam. The beam carries a load of 42 kN.

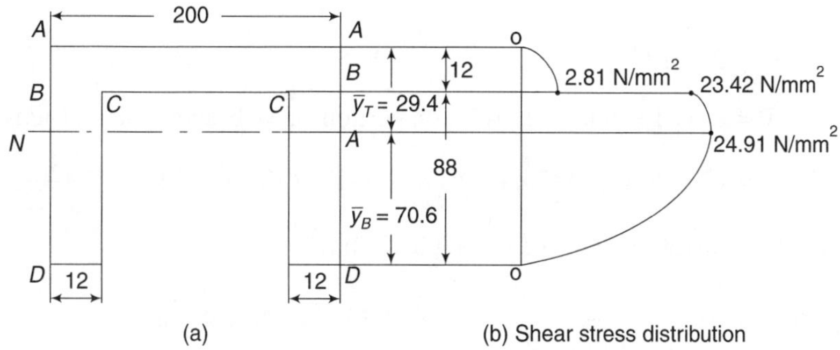

Fig. 5.87

Solution: Load $F = 42000$ N

$$\text{Centre of gravity } \bar{y}_B = \frac{A_1y_1 - A_2y_2}{A_1 - A_2}$$

$$= \frac{200 \times 100 \times 50 - (200 - 24) \times 88 \times 44}{200 \times 100 - (200 - 24) \times 88}$$

$$\bar{y}_B = 70.60 \text{ mm}$$

$$\therefore \quad \bar{y}_T = 100 - 70.60 = 29.40 \text{ mm}$$

Moment of inertia of the channel about neutral axis

$$I = I_1 - I_2 = (I_{cg} + AK^2)_1 - (I_{cg} + AK^2)_2$$

$$= \left\{\frac{200 \times 100^3}{12} + 200 \times 100\,(50 - 70.6)^2\right\}$$

$$- \left\{\frac{(200 - 24) \times 88^3}{12} + (200 - 24)\,88 \times (44 - 70.6)^2\right\}$$

$$= 4.20 \times 10^6 \text{ mm}^4$$

Shear stress at AA, $\tau_{AA} = 0$

Shear stress at the bottom of flange at BB,

$$\tau_{BB} = \frac{F}{Ib}.a\bar{y}$$

$$\tau_{BB} = \frac{42000}{4.2 \times 10^6 \times 200} \times (200 \times 12)\,(29.4 - 6) = 2.81 \text{ N/mm}^2$$

Shear stress at the neck of web at CC,

$$\tau_{CC} = \frac{F}{Ib}.a\bar{y}$$

$$\tau = \tau_{BB}\frac{B}{b} = 2.81 \times \frac{200}{(2 \times 12)} = 23.42 \text{ N/mm}^2$$

At neutral axis at NA,

$$\tau_{NA} = \frac{F}{Ib}\, a\bar{y}$$

$$b = 2 \times 12$$

$$a\bar{y} = a_1\bar{y}_1 \text{ for flange} + a_2\bar{y}_2 \text{ for portion of web above neutral axis}$$

$$= (200 \times 12) \times (29.4 - 6) + 2 \times 12 \times (29.4 - 12) \times \left(\frac{29.4 - 12}{2}\right)$$

$$= 56160 + 3633.12 = 59793.12 \text{ mm}^3.$$

$$\therefore \quad \tau_{\text{NA}} = \frac{42000}{4.2 \times 10^6 \times (2 \times 12)} \times 59793.12 = 24.91 \text{ N/mm}^2$$

At DD at the bottom $\tau_{DD} = 0$

The shear stress distribution is shown in Fig. 5.87(b)

Problem 68 A steel beam 120 mm × 180 mm is simply supported at the ends at 6 m apart. It has a central bore of 60 mm and carries a central load of 300 kN. Find the shear stress induced at the beginning of the bore and at the neutral axis.

Solution: The cross section is shown in Fig. 5.88. For simply supported beam with central load.

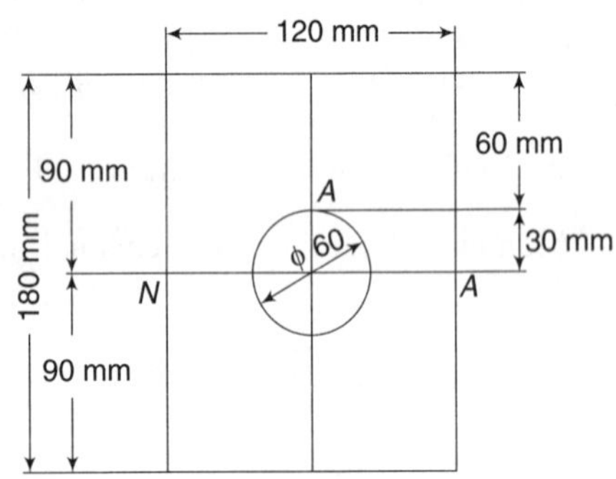

Fig. 5.88

Maximum shear force, $F = \dfrac{W}{2} = \dfrac{300000}{2}$

Due to symmetry, the centre of gravity lies in the middle.

$$\bar{y} = \frac{180}{2} = 90 \text{ mm}$$

Moment of inertia

$$I = I_1 - I_2$$

$$I = \frac{120 \times 180^3}{12} - \frac{\pi}{64} \times 60^4$$

$$= 57.684 \times 10^6 \text{ mm}^4$$

Shear stress at the beginning of circle at A,

$$\tau_A = \frac{F}{Ib}.a\bar{y}$$

b (Full width) 120 mm at A

$$a\bar{y} = 60 \times 120\left(90 - \frac{60}{2}\right) = 432000 \text{ mm}^3$$

$$\therefore \qquad \tau_A = \frac{150000}{120 \times 57.684 \times 10^6} \times 432000 = 9.36 \text{ mm}^2$$

Shear stress at the neutral axis

$$\tau_{NA} = \frac{F}{Ib} \cdot a\bar{y}$$

$$b = 120 - 60 = 60 \text{ mm}$$

$$a\bar{y} = a\bar{y} \text{ for the rectangle above } NA \ (120 \times 90) - a\bar{y} \text{ for semicircle.}$$

$$= 120 \times 90 \times \frac{90}{2} - \frac{\pi}{8} \times 60^2 \times \left(\frac{4 \times 30}{3\pi}\right) = 450000 \text{ mm}^3$$

(For semicircle, CG is at $\frac{4r}{3\pi}$ from diameter line)

$$\therefore \qquad \tau_{NA} = \frac{150000}{57.684 \times 10^6 \times 60} \times 450000 = 19.50 \text{ N/mm}^2$$

Shear stress distribution for beams of various cross sections

Figure. 5.89 shows the shear stress distribution of beams of some standard cross sections.

Fig. 5.89 Shear stress distribution for standard cross sections

PROBLEMS ON BENDING AND SHEAR STRESS AS IN BEAMS

Problem 69 A 100 mm × 200 mm simply supported beam carries a central load W. The permissible stresses in bending and shear are 15 N/mm^2 and 1.2 N/mm^2 respectively. Determine the safe load W, if the span of the beam is 3 m.

Solution: Allowable stress in bending

$$\sigma = 15 \text{ N/mm}^2$$

In shear $\tau = 1.2 \text{ N/mm}^2$

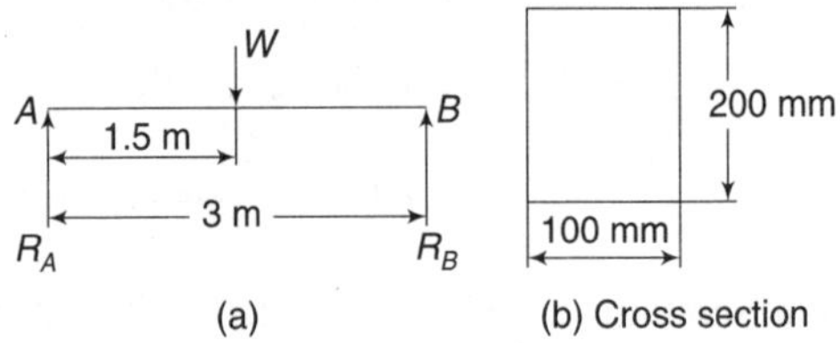

Fig. 5.90

For simply supported beam with central load

$$\text{Reactions: } R_A = R_B = \frac{W}{2}$$

Maximum bending moment

$$M = \frac{Wl}{4} = \frac{W \times 3000}{4} = 750\ W$$

$$I = \frac{bd^3}{12} = \frac{100 \times 200^3}{12} = 66.6667 \times 10^6 \text{ mm}^4$$

From bending

$$\frac{M}{I} = \frac{\sigma}{y}$$

$$\frac{750\ W}{66.6667 \times 10^6} = \frac{15}{\left(\frac{200}{2}\right)}$$

$\therefore$ Load $W = 13333.33$ N

For rectangle, the maximum shear stress occurs at the neutral axis and is equal to 1.5 τ_{avg}.

$$\tau_{max} = 1.5\ \tau_{avg}$$

$$1.2 = 1.5 \cdot \frac{F}{A}$$

Maximum shear force

$$F = R_A = R_B = \frac{W}{2}$$

$$1.2 = 1.5 \times \frac{\left(\frac{W}{2}\right)}{100 \times 200}$$

$\therefore$ Load $W = 32000$ N

$\therefore$ Safe central load $W = 13333.33$ N

(Smaller of the above two values)

Problem 70 A beam of an I-section consists of 180 mm × 15 mm flanges and a web of 280 mm × 15 mm. It is subjected to a bending moment of 120 kN-m and a shear force of 60 kN. Sketch the bending and shear stress distribution along the depth of the section.

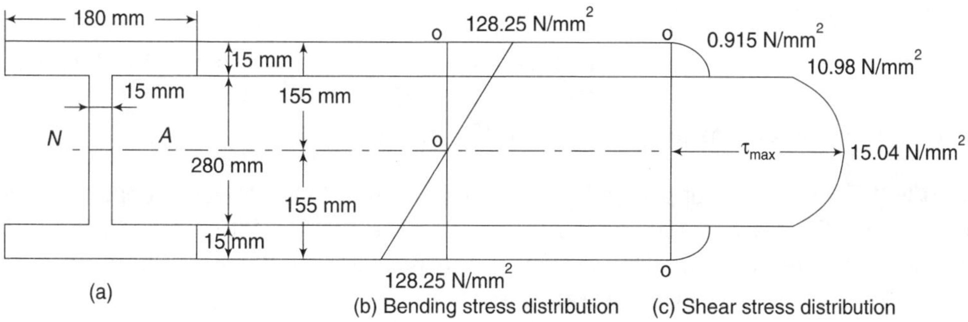

Fig. 5.91

Solution: Due to symmetry, CG lies in the middle

$$\bar{y} = \left(\frac{280 + 2 \times 15}{2}\right) = 155 \text{ mm.}$$

Moment of inertia about neutral axis

$$I = \frac{BH^3 - bh^3}{12} = \frac{180 \times 310^3 - (180 - 15) \times 280^3}{12}$$

$$= 145.025 \times 10^6 \text{ mm}^4$$

Bending moment $M = 120 \text{ kN-m} = 120 \times 10^6 \text{ N-mm}$

Shear force $F = 60 \text{ kN} = 60 \times 10^3 \text{ N}$

From bending equation $\frac{M}{I} = \frac{\sigma}{y}$

Bending stress $\sigma = \frac{M}{I}.y = \frac{120 \times 10^6 \times 155}{145.025 \times 10^6} = 128.25 \text{ N/mm}^2$

This is tensile at one side and compressive at the other side of neutral axis.

Bending stress at neutral axis = 0

The bending stress distribution is shown in Fig. 5.91(b)

Shear stress at the top of flange = 0

Shear stress at the bottom of flange at the neck

$$= \frac{F}{8I}(D^2 - d^2)$$

$$= \frac{60000\,(310^2 - 280^2)}{8 \times 145.025 \times 10^6} = 0.915 \text{ N/mm}^2$$

Shear stress at the neck of web

$$= \frac{B}{b}\left[\frac{F}{8I}(D^2 - d^2)\right] = \frac{180}{15} \times 0.915 = 10.98 \text{ N/mm}^2$$

At the neutral axis,

$$= \frac{B}{b}\frac{F}{8I}(D^2 - d^2) + \frac{F}{8I}d^2$$

$$= 10.98 + \frac{60000}{8 \times 145.025 \times 10^6} \times 280^2 = 15.04 \text{ N/mm}^2$$

The shear stress distribution is shown in Fig. 5.91(c)

Problem 71 A simply supported beam of I-section 6 m long between supports carries a UDL. The flanges are 100 mm × 20 mm and the web is 150 mm × 12 mm. The allowable tensile, compressive and shear stresses are 90 N/mm^2, 120 N/mm^2 and 30 N/mm^2 respectively. Find the safe intensity of UDL that can be applied.

Solution:

$$\sigma_T = 90 \text{ N/mm}^2$$

$$\sigma_C = 120 \text{ N/mm}^2$$

$$\tau_{max} = 30 \text{ N/mm}^2$$

Let UDL = w N/m run.

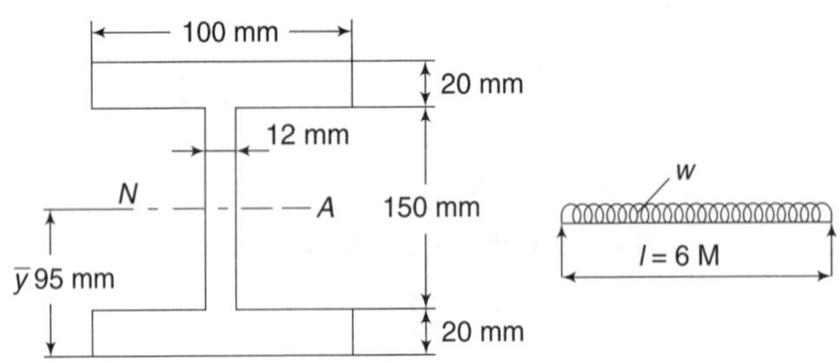

Fig. 5.92

The I-section is as shown in Fig. 5.92

$$\bar{y} = \frac{190}{2} = 95 \text{ mm}$$

Moment of inertia $\quad I = \frac{BH^3 - bh^3}{12} = \frac{100 \times 190^3 - (100 - 12) \times 150^3}{12}$

$$= 32.408 \times 10^6 \text{ mm}^4$$

Consider bending $\quad \frac{M}{I} = \frac{\sigma}{y}$

For simply supported beam with full UDL, the maximum bending moment

$$M = \frac{wl^2}{8} = \frac{w \times 6^2}{8} \text{ N-m}$$

$$= 4.5\ w \times 1000 \text{ N-mm} = 4500 \text{ w.}$$

Considering tensile stress condition.

$$\frac{4500\ w}{32.408 \times 10^6} = \frac{60}{95}$$

$$\therefore \qquad w = 4548.49 \text{ N/m}$$

Considering compressive stress condition

$$\frac{4500\ w}{32.408 \times 10^6} = \frac{90}{95}$$

$$\therefore \qquad w = 6822.74 \text{ N/m}$$

For simply supported beam with full UDL.

The maximum shear stress $F = \frac{wl}{2}$ at the supports

$$F = \frac{w \times 6}{2} = 3\ w \text{ N}$$

For I-section maximum shear stress occurs at neutral axis and equal to

$$= \frac{B}{b}\frac{F}{8I}(D^2 - d^2) + \frac{F}{8I}d^2$$

$$30 = \frac{100}{12} \times \frac{F}{8 \times 32.408 \times 10^6}(190^2 - 150^2) + \frac{F}{8 \times 32.408 \times 10^6} \times 150^2$$

$$\therefore \qquad F = 3\ w = 57260.8 \text{ N}$$

$$\therefore \qquad w = 19086.92 \text{ N/m}$$

$\therefore$ Maximum intensity of UDL = Minimum of the above three values

$$w = 4548.49 \text{ N/m.}$$

Problem 72 A T-section is formed by cutting the bottom *F* lange of a I-section. The *F* lange are 100 mm × 20 mm and the web is 150 mm × 20 mm. Draw the bending stress and shear stress diagrams if the bending moment and shear force at a section are 10 kN-m (hogging) and 50 kN respectively.

Solution: Bending moment

$$M = 10 \text{ kN-m} = 10 \times 10^6 \text{ N-mm}$$

Shear force $\qquad F = 50 \text{ kN} = 50 \times 10^3 \text{ N}$

For hogging moment, top fibres experience tensile stress and bottom compressive.

The T-section is shown in Fig. 5.93 (a)

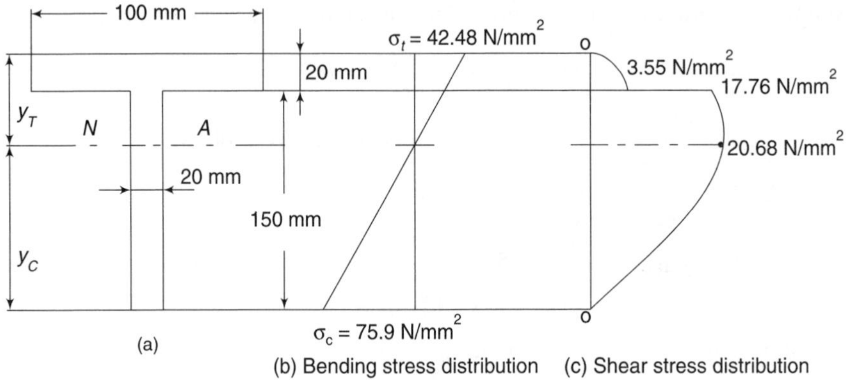

Fig. 5.93

Centre of gravity

$$y_C = \frac{A_1y_1 + A_2y_2}{A_1 + A_2} = \frac{150 \times 20 \times 75 + 100 \times 20 \times (150 + 10)}{150 \times 20 + 100 \times 20}$$

$$y_C = 109 \text{ mm (from bottom)}$$

$$\therefore \quad y_T = 170 - 109 = 61 \text{ mm}$$

Bending stress: Equation of bending, $\dfrac{M}{I} = \dfrac{\sigma}{y}$

$$\therefore \quad \text{Bending stress } \sigma = \frac{M}{I}\cdot y$$

Moment of inertia $\quad I = I_1 + I_2 = (I_{CG} + AK^2)_1 + (I_{CG} + AK^2)_2$

$$= \left\{\frac{100 \times 20^3}{12} + 100 \times 20\,(61 - 10)^2\right\} + \left\{\frac{20 \times 150^3}{12} + 20 \times 150\,(109.75)^2\right\}$$

$$I = 14.36 \times 10^6 \text{ mm}^4$$

Tensile stress at the top due to bending

$$\sigma_t = \frac{M}{I}\cdot y_t = \frac{10 \times 10^6}{14.36 \times 10^6} \times 61 = 42.48 \text{ N/mm}^2$$

Compressive stress at the bottom due to bending

$$\sigma_C = \frac{M}{I}\cdot y_C = \frac{10 \times 10^6}{14.36 \times 10^6} \times 109 = 75.9 \text{ N/mm}^2$$

The bending stress distribution is shown in Fig. 5.93 (b)

Shear stress: Shear stress at the top = 0

Shear stress at the bottom of the *F* lange

$$\tau = \frac{F}{IB} \cdot a\bar{y}$$

$$= \frac{50 \times 10^3}{14.36 \times 10^6 \times 100} (100 \times 20) \times (61 - 10) = 3.55 \text{ N/mm}^2$$

Shear stress at the neck of web

$$\tau = \frac{F}{Ib} a\bar{y} = \frac{F}{Ib} \cdot a\bar{y} \times \frac{B}{b}$$

$$= 3.55 \times \frac{100}{20} = 17.76 \text{ N/mm}^2$$

Maximum shear stress at neutral axis

$$\tau = \frac{F}{Ib} a\bar{y}$$

$$= \frac{50 \times 10^3}{14.36 \times 10^6 \times 20} \left[100 \times 20 \times (61 - 10) + 20 (61 - 20) \left(\frac{61 - 20}{2}\right)\right]$$

$$= 20.68 \text{ N/mm}^2$$

Shear stress at the bottom = 0

The shear stress distribution is shown in Fig. 5.93 (c).

Problem 73 A simply supported beam of circular cross section of length l carries a concentrated load W at one fourth of its span from the support. Find the ratio of length to its depth such that both bending and shear stresses read maximum allowable values simultaneously. The maximum bending is 5 times that maximum shear stress.

Solution: The reaction are

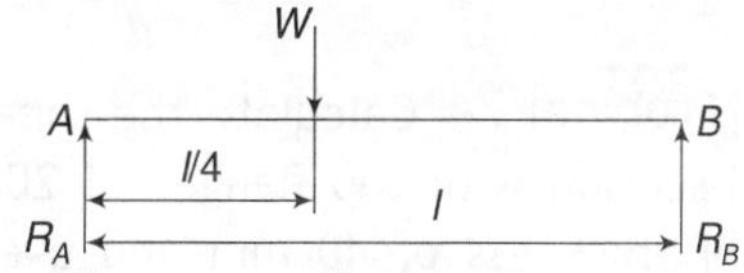

Fig. 5.94

$$R_A = \frac{W \cdot \left(\frac{3}{4} l\right)}{l} = 0.75 \ W$$

$$R_B = \frac{W \cdot \left(\frac{1}{4} \cdot l\right)}{l} = 0.25 \ W$$

The maximum shear force $F = 0.75 \ W$

(Maximum of the reactions)

Bending moment

$$M = R_A \cdot \frac{l}{4} = 0.75 \ W \times \frac{l}{4}$$

$$M = 0.1875 \ Wl$$

$$\text{Bending equation} = \frac{M}{I} = \frac{\sigma}{y}$$

$$I = \frac{\pi d^4}{64} \text{ and } y = \frac{\text{Depth}}{2} = \frac{d}{2}$$

$\therefore$ $$\text{Bending stress } \sigma = \frac{M \cdot y}{I}$$

$$\sigma = \frac{0.1875\ Wl}{\left(\frac{\pi d^4}{64}\right)} \cdot \frac{d}{2} = \frac{1.91\ Wl}{d^3}$$

For circular cross section, maximum shear stress

$$\tau = \frac{4}{3}\frac{F}{A} = \frac{4}{3} \times \frac{0.75\ W}{\left(\frac{\pi d^2}{4}\right)}$$

$$\tau = 1.273\ \frac{W}{d^2}$$

$\therefore$ $$\frac{\sigma}{\tau} = \frac{\left(\frac{191\ Wl}{d^3}\right)}{\left(\frac{1.273\ W}{d^2}\right)} = \frac{1.5\ l}{d}$$

But it is given that bending stress is 5 times the shear stress, i.e., $\sigma = 5\tau$

$$\frac{\sigma}{\tau} = 5$$

$\therefore$ $$\frac{\sigma}{\tau} = 5 = 1.5\ \frac{l}{d}$$

$\therefore$ $$\frac{l}{d} = 7.5.$$

Problem 74 Calculate the shear stress and bending stress values for an unsymmetric I-section with top flange of 200 × 30 mm, bottom flange of 150 × 24 mm, the web of thickness of 40 mm and over all height of section of 140 mm when it carries a shear force of 42 kN and a bending moment of 69 kNm. Draw the distribution of the stresses.

Solution: Figure 5.95 (a) shows the across section.

Centre of gravity

$$y_B = \frac{A_1X_1 + A_2X_2 + A_3X_3}{A_1 + A_2 + A_3}$$

$$= \frac{200 \times 30(140 - 15) + (86 \times 40)\left(24 + \frac{86}{2}\right) + 150 \times 24 \times (12)}{(200 \times 30 + 86 \times 40 + 150 \times 24)}$$

$$= 78.50 \text{ mm}$$

$\therefore$ $$y_T = 140 - yB = 140 - 78.50 = 61.50 \text{ mm}$$

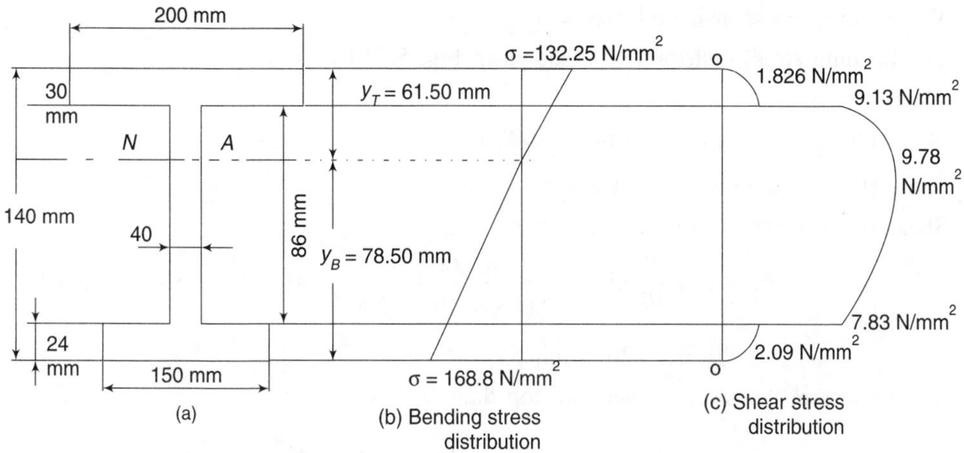

Fig. 5.95

Moment of inertia

$$I = I_1 + I_2 + I_3$$

$$I_1 = \frac{bh^3}{12} + AR^2 = \frac{200 \times 30^3}{12} + 200 \times 30\,(78.5 - 125)^2$$

$$= 13.42 \times 10^6 \text{ mm}^4$$

$$I_2 = \frac{40 \times 80^3}{12} + 40 \times 86\,(78.5 - 67)^2 = 2.575 \times 10^6 \text{ mm}^4$$

$$I_3 = \frac{150 \times 24^3}{12} + 150 \times 24\,(78.5 - 12)^2 = 16.093 \times 10^6 \text{ mm}^4$$

$$\therefore \quad I = I_1 + I_2 + I_3 = 32.088 \times 10^6 \text{ mm}^4$$

Bending stress

WKT

$$\frac{M}{I} = \frac{\sigma}{y}$$

$$\therefore \quad \text{Bending stress } \sigma = \frac{M \cdot y}{I}$$

$$M = \text{Bending moment} = 69 \text{ kN-m} = 69 \times 10^6 \text{ N-mm}$$

For top fibres

$$\sigma_T = \frac{M \cdot y_T}{I} = \frac{69 \times 10^6 \times 61.5}{32.088 \times 10^6} = 132.25 \text{ N/mm}^2$$

For top bottom fibres

$$\sigma_B = \frac{M \cdot y_B}{I} = \frac{69 \times 10^6 \times 78.5}{32.088 \times 10^6} = 168.8 \text{ N/mm}^2$$

The bending stress at neutal axis = 0

The bending stress distribution is shown in Fig. 5.94(b).

Shear stresses

Shear Force $F = 42 \text{ kN} = 42000 \text{ N}$

Shear stress at the top of flange = 0

Shear stress at the bottom of top flange

$$\tau = \frac{F}{IB} a\bar{y} = \frac{42000}{32.088 \times 10^6 \times 200} (200 \times 30) \times (61.5 - 15)$$

$$= 1.826 \text{ N/mm}^2$$

Shear stress at the neck of web and top flange.

$$= \frac{F}{Ib} \cdot a\bar{y} = \left(\frac{F}{IB} \cdot a\bar{y}\right) \frac{B}{b} = 1.826 \times \frac{200}{40} = 9.13 \text{ N/mm}^2$$

Shear stress at neutral axis $= \frac{F}{Ib} \cdot a\bar{y}$

$a\bar{y} = a_1 \bar{y}_1$ for top flange + $a_2 \bar{y}_2$ for portion of web above neutral axis

$$= 200 \times 30 \times (61.5 - 15) + 40(61.5 - 30) \frac{(61.5 - 30)}{2}$$

$$= 298845 \text{ mm}^3$$

$$\therefore \quad \tau = \frac{42000}{32.088 \times 10^6 \times 40} \times 298845 = 9.78 \text{ N/mm}^2$$

Shear stress at the neck of web and bottom flange

$$\tau = \frac{F}{Ib} \cdot a\bar{y}$$

$$= \frac{42000}{32.088 \times 10^6 \times 40} (150 \times 24)(78.5 - 12) = 7.83 \text{ N/mm}^2$$

Shear stress at the top of bottom flange

$$\tau = \frac{F}{IB} \cdot a\bar{y} = \frac{F}{Ib} . a\bar{y} \left(\frac{b}{B}\right)$$

$$= 7.83 \times \frac{40}{150} = 2.09 \text{ N/mm}^2$$

Shear stress at the bottom of bottom flange = 0

Shear stress distribution is shown in Fig. 5.95 (c).

Problem 75 A cantilever beam of span 1.8 m has rectangular cross section 120 mm wide and 200 mm depth. Find the safe point load at the free end if the maximum shear stress is not to exceed 3 N/mm^2. Also find the maximum bending stresses induced in the beam.

Solution: **Given data:**

$$\text{Maximum shear stress} = 3 \text{ N/mm}^2$$
$$\text{Length of beam } l = 1.8 \text{ m} = 1800 \text{ mm}$$
$$\text{Cross section width} = b = 120 \text{ mm}$$
$$\text{Depth} = d = 200 \text{ mm.}$$

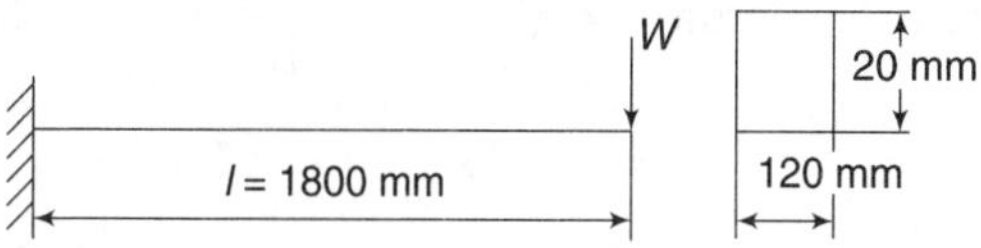

Fig. 5.96

For rectangular cross section

$$\bar{y} = \frac{200}{2} = 100 \text{ mm}$$

$$I = \frac{bd^3}{12} = \frac{120 \times 200^3}{12} = 80 \times 10^6 \text{ mm}^4$$

$$\text{Maximum shear stress } \tau_{max} = 1.5\ \tau_{avg}$$

$$3 = 1.5 \frac{F}{A}$$

$$\therefore \quad F = \frac{3A}{1.5} = \frac{3 \times 120 \times 200}{1.5} = 48000 \text{ N}$$

For cantilever max shear force

$$F = W = 48000 \text{ N}$$

Maximum bending moment

$$M = W \cdot l = 48000 \times 1800$$

$$M = 86.4 \times 10^6 \text{ Nmm}$$

$$\text{For bending } \frac{M}{I} = \frac{\sigma}{C}$$

$$\therefore \quad \text{Bending stress } \sigma = \frac{M \cdot y}{I} = \frac{86.4 \times 10^6}{80 \times 10^6} \times 100$$

$$\sigma = 108 \text{ N/mm}^2.$$

Problem 76 A beam of rectangular cross section 150 mm wide, 240 mm deep is simply supported at the ends over a span of 9 m. Find the intensity of uniformly distributed load that the beam can carry if the bending and shear stresses are limited to 90 N/mm^2 and 18 N/mm^2 respectively.

Solution:

$$\text{Allowable bending stress } \sigma_{max} = 90 \text{ N/mm}^2$$

$$\text{Allowable shear stress } \tau_{max} = 18 \text{ N/mm}^2$$

$$\text{For rectangular cross section, } \bar{y} = \frac{d}{2} = \frac{240}{2} = 120 \text{ mm}$$

$$\text{Moment of inertia } I = \frac{bd^3}{12} = \frac{150 \times 240^3}{12} = 172.8 \times 10^6 \text{ mm}^4$$

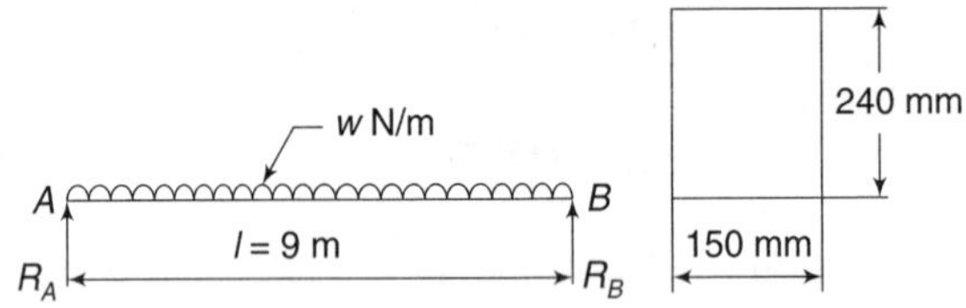

Fig. 5.97

For simply supported beams with full UDL as shown in Fig. 5.97, maximum shear force

$$F = R_A = R_B = \frac{wl}{2}$$

$$F = \frac{w \times 9}{2} = 4.5 \ w \text{ N}$$

For rectangular cross section × Maximum shear

$$\tau_{max} = 1.5 \ \tau_{avg} = 1.5 \ \frac{F}{A}$$

$$18 = 1.5 \times \frac{4.5 \ w}{(bd)} = 1.5 \times \frac{4.5 \ w}{(150 \times 240)}$$

∴ Intensity of UDL $w = 96000$ N/m

For simply supported beam with full UDL.

$$\text{Maximum bending moment } M = \frac{wl^2}{8} = \frac{w \times 9^2}{8} \text{ Nm}$$

$$M = \frac{w \times 81 \times 10^3}{8} \text{ N-mm}$$

$$\text{Equation of bending } \frac{M}{I} = \frac{\sigma}{y}$$

∴
$$\frac{\left(\dfrac{w \times 81 \times 10^3}{8}\right)}{172.8 \times 10^6} = \frac{90}{120}$$

∴ Intensity of UDL $w = 12800$ N/m

∴ Safe UDL that can be applied on the beam,

$$w = 12800 \text{ N/m} = 12.8 \text{ kN/m. (smaller one)}$$

Problem 77 A I-section has flange of width b and over all depth $2b$. The thickness of flange and web and each t. Find the ratio of maximum shear stress intensity to mean shear stress intensity. t is small with respect to b.

Solution:

Figure 5.98 shows the given I-section.

Overall width $B = b$

Overall depth $D = 2b$

Height of web $d = 2b - 2t = 2(b - t)$

Thickness of flange $= t$

Thickness of web $= t$

CG lies in the middle $\bar{y} = \frac{2b}{2} = b$

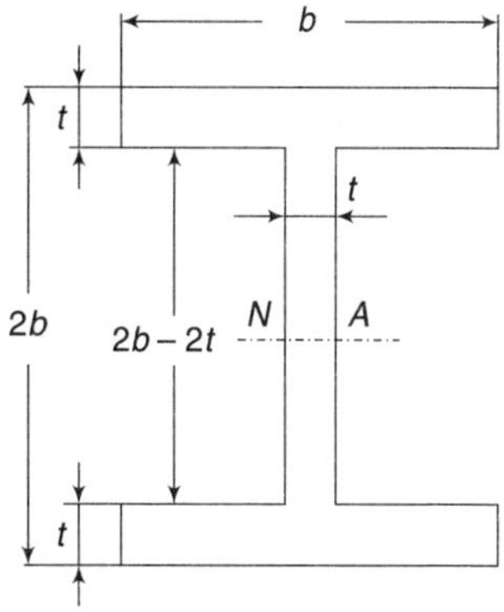

Fig. 5.98

For I section, moment of inertia $I = \frac{BD^3 - bd^3}{12}$

$$= \frac{b(2b)^3 - (b - t)(2b - 2t)^3}{12} = \frac{8b^4 - 8(b - t)^4}{12}$$

$$= \frac{8b^4 - 8(b - t)^2 (b - t)^2}{12}$$

$$= \frac{8b^4 - 8(b^2 + t^2 - 2bt)(b^2 + t^2 - 2bt)}{12}$$

$$= \frac{8b^4 - 8(b^4 + b^2t^2 - 2b^3t + b^2t^2 + t^4 - 2bt^3 - 2b^3t - 2bt^3 + 4b^2t^2)}{12}$$

Since t is small with respect to b, neglect the higher power of t,

$$I = \frac{8b^4 - 8(b^4 - 4b^3t)}{12} = \frac{8b^4 - 8b^4 + 32b^3t}{12}$$

$$I = \frac{32}{12} b^3t = \frac{8}{3} b^3t$$

Maximum shear stress at neutral axis

$$\tau_{max} = \frac{F}{Ib} \cdot a\bar{y}$$

$a\bar{y} = a_1\bar{y}_1$ for flange + $a_2\bar{y}_2$ for web above neutral axis

$$= bt\left(b - \frac{t}{2}\right) + (b - t)\, t\frac{(b - t)}{2}$$

$$= b^2t - \frac{bt^2}{2} + \frac{t}{2}(b^2 + t^2 - 2bt)$$

$$= b^2t - \frac{bt^2}{2} + \frac{b^2t}{2} + \frac{t^3}{2} - bt^2$$

Neglecting the higher powers of t

$$a\bar{y} = \frac{3b^2t}{2} = 1.5\ b^2t$$

$$\therefore \qquad \tau_{max} = \frac{F}{\left(\frac{8}{3}\ b^3t\right) \times t} \times 1.5b^2t = \frac{9}{16}\left(\frac{F}{bt}\right)$$

Mean shear stress $\tau_{mean} = \dfrac{F}{A}$

$$A = \text{Area of I-section} = 2 \times b \times t + (2b - 2t)\ t$$

$$= 2bt + 2bt - 2t^2 = 4bt$$

(Neglecting higher power of t)

$$\therefore \qquad \tau_{mean} = \frac{F}{4bt}$$

$$\therefore \qquad \text{Ratio } \frac{\tau_{max}}{\tau_{mean}} = \frac{\left(\frac{9}{16}\frac{F}{bt}\right)}{\left(\frac{F}{4bt}\right)} = \frac{9}{4} = 2.25$$

The maximum shear stress is 2.25 times the mean shear stress.

THINGS TO REMEMBER

- Simple bending equation is $\dfrac{M}{I} = \dfrac{\sigma}{y} = \dfrac{E}{R}$
- Moment of resistance of a section $M = \dfrac{\sigma.I}{y}$ or $= \sigma Z$
- Section modulus $Z = \dfrac{I}{y}$
- Section modulus of rectangular section $= \dfrac{bd^2}{6}$
- Hollow rectangular section $= \dfrac{BD^2 - bd^3}{6D}$
- Circular section $= \dfrac{\pi}{32}\ d^3$
- Hollow circular section $= \dfrac{\pi}{32}\ \dfrac{(D^4 - d^4)}{D}$
- Triangular section $= \dfrac{bh^2}{24}$
- The dimensions of the strongest rectangular beam that can be cut from a circular log are $\sqrt{\dfrac{2}{3}}\ d$ and $\dfrac{d}{\sqrt{3}}$

- For a flitcher beam made of two materials having modulus of elasticity as E_1 and E_2 may be replaced by an equivalent beam of material 1 for which the width of 2nd materials is replaced by $\frac{E_2}{E_1}$ times the width of 2nd material

$$b_{1eq} = b_2 = \frac{E_2}{E_1}$$

- Modular ratio is the ratio of modulii of elasticity of the two materials

$$m = \frac{E_1}{E_2} \text{ or } \frac{E_2}{E_1}$$

- Moment of resistance of a flitcher beam is the sum of moment of resistance of individual beams
- Shear stress at a distance y from neutral axis is given by $\tau = \frac{F}{bI} a\bar{y}$
  where, F = shear force, b = width section at neutral axis.
  I = moment of inertia of entire section and $a\bar{y}$ is the moment of area of the portion above neutral axis
- Maximum shear stress is standard section and the location of maximum shear stress are:

Section	τ_{max}	*Location*
Rectangular	$1.5\ \tau_{avg} = 1.5\ \frac{F}{A}$	At neutral axis
Circular	$\frac{4}{3}\tau_{avg} = \frac{4}{3}\frac{F}{A}$	At neutral axis
Triangular	$\frac{4}{3}\tau_{avg} = \frac{4}{3}\frac{F}{A}$	At neutral axis
	$1.5\ \tau_{avg} = 1.5\ \frac{F}{A}$	At the middle of height
Square section with one of the diagonals horizontal	$\frac{9}{8}\tau_{avg} = \frac{9}{8}\frac{F}{A}$	At a height of $\frac{3a}{4\sqrt{2}}$ or $\frac{3}{8}$ height
Hexagonal section with two sides horizon-tal	$1.2\ \tau_{avg}$	At neutral axis

- For symmetric I-section of width B, total depth D, width of web b and height of web d
- Shear stress at bottom of flange, $\tau = \frac{F}{8I}(D^2 - d^2)$
- Shear stress at the neck of web and flange $\tau = \frac{B}{b} \cdot \frac{F}{8I}(D^2 - d^2)$
- Shear stress at the neutral axis $\tau = \frac{B}{b}\frac{F}{8I}(D^2 - d^2) + \frac{F}{8I}d^2$

In I-section, majority of shear stress is taken by web.

EXERCISES

1. A simply supported beam of span 6 m has a rectangular cross section 200 mm wide and 300 mm deep. The allowable stress is 90 MPa. Find
   (a) The safe central load
   (b) The intensity of UDL that can be carried.
2. A hollow beam of rectangular cross section 100 mm wide 150 mm deep has a thickness of 10 mm. It is fixed at one end carries a point load of 180 kN at a distance of 3 m from the fixed end. Find the maximum tensile and compressive stresses induced. Sketch the stress distribution.
3. A symmetric I-section of 100 mm × 150 mm × 10 mm of steel having allowable stress of 120 MPa is simply supported at the ends over a span of 3 m. Find the safe UDL it can carry and the moment of resistance.
4. A cantilever beam of unsymmetric I-section with 60 × 20 top flange, 100 × 25 bottom flange and 150 × 20 web is 3 m long. If the stress is not to exceed 120 MPa, find the safe point load that can be applied at its free end.
5. A simply supported 3 m long beam of has a hollow rectangular cross section as shown in Fig. 5.99. Find the safe UDL over the entire span that can be carried, if the allowable stress is 90 MPa.

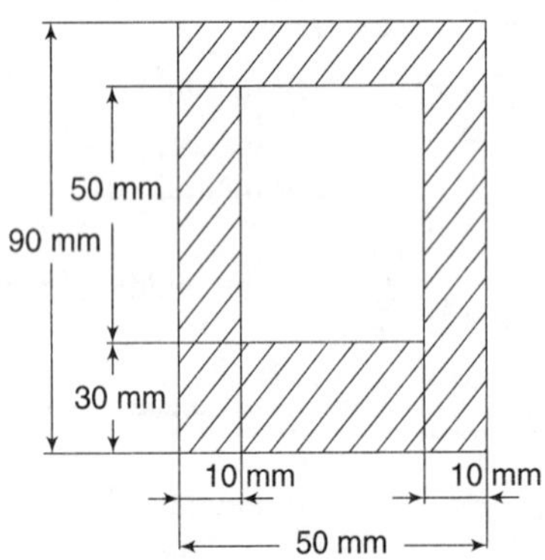

Fig. 5.99

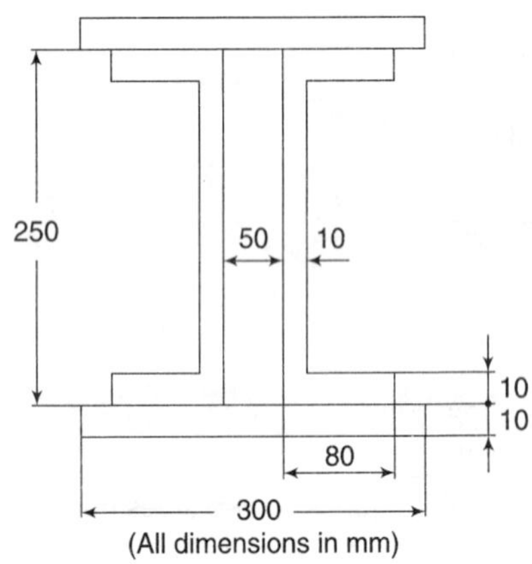

Fig. 5.100

6. Figure 5.100 shows a beam consisting of two identical channels connected with two equal size cover plates. Find the moment of resistance if the allowable stress is 120 MPa.
7. A channel section has an overall depth of 200 mm with each flange of 100 mm, web thickness 10 mm and flange thickness of 12 mm is used as a simply supported beam of 5 m length with the web at the top. Find the maximum central load that can be applied limiting the allowable tensile and compressive stresses to 120 MPa and 90 MPa respectively.
8. Three beams of circular, square and rectangular cross section, have same length, same allowable bending stress and carry same bending moment. The depth of rectangle is twice

its width. Compare the weight of circular and rectangular cross section with square section beam.

9. Determine the ratio of moment of resistance in planes *YY* and *XX* for a section of beam shown in Fig. 5.101

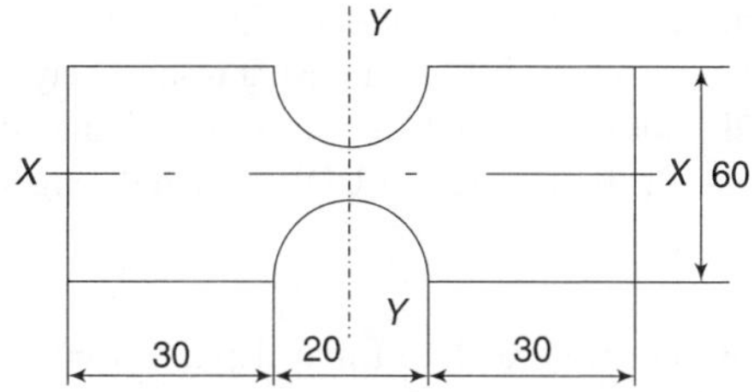

Fig. 5.101

10. A wooden beam 40 mm wide and 150 mm deep is to be reinforced with two plates each 12 mm thick and 90 mm wide. Determine the moment of resistance of the flitched beam when

(a) the plates are attached on either side, and

(b) the plates are attached at the top and bottom.

11. A flitcher beam is made of a wooden section 90 mm wide and 150 mm deep, reinforced at the bottom by a steel plate 90 mm wide and 12 mm thick. The beam is simply supported at the ends over a span of 5 m and carries a UDL of 5 kN/m over the entire span. Find the bending stresses developed in wood and steel. Take $E_S = 2 \times 10^5$ MPa and $E_W = 1 \times 10^4$ MPa.

12. A flitcher beam is made up of two timber sections each 90 mm wide 150 mm deep with a steel plate of 15 mm thick and 110 mm deep placed centrally between them and is firmly attached to them. Calculate the moment of resistance of the flitcher beam if the stresses in timber is to be limited to 7 N/mm^2. Find the corresponding stress in steel. Take $E_T = 1.1 \times 10^4$ N/mm^2, $E_S = 2.1 \times 10^5$ N/mm^2.

13. A timber beam of rectangular cross section 90 mm × 150 mm is strengthened all around with steel plates of 10 mm thick. Determine the safe UDL this beam carry if it is 6 m long and is simply supported at the ends. The stress in steel is not to exceed 120 N/mm^2. Take $\frac{E_S}{E_T} = 20$.

14. A flitcher beam consists of a wooden beam 90 mm wide and 150 mm deep strengthened by steel plates of 10 mm thick and 110 mm deep on either side of wood and placed symmetrically. The stresses in wood and steel are not to exceed 6 N/mm^2 and 100 N/mm^2, find the moment of resistance of the combined beam. Take $E_S = 20\ E_W$.

15. A wooden round bar of diameter 150 mm is surrounded by fitting a steel tube of external diameter 200 mm. If the bending stresses in wood and steel are not to exceed 7 N/mm^2 and 120 N/mm^2, find the moment of resistance. Take $E_S = 20\ E_W$

16. In a I-section the width of each flange is *b* and total depth is also *b*. The thickness of flange and web is *t* which is small with respect to *b*. Determine the ratio of maximum shear stress to the average shear stress.

$$\left(\textbf{\textit{Ans.}}\ \frac{\tau_{\max}}{\tau_{\text{avg}}} = 3.2\right)$$

17. A T-section has flange of 100 mm × 12 mm and the web of 88 mm × 12 mm. Draw the shear stress distribution across the depth indicating values. The beam is subjected to a shear force of 36 kN.

(***Ans.*** τ_f = 482 N/mm^2, τ_w = 40.12 N/mm^2, τ_{NA} = 42.71 N/mm^2)

18. A beam having rectangular cross section is simply supported over a span of l. If the maximum bending and shear stresses are 30 MPa and 2 MPa respectively. Determine the ratio of length of depth assuming the maximum bending stresses and maximum shear stress will occur simultaneously. The beam is subjected to UDL over full length.

$\left(\textbf{\textit{Ans.}}\ \frac{l}{d} = 15\right)$

19. A I-section beam 150 mm wide and 350 mm depth has both flanges of 25 mm thick and web of 12 mm thick. It is simply supported over a span of 4 metres and carries a UDL of 80 kN/m. Determine the magnitude of bending and shear stress in the flange at the junction of top flange and web at 1/4th of its span.

(***Ans.*** σ = 79.85 N/mm^2 (Comp) and τ 1.44 N/mm^2)

20. An I-section beam of each flange as 190 mm × 25 mm and the web of 450 mm × 15 mm carries a concentrated load of 800 kN at its middle. It is simply supported at the ends over a span of 3 m. Calculate the maximum shear stress induced.

(***Ans.*** = τ_{max} = 62.36 N/mm^2)

21. A cantilever beam of rectangular cross section 100 mm wide and 200 mm deep is 2 m long. Determine the concentrated load at the free end if the shear stress is not to exceed 15 N/mm^2. Also find the bending stresses induced in the section at the fixed end.

(***Ans.*** W = 200000 N, σ_{max} = 600 N/mm^2)

22. A simply supported beam is 3 m long and of rectangular cross section 120 mm wide 240 mm deep. Find the intensity of UDL that can be safely applied if the bending and shear stresses are not to exceed 30 N/mm^2 and 2.7 N/mm^2 respectively.

(***Ans.*** UDL w = 30.72 kN/m)

23. A symmetrical I-section of 200 mm wide and 500 mm deep has each flanges of 25 mm thick and web of 20 mm thick. It is subjected to a shear force of 90 kN. Find the shear force resisted by the flange and web. Also express the forces as percentage of total force.

(***Ans.*** F_F = 3.8 kN and 4.2%, F_N = 86.2 kN and 95.8%)

24. The unsymmetric I-section has the top flange of 160 mm × 20 mm, web of 200 mm × 20 mm and bottom flange of 80 mm × 20 mm carries a shear force of 100 kN. Find the shear stress induced at critical points across the depth.

(***Ans.*** $\tau_{\text{top flange}}$ = 2.65 N/mm^2, $\tau_{\text{Top flangeneck}}$ = 21.16 N/mm^2, τ_{NA} = 25.86 N/mm^2,
$\tau_{\text{bottom flangeneck}}$ = 15.26 N/mm^2, $\tau_{\text{bottom flange}}$ = 3.81 N/mm^2)

25. A steel beam of I-section of width 200 mm and total depth of 600 mm has the thickness of each flange and web of 20 mm. What percentage of moment and shear forces are resisted by flanges and web.

(***Ans.*** M_{Flanges} = 69.75% and M_{web} = 30.25% F_{Flange} = 2.5% and F_{web} = 97.5%)

26. A beam of square section of side 90 mm is placed with one of its diagonal vertical. If the beam is subjected to a vertical force of 90 kN, find

(a) the maximum shear stress,

(b) the average shear stress, and

(c) the shear stress at neutral axis.

(***Ans.*** (a) 12.5 N/mm^2, (b) and (c) = 11.11 N/mm^2)

27. A beam of triangular cross section has a base of 120 mm and height of 75 mm. Find the maximum shear stress and shear stress at the neutral axis if the beam is subjected to shear force of 60 kN.

(***Ans.*** τ_{max} = 20 N/mm^2, τ_{NA} = 17.78 N/mm^2)

Objective Type Questions

1. Equation of bending is

(a) $\frac{M}{I} = \frac{\sigma}{E} = \frac{R}{y}$ (b) $\frac{M}{I} = \frac{\sigma}{y} = \frac{E}{R}$

(c) $\frac{M}{E} = \frac{\sigma}{y} = \frac{R}{I}$ (d) $\frac{M}{y} = \frac{\sigma}{I} = \frac{E}{R}$

2. Section modulus is the ratio of

(a) moment of inertia to farthest distance from CG $\left(\frac{I}{y}\right)$

(b) bending moment to moment of inertia $\left(\frac{M}{I}\right)$

(c) moment of inertia to radius of curvature $\left(\frac{I}{R}\right)$

(d) farthest distance from CG to moment of inertia $\left(\frac{y}{I}\right)$

3. Section modulus for rectangular section is

(a) $\frac{bd^3}{12}$ (b) $\frac{bd^2}{6}$

(c) $\frac{bd^2}{12}$ (d) $\frac{bd^3}{6}$

4. Section modulus for hollow rectangular section is

(a) $\frac{BD^3 - bd^3}{6D}$ (b) $\frac{BD^2 - bd^2}{6D}$

(c) $\frac{BD^3 - bd^3}{12D}$ (d) $\frac{BD^2 - bd^2}{6d}$

5. Section modulus for circular section is

(a) $\frac{\pi d^4}{64}$ (b) $\frac{\pi d^4}{32}$

(c) $\frac{\pi d^3}{16}$ (d) $\frac{\pi d^3}{32}$

6. Section modulus for hollow circular section is

(a) $\frac{\pi(D^4 - d^4)}{32\ D}$ (b) $\frac{\pi(D^4 - d^4)}{32\ d}$

(c) $\frac{\pi(D^4 - d^4)}{64\ d}$ (d) $\frac{\pi(D^3 - d^3)}{32\ D}$

7. Section modulus for triangular section is

(a) $\frac{bh^3}{12}$ (b) $\frac{bh^3}{36}$

(c) $\frac{bh^2}{24}$ (d) $\frac{bh^2}{12}$

8. The width of equivalent beam of material 2 to be used as material 1 in a flitcher beam is

(a) $b_2 \frac{E_2}{E_1}$ (b) $b_2 \frac{E_1}{E_2}$

(c) $\frac{b_2}{b_1} E_2$ (d) $\frac{b_2}{b_1} E_1$

9. Moment of resistance of a flitcher beam made of two different materials is given by $M =$

(a) $M_1 - M_2$ (b) $\frac{M_1}{M_2}$

(c) $\frac{M_2}{M_1}$ (d) $M_1 + M_2$

10. Bending stress is maximum at

(a) Neutral axis (b) CG axis

(c) outer fibre (d) None

11. At neutral axis, the bending stress in pure bending is

(a) maximum (b) zero

(c) tensile (d) compressive

12. For a symmetric beam about neutral axis, the ratio of tensile and compressive stresses due to bending is

(a) < 1 (b) > 1

(c) 1 (d) zero

13. For a rectangular cross section, the ratio of maximum shear stress to average shear stress is

(a) 1 (b) $\frac{4}{3}$

(c) 1.5 (d) 2

14. For a circular cross section, the ratio of maximum shear stress to average shear stress is

(a) 1 (b) $\frac{4}{3}$

(c) 1.5 (d) $\frac{\pi}{4}$

15. For a triangular cross section, the ratio of maximum shear stress to average shear stress is

(a) 1 (b) $\frac{4}{3}$

(c) 1.5 (d) $\frac{1}{2}$

16. The general equation for shear stress in a beam of width b is

(a) $\frac{F}{8I} a\bar{y}$ (b) $\frac{F}{bI} (a\bar{y})^2$

(c) $\frac{F}{I} a\bar{y}$ (d) $\frac{F}{bI} a\bar{y}$

17. For rectangular cross section, the maximum shear stress occurs at

(a) top (b) bottom

(c) neutral axis (d) at $\frac{1}{4}$ of depth

18. For circular cross section, the maximum shear stress occurs at

(a) top (b) bottom

(c) neutral axis (d) at $\frac{1}{2}$ of radius

19. For triangular cross section, the maximum shear stress occurs at

(a) 1/2 of height (b) neutral axis

(c) top (d) $\frac{1}{3}$ from top

20. For I-section maximum shear stress occurs

(a) at top (b) at the bottom of top flange

(c) at the neck of web (d) at the neutral axis

21. For T-section, maximum shear stress occurs at

(a) at top (b) at the bottom of flange

(c) at the neck of web (d) at the neutral axis

22. The ratio of maximum shear stress for a rectangular cross section and a circular cross section beam subjected to same force and having some area of cross section is

(a) 1 (b) 1.5

(c) $\frac{4}{3}$ (d) 1.125

23. The maximum shear stress in a beam of rectangular cross section 100 mm × 200 mm subjected to a shear force of 60 kN is

(a) 3 N/mm^2 (b) 6 N/mm^2

(c) 2 N/mm^2 (d) 4.5 N/mm^2

24. The maximum shear stress in a beam of circular cross section of diameter 60 mm subjected to a shear force of 30 kN is

(a) 14.15 N/mm^2 (b) 5.305 N/mm^2

(c) 15.92 N/mm^2 (d) 10.61 N/mm^2

25. The maximum shear stress in a beam of triangular cross section of base 90 mm and height 60 mm, subjected to a shear force of 45 kN is

(a) 25 N/mm^2 (b) 16.67 N/mm^2

(c) 22.22 N/mm^2 (d) 8.33 N/mm^2

26. The shear stress at the neutral axis for a triangular cross section beam of base 100 mm and height 72 mm subjected to a shear force of 60 kN is.

(a) 16.67 N/mm^2 (b) 25 N/mm^2

(c) 11.11 N/mm^2 (d) 22.22 N/mm^2

27. In an I-section, majority of shear force is taken by

(a) Top flange (b) Bottom flange

(c) Web (d) Both flanges

28. Circular cross section beam having same area as of rectangular cross section beam is _______ stronger than rectangular cross section:

(a) more (b) less

(c) equally (d) None of the above

29. In a rectangular cross sectional beam, the variation of shear stress across depth is

(a) straight line (b) parabolic

(c) cubic (d) linear

30. In a circular cross sectional beam, the variation of shear stress across depth is

(a) linear (b) straight line

(c) cubic (d) parabolic

31. For a square cross section beam with one diagonal horizontal, the maximum shear stress occurs at

(a) top (b) $\frac{3}{8}$ of depth

(c) neutral axis (d) $\frac{1}{2}$ of depth

32. For a square cross-sectional beam the shear stress at neutral axis is equal to

(a) $\frac{F}{A}$ (b) $1.5\,\frac{F}{A}$

(c) $\frac{4}{3}\frac{F}{A}$ (d) $\frac{9}{8}\frac{F}{A}$

Key to Answers

1. (b)	2. (a)	3. (b)	4. (a)	5. (d)	6. (a)	7. (c)	8. (a)
9. (d)	10. (c)	11. (b)	12. (c)	13. (c)	14. (b)	15. (c)	16. (d)
17. (c)	18. (c)	19. (a)	20. (d)	21. (d)	22. (d)	23. (d)	24. (a)
25. (a)	26. (d)	27. (c)	28. (a)	29. (b)	30. (d)	31. (b)	32. (b)

Principal Stresses and Strains

CHAPTER 6

6.1 INTRODUCTION

A structural member may be subjected to different types of stresses like normal stress, shear stress, etc., simultaneously. The member may be subjected to direct stresses in the three different directions x, y and z. The shear stress either direct or due to torsion may also act. For example, a beam is subjected to bending and shear stresses. A shaft may be subjected to shear stress due to torque transmission, bending stresses due to belt tensions or gear forces and also axial stresses. For designing a member, it is necessary to find the combined effect of these stresses. For such type of loading, there always exist three orthogonal planes carrying only normal stresses and there will be no shear stress. Such planes where only normal stresses exist and no shear stress are called principal planes and normal stresses acting on them are called normal stresses or principal stresses. The greatest normal stress on any plane is called maximum principal stress and the least one is called minimum principal stress. For safe design, the maximum principal stress is to be within the permissible stress. In this chapter we will see the effect of such combined or principal stresses and strains.

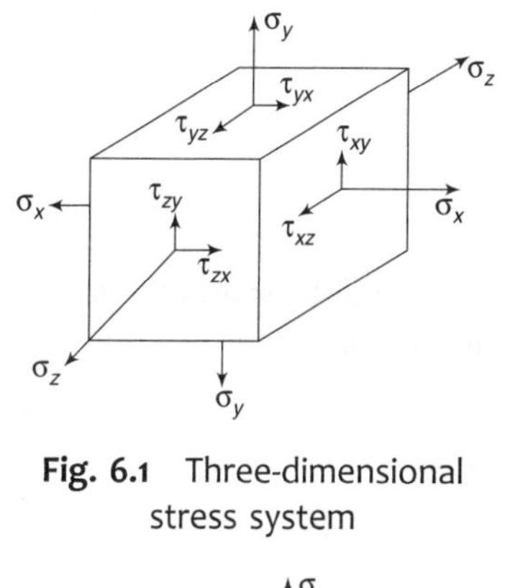

Fig. 6.1 Three-dimensional stress system

Figure 6.1 shows a three-dimensional stress system with various stresses acting on it.

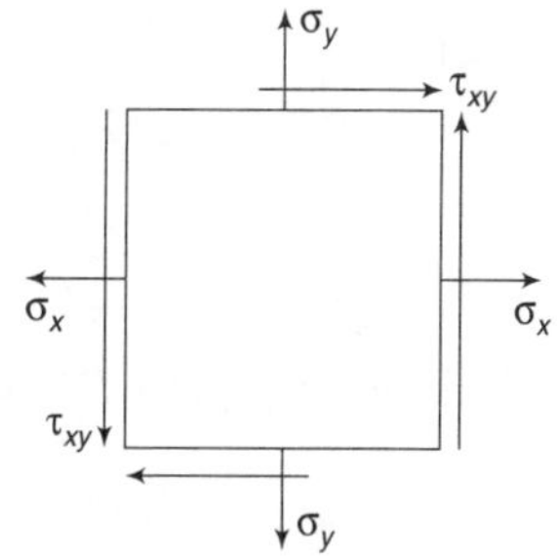

Fig. 6.2 Two-dimensional stress system

In may situations two-dimensional stress system may be sufficient and the general two-dimensional stress system is shown in Fig. 6.2.

6.2 NORMAL AND TANGENTIAL OR SHEAR STRESS ON AN INCLINED PLANE

The following stressed conditions are considered for analysis of normal and shear stresses and a plane inclined at an angle θ:

(a) Unidirectional stress
(b) Pure shear stress
(c) Biaxial stresses
(d) Biaxial stresses combined with shear stress (general two-dimensional stress system)

6.2.1 Member Subjected to Unidirectional Stress

Consider an element subjected to a direct uniaxial stress σ_x as shown in Fig. 6.3. Consider an inclined plane *DE* which is inclined at an angle θ to the plane of stress σ_x, measured in anticlockwise direction.

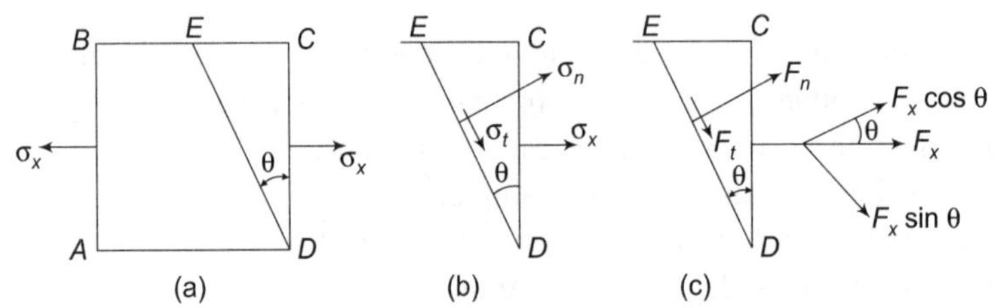

Fig. 6.3

Figure 6.3(b) shows the stress acting on the element and Fig. 6.3(c) shows the forces acting on the element.

$$\sigma_x = \text{Uniaxial stress acting} = \frac{F_x}{A}$$

$$\sigma_n = \text{Normal stress on inclined plane } DE$$

$$\sigma_t = \text{Tangential or shear stress on inclined plane } DE.$$

Let t = Thickness of the member

$$F_x = \text{Load on plane } CD = \sigma_x \cdot A = \sigma_x \cdot CD.t$$

$$F_n = \text{Normal load on plane } DE = \sigma_n \cdot A = \sigma_n \cdot DE \cdot t$$

$$F_t = \text{Tangential load on plane } DE = \sigma_t \cdot A = \sigma_t \cdot DE \cdot t$$

From triangle *CDE*, $\cos\theta = \dfrac{CD}{DE}$

$$\therefore \quad DE = \frac{CD}{\cos\theta}$$

$$\therefore \quad F_n = \sigma_n \cdot \frac{CD}{\cos\theta} \cdot t$$

$$F_t = \sigma_t \frac{CD}{\cos\theta} \cdot t$$

Resolve the given force F_x perpendicular to and along inclined plane *DE*.

Normal to plane $F_{xn} = F_x \cdot \cos\theta = \sigma_x \cdot CD \cdot t \cdot \cos\theta$

Along plane, $F_{xt} = F_x \cdot \sin\theta = \sigma_x\, CD \cdot t \cdot \sin\theta$.

Consider forces perpendicular to plane (normal to plane)

$$\sigma_x \cdot CD \cdot t \cdot \cos\theta = \sigma_n \cdot \frac{CD}{\cos\theta} \cdot t.$$

$$\sigma_n = \sigma_x \cdot \cos^2\theta \qquad \text{...(i)}$$

Forces along plane *DE*

$$\sigma_x \cdot CD \cdot t \cdot \sin\theta = \sigma_t \cdot \frac{CD}{\cos\theta} \cdot t$$

$$\therefore \quad \sigma_t = \sigma_x \cdot \sin\theta \cdot \cos\theta$$

$$\sigma_t = \frac{\sigma_x}{2} \sin 2\theta \qquad \text{...(ii)}$$

Resultant stress on plane *DE*

$$\sigma_R = \sqrt{\sigma_n^2 + \sigma_t^2} = \sqrt{(\sigma_x \cos^2\theta)^2 + (\sigma_x \sin\theta \cos\theta)^2}$$

$$= \sqrt{\sigma_x^2 \cos^4\theta + \sigma_x^2 \cos^2\theta \sin^2\theta}$$

$$\sigma_R = \sqrt{\sigma_x^2 \cos^2\theta (\cos^2\theta + \sin^2\theta)} = \sigma_x \cos\theta$$

$$\sigma_R = \sigma_x \cos\theta \qquad \text{...(iii)}$$

From Fig. 6.3 it can be seen that θ varies from 0 to 90°

From Eq. (i), $\sigma_n = \sigma_x \cos^2\theta$

This will be maximum when $\cos^2\theta = 1$

$$\cos\theta = 1 \text{ or } \theta = 0$$

$\therefore$ Maximum value of $\sigma_n = \sigma_x$ at $\theta = 0$

σ_n is minimum at $\theta = 90°$

Minimum value of $\sigma_n = \sigma_x \cdot \cos^2 90 = 0$.

Also $$\sigma_t = \frac{\sigma_x}{2} \cdot \sin 2\theta$$

This is maximum when $\sin 2\theta = 1$

$$\therefore \quad 2\theta = 90° \text{ or } \theta = 45°$$

Maximum value of $\sigma_t = \frac{\sigma_x}{2} \sin(2 \times 45) = \frac{\sigma_x}{2}$ at $\theta = 45°$

where, $\theta = 0$ or $90°$, $\sin 2\theta = 0$

$\therefore$ $\sigma_t = 0$ at $\theta = 0$ or $90°$

The maximum shear stress occurs at a plane 45° to the plane of uniaxial stress.

6.2.2 Member Subjected to Pure Shear Stress (τ)

Figure 6.4 shows a member *ABCD* of thickness *t* subjected to pure shear stress τ. Consider any sectional plane *DE* at angle θ with plane *CD* in counterclockwise direction.

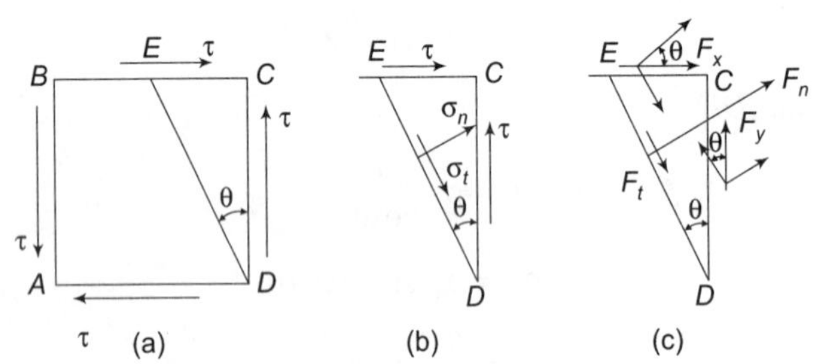

Fig. 6.4 Member subjected to pure shear

τ = Shear stress

σ_n = Normal stress on inclined plane *DE*

σ_t = Tangential or shear stress on inclined plane *DE*

Let t = Thickness of the member

F_x = Load in *x*-direction = $\tau \cdot A = \tau \cdot EC \cdot t$

F_y = Load in *y*-direction = $\tau \cdot A = \tau \cdot CD \cdot t$

F_n = Normal load on plane *DE* = $\sigma_n \cdot DE \cdot t$

F_t = Tangential load on plane *DE* = $\sigma_t \cdot DE \cdot t$

From triangle *CDE*, $\cos\theta = \dfrac{CD}{DE}$ and $\sin\theta = \dfrac{EC}{DE}$

Resolve the forces F_x and F_y perpendicular to the plane *DE* and along the plane.

$$\text{Normal to plane, } F_n = F_y \sin\theta + F_x \cos\theta$$

$$\sigma_n \cdot DE.t = \tau \cdot CD \cdot t \cdot \sin\theta + \tau \cdot EC \cdot t \cdot \cos\theta$$

$$\sigma_n = \tau \cdot \frac{CD}{DE} \cdot \sin\theta + \tau \cdot \frac{EC}{DE} \cdot \cos\theta$$

$$= \tau \cdot \cos\theta \cdot \sin\theta + \tau \cdot \sin\theta \cdot \cos\theta$$

$$\sigma_n = 2\tau \cos\theta \cdot \sin\theta = \tau \cdot \sin 2\theta$$

$$\text{Normal stress} = \sigma_n = \tau \cdot \sin 2\theta.$$

$$\text{Along plane, } F_t = F_y \cos\theta - F_x \sin\theta$$

$$\sigma_t \cdot DE \cdot t = \tau \cdot CD \cdot t \cdot \cos\theta - \tau \cdot EC \cdot t \cdot \sin\theta$$

$$\sigma_E = \tau \cdot \frac{DC}{DE} \cdot \cos\theta - \tau \cdot \frac{EC}{DE} \cdot \sin\theta$$

$$= \tau \cdot \cos^2\theta - \tau \sin^2\theta = \tau (\cos^2\theta - \sin\theta^2)$$

$$\text{Tangential stress } \sigma_t = \tau \cos 2\theta$$

The principal planes are planes on which tangential stress $\sigma_t = 0$

$$\text{Equating } \sigma_t = 0 \text{ for principal planes}$$

$$\sigma_t = \tau \cdot \cos 2\theta = 0$$

$$\cos 2\theta = 0$$

$$\therefore \quad 2\theta = 90° \text{ or } 270°$$

$$\theta = 45° \text{ or } 135°$$

Hence, the two principal planes are at 45° and 135° with the plane *CD*. The principal planes are two right angles to each other.

The two principal stresses are

$$\sigma_1 = \tau \sin 2\theta = \tau \sin (2 \times 45) = \tau$$

$$\sigma_2 = \tau \sin 2\theta = \tau \sin (2 \times 135) = -\tau$$

It is seen that on one principal plane there will be tensile stress $+\tau$ and on the other principal plane, there will be compressive stress $-\tau$. These principal stresses are also called diagonal tensile and diagonal compressive stresses.

For the tangential stress to be maximum,

$$\sigma_t = \tau \cos 2\theta \text{ to be maximum}$$

$$\text{Differentiating } \frac{d\sigma_t}{d\theta} = 0 \Rightarrow 2 \sin 2\theta = 0$$

$$\therefore \quad \sin 2\theta = 0$$

$$2\theta = 0 \text{ or } 180°$$

The two maximum values are

$$\sigma_{t\,max} = \tau \cdot \cos (2 \times 0) = \tau$$

$$\sigma_{t\,max} = \tau \cos (2 \times 90) = -\tau$$

Hence, maximum shear stress of intensity τ is carried on plane *CD* and *BC*. From the above, it may be noted that

(i) The two principal planes are normal to each other

(ii) The planes of maximum shear stress are normal to each other

(iii) The planes of shear stress are at 45° to the principal planes.

6.2.3 Member Subjected to Biaxial Stresses (σ_x and σ_y)

Figure 6.5 shows a member *ABCD* of thickness *t* subjected to two mutually perpendicular stresses, σ_x and σ_y. Consider any sectional plane *DE* inclined at an angle θ with plane of σ_x, i.e., plane *CD* is counterclockwise direction.

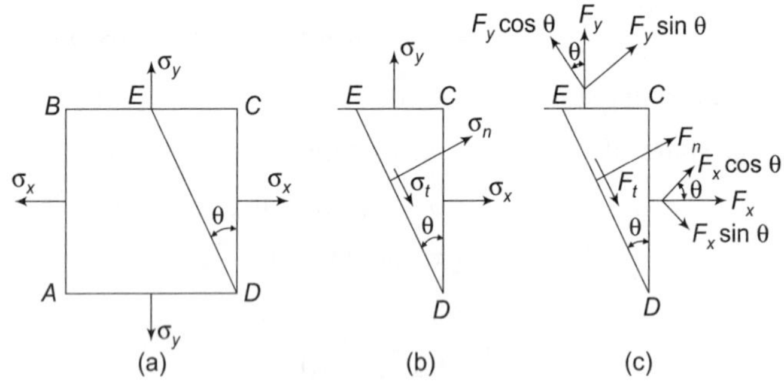

Fig. 6.5 Member subjected to biaxial stresses

F_x = Load in *x*-direction $= \sigma_x \cdot A = \sigma_x \cdot CD \cdot t$

F_y = Load in *y*-direction $= \sigma_y\ A = \sigma_y \cdot EC \cdot t$

F_n = Normal load on plane $DE = \sigma_n \cdot A = \sigma_n \cdot DE \cdot t$

F_t = Tangential or shear load on plane $DE = \sigma_E\ A = \sigma_t \cdot DE \cdot t$

From triangle *CDE*, $\sin\theta = \dfrac{EC}{DE}$

$$\cos\theta = \frac{CD}{DE}$$

Resolve the forces along and perpendicular to plane *DE*.

Consider forces perpendicular to plane *DE*.

$$F_n = F_x \cdot \cos\theta + F_y \cdot \sin Q$$

$$\sigma_n\ DE \cdot t = \sigma_x \cdot CD \cdot t \cdot \cos\theta + \sigma_y \cdot EC \cdot t \cdot \sin\theta$$

Divide by $DE \times t$

$$\sigma_n = \sigma_x \cdot \frac{CD}{DE} \cdot \cos\theta + \sigma_y \frac{EC}{DE} \cdot \sin\theta = \sigma_x \cos^2\theta + \sigma_y \sin^2\theta$$

We know that $\cos 2\theta = \cos^2\theta - \sin^2\theta$

$\therefore$ $$\cos 2\theta = (1 - \sin^2\theta) - \sin^2\theta = 1 - 2\sin^2\theta$$

$\therefore$ $$\sin^2\theta = \frac{1 - \cos 2\theta}{2}$$

and $$\cos^2\theta = \frac{1 + \cos 2\theta}{2}$$

$$\sigma_n = \sigma_x\left(\frac{1+\cos 2\theta}{2}\right) + \sigma_y\left(\frac{1-\cos 2\theta}{2}\right)$$

$$= \frac{\sigma_x}{2} + \frac{\sigma_x \cos 2\theta}{2} + \frac{\sigma_y}{2} - \frac{\sigma_y \cos 2\theta}{2}$$

$\therefore$ Normal stress $\sigma_n = \dfrac{(\sigma_x + \sigma_y)}{2} + \left(\dfrac{\sigma_x - \sigma_y}{2}\right)\cos 2\theta$

Consider forces parallel to the plane DE

$$F_t = F_x \sin\theta - F_y \cos\theta$$

$$\sigma_t \cdot DE \cdot t = \sigma_x \cdot CD \cdot t \cdot \sin\theta - \sigma_y \cdot EC \cdot t \cdot \cos\theta$$

Divide by $DE \cdot t$

$$\sigma_t = \sigma_x \frac{CD}{DE}\sin\theta - \sigma_y \cdot \frac{EC}{DE} \cdot \cos\theta$$

Substitute $\dfrac{CD}{DE} = \cos\theta$ and $\dfrac{EC}{DE} = \sin\theta$

$$\sigma_t = \sigma_x \cdot \cos\theta \sin\theta - \sigma_y \cdot \sin\theta \cos\theta$$

$$= (\sigma_x - \sigma_y)\sin\theta\cos\theta$$

$$\sigma_t = (\sigma_x - \sigma_y)\frac{\sin 2\theta}{2}$$

Tangential stress or shear stress $\sigma_t = \left(\dfrac{\sigma_x - \sigma_y}{2}\right)\sin 2\theta$

The resultant stress on plane $DE = \sigma_R = \sqrt{\sigma_n^2 + \sigma_t^2}$

$$= \sqrt{(\sigma_x \cos^2\theta + \sigma_y \sin^2\theta)^2 + \left[(\sigma_x - \sigma_y)^2 \sin\theta\cos\theta\right]^2}$$

$$= \sqrt{\begin{array}{l}\sigma_x^2\cos^4\theta + \sigma_y^2 \sin^4\theta + 2\,\sigma_x\sigma_y \cos^2\theta \sin^2\theta + \sigma_x^2 \sin^2\theta.\cos^2\theta \\ \quad + \sigma_y^2 \sin^2\theta \cos^2\theta - 2\sigma_x\sigma_y \sin^2\theta \cos^2\theta\end{array}}$$

$$= \sqrt{\sigma_x^2 \cos^2\theta(\cos^2\theta + \sin^2\theta) + \sigma_y^2 \sin^2\theta(\sin^2\theta + \cos^2\theta)}$$

$$\sigma_R = \sqrt{\sigma_x^2 \cos^2\theta + \sigma_y^2 \sin^2\theta}$$

Consider $\sigma_n = \dfrac{\sigma_x + \sigma_y}{2} + \dfrac{\sigma_x - \sigma_y}{2}\cos^2\theta$ and $\sigma_t = \dfrac{\sigma_x - \sigma_y}{2}\sin 2\theta$

For principal planes, the shear stress, $\sigma_t = 0$

$$\frac{\sigma_x - \sigma_y}{2}\sin 2\theta = 0$$

$$\therefore \qquad 2\theta = 0 \text{ or } 180°$$

$$\theta = 0 \text{ or } 90°$$

when $\theta = 0$, maximum principal stress $\sigma_1 = \dfrac{\sigma_x + \sigma_y}{2} + \dfrac{\sigma_x - \sigma_y}{2} \cos\theta$

$$\sigma_1 = \frac{\sigma_x + \sigma_y}{2} + \left(\frac{\sigma_x - \sigma_y}{2}\right) \times 1 = \sigma_x$$

and at $\theta = 90$, minimum principal stress $\sigma_2 = \dfrac{\sigma_x + \sigma_y}{2} + \dfrac{\sigma_x - \sigma_y}{2} . \cos(2 \times 90)$

$$\sigma_2 = \frac{\sigma_x + \sigma_y}{2} + \frac{\sigma_x - \sigma_y}{2}(-1) = \sigma_y$$

The two principal stresses σ_1 and σ_2 are at 90° to each other.

Now for the shear stresses to be maximum

$$\sigma_t = \left(\frac{\sigma_x - \sigma_y}{2}\right) \sin 2\theta \text{ is to be maximum.}$$

Hence, sin 2 θ is to be maximum

$$\sin 2\theta = 1$$

$$\therefore \qquad 2\theta = 90° \text{ or } 270°$$

$$\theta = 45° \text{ or } 135°$$

Maximum shear stress $\sigma_{t\max} = \dfrac{\sigma_x - \sigma_y}{2}$

Maximum shear stress occurs on two mutually perpendicular planes at θ = 45° or 135° with the plane carrying σ_x or maximum principal stress σ_1.

6.2.4 Member Subjected to Biaxial Stresses σ_x and σ_y Combined with Shear Stress τ. (Generalised Two-Dimensional Stress)

Figure 6.6 shows a member *ABCD* of thickness *t* subjected to biaxial stresses σ_x and σ_y combined with shear stress τ. Consider any sectional plane *DE* inclined at an angle θ with plane of σ_x, taken in counterclockwise direction.

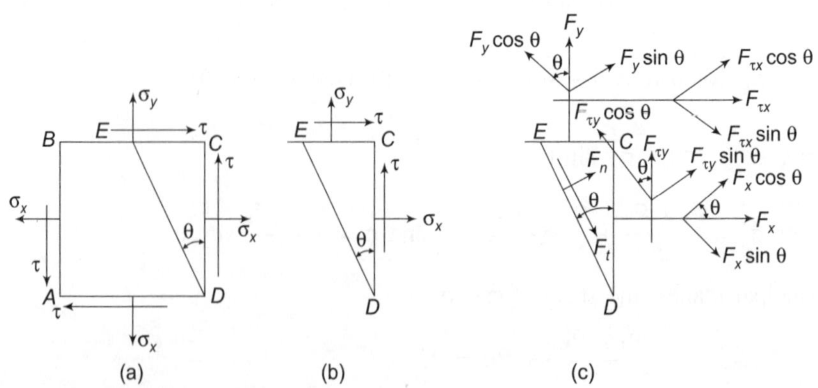

Fig. 6.6 Member subjected to biaxial stresses combined with shear stress.

$$F_x = \text{Load in } x\text{-direction} = \sigma_x \cdot A = \sigma_x \cdot CD \cdot t$$
$$F_y = \text{Load in } y\text{-direction} = \sigma_y\ A = \sigma_y \cdot EC \cdot t$$
$$F_{\tau x} = \text{Shear load on } CE = \tau \cdot A = \tau \cdot EC \cdot t$$
$$F_{\tau y} = \text{Shear load on } CD = \tau \cdot A = \tau \cdot CD \cdot t$$
$$F_n = \text{Normal load on } DE = \sigma_t A = \sigma \cdot DE \cdot t$$
$$F_t = \text{Tangential or shear load on } DE = \sigma_t \cdot DE \cdot t$$

Resolving the forces along and perpendicular to the plane.

Forces perpendicular to the plane:

$$F_n = F_x \cos\theta + F_y \sin\theta + F_{\tau x} \cdot \cos\theta + F_{\tau y} \sin\theta$$

$$\sigma_n \cdot DE \cdot t = \sigma_x \cdot CD \cdot t \cos\theta + \sigma_y \cdot CE \cdot t \cdot \sin\theta + \tau \cdot CE \cdot t \cdot \cos\theta + \tau \cdot CD \cdot t \cdot \sin\theta$$

Dividing by $DE \cdot t$, $\sigma_n = \sigma_x \cdot \dfrac{CD}{DE} \cdot \cos\theta + \sigma_y \cdot \dfrac{CE}{DE} \sin\theta + \tau \dfrac{CE}{DE} \cos\theta + \tau \dfrac{CD}{DE} \cdot \sin\theta$

From triangle CDE, $\sin\theta = \dfrac{EC}{DE}$ and $\cos\theta = \dfrac{CD}{DE}$

Substituting

$$\sigma_n = \sigma_x \cdot \cos^2\theta + \sigma_y \sin^2\theta + \tau \cdot \sin\theta \cos\theta + \tau \cdot \cos\theta \sin\theta$$

Substitute $\cos^2\theta = \dfrac{1 + \cos^2\theta}{2}$ and $\sin^2\theta = \dfrac{1 - \cos 2\theta}{2}$

$$\sigma_n = \sigma_x \left(\frac{1 + \cos 2\theta}{2}\right) + \sigma_y \left(\frac{1 - \cos 2\theta}{2}\right) + \tau \cdot (2 \sin\theta \cos\theta)$$

Normal stress

$$\sigma_n = \frac{\sigma_x + \sigma_y}{2} + \frac{\sigma_x - \sigma_y}{2} \cdot \cos 2\theta + \tau \cdot \sin 2\theta$$

Forces parallel to the palne

$$F_t = F_x \cdot \sin\theta - F_y \cdot \cos\theta + F_{\tau x} \cdot \sin\theta - F_{\tau y} \cdot \cos\theta$$

$$\sigma_t \cdot DE \cdot t = \sigma_x \cdot CD \cdot t \cdot \sin\theta - \sigma_y\ CE \cdot t \cos\theta + \tau \cdot CE \cdot t \cdot \sin\theta - \tau \cdot CD \cdot t \cdot \cos\theta.$$

Dividing by $DE \cdot t$, $\sigma_t = \sigma_x \cdot \dfrac{CD}{DE} \sin\theta - \sigma_y \cdot \dfrac{CE}{DE} \cdot \cos\theta + \tau \dfrac{CE}{DE} \cdot \sin\theta - \tau \dfrac{CD}{DE} \cos\theta$

Substituting $\dfrac{CD}{DE} = \cos\theta$ and $\dfrac{CE}{DE} = \sin\theta$

$$\sigma_t = \sigma_x \cdot \sin\theta \cos\theta - \sigma_y \cdot \sin\theta \cdot \cos\theta + \tau \sin^2\theta - \tau \cos^2\theta$$
$$= (\sigma_x - \sigma_y) \sin\theta \cos\theta - \tau (\cos^2\theta - \sin^2\theta)$$
$$= (\sigma_x - \sigma_y) \frac{\sin 2\theta}{2} - \tau \cos 2\theta$$

Tangential or shear stress

$$\sigma_t = \left(\frac{\sigma_x - \sigma_y}{2}\right)\sin 2\theta - \tau\cos 2\theta$$

(i) Direction of principal planes

We have seen earlier that the planes on which shear stress is zero are called principal planes.

For principal planes, $\sigma_t = \dfrac{\sigma_x - \sigma_y}{2}\sin 2\theta - \tau\cos 2\theta = 0$

$$\frac{\sigma_x - \sigma_y}{2}\sin 2\theta = \tau\cos 2\theta$$

$$\therefore \quad \frac{\sin 2\theta}{\cos 2\theta} = \frac{\tau}{\left(\dfrac{\sigma_x - \sigma_y}{2}\right)}$$

$$\tan 2\theta = \frac{2\tau}{\sigma_x - \sigma_y}$$

Represent this using a triangle as shown in Fig. 6.7.

From Fig. 6.7 $\quad \tan 2\theta = \dfrac{BC}{AB} = \dfrac{2\tau}{\sigma_x - \sigma_y}$

$$\therefore \quad AC = \sqrt{AB^2 + BC^2}$$

$$AC = \sqrt{(\sigma_x - \sigma_y)^2 + 4\tau^2}$$

$$2\theta = \tan^{-1}\left(\frac{2\tau}{\sigma_x - \sigma_y}\right)$$

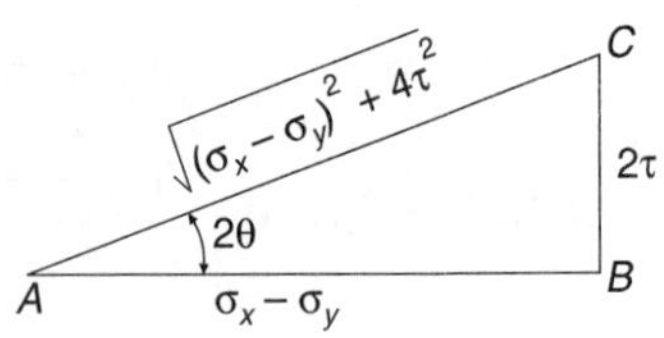

Fig. 6.7

This will have two values that are at 180° to each other. Let $2\theta_1$ and $2\theta_2$ be the angles.

From Fig. 6.7, $\quad \sin 2\theta_1 = \dfrac{2\tau}{\sqrt{(\sigma_x - \sigma_y)^2 + 4\tau^2}}$

$$\cos 2\theta_1 = \frac{(\sigma_x - \sigma_y)}{\sqrt{(\sigma_x - \sigma_y)^2 + 4\tau^2}}$$

$$\sin 2\theta_2 = -\frac{-2\tau}{\sqrt{(\sigma_x - \sigma_y)^2 + 4\tau^2}}$$

$$\cos 2\theta_2 = \frac{-(\sigma_x - \sigma_y)}{\sqrt{(\sigma_x - \sigma_y)^2 + 4\tau^2}}$$

Since $\quad 2\theta_2 = 180 + 2\theta_1$

$$\theta_2 = 90 + \theta_1$$

Hence θ_1 and θ_2 are at 90° apart and are the directions of two principal planes. It may be observed that the two principal planes are at 90° away.

(ii) Magnitude of principal stress

To get the magnitude of two principal stresses σ_1 and σ_2, the values of $2\theta_1$ and $2\theta_2$ are to be substituted in the expression for σ_n.

$$\sigma_1 = \frac{\sigma_x + \sigma_y}{2}\left(\frac{\sigma_x - \sigma_y}{2}\right)\left(\frac{\sigma_x - \sigma_y}{\sqrt{(\sigma_x - \sigma y)^2 + 4\tau^2}}\right) + \tau\left(\frac{2\tau}{\sqrt{(\sigma_x - \sigma_y)^2 + 4\tau^2}}\right)$$

$$= \frac{\sigma_x + \sigma_y}{2} + \frac{(\sigma_x - \sigma_y)^2 + 4\tau^2}{2\sqrt{(\sigma_x - \sigma_y)^2 + 4\tau^2}} = \frac{\sigma_x + \sigma_y}{2} + \frac{1}{2}\cdot\sqrt{(\sigma_x - \sigma_y)^2 + 4\tau^2}$$

$$= \frac{\sigma_x + \sigma_y}{2} + \sqrt{\left(\frac{\sigma_x - \sigma_y}{2}\right)^2 + \tau^2}$$

or

$$\sigma_1 = \frac{1}{2}\left[(\sigma_x + \sigma_y) + \sqrt{(\sigma_x - \sigma_y)^2 + 4\tau^2}\right]$$

Similarly,

$$\sigma_2 = \frac{\sigma_x + \sigma_y}{2} + \left(\frac{\sigma_x - \sigma_y}{2}\right)\left[\frac{-(\sigma_x + \sigma_y)}{\sqrt{(\sigma_x - \sigma_y)^2 + 4\tau^2}}\right] + \tau\left[\frac{-2\tau}{\sqrt{(\sigma_x - \sigma_y)^2 + 4\tau^2}}\right]$$

$$= \frac{\sigma_x + \sigma_y}{2} - \frac{(\sigma_x - \sigma_y)^2 + 4\tau^2}{2\sqrt{(\sigma_x - \sigma_y)^2 + 4\tau^2}} = \frac{\sigma_x + \sigma_y}{2} - \frac{1}{2}\sqrt{(\sigma_x - \sigma_y)^2 + 4\tau^2}$$

$$= \frac{\sigma_x + \sigma_y}{2} - \sqrt{(\sigma_x - \sigma_y)^2 + \tau^2}$$

or $$\sigma_2 = \frac{1}{2}\left[(\sigma_x + \sigma_y) - \sqrt{(\sigma_x - \sigma_y)^2 + 4\tau^2}\right]$$

For planes of maximum shear stress, differentiate σ_t w.r.t. θ and equate to zero

$$\frac{d\sigma_t}{d\theta} = \frac{\sigma_x - \sigma_y}{2}\cdot 2\cdot\cos 2\theta_t - \tau\cdot 2\,(-\sin 2\theta_t) = 0$$

$$(\sigma_x - \sigma_y)\cos 2\theta_t = -2\tau\cdot\sin 2\theta_t$$

$$\therefore \quad \frac{\sin 2\theta_t}{\cos 2\theta_t} = \left(\frac{\sigma_x - \sigma_y}{-2\tau}\right)$$

$$\tan 2\theta_t = -\frac{(\sigma_x - \sigma_y)}{2\tau}$$

where θ_t = Angle of maximum shear stress plane.

Representing this by a triangle as shown in Fig. 6.8

$$\sin 2\theta_t = \frac{(\sigma_x - \sigma_y)}{\sqrt{(\sigma_x - \sigma_y)^2 + 4\tau^2}}$$

$$\cos 2\theta = \frac{-2\tau}{\sqrt{(\sigma_x - \sigma_y)^2 + 4\tau^2}}$$

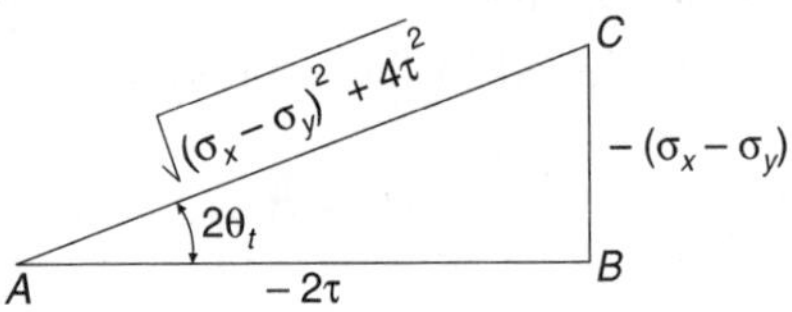

Fig. 6.8

Substituting these values in tangential stress equation

$$\sigma_{t\,max} = \left(\frac{\sigma_x - \sigma_y}{2}\right)\left[\frac{(\sigma_x - \sigma_y)}{\sqrt{(\sigma_x - \sigma_y)^2 + 4\tau^2}}\right] - \tau\left[\frac{-2\tau}{\sqrt{(\sigma_x - \sigma_y)^2 + 4\tau^2}}\right]$$

$$= \frac{(\sigma_x - \sigma_y)^2 + 4\tau^2}{2\sqrt{(\sigma_x - \sigma_y)^2 + 4\tau^2}} = \frac{1}{2}\sqrt{(\sigma_x - \sigma_y)^2 + 4\tau^2} = \sqrt{\left(\frac{\sigma_x - \sigma_y}{2}\right)^2 + 4\tau^2}$$

$$\sigma_{t\,max} = \frac{1}{2}(\sigma_1 - \sigma_2)$$

Maximum shear stress $= \frac{1}{2}$ the difference of the principal stresses σ_1 and σ_2

$$\tan 2\theta = \frac{2\tau}{(\sigma_x - \sigma_y)} \text{ and } \tan 2\theta_t = \frac{(\sigma_x - \sigma_y)}{-2\tau}$$

$\therefore \quad \tan 2\theta \times \tan 2\theta_t = -1$

The values of 2θ and $2\theta_t$ differ by 90°. Hence, the principal planes and shear planes differ by 45°.

SOLVED PROBLEMS

Problem 1 A round bar of diameter 42 mm is subjected to an axial force of 60 kN. Find the stresses on a plane making 30° to the plane of axial stress. Also find maximum shear stress.

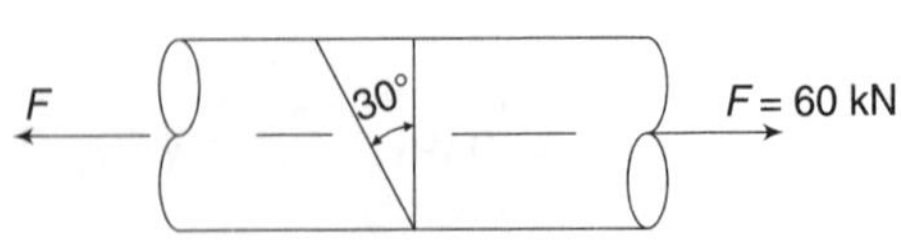

Fig. 6.9

Solution:

$$\text{Axial load } F = 60 \text{ kN} = 60000 \text{ N}$$

$$\text{Diameter } d = 42 \text{ mm}$$

$$\text{Angle } \theta = 30°$$

$$\text{Axial stress } \sigma_x = \frac{F}{A} = \frac{60000}{\left(\frac{\pi}{4} \times 42^2\right)} = 43.31 \text{ N/mm}^2$$

Normal stress on plane inclined at θ, $\sigma_n = \sigma_x \cos^2\theta$

$$\sigma_n = 43.31 \cos^2 30 = 32.48 \text{ N/mm}^2$$

$$\text{Shear stress } \sigma_t = \frac{\sigma_x}{2} \sin 2\theta = \frac{43.31}{2} \sin(2 \times 30) = 18.75 \text{ N/mm}^2$$

Maximum shear stress occurs at θ = 45°

$$\text{and} \quad \text{Maximum shear stress } \sigma_{t\,max} = \frac{\sigma_x}{2} = \frac{43.31}{2} = 21.655 \text{ N/mm}^2.$$

Problem 2 Find the safe axial load that can be applied on to a bar of rectangular cross section 30 mm wide, 60 mm depth if the normal and shear stresses on a plane inclined at 25° are to be limited to 15 N/mm² and 9 N/mm² respectively.

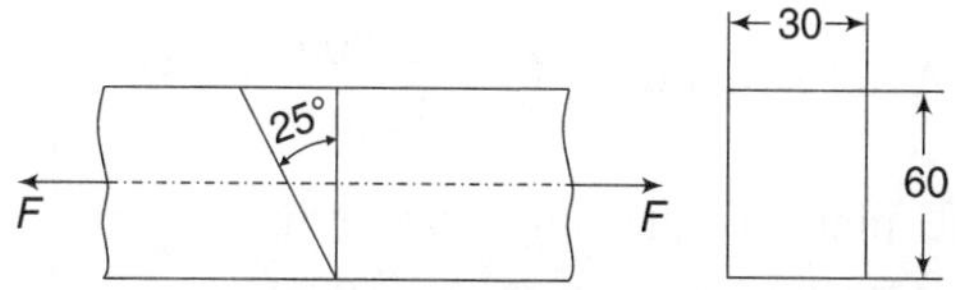

Fig. 6.10

Solution:

$$\text{Width } b = 30 \text{ mm, depth } h = 60 \text{ mm}$$

$$\therefore \quad \text{Area of cross section } A = bh = 30 \times 60 = 1800 \text{ mm}^2$$

$$\text{Angle of plane } \theta = 25°$$

$$\text{Normal stress } \sigma_n = 15 \text{ N/mm}^2$$

$$\text{Shear stress } \sigma_t = 9 \text{ N/mm}^2$$

$$\text{Axial stress } \sigma_x = \frac{F}{A}$$

Normal stress on a plane inclined at θ, $\sigma_n = \sigma_x \cdot \cos^2\theta$

$$15 = \sigma_x . \cos^2 25$$

$$\therefore \quad \text{Axial stress } \sigma_x = \frac{15}{\cos^2 25} = 18.26 \text{ N/mm}^2$$

$$\text{Shear stress } \sigma_t = \frac{\sigma_x}{2} \sin 2\theta$$

$$9 = \frac{\sigma_x}{2} \sin (2 \times 25)$$

$$\therefore \quad \sigma_x = 23.497 \text{ N/mm}^2$$

∴ Possible axial stress $\sigma_x = 18.26 \text{ N/mm}^2$ (minimum of the above two values)

$$\therefore \quad \text{Axial load } F = \sigma_x . A = 18.26 \times 1800 = 32.868 \text{ N.}$$

Problem 3 Find the normal and shear stresses on the oblique plane through point *P* in a bar *AB* which has a diameter of 40 mm as shown in Fig. 6.11.

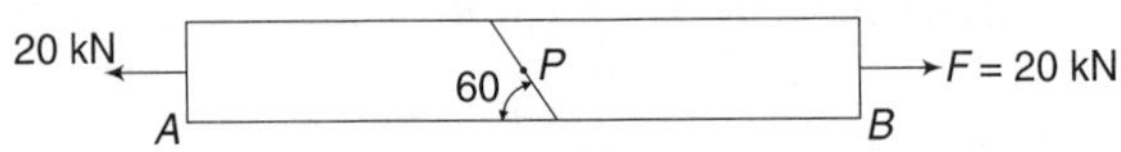

Fig. 6.11

Solution:

$$\text{Axial load } F = 20 \text{ kN} = 20 \times 10^3 \text{ N}$$

$$\text{Diameter } d = 40 \text{ mm}$$

$$\text{Area of cross section } A = \frac{\pi}{4}d^2 = \frac{\pi}{4} \times 40^2 = 1256.6 \text{ mm}^2$$

$$\therefore \quad \text{Axial stress } \sigma_x = \frac{F}{A} = \frac{20000}{1256.6} = 15.92 \text{ N/mm}^2$$

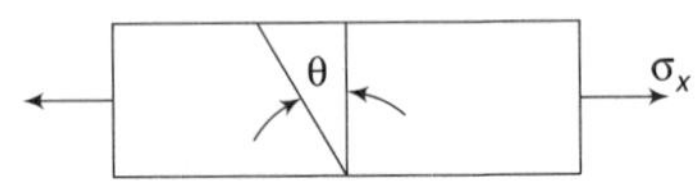

Fig. 6.12

Angle of the oblique plane is the angle taken in anticlockwise direction from the plane of σ_x as shown in adjacent Fig. 6.12.

$$\therefore \quad \theta = 90 - 60 = 30°$$

$$\text{Normal stress on the oblique plane } \sigma_n = \sigma_x \cdot \cos^2\theta = 15.92 \cos^2 30 = 11.94 \text{ N/mm}^2$$

$$\text{Shear stress on the oblique plane } \sigma_t = \frac{\sigma_x}{2} \sin 2\theta = \frac{15.92}{2} \sin (2 \times 30)$$

$$= 6.849 \text{ N/mm}^2.$$

Problem 4 An element is subjected to pure shear stress of 60 MPa as shown in Fig. 6.13. Find the normal and shear stress on a plane inclined at 20°.

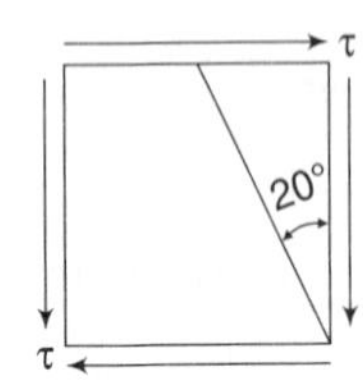

Fig. 6.13

Solution:

$$\text{Shear stress } \tau = 60 \text{ MPa}$$

$$\text{Angle } \theta = 20°$$

$$\text{Normal stress on inclined plane } \sigma_n = \tau . \sin 2\,\theta = 60 \cdot \sin (2 \times 20)$$

$$= 38.57 \text{ N/mm}^2$$

$$\text{Shear stress on inclined plane } \sigma_t = \tau \cos 2\theta = 60 \cdot \cos (2 \times 20)$$

$$= 54.96 \text{ N/mm}^2.$$

Problem 5 The direct stresses acting at a point in a strained material are shown in Fig. 6.14. Find the normal and shear stresses acting on a plane inclined at 30° to the plane of major principal stress. Also find the resultant stress and its obliquity.

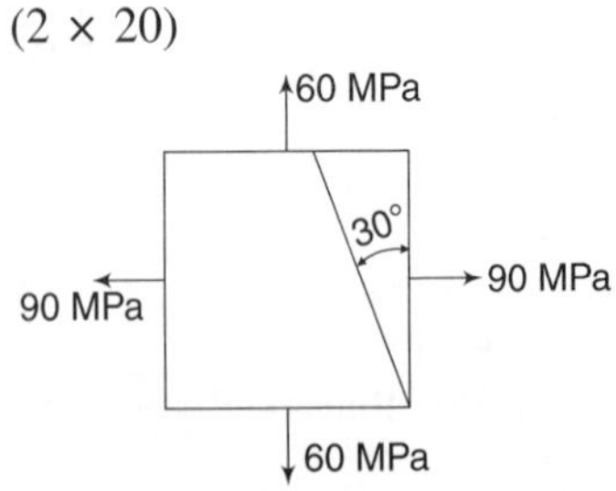

Fig. 6.14

Solution: $\sigma_x = 90\text{ MPa}, \quad \sigma_y = 60\text{ MPa}, \quad \theta = 30°$

Normal stress on inclined plane $\sigma_n = \left(\frac{\sigma_x + \sigma_y}{2}\right) + \left(\frac{\sigma_x - \sigma_y}{2}\right)\cos 2\theta$

$$= \left(\frac{90 + 60}{2}\right) + \left(\frac{90 - 60}{2}\right)\cos(2 \times 30) = 82.5\text{ MPa}$$

Shear stress on inclined plane $\sigma_t = \left(\frac{\sigma_x - \sigma_y}{2}\right)\sin 2\theta$

$$= \left(\frac{90 - 60}{2}\right)\cdot \sin(2 \times 30) = 12.99\text{ MPa}$$

Resultant stress $\sigma_R = \sqrt{\sigma_n^2 + \sigma_t^2} = \sqrt{82.5^2 + 12.99^2} = 83.52\text{ MPa}$

Inclination of resultant stress $\phi = \tan^{-1}\left(\frac{\sigma_n}{\sigma_t}\right)$

$$\phi = \tan^{-1}\left(\frac{82.5}{12.99}\right) = 81.05°$$

$$= 30° + 81.05°$$

$$= 111.05° \text{ from plane of } \sigma_x$$

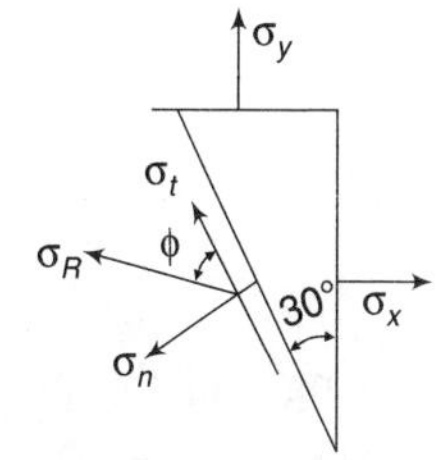

Fig. 6.14(a)

The stresses angle and angles are shown in Fig. 6.14(a).

Problem 6 Figure 6.15 shows an element subjected to two direct stresses. Find the normal and shear stresses along a plane inclined at 25°.

Solution: $\sigma_x = 100\text{ MPa}, \sigma_y = -60\text{ MPa}$ (compressive)

$\theta = 25°$

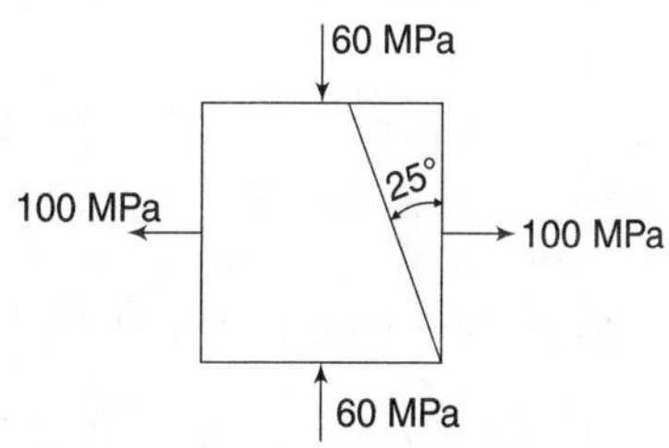

Fig. 6.15

Normal stress on inclined plane

$$\sigma_n = \left(\frac{\sigma_x + \sigma_y}{2}\right) + \left(\frac{\sigma_x - \sigma_y}{2}\right)\cos 2\theta$$

$$= \frac{100 + (-60)}{2} + \left[\frac{100 - (-60)}{2}\right]\cos(2 \times 25) = 71.42\text{ MPa}$$

Shear stress on inclined plane, $\sigma_t = \frac{\sigma_x - \sigma_y}{2}\cdot \sin 2\theta$

$$= \left[\frac{100 - (-60)}{2}\right]\cdot \sin(2 \times 25) = 61.28\text{ MPa}$$

Resultant stress $\sigma_R = \sqrt{\sigma_n^2 + \sigma_E^2} = \sqrt{71.42^2 + 61.28^2} = 94.11\text{ MPa.}$

Problem 7 At a point of a strained material, the principal stresses are 100 MPa compressive and 40 MPa tensile. Determine the resultant stress in magnitude and direction on a plane inclined at 65° to the axis of the major principal stress. What is the maximum shear stress in the material?

Solution:

$$\sigma_x = -100 \text{ MPa } (-\text{ve because compressive})$$

$$\sigma_y = 40 \text{ MPa}$$

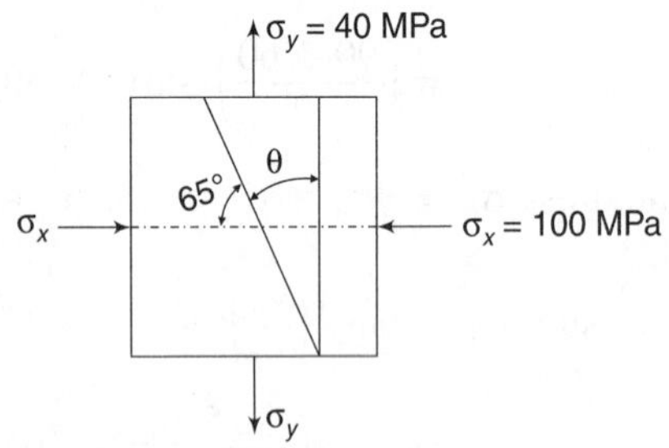

Fig. 6.16

Angle of the plane is at 65° with the axis.

Hence, θ = 90 – 65 = 25° from the plane of maximum principal stress.

$$\text{Normal stress on inclined plane } \sigma_n = \frac{\sigma_x + \sigma_y}{2} + \left(\frac{\sigma_x - \sigma_y}{2}\right) \cos 2\theta$$

$$= \frac{-100 + 40}{2} + \left(\frac{-100 - 40}{2}\right) \cos (2 \times 25) = -\ 74.995 \text{ MPa}$$

$$\text{Shear stress on inclined plane } \sigma_t = \left(\frac{\sigma_x - \sigma_y}{2}\right) \sin 2\theta$$

$$= \left(\frac{-\ 100 - 40}{2}\right) \sin 2 \times 25 = -\ 53.62 \text{ MPa}$$

$$\text{Resultant stress } \sigma_R = \sqrt{\sigma_n{}^2 + \sigma_t{}^2} = \sqrt{(-74.995)^2 + (-53.62)^2} = 92.19 \text{ MPa}$$

Inclination of resultant stress from the inclined plane

$$\phi = \tan^{-1} \frac{\sigma_n}{\sigma_t} = \tan^{-1} \left(\frac{-74.995}{-53.62}\right) = 54.44°$$

$$\text{Maximum shear stress } \tau_{\max} = \frac{\sigma_1 - \sigma_2}{2} = \frac{\sigma_x - \sigma_y}{2}$$

$$\tau_{\max} = \left|\frac{-100 - 40}{2}\right| = \pm\ 70 \text{ MPa.}$$

Problem 8 For a general two-dimensional stress system, show that (i) sum of normal stresses in any two mutually perpendicular direction is constant, and (ii) principal planes are also the normal planes.

Solution:

(i) For generating two-dimensional stress system,

$$\text{Normal stress } \sigma_n = \frac{\sigma_x + \sigma_y}{2} + \frac{\sigma_x - \sigma_y}{2} \cos 2\theta + \tau \sin 2\theta \qquad \text{...(i)}$$

$$\text{and shear stress } \sigma_t = \frac{\sigma_x - \sigma_y}{2} \sin 2\theta - \tau \cos 2\theta \qquad \text{...(ii)}$$

Let σ_n^1 be the stress on a plane at $\theta + 90°$

Substituting $\theta = \theta + 90°$ for the stress on perpendicular plane

$$\sigma_n^1 = \frac{\sigma_x + \sigma_y}{2} = \left(\frac{\sigma_x - \sigma_y}{2}\right) \cos 2\,(90 + \theta) + \tau \sin 2(90 + \theta)$$

$$\sigma_n^1 = \frac{\sigma_x + \sigma_y}{2} + \frac{\sigma_x - \sigma_y}{2} \cos (180 + 2\theta) + \tau \sin (180 + 2\theta)$$

$$= \frac{\sigma_x + \sigma_y}{2} - \left(\frac{\sigma_x - \sigma_y}{2}\right) \cos 2\theta - \tau \sin 2\theta \qquad \text{...(iii)}$$

Adding σ_n and σ_n^1 [(adding Eqs. (i) and (ii)]

$$\sigma_n + \sigma_n^1 = \sigma_x + \sigma_y$$

Hence, the sum of normal stresses on any two mutually perpendicular planes is constant and is equal to $(\sigma_x + \sigma_y)$.

(ii) For the maximum normal stress, $\dfrac{d\sigma_n}{d\theta} = 0$

$$\frac{d\sigma_n}{d\theta} = \left(\frac{\sigma_x - \sigma_y}{2}\right) 2\,(-\sin 2\theta) + \tau.2\,(\cos 2\theta) = 0$$

$$\frac{\sigma_x - \sigma_y}{2} \sin 2\theta - \tau \cos 2\theta = 0$$

$$\sigma_t = 0$$

For principal planes, the shear stress on those plane is zero. This is the condition for principal planes.

Thus, principal planes are also the normal planes.

Problem 9 A rectangular block is subjected to stresses as shown in Fig. 6.17. Find the direction and magnitude of principal stresses and maximum shear stress. Sketch the planes and mark the stresses on planes.

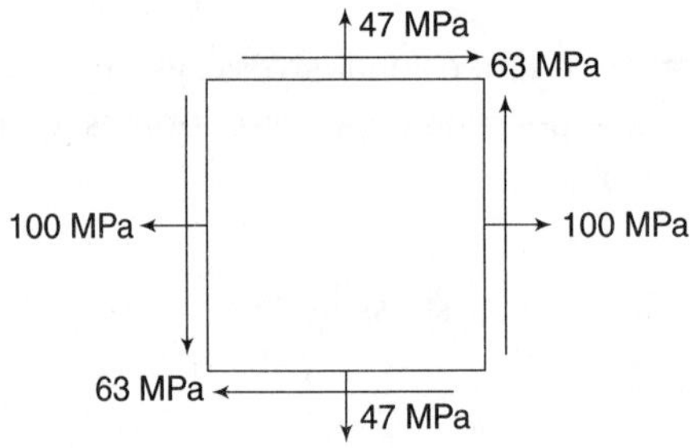

Fig. 6.17

Solution: $\sigma_x = 100$ MPa, $\sigma_y = 47$ MPa, $\tau = 63$ MPa

(i) Maximum principal stress $\sigma_1 = \dfrac{\sigma_x + \sigma_y}{2} + \sqrt{(\sigma_x - \sigma_y)^2 + \tau^2}$

$$= \frac{100 + 47}{2} + \sqrt{\left(\frac{100 - 47}{2}\right)^2 + 63^2} = 141.847 \text{ MPa}$$

(ii) Minimum principal stress $\sigma_z = \dfrac{\sigma_x + \sigma_y}{2} - \sqrt{\left(\dfrac{\sigma_x - \sigma_y}{2}\right)^2 + \tau^2}$

$$= \frac{100 + 47}{2} - \sqrt{\left(\frac{100 - 47}{2}\right)^2 + 63^2} = 5.153 \text{ MPa}$$

(iii) Direction of principal stresses

$$\tan 2\theta_1 = \frac{2\tau}{\sigma_x - \sigma_y} = \frac{2 \times 63}{100 - 47}$$

$\therefore$ $2\theta_1 = 67.1867°$ and $2\theta_2 = 180 + 2\theta_1 = 247.1867°$

$\therefore$ $\theta_1 = 33.5933°$ and $\theta_2 = 123.5933°$

(iv) Maximum shear stress $\tau_{max} = \sqrt{\left(\dfrac{\sigma_x - \sigma_z}{2}\right)^2 + \tau^2}$

$$\tau_{max} = \sqrt{\left(\frac{100 - 47}{2}\right)^2 + 63^2} = \pm\ 68.347 \text{ MPa}$$

(v) Directing of shear plane

$$\tan 2\theta_t = -\frac{(\sigma_x - \sigma_y)}{2\tau} = \frac{-(100 - 47)}{2 \times 63}$$

$$2\theta_t = -\ 22.813 \text{ or } -\ 22.813 + 180 = 157.186°$$

$\therefore$ $\theta_t = -\ 11.407$ or $78.593°$

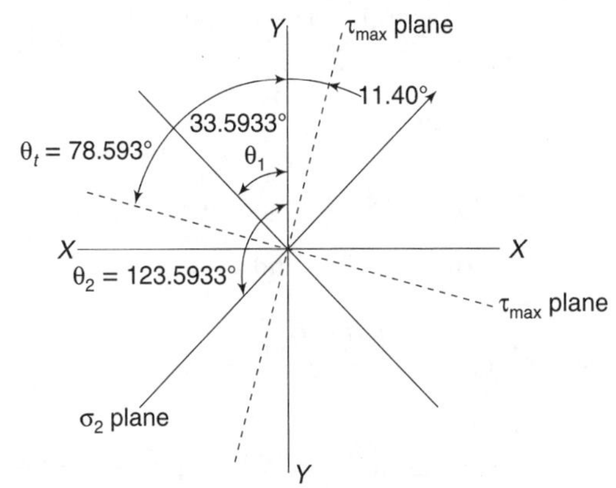

Fig. 6.17(a)

Problem 10 A plane element is subjected to 60 N/mm^2 tensile and 45 N/mm^2 compressive on two mutually perpendicular planes and shear stress of 15 N/mm^2. Determine the principal stress, maximum shear stress and their planes.

Solution: σ_x = 60 MPa, σ_y = – 45 MPa (Compressive), τ = 15 MPa

The stressed element is shown in Fig. 6.18(a)

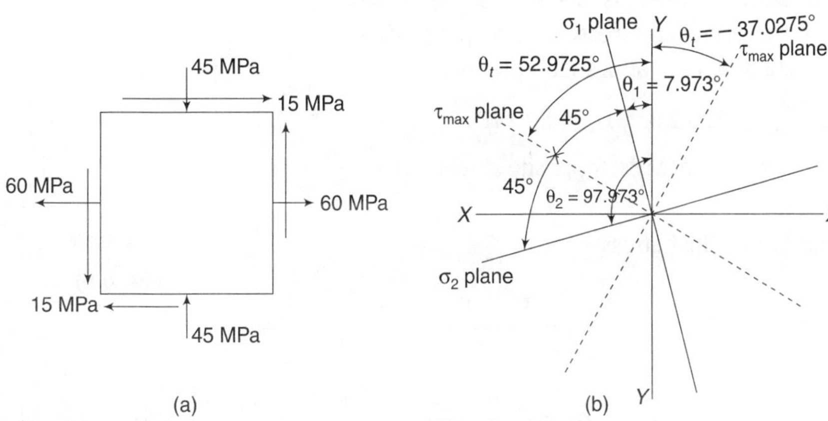

Fig. 6.18

(i) Maximum principal stress $\sigma_1 = \dfrac{\sigma_x + \sigma_y}{2} + \sqrt{\left(\dfrac{\sigma_x - \sigma_y}{2}\right)^2 + \tau^2}$

$$= \frac{60 - 45}{2} + \sqrt{\left(\frac{60 + 45}{2}\right)^2 + 15^2} = -\ 62.1 \text{ MPa}$$

(ii) Minimum principal stress $\sigma_2 = \dfrac{\sigma_x + \sigma_y}{2} - \sqrt{\left(\dfrac{\sigma_x - \sigma_y}{2}\right)^2 + \tau^2}$

$$= \frac{60 - 45}{2} - \sqrt{\left(\frac{60 + 45}{2}\right)^2 + 15^2} = -47.1 \text{ MPa}$$

(iii) Angles of principal planes $\tan 2\theta_1 = \dfrac{2\tau}{\sigma_x - \sigma_y} = \dfrac{2 \times 15}{60 + 45}$

$$\therefore \quad 2\theta_1 = 15.945° \text{ and } 2\theta_2 = 15.945 + 180 = 195.945°$$

$$\theta_1 = 7.973° \text{ and } \theta_2 = 97.973°$$

(iv) Maximum shear stress $\tau_{max} = \sqrt{\left(\dfrac{\sigma_x - \sigma_y}{2}\right)^2 + \tau^2}$

$$\tau_{max} = \sqrt{\left(\frac{60 + 45}{2}\right)^2 + 15^2} = \pm 54.6 \text{ MPa}$$

(v) Angle of shear planes $\tan 2\theta_t = \dfrac{-(\sigma_x - \sigma_y)}{2\tau} = \dfrac{-(60 + 45)}{2 \times 15}$

$$2\theta_t = -74.055 \text{ and } -74.055 + 180° = 105.945°$$

$$\therefore \quad \theta_t = -37.0275° \text{ and } 52.9725°$$

The principal and shear planes are shown in Fig. 6.18(b).

Problem 11 The state of stress in a two dimensional stressed member is shown in Fig. 6.19. Determine the principal stresses, principal planes, maximum shear stress and shear planes.

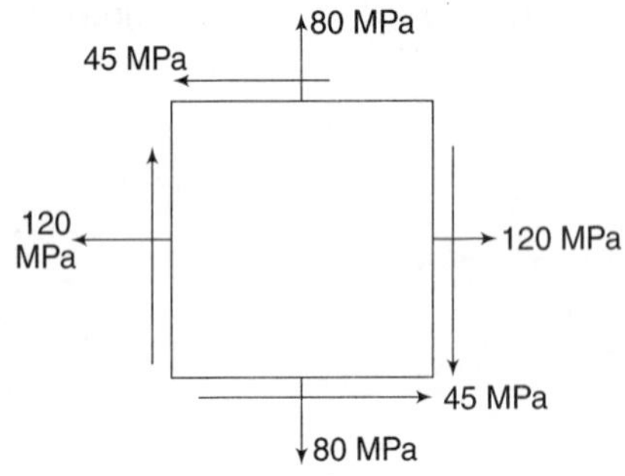

Fig. 6.19

Solution: $\sigma_x = 120$ MPa, $\sigma_y = 80$ MPa

$\tau = -45$ MPa (opposite direction)

Maximum principal stress

$$\sigma_1 = \frac{\sigma_x + \sigma_y}{2} + \sqrt{\left(\frac{\sigma_x - \sigma_y}{2}\right)^2 + \tau^2}$$

$$\sigma_1 = \frac{120 + 80}{2} + \sqrt{\left(\frac{120 - 80}{2}\right)^2 + (-45)^2} = 149.24 \text{ MPa}$$

Minimum principal stress $\sigma_z = \dfrac{\sigma_x + \sigma_y}{2} - \sqrt{\left(\dfrac{\sigma_x - \sigma_y}{2}\right)^2 + \tau^2}$

$$= \frac{120 + 80}{2} - \sqrt{\left(\frac{120 - 80}{2}\right)^2 + (-45)^2} = 50.76 \text{ MPa}$$

Angles of principal planes with the plane of σ_x.

$$\tan 2\theta = \frac{2\tau}{\sigma_x - \sigma_y} = \frac{2(-45)}{120-80}$$

$\therefore$ $$2\theta_1 = -66.04 \text{ and } 2\theta_2 = -60.04 + 180 = 113.96°$$

$\therefore$ $$\theta_1 = -33.02 \text{ and } \theta_2 = 56.98°$$

Maximum shear stress $$\tau_{max} = \sqrt{\left(\frac{\sigma_x - \sigma_y}{2}\right)^2 + \tau^2}$$

$$= \sqrt{\left(\frac{120-80}{2}\right)^2 + (-45)^2} = \pm 49.24 \text{ MPa}$$

Angles of shear planes with the plane of σ_x

$$\tan 2\theta_t = \frac{\sigma_x - \sigma_y}{-2\tau} = \frac{120-80}{-2(-45)}$$

$\therefore$ $$2\theta_t = 23.96° \text{ or } 23.96 + 180 = 203.96°$$

$\therefore$ $$\theta_t = 11.98° \text{ or } 101.98°$$

The planes of principal stresses and shear stress are shown in Fig. 6.19(a).

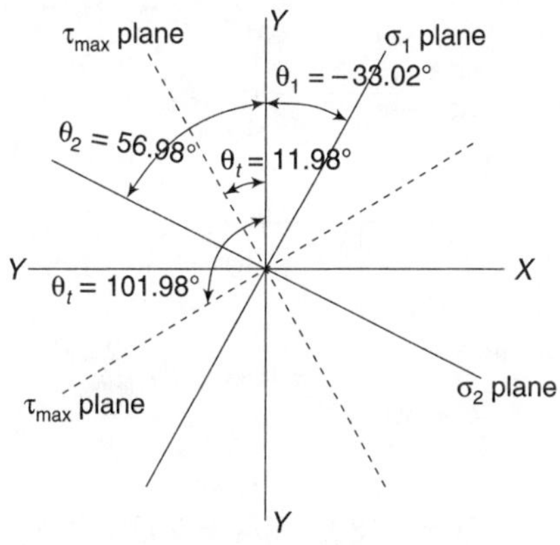

Fig. 6.19(a)

Problem 12 An element is subjected to stresses as shown in Fig. 6.20. Determine the tangential and normal stress along plane inclined at 30°. Also find principal stresses, maximum shear stress and their directions.

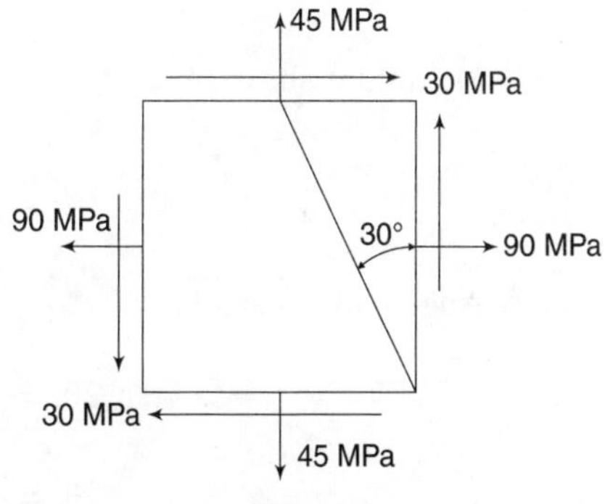

Fig. 6.20

Solution: $$\sigma_x = 90 \text{ MPa}$$

$$\sigma_y = 45 \text{ MPa}$$

$$\tau = 30 \text{ MPa}$$

Angle of plane $\theta = 30°$

(i) Normal stress along inclined plane

$$\sigma_n = \frac{\sigma_x + \sigma_y}{2} + \frac{\sigma_x - \sigma_y}{2}\cos 2\theta + \tau \sin 2\theta$$

$$= \frac{90 + 45}{2} + \frac{90 - 45}{2}\cos(2 \times 30) + 30 \sin(2 \times 30) = 104.73 \text{ MPa}$$

(ii) Shear stress or tangential stress along inclined plane

$$\sigma_t = \frac{\sigma_x - \sigma_y}{2}\sin 2\theta - \tau \cos 2\theta$$

$$= \frac{90 - 45}{2}\sin(2 \times 30) - 30 \cos(2 \times 30) = 4.49 \text{ MPa}$$

(iii) Maximum principal stress

$$\sigma_1 = \frac{\sigma_x + \sigma_y}{2} + \sqrt{\left(\frac{\sigma_x - \sigma_y}{2}\right)^2 + \tau^2}$$

$$\sigma_1 = \frac{90 + 45}{2} + \sqrt{\left(\frac{90 - 45}{2}\right)^2 + 30^2} = 105 \text{ MPa}$$

(iv) Minimum principal stress

$$\sigma_2 = \frac{\sigma_x + \sigma_y}{2} - \sqrt{\left(\frac{\sigma_x - \sigma_y}{2}\right)^2 + \tau^2}$$

$$= \frac{90 + 45}{2} - \sqrt{\left(\frac{90 - 45}{2}\right)^2 + 30^2} = 30 \text{ MPa}$$

(v) Angles of principal planes

$$\tan 2\theta_1 = \frac{2\tau}{\sigma_x - \sigma_y} = \frac{2 \times 30}{90 - 45}$$

$$\therefore \quad 2\theta_1 = 53.13° \text{ and } 2\theta_2 = 180 + 2\theta_1 = 233.13°$$

$$\therefore \quad \theta_1 = 26.57° \text{ and } \theta_2 = 116.57°$$

(vi) Maximum shear stress $\tau_{max} = \sqrt{\left(\frac{\sigma_x - \sigma_y}{2}\right)^2 + \tau^2}$

$$= \sqrt{\left(\frac{90 - 45}{2}\right)^2 + 36^2} = \pm 37.5 \text{ MPa}$$

(vii) Angles of shear planes

$$\tan 2\theta_t = \frac{\sigma_x - \sigma_y}{-2\tau} = \frac{90 - 45}{-2 \times 30}$$

$$\therefore \quad 2\theta_t = -36.87° \text{ or } 143.13°$$

$$\theta_t = -18.43° \text{ or } 71.57°.$$

Problem 13 An element is subjected to stresses as shown in Fig. 6.21 Determine

(i) Principal stresses and their planes

(ii) Normal and tangential stresses along plane *BC*

(iii) Maximum shear stress and its location.

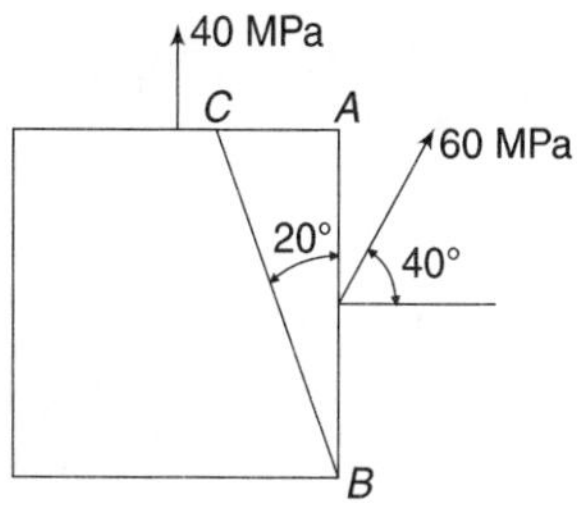

Fig. 6.21

Solution: Resolving the inclined stress 60 MPa into horizontal and vertical components

Horizontal component = 60 cos 40 = 45.96 MPa

Vertical component = 60 sin 40 = 38.57 MPa

The horizontal component is stress in *x*-direction and the vertical component is shear stress on plane.

The element with equivalent stresses is shown in Fig. 6.21(a).

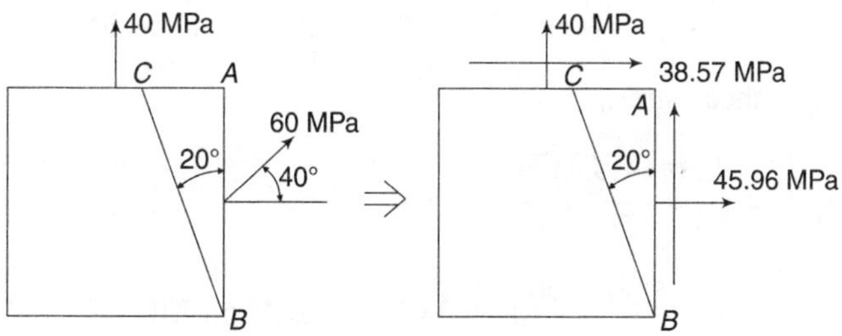

Fig. 6.21(a)

Now σ_x = 45.96 MPa, σ_y = 40 MPa and τ = 38.57 MPa

(i) Principal stresses and their planes

(a) Maximum principal stress

$$\sigma_1 = \frac{\sigma_x + \sigma_y}{2} + \sqrt{\left(\frac{\sigma_x - \sigma_y}{2}\right)^2 + \tau^2}$$

$$= \frac{45.96 + 40}{2} + \sqrt{\left(\frac{45.96 - 40}{2}\right)^2 + 38.57^2} = 81.66 \text{ MPa}$$

(b) Minimum principal stress

$$\sigma_2 = \frac{\sigma_x + \sigma_y}{2} - \sqrt{\left(\frac{\sigma_x - \sigma_y}{2}\right)^2 + \tau^2}$$

$$= \frac{45.96 + 40}{2} - \sqrt{\left(\frac{45.96 - 40}{2}\right)^2 + 38.57^2} = 4.295 \text{ MPa}$$

(c) Angles of principal planes

$$\tan 2\theta = \frac{2\tau}{\sigma_x - \sigma_y} = \frac{2 \times 38.57}{45.96 - 40} = 12.943$$

$\therefore \quad 2\theta_1 = 85.58 \text{ or } 2\theta_2 = 2\theta_1 + 180 = 265.58°$

$\therefore \quad \theta_1 = 42.79° \text{ and } \theta_2 = 132.79°$

(ii) Normal and shear stress along inclined plane *BC*

(a) Normal stress on inclined plane *BC*

$$\sigma_n = \frac{\sigma_x + \sigma_y}{2} + \frac{\sigma_x - \sigma_y}{2}\cos 2\theta + \tau \sin 2\theta$$

$$= \frac{45.96 + 40}{2} + \frac{45.96 - 40}{2}\cos 2 \times 20 + 38.57 \sin 2 \times 20$$

$$= 70.06 \text{ MPa}$$

(b) Shear stress on inclined plane *BC*

$$\sigma_t = \frac{\sigma_x - \sigma_y}{2}\sin 2\theta - \tau \cos 2\theta$$

$$= \frac{45.96 - 40}{2}\sin (2 \times 20) - 38.57 \cos (2 \times 20) = -27.63 \text{ MPa}$$

(iii) (a) Maximum shear stress

$$\tau_{max} = \sqrt{\left(\frac{\sigma_x - \sigma_y}{2}\right)^2 + \tau^2}$$

$$= \sqrt{\left(\frac{45.96 - 40}{2}\right)^2 + 38.57^2} = \pm 38.68 \text{ MPa}$$

(b) Angles of shear planes

$$\tan 2\theta_t = \frac{\sigma_x - \sigma_y}{-2\tau} = \frac{45.96 - 40}{-2 \times 38.57}$$

$\therefore \quad 2\theta_t = -4.418° \text{ or } 175.58°$

$\therefore \quad \theta_t = -2.209° \text{ or } 87.79°.$

Problem 14 An element is subjected to stresses across two perpendicular planes as shown in Fig. 6.22. Determine

(i) Normal and shear stresses and resultant stress on plane *EC*.

(ii) Maximum and minimum principal stresses and their obliquity.

(iii) Maximum shear stress and its obliquity.

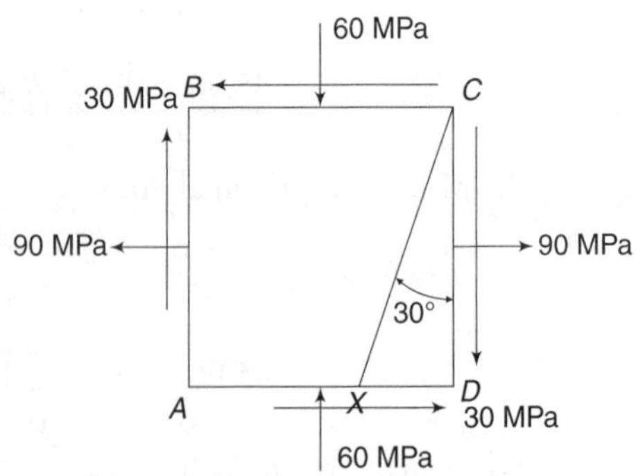

Fig. 6.22

Solution: $\sigma_x = 90$ MPa,

$\sigma_y = -60$ MPa (Compressive)

$\tau = 30$ MPa, $\theta = 30°$

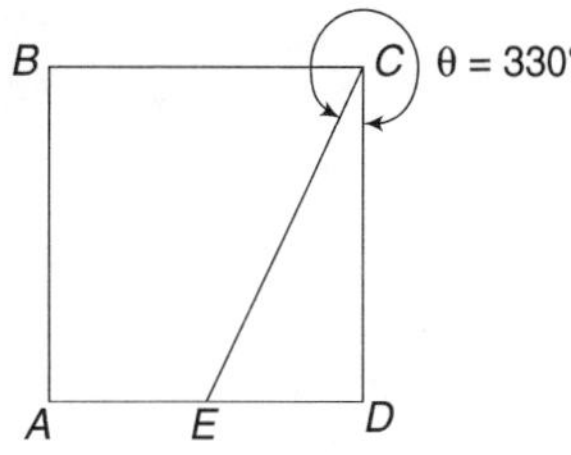

Fig. 6.22(a)

But θ is the angle to the plane from plane of σ_x in counter-clockwise direction.

$$\therefore \quad \theta = 360 - \theta = 360 - 30 = 330°$$

(i) Normal and shear stresses on plane *CE*

(a) Normal stress $\sigma_n = \dfrac{\sigma_x + \sigma_y}{2} + \dfrac{\sigma_x - \sigma_y}{2} \cos 2\theta + \tau \sin 2\theta$

$$= \frac{90 + (-60)}{2} + \frac{90 - (-60)}{2} \cos (2 \times 330) + 30 \sin (2 \times 330)$$

$$\sigma_n = 26.52 \text{ MPa}$$

(b) Shear stress $\sigma_t = \dfrac{\sigma_x - \sigma_y}{2} \sin 2\,\theta - \tau \cos 2\,\theta$

$$= \frac{90 + 60}{2} \sin (2 \times 330) - 30 \cos (2 \times 330) = -79.95 \text{ MPa}$$

(c) Resultant stress on plane $CE = \sqrt{\sigma_n^2 + \sigma_t^2}$

$$= \sqrt{26.52^2 + (-79.95)^2} = 84.235 \text{ MPa}$$

(d) Obliquity $\tan \phi = \left|\dfrac{\sigma_t}{\sigma_n}\right| \Rightarrow = \left(\dfrac{79.95}{26.52}\right)$

$\therefore \quad \phi = 71.65°$.

(**Note**: Given angle of inclined plane *CE* can be taken as θ = – 30° in the above σ_n and σ_t calculating)

(ii) Principal stresses and obliquity

(a) Maximum principal stress

$$\sigma_1 = \frac{\sigma_x + \sigma_y}{2} - \sqrt{\left(\frac{\sigma_x - \sigma_y}{2}\right)^2 + \tau^2}$$

$$= \frac{90 - 60}{2} + \sqrt{\left(\frac{90 + 60}{2}\right)^2 + 30^2} = 95.78 \text{ MPa}$$

(b) Minimum principal stress

$$\sigma_2 = \frac{\sigma_x + \sigma_y}{2} - \sqrt{\left(\frac{\sigma_x - \sigma_y}{2}\right)^2 + \tau^2}$$

$$= \frac{90 - 60}{2} - \sqrt{\left(\frac{90 + 60}{2}\right)^2 + 30^2} = -65.78 \text{ MPa}$$

(c) Angle of principal planes (obliquity)

$$\tan 2\theta = \frac{2\tau}{\sigma_x - \sigma_y} = \frac{2 \times 30}{90 - (-60)} = 0.4$$

$$2\theta = 21.8° \text{ or } 21.8 + 180 = 201.8$$

$$\therefore \quad \theta = 10.9° \text{ or } 100.9°$$

(iii) (a) Maximum shear stress $\tau_{max} = \sqrt{\left(\frac{\sigma_x - \sigma_y}{2}\right)^2 + \tau^2}$

$$= \sqrt{\left(\frac{90 + 60}{2}\right)^2 + 30^2} = \pm\ 80.78 \text{ MPa}$$

(b) Angle of shear planes $\tan 2\theta_t = \frac{\sigma_x - \sigma_y}{-2\tau} = \frac{90 + 60°}{-2 \times 30}$

$$\therefore \quad 2\theta_t = -68.19° \text{ or } -68.19 + 180 = 111.8°$$

$$\therefore \quad \theta_t = -34.1° \text{ or } 55.9°.$$

Problem 15 Figure 6.23 shows an element subjected to stresses. Determine

(i) Normal and shear stresses along plane *DE* inclined at 30°

(ii) Principal stresses and their location

(iii) Maximum shear stress and its location

(iv) Sketch the principal planes.

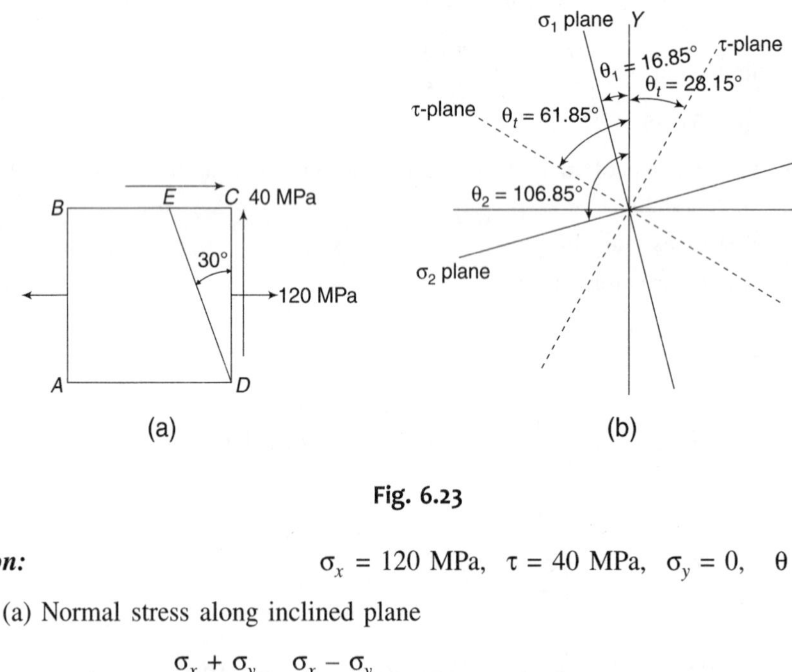

Fig. 6.23

Solution: $\sigma_x = 120 \text{ MPa}, \quad \tau = 40 \text{ MPa}, \quad \sigma_y = 0, \quad \theta = 30°$

(i) (a) Normal stress along inclined plane

$$\sigma_n = \frac{\sigma_x + \sigma_y}{2} + \frac{\sigma_x - \sigma_y}{2} \cos 2\theta + \tau \sin 2\theta$$

$$= \frac{120 + 0}{2} + \frac{120 - 0}{2} \cos (2 \times 30) + 40 \sin 2\,(30) = 124.64 \text{ MPa}$$

(b) Shear stress along plane *DE*

$$\sigma_t = \frac{\sigma_x - \sigma_y}{2} \sin 2\theta - \tau \cos 2\theta$$

$$= \frac{120 - 0}{2} \sin (2 \times 30) - 40 \cos (2 \times 30) = 31.96 \text{ MPa}$$

(ii) (a) Maximum principal stress

$$\sigma_1 = \frac{\sigma_x + \sigma_y}{2} + \sqrt{\left(\frac{\sigma_x - \sigma_y}{2}\right)^2 + \tau^2}$$

$$= \frac{120 + 0}{2} + \sqrt{\left(\frac{120 - 0}{2}\right)^2 + 40^2} = 132.11 \text{ MPa}$$

(b) Minimum principal stress

$$\sigma_2 = \frac{\sigma_x + \sigma_y}{2} - \sqrt{\left(\frac{\sigma_x - \sigma_y}{2}\right)^2 + \tau^2}$$

$$= \frac{120 + 0}{2} - \sqrt{\left(\frac{120 - 0}{2}\right)^2 + 40^2} = -12.11 \text{ MPa}$$

(c) Planes of principal stresses

$$\tan 2\theta = \frac{2\tau}{\sigma_x - \sigma_y} = \frac{2 \times 40}{120 - 0} = 0.667$$

$$2\theta = 33.69° \text{ or } 180 + 33.69 = 213.69°$$

$$\therefore \quad \theta_1 = 16.85° \text{ and } \theta_2 = 106.85°$$

(iii) (a) Maximum shear stress

$$\tau_{max} = \sqrt{\left(\frac{\sigma_x - \sigma_y}{2}\right)^2 + \tau^2} = \sqrt{\left(\frac{120 - 0}{2}\right)^2 + 40^2} = \pm 72.11 \text{ MPa}$$

(b) Angle of shear plane $\tan 2\theta_t = \dfrac{\sigma_x - \sigma_y}{-2\tau} = \dfrac{120 - 0}{-2 \times 40} = -1.5$

$$2\theta_t = -56.3° \text{ or } -56.3 + 180 = 123.69°$$

$$\theta_t = -28.15° \text{ or } 61.85°$$

The planes are shown in Fig. 6.23(b).

Problem 16 An element is subjected to 100 N/mm² tensile in x-direction and 80 N/mm² compression in *y*-direction perpendicular to *x* and a shear stress τ. The maximum principal stress is to be limited to 110 N/mm². Find

(i) The magnitude of shear stress

(ii) Maximum shear stress.

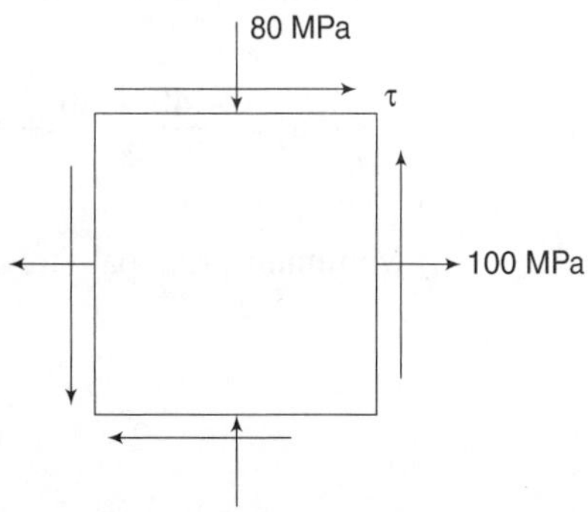

Fig. 6.24

Solution: $\sigma_x = 100$ MPa. $\sigma_y = -80$ MPa (Compressive)

Maximum principal stress $\sigma_1 = 110$ MPa

(i) Magnitude of shear stress τ

We know that maximum principal stress $\sigma_1 = \dfrac{\sigma_x + \sigma_y}{2} + \sqrt{\left(\dfrac{\sigma_x - \sigma_y}{2}\right)^2 + \tau^2}$

$$110 = \frac{100 - 80}{2} + \sqrt{\left(\frac{100 + 80}{2}\right)^2 + \tau^2}$$

$\therefore$ Shear stress $\tau = 43.59$ MPa

(ii) Maximum shear stress $\tau_{max} = \sqrt{\left(\dfrac{\sigma_x - \sigma_y}{2}\right)^2 + \tau^2}$

$$\tau_{max} = \sqrt{\left(\frac{100 + 80}{2}\right)^2 + 43.59^2} = \pm 100 \text{ MPa.}$$

Problem 17 An element is subjected to biaxial stresses accompanied with the shear stress as shown in Fig. 6.25. Determine

(a) The principal stresses and their directions

(b) Maximum shear stress and its direction

(c) Sketch the planes.

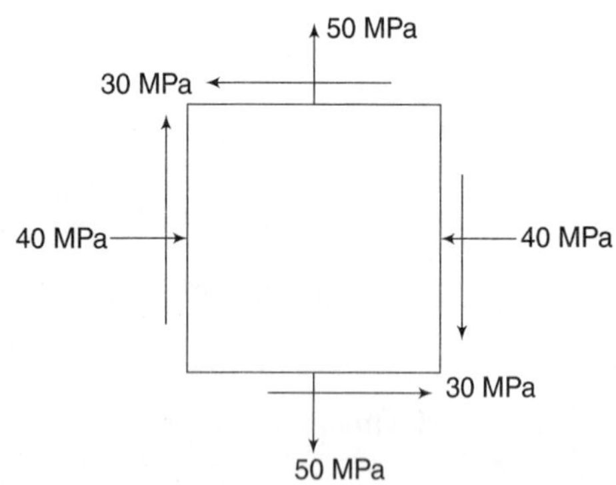

Fig. 6.25

Solution:

$\sigma_x = -40$ MPa (Compressive)

$\sigma_y = 50$ MPa, $\tau = -30$ MPa (Opposite direction)

(a) (i) Maximum principal stresses $\sigma_1 = \dfrac{\sigma_x + \sigma_y}{2} + \sqrt{\left(\dfrac{\sigma_x - \sigma_y}{2}\right)^2 + \tau^2}$

$$\sigma_1 = \frac{-40 + 50}{2} + \sqrt{\left(\frac{-40 - 50}{2}\right)^2 + (-30)^2} = 59.08 \text{ MPa}$$

(ii) Minimum principal stress $\sigma_2 = \dfrac{\sigma_x + \sigma_y}{2} - \sqrt{\left(\dfrac{\sigma_x - \sigma_y}{2}\right)^2 + \tau^2}$

$$\sigma_2 = \frac{-40 + 50}{2} - \sqrt{\left(\frac{-40 - 50}{2}\right)^2 + (-30)^2} = -49.08 \text{ MPa}$$

(iii) Angle of principal planes $\tan 2\theta = \left(\dfrac{2\tau}{\sigma_x - \sigma_y}\right) = \left(\dfrac{2 \times -30}{-40 - 50}\right)$

$$2\theta = 33.69° \text{ or } 180 + 33.69 = 213.69°$$

$$\therefore \quad \theta_1 = 16.85° \text{ and } \theta_2 = 106.85°$$

(b) (i) Maximum shear stress $\tau_{max} = \sqrt{\left(\frac{\sigma_x - \sigma_y}{2}\right)^2 + \tau^2}$

$$\tau_{max} = \sqrt{\left(\frac{-40-50}{2}\right)^2 + (-30)^2} = 54.08 \text{ MPa}$$

(ii) Angle of shear plane $\tan 2\theta_t = \left(\frac{\sigma_x - \sigma_y}{-2\tau}\right)$

$$\tan 2\theta_t = \left(\frac{-40-50}{-2 \times 30}\right)$$

$$\therefore \quad 2\theta_t = 56.31 \text{ or } 56.31 + 180 = -236.31°$$

$$\therefore \quad \theta_t = 28.155° \text{ or } 118.155° \text{ (Both –ve)}$$

(c) Location of planes

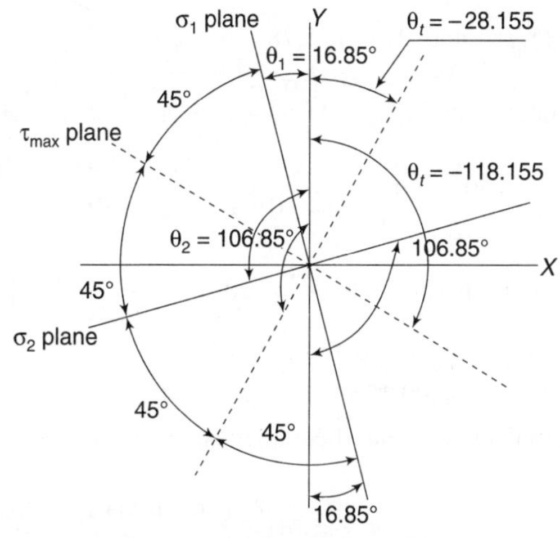

Fig. 6.25(a)

The principal and shear planes are shown in Fig. 6.25.

Problem 18 The state of stress in two-dimensionally stressed body is shown in Fig. 6.26. Determine the principal stresses and planes, maximum shear stresses and their planes. Indicate the plane by a sketch.

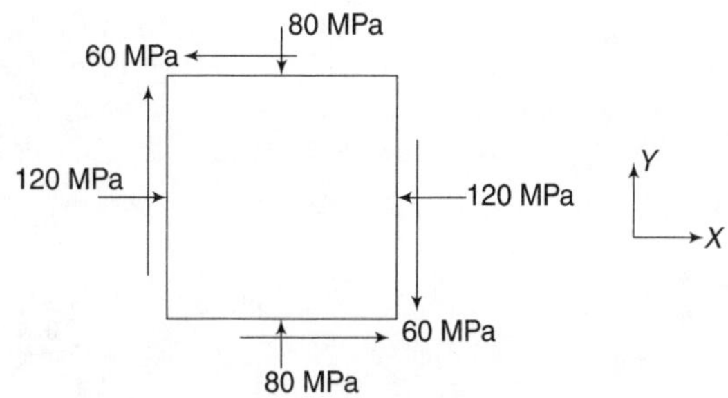

Fig. 6.26

Solution:

$$\sigma_x = -120 \text{ MPa (Compressive)}$$

$\sigma_y = -\ 80$ MPa (Compressive)

$\tau = -\ 60$ MPa (Opposite direction)

(i) (a) Maximum principal stress $\sigma_1 = \dfrac{\sigma_x + \sigma_y}{2} + \sqrt{\left(\dfrac{\sigma_x - \sigma_y}{2}\right)^2 + \tau^2}$

$$\sigma_1 = \frac{-\ 120 - 80}{2} + \sqrt{\left(\frac{-\ 120 + 80}{2}\right)^2 + (-\ 60)^2} = -\ 36.75 \text{ MPa}$$

(b) Minimum principal stress $\sigma_2 = \dfrac{\sigma_x + \sigma_y}{2} - \sqrt{\left(\dfrac{\sigma_x - \sigma_y}{2}\right)^2 + \tau^2}$

$$\sigma_2 = \frac{-120 - 80}{2} - \sqrt{\left(\frac{-120 + 80}{2}\right)^2 + (-\ 60)^2} = -\ 163.25 \text{ MPa}$$

(c) Planes of principal stresses $\tan 2\theta = \dfrac{2\tau}{\sigma_x - \sigma_y} \left[\dfrac{2(-60)}{-120 + 80}\right]$

$\therefore \quad 2\theta = 71.565$ or $71.565 + 180 = 251.565$

$\therefore \quad \theta_1 = 35.78°$ and $\theta_2 = 125.78°$

(ii) (a) Maximum shear stress $\tau_{max} = \sqrt{\left(\dfrac{\sigma_x - \sigma_y}{2}\right)^2 + \tau^2}$

$$\tau_{max} = \sqrt{\left(\frac{-120 + 80}{2}\right)^2 + (-\ 60)^2} = 63.25 \text{ MPa}$$

(b) Planes of maximum shear stress $\tan 2\theta_t = \left(\dfrac{\sigma_x - \sigma_y}{-\ 2\tau}\right) = \left(-\dfrac{120 + 80}{-\ 2 \times 60}\right)$

$\therefore \quad 2\theta_t = 18.43°$ or $180 + 18.43 = 198.43°$

$\therefore \quad \theta_t = 9.215°$ or $99.125°$

(iii) Position of principal planes and shear planes.

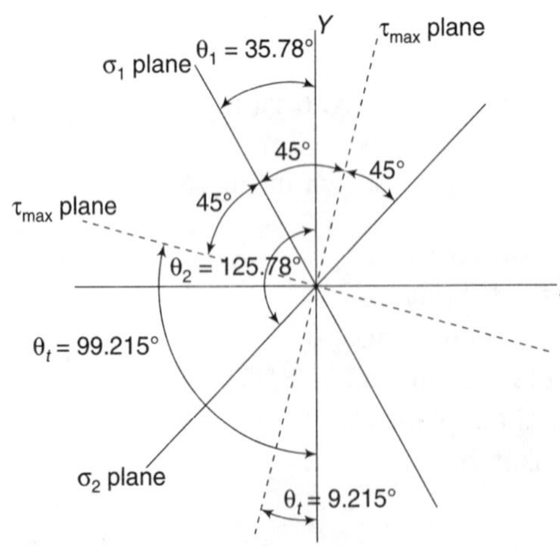

Fig. 6.26(a)

Problem 19 A block is 90 mm long, 60 mm high and 30 mm thick. It is subjected to uniformly distribute forces of 30 kN and 18 kN as shown in Fig. 6.27. Calculate the normal and shear stresses and the resultant stress along the diagonal.

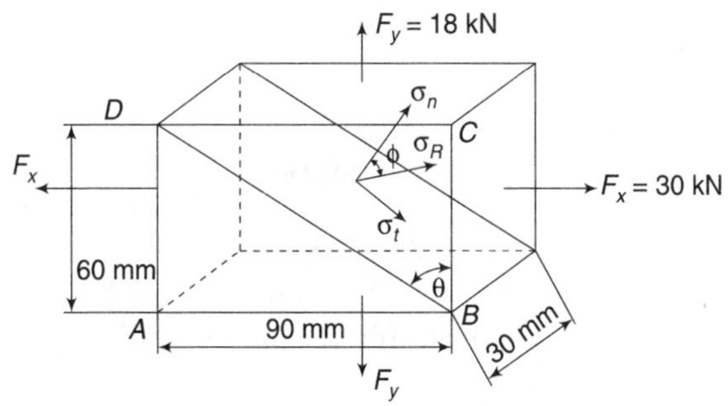

Fig. 6.27

Solution:

$$F_x = 30000 \text{ N acting an area } 60 \times 30$$

$$\therefore \quad \text{Stress } \sigma_x = \frac{F_x}{A_x} = \frac{30000}{60 \times 30} = 16.67 \text{ MPa}$$

$$F_y = 18000 \text{ N acting an area } 90 \times 30.$$

$$\therefore \quad \text{Stress } \sigma_y = \frac{F_y}{A_y} = \frac{18000}{90 \times 30} = 6.67 \text{ MPa}$$

Inclined plane angle $\theta = \tan^{-1}\left(\frac{90}{60}\right) = 56.31°$

Stress on inclined plane:

(i) Normal stress $\sigma_n = \frac{\sigma_x + \sigma_y}{2} + \left(\frac{\sigma_x - \sigma_y}{2}\right) \cos 2\theta = 11.28$ MPa

(ii) Shear stress $\sigma_t = \frac{\sigma_x - \sigma_y}{2} \sin 2\theta = 9.23$ MPa

(iii) Resultant stress $\sigma_R = \sqrt{\sigma_n^2 + \sigma_t^2} = \sqrt{11.28^2 + 9.23^2} = 14.58$ MPa

(iv) Angle of resultant stress $\phi = \tan^{-1}\left(\frac{\sigma_t}{\sigma_n}\right) = 39.29°$.

Problem 20 An element is subjected to stresses as shown in Fig. 6.28. A circle of diameter 200 mm on the element is converted into ellipse after the application stresses. Determine the major and minor axis of the ellipse and their orientation. Taken $E = 2 \times 10^5$ MPa and Poisson's ratio as 0.3.

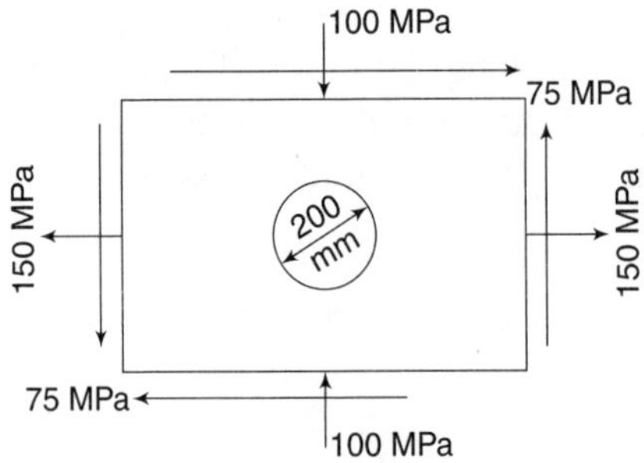

Fig. 6.28

Solution: $\sigma_x = 150$ MPa, $\sigma_y = -100$ MPa (Comp.) $\tau = 75$ MPa.

$E = 2 \times 10^5$ MPa. Poisson's ratio $\mu = 0.3$

The circle elongates due to maximum principal stress and compresses due to minimum principal stress and thus becomes an ellipse.

(i) Maximum principal stress $\sigma_1 = \dfrac{\sigma_x + \sigma_y}{2} + \sqrt{\left(\dfrac{\sigma_x - \sigma_y}{2}\right)^2 + \tau^2}$

$$= \frac{150 - 100}{2} + \sqrt{\left[\frac{150 - (-100)}{2}\right]^2 + 75^2} = 170.77 \text{ MPa (Tensile)}$$

(ii) Minimum principal stress $\sigma_2 = \dfrac{\sigma_x + \sigma_y}{2} - \sqrt{\left(\dfrac{\sigma_x - \sigma_y}{2}\right)^2 + \tau^2}$

$$= \frac{150 - 100}{2} - \sqrt{\left(\frac{150 + 100}{2}\right)^2 + 75^2} = -120.77 \text{ MPa (Compressive)}$$

(iii) Major principal strain $\in_1 = \dfrac{\sigma_1}{E} - \dfrac{\mu\,\sigma_2}{E}$

$$= \frac{170.77}{2 \times 10^5} - 0.3\left(\frac{-120.77}{2 \times 10^5}\right) = 1.035 \times 10^{-3} \text{ (Tensile)}$$

(iv) Minor principal strain $\in_2 = \dfrac{\sigma_2}{E} - \dfrac{\mu.\sigma_1}{E}$

$$= \frac{-120.77}{2 \times 10^5} - 0.3 \times \frac{170.77}{2 \times 10^5} = -0.86 \times 10^{-3} \text{ (Compressive)}$$

Stain $\in = \dfrac{\delta d}{d}$

$\therefore$ Change in diameter $\delta d = \in . d$

Major axis of ellipse $= d = \in_1 d$

$$= 200 + 1.035 \times 10^{-3} \times 200 = 200.207 \text{ mm}$$

Minor axis of ellipse = $d = \in_2 d$

$$= 200 - 0.86 \times 10^{-3} \times 200 = 199.828 \text{ mm}$$

Orientation of major axis $\theta_1 = \dfrac{1}{2}\tan^{-1}\left(\dfrac{2\tau}{\sigma_x - \sigma_y}\right)$

$$\theta_1 = \frac{1}{2}\tan^{-1}\left(\frac{2 \times 75}{150 - (-100)}\right) = 15.48°$$

$$\theta_2 = \theta_1 + 90 = 15.48 + 90 = 105.98$$

The ellipse is shown in Fig. 6.28(a).

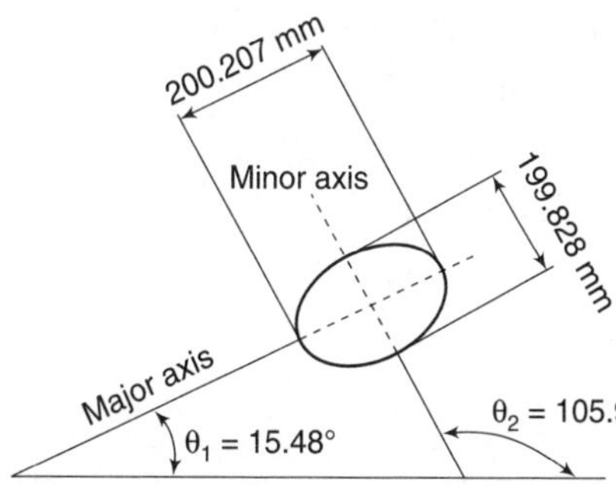

Fig. 6.28(a)

6.3 MOHR'S CIRCLE FOR PLANE STRESSES

Mohr's circle is a graphical representation of direct stresses along one axis and shear stress along another axis. By drawing Mohr's circle, principal stresses and planes, maximum shear stress and planes can be found. From this, normal and tangential stresses on any inclined plane can also be found easily.

Mohr's circle can be drawn for the following cases:

(i) A body subjected to unidirectional stress.

(ii) A body subjected to two mutually perpendicular like stresses of unequal magnitude (both tensile or both compressive).

(iii) A body subjected to two mutually perpendicular unlike and unequal stresses (one tensile and one compressive).

(iv) A body subjected to a direct stress (tensile or compressive) accompanied by a simple shear stress.

(v) A body subjected to two mutually perpendicular stresses accompanied by a simple shear stress.

6.4 SIGN CONVENTIONS FOR MOHR'S CIRCLE

Mohr's circles are easily plotted on a graph sheet. Tensile stresses are considered positive and are plotted on to the right side of origin '*O*' and compressive stresses are considered –ve and are plotted on to the left side of origin '*O*'

Tangential or shear stress (τ) is considered positive that produces clockwise moment and is drawn below x-axis.

θ is considered positive when it is in the anticlockwise direction and negative when it is in the clockwise direction. Angles are taken as 2θ on Mohr's circle.

6.5 MOHR'S CIRCLES FOR VARIOUS STRESS CONDITIONS

6.5.1 Mohr's Circles for Unidirectional Stress, σ_x

Figure 6.29 shows an element subjected to a stress σ_x.

Procedure of drawing Mohr's circle:

1. Draw x and y axes as shown in Fig. 6.29
2. Select suitable scale for representing stresses
3. From origin O take the scaled length of stress σ_x and mark point A.

   $$\text{Length } OA = \frac{\sigma_x}{\text{Scale}}$$

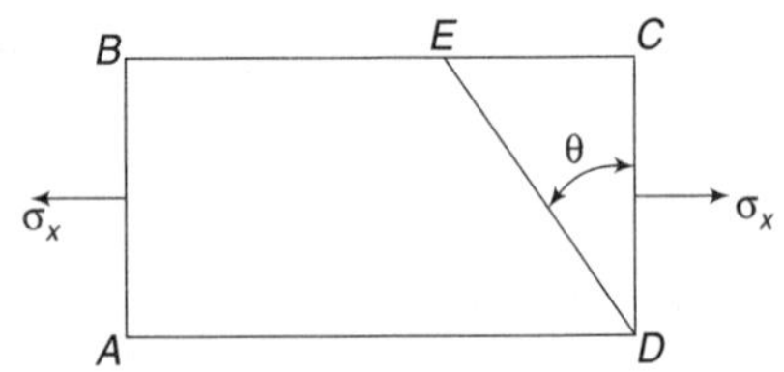

Fig. 6.29

4. Locate midpoint of OA as C. Draw a circle of radius CA from centre C.
5. This circle cuts the x-axis at two point A and O.
6. OA represents maximum principal stress.
7. From OA, measure angle 2θ (twice the given angle in counterclockwise direction) and mark point P on Mohr's circle. Drop a perpendicular PE from P.

   OE is the normal stress on inclined plane and PE is the shear stress. OP is the resultant stress.

   $$\phi = \frac{1}{2}\angle EOP$$

   $$\sigma_n = OE,\ \sigma_t = PE,\ \sigma_R = OP$$

   (**Note:** Remember to multiply these with scale)
8. Angle of obliquity of resultant stress is given by angle EOP and is equal to 2ϕ
9. *Mohr's circle:* Radius of Mohr's circle is the maximum shear stress $\tau_{max} = CP = CA = OC$

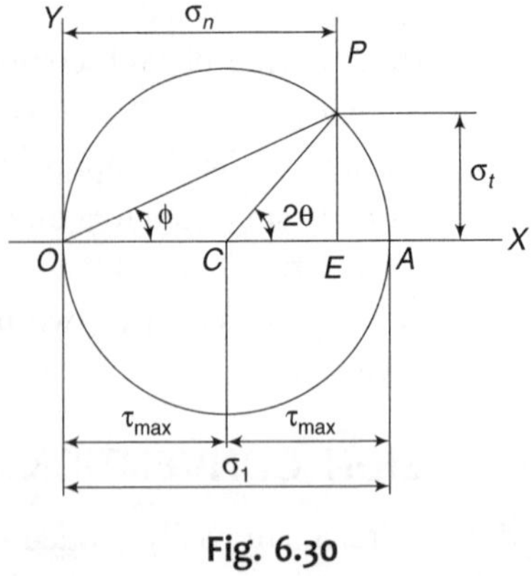

Fig. 6.30

Proof:

(i) Maximum principal stress

$$\sigma_1 = OA = \sigma_x$$

(ii) Maximum shear stress $\tau_{max} = \frac{1}{2}\sigma_1 = \frac{1}{2}OA$

$$= OC = CA = CP \text{ (Radius of circle)}$$

(iii) Normal stress $\sigma_n = OP = OC + CE = \frac{\sigma_x}{2} + CP\cos 2\theta$

$$\sigma_n = \frac{\sigma_x}{2} + \frac{\sigma_x}{2}\cos 2\theta = \frac{\sigma_x}{2}(1 + \cos 2\theta)$$

Put $1 = \cos^2\theta + \sin^2\theta$ and $\cos 2\theta = \cos^2\theta - \sin^2\theta$

$$= \frac{\sigma_x}{2}(\cos^2\theta + \sin^2\theta + \cos^2\theta - \sin^2\theta)$$

$$\sigma_n = \sigma_x \cdot \cos^2\theta = OE$$

(iv) Shear stress $\sigma_t = PE$

$$= CP \sin 2\theta = \frac{\sigma_x}{2}\sin 2\theta$$

(v) Resultant stress $\sigma_R = OP = \sqrt{OE^2 + PE^2} = \sqrt{\sigma_n^2 + \sigma_t^2}$.

6.5.2 Mohr's Circle for Two Mutually Perpendicular Unequal like Stresses (σ_x and σ_y)

Figure 6.31 shown an element subjected to two unequal like stress σ_x and σ_y on two mutually perpendicular planes. *ED* is the inclined plane on which normal and stresses are to be found.

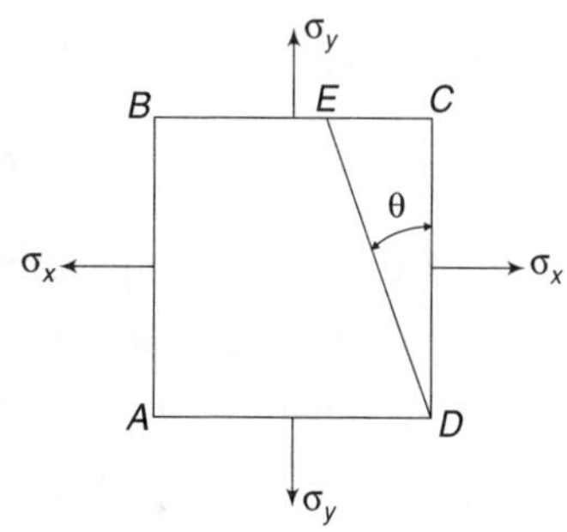

Fig. 6.31

Mohr's Circle:

(i) Draw two mutually perpendicular lines *x* and *y* as shown in Fig. 6.32.

(ii) Mark σ_x on *X*-axis at *A*, *OA* = Scaled length of σ_x

(iii) Mark σ_y on *X*-axis at *B*, *OB* = Scaled length of σ_y

If both are tensile, mark on position side of *X*. If both are compressive, mark on negative side of *x*-axis from '*O*'

(iv) Mark midpoint of *AB* as *C*. With *C* as centre *CA* or *CB* as radius, draw a circle. This cuts *X*-axis at *A* and *B*.

Maximum principal stresses $\sigma_1 = (OA \times \text{Scale})$

Minimum principal stress $\sigma_2 = OB \times \text{Scale}$

Maximum shear stress $\tau_{max} = CA = BC =$ radius of circle

(v) To find normal and shear stresses on inclined plane *DE*, inclined at angle θ

Mark angle = 2θ from *X* in counterclockwise direction from *C* to meet the circle at *P*. Drop a perpendicular *PE* from *P* to *X*-axis (σ_x line)

Normal stress $\sigma_n = OE \times \text{Scale}$

Shear stress $\sigma_t = PE \times \text{Scale}$

Resultant stress $\sigma_R = OP \times \text{Scale}$

Angle of resultant stress $\phi = \frac{1}{2}\tan^{-1}\left(\frac{PE}{OP}\right)$

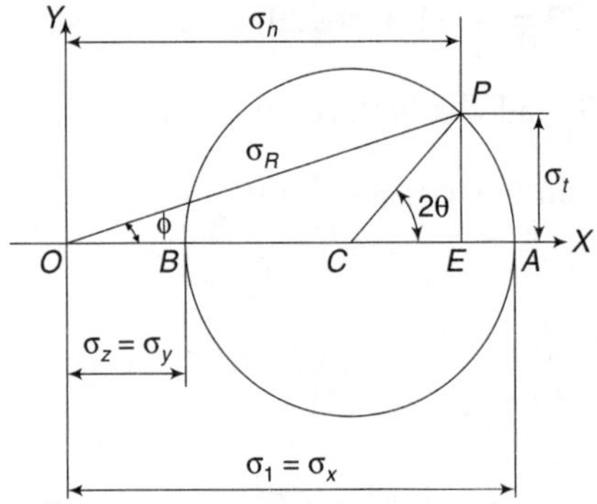

Fig. 6.32

Proof: $OA = \sigma_x$ and $OB = \sigma_y$

$$OC = \frac{\sigma_x + \sigma_y}{2}$$

$$BC = AC = \frac{AB}{2} = \frac{\sigma_x - \sigma_y}{2}$$

(i) Maximum principal stress $\sigma_1 = OA = \sigma_x$

(ii) Minimum principal stress $\sigma_2 = OB = \sigma_y$

(iii) Maximum shear stress $\tau_{max} = BC = CA = CP$ (Radius) $= \dfrac{\sigma_x - \sigma_y}{2} = \dfrac{\sigma_1 - \sigma_2}{2}$

(iv) Normal stress on plane inclined at θ

$$\sigma_n = OE = OC + CE$$

$$OC = \frac{\sigma_x + \sigma_y}{2}$$

From triangle *PEC*, $\cos 2\theta = \dfrac{CE}{PC}$

$$\therefore \quad CE = PC \cos 2\theta = \left(\frac{\sigma_x - \sigma_y}{2}\right) \cos 2\theta$$

$$\therefore \quad \sigma_n = \frac{\sigma_x + \sigma_y}{2} + \left(\frac{\sigma_x - \sigma_y}{2}\right) \cos 2\theta$$

(v) Shear stress on inclined plane

$$\sigma_t = CE$$

From triangle *PEC*, $\sin 2\theta = \dfrac{EC}{PC}$

$$\therefore \quad CE = PC \sin 2\theta = \frac{\sigma_x - \sigma_y}{2} \sin 2\theta$$

(vi) Resultant stress $\sigma_R = OP = \sqrt{OE^2 + PE^2}$

$$\sigma_R = OP = \sqrt{\sigma_n^{\,2} + \sigma_t^{\,2}}$$

$$\text{Obliquity of resultant stress } \tan\phi = \frac{PE}{OE}$$

$$\phi = \tan^{-1}\left(\frac{PE}{OE}\right).$$

6.5.3 Mohr's Circle for Mutually Perpendicular Unequal Unlike Stresses

Let σ_x = stress in x-direction (Tensile)

σ_y = stress in y-direction (Compressive)

Figure 6.33 shows an element subjected to two unlike stresses.

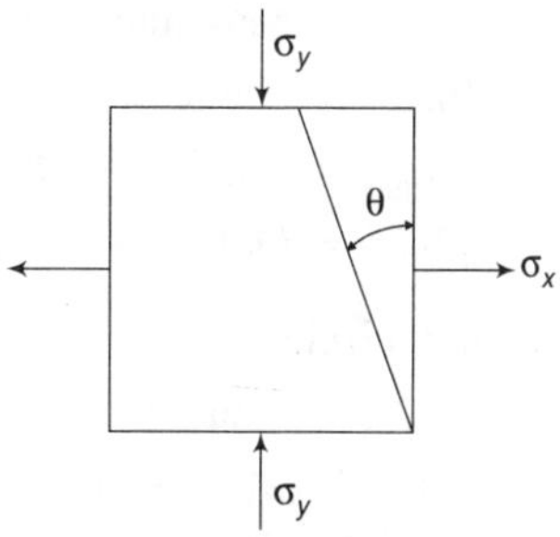

Fig. 6.33

Mohr's circle

1. Draw X- and Y-axes as shown in Fig. 6.34. Take σ_x values on +ve side of X-axis and mark point A. OA = Scaled length of σ_x.
2. Mark OB = Scaled length of σ_y on negative side of X-axis.
3. Mark centre of AB as C with C as centre, and CA or CB as radius, draw a circle.
4. $OA = \sigma_1$ = Major principal stress

   $OB = \sigma_2$ = Major principal stress

   $BC = AC = \tau_{max}$ = Maximum shear stress (radius of the circle)
5. Mark angle of 2θ from CA in counterclockwise direction from point C to meet circle at P. Drop a perpendicular PE.

   Normal stress on inclined plane $\sigma_n = OE$

   Shear stress an inclined plane $\sigma_t = PE$

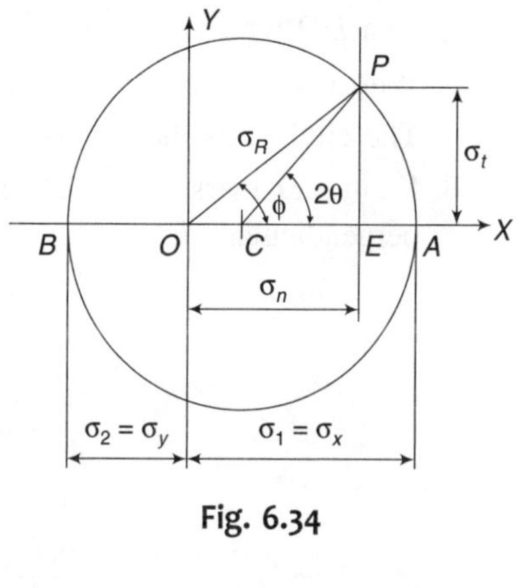

Fig. 6.34

The proof is similar to two mutually perpendicular like stresses as discussed in the previous case.

6.5.4 Mohr's Circle for an Element Subjected to two Mutually Perpendicular Stresses Combined with Shear Stress (Generalized Two-Dimensional Stress Condition)

Figure 6.35 shows an element subjected to two mutually perpendicular stresses accompanied by a shear stress

σ_x = Stress in X-direction

σ_y = Stress in Y-direction

τ = Shear stress

θ = Angle of inclined plane *ED* with plane of σ_x in counterclockwise direction

σ_n = Normal stress on inclined plane *ED*

σ_t = Shear stress in inclined plane *ED*

σ_R = Resultant stress on inclined plane *ED*

σ_1, σ_2 = Maximum and minimum principal stresses

θ_1, θ_2 = Angles of principle planes.

τ_{max} = Maximum shear stress

ϕ = Obliquity of resultant stress σ_R with σ_n

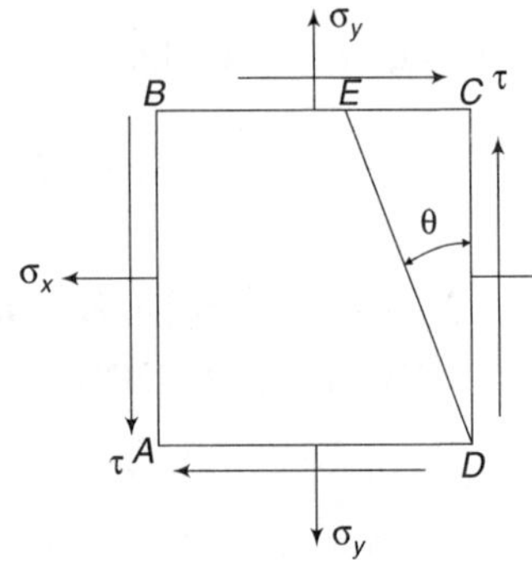

Fig. 6.35

Mohr's circle

1. Draw *X* and *Y* axes at '*O*' as shown in Fig. 6.36
2. Mark $OA = \sigma_x$, $OB = \sigma_y$ to the positive side of *X* if positive and negative side of *X* if any stress is negative. Select suitable scale.
3. Mark centre of *AB* as *C*. Mark shear stress on *Y*-axis from *A* and *B*. If the shear stress is positive, then mark *AD* downward from *A* and *BD* upward from *B*, forming clockwise. If the shear stress is negative, then mark *AD* upward from *A*, and *BD* downward from *B*, forming counterclockwise.
4. Join *DD* which will pass through the centre of Mohr's circle *C*.
5. With *C* as centre, *CD* as radius, draw a circle.

   The circle cuts the *X*-axis at *P* and *Q*.
6. From *CD* mark angle 2θ and draw a line *CT* in anticlockwise direction. Drop a perpendicular *TE*.

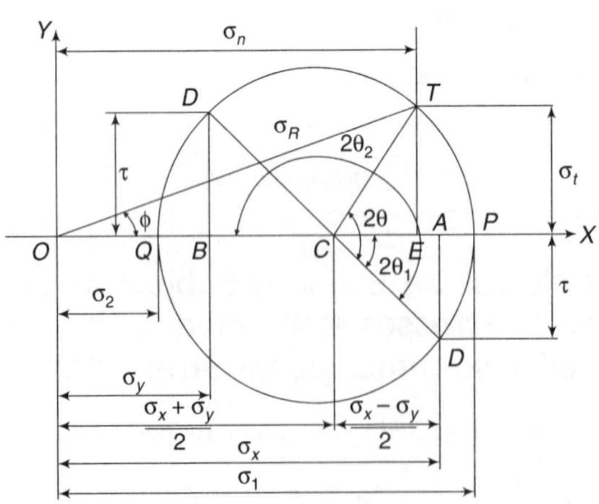

Fig. 6.36 Mohr's circle

From Mohr's circle measure the following and multiply with scale to get the various stresses.

- Maximum principal stress $\sigma_1 = OP$
- Minimum principal stress $\sigma_2 = OQ$
- Maximum shear stress $\tau_{max} = CP = CQ = CT$ (Radius of Mohr's circle)
- Normal stress on inclined plane $\sigma_n = OE$
- Shear stress on inclined plane $\sigma_t = TE$
- Resultant stress on inclined plane $\sigma_R = OP$
- Angle of obliquity $\phi = \angle EOP$

Proof:

(i) Maximum principal stress $\sigma_1 = OP = OC + CP$

$$OC = \frac{OA + OB}{2} = \frac{\sigma_x + \sigma_y}{2}$$

$$CP = \text{Radius of circle} = CD$$

From triangle CAD, $CD = \sqrt{AC^2 + AD^2}$

$$AC = \frac{\sigma_x - \sigma_y}{2} \text{ and } AD = \tau$$

$$\therefore \quad CP = \sqrt{\left(\frac{\sigma_x - \sigma_y}{2}\right)^2 + \tau^2}$$

$$\therefore \quad \sigma_1 = OP = OC + CP = \frac{\sigma_x + \sigma_y}{2} + \sqrt{\left(\frac{\sigma_x - \sigma_y}{2}\right)^2 + \tau^2}$$

(ii) Minimum principal stress $\sigma_2 = OQ = OC - CQ$

$$OC = \frac{\sigma_x + \sigma_y}{2} \text{ and } CQ = CP = \sqrt{\left(\frac{\sigma_x - \sigma_y}{2}\right)^2 + \tau^2}$$

$$\therefore \quad \sigma_2 = OQ = \frac{\sigma_x + \sigma_y}{2} - \sqrt{\left(\frac{\sigma_x - \sigma_y}{2}\right)^2 + \tau^2}$$

(iii) Angles of principal planes

From triangle CAD, $\tan 2\theta_1 = \dfrac{AD}{AC}$

$$AD = \tau,\ AC = \frac{\sigma_x - \sigma_y}{2}$$

$$\therefore \quad \tan 2\theta_1 = \frac{\tau}{\left(\dfrac{\sigma_x - \sigma_y}{2}\right)} = \frac{2\tau}{\sigma_x - \sigma_y}$$

$$2\theta_1 = \tan^{-1} = \frac{2\tau}{\sigma_x - \sigma_y}$$

$$2\theta_2 = 180 + 2\,\theta_1$$

$$\therefore \qquad \theta_1 = \frac{1}{2}\tan^{-1}\left(\frac{2\tau}{\sigma_x - \sigma_y}\right) \text{ or } \theta_2 = 90 + \theta_1$$

(iv) Maximum shear stress $\tau_{max} = CT = CA = CB$ (Radius of Mohr's circle)

$$\tau_{max} = \sqrt{\left(\frac{\sigma_x - \sigma_y}{2}\right)^2 + \tau^2}$$

(v) Angle of shear planes,

From triangle *CAD* $\quad \tan 2\theta_t = \dfrac{AC}{AD} = \left(\dfrac{\sigma_x - \sigma_y}{2\tau}\right)$

This angle is clockwise, hence negative

$$2\theta_t = \tan^{-1}\left(\frac{\sigma_x - \sigma_y}{-2\tau}\right)$$

(vi) Normal stress on inclined plane $\sigma_n = OE$

$$\sigma_n = OC + CE$$

$$OC = \frac{\sigma_x + \sigma_y}{2}$$

From triangle *TCE*, $\angle TCE = 2\theta - 2\theta_t$

$$\cos(2\theta - 2\theta_1) = \frac{CE}{CT}$$

$$CE = CT\cos(2\theta - 2\theta_1)$$

$$\cos(2\theta - 2\theta_1) = \cos 2\theta \cos 2\theta_1 + \sin 2\theta \cdot \sin 2\theta_1$$

From triangle *CAD*, $\cos 2\theta_1 = \dfrac{CA}{CD}$ and $\sin 2\theta_1 = \dfrac{AD}{CD}$

But $\qquad CD = CT =$ Radius of circle

$$\therefore \qquad \cos 2\theta_1 = \frac{CA}{CT} \text{ and } \sin 2\theta_1 = \frac{AD}{CT}$$

$$\therefore \qquad CE = CT\left(\cos 2\theta \cdot \frac{CA}{CT} + \sin 2\theta \cdot \frac{AD}{CT}\right)$$

$$CE = CA\cos 2\theta + AD\sin 2\theta$$

But $\qquad CA = \dfrac{\sigma_x - \sigma_y}{2}$ and $AD = \tau$

$$CE = \frac{\sigma_x - \sigma_y}{2}\cos 2\theta + \tau\sin 2\theta$$

$$\therefore \qquad \sigma_n = OE = OC + CE = \frac{\sigma_x + \sigma_y}{2} + \frac{\sigma_x - \sigma_y}{2}\cos 2\theta + \tau\sin 2\theta$$

(vii) Shear stress on inclined plane $\sigma_t = TE$

From triangle TCE, $\angle TCE = 2\theta - 2\theta_1$

$$\sin (2\theta - 2\theta_1) = \frac{TE}{CT}$$

$$\therefore \quad TE = CT \sin (2\theta - 2\theta_1)$$

$$\sigma_t = CT (\sin 2\theta \cdot \cos 2\theta_1 - \cos 2\theta \cdot \sin 2\theta_1)$$

$$= CT \left(\sin 2\theta \cdot \frac{CA}{CT} - \cos 2\theta \cdot \frac{AD}{CT}\right)$$

$$= CA \sin 2\theta - AD \cos 2\theta$$

$$\therefore \quad \sigma_t = \frac{\sigma_x - \sigma_y}{2} . \sin 2\theta - \tau \cos 2\theta.$$

SOLVED PROBLEMS

Problem 21 A point is subjected to two mutually perpendicular stresses of 60 N/mm^2 and 30 N/mm^2. Determine the normal tangential and resultant stresses on a plane inclined at 30°. Also find maximum shear stress. Use Mohr's circle method.

Solution: $\sigma_x = 60$ MPa, $\sigma_y = 30$ MPa, $\theta = 30°$

Draw Mohr's circle as explained below (Fig. 6.37)

(i) Take any point O as origin and draw X- and Y-axes Fig. 6.37(a)

(ii) Select scale 1 cm = 5 MPa

Draw $OA = \frac{50}{5} = 10$ cm to represent 50 N/mm^2.

Draw $OD = \frac{30}{5} = 6$ cm to represent 30 N/mm^2

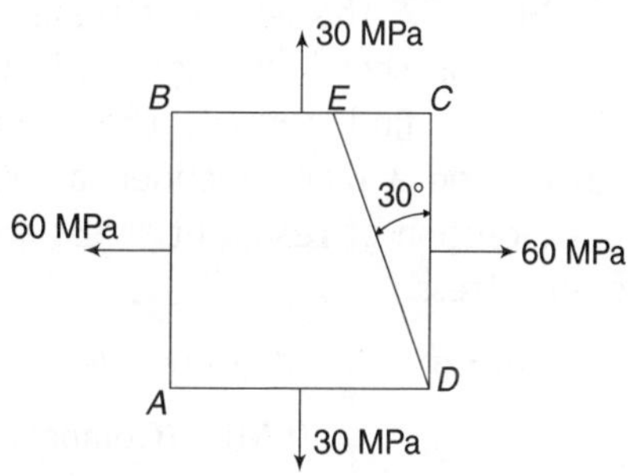

Fig. 6.37

on X-axis. (both are tensile and hence both are towards +ve side of X-axis)

(iii) Bisect AB at C. (midpoint of AB). With C as centre and CA as radius draw one circle, called Mohr's circle. This passes through B.

(iv) From C, draw a line CP at angle of $2\theta = 2 \times 30°$ from CA in counterclockwise direction.

(v) Draw a perpendicular PE from P. Join OP.

Maximum principal stress $\sigma_1 = OA = \sigma_x = 50$ MPa

Minimum principal stress $\sigma_2 = OB = \sigma_y = 30$ MPa

Angle of principal planes $\theta_1 = 0$ and $\theta_2 = 180°$

Normal stress on inclined plane = OE × Scale

$$\sigma_n = 9 \times 5 = 45 \text{ MPa}$$

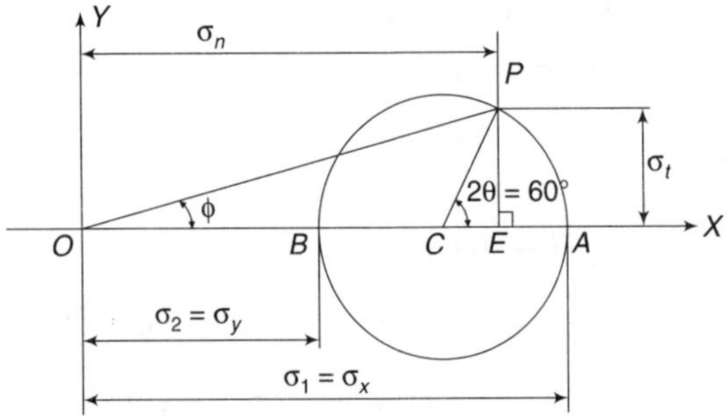

Fig. 6.37(a)

Shear stress on inclined plane = $PE \times$ Scale

$$\sigma_t = 1.75 \times 5 = 8.75 \text{ MPa}$$

Resultant stress $\sigma_R = OP \times \text{Scale} = 9.2 \times 5 = 46$ MPa

Obliquity of resultant stress = $\phi = \angle POE = 11°$

Maximum shear stress τ_{max} = Radius of Mohr's circle

$$= CP = AC = BC = \pm 10 \text{ MPa.}$$

Problem 22 The state of stresses in a stressed element is shown in Fig. 6.38. Draw Mohr's circle and find normal, shear and resultant stresses on a plane inclined at 50°. Also find the direction of resultant stress and maximum shear stress.

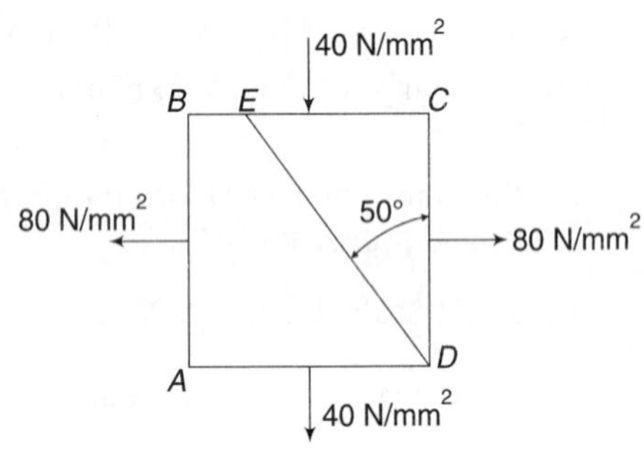

Fig. 6.38

Solution: $\sigma_x = 80$ MPa (Tensile)

$\sigma_y = -40$ MPa (Compressive)

$\theta = 50°$

Draw Mohr's circle as explained below (Fig. 6.39)

(i) Select a point '*O*' as origin and draw *X* and *Y* axes

(ii) Select scale to represent stresses, 1 cm = 10 MPa.

Take $OA = \dfrac{80}{10} = 8$ cm and mark on *X*-axis towards positive side to represent σ_x.

Take $OB = \dfrac{-40}{10} = 4$ cm and mark on *X*-axis towards negative side (Left side)

(iii) Bisect *AB* at *C* (Midpoint).

With *C* as centre *CA* as radius draw one circle. This is Mohr's circle.

(iv) From C, draw a line CP at an angle of $2\theta = 2 \times 50 = 100°$ from CA in counter-clockwise direction.

(v) Drop a perpendicular PE from P. Join OP.

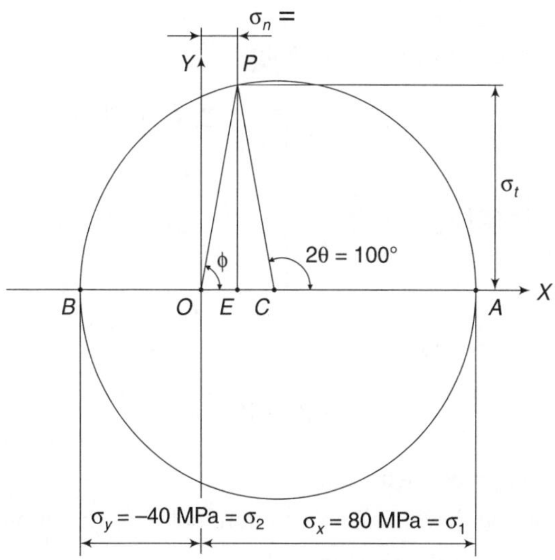

Fig. 6.39 Mohr's circle

From Mohr's circle

Normal stress along inclined plane $\sigma_n = OE \times \text{Scale} = 0.96 \times 10 = 9.6$ MPa

Shear stress along inclined plane $\sigma_t = PE \times \text{Scale} = 5.9 \times 10 = 59$ MPa

Resultant stress along inclined plane $\sigma_R = OP \times \text{Scale} = 6 \times 10 = 60$ MPa

Angle of resultant stress $= \phi = \angle POE = 81°$

Maximum shear stress τ_{max} = Radius of Mohr's circle

$= CA = CP = BC = 6 \times 10 = 60$ MPa.

Problem 23 A shaft of 30 mm diameter is subjected to a direct tensile force of 18 kN and a sharing force of 12 kN. Determine, drawing Mohr's circle:

(i) Normal and shear stresses on a plane at an angle 60° to the longitudinal axis

(ii) Resultant stress on the above plane and is obliquity

(iii) Principal stresses and their location

(iv) Maximum shear stress and its location.

Solution: Diameter $d = 30$ mm

Direct tensile load $F_T = 18$ kN $= 18000$ N

$$\therefore \text{ Direct tensile stress } \sigma_x = \frac{F_T}{A} = \frac{18000}{\left(\frac{\pi}{4} \times 30^2\right)} = 25.46 \text{ MPa}$$

Shear load F_S = 12 kN = 12000 N

$$\therefore \quad \text{Shear stress } \tau = \frac{F_S}{A} = \frac{12000}{\left(\frac{\pi}{4} \times 30^2\right)} = 16.98 \text{ MPa}$$

The stressed shaft is represented in Fig. 6.40

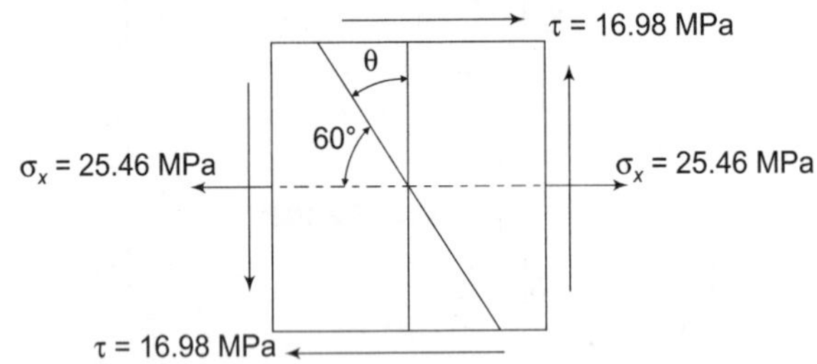

Fig. 6.40

Angle of inclined plane = 60° with axis

$\therefore$ θ = 90 – 60 = 30° with vertical in counterclockwise direction.

Draw the Mohr's circle as follows (Fig. 6.41)

(i) Select origin O and draw X- and Y-axes at o

(ii) Select scale to represent stresses. 1 cm = 5 MPa

Take $OA = \frac{25.46}{5}$ = 5.09 cm and mark point A on +ve side of x-axis representing σ_x.

Since $\sigma_y = 0$, point B lies at O itself.

(iii) Bisect AB at C. (Midpoint). Draw perpendicular line at A downward as $AD = \frac{\tau}{\text{Scale}}$ $= \frac{16.98}{5}$ = 3.396 cm.

Also mark D at B upward of the same length.

$$BD = AD = \tau = 3.396 \text{ cm}$$

(iv) With C as centre, CD as radius, draw a circle that cuts X-axis at P and Q. This, is the Mohr's circle

(v) From C, draw a line CT at angle of $2\theta = 2 \times 30 = 60°$ counterclockwise to line CD.

Drop a perpendicular from T as TE. Join OT.

From Mohr's circle:

Normal stress along inclined plane = $\sigma_n = OE \times \text{Scale} = 6.8 \times 5 = 34$ MPa

Shear stress along inclined plane = $\sigma_t = TE \times \text{Scale} = 0.5 \times 5 = 2.5$ MPa

Resultant stress = $\sigma_R = OT \times \text{Scale} = 6.81 \times 5 = 34.05$ MPa

Obliquity of resultant stress = $\varphi = \angle TOE = 4°$

Maximum principal stress $\sigma_1 = OA \times \text{Scale} = 6.8 \times 5 = 34$ MPa

Minimum principal stress $\sigma_2 = OB \times \text{Scale} = 1.7 \times 5 = -8.5$ MPa

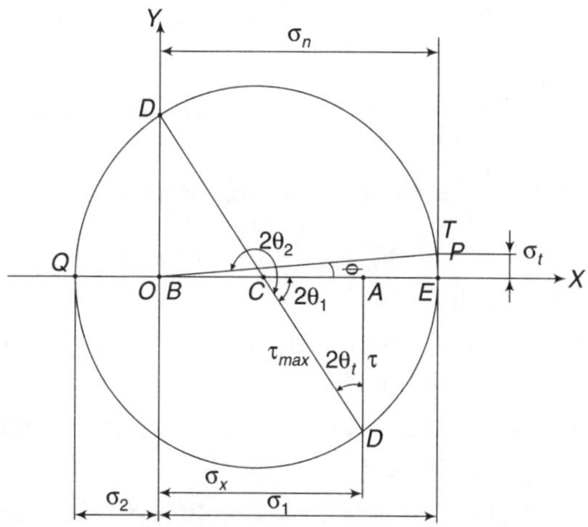

Fig. 6.41 Mohr's circle

Principal planes, $2\theta_1 = \angle DCP = 53°$

$\therefore \quad \theta_1 = 26.5°$

$2\theta_2 = \angle DCB = 53 + 180 = 233°$

$\theta_2 = 116.5°$

Maximum shear stress τ_{max} = Radius of Mohr's circle

$= CD = CP = 4.2 \times \text{Scale} = 4.2 \times 5 = 21$ MPa

Angle of shear planes, $2\theta_t = \angle CDA = 37°$

$\theta_t = 18.5$ or $108.5°$.

Problem 24 A point in a strained material is subjected to stresses as shown in Fig. 6.42. Find graphically the normal, tangential and resultant stresses on the plane *DE*. Also find the maximum and minimum principal stresses and their location and maximum shear stress and its location. Verify the answers analytically.

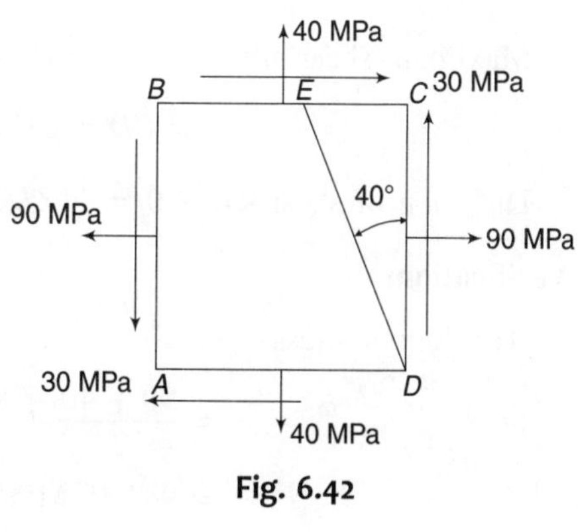

Fig. 6.42

Solution: $\sigma_x = 90$ MPa, $\sigma_y = 40$ MPa

$\tau = 30$ MPa, $\theta = 40°$

Mohr's circle:

Draw Mohr's circle as explained in Fig. 6.43

(i) Take origin O and mark X and Y axes lines.

(ii) Select scale to represent the stress.

1 cm = 10 MPa

Take $\sigma_x = 90 \text{ MPa} = \frac{90}{\text{Scale}} = \frac{90}{10} = 9$ cm as OA and mark A an X-axis

Take $\sigma_y = 40 \text{ MPa} = \frac{40}{10} = 4$ cm as OB and Mark B an X-axis (Both positive)

(iii) Take $\tau = 30 \text{ MPa} = \frac{30}{\text{Scale}} = \frac{30}{10} = 3$ cm

as $AD = BD$ in clockwise direction, perpendicular to X-axis. Join DD and that passes through C which is midpoint of AB.

(iv) With C as centre CD as radius, draw a circle which is called Mohr's circle. This cuts X axis at two point P and Q.

(v) From point C, mark angle $DCT = 2\theta = 2 \times 40 = 80°$ in counterclockwise direction from line CD.

Drop a perpendicular from T to X-axis as TE join OT.

From Mohr's circle:

Normal stresses $\sigma_n = OE \times \text{Scale} = 9.9 \times 10 = 99$ MPa

Shear stress $\sigma_t = TE \times \text{Scale} = 1.9 \times 10 = 19$ MPa

Resultant stress $\sigma_R = OT \times \text{Scale} = 10.1 \times 10 = 101$ MPa

Obliquity of resultant stress $\phi = \angle EOT = 11°$

Major principal stress $\sigma_1 = OP \times \text{Scale} = 10.4 \times 10 = 104$ MPa

Minor principal stress $\sigma_2 = OQ \times \text{Scale} = 2.6 \times 10 = 26$ MPa

Direction of principal stress $2\theta_1 = 50$ or $\theta_1 = 25°$

$$2\theta_2 = 2\theta_1 + 180 = 50 + 180 = 230$$

$$\therefore \quad \theta_2 = 115°$$

Maximum shear stress $= \tau_{max}$ = Radius of Mohr's circle

$$= CD = CP = CQ = 3.9 \times \text{Scale} = 39 \text{ MPa}$$

Direction of shear stress $\theta_t = \frac{1}{2} \angle DCA = \frac{1}{2} \times 40 = 20°$ (–ve)

Verification:

(i) Normal stress
$$\sigma_n = \frac{\sigma_x + \sigma_y}{2} + \frac{\sigma_x - \sigma_y}{2} \cos 2\theta + \tau \sin 2\theta$$
$$= \frac{90 + 40}{2} + \frac{90 - 40}{2} \cos (2 \times 40) + 30 \sin 2 \times 40$$
$$= 98.885 \text{ MPa}$$

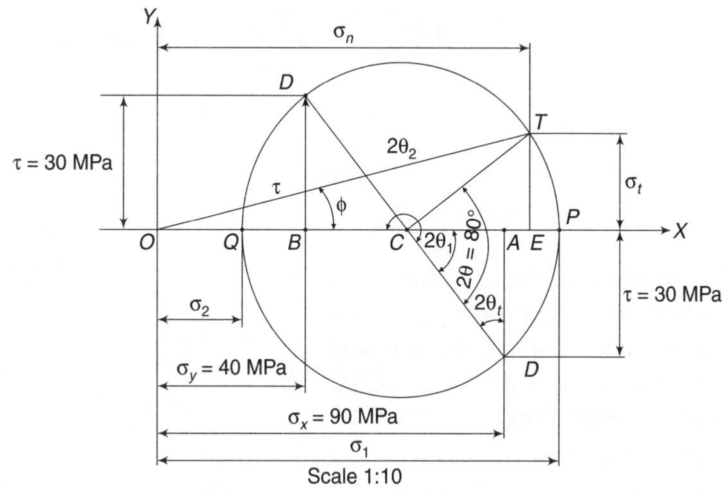

Fig. 6.43 Mohr's circle

(ii) Shear stress $\sigma_t = \dfrac{\sigma_x - \sigma_y}{2} \sin 2\theta - \tau \cos 2\theta = 19.41$ MPa

(iii) Resultant stress $\sigma_R = \sqrt{\sigma_n^{\,2} + \sigma_t^{\,2}} = \sqrt{98.885^2 + 19.41^2} = 100.77$ MPa

(iv) Obliquity $\phi = \tan^{-1}\left(\dfrac{\sigma_t}{\sigma_n}\right) = \tan^{-1}\left(\dfrac{19.41}{98.885}\right) = 11.1°$

(v) Major principal stress

$$\sigma_1^2 = \frac{\sigma_x + \sigma_y}{2} + \sqrt{\left(\frac{\sigma_x - \sigma_y}{2}\right)^2 + \tau^2}$$

$$= \frac{90 + 40}{2} + \sqrt{\left(\frac{90 - 40}{2}\right)^2 + 30^2} + 104.05 \text{ MPa}$$

(vi) Minor principal stress $\sigma_2 = \dfrac{\sigma_x + \sigma_y}{2} - \sqrt{\left(\dfrac{\sigma_x - \sigma_y}{2}\right)^2 + \tau^2}$

$$= \frac{90 + 40}{2} - \sqrt{\left(\frac{90 - 40}{2}\right)^2 + 30^2} = 25.95 \text{ MPa}$$

(vii) Principal planes $\tan 2\theta_1 = \dfrac{2\tau}{\sigma_x - \sigma_y} = \dfrac{2 \times 30}{90 - 40}$

$$2\theta_1 = 50.19 \text{ or } 2\theta_2 = 50.19 + 180 = 230.19°$$
$$\theta_1 = 25.095 \text{ or } \theta_2 = 115.095°$$

(viii) Maximum shear stress $\tau_{max} = \sqrt{\left(\dfrac{\sigma_x - \sigma_y}{2}\right)^2 + \tau^2}$

$$= \sqrt{\left(\frac{90 - 40}{2}\right)^2 + 30^2} = 39.05 \text{ MPa}$$

(ix) Shear plane $\tan 2\theta_t = \dfrac{\sigma_x - \sigma_y}{-2\tau} = \dfrac{90 - 40}{-2 \times 30}$

$$2\theta_t = -39.806° \text{ or } 140.194°$$

$$\theta_t = -19.90° \text{ or } 70.097°.$$

Problem 25 A stressed element is subjected to 60 MPa tensile and 40 MPa compressive stresses on two mutually perpendicular planes and a shear stress of 10 MPa as shown in Fig. 6.44. Determine by Mohr's circle method:

(i) Principal stresses and planes

(ii) Maximum shear stress and planes.

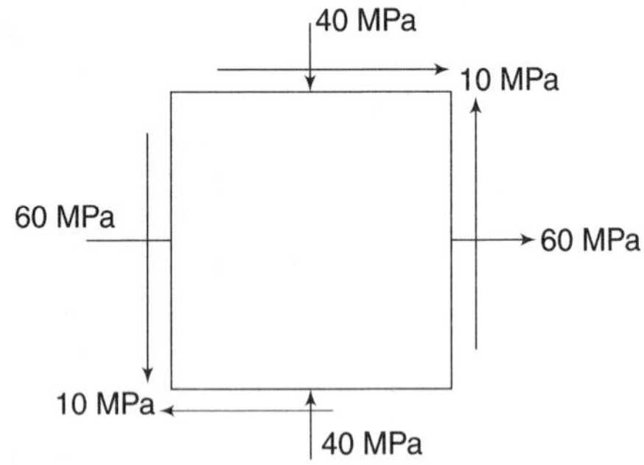

Fig. 6.44

Solution: $\sigma_x = 60$ MPa, $\sigma_y = -40$ MPa (Compressive)

$$\tau = 10 \text{ MPa}$$

Mohr's Circle:

Draw Mohr's circle as explained in Fig. 6.45.

(i) Draw *X*- and *Y*-axes and mark origin '*O*'

(ii) Scale 1 cm = 10 MPa. Draw $\sigma_x = \dfrac{60}{10} = 6$ cm on to the positive side of *X*-axis as *OA*.

Draw $\sigma_y = \dfrac{-40}{10} = -4$ cm on to the left side (negative side) of *X*-axis as *OB*.

(iii) Draw $\tau = \dfrac{10}{10} = 1$ cm in *y*-direction from *A* towards downward as *AD* and from *B* upward as *BD*. Join *DD* that passes through *C*, the mid point of *AB*.

(iv) With *C* as centre *CD* as radius draw a circle. This is Mohr's circle. This circle cuts *X*-axis at two points *P* and *Q*.

From Mohr's circle:

Maximum principal stress $\sigma_1 = OP \times$ Scale: $6.1 \times 10 = 61$ MPa

Minimum principal stress $\sigma_2 = OQ \times$ Scale: $4.1 \times 10 = 41$ MPa

Angle of principal planes, $2\theta_1 = \angle DCP = 11°$

$$\therefore \quad \theta_1 = \frac{1}{2} 11 = 5.5°$$

$$2\theta_2 = \angle CCQ \text{ (in CCW direction)}$$

$$2\theta_2 = 2\theta_1 + 180 = 191$$

$$\therefore \quad \theta_2 = 95.5°$$

Maximum shear stress τ_{max} = Radius of Moh's circle

$= CD = 5.1 \times 10 = 51$ MPa

Angles of shear planes $2\theta_t = \angle CDA = 79°$

$\theta_t = 39.5°$ or $39.5 + 90 = 129.5°$

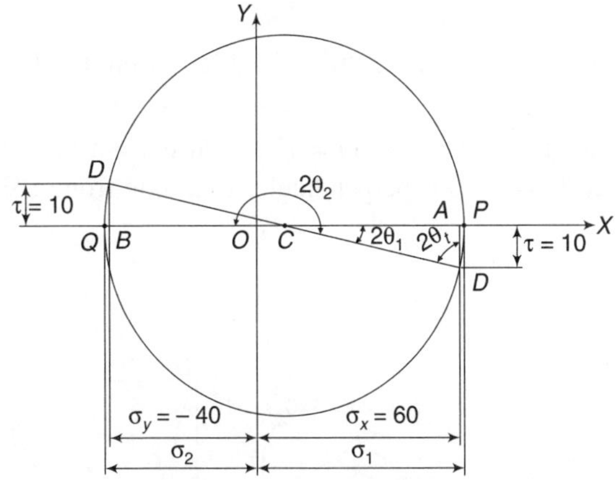

Fig. 6.45 Mohr's circle

Problem 26 A point in a machine member is stressed as shown in Fig. 6.46. Determine, drawing Mohr's circle,

(i) The stress on plane inclined at 60° with respect to the plane of stress 80 MPa.

(ii) Magnitude and location of principal stresses

(iii) Maximum shear stress and location.

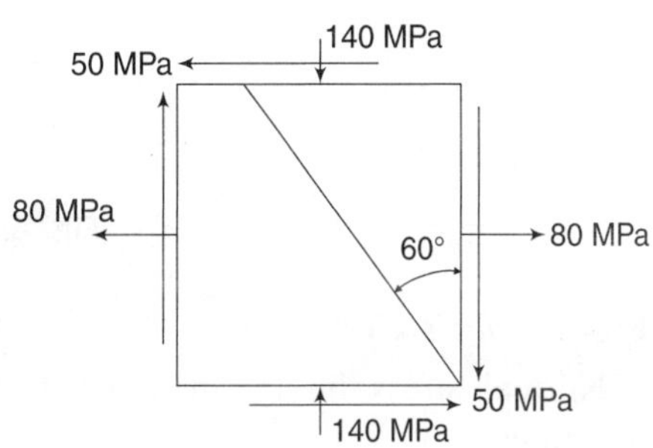

Fig. 6.46

Solution: $\sigma_x = 80$ MPa $\sigma_y = -140$ MPa (Compressive)

$\tau = -50$ MPa (Opposite direction)

$$\theta = 60°$$

Mohr's circle:

Draw Mohr's circle as explained in Fig. 6.47.

(i) Draw *X*- and *Y*-axes and mark origin *O*

(ii) Select scale 1 cm = 20 MPa.

Take $\sigma_x = \dfrac{80}{20} = 4$ cm on *X*-axis positive direction and mark *OA* from *O*.

(iii) Take $\sigma_y = \dfrac{-140}{20} = -7$ cm on an X-axis negative direction and mark OB from O

(iv) Take $\tau = \dfrac{-50}{20} = -2.5$ cm

Mark AD perpendicular to OA at A upwards since τ is negative. Also mark BD perpendicular to OB downward for negative τ. (CCW) Join D and D, that passes through midpoint of AB at C.

(v) With C as center, CD as radius, draw the Mohr's circle. This cuts X-axis at two points P and Q.

(vi) From CD, take angle = $2\theta = 2 \times 60 = 120°$ and mark a line in CCW direction that cuts the circle at T. Draw TE perpendicular to x-axis from τ. Join OT.

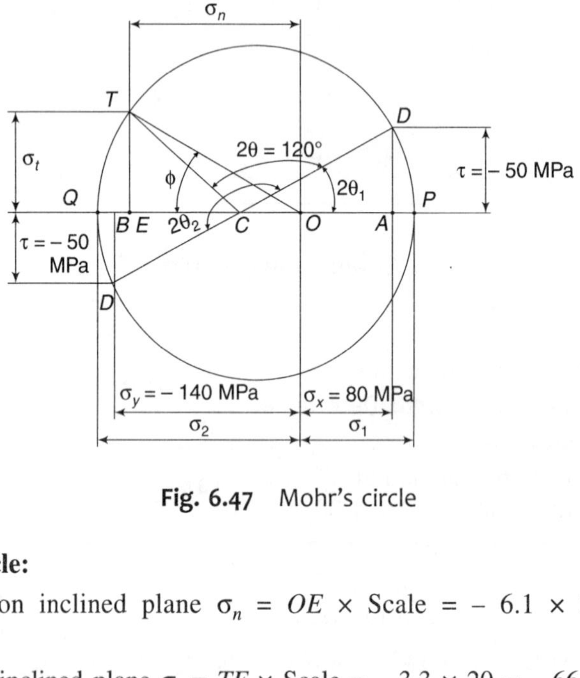

Fig. 6.47 Mohr's circle

From Mohr's circle:

Normal stress on inclined plane $\sigma_n = OE \times \text{Scale} = -6.1 \times 20 = -122$ MPa (Compressive)

Shear stress on inclined plane $\sigma_t = TE \times \text{Scale} = -3.3 \times 20 = -66$ MPa

Resultant stress $\sigma_R = OT \times \text{Scale} = 6.9 \times 20 = 138$ MPa

Angle of obliquity $\phi = \angle EOT = 28.5°$

Maximum principal stress $\sigma_1 = OP \times \text{Scale} = 4.5 \times 20 = 90$ MPa

Minimum principal stress $\sigma_2 = OQ \times \text{Scale} = -7.5 \times 20 = -150$ MPa (Compressive)

Direction of principal stresses $2\theta_1 = \angle DOP = -25°$

$$\therefore \qquad \theta_1 = \frac{-25°}{2} = -12.5°$$

$$2\theta_2 = \angle DOQ = 2\theta_1 + 180 = -25 + 180 = 155°$$

$$\theta_2 = 77.5°$$

Maximum shear stress: τ_{max} = Radius of Mohr's circle

$= CD \times \text{Scale} = 5.8 \times 20 = 116$ MPa

Angle of shear planes, $2\theta_t = \angle CDA = 65°$

$\therefore \quad \theta_t = 32.5°$ or $32.5 + 90 = 112.5°$.

Problem 27 A point in a strained material is subjected to pure shear stress of 50 MPa. Determine the maximum and minimum principal stresses, maximum shear stress and their location using Mohr's circle method.

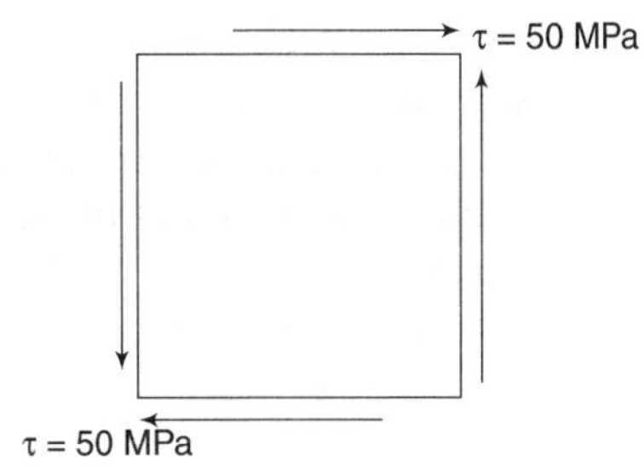

Fig. 6.48

Solution: $\tau = 50$ MPa

Mohr's circle: Fig. 6.49

(i) Take origin O and draw X- and Y-axes

(ii) Since there are no normal stresses, $\sigma_x = 0$ and $\sigma_y = 0$, hence points A and B lie at O itself.

(iii) From O take τ above and below on Y-axis equal to 50 MPa to scale 1 cm = 10 MPa. Mark points D, D.

(iv) With O as centre (which is also midpoint of AB) OD as radius draw Mohr's circle, that cuts X-axis at P and Q.

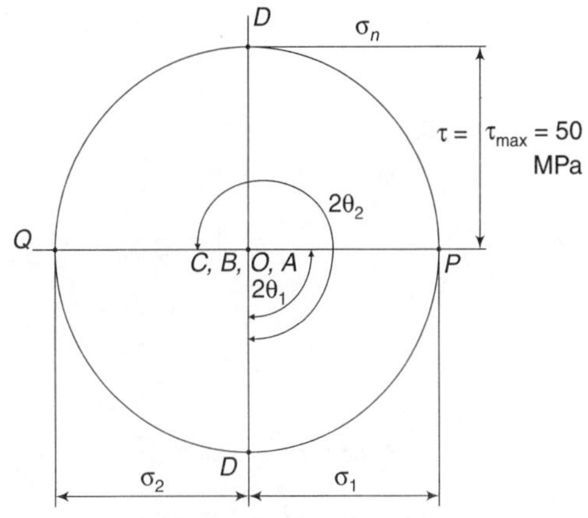

Fig. 6.49 Mohr's circle

Maximum principal stress $\sigma_1 = OP \times \text{Scale} = 5 \times 10 = 50 \text{ MPa} = \tau$

Minimum principal stress $\sigma_2 = OQ \times \text{Scale} = -5 \times 10 = -50 \text{ MPa} = \tau$

Maximum shear stress $\tau_{max} = OD \times \text{Scale} = 5 \times 10 = 50 \text{ MPa} = \tau$

(Radius Mohr's circle)

Angles of principal planes $2\theta_1 = \angle DCP = 90 \Rightarrow \theta_1 = 45°$

and $2\theta_2 = \angle DCQ = 270° \Rightarrow \theta_2 = 135°$

Angles of shear planes $2\theta_t = \angle DOD = 180° \Rightarrow \theta_t = 90°$

or $\theta_t = 90 + 90 = 180°$.

Problem 28 An element is subjected to a state of stresses as shown in Fig. 6.50. Using Mohr's circle method determine:

(i) Normal and tangential stresses on a plane inclined at 110° with respect to axis

(ii) Principal stresses and their location

(iii) Maximum shear stress and its location.

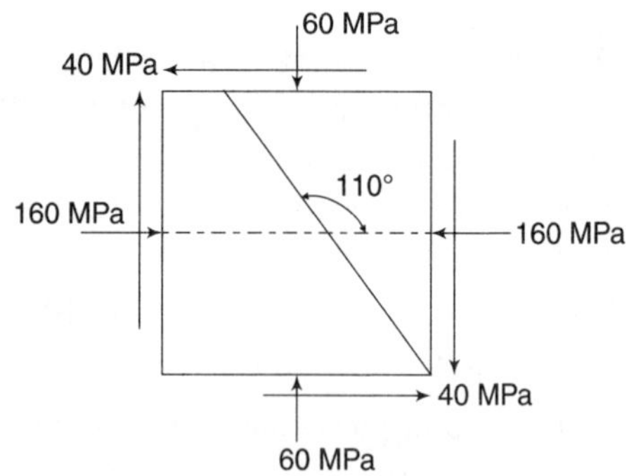

Fig. 6.50

Solution: $\sigma_x = -160$ MPa (Compressive)

$\sigma_y = -60$ MPa (Compressive)

$\tau = -40$ MPa (Direction is opposite)

$\theta = 110 - 90 = 20°$ (Counterclockwise)

Draw the Mohr's circle as explained in the previous problems.

Scale 1 cm = 20 MPa.

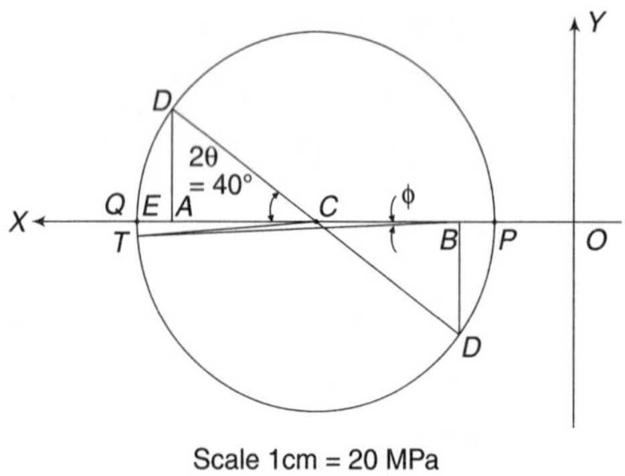

Fig. 6.51 Mohr's circle

From Mohr's circle:

Normal stress on inclined plane $\sigma_n = OE \times \text{Scale} = -8.7 \times 20 = -174$ MPa

(i) Shear stress on inclined plane $\sigma_t = TE \times \text{Scale} = -0.1 \times 20 = -2$ MPa

Resultant stress $\sigma_R = OT \times \text{Scale} = 8.7 \times 20 = 174$ MPa

Obliquity $\phi = \angle OTE = -0.5°$

(ii) Maximum principal stress $\sigma_1 = OP \times \text{Scale} = -2.3 \times 20 = -46$ MPa

Minimum principal stress $\sigma_2 = OQ \times \text{Scale} = -8.7 \times 20 = -174$ MPa

Angles of principal planes: $2\theta = \angle CAD = -38.66°$ or $-218.66°$

$$\theta_1 = -19.33° \text{ and } \theta_2 = -109.33°$$

(measured in CW direction, hence – ve)

(iii) Maximum shear stress $\tau_{max} = CD \times \text{Scale}$

(Radius of Mohr's circle)

$$\tau_{max} = 3.2 \times 20 = 64 \text{ MPa}$$

Angles of shear planes $2\theta_t = \angle CDA = +52°$

$$\theta_t = +26°.$$

Problem 29 A certain point in a strained material the state of stress is shown in Fig. 6.52. Find by drawing Mohr's circle.

(i) The normal and shear stresses on inclined plane *EF*

(ii) Principal stresses and planes

(iii) Maximum shear stress and their planes.

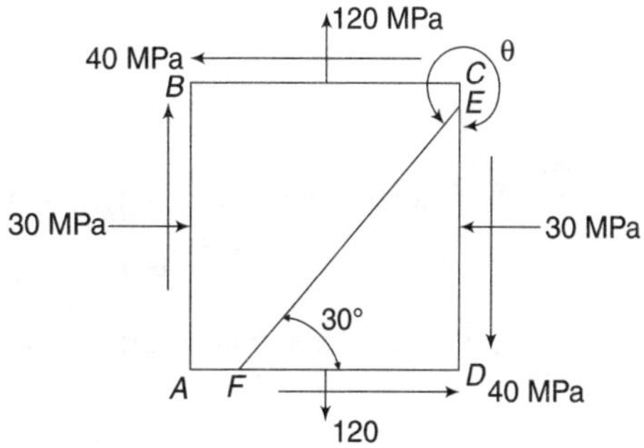

Fig. 6.52

Solution:

$\sigma_x = -30$ MPa (Compressive)

$\sigma_y = 120$ MPa (Tensile)

$\tau = -40$ MPa (Opposite direction)

$\theta = 30° = 270° + 30° = 300°$ or $\theta = -60°$

(Angle is measured in CCW direction from *CD*)

Mohr's Circle: Figure 6.53

Draw Mohr's circle as explained in earlier problems.

From Mohr's circle:

(i) Normal stress on inclined plane, $\sigma_n = OE \times \text{Scale} = 11.7 \times 10 = 117$ MPa

Shear stress an inclined plane $\sigma_t = TE \times \text{Scale} = -4.5 \times 10 = -45$ MPa (Upward)

Resultant stress $\sigma_R = OT \times \text{Scale} = 12.6 \times 10 = 126$ MPa

Obliquity $\phi = \angle EOT = 21°$

(ii) Maximum principal stress $\sigma_1 = OP \times \text{Scale} = 13 \times 10 = 130$ MPa

- Minimum principal stress $\sigma_2 = OQ \times \text{scale} = -4 \times 10 = -40$ MPa
- Angles of principal planes $2\theta_1 = \angle DCA = 28°$

And $2\theta_2 = \angle DCA = 208°$

$\therefore$ $\theta_1 = 14°$ and $\theta 2 = 104°$

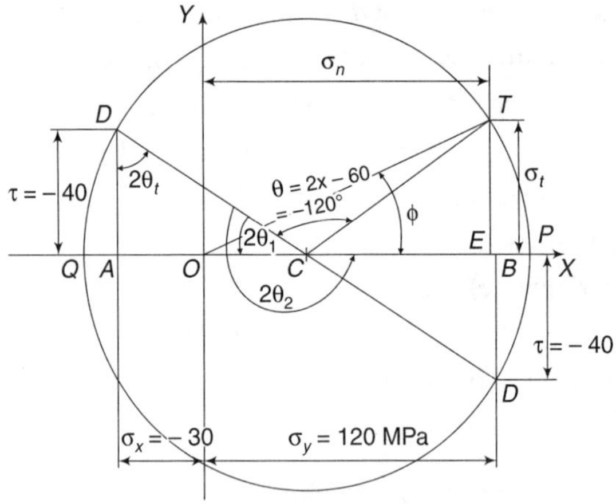

Fig. 6.53 Mohr's circle

(iii) Maximum shear stress = τ_{max} = Radius of Mohr's circle

$$= CD \times \text{Scale} = 8.5 \times 10 = 85 \text{ MPa}$$

Angles of shear planes $2\theta_t = \angle CDA = -62°$

$$\therefore \qquad \theta_t = \frac{-62}{2} = -31° \text{ or } -31 + 90 = 59°$$

$$\theta_t = 180 - 31 = 149° \text{ or } 59°.$$

Problem 30 The state of stresses at a point is shown in Fig. 6.54. Draw Mohr's circle and determine

(i) Normal and shear stresses along plane *EF* that cuts the element at 50° with axis. Also find resultant stress and its obliquity

(ii) Principal stresses and planes

(iii) Maximum shear stress and their planes

(iv) Verify the answers analytically.

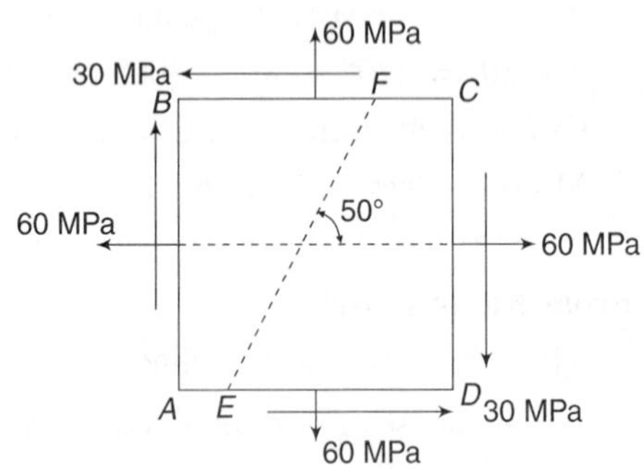

Fig. 6.54

Solution: $\sigma_x = 60$ MPa

$\sigma_y = 60$ MPa

$\tau = -30$ MPa (Opposite direction)

θ = Angle from plane *CD* to *EF* in CCW direction

$\theta = 270 + 50 = 320°$ or $90 - 50 = 40°$ (CW)

Part A: Mohr's circle (Fig. 6.55)

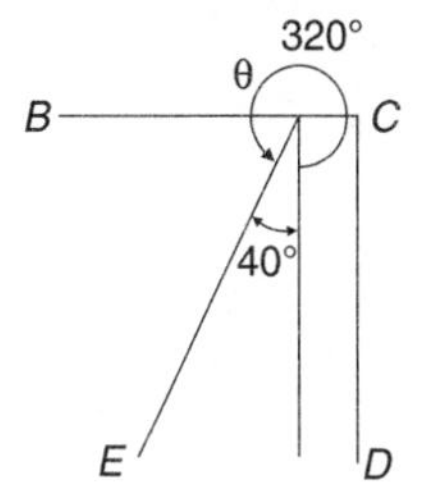

Draw Mohr's circle for the above stresses. Select a scale of 1 cm = 10 MPa

From Mohr's circle:

(i) Normal stress on inclined plane $\sigma_n = OE \times \text{Scale} = 8.95 \times 10 = 89.5$ MPa

Shear stress on inclined planes $\sigma_t = TE \times \text{Scale} = 0.5 \times 10 = 5$ MPa

Resultant stress $\sigma_R = OT \times \text{Scale} = 9 \times 10 = 90$ MPa

Obliquity $\phi = \angle EOT = 3.5°$

(ii) Major principal stress $\sigma_1 = OP \times \text{Scale} = 9 \times 10 = 90$ MPa

Minor principal stress $\sigma_2 = OQ \times \text{Scale} = 3 \times 10 = 30$ MPa

Principal planes $2\,\theta_1 = \angle DCP = 90° \;\therefore\; \theta_1 = 45°$

$2\theta_2 = \angle DCQ = 270°, \;\; \theta_2 = 135°$

(iii) Maximum shear stresss τ_{max} = Radius of Mohr's circle

$$\tau_{max} = CD \times \text{Scale} = 3 \times 10 = 30 \text{ MPa}$$

Angles of shear planes $\quad 2\theta_t = 0$ or 180

$\therefore \qquad \theta_t = 0$ or 90°

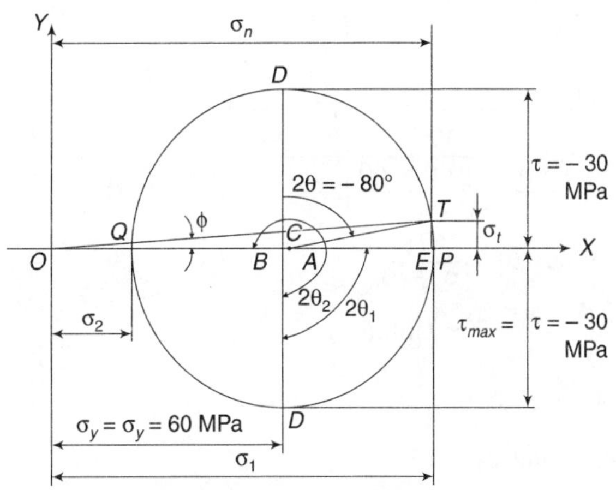

Fig. 6.55 Mohr's circle.

Part B: Analytical verification

(i) Noraml stress on inclined plane

$$\sigma_n = \frac{\sigma_x + \sigma_y}{2} + \frac{\sigma_x - \sigma_y}{2} \cos 2\theta + \tau \sin 2\theta$$

$$= \frac{60 + 60}{2} + \frac{60 - 60}{2} \cos (2 \times - 40) - 30 \sin (2 \times 40)$$

$$= 89.544 \text{ MPa}$$

Shear stress or tangential stress on inclined plane

$$\sigma_t = \frac{\sigma_x - \sigma_y}{2} \sin 2\theta - \tau \cos 2\theta$$

$$= \frac{60 - 60}{2} \sin (2 \times - 40) - (- 30) \cos (2 \times - 40)$$

$$\sigma_t = 5.209 \text{ MPa}$$

Resultant stress on inclined plane

$$\sigma_R = \sqrt{\sigma_n^2 + \sigma_t^2} = \sqrt{89.544^2 + 5.209^2} = 89.695 \text{ MPa}$$

Obliquity $\tan \phi = \frac{\sigma_t}{\sigma_n} = \frac{5.209}{89.544}$

$\therefore \quad \phi = 3.33°$

(ii) Maximum principal stress

$$\sigma_1 = \frac{\sigma_x + \sigma_y}{2} + \sqrt{\left(\frac{\sigma_x + \sigma_y}{2}\right)^2 + \tau^2}$$

$$= \frac{60 + 60}{2} + \sqrt{\left(\frac{60 - 60}{2}\right)^2 + (- 30)^2} + 90 \text{ MPa}$$

Minimum principal stress $\sigma_2 = \frac{\sigma_x + \sigma_y}{2} - \sqrt{\left(\frac{\sigma_x + \sigma_y}{2}\right)^2 + \tau^2}$

$$= \frac{60 + 60}{2} - \sqrt{\left(\frac{60 - 60}{2}\right)^2 + (- 30)^2} = 30 \text{ MPa}$$

Principal plane, $\tan 2\theta = \frac{2\tau}{\sigma_x - \sigma_y}$

$$\tan 2\theta = \frac{2\ X -30}{60 - 60} = \propto \text{ (Infinity)}$$

$\therefore \quad 2\theta = 90 \text{ or } 270°$

$\therefore \quad \theta_1 = 45° \text{ and } \theta_2 = 135°$

(iii) Maximum shear stress $\tau_{max} = \sqrt{\left(\frac{\sigma_x - \sigma_y}{2}\right)^2 + \tau^2}$

$$= \sqrt{\left(\frac{60 - 60}{2}\right)^2 + (- 30)^2} = \pm\ 30 \text{ MPa}$$

Angles shear planes, $\tan 2\theta_t = \frac{\sigma_x - \sigma_y}{-2\tau} = \frac{60 - 60}{- 2 \times (-30)} = 0$

$\therefore \quad 2\theta_t = 0 \text{ or } 180°$

$\theta_t = 0 \text{ or } 90°$

THINGS TO REMEMBER

- In a strained material, the planes on which shear stresses do not exist are defined as principal planes.
- The normal stresses on the principal planes are called principal stresses or maximum normal stresses.
- The planes on which maximum shear stress occurs are called shear planes.
- The major principal stress $\sigma_1 = \dfrac{\sigma_x + \sigma_y}{2} + \sqrt{\left(\dfrac{\sigma_x - \sigma_y}{2}\right)^2 + \tau^2}$.
- Minor principal stress $\sigma_2 = \dfrac{\sigma_x + \sigma_y}{2} - \sqrt{\left(\dfrac{\sigma_x - \sigma_y}{2}\right)^2 + \tau^2}$.
- Maximum shear stress $\tau_{max} = \dfrac{\sigma_1 - \sigma_2}{2} = \sqrt{\left(\dfrac{\sigma_x - \sigma_y}{2}\right)^2 + \tau^2}$.
- Normal stress on any plane inclined at θ with σ_x plane is
  $\sigma_n = \dfrac{\sigma_x + \sigma_y}{2} + \dfrac{\sigma_x - \sigma_y}{2}\cos 2\theta + \tau \sin 2\theta$.
- Shear stresses or tangential stress on any plane inclined at θ with σ_x plane is
  $\sigma_t = \dfrac{\sigma_x - \sigma_y}{2}\sin 2\theta - \tau \cos 2\theta$.
- Resultant stress on inclined plane $\sigma_R = \sqrt{\sigma_n^2 + \sigma_t^2}$.
- Obliquity of resultant stress with plane of σ_x, $\tan\phi = \dfrac{\sigma_t}{\sigma_n}$.
- The angles of principal planes $\tan 2\theta = \dfrac{2\tau}{\sigma_x - \sigma_y}$.
- θ_1 and θ_2 are 90° apart
- The angles of shear planes $\tan 2\theta_t = \dfrac{\sigma_x - \sigma_y}{-2\tau}$.
- The principal planes and shear planes are separated by 45° apart.

EXERCISES

1. A tension bar is prepared by joining two wooden pieces 60 mm × 120 mm at their ends by cutting at an angle of 60° as shown in Fig. 6.56. The member is subjected to a pull of 18 kN. Find the normal and shear stresses at the joint.

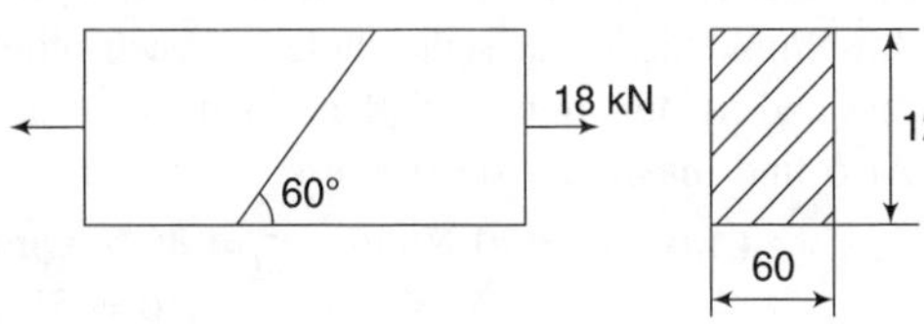

Fig. 6.56

(*Hint:* $\sigma_x = \frac{F}{A} = \frac{18000}{60 \times 120} = 2.5$ MPa, $\theta = 90 - 60 = -30°$)

(***Ans.*** $\sigma_n = 1.875$ MPa, $\sigma_t = -1.083$ MPa.)

2. A circular bar of diameter 30 mm is subjected to an axial load. If the normal and tangential stresses on a 30° inclined plane are limited to 90 MPa and 60 MPa respectively, determine the safe load that can be applied.

($F = 84.823$ N)

3. The direct stresses at a point on a stressed material are 100 N/mm^2 compressive and 60 N/mm^2 tensile. Find the stresses on a plane inclined at 25° to the plane of 100 N/mm^2 stress. Also find resultant stress and its inclination.

($\sigma_n = 71.423$ MPa, $\sigma_t = 61.284$ MPa, $\sigma_R = 94.111$ MPa, $\phi = 40.631°$)

4. A point in a stressed material is subjected to tensile stress of 50 N/mm^2 and 30 N/mm^2 on two mutually perpendicular planes. Calculate the normal, tangential and resultant stresses and obliquity of resultant stress, on a plane making an angle of 30° with the axis of the second stress. Also find maximum shear stress.

(***Ans.*** $\sigma_n = 45$ N/mm^2, $\sigma_t = 8.66$ N/mm^2, $\sigma_R = 45.83$ N/mm^2, $\phi = 10.89°$, $\tau_{max} = \pm 10$ N/mm^2)

5. The state of stress at a point in a stressed material is shown in Fig. 6.57. Determine using Mohr's circle:

(i) Normal, shear stresses and resultant stress and its obliquity on inclined plane *AB*

(ii) Principal stresses and their location

(iii) Maximum shear stresses and their location

(iv) Verify using analytical method.

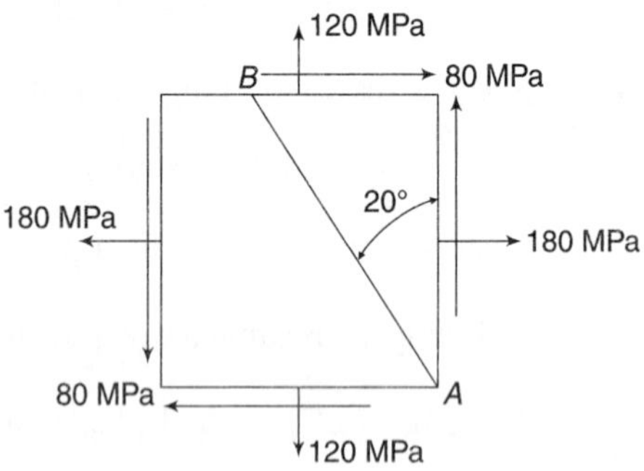

Fig. 6.57

(***Ans.*** (i) $\sigma_n = 224.404$ MPa, $\sigma_t = -42$ MPa, $\sigma_R = 228.3$ MPa, $\phi = 10.6°$,
(ii) $\sigma_1 = 235.44$ MPa, $\sigma_2 = 65.56$ MPa, $\theta_1 = 34.722°$ and $\theta_2 = 124.722°$,
(iii) $\tau_{max} = 85.44$ MPa, $\theta_t = -10.278°$ and 79.72°)

6. In a stressed material, at a certain point, direct stresses of 120 N/mm^2 tensile and 100 N/mm^2 on two mutually perpendicular planes. The major principal stress in the material is to be limited to 160 N/mm^2. What is the safe shear stress that the material may be subjected on given planes. Also find minor principal stresses and maximum shear stress at that point.

(***Ans.*** $\tau = 101.98$ N/mm^2, $\sigma_2 = -140$ N/mm^2, $\tau_{max} = 150$ N/mm^2)

7. A point in a strained material is subjected to a tensile stress of 120 N/mm^2 and a compressive stresses of 80 N/mm^2 acting at right angles to each other. Find normal and shear stresses, resultant stress and its direction on a plane inclined at an angle of 30° with axis of compressive stress. Also find maximum shear stress.

(***Ans.*** $\sigma_n = 70$ N/mm^2, $\sigma_t = 86.6$ N/mm^2, $\sigma_R = 111.35$ N/mm^2, $\phi = 51.05°$ and $\tau_{max} = 100$ MPa)

8. Explain the step by step procedure of drawing Mohr's circle diagram for an element under combined stresses on two mutually perpendicular planes which are unlike and unequal (σ_x is tensile σ_y is compressive).

9. An element is subjected to in Fig. 6.58 Determine

(i) Principal stresses and their direction's analytically

(ii) Find normal and tangential stresses on plane *BC* graphically

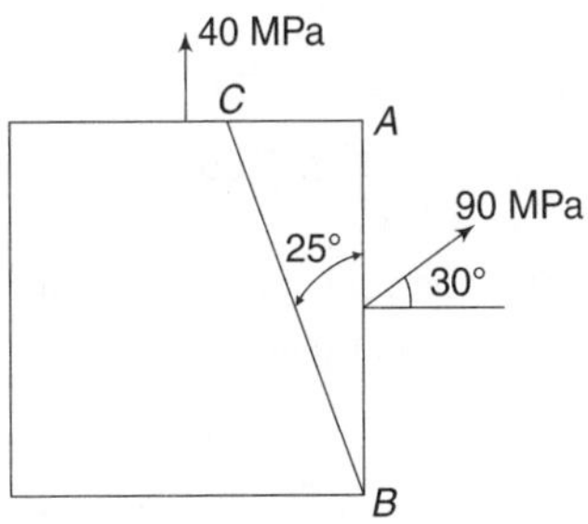

Fig. 6.58

(*Ans.* (i) σ_1 = 109.2 MPa, σ_2 = – 29.25 MPa, θ_1 = 33.03°, θ_2 = 123.03°, (ii) σ_n = 107.29 MPa and σ_t = –13.63 MPa)

10. An element is subjected to combined stresses as shown in Fig. 6.59. Derive an expression for normal, tangential and resultant stresses on an oblique plane *BE*. Also find the maximum shear stress and the plane on which they act.

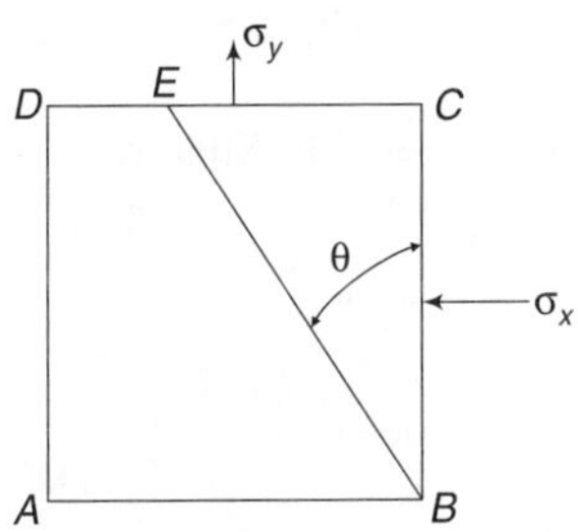

Fig. 6.59

11. At a point in a stressed plate there is a tensile stress of 100 N/mm^2 in the horizontal direction, and a compressive stress of 50 N/mm^2 along the vertical axis along with shear stress of 50 N/mm^2 on each plane. Determine the principal stresses, plane of principal stresses, maximum shear stress and its obliquity, verify the values by Mohr's circle method.

(*Ans.* σ_1 = 115.14 MPa, σ_2 = –65.14 MPa, θ_1 = 16.85°, θ_2 = 106.85°, τ_{max} = 90.14 MPa, θ_t = –28.15° and 61.85°)

12. Prove that in case of a bar subjected to uniaxial stress, maximum shear stress occurs at 45° of the axis of the bar.

13. A rectangular bar of cross-sectional area 11000 mm^2 is subjected to a tensile load P as shown in Fig. 6.60. The permissible normal and shear stresses on the oblique plane BC are 7 N/mm^2 and 3.5 N/mm^2, respectively. Determine the safe value of force P.

(***Ans.*** P = 88.912 kN)

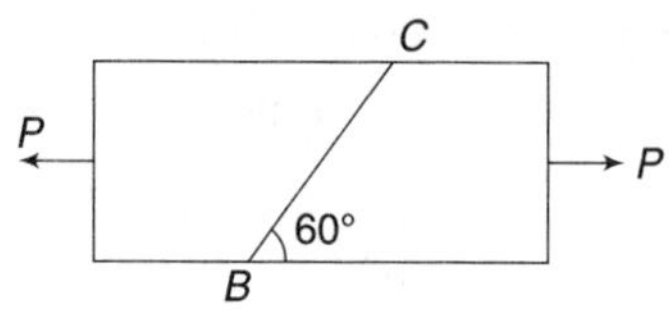

Fig. 6.60

14. Using Mohr's circle, determine the principal stresses and planes, maximum shear stress and planes and maximum shear stress and planes.

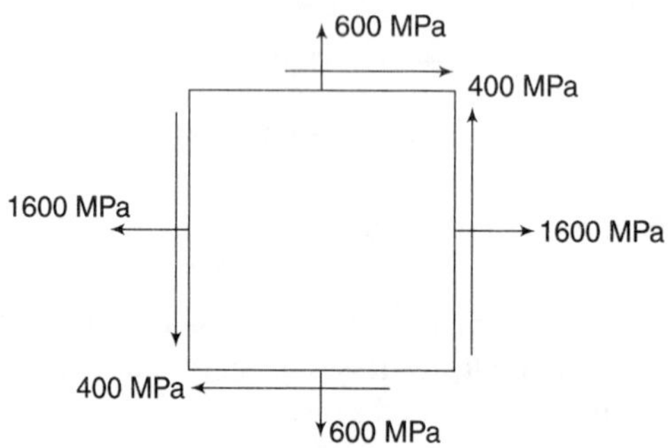

Fig. 6.61

(***Ans.*** σ_1 = 1740 MPa, σ_2 = 460 MPa, θ_1 = 19.25°, θ_2 = 109.25°, τ_{max} = 640 MPa, θ_t = –25.67° and 64.33°)

15. An element with the stresses acting on it, is shown in Fig. 6.62.

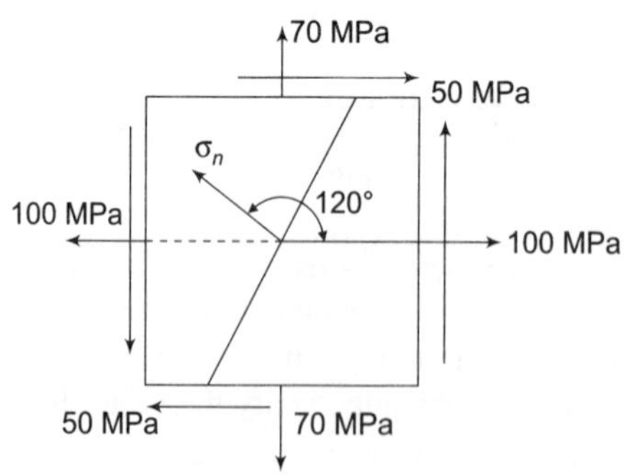

Fig. 6.62

(i) By Mohr's circle method determine

(a) Normal and shear stresses acting on plane EF.

(b) Maximum shear stress and their location.

(ii) Verify answers analytically

(*Ans.* (i) σ_n = 34.2 MPa, σ_t = 12 MPa,
(ii) σ_1 = 137.2 MPa, σ_2 = 32.8 MPa, θ_1 = 36.7°, θ_2 = 126.7°
(iii) τ_{max} = ± 52.2 MPa, θ_t = 81.7° and 171.7°)

Objective Type Questions

1. The normal stress on an inclined plane at an angle θ to the cross section of a body which is subjected to an axial stress σ is given by

(a) $\frac{\sigma}{2}\sin\theta$ (b) $\frac{\sigma}{2}\sin 2\theta$

(c) $\sigma\cos^2\theta$ (d) $\frac{\sigma}{2}\cos 2\theta$

2. The normal stress on an oblique plane at an angle θ for a body subjected to axial stress will be maximum at θ =

(a) 0 (b) 90°

(c) 45° (d) 180°

3. The shear stress on an oblique plane at an angle θ to the cross section of a body subjected to an axial stress σ is given by

(a) $\frac{\sigma}{2}\sin\theta$ (b) $\frac{\sigma}{2}\sin 2\theta$

(c) σ cios^{2}θ (d) $\frac{\sigma}{2}\cos 2\theta$

4. The shear stress on an oblique plane at an angle θ for a body subjected to axial of stress will be maximum at θ =

(a) 0° (b) 90°

(c) 45° (d) 180°

5. If a member is subjected to an axial tensile load, the plane normal to the axis of loading carries

(a) Minimum normal stress (b) Maximum normal stress

(c) Maximum shear stress (d) Minimum shear stress

6. The resultant stress on an oblique plane for a body subjected to axial stress is equal to

(a) $\frac{\sigma}{2}\sin 2\theta$ (b) σ cos2θ

(c) σ cos θ (d) σ sin θ

7. If a member is subjected to an axial tensile load, the plane inclined at 45° to axis of loading carries

(a) Maximum principal stress (b) Minimum principal stress

(c) Resultant stress (d) Maximum shear stress

8. For a member subjected to axial stress σ, the maximum shear stress induced is

(a) σ (b) $\frac{\sigma}{2}$

(c) 2σ (d) $\frac{\sigma}{\sqrt{2}}$

9. For a biaxial stressed member, maximum shear stress is given by

(a) $\frac{\sigma_1 + \sigma_2}{2}$ (b) $\frac{\sigma_1 - \sigma_2}{2}$

(c) $\sigma_1 - \sigma_2$ (d) $\sigma_1 + \sigma_2$

10. For a generalised biaxial stress combined with shear stress, maximum principal stress is given by

(a) $\frac{\sigma_x + \sigma_y}{2} + \sqrt{\left(\frac{\sigma_x - \sigma_y}{2}\right)^2 + \tau^2}$ (b) $\frac{\sigma_x + \sigma_y}{2} + \sqrt{\left(\frac{\sigma_x + \sigma_y}{2}\right)^2 + \tau^2}$

(c) $\frac{\sigma_x - \sigma_y}{2} + \sqrt{\left(\frac{\sigma_x - \sigma_y}{2}\right)^2 + \tau^2}$ (d) $\frac{\sigma_x - \sigma_y}{2} - \sqrt{\left(\frac{\sigma_x - \sigma_y}{2}\right)^2 + \tau^2}$

11. For a generalised biaxial stress combined with shear stress, minimum principal stress is given by

(a) $\frac{\sigma_x + \sigma_y}{2} + \sqrt{\left(\frac{\sigma_x - \sigma_y}{2}\right)^2 + \tau^2}$ (b) $\frac{\sigma_x - \sigma_y}{2} + \sqrt{\left(\frac{\sigma_x + \sigma_y}{2}\right)^2 + \tau^2}$

(c) $\frac{\sigma_x + \sigma_y}{2} - \sqrt{\left(\frac{\sigma_x - \sigma_y}{2}\right)^2 + \tau^2}$ (d) $\frac{\sigma_x - \sigma_y}{2} - \sqrt{\left(\frac{\sigma_x + \sigma_y}{2}\right)^2 + \tau^2}$

12. For a generalised biaxial stress combined with shear stress, the maximum shear stress is given by

(a) $\sqrt{\left(\frac{\sigma_x - \sigma_y}{2}\right)^2 - \tau^2}$ (b) $\sqrt{\left(\frac{\sigma_x + \sigma_y}{2}\right)^2 + \tau^2}$

(c) $\sqrt{\left(\frac{\sigma_x - \sigma_y}{2}\right)^2 + \tau^2}$ (d) $\sqrt{\left(\frac{\sigma_x + \sigma_y}{2}\right)^2 - \tau^2}$

13. For general biaxial stresses combined with shear stress, the angle of principal planes are given by tan 2θ =

(a) $\frac{\sigma_x + \sigma_y}{2\tau}$ (b) $\frac{2\tau}{\sigma_x + \sigma_y}$

(c) $\frac{\sigma_x + \sigma_y}{-2\,\tau}$ (d) $\frac{2\tau}{\sigma_x - \sigma_y}$

14. For general biaxial stresses combined with shear stress, the angle of shear planes are given by tan $2\,\theta_t$ =

(a) $\frac{\sigma_x + \sigma_y}{2\tau}$ (b) $\frac{2\tau}{\sigma_x + \sigma_y}$

(c) $\frac{\sigma_x - \sigma_y}{-\,2\tau}$ (d) $\frac{2\tau}{\sigma_x - \sigma_y}$

15. The normal stress on a plane inclined at θ with the plane of σ_x for a biaxial stress combined with shear stress is given by

(a) $\dfrac{\sigma_x + \sigma_y}{2} + \dfrac{\sigma_x - \sigma_y}{2} \sin 2\theta + \tau \cos 2\theta$

(b) $\dfrac{\sigma_x + \sigma_y}{2} + \dfrac{\sigma_x - \sigma_y}{2} \cos 2\theta + \tau \sin 2\theta$

(c) $\dfrac{\sigma_x - \sigma_y}{2} + \dfrac{\sigma_x + \sigma_y}{2} \cos 2\theta + \tau \sin 2\theta$

(d) $\dfrac{\sigma_x - \sigma_y}{2} + \dfrac{\sigma_x + \sigma_y}{2} \sin 2\theta + \tau \cos 2\theta$

16. The shear stress on a plane inclined at θ with the plane of σ_x for a biaxial stress combined with shear stress is given by

(a) $\dfrac{\sigma_x - \sigma_y}{2} \sin 2\theta + \tau \cos 2\theta$

(b) $\dfrac{\sigma_x - \sigma_y}{2} \sin 2\theta - \tau \cos 2\theta$

(c) $\dfrac{\sigma_x + \sigma_y}{2} \sin 2\theta + \tau \cos 2\theta$

(d) $\dfrac{\sigma_x + \sigma_y}{2} \cos 2\theta + \tau \sin 2\theta$

Key to Answers

1. (c)	2. (a)	3. (b)	4. (c)	5. (b)	6. (c)	7. (d)	8. (b)
9. (b)	10. (a)	11. (c)	12. (c)	13. (d)	14. (c)	15. (b)	16. (b)

CHAPTER 7

Deflection of Beams

7.1 INTRODUCTION

Most of the engineering materials are elastic in nature and they deflect under load. In some cases, the deflection is undesirable and is to be kept minimum. While designing beams the decision is not only based on the strength but also on the point of deflection. Sometimes excessive deflection leads to misalignment and non-performance of machine elements. Hence, it is necessary to calculate the deflection of beams.

There are several methods to calculate beam deflection. Some of them are double integration or Macaulay's method, Castigliano theorem or strain energy method, area moment method and method of superpositioning. Double integration or Macaulay's method is widely used because of its simiplicity and deflection at all points of the beam can be easily calculated and the elastic curve for deflection can be drawn. However, it is simple and better used for determinate beams only.

In the earlier chapter for bending stresses, it is seen that the bending equation is

$$\frac{M}{I} = \frac{\sigma}{y} = \frac{E}{R}.$$

The beam subjected to bending moment M will result in deflecting it into circular arc of radius R.

7.2 DIFFERENTIAL EQUATION FOR DEFLECTION AND SLOPE

Consider a beam bent into a circular arc as shown in Fig. 7.1.

Consider an elemental length $AB = ds$.

Draw tangents AD and BE at A and B.

P is the point of intersection of DA extended with BE.

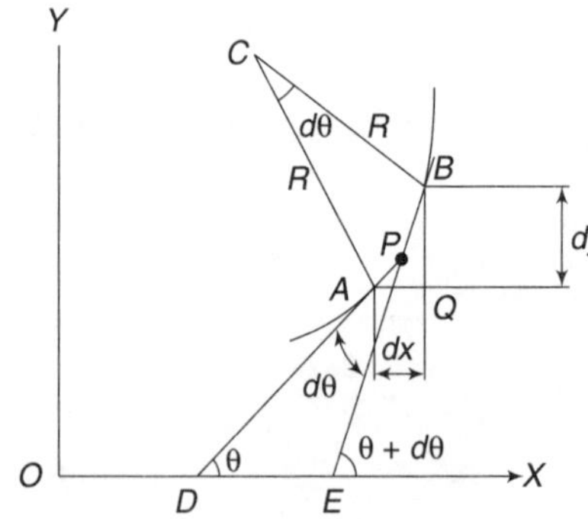

Fig. 7.1

C is the centre of curvature of the Arc AB.

$$\therefore \quad CA = CB = \text{Radius} = R$$

$$\text{Let the angle } ACB = d\theta$$

$$\text{Let angle } PDE = \theta$$

Since AD is $\perp$ to AC and BE is $\perp$ to BC, $\angle DPE = d\theta$

From triangle DPE, $\angle EDP + \angle DPE$ = External angle PEX

$$\therefore \quad \angle PEX = \theta + d\theta$$

$$\text{Length of sector } AB = ds = R \cdot d\theta$$

$$\therefore \quad R = \frac{ds}{d\theta}$$

Draw horizontal line from A and vertical line from B to meet at Q.

Since AB is very small, consider $AB\,Q$ as a triangle,

$$AQ = dx \text{ and } BQ = dy$$

$$\tan\theta = \frac{dy}{dx}, \cos\theta = \frac{dx}{ds} \text{ or } \frac{ds}{dx} = \frac{1}{\cos\theta} = \sec\theta$$

But
$$R = \frac{ds}{d\theta} = \frac{\left(\frac{ds}{dx}\right)}{\left(\frac{d\theta}{dx}\right)} = \frac{\sec\theta}{\left(\frac{d\theta}{dx}\right)} \qquad \text{...(i)}$$

$$\tan\theta = \frac{dy}{dx}$$

$$\text{Differentiating w.r.t. } x, \quad \sec^2\theta \frac{d\theta}{dx} = \frac{d^2y}{dx^2}$$

$$\therefore \quad \frac{d\theta}{dx} = \frac{d^2y}{dx^2} \cdot \frac{1}{\sec^2\theta}$$

$$\text{Substituting in Eq. (i), } R = \frac{\sec\theta}{\left(\frac{d^2y}{dx^2} \cdot \frac{1}{\sec^2\theta}\right)}$$

$$R = \frac{\sec^3\theta}{\left(\frac{d^2y}{dx^2}\right)}$$

or
$$\frac{1}{R} = \frac{d^2y}{dx^2} \cdot \frac{1}{\sec^3\theta}$$

But
$$\sec^2\theta = 1 + \tan^2\theta$$

$$\sec^3\theta = (1 + \tan^2\theta)^{3/2}$$

$$\therefore \quad \frac{1}{R} \cdot \frac{d^2y}{dx^2} \cdot \frac{1}{(1 + \tan^2\theta)^{3/2}}$$

tan θ is the slope of the bent beam at *A*. For a beam bent, the slope tan θ will be very small and $\tan^2\theta$ will be further small. Hence, neglected.

$$\frac{1}{R} = \frac{d^2y}{dx^2} \quad \text{...(ii)}$$

From equation of bending $\frac{M}{I} = \frac{E}{R}$

$$\therefore \quad \frac{1}{R} = \frac{M}{EI} \quad \text{...(iii)}$$

Equating Eqs. (ii) and (iii)

$$\frac{M}{EI} = \frac{d^2y}{dx^2}$$

$$M = EI\frac{d^2y}{dx^2}$$

EI is known as flexural rigidity.

It is the resistance offered for flexing or bending.

$$EI = \frac{d^2y}{dx^2} = M$$

is called differential equation for deflection.

7.3 DOUBLE INTEGRATION METHOD OF FINDING SLOPE AND DEFLECTION

The deflection equation is $EI\frac{d^2y}{dx^2} = M$

M is the bending moment at a distance *x* from one end.

Integrating once, $EI\frac{dy}{dx} = \int_0^x M \cdot dx + C_1$

From this equation the slope $\frac{dy}{dx}$ can be calculated.

Integrating again, $EIY = \iint Mdx + C_1 x + C_2$

where C_1 and C_2 are constants of integration and can be found by applying boundaries conditions like

(i) at the supports deflection $y = 0$

(ii) at the point of symmetry and at maximum deflection point

$$\text{Slope } \frac{dy}{dx} = 0.$$

7.4 DEFLECTION OF CANTILEVERS: SOME GENERAL CASES

7.4.1 Cantilever Subjected to Moment at Free End

Figure 7.2 shows a cantilever of length l subjected to a moment M at the end. Let E be the modulus of elasticity and I the moment of inertia of the cross section.

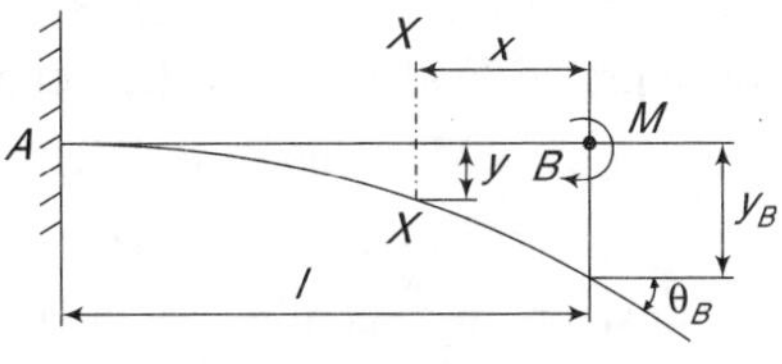

Fig. 7.2

$$\text{Moment at any section } XX = M_x = M$$

$$EI\frac{d^2y}{dx^2} = M_x = M$$

$$\text{Integrating once, } EI\frac{dy}{dx} = M \cdot x + C_1$$

$$\text{Integrating again, } EIY = \frac{M \cdot x^2}{2} + C_1 x + C_2$$

Applying boundary conditions:

$$\text{At the fixed end, } x = l, \text{ slope } \frac{dy}{dx} = 0$$

$$O = Ml + C_1$$

$$\therefore \quad C_1 = -Ml$$

$$\therefore \quad EI\frac{dy}{dx} = Mx - Ml.$$

At the free end, at $x = 0$, $EI\left(\frac{dy}{dx}\right)_B = -Ml$

$$\therefore \quad \text{Slope } \frac{dy}{dx} = \theta_B = \frac{-Ml}{EI}$$

$$EIY = \frac{Mx^2}{2} - Mlx + C_2$$

At $x = l$, at the fixed end, deflection $y = 0$

$$O = \frac{Ml^2}{2} - Ml^2 + C_2$$

$$\therefore \quad C_2 = \frac{Ml^2}{2}$$

$$\therefore \quad EIY = \frac{Mx^2}{2} + Mlx + \frac{Ml^2}{2}$$

$$\text{At free end, at } x = 0, \text{ deflection } y = \frac{Ml^2}{2EI}$$

(the deflection is downward)

7.4.2 Cantilever Subjected to Concentrated Load *W* at the Free End

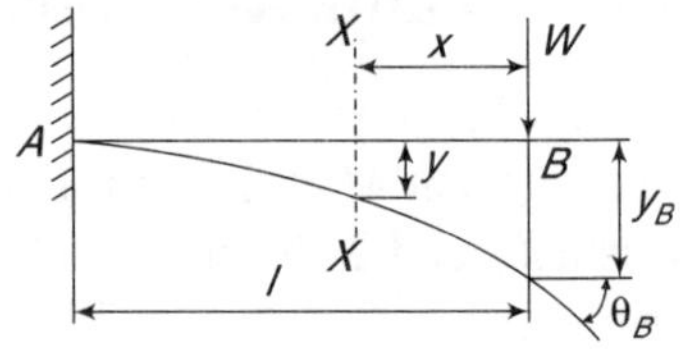

Fig. 7.3

Figure 7.3 shows a cantilever of length l subjected to a concentrated load W at the free end.

E = Modulus of elasticity

I = Moment of inertia of the section

Consider any section XX at a distance x from free end.

Bending moment at section XX, $Mx = W.x$ (hogging)

$$EI\frac{d^2y}{dx^2} = M_x = Wx$$

$$\text{Integrating once } EI\frac{dy}{dx} = W\frac{x^2}{2} + C_1 \qquad \text{...(i)}$$

Applying the boundary conditions,

$$\text{At the fixed end } x = l, \text{ slope } \frac{dy}{dx} = 0$$

$$O = W\frac{l^2}{2} + C_1$$

$$\therefore \quad C_1 = -W\frac{l^2}{2}$$

$$\text{At the free end at } x = 0, \text{ Slope } \frac{dy}{dx} = -\frac{Wl^2}{2EI}$$

$$\text{Integrating again, } EIY = \frac{Wx^3}{6} + C_1x + C_2$$

$$EIY = \frac{Wx^3}{6} + \frac{Wl^2}{2}x + C_2$$

At the fixed end at $x = l$, deflection $y = 0$

$$O = \frac{Wl^3}{6} + \frac{Wl^3}{2} + C_2$$

$$\therefore \quad C_2 = \frac{Wl^3}{2} + \frac{Wl^3}{6} = \frac{3Wl^3 - Wl^3}{6} = \frac{Wl^3}{3}$$

$$EIY = \frac{Wx^3}{6} + \frac{Wl^2}{2}x + \frac{Wl^3}{3}$$

At free end, at $x = 0$, $y = y_B$

$$EIY_B = -O + O + \frac{Wl^3}{3}$$

$$y_B = \frac{Wl^3}{3EI} \quad (y \text{ is downward})$$

7.4.3 Cantilever of Length *l* Carrying a Concentrated Load *W* at a Distance *a* from Fixed End

Load W at a distance a from fixed end. Figure 7.4 shows a cantilever of length l subjected to a concentrated load W at a distance a from the fixed end.

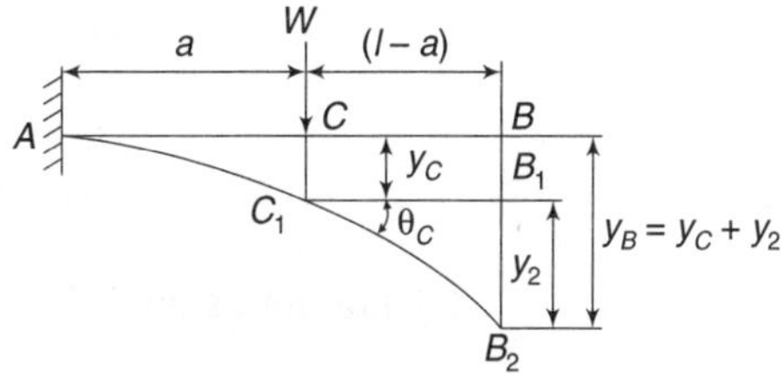

Fig. 7.4

Let y_c be the deflection and Q_C be the slope at C.

$$\text{WKT deflection } y_C = \frac{Wa^3}{3EI}$$

and

$$\text{slope } \frac{dy}{dx} = \theta_C = \frac{Wa^2}{2EI}$$

Since the portion of beam BC is not subjected to any load or moment, it remains straight and its slope

$$\theta_C = \frac{Wa^2}{2EI}$$

$$\text{From } \Delta\, B_1\, B_2\, C_1, \text{ slope } \tan\theta_C = \frac{BB_2}{B_1C_1}$$

$$\text{For small angles } \tan\theta_C = \theta_C = \frac{y_2}{l-a}$$

$$\therefore \quad y_2 = (l-a)\,\theta_C = (l-a)\frac{Wa^2}{2EI}$$

$$\text{Total deflection at } B,\ y_B = y_C + y_2$$

$$y_B = \frac{Wa^3}{3EI} + \frac{Wa^2}{2EI}(l-a).$$

7.4.4 Cantilever of Length *l* Carrying a Uniformly Distributed Load *w* per Unit Length Over Full Length

Figure 7.5 shows a cantilever of length l carrying a UDL of w/unit length over full length.

Consider any section XX at a distance x from free end.

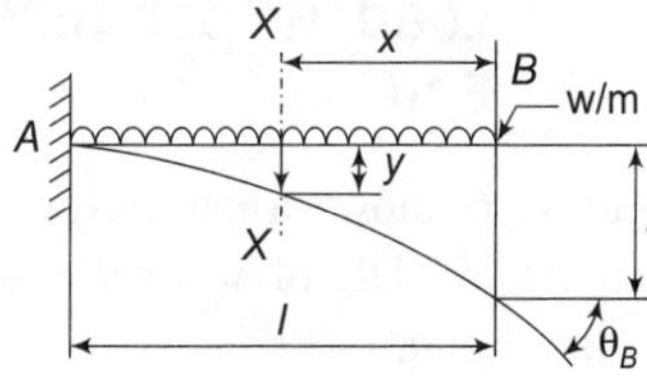

Fig. 7.5

E = Modulus of elasticity

I = Moment of inertia of the section

$$\text{Bending moment at } XX,\ M_X = \frac{wx^2}{2}$$

$$\text{Integrating } EI\frac{dy}{dx} = \frac{wx^3}{6} + C_1$$

$$\text{At the fixed end at } A,\ x = l \text{ and slope } \frac{dy}{dx} = 0$$

$$O = \frac{wl^3}{6} + C_1$$

$$\therefore \quad C_1 = \frac{-wl^3}{6}$$

$$\therefore \quad EI\frac{dy}{dx} = \frac{wx^3}{6} - \frac{wl^3}{6}$$

$$\text{Integrating again } EIY = \frac{wx^4}{24} - \frac{wl^3x}{6} + C_2$$

At the fixed end at A, deflection $y = 0$

$$O = \frac{wl^4}{24} - \frac{wl^4}{6} + C_2$$

$$\therefore \quad C_2 = \frac{4wl^4 - wl^4}{24} = \frac{3wl^4}{24} = \frac{wl^4}{8}$$

$$\therefore \quad EIY = \frac{wx^4}{24} - \frac{wl^3x}{6} + \frac{wl^4}{8}$$

At B, at $x = 0$, deflection is maximum

$$EIY_B = \frac{wl^4}{8}$$

$$y_{max} = y_B = \frac{wl^4}{8EI} \text{ (downward deflection)}$$

Slope at the end at $x = 0$

$$EI\left(\frac{dy}{dx}\right)_B = \frac{wl^3}{6}$$

or

$$\theta_B = \frac{dy}{dx} = \frac{wl^3}{6EI}.$$

7.4.5 Cantilever of Length *l* Carrying a Uniformly Distributed Load *w* per unit Length Over a Distance of *a* from Fixed End

Figure 7.6 shows a cantilever beam of length l carrying a UDL of w over a length of a from the fixed end.

E = Modulus of elasticity

I = Moment of inertia of the section

Let y_C be the deflection and θ_C be the slope at point C.

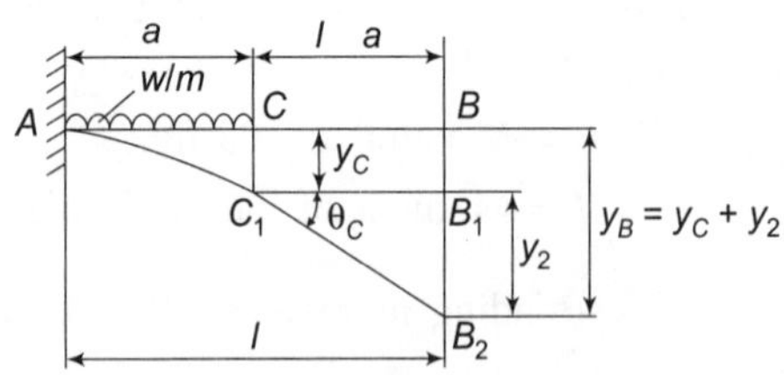

Fig. 7.6

For cantilever with UDL, deflection $y_C = \frac{wa^4}{8EI}$

$$\text{and slope } \theta_C = \frac{wa^3}{6EI}$$

Since the portion of the beam BC is not subjected to any load or moment, it remains straight and its slope

$$\tan \theta_C = \frac{wa^3}{6EI}$$

$$\text{For small angle } \tan \theta_C = Q_C$$

$$\text{From } \Delta\, B_1\, B_2\, C_1, \text{ slope } \theta_C = \frac{B_1B_2}{B_1C_1} = \frac{y_2}{l - a}$$

$$\therefore \quad y_2 = (l - a)\, \theta_C = (l - a)\, \frac{wa^3}{6EI}$$

$$\therefore \quad \text{Total deflection at } B,\ y_B = y_C + y_2$$

$$y_{max} = y_B = \frac{wa^4}{8EI} + \frac{wa^3}{6EI}(l - a).$$

7.4.6 Cantilever of Length *l* Carrying a Uniformly Distributed Load *w* Over a Length of *a* from the Free End

Figure 7.7 (a) shows a cantilever of length l subjected to a UDL of w/unit length over a span of a from the fixed end.

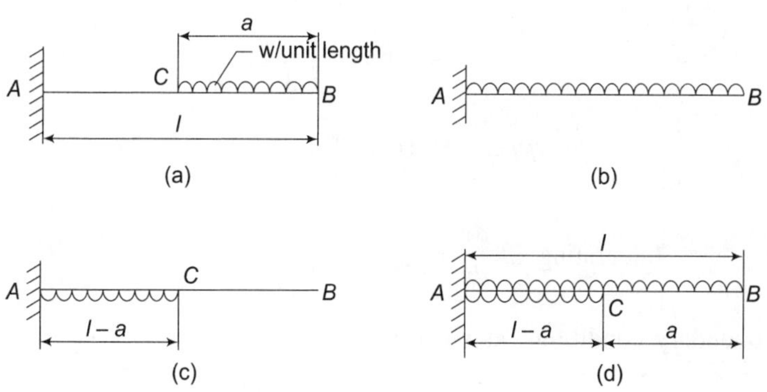

Fig. 7.7

E = Modulus of elasticity

I = Moment of inertia of the section.

To find the deflection at free end B, the following effects may be considered.

(i) First apply UDL to the entire length l as shown in Fig. 7.7 (b)

(ii) Apply an upward UDL at the portion AC over a span of $(l - a)$ as shown in Fig. 7.7 (c)

The net effect is the downward deflection due to full UDL – upward deflection due to UDL over the remaining span $(l - a)$ as shown in Fig 7.7 (d)

$$\therefore \quad y_B = y_{B\ (\text{FULL})} - y_B\ (\text{for } l - a)$$

$$y = \frac{wl^4}{8EI} - \left\{ \frac{w(l-a)^4}{8EI} + \frac{w(l-a)^3}{6EI} \cdot a \right\}$$

7.4.7 Cantilever of Length *l* Carrying a Uniformly Varying Load UDL whose Intensity Varies Uniformly from *O* at Free End to *w* per unit Length at Fixed End

Figure 7.8 shows a cantilever of length l subjected to a load that varies uniformly from O at free end to $w/$ unit length at fixed end.

E = Modulus of elasticity

I = Moment of inertia of the section

Consider a section XX at a distance x from the free end.

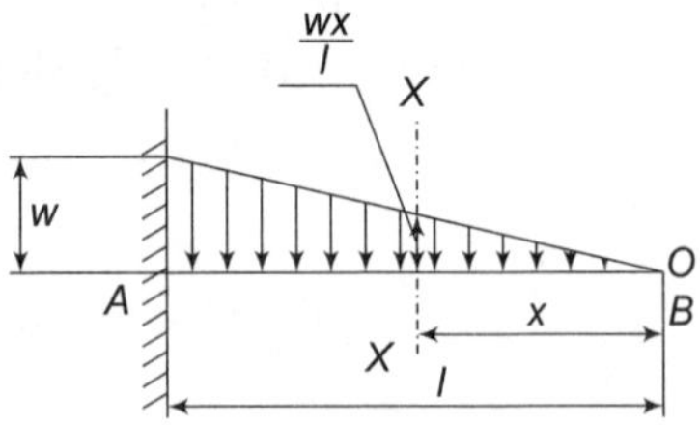

Fig. 7.8

$$\text{Intensity of load at } XX,\ w_x = \frac{w}{l} \cdot x \text{ per unit length}$$

The cg for the UVL lies at $\frac{1}{3}$ of its base $= \frac{x}{3}$

$$\text{Bending moment at } XX,\ M_x = \frac{1}{2}(w_x . x)\frac{x}{3} \text{ (Hogging)}$$

$$= \frac{1}{2}\left(\frac{w_x}{l} . x\right)\frac{x}{3} = \frac{wx^3}{6l}$$

$$\therefore \quad EI\frac{d^2y}{dx^2} = M_x = \frac{wx^3}{6l}$$

$$\text{Integrating } EI\frac{dy}{dx} = \frac{wx^4}{24l} + C_1$$

Applying boundary conditions, at $x = l$, slope $\frac{dy}{dx} = 0$

$$O = \frac{wl^4}{24l} + C_1$$

$$\therefore \quad C_1 = \frac{-wl^3}{24}$$

$$\therefore \quad EI\frac{dy}{dx} = \frac{wx^4}{24l} - \frac{wl^3}{24}$$

At the free end at B, $x = 0$

$$\therefore \quad \text{Slope} = \frac{dy}{dx} = \frac{-wl^3}{24}$$

$$\text{Integrating again, } EIY = \frac{wx^5}{120l} - \frac{wl^3x}{24} + C_2$$

Applying boundary conditions, at the fixed end at A,

$$x = l, \text{ deflection } y = 0$$

$$O = \frac{wl^5}{120l} - \frac{wl^3l}{24} + C_2$$

$$\therefore \quad C_2 = \frac{wl^4}{24} - \frac{wl^4}{120} = \frac{5wl^4 - wl^4}{120} = \frac{wl^4}{30}$$

$$\therefore \quad EIY = \frac{wx^5}{120l} - \frac{wl^3x}{24} + \frac{wl^4}{30}$$

At the free end, at B, $x = 0$. The deflection will be maximum

$$y_B = y = \frac{wl^4}{30EI}.$$

7.4.8 Cantilever of Length *l* Carrying a Uniformly Varying Load that Varies Uniformly from *O* at Fixed End to *w*/Unit Length of Free End

Figure 7.9 (a) shows a cantilever of length l carrying a UVL that varies uniformly form O at fixed end to w/unit length at free end.

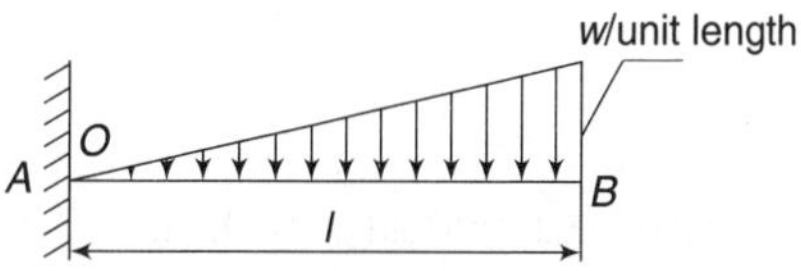

Fig. 7.9 (a)

This UVL can be treated as a full UDL of w/unit length over the entire span acting downward and UVL of w/unit run at fixed end varying to O at free end as shown in Fig 7.9 (b).

Deflection at B = Downward deflection due to UDL – Upward deflection due to UVL.

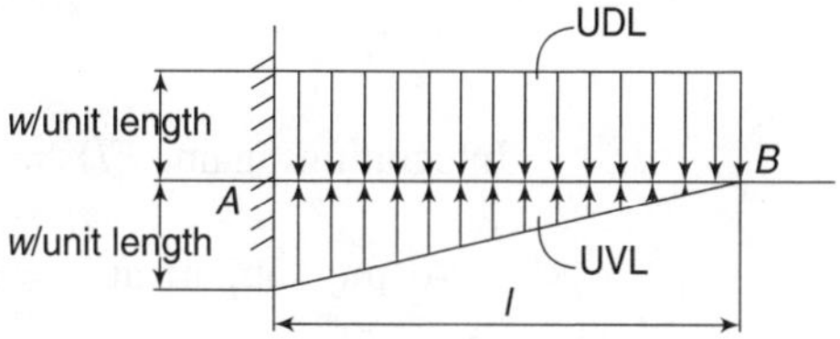

Fig. 7.9 (b)

$$y = \frac{wl^4}{8EI} - \frac{wl^4}{30EI} = \frac{15\ wl^4 - 4wl^4}{120\ EI}$$

$$y = \frac{11}{120}\frac{wl^4}{EI}.$$

7.5 DEFLECTION OF SIMPLY SUPPORTED BEAMS: SOME GENERAL CASES

7.5.1 Simply Supported Beam of Span *l* Carrying a Central Concentrated Load, *W*

Figure 7.10 shows a simply supported beam of length l carrying a concentrated load W at its middle.

Since the load is in the middle, both reactions will be equal.

$$R_A = R_B = \frac{W}{2}$$

Consider any section XX at a distance x from left side support A.

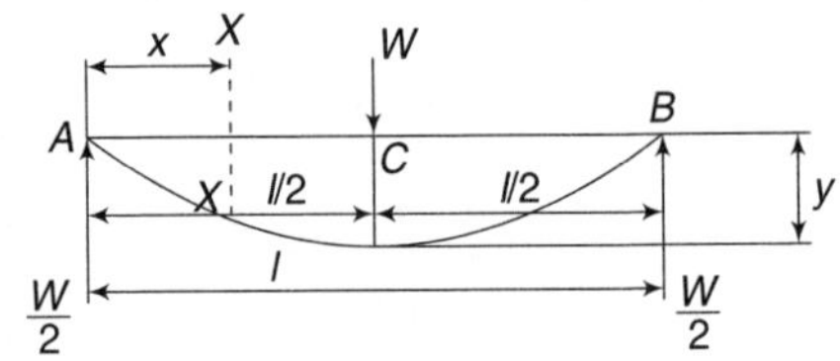

Fig. 7.10

$$\text{Bending moment at } XX,\ M_x = \frac{W}{2}.x$$

Substituting in the deflection equation

$$EI\frac{d^2y}{dx^2} = M_x = \frac{W}{2}.t$$

$$\text{Integrating, } EI\frac{dy}{dx} = \frac{Wx^2}{4} + C_1$$

Since the maximum deflection occurs at the mid span, the slope $\frac{dy}{dx} = 0$ at the middle, at $x = \frac{l}{2}$

$$O = \frac{W}{4}\left(\frac{l}{2}\right)^2 + C_1$$

$$C_1 = \frac{-Wl^2}{16}$$

$$\therefore \quad EI\frac{dy}{dx} = \frac{Wx^2}{4} - \frac{Wl^2}{16}$$

At the support at $x = 0$, the slope $\frac{dy}{dx} = \frac{-Wl^2}{16EI}$

$$\theta = \frac{dy}{dx} = \frac{-Wl^2}{16EI}$$

$$\text{Integrating again, } EIY = \frac{Wx^3}{12} - \frac{Wl^2x}{16} + C_2$$

At the support at $x = 0$, deflection $y = 0$

$$0 = 0 - 0 + C_2$$

$$\therefore \quad C_2 = 0$$

$$\therefore \quad EIY = \frac{Wx^3}{12} - \frac{Wl^2x}{16}$$

For deflection at C $\quad x = \frac{l}{2}$

$$EIY = \frac{W}{12}\left(\frac{l}{2}\right)^3 - \frac{Wl^2}{16}\frac{l}{2} = \frac{Wl^3}{96} - \frac{Wl^3}{32} = \frac{Wl^3 - 3Wl^3}{96}$$

$$\therefore \quad y = \frac{-Wl^3}{48EI} \quad \text{(Downward)}$$

7.5.2 Simply Supported Beam of Length *l* Carrying a Uniformly Distributed Load of *w*/Unit Length Over the Entire Span

Figure 7.11 shows a simply supported beam of length l subjected to a uniformly distributed load w/unit length over the entire span.

E = Modulus of elasticity

I = Moment of inertia of the section

$$\text{Reactions } R_A = R_B = \frac{wl}{2}$$

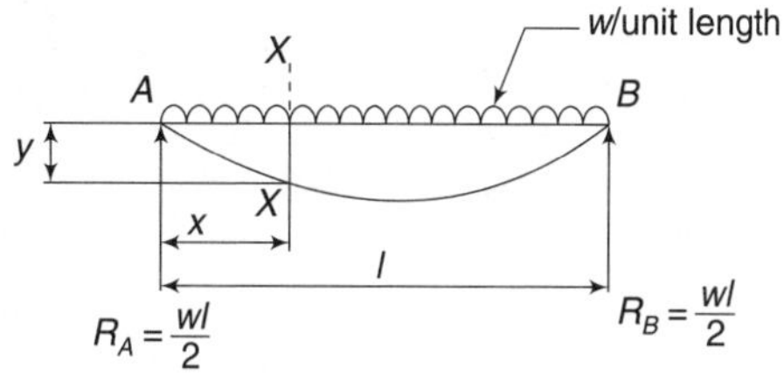

Fig. 7.11

Consider any section XX at distance x from support A.

$$\text{Bending moment at section } XX,\ M_x = \frac{wl}{2}x - wx\cdot\left(\frac{x}{2}\right)$$

$$M_x = \frac{wlx}{2} - \frac{wx^2}{2}$$

$$\text{Substituting in the deflection equation, } EI\frac{d^2y}{dx^2} = M_x = \frac{wlx}{2} - \frac{wx^2}{2}$$

$$\text{Integrating } EI\frac{dy}{dx} = \frac{wlx^2}{4} - \frac{wx^3}{6} + C_1$$

Due to symmetry, the deflection will be maximum at the middle of span, at $x = \frac{l}{2}$. Hence, slope $\frac{dy}{dx} = 0$ at $x = \frac{l}{2}$

$$0 = \frac{wl}{4}\left(\frac{l}{2}\right)^2 - \frac{w}{6}\left(\frac{l}{2}\right)^3 + C_1 = \frac{wl^3}{16} - \frac{wl^3}{48} + C_1$$

$$C_1 = \frac{-wl^3}{16} + \frac{wl^3}{48} = \frac{-3wl^3 + wl^3}{48} = \frac{-wl^3}{24}$$

$$\therefore \quad EI\frac{dy}{dx} = \frac{wl}{4}x^2 - \frac{wx^3}{6} - \frac{wl^3}{24}$$

The slope at support, at $x = 0$,

$$\frac{dy}{dx} = \frac{-wl^3}{24EI}$$

$$\text{Integrating again, } EIY = \frac{wlx^3}{12} - \frac{wx^4}{24} - \frac{wl^3}{24}x + C_2$$

At the fixed end at A, $x = 0$ deflection $y = 0$

$$O = O - O - O + C_2$$

$$\therefore \quad C_2 = 0$$

$$\therefore \quad EIY = \frac{wl}{12}x^3 - \frac{w}{24}x^4 - \frac{wl^3}{24}x$$

Deflection will be maximum in the middle at $x = \frac{l}{2}$

$$EIY = \frac{wl}{12}\left(\frac{l}{2}\right)^3 - \frac{w}{24}\left(\frac{l}{2}\right)^4 - \frac{wl^3}{24}\left(\frac{l}{2}\right)$$

$$= \frac{wl^4}{96} - \frac{wl^4}{384} - \frac{wl^4}{48}$$

$$= \frac{4wl^4 - wl^4 - 8\,wl^4}{384} = \frac{-5wl^4}{354}$$

$$\therefore \qquad y = \frac{-5\,wl^4}{384\,EI} \text{ (downward).}$$

7.5.3 A Simply Supported Beam of Length *l* Subjected to a Uniformly Varying Load Varying from *O* at One End to *w*/Unit Length at the Other End

Figure 7.12 shows a simply supported beam of length l carrying a triangular load that varies uniformly from O at one end to w/unit length at the other end.

E = Modulus of elasticity

I = Moment of inertia of the section

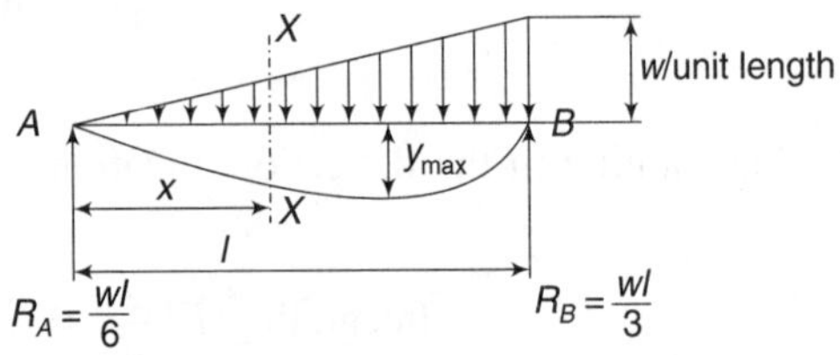

Fig. 7.12

$$R_A + R_B = \frac{1}{2}\cdot wl$$

Taking moment about A

$$R_B \cdot l = \frac{1}{2}\,wl\cdot\left(\frac{2l}{3}\right) \quad \left(\because \text{ cg of triangle is at } \frac{2}{3}\,l \text{ from } A\right)$$

$$R_B = \frac{wl}{3}$$

$$R_A = \frac{wl}{2} - R_B = \frac{wl}{2} - \frac{wl}{3} = \frac{wl}{6}$$

Consider any section XX at a distance x from A.

Bending moment at section XX, $M_x = R_A\,x - \left(\frac{1}{2}\,w_x\cdot x\right)\frac{x}{3}$

$$M_x = \frac{wl}{6}\,x - \frac{1}{6}\,w_x\cdot x^2$$

where, $\qquad w_x$ = intensity of load at x.

From similar triangles $\frac{w}{l} = \frac{w_x}{x}$

$\therefore \qquad$ Intensity of load at x, $w_x = \frac{w}{l}.x$

$$\therefore \qquad M_x = \frac{wl}{6}.x - \frac{1}{6_l} w.x^3$$

Substituting in deflection equation

$$EI \frac{d^2y}{dx^2} = M_x = \frac{wlx}{6} - \frac{wx^3}{6l}$$

$$\text{Integrating, } EI \frac{dy}{dx} = \frac{wlx^2}{12} - \frac{wx^4}{24l} + C_1$$

$$\text{Integrating again, } EIY = \frac{wlx^3}{36} - \frac{wx^5}{120l} + C_1 x + C_2$$

Applying boundary conditions

At the support A, at $x = 0$, deflection $y = 0$

$$0 = 0 - 0 + 0 + C_2$$

$$\therefore \qquad C_2 = 0$$

$$\therefore \qquad EIY = \frac{wlx^3}{36} - \frac{wx^5}{120l} + C_1x$$

At support B, at $x = l$, deflection $y = 0$

$$0 = \frac{wl}{36} l^3 - \frac{wl^5}{120l} + C_1l$$

$$C_1l = \frac{-wl^4}{36} + \frac{wl^4}{120} = \frac{-10wl^4 + 3wl^4}{360} = \frac{-7wl^4}{360}$$

$$C_1 = \frac{-7wl^3}{360}$$

$$\therefore \qquad EI \frac{dy}{dx} = \frac{wlx^2}{12} - \frac{wx^4}{24l} - \frac{7wl^3}{360}$$

$$\text{and} \qquad EIY = \frac{wlx^3}{36} - \frac{wx^5}{120l} - \frac{7wl^3x}{360}$$

At the point of maximum deflection, $\frac{dy}{dx} = 0$

Substituting in the above equation

$$O = \frac{wlx^2}{12} - \frac{wx^4}{24l} - \frac{7wl^3}{360}$$

$$\frac{wx^4}{24l} - \frac{wlx^2}{12} + \frac{7wl^3}{360} = 0$$

Divide by $\frac{w}{24l}$

$$x^4 - 2l^2 x^2 + \frac{7}{15} l^4 = 0$$

$$\therefore \qquad x^2 = \frac{-b \pm \sqrt{b^2 - 4ac}}{2a}$$

$$= \frac{2l^2 \pm \sqrt{(2l)^2 - 4(1)\left(\frac{7}{15}l^4\right)}}{2 \times 1}$$

$$= \frac{2l^2 \pm \sqrt{(2l)^2 - (2l)^2 \frac{7}{15}}}{2} = \frac{2l^2 \pm 2l^2\sqrt{1 - \frac{7}{15}}}{2}$$

$$= \frac{2l^2\left[1 \pm \sqrt{1 - \frac{7}{15}}\right]}{2} = l(1 \pm 0.7303)$$

$$x^2 = 1.7303\, l \quad \text{or} \quad 0.2697\, l$$

Considering feasible value, $x^2 = 0.2697\, l$

$\therefore \qquad x = 0.5193\, l$

Substituting this in y equation.

$$EIY_{max} = \frac{wl}{36}(0.5193l)^3 - \frac{w}{120l}(0.5193l)^5 - \frac{7}{360}wl^3\,(0.5193\, l)$$

$$= wl^4\left[\frac{0.5193^3}{36} - \frac{0.5193^5}{120} - \frac{7}{360} \times 0.5193\right]$$

$$= wl^4\,(-0.00652)$$

$\therefore$ Maximum deflection $y = \dfrac{-0.00652\, wl^4}{EI}$ (Downward)

This maximum deflection occurs at a distance of $0.5193\, l$ from the end with load intensity of zero.

To find the deflection at the mid span at $x = \dfrac{l}{2}$

$$EIY = \frac{wl}{36}\left(\frac{l}{2}\right)^3 - \frac{w}{120l}\left(\frac{l}{2}\right)^5 - \frac{7}{360}l^3\left(\frac{l}{2}\right)$$

$$= \frac{wl^4}{36 \times 8} - \frac{wl^4}{120 \times 32} - \frac{7}{720} \times wl^4$$

$$= wl^4\left[\frac{1}{288} - \frac{1}{3840} - \frac{7}{720}\right]$$

$$y = \frac{-5}{768}\frac{wl^4}{EI} \text{ (Downward)}$$

$$y_{UVL} = \frac{1}{2}\left(\frac{-5\, wl^4}{384\, EI}\right) = \frac{1}{2}\,(y_{\text{for UDL}})$$

The deflection due to UVL at the mid span is $\frac{1}{2}$ the deflection of UDL of same intensity.

7.5.4 A Simply Supported Beam of Span *l* Carries a Uniformly Distributed Over a Length of Half of Its Span from One End Support

Figure 7.13 (a) shows a simply supported beam of length l carrying a UDL of intensity w/unit length over $\frac{1}{2}$ of its span. Let y be the deflection due to this

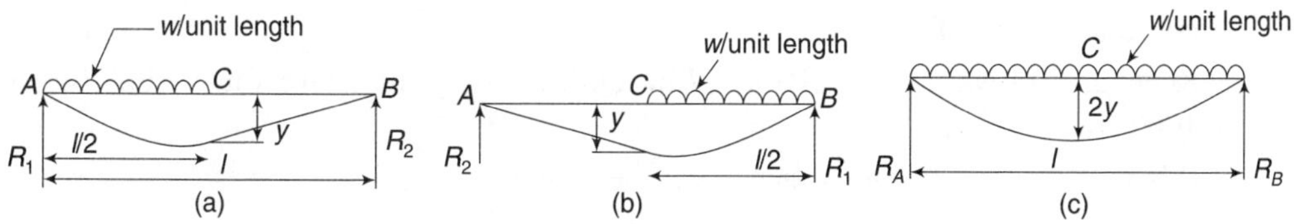

Fig. 7.13

Consider that if only the right half span is loaded with same load intensity, then also the same deflection y will occur at the middle as shown in Fig. 7.13 (b). If both left half and right half spans of the beam are loaded as shown in Fig. 7.13 (c) then the total deflection at the middle will be $= y + y = 2y$.

We know that for a simply supported beam with UDL over the entire span, the deflection in the middle,

$$= 2y = \frac{5wl^4}{384\ EI}$$

Hence, when only one half of the span is loaded, then the deflection at the centre,

$$y = \frac{1}{2}\left[\frac{5}{384}\frac{wl^4}{EI}\right] = \frac{5}{768}\frac{wl^4}{EI}.$$

SOLVED PROBLEMS (CANTILEVERS)

Problem 1 Find the deflection at the free end of the cantilever loaded as shown in Fig. 7.14. Take moment of inertia as 3×10^9 mm^4 and modulus of elasticity as 120 GPa.

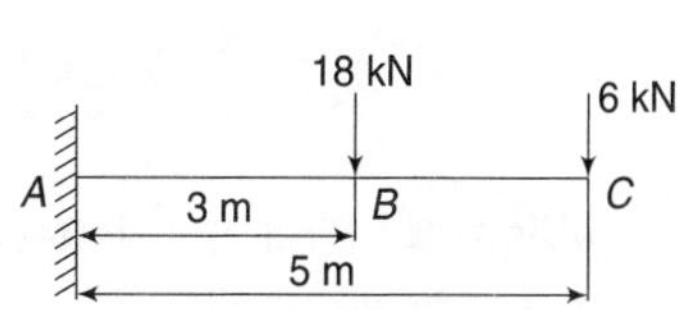

Fig. 7.14

Solution: **Given data:**

Moment of inertia $I = 3 \times 10^9$ mm^4

Modulus of elasticity $E = 210$ GPa $= 210 \times 10^3$ MPa.

Deflection at the free end C = Deflection at C due to load at end C + Deflection at C due to load at B

$$y_C = \frac{W_1 l^3}{3EI} + \left(\frac{W_2 a^3}{3EI} + \frac{W_2 a^2(l-a)}{2EI}\right)$$

where, $W_1 = 6$ kN, $W_2 = 18$ kN, $l = 5$ m and $a = 3$ m

$$y_C = \frac{6 \times 5^3}{3EI} + \frac{18 \times 3^3}{3EI} + \frac{18 \times 3^2(5-3)}{2EI} = \frac{574}{EI} \text{ m}$$

Note: To balance the units either

(i) The load in kN to be converted to Newtons and length in metres to be converted to mm

OR

(ii) E in N/mm^2 is to be converted to kN/m^2 and moment of inertia I in mm^2 is to be converted to m^4. That is,

Keeping E in N/mm^2 or MPa and I in mm^4, multiply the above value with $1000 \times 1000^3 = 10^{12}$

(kN to N and m^3 to mm^3) then y will be in mm

OR

Keeping W in kN and length in metres, multiply denominator EI with 10^{-9}, then y will be in metres

$$E = \frac{\text{N}}{\text{mm}^2} = \left(\frac{\text{kN}}{\text{mm}^2}\right) \times 10^{-3} \frac{\text{kN}}{\text{mm}^2} = \frac{\text{N}}{\text{mm}^2} \times \frac{10^{-3}}{10^{-6}} \frac{\text{kN}_2}{\text{m}}$$

$$I = \text{mm}^4 = (\text{mm}^4 \times 10^{12}) \text{ m}^4$$

$$EI = \left(\frac{\text{N}}{\text{mm}^2} \times \text{mm}^4\right) \frac{10^{-3}}{10^{-6}} \times 10^{-12} = 10^{-9}$$

We will multiply the above equation with 10^{12} to convert to N and mm.

$$y_C = \frac{574}{EI} \times 10^{12} \text{ mm} = \frac{574 \times 10^{12}}{210 \times 10^3 \times 3 \times 10^9} = 0.911 \text{ mm}$$

Alternatively, multiply EI with 10^{-9}

$$y_C = \frac{574}{EI} = \frac{574}{(210 \times 10^3 \times 3 \times 10^9) \times 10^{-9}} = 9.11 \times 10^{-4} \text{ m}$$

$$y_C = 0.911 \times 10^{-3} \text{ m} = 9.11 \text{ mm}.$$

Problem 2 Find the deflection under the load and at the free end of a cantilever beam as shown in Fig. 7.15. Take $E = 2 \times 10^5$ N/mm^2 and $I = 2 \times 10^8$ mm^4.

Solution: **Given data:**

Modulus of elasticity $E = 2 \times 10^5$ N/mm^2

Moment of inertia $I = 2 \times 10^8$ N/mm^2

Load $W = 12$ kN

$l = 6$ m

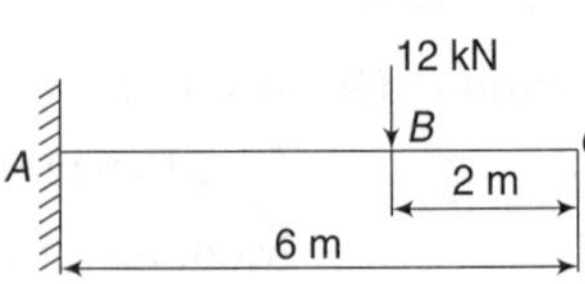

Fig. 7.15

$a = 6 - 2 = 4$ m from fixed end.

Deflection under load $y_B = \dfrac{Wa^3}{3EI}$

$$y_B = \frac{12 \times 4^3 \times 10^{12}}{3 \times 2 \times 10^5 \times 2 \times 10^8} \times 10^{12}$$

(**Note:** 10^{12} is to convert kN to N and m^3 to mm^3)

$$= 6.4 \text{ mm}$$

Deflection at the free end $y_C = \dfrac{wa^3}{3EI} + \dfrac{wa^2}{2EI}(l - a)$

$$y_C = \left[\frac{12 \times 4^3}{3 \times 2 \times 10^5 \times 2 \times 10^8} + \frac{12 \times 4^2}{2 \times 2 \times 10^5 \times 2 \times 10^8}(6 - 4)\right] \times 10^{12}$$

$$y_C = 6.4 + 4.8 = 11.2 \text{ mm.}$$

Problem 3 Find the maximum deflection at the free end of a cantilever beam carrying a UDL of 20 kN/m over the entire span of 3 m. The modulus of elasticity of the material is 200 kN/mm^2 and moment of inertia of the section is 5×10^8 mm^4.

Solution: Figure 7.16 shows the cantilever of a length 3 m with full UDL.

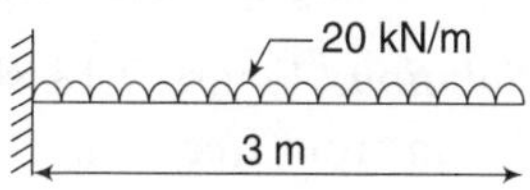

Fig. 7.16

$$\text{Intensity of UDL } w = 20 \text{ kNm}$$

$$\text{Length } l = 3 \text{ m}$$

$$E = 200 \text{ kN/mm}^2 = 200 \times 10^3 \text{ N/mm}^2$$

$$I = 5 \times 10^8 \text{ mm}^4$$

$$\therefore \quad EI = (200 \times 10^3) \times 5 \times 10^8 = 10 \times 10^{13} \text{ N-mm}^2$$

$$EI = (10 \times 10^{13}) \times 10^{-9} \text{ kN-m}^2$$

$$= 10^5 \text{ kN-m}^2$$

For cantilever with full UDL, the deflection at the free end $y = \dfrac{wl^4}{8EI} = \dfrac{20 \times 3^4}{8 \times 10^5} = 0.0162$ m

$$= 16.2 \text{ mm.}$$

Problem 4 Find the deflection at the free end of a cantilever beam located as shown in Fig. 7.17 Take $E = 2 \times 10^5$ N/mm^2 and $I = 2 \times 10^9$ mm^4.

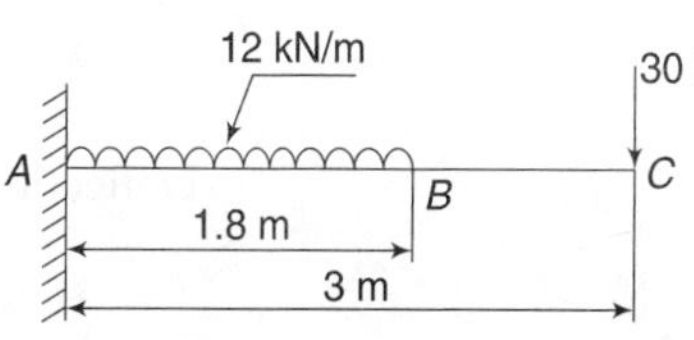

Fig. 7.17

Solution: Deflection at the free end $y_C = y_{C1} + y_{C2}$

where y_{C1} = Deflection at C due to concentrated load of 30 kN.

and y_{C2} = deflection at C due to UDL at AB.

$$y_{C1} = \frac{Wl^3}{3EI} = \frac{30 \times 3^3}{3 \times 2 \times 10^5 \times 2 \times 10^9 \times 10^{12}}$$

$$= 0.675 \text{ mm}$$

($\times 10^{12}$ is to convert kN to N and l in m to mm)

$$y_{C2} = \frac{wa^4}{8EI} + \frac{wa^3}{6EI}(l - a)$$

$$= \left[\frac{20 \times 1.8^4}{8 \times 2 \times 10^5 \times 2 \times 10^9} + \frac{20 \times 1.8^3}{6 \times 2 \times 10^5 \times 2 \times 10^9}(3 - 1.8)\right] \times 10^{12}$$

$$y_{C2} = 1.2393 \text{ mm}$$

$\therefore$ Total deflection at C $y_C = y_{C1} + y_{C2} = 0.675 + 1.2393$

$$= 1.9143 \text{ mm.}$$

Problem 5 A cantilever of rectangular cross section 100 mm wide and 200 mm deep is 3 m long and carries a UDL of 60 kN/m over a length of 1.8 m from the free end. Find the deflection and slope at the free end taking modulus of elasticity as 206 kN/mm^2.

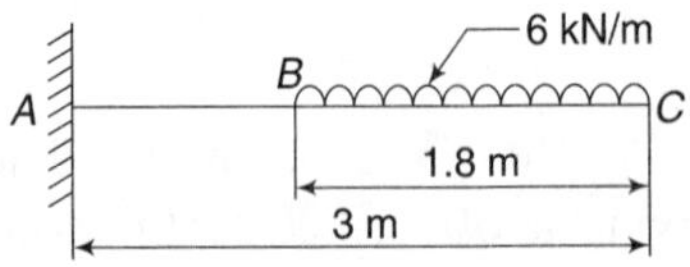

Fig. 7.18

Solution: Figure 7.18 shows the given cantilever with UDL of 6 kN/m over a span of 1.8 m from free end.

$$\text{Moment of inertia } I = \frac{bd^3}{12} = \frac{100 \times 200^3}{12} = 66.667 \times 10^6 \text{ mm}^4$$

This can be treated as a cantilever with UDL over the entire span acting downward and upward UDL over the span AB.

$$\text{The deflection } C\ y_C = y_{C1} - y_{C2}$$

$$= \frac{wl^4}{8EI} - \left[\frac{wa^4}{8EI} + \frac{wa^3}{6EI}(l - a)\right]$$

$$= \frac{6 \times 3^4}{8EI} - \left[\frac{6 \times 1.2^4}{8EI} + \frac{6 \times 1.2^3(3 - 1.2)}{6EI}\right]$$

$$= \frac{1}{EI}\left[\frac{6 \times 3^4}{8} - \frac{6 \times 1.2^4}{8} - \frac{6 \times 1.2^3 \times 1.8}{6}\right]$$

$$\text{Deflection } y_C = \frac{1}{206 \times 10^3 \times 66.667 \times 10^6}[56.0844] \times 10^{12}$$

$$= 4.084 \text{ mm}$$

Slope at the free end is given by

$$\theta_B = \frac{wl^3}{6EI} - \frac{w(l - a)^3}{6EI}$$

$$\theta_B = \left[\frac{6 \times 3^3}{6 \times 206 \times 10^3 \times 66.667 \times 10^6} - \frac{6 \times (3 - 1.2)^3}{6 \times 200 \times 10^3 \times 66.667 \times 10^6}\right] \times 10^9$$

(Since the load is in kN and lengths are in m, multiply the above slope expression with $1000 \times 1000^2 = 10^9$ to convert to N and mm)

$$\theta_B = 1.54135 \times 10^{-3} \text{ radius}$$

$$\theta_B = 1.54135 \times 10^{-3} \times \frac{180}{\pi} = 0.0883 \text{ degree.}$$

Problem 6 A cantilever beam of circular cross section of diameter 120 mm is 3 m long. It carries a UDL of 3 kN/m run over a span of 1 m from the free end and a point load of 1.8 kN at its free end. Find the slope and deflection at the free end. Modulus of elasticity for the material may be taken as 200 GPa.

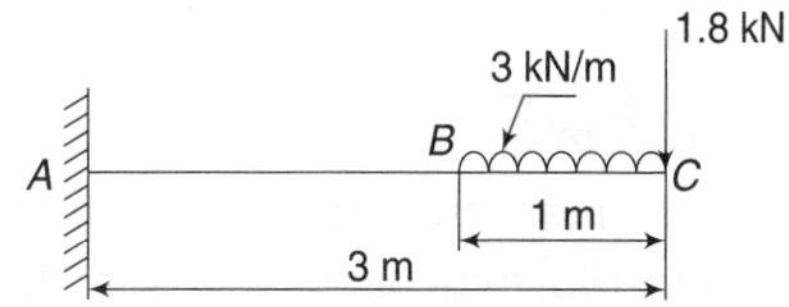

Fig. 7.19

Solution: Diameter $d = 120$ mm

$\therefore$ Moment of inertia of the section $I = \dfrac{\pi d^4}{64}$

$$I = \frac{\pi \times 120^4}{64} = 10.179 \times 10^6 \text{ mm}^4$$

$$\text{Modulus of elasticity} = 200 \text{ GPa} = 200 \times 10^3 \text{ N/mm}^2 = 2 \times 10^5 \text{ N/mm}^2$$

$$l = AC = 3 \text{ m}$$

$$a = AB = 3 - 1 = 2 \text{ m}$$

$$\text{Slope at the free end } \theta_C = \theta_{C1} + [\theta_{C_2} - \theta_{C_3}]$$

θ_{C_1} = Slope at C due to point load 1.8 kN

θ_{C_2} = Slope at C due to UDL of 3 kN/m over full span AC.

θ_{C_3} = Slope at C due to upward UDL of 3 kN/m over the span AB.

$$\theta_B = \frac{wl^2}{2EI} + \left[\frac{wl^3}{6EI} - \frac{wa^3}{6EI}\right]$$

$$= \frac{1}{EI}\left[\frac{1.8 \times 3^2}{2} + \frac{3 \times 3^3}{6} - \frac{3 \times 2^3}{6}\right] = \frac{17.6}{EI}$$

$$\theta_B = \frac{17.6}{2 \times 10^5 \times 10.179 \times 10^6} \times 10^9 = 0.008645 \text{ radians}$$

$$\text{Slope } \theta_B = 0.008645 \times \frac{180}{\pi} = 0.495336 \text{ degree}$$

Similarly, deflection at free end $y_C = y_{C1} + y_{C2} - y_{C3}$

y_{C1} = Deflection at C due to point load

y_{C2} = Deflection at C due to UDL over full span

y_{C3} = Upward deflection at C due to upward UDL over the span AB

$$y_C = \frac{wl^3}{3EI} + \frac{wl^4}{8EI} - \left[\frac{wa^4}{8EI} + \frac{wa^3}{6EI}(l - a)\right]$$

$$= \frac{1.8 \times 3^3}{3EI} + \frac{3 \times 3^4}{8EI} - \frac{3 \times 2^4}{8EI} - \frac{3 \times 2^3}{6EI}(3 - 2)$$

$$y_C = \frac{36.575}{EI} = \frac{36.575 \times 10^{12}}{2 \times 10^5 \times 10.179 \times 10^6} \times 10^{12} = 17.966 \text{ mm.}$$

Problem 7 A cantilever beam of rectangular cross section 120 mm wide, 180 mm deep and 3 m long carries a uniformly varying load varying from 30 kN/m at the free end to 75 kN/m at the fixed end. Find the slope and deflection of cantilever at its free end, modulus of elasticity may be taken as 2×10^5 MPa.

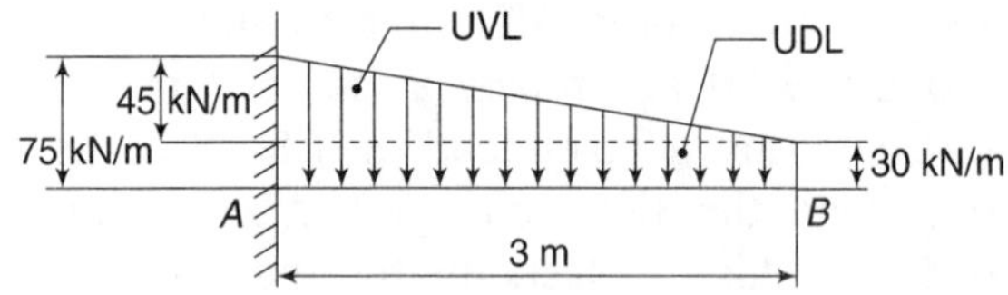

Fig. 7.20

Solution: The given cantilever with loading is shown in Fig. 7.20.

The given trapezoidal load can be converted into a UDL of 30 kN/m and a UVL that varies from zero at B to 75 – 30 = 45 kN/m at A as shown in Fig. 7.20.

$\therefore$ Span $l = 3$ m

$$\text{Moment of inertia } I = \frac{bd^3}{12} = \frac{120 \times 180^3}{12} = 58.32 \times 10^6 \text{ mm}^4$$

$$E = 2 \times 10^5 \text{ MPa}$$

$$\text{Slope at } B \ \theta_B = \theta_{B_1} + \theta_{B_2}$$

where, θ_{B_1} = Slope at B due to UDL

θ_{B_2} = Slope at B due to UVL

$$\theta_B = \frac{w_1 l^3}{6EI} + \frac{w_2 l^3}{24EI}$$

$$\therefore \quad \theta_B = \left[\frac{30 \times 3^3}{6 \times 2 \times 10^5 \times 58.32 \times 10^6} + \frac{45 \times 3^3}{24 \times 2 \times 10^5 \times 58.32 \times 10^6}\right] \times 10^9$$

$$\text{Slope } \theta_B = 0.015914 \text{ radians} = 0.015914 \times \frac{180}{\pi} = 0.9118 \text{ degree}$$

Deflection at the free end $y_B = y_{B1} + y_{B2}$

where, y_{B1} = Deflection at B due to UDL

and y_{B2} = Deflection at B due to UVL

$$y_B = \frac{w_1 l^4}{8EI} + \frac{w_2 l^4}{30EI}$$

$$= \left(\frac{30 \times 3^4}{8 \times 2 \times 10^5 \times 58.32 \times 10^6} + \frac{45 \times 3^4}{30 \times 2 \times 10^5 \times 58.32 \times 10^6} \right) \times 10^{12}$$

$$y_B = 36.458 \text{ mm.}$$

Problem 8 For the cantilever beam shown in Fig. 7.21, determine the deflection at the free end. The modulus of elasticity for the material is 2×10^5 N/mm^2 and moment of inertia is 3×10^8 mm^4

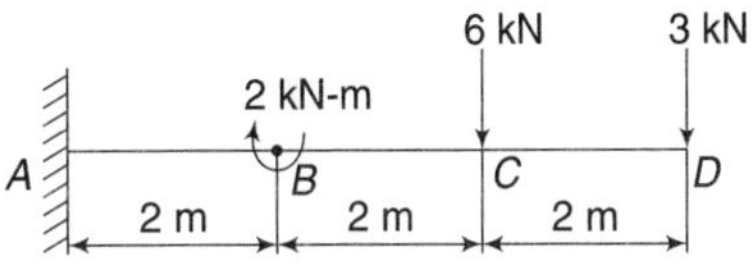

Fig. 7.21

Solution: $E = 2 \times 10^5$ N/mm^2

$I = 3 \times 10^8$ mm^4

$W_1 = 3$ kN, $l = 6$ m

$W_2 = 6$ kN, $a_2 = 4$ m

$M = 2$ kN M, $a_1 = 2$ m

Total deflection at D $y_D = y_{D1} + y_{D2} + y_{D3}$

y_{D1} = Deflection due to point load W_1

y_{D2} = Deflection due to point load W_2

and y_{D3} = Deflection due to moment M

$$\therefore \quad y_D = \frac{W_1 l^3}{3EI} + \left[\frac{W_2 a_2^3}{3EI} + \frac{W_2 a_2^2}{2EI}(l - a_2) \right] + \left[\frac{M a_1^2}{2EI} + \frac{M a_1}{EI}(l - a_1) \right]$$

$$= \frac{3 \times 6^3}{3EI} + \left[\frac{6 \times 4^3}{3EI} + \frac{6 \times 4^2(6 - 4)}{2EI} \right] + \left[\frac{2 \times 2^2}{2EI} + \frac{2 \times 2(6 - 2)}{EI} \right]$$

$$y_D = \frac{460}{EI} = \frac{460 \times 10^{12}}{2 \times 10^5 \times 3 \times 10^8} = 7.667 \text{ mm.}$$

Problem 9 Find the slope and deflection at the free end of the cantilever shown in Fig. 7.22. Take $EI = 2 \times 10^{13}$ N-mm^2.

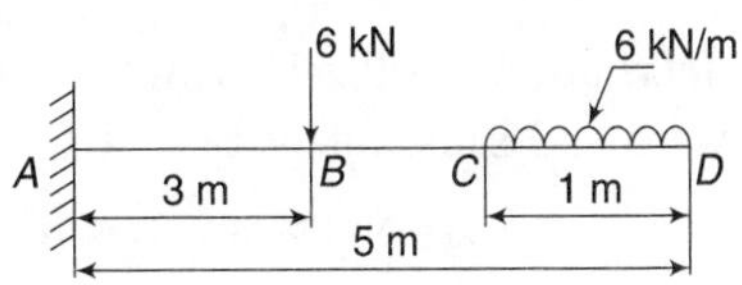

Fig. 7.22

Solution: $EI = 2 \times 10^{13}$ N-mm^2

Point load $W_1 = 6$ kN

UDL $w = 6$ kN/m

Slope at D, $\theta_D = \theta_{D_1} + \theta_{D_2} - \theta_{D_3}$

where θ_{D_1} = Slope due to point load

θ_{D_2} = Slope due to UDL over full span

θ_{D_3} = Slope due to upward UDL over AC

$$\theta_D = \frac{Wa_1^2}{2EI} + \frac{wl^3}{6EI} - \frac{wa_2^3}{6EI}$$

$$= \frac{6 \times 3^2}{2EI} + \frac{6 \times 5^3}{6EI} - \frac{6 \times 4^3}{6EI} = \frac{88}{EI}$$

$$\theta_B = \left(\frac{88}{2 \times 10^{13}}\right) \times 10^9 = 0.0044 \text{ radian}$$

$$\theta_B = 0.0044 \times \frac{180}{\pi} = 0.252 \text{ degree}$$

Deflection at free end D, $y_D = y_{D_1} + y_{D_2} - y_{D_3}$

where y_{D_1} = Deflection at D due to point load

y_{D_2} = Deflection at D due to UDL over full span AD

y_{D_3} = Deflection at D due to upward UDL over AC

$$y_D = \left\{\frac{Wa_1^3}{3EI} + \frac{Wa_1^2(l - a_1)}{2EI}\right\} + \frac{wl^4}{8EI} - \left\{\frac{wa_2^4}{8EI} + \frac{wa_2^3}{6EI}(l - a_2)\right\}$$

$$= \frac{1}{EI}\left[\frac{6 \times 3^3}{2} + \frac{6 \times 3^2}{2}(5 - 3) + \frac{6 \times 5^4}{8} - \left\{\frac{6 \times 4^4}{8} + \frac{6 \times 4^3}{6}(5 - 4)\right\}\right]$$

$$y_D = \frac{320.75}{EI} = \frac{320.75}{2 \times 10^{13} \times 10^{13}} = 16.0375 \text{ mm.}$$

Problem 10 Determine the slope and deflection at the free end of the cantilever shown in Fig. 7.23. The modulus of elasticity is 200 kN/mm^2. The beam is of rectangular cross section of width 120 mm and depth 270 mm.

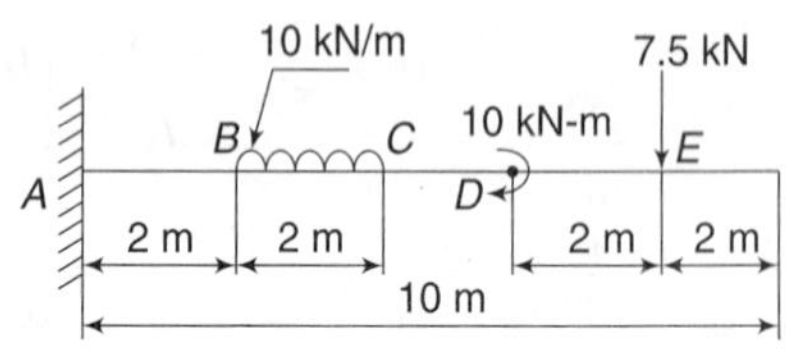

Fig. 7.23

Solution: Consider UDL from A to C acting downward and same intensity of UDL acting upward from A to B

Slope at free end F, $\theta_F = \theta_{F_1} + \theta_{F_2} + \theta_{F_3} + \theta_{F_4}$

where, θ_{F_1} = Slope at F due to point load at E

θ_{F_2} = Slope at F due to moment at D

θ_{F_3} = Slope at F due to UDL from A to C

θ_{F_4} = Slope at F due to UDL from A to B (Upward)

$$\text{Moment of inertia } I = \frac{bd^3}{12} = \frac{120 \times 270^3}{12} = 196.83 \times 10^6 \text{ mm}^4$$

Modulus of elasticity $E = 200 \text{ GPa} = 200 \times 10^3 \text{ MPa} = 2 \times 10^5 \text{ MPa}.$

$$\theta_F = \frac{Wa_1^2}{2EI} + \frac{Ma_2}{EI} + \frac{wa_3{}^3}{6EI} - \frac{wa_4{}^3}{6EI}$$

$$W = 7.5 \text{ kN and } a_1 = 8 \text{ m}$$

$$M = 10 \text{ kN-m and } a_2 = 6 \text{ m}$$

$$w = 10 \text{ kN/m, } d^3 = 4 \text{ m and } a_4 = 2 \text{ m}$$

$$\therefore \quad \theta_F = \frac{7.5 \times 8^2}{2EI} + \frac{10 \times 6}{EI} + \frac{10 \times 4^3}{6EI} - \frac{10 \times 2^3}{6EI} = \frac{393.333}{EI}$$

$$= \frac{393.333 \times 10^9}{2 \times 10^5 \times 196.83 \times 10^6} = 0.009992 \text{ radians}$$

$$\theta_F = 0.009992 \times \frac{180}{\pi} = 0.5725 \text{ degrees}$$

Deflection at the free end F $y_F = y_{F_1} + y_{F_2} + y_{F_3} + y_{F_4}$

where, y_{F_1} = Deflection at F due to point load at E

$$= \frac{Wa_1{}^3}{3EI} + \frac{Wa_1{}^2}{2EI}(l - a_1)$$

$$y_{F_1} = \frac{7.5 \times 8^3}{3EI} + \frac{7.5 \times 8^2}{2EI}(10 - 8) = \frac{1760}{EI}$$

y_{F_2} = Deflection at F due to moment at D

$$= \frac{Ma_2^2}{2EI} + \frac{Ma_2}{EI}(l - a_2)$$

$$y_{F_2} = \frac{10 \times 6^2}{2EI} + \frac{10 \times 6(10 - 6)}{EI} = \frac{420}{EI}$$

y_{F_3} = Deflection at F due to UDL at AC

$$= \frac{wa_3{}^4}{8EI} + \frac{wa_3{}^3}{6EI}(l - a_3)$$

$$y_{F_3} = \frac{10 \times 4^4}{8EI} + \frac{10 \times 4^3}{6EI}(10 - 4) = \frac{960}{EI}$$

y_{F_4} = Deflection at F due to upward UDL at AB

$$= \frac{wa_4^4}{8EI} + \frac{wa_4^3}{6EI}(l - a_4)$$

$$= \frac{10 \times 2^4}{8EI} + \frac{10 \times 2^3}{6EI}(10 - 2) = 126.667/EI$$

$$\therefore \quad \text{Total deflection at } F,\ y_F = \frac{1760}{EI} + \frac{420}{EI} + \frac{960}{EI} - \frac{126.667}{EI}$$

$$= \frac{3013.333}{EI}$$

$$= \frac{3013.333 \times 10^{12}}{2 \times 10^5 \times 196.83 \times 10^6}$$

$$y_F = 76.547 \text{ mm.}$$

Problem 11 A cantilever of circular section tapers uniformly from a diameter d at the free end to a diameter $2d$ at the fixed end. It is subjected to a concentrated load W at its free end. Find the deflection at the free end. Also calculate the diameter of the cantilever of same length and of uniform cross section that would have the same end deflection.

Solution: Figure 7.24 shows the given cantilever. Extend the top and bottom lines to meet at C.

From similar triangles APC and BQC,

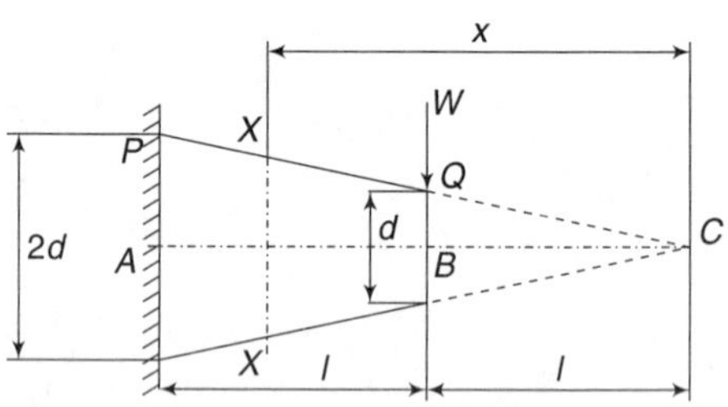

Fig. 7.24

$$\frac{AP}{BQ} = \frac{AC}{BC}$$

$$\frac{d}{d/2} = \frac{AC}{BC}$$

$$\therefore \quad \frac{AC}{BC} = 2$$

$$\therefore \quad AC = 2\ BC = 2\ AB = 2l$$

$$\therefore \quad BC = l$$

Consider any section XX at a distance x from C.

From similar triangles,
$$\frac{d_x}{x} = \frac{2d}{2l}$$

$$\therefore \quad \text{Diameter at } XX,\ dx = \frac{d}{l}.x$$

$$\text{Moment of inertia } I = \frac{\pi d_x^4}{64} = \frac{\pi d^4\ x^4}{64\ l^4}$$

$$\text{Bending moment at } XX,\ M_x = W(x - l)$$

Substituting in the deflection equation.

$$EI\frac{d^2y}{dx^2} = W(x - l)$$

$$\therefore \qquad \frac{d^2y}{dx^2} = \frac{W}{EI}(x-l) = \frac{W}{E\left(\frac{\pi d^4 x^4}{64 l^4}\right)}.(x-l)$$

$$\frac{d^2y}{dx^2} = \frac{64Wl^4}{\pi E d^4}\left(\frac{x-l}{x^4}\right)$$

For the given problem W, l, E and d are constants

Put $$\frac{64Wl^4}{\pi E d^4} = K \text{ (Some constant)}$$

$$\therefore \qquad \frac{d^2y}{dx^2} = K\left(\frac{x-l}{x^4}\right) = K\left(\frac{1}{x^3} - \frac{l}{x^4}\right)$$

$$= K(x^{-3} - l\,x^{-4})$$

Integrating, $$\frac{dy}{dx} = K\left(\frac{x^{-2}}{-2} - \frac{lx^{-3}}{-3}\right) + C_1$$

$$= K\left(\frac{-x^{-2}}{2} + \frac{lx^{-3}}{3}\right) + C_1$$

Applying boundary conditions, at fixed end,

$$x = 2l, \text{ slope } \frac{dy}{dx} = 0$$

$$0 = K\left(\frac{-(2l)^{-2}}{2} + \frac{l(2l)^{-3}}{3}\right) + C_1$$

$$0 = K\left(\frac{-1}{8l^2} + \frac{1}{24l^2}\right) + C_1$$

$$C_1 = K\left(\frac{1}{8l^2} - \frac{1}{24l^2}\right) = K\left(\frac{3-1}{24l^2}\right) = \frac{K}{12l^2}$$

$$C_1 = \frac{Kl^{-2}}{12}$$

Substituting in slope equation

$$\frac{dy}{dx} = K\left(\frac{-x^{-2}}{2} + \frac{lx^{-3}}{3}\right) + \frac{K}{12}l^{-2}$$

$$= K\left(\frac{-x^{-2}}{2} + \frac{lx^{-3}}{3} + \frac{l^{-2}}{12}\right)$$

Integrating

$$y = K\left(\frac{-x^{-1}}{2(-1)} + \frac{lx^{-2}}{3(-2)} + \frac{l^{-2}x}{12}\right) + C_2$$

$$y = K\left(\frac{x^{-1}}{2} - \frac{lx^{-2}}{6} + \frac{l^{-2}x}{12}\right) + C_2$$

Applying boundary conditions, at the fixed end, $x = 2l$, deflection $y = 0$

$$0 = K\left(\frac{(2l)^{-1}}{2} - \frac{l\,(2l)^{-2}}{6} + \frac{l^{-2}(2l)}{12}\right) + C_2$$

$$= K\left(\frac{1}{4l} - \frac{1}{24l} + \frac{1}{6l}\right) + C_2$$

$$C_2 = -K\left(\frac{6 - 1 + 4}{24l}\right) = \frac{-9K}{24l}$$

$$C_2 = \frac{-3K}{8l} = \frac{-3Kl^{-1}}{8}$$

Substituting in the deflection equation

$$y = K\left(\frac{x^{-1}}{2} - \frac{lx^{-2}}{6} + \frac{l^{-2}\cdot x}{12}\right) - \frac{3Kl^{-1}}{8}$$

$$y = K\left(\frac{x^{-1}}{2} - \frac{l\,x^{-2}}{6} + \frac{l^{-2}x}{12} - \frac{3l^{-1}}{8}\right)$$

At $x = l$, at free end, deflection

$$y = K\left(\frac{l^{-1}}{2}\,\frac{l\cdot l^{-2}}{6} + \frac{l^{-2}.l}{12} - \frac{3l^{-1}}{8}\right)$$

$$= K\left(\frac{1}{2l} - \frac{1}{6l} + \frac{1}{12l} - \frac{3}{8l}\right)$$

$$y = K\left(\frac{12 - 4 + 2 - 9}{24\,l}\right) = \frac{K}{24l}$$

Substituting for K value, deflection of taper bar

$$y = \frac{64\,Wl^4}{\pi Ed^4}\cdot\frac{1}{24\,l} = \frac{8}{3}\,\frac{Wl^3}{\pi Ed^4} \qquad \text{...(1)}$$

For beam of uniform diameter D,

$$\text{Moment of inertia } I = \frac{\pi D^4}{64}$$

Deflection for cantilever with point load at the free end,

$$y = \frac{Wl^3}{3EI} = \frac{Wl^3}{3E\left(\frac{\pi D^4}{64}\right)} = \frac{64Wl^3}{3E\pi D^4} \qquad \text{...(2)}$$

For the deflection to be same, equating Eqs. (i) and (ii)

$$\frac{64\,Wl^3}{3\pi ED^4} = \frac{8}{3}\,\frac{Wl^3}{\pi Ed^4}$$

$$D^4 = \frac{64\ Wl^3}{3\pi E} \times \frac{3\pi E d^4}{8wl^3} = 8d^4$$

$$\therefore \qquad D = d(8)^{1/4} = 1.6818\ d$$

Equivalent dia $D = 1.6818\ d$.

7.6 MACAULAY'S METHOD OF DOUBLE INTEGRATION FOR DEFLECTION

This is mainly a double integration method for determining the deflection of beams. This is a convenient method of determining the deflections of beams at any point on the beam. This method involves in systematically expressing the bending moments due to various loads, equating to deflection equation, integrating them in a specific manner for arriving at slope and deflection at various points on the beam.

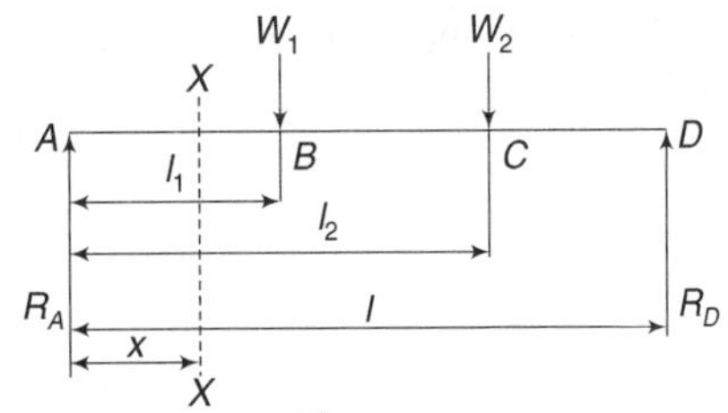

Fig. 7.25

For simplicity and to understand this method, let us consider a simply supported beam subjected to point loads as shown in Fig. 7.25.

W_1 and W_2 are two point loads at distances l_1 and l_2 from support A. Let R_A and R_D be the reactions at A and D.

For a section XX at a distance x from A lying between A and B, bending moment

$$M_x = R_A \cdot x$$

For a section XX at a distance x from A lying between B and C, bending moment

$$M_x = R_A \cdot x - W_1\ (x - l_1)$$

Similarly for a section XX at a distance x from A lying between C and D, bending moment

$$M_x = R_A \cdot x - W_1\ (x - l_1) - W\ (x - l_2)$$

In general, the bending moment at any section is given by

$$M_x = R_A \cdot x \;\vdots\; -W(x - l_1) \;\vdots\; -W(x - l_2) \;\vdots$$

The way this expression is written is something special.

For the section between A and B, $x \le l_1$.

By substituting $x \le l_1$ the terms $(x - l_1)$ and $(x - l_2)$ will be either zero or negative and are not applicable. Hence, such –ve or zero terms will be ignored.

It is also clear that for section between A and B, $M_x = R_A \cdot x$ and hence only the first term will be existing.

Similarly for a section between B and C, $x \le l_2$ and only first two terms will be existing. The third term will be –ve or zero and will be ignored.

For a section between C and D, $x > l_2$ and hence all the three terms will be existing.

It may be noted that as the magnitude of x goes on increasing, additional terms are appearing. It may also be noted that if any term $(x - l_1)$, $(x - l_2)$ etc., is zero or –ve, such terms are not applicable for that point and are to be neglected.

As such, for values of x between 0 and l_1 the only first term of the expression should be considered.

For values of x between l_1 and l_2, first two terms should be considered.

For values of x between l_2 and l, all the three terms should be considered.

Equating the moment to deflection equation,

$$EI \frac{d^2y}{dx^2} = R_A \cdot x + W_1 (x - l_1) + W_2 (x - l_2)$$

$$\text{Integrating, } EI \frac{dy}{dx} = R_A \frac{x^2}{2} + C_1 + \frac{W_1 (x - l_1)^2}{2} + \frac{W_2 (x - l_2)^2}{2}$$

There are very important points to be noted during integrating using Macaulay's method:

(i) The constant of integrating C_1 is written first soon after the first term, since it is applicable to all points.

(ii) Expression $(x - l_1)$ or $(x - l_2)$ may be integrated as $\frac{(x - l_1)^2}{2}$ and $\frac{(x - l_2)^2}{2}$, etc., and not as $\frac{x^2}{2} - l_1x$ and $\frac{x^2}{2} - l_2x$, etc. $(x - l_1)$ or $(x - l_2)$ are to be treated as single term.

(iii) If any of the terms $(x - l_1)$ or $(x - l_2)$, etc., are negative, and such as terms are neglected in the expression.

Integrating again for deflection

$$\therefore \qquad EIY = R_A \frac{x^3}{6} + C_1 x + C_2 - \frac{W_1(x - l_1)^3}{6} - \frac{W_2 (x - l_2)^3}{6}$$

It may be noted that C_2 is written after the first term as C_1 since it is common for all points.

Hence, again $(x - l_1)^2$ is integrated as $\frac{(x - l_1)^3}{3}$ and $(x - l_2)^2$ is integrated as $\frac{(x - l_2)^3}{3}$ as discussed above.

The constants C_1 and C_2 are valid for all values of x. C_1 and C_2 can be evaluated from the boundary conditions. For this example of simply supported beam, at the supports the deflection is zero.

At $x = 0$, $y = 0$ and hence $C_2 = 0$.

At $x = l$, $y = 0$, the constant C_1 can be arrived at.

Once C_1 and C_2 are found, the slope $\frac{dy}{dx}$ and deflection y at any point on the beam can be calculated.

This method of double integration for determining the deflection is called Macaulay's method.

Problem 12 A simply supported beam of length l carries a point load W at a distance a from the left support. Find the deflection under load and position and value of maximum deflection of the beam. Also find the deflection at the centre of the beam.

Solution: Figure 7.26 shows a simply supported beam of length l carrying a concentrated point load W at a distance a from left side support A.

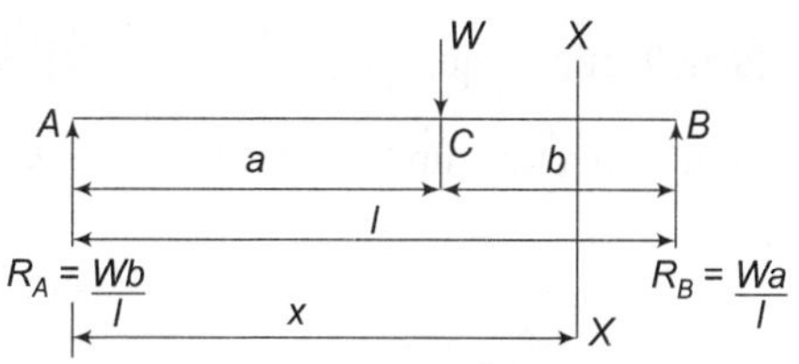

Fig. 7.26

$$E = \text{Modulus of elasticity}$$

$$I = \text{Moment of inertia of the section}$$

$$AB = l,\ AC = a,\ BC = l - a = b$$

$$\text{Reaction at } A,\ R_A = \frac{W \cdot b}{l}$$

and

$$\text{Reaction at } B,\ R_B = \frac{W \cdot a}{l}$$

Considering Macaulay's method,

$$\text{Moment } M_x = R_A \cdot x - W(x - a) = \frac{Wb}{l}x - W(x - a)$$

$$\therefore \quad EI\frac{d^2y}{dx^2} = \frac{Wb}{l}x - W(x - a)$$

$$\text{Integrating, } EI\frac{dy}{dx} = \frac{Wb}{l}\frac{x^2}{2} + C_1 - \frac{W(x - a)^2}{2}$$

$$\text{Integrating again, } EIY = \frac{Wb}{l} \cdot \frac{x^3}{6} + C_1x + C_2 - \frac{W(x - a)^3}{6}$$

Applying boundary conditions, at supports A and B, the deflection = 0.

$$\text{At } A,\ x = 0,\ y = 0$$

$$0 = 0 + 0 + C_2$$

$$\therefore \quad C_2 = 0$$

$$\text{At } B,\ x = l,\ y = 0$$

$$0 = \frac{Wb}{l}\frac{l^3}{6} + C_1l - \frac{W(l - a)^3}{6}$$

$$0 = \frac{Wbl^2}{6} + C_1 l - \frac{Wb^3}{6}$$

$$\therefore \quad C_1 l = \frac{Wb^3}{6} - \frac{Wbl^2}{6} = \frac{-Wb}{6}(l^2 - b^2)$$

$$\therefore \quad C_1 = \frac{-Wb}{6l}(l^2 - b^2)$$

$\therefore$ Slope equation is $EI\dfrac{dy}{dx} = \dfrac{Wb}{l}\cdot\dfrac{x^2}{2} - \dfrac{Wb}{6l}(l^2 - b^2) - \dfrac{W(x-a)^2}{2}$ ...(1)

and deflection equation, $EIY = \dfrac{Wb}{l}\dfrac{x^3}{6} - \dfrac{Wb}{6l}(l^2 - b^2)\,x - \dfrac{W(x-a)^3}{6}$ ...(2)

For deflection under load at C_1, $x = a$

$$EIy_C = \frac{Wb}{l}\frac{a^3}{6} - \frac{Wb}{6l}(l^2 - b^2)a - 0$$

$$EIy_C = \frac{Wba}{6l}(a^2 - l^2 + b^2)$$

$$= \frac{Wab}{6l}(a^2 - (a + b)^2 + b^2) \quad (\because l = a + b)$$

$$= \frac{Wab}{6l}(a^2 - a^2 - b^2 - 2ab + b^2) = \frac{Wab}{6l}(-2ab)$$

$$EIy_C = \frac{-Wa^2b^2}{3l}$$

$$\therefore \quad y_C = \frac{-Wa^2b^2}{3EIl} \text{ (downward)}$$

To find maximum deflection:

For maximum deflection, slope $\dfrac{dy}{dx} = 0$

It can be seen that the deflection will be maximum between A and C. From Eqn. (1)

$$0 = \frac{Wb}{l}\frac{x^2}{2} - \frac{Wb}{6l}(l^2 - b^2)$$

$$\therefore \quad x^2 = \frac{l^2 - b^2}{3}$$

$$\therefore \quad x = \frac{\sqrt{l^2 - b^2}}{3}$$

Substituting in deflection equation

$$EIY_{max} = \frac{Wb}{6l}\left(\frac{\sqrt{l^2 - b^2}}{3}\right)^3 - \frac{Wb}{6l}(l^2 - b^2)\left(\frac{\sqrt{l^2 - b^2}}{3}\right)$$

$$= \frac{Wb}{6l}\left(\frac{l^2 - b^2}{3}\right)^{3/2} - \frac{Wb}{6l}\frac{(l^2 - b^2)^{3/2}}{\sqrt{3}}$$

$$= \frac{Wb}{6l}(l^2 - b^2)^{3/2}\left[\frac{1}{3\sqrt{3}} - \frac{1}{\sqrt{3}}\right]$$

$$= \frac{Wb}{6l}(l^2 - b^2)^{3/2}\left[\frac{1-3}{3\sqrt{3}}\right]$$

$$= \frac{-Wb\,(l^2 - b^2)^{3/2}}{9\sqrt{3}\,l}$$

$$\therefore \quad y_{max} = \frac{-Wb(l^2 - b^2)^{3/2}}{9\sqrt{3}\,.EIl}$$

To find the deflection at the centre of beam.

$$x = \frac{l}{2} < a$$

Substituting in the deflection equation, [Eqn. (2)]

$$EIY = \frac{Wb}{6l}\left(\frac{l}{2}\right)^3 - \frac{Wb}{6l}(l^2 - b^2)\frac{l}{2}$$

$$= \frac{Wbl^2}{48} - \frac{Wb}{12}\cdot(l^2 - b^2)$$

$$= \frac{Wbl^2}{48} - \frac{Wbl^2}{12} + \frac{Wb^3}{12}$$

$$= \frac{Wb}{48}(l^2 - 4l^2 + 4b^2)$$

$$EIY = \frac{-Wb}{48}(3l^2 - 4b^2)$$

$$\therefore \quad y_{centre} = \frac{-Wb}{48EI}(3l^2 - 4b^2) \text{ (Downward)}$$

Problem 13 A simply supported beam of length 8 m carries a point load of 21 kN at a distance of 5 m from the left side support. Calculate the maximum deflection, deflection at the load point and deflection at the mid-span. Take the flexural rigidity as 3×10^{12} N-mm^2.

Solution:

$$W = 21 \text{ kN } a = 5 \text{ m}, l = 8 \text{ m}, b = l - a\ 3 \text{ m}$$

$$\text{Flexural rigidity } EI = 3 \times 10^{12} \text{ N-mm}^2$$

$$R_A + R_B = 21$$

$$\text{Taking moments about } A,\ 21 \times 5 = R_B \times 8$$

$$\therefore \quad R_B = \frac{21 \times 5}{8} = 13.125 \text{ kN}$$

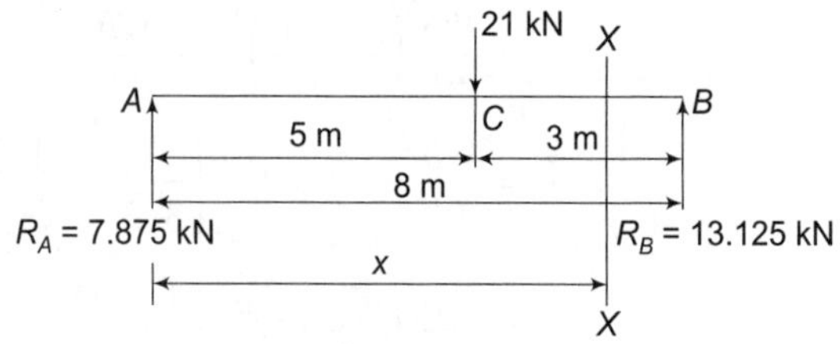

Fig. 7.27

$$R_A = 21 - R_B = 7.875 \text{ kN}$$

Bending moment at any section XX at a distance x from A

$$M_x = R_A \cdot x - W(x - a)$$

$$= 7.875\, x - 21(x - 5)$$

$$\therefore \quad EI \frac{d^2y}{dx^2} = 7.875\, x - 21(x - 5) \text{ kN-m}$$

$$\text{Integrating, } EI \frac{dy}{dx} = 7.875 \frac{x^2}{2} + C_1 - \frac{21(x-5)^2}{2} \text{ kN-m}^2$$

Integrating again

$$EIy = 7.875 \frac{x^3}{6} + C_1\, x + C_2 - \frac{21\,(x-5)^3}{6} \text{ kN-m}^3$$

Applying boundary conditions,

At support A, $x = 0$, deflection $y = 0$

$$0 = 0 + 0 + C_2$$

$$\therefore \quad C_2 = 0$$

At support B, $x = l = 8$, deflection $y = 0$

$$0 = 7.875 \frac{8^3}{6} + C_1 8 - \frac{21(8-5)^3}{6}$$

$$\therefore \quad C_1 = 72.1875$$

$$\therefore \quad EI \frac{dy}{dx} = 7.875 \frac{x^2}{2} - 72.1875 - \frac{21(x-5)^2}{2}$$

and

$$EIy = 7.875 \frac{x^3}{6} - 72.1875\, x - \frac{21(x-5)^3}{6}$$

$$\text{For maximum deflection, } \frac{dy}{dx} = 0$$

It is seen that maximum deflection occurs in the portion AC. Hence, $x < a$

$$0 = 7.875 \frac{x^2}{2} - 72.1875$$

$$\therefore \quad x^2 = 18.333 \text{ or } x = 4.2817 \text{ m from } A.$$

Since this is less than $a = 5$, the maximum deflection occurs in the portion AC.

Substituting in deflection equation,

$$EIy_{max} = 7.875.\frac{4.2817^3}{6} - 72.1875 \times 4.2817$$

$$EIy_{max} = -206.0557 \text{ kN-m}^3$$

$$\therefore \quad y_{max} = \frac{-206.0557}{EI} = \frac{-206.0557}{3 \times 10^{12}} \times 10^{12}$$

$$y_{max} = -68.685 \text{ mm (Downward) At } x = 4.2817 \text{ m from } A.$$

Note: From the formulae,

$$y_{max} = \frac{-Wb\,(l^2 - b^2)^{3/2}}{9\sqrt{3}\;EIl}$$

$$\text{At } x = \sqrt{\frac{l^2 - b^2}{3}}$$

$$x = \sqrt{\frac{8^2 - 3^2}{3}} = 4.2817 \text{ mm}$$

and

$$y_{max} = \frac{-21 \times 3\,(8^2 - 3^2)^{3/2}}{9\sqrt{3} \times 3 \times 10^{12} \times 8} \times 10^{12}$$

$$y_{max} = -68.686 \text{ mm (Downward)}$$

To find deflection under load at $x = 5$,

$$EIy_c = 7.875 \times \frac{5^3}{6} - 72.1875 \times 5$$

$$= -196.875 \text{ kN-m}^3$$

$$\therefore \quad y_c = \frac{-196.875}{EI} = \frac{-196.875 \times 10^{12}}{3 \times 10^{12}} = -65.625 \text{ mm}$$

To find deflection at the mid-span, $x = \frac{l}{2} = 4\text{m}$,

$$EIy_{mid} = 7.875 \times \frac{4^3}{6} - 72.1875 \times 4$$

$$EIy_{mid} = -204.75 \text{ kN-m}^3$$

$$\therefore \quad y_{mid} = \frac{-204.75}{EI} = \frac{-204.75 \times 10^{12}}{3 \times 10^{12}} = -68.25 \text{ mm.}$$

Problem 14 A beam 12 m is simply supported at the ends. It carries two concentrated loads of 90 kN and 60 kN at distance 3 m and 8 m respectively from the left end support. Calculate the deflection under each load and maximum deflection. The modulus of elasticity is 200 kN/mm^2 and moment of inertia is 3×10^9 mm^4.

Solution: $E = 200 \text{ kN/mm}^2 = 2 \times 10^5 \text{ N/mm}^2$

$I = 3 \times 10^9 \text{ mm}^4$

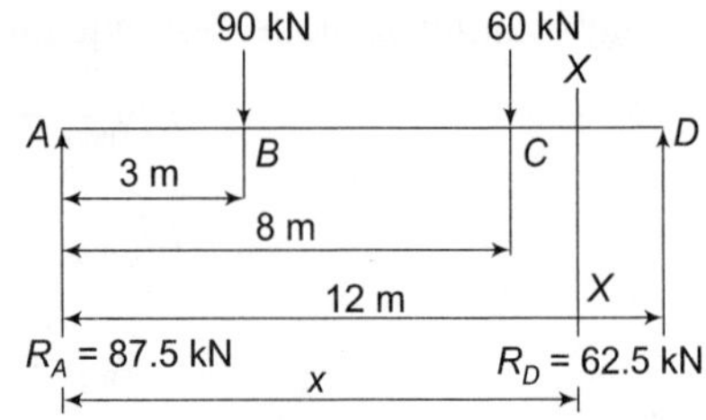

Fig. 7.28

To find reactions; $R_A + R_D = 90 + 60 = 150$ kN

Taking moments about A,

$$90 \times 3 + 60 \times 8 - R_D \times 12 = 0$$

$\therefore$ $R_D = 62.5$ kN

$$R_A = 150 - R_D = 87.5 \text{ kN}$$

Considering a section between CD, moment at XX,

$$M_x = R_A \cdot x - 90\,(x - 3) - 60\,(x - 8) \text{ kN-m}$$

$$\text{Deflection equation, } EI\,\frac{d^2y}{dx^2} = 87.5\,x - 90\,(x - 3) - 60\,(x - 8)$$

$$\text{Integrating } EI\,\frac{dy}{dx} = 87.5\,\frac{x^2}{2} + C_1 - \frac{90(x - 30)^2}{2} - \frac{60\,(x - 8)^2}{2}$$

$$EI\,\frac{dy}{dx} = 43.75\,x^2 + C_1 - 45\,(x - 3)^2 - 30\,(x - 8)^2 \text{ kN-m}^2$$

$$\text{Integrating again, } EIy = 43.75\,\frac{x^3}{3} + C_1x + C_2 - 45\,\frac{(x - 3)^3}{3} - 30\frac{(x - 8)^3}{3}$$

$$EIy = 14.583\,x^3 - C_1x + C_2 - 15\,(x - 3)^3 - 10(x - 8)^3 \text{ kN-m}^3$$

Applying boundary conditions, at A, $x = 0$, $y = 0$

$$0 = 0 - 0 + C_2$$

$\therefore$ $C_2 = 0$

At $x = l = 12$ m at B, $y = 0$

$$0 = 14.583 \times 12^3 - C_1 \times 12 - 15(12 - 3)^3 - 10\,(12 - 8)^3$$

$$C_1 = -\,819.368$$

$\therefore$ $$EI\,\frac{dy}{dx} = 43.75\,x^2 - 819.368 - 45(x - 3)^3 - 30(x - 8)^2$$

and $$EIy = 14.583\,x^3 - 819.368\,x - 15(x - 3)^3 - 10(x - 8)^3$$

To find deflection under load.

At B, $x = 3$ m

$$EIy_B = 14.583 \times 3^3 - 819.368 \times 3 - 0 - 0 \text{ kN-m}^3$$

$$y_B = \frac{-\,2064.363}{EI} = \frac{-\,2064.363 \times 10^{12}}{2 \times 10^5 \times 3 \times 10^9} = -3.44 \text{ mm}$$

At C, $x = 8$ m

$$EIy_c = 14.583 \times 8^3 - 819.368 \times 8 - 15(8 - 3)^3 \text{ kN-m}^3$$

$$\therefore \qquad y_C = \frac{-963.448}{EI} = \frac{-963.448}{2 \times 10^5 \times 3 \times 10^9} \times 10^{12} = -1.6057 \text{ mm}$$

To find maximum deflection,

For maximum deflection, $\frac{dy}{dx} = 0$

It may be assumed that maximum deflection occurs between B and C.

$$0 = 43.75\, x^2 - 819.368 - 45(x - 3)$$

$$43.75\, x^2 - 45x - 684.368 = 0$$

$$x = 4.503 \text{ or } -3.47$$

Taking feasible value, $x = 4.503$

Since this is between B and C, our assumption is correct.

Substituting in deflection equation,

$$EIy_{max} = 14.583\,(4.503)^3 - 819.368\,(4.503) - 15(4.503 - 3)^3 - 0$$

$$y_{max} = \frac{-2409.01}{EI}$$

$$y_{max} = \frac{-2409.01 \times 10^{12}}{2 \times 10^5 \times 3 \times 10^9} = -4.015 \text{ mm.}$$

Problem 15 Find the maximum deflection and maximum slope for a simply supported beam loaded as shown in Fig. 7.29. The moment of inertia is 9×10^8 mm^4 and modulus of rigidity is 2×10^5 N/mm^2.

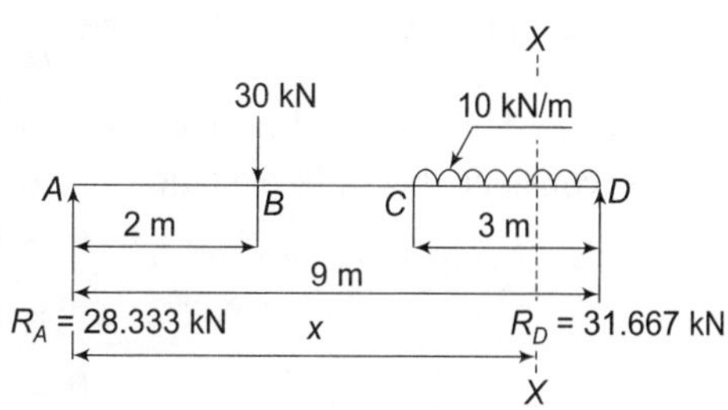

Fig. 7.29

Solution: $I = 9 \times 10^8$ mm^4

$E = 2 \times 10^5$ N/mm^2

Reaction, $R_A + R_D = 30 + 10 \times 3 = 60$ kN

Taking moment about A, $30 \times 2 + (10 \times 3)(7.5) - R_D \times 9 = 0$

$$\therefore \qquad R_D = 31.667 \text{ kN}$$

$$R_A = 60 - R_D = 28.333 \text{ kN}$$

Consider a section XX between CD.

$$\text{Moment } M_x = R_A.x - 30\,(x - 2) - 10.\,(x - 6)\,\frac{(x - 6)}{2}$$

$$M_x = 28.333\,x - 30\,(x - 2) - 5\,(x - 6)^2 \text{ kN-m}$$

$$\therefore \qquad EI\,\frac{d^2y}{dx} = M_x = 28.333\,x - 30(x - 2) - 5(x - 6)^2$$

$$\text{Integrating, } EI\,\frac{dy}{dx} = 28.333\,\frac{x^2}{2} + C_1 - \frac{30\,(x - 2)^2}{2} - \frac{5(x - 6)^3}{3} \text{ kN-m}^2$$

Integrating again, $EIy = 28.333\,\dfrac{x^3}{6} + C_1x + C_2 - \dfrac{30(x-2)^3}{6} - \dfrac{5(x-6)^4}{12}$ kN-m^3

Applying boundary conditions,

At support A, $x = 0$, deflection $y = 0$

$$0 = 0 + 0 + C_2 - 0 - 0$$

$$\therefore \quad C_2 = 0$$

At supports B, $x = l = 9$ m, deflection $y = 0$

$$0 = 28.333\,\frac{9^3}{6} + C_1 9 - \frac{30(9-2)^3}{6} - \frac{5}{12}(9-6)^4$$

$$\therefore \quad C_1 = -188.19$$

$$\therefore \quad \text{The slope equation, } EI\frac{dy}{dx} = 28.333\,\frac{x^2}{2} + C_1 - \frac{30(x-2)^2}{2} - \frac{5}{3}(x-6)^3$$

$$EI\,\frac{dy}{dx} = 14.167\,x^2 - 188.19 - 15(x-2)^2 - 1.667(x-6)^3$$

$$\text{Deflection equation, } EIy = 28.333\,\frac{x^3}{6} - 188.19\,x - \frac{30}{6}(x-2)^3 - \frac{5}{12}(x-6)^3$$

$$EIy = 4.722\,x^3 - 188.19x - 5(x-2)^2 - 0.417\,(x-6)^3$$

For maximum deflection, the slope $\dfrac{dy}{dx} = 0$

From the loading pattern given, assume that the maximum deflection occurs between C and D, consider the slope equation up to D,

$$0 = 14.167\,x^2 - 188.19 - 15(x-2)^2$$

$$= 14.167\,x^2 - 188.19 - 15(x^2 + 4 - 4x)$$

$$0 = 14.167\,x^2 - 188.19 - 15x^2 - 60 + 60\,x$$

$$0.833\,x^2 - 60\,x + 248.19 = 0$$

$$\text{Solving } x = 67.622 \text{ and } 4.406$$

Adopting the feasible value, $x = 4.406$ mm which lies between B and C, hence our assumption is correct.

Substituting in deflection equation and considering up to C,

$$EIy_{\max} = 4.722\,(4.406)^3 - 188.19 \times 4.406 - 5(4.406 - 2)^3 - 0$$

$$y_{\max} = \frac{-494.918}{EI}$$

$$y_{\max} = \frac{-494.918 \times 10^{12}}{2 \times 10^5 \times 9 \times 10^8} = -2.7495 \text{ mm.}$$

Maximum slope

Slope will be maximum at the supports A or B.

$$\text{At } A, x = 0, EI\left(\frac{dy}{dx}\right)_A = -188.19 \text{ kNm}^2$$

$$\text{At } B, x = 9, EI\left(\frac{dy}{dx}\right)_B = 14.167 \times 9^2 - 188.19 - 15(9-2)^2 - 1.667\,(9-6)^3$$

$$EI\left(\frac{dy}{dx}\right)_B = 332.658 \text{ kN-m}^3$$

It is seen that the slope is maximum at B.

$$\left(\frac{dy}{dx}\right)_B = \frac{332.658}{EI}$$

$$= \frac{332.658 \times 10^9}{2 \times 10^5 \times 9 \times 10^8} = 0.001848 \text{ radian}$$

$$\therefore \qquad \text{Maximum slope} = \left(\frac{dy}{dx}\right)_B = 0.001848 \times \frac{180}{\pi} = 0.1059 \text{ degree.}$$

Problem 16 Figure 7.30 shows a simply supported beam of uniform section. Determine the position and magnitude of maximum deflection. Modulus of elasticity = 200 GPa. Moment of inertia of the section = 3.6 × 10^8 mm^4.

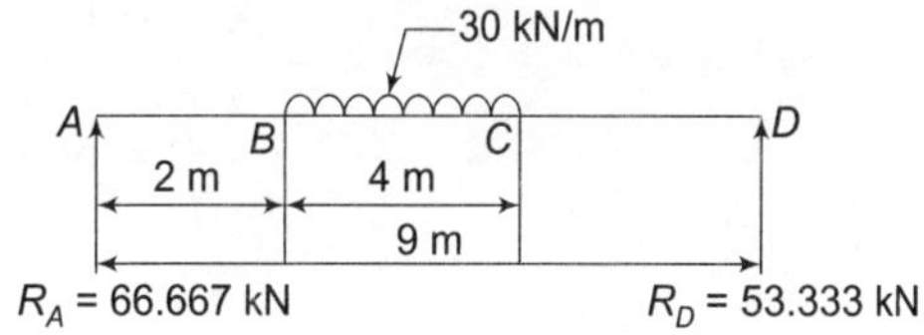

Fig. 7.30

Solution:

$$E = 200 \text{ GPa} = 2 \times 10^5 \text{ MPa}$$

$$I = 3.6 \times 10^8 \text{ mm}^4$$

For bending moment at any section XX, the UDL is to be extended up to right side support and same intensity of UDL is to be placed upward in the portion CD. Then the general expression for the bending moment can be written.

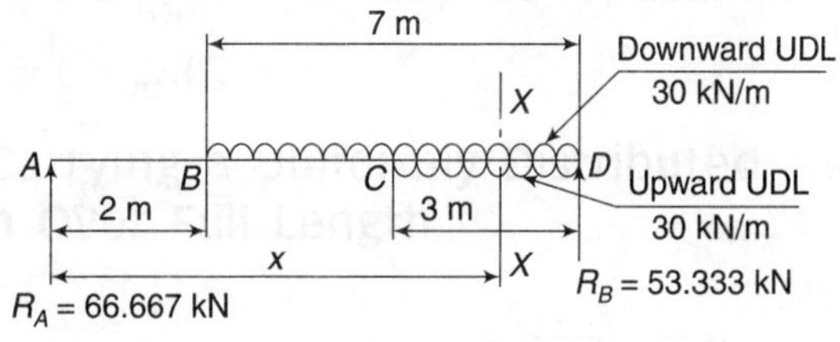

Fig. 7.31

The rearranged loading on the beam is shown in Fig. 7.31.

Reactions: $\qquad R_A + R_D = 30 \times 4 = 120$ kN

Taking moments about A, $R_D \times 9 = 30 \times 4\left(2 + \frac{4}{2}\right)$

$$\therefore \qquad R_D = 53.333 \text{ kN}$$

$$R_A = 120 - R_D = 66.667 \text{ kN}$$

Bending moment at XX,

$$M_x = 66.667\ x - 30\ (x - 2)\ \frac{(x-2)}{2} + 30\ (x - 6)\ \frac{(x-6)}{2}$$

$$M_x = 66.667\ x - 15\ (x - 2)^2 + 15(x - 6)^2 \text{ kN-m}$$

$$\therefore \qquad EI\ \frac{d^2y}{dx^2} = 66.667\ x - 15\ (x - 2)^2 + 15\ (x - 6)^2 \text{ kN-m}$$

Integrating, $EI\ \frac{dy}{dx} = 66.667\ \frac{x^2}{2} + C_1 - \frac{15\ (x-2)^3}{3} + \frac{15\ (x-6)^3}{3}$ kN-m^2 ...(1)

Integrating again, $EIy = 66.667\ \frac{x^3}{6} + C_1x + C_2 - \frac{15\ (x-2)^4}{12} + \frac{15(x-6)^4}{12}$ kN-m^3 ...(2)

At $x = 0$, $y = 0$ at support A

$$0 = 0 + 0 + C_2$$

$$\therefore \qquad C_2 = 0$$

At $x = l = 9$ m at support B, $y = 0$

$$0 = 66.667 \times \frac{93}{6} + C_1 \times 9 - \frac{15}{12}\ (9 - 2)^4 + \frac{15}{12}\ (9 - 6)^4$$

$$C_1 = -\ 577.782$$

For maximum deflection, $\frac{dy}{dx} = 0$

The maximum deflection occurs between B and C. From equation (1)

$$0 = 6.667\frac{x^2}{2} - 577.782\ \frac{15(x-6)^3}{3}$$

$$= 33.333\ x^2 - 577.782 - 5(x^3 - 63 - 18x^2 + 108x)$$

$$0 = 33.333\ x^2 - 577.782 - 5x^3 + 1080 + 90x^2 - 540x$$

$$5x^3 - 123.333\ x^2 + 540\ x - 1080 = 0$$

$$x = 19.7525,\ 2.457 \text{ and } 2.457$$

Taking the feasible value of x, $x = 2.457$ m

This value of $x = 2.457$ is between B and C and hence our assumption of y_{max} is correct.

Substitute $x = 2.457$ in deflection equation (2)

$$EIy_{max} = 66.667 \times \frac{2.457^3}{6} - 577.782 \times 2.457 - \frac{5}{12}(2.457 - 2)^4 + 0$$

$$EIy_{max} = -1254.858 \text{ kNm}^3$$

$$\therefore \qquad y_{max} = \frac{-1254.858}{2 \times 10^5 \times 3.6 \times 10^8} \times 10^{12} = -17.429 \text{ mm.}$$

Problem 17 Figure 7.32 shows a simply supported beam of length 7 m, subjected to two point loads and a UDL. Determine the deflection at *B*, *C* and *D* on the beam. The modulus of elasticity for beam material is 210 GPa and moment of inertia of the section is 9 × 10^7 mm^4.

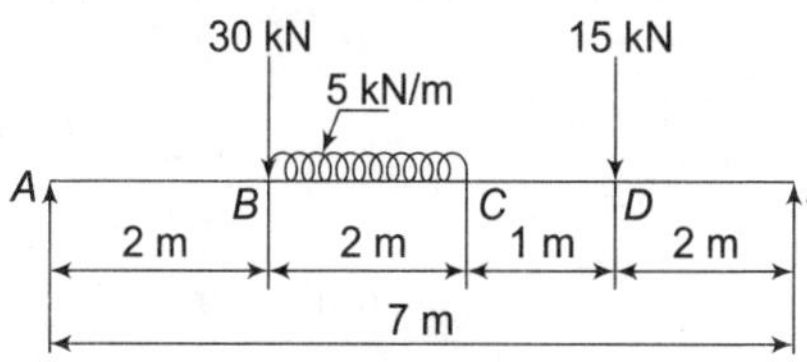

Fig. 7.32

Solution:

$$E = 210 \text{ GPa} = 210 \times 10^3 \text{ MPa}$$

$$I = 9 \times 10^7 \text{ mm}^4$$

$$\text{Reactions: } R_A + R_E = 30 + (5 \times 2) + 15 = 55 \text{ kN}$$

Taking moments about *A*, $30 \times 2 + (5 \times 2)(2 + 1) + 15 \times 5 - R_E \times 7 = 0$

$$\therefore \qquad R_E = 23.57 \text{ kN}$$

$$R_A = 31.43 \text{ kN}$$

Consider a section *XX* between *D* and *E* covering all forces, at a distance *x* from *A*.

For bending moment at *XX*, the UDL is to be extended from *C* to *E* up to the right side support end is to be compensated by an upward UDL between *C* and *E* as shown in Fig. 7.33.

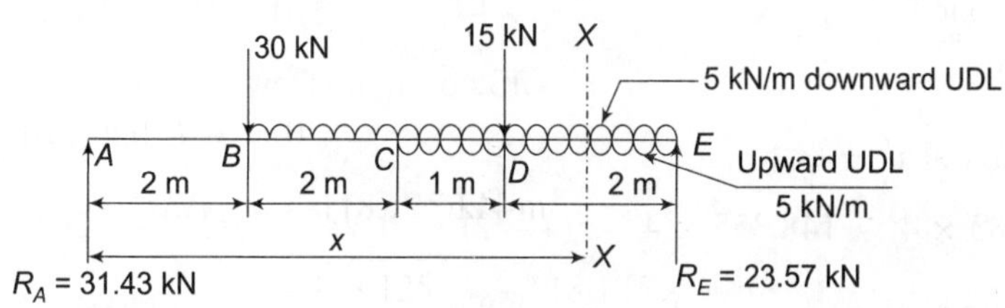

Fig. 7.33

Bending moment at *XX*,

$$M_x = R_A \cdot x - 30(x - 2) - \frac{5(x-2)^2}{2} + \frac{5(x-4)^2}{2} - 15(x - 5)$$

Substituting in the deflection equation

$$EI\frac{d^2y}{dx^2} = 31.43\,x - 30\,(x-2) - 2.5(x-2)^2 + 2.5\,(x-4)^2 - 15\,(x-5)$$

Integrating,

$$EI\frac{dy}{dx} = 31.43\,\frac{x^2}{2} + C_1 - \frac{30(x-2)^2}{2} - \frac{2.5\,(x-2)^3}{3} + \frac{2.5(x-4)^3}{3} - \frac{15(x-5)^2}{2}$$

$$EI\frac{dy}{dx} = 16.715\,x^2 + C_1 - 15(x-2)^2 - 0.833\,(x-2)^3 + 0.833\,(x-4)^3 - 7.5\,(x-5)^2 \text{ kN-m}^2$$

Integrating again,

$$EIy = 16.715\,\frac{x^3}{3} + C_1\,x + C_2 - \frac{15(x-2)^3}{3} - \frac{0.833(x-2)^4}{4} + \frac{0.833\,(x-4)^4}{4} - \frac{7.5(x-5)^3}{3}$$

$$EIy = 5.238\,x^3 + C_1x + C_2 - 5(x-2)^3 - 0.208\,(x-2)^4 + 0.208(x-4)^4 - 2.5\,(x-5)^3 \text{ kN-m}^3$$

Applying boundary conditions,

At the support A, zero, deflection $y = 0$

$$0 = 0 + 0 + C_2$$

$$\therefore \quad C_2 = 0$$

At the support B, $x = l = 7$, deflection $y = 0$

$$O = 5.238 \times 7^3 + C_1 \times 7 - 5(7-2)^3 - 0.208\,(7-2)^4 + 0.28(7-4)^4 - 2.5\,(7-5)^3$$

$$\therefore \quad C_1 = -\,148.355$$

To find deflection at B, $x = 2$ m

$$EIy_B = 5.283 \times 2^3 - 148.355 \times 2 \text{ kN m}^3$$

$$\therefore \quad y_B = \frac{-\,254.446}{EI} = \frac{-\,254.446 \times 10^{12}}{210 \times 10^3 \times 9 \times 10^7}$$

$$= -\,13.463 \text{ mm (Downward)}$$

For deflection at C, $x = 4$ m

$$EIY_C = 5.283 \times 4^3 - 148.355 \times 4 - 5(4-2)^3 - 0.208\,(4-2)^4 + 0.208(4-4)^4 - 0$$

$$EIy_C = -298.636 \text{ kNm}^3$$

$$\therefore \quad y_C = \frac{-\,298.636}{210 \times 10^3 \times 9 \times 10^7} \times 10^{12}$$

$$= 15.8 \text{ mm (Downward)}$$

For deflection at D, $x = 5$ m

$$EIY_D = 5.283 \times 5^3 - 148.355 \times 5 - 5(5-2)^3 - 0.208(5-2)^4 + 0.208\,(5-4)^4 - 0$$

$$y_D = \frac{-238.872 \times 10^{12}}{210 \times 10^3 \times 9 \times 10^7} = -13.205 \text{ mm.}$$

Problem 18 A simply supported beam of length l carries a uniformly distributed load of w/unit length over a span of a as shown in Fig. 7.34. Determine the value of a for which maximum deflection will occur at the left end of the UDL.

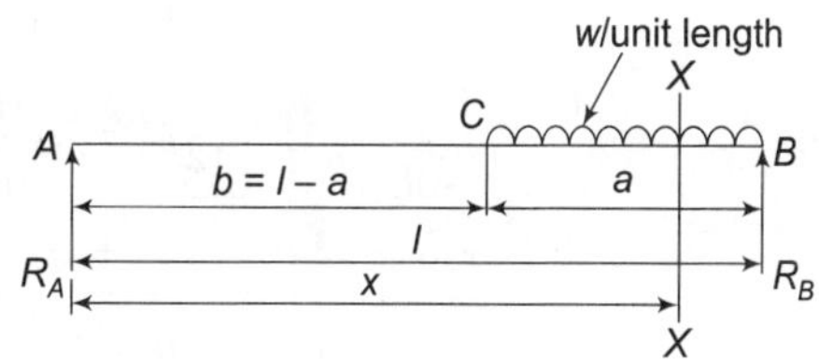

Fig. 7.34

Solution: Reaction at A, $R_A = \dfrac{w \cdot a^2}{2l}$

$\therefore$ Reaction at B, $R_B = w \cdot a - \dfrac{wa^2}{2l} = \dfrac{wa}{2l}(2l - a)$

Bending moment at any section XX at a distance x from left support A,

$$M_x = R_A \cdot x - \frac{w(x-b)^2}{2}$$

$$M_x = \frac{wa^2x}{2l} - \frac{w\,(x-b)^2}{2}$$

$$\text{Equating to } EI\frac{d^2y}{dx^2} = M_x = \frac{wa^2x^2}{2l} - \frac{w\,(x-b)^2}{2}$$

$$\text{Integrating, } EI\frac{dy}{dx} = \frac{wa^2x^2}{4l} + C_1 - \frac{w(x-b)^3}{6}$$

$$\text{Integrating again, } EIY = \frac{wa^2x^3}{12l} + C_1x + C_2 - \frac{w\,(x-b)^4}{24}$$

Applying boundary conditions

At $x = 0$ at support A, deflection $y = 0$

$$0 = 0 + 0 + C_2$$

$$\therefore \quad C_2 = 0$$

At $x = l$ at support B, deflection $y = 0$

$$0 = \frac{wa^2l^3}{12} + C_1l - \frac{w\,(l-b)^4}{24}$$

$$\therefore \quad C_1 = \frac{-wa^2l^3}{12l} + \frac{wa^4}{24l} \quad (\because l - b = a)$$

$$C_1 = \frac{-wa^2}{24l}(2l^2 - a^2)$$

For maximum deflection to occur at left end of UDL,

i.e., at C, $x = b = l - a$, $\frac{dy}{dx} = 0$

$$0 = \frac{wa^2}{4l}(l - a)^2 - \frac{wa^2}{24l}(2l^2 - a^2)$$

$$0 = \frac{wa^2}{24l}\left[6(l - a)^2 - (2l^2 - a^2)\right]$$

or $$6(l - a^2) - (2l^2 - a^2) = 0$$

$$6(l^2 + a^2 - 2la) - 2l^2 + a^2 = 0$$

$$7a^2 - 12la + 4l^2 = 0$$

$$a = \frac{12l \pm \sqrt{(12l)^2 - 4 \times 7 - 4l^2}}{2 \times 7}$$

$$a = \frac{12l \pm 5.65685\, l}{14} = 1.2612\, l \text{ or } 0.4531\, l$$

Take the possible value of $a = 0.4531\, l$,

For maximum deflection at $a = 0.4531\, l$,

At $a = 0.4531\, l$, $x = l - a = l - 0.4531\, l$

$$EIy_{max} = \frac{w(0.4531l)^2}{12}(l - 0.4531l^3) - \frac{w(0.4531l)^2}{24l}\left[2l^2 - (0.4531)^2\right] \times (l - 0.4531\, l)$$

$$EIy_{max} = -\,0.005598\, wl^4$$

$\therefore$ $$y_{max} = \frac{-\,0.005598\, wl^4}{EI} \text{ at } a = 0.4531\, l.$$

Problem 19 A simply supported beam of length l carries two concentrated loads of each W at a distance of a from each end. Find the deflection under each load and maximum deflection.

Solution: The loading of beam is symmetrical and hence both the reactions will be equal $R_A = R_D = W$

Also, maximum deflection occurs in the middle of beam at $x = \frac{l}{2}$

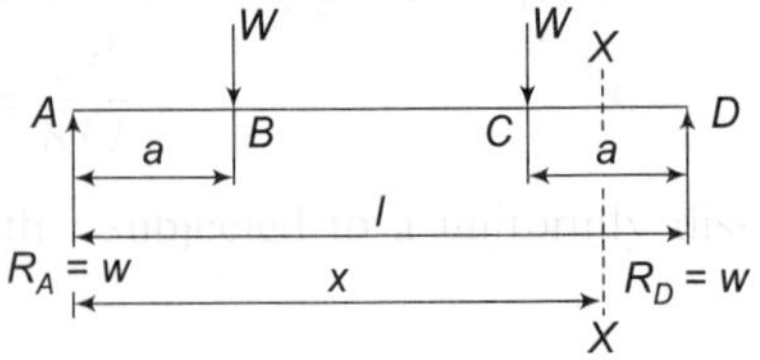

Fig. 7.35

Bending moment up to the mid-span, where the deflection will be maximum.

$$M_x = W \cdot x - W(x - a).$$

$$\therefore \qquad EI\frac{d^2y}{dx^2} = Wx - W(x-a)$$

$$\text{Integrating } EI\frac{dy}{dx} = \frac{Wx^2}{2} + C_1 \frac{-W(x-a)^2}{2}$$

The deflection will be maximum at the centre at $x = \frac{l}{2}$, the slope $\frac{dy}{dx} = 0$

$$0 = \frac{W}{2}\left(\frac{l}{2}\right)^2 + C_1 - \frac{W}{2}\left(\frac{l}{2} - a\right)^2$$

$$\therefore \qquad C_1 = \frac{-Wl^2}{8} + \frac{W}{2}\left(\frac{l^2}{4} + a^2 - la\right)$$

$$= \frac{-Wl^2}{8} + \frac{Wl^2}{8} + \frac{Wa^2}{2} - \frac{Wla}{2}$$

$$C_1 = \frac{-Wa}{2}(l-a)$$

$$\therefore \qquad EI\frac{dy}{dx} = \frac{Wx^2}{2} - \frac{Wa}{2}(l-a) - \frac{W}{2}(x-a)^2$$

Integrating again

$$EIy = \frac{Wx^3}{6} - \frac{Wa}{2}(l-a)\,x - \frac{W}{2}\frac{(x-a)^3}{3} + C_2$$

$$EIy = \frac{Wx^3}{6} - \frac{Wa}{2}(l-a)\,x - \frac{W}{6}(x-a)^3 + C_2$$

At $x = 0$, at the support, deflection $y = 0$

$$0 = 0 - 0 - 0 + C_2$$

$$\therefore \qquad C_2 = 0$$

$$EIy = \frac{Wx^3}{6} - \frac{Wa}{2}(l-a)\,x - \frac{W}{6}(x-a)^3$$

For deflection under load at B, $x = a$

$$EIy_B = \frac{Wa^3}{6} - \frac{Wa}{2}(l-a)a = \frac{Wa^3}{6} - \frac{Wa^2l}{2} + \frac{Wa^3}{2}$$

$$= \frac{Wa^3 - 3\,Wa^2l + 3\,Wa^3}{6} = -\frac{Wa^2}{6}(3l - 4a)$$

$$\therefore \qquad y_B = \frac{-Wa^2}{6EI}(3l - 4a)$$

For maximum deflection at the middle, $x = \frac{l}{2}$

$$EIy_{max} = \frac{W}{6}\left(\frac{l}{2}\right)^3 - \frac{Wa}{2}(l-a)\frac{l}{2} - \frac{W}{6}\left(\frac{l}{2}-a\right)^3$$

$$= \frac{Wl^3}{48} - \frac{Wal}{4}(l-a) - \frac{W}{6}\left(\frac{l-2a}{2}\right)^3$$

$$= \frac{Wl^3}{48} - \frac{Wal^2}{4} + \frac{Wa^2l}{4} - \frac{W}{6}\left(\frac{l^3 - 8a^3 - 6l^2a + 12la^2}{8}\right)$$

$$= \frac{Wl^3}{48} - \frac{Wal^2}{4} + \frac{Wa^2l}{4} - \frac{W}{48}(l^3 - 8a^3 - 6l^2\,a + 12\,la^2)$$

$$= \frac{W}{48}(l^3 - 12a\,l^2 + 12\,a^2l - l^3 + 8a^3 + 6l^2a - 12\,la^2)$$

$$= \frac{W}{48}(8a^3 - 6al^2) = \frac{-Wa}{48} \times 2\,(3l^2 - 4a^2)$$

$$EIy_{max} = \frac{-Wa}{24}(3l^2 - 4a^2)$$

$$\therefore \quad y_{max} = \frac{-Wa}{24EI}(3l^2 - 4a^2).$$

Problem 20 A beam of uniform cross section of length l is simply supported at its ends and carries a symmetrical triangular load whose intensity varies from zero at the supports to w/unit length at the middle. Find the slope and deflection at the middle.

Solution: Figure 7.36 shows a simply supported beam of length l subjected to a symmetric triangular load.

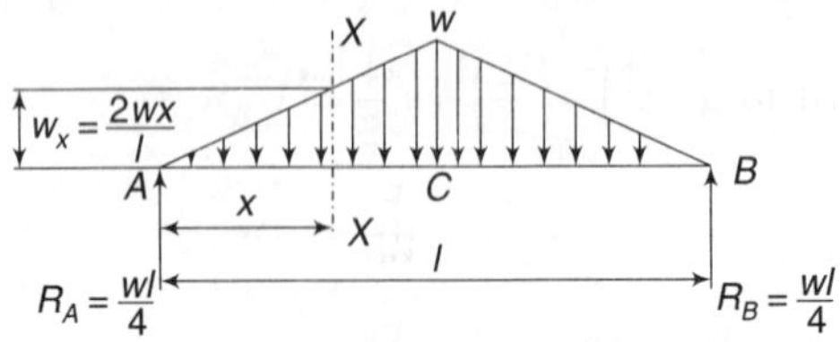

Fig. 7.36

E = Modulus of elasticity

I = Moment of inertia of the section

Due to symmetry, both reactions will be equal,

$$R_A = R_B = \frac{W}{2} = \frac{l}{2}\left(\frac{w.l}{2}\right) = \frac{wl}{4}$$

Consider any section XX at a distance x from A between A and C, such that $x < \frac{l}{2}$.

Due to symmetrical loading maximum deflection occurs in the middle at C, hence consider one half for simplicity.

Let the load intensity at XX be w_x

From similar triangles, $\frac{w_x}{x} = \frac{w}{l/2}$

$\therefore$ $$w_x = \frac{2wx}{l}$$

Bending moment at section XX

$$M_x = \frac{wl}{4} \cdot x - \left(\frac{1}{2} w_x . x\right) \frac{x}{3}$$

$$M_x = \frac{wlx}{4} - \frac{1}{2}\frac{2wx}{l} x \frac{x}{3} = \frac{wlx}{4} - \frac{wx^3}{3l}$$

Substituting in deflection equation,

$$EI \frac{d^2y}{dx^2} = \frac{wlx}{4} - \frac{wx^3}{3l}$$

Integrating, we get $EI \frac{dy}{dx} = \frac{wlx^2}{8} + C_1 - \frac{wx^4}{12l}$

At C, at $x = \frac{l}{2}$, the deflection is maximum and the slope $\frac{dy}{dx} = 0$

$$0 = \frac{wl}{8}\left(\frac{l}{2}\right)^2 + C_1 - \frac{w}{12l}\left(\frac{l}{2}\right)^4$$

$$C_1 = \frac{-wl^3}{32} + \frac{wl^3}{192}$$

$$C_1 = \frac{-wl^3 (6 - 1)}{192} = \frac{-5}{192} wl^3$$

$\therefore$ $$EI \frac{dy}{dx} = \frac{wlx^2}{8} - \frac{5}{192} wl^3 - \frac{wx^4}{12l} \text{ (Slope equation)}$$

The slope will be maximum at the supports A or B put $x = 0$ for slope at A,

$$EI \left(\frac{dy}{dx}\right)_A = \frac{-5}{192} wl^3 \text{ (Max. slope)}$$

Integrating the slope equation,

$$EIy = \frac{wlx^3}{24} - \frac{5}{192} wl^3x + C_2 - \frac{wx^5}{60l}$$

At $x = 0$ at support A, deflection $y = 0$

$$0 = 0 - 0 + C_2 - 0$$

$\therefore$ $$C_2 = 0$$

The deflection equation is $EIy = \frac{wlx^3}{24} - \frac{5}{192} wl^3x - \frac{wx^5}{60l}$

The deflection will be maximum at C at $x = \frac{l}{2}$

$$EIy_{max} = \frac{wl}{24}\left(\frac{l}{2}\right)^3 - \frac{5wl^3}{192}\left(\frac{l}{2}\right) - \frac{w}{60l}\left(\frac{l}{2}\right)^5$$

$$= \frac{wl^4}{192} - \frac{5wl^4}{384} - \frac{wl^4}{1920} = \frac{wl^4(10 - 25 - 1)}{1920}$$

$$= \frac{-16wl^4}{1920} = \frac{-wl^4}{120}$$

$$\therefore \quad y_{max} = \frac{-wl^4}{120\,EI} \text{ (at middle at } C)$$

Problem 21 A simply supported beam of length 6 m carries a trapezoidal load whose intensity varies from 12 kN/m from one end to 48 kN/m at the other end. The modulus of elasticity is 210 kN/mm^2. It is made of uniform rectangular cross section whose width is 150 mm. The allowable bending stress for the materials is 120 N/mm^2. Determine the depth of the beam and the deflection at the mid-span of the beam.

Solution: The given beam with trapezoidal loading is shown in Fig. 7.37.

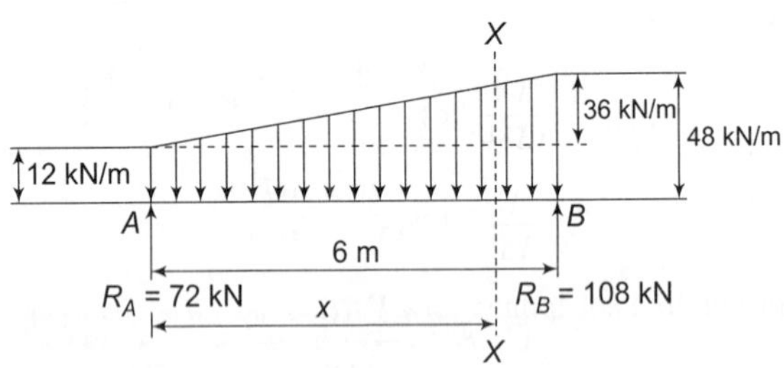

Fig. 7.37

$$E = 210 \times 10^3 \text{ N/mm}^2$$

$$I = \frac{bd^3}{12} = \frac{150\,d^3}{12}$$

Allowable bending stress σ = 120 MPa.

Reactions:

Consider the given load as one UDL of intensity 12 kN/m over full span and a UVL that varies from 0 at A to 48 – 12 = 36 kN/m at B.

$$R_A + R_B = 12 \times 6 + \frac{1}{2} \times 36 \times 6 = 180 \text{ kN}$$

Taking moments about A, $(12 \times 6) \times \frac{6}{2} + \left(\frac{1}{2} \times 36 \times 6\right)\frac{2 \times 6}{3} - R_B \times 6 = 0$

$$\therefore \quad R_B = 108 \text{ kN}$$

$$R_A = 180 - 108 = 72 \text{ kN}$$

To find I

$$\text{Bending equation, } \frac{M}{I} = \frac{\sigma}{y}$$

$$M = \text{Maximum bending moment.}$$

Bending moment is maximum whose shear force is zero, changing its sign between AB.

Let shear force be zero at a distance x from left end A

$$SF_x = 72 - 12x - \frac{1}{2} \times x \cdot w_x$$

$$w_x = \text{Intensity of UVL.}$$

From similar triangles of UVL, $\dfrac{w_x}{x} = \dfrac{wl}{6}$

$$w_x = \frac{w \cdot x}{6} = \frac{36 \times x}{6} = 6x$$

$$SF_x = 72 - 12x - \frac{1}{2} \cdot 6x \cdot x$$

$$0 = 72 - 12x - 3x^2$$

$$3x^2 + 12x - 72 = 0$$

Solving $$x = 3.2915 \text{ m}$$

Bending moment at XX, $M_x = 72\,x - 12.\dfrac{x^2}{2} - \left(\dfrac{1}{2} \cdot w_x \cdot x\right)\dfrac{x}{3}$

$$= 72\,x - 6x^2 - \left(\frac{1}{2} \times 6x \cdot x\right)\frac{x}{3} = 72\,x - 6x^2 - x^3$$

Substituting $x = 3.2915$ for maximum bending moment,

$M_{max} = 72 \times 3.2915 - 6(3.2915)^2 - 3.2915^3 = 136.324 \text{ kN-m} = 136.324 \times 10^6 \text{ Nmm}$

Substituting in the bending moment equation,

$$\frac{136.324 \times 10^6}{I} = \frac{120}{\left(\frac{d}{2}\right)}$$

$$\frac{136.324 \times 10^6}{\left(\frac{bd^3}{12}\right)} = \frac{120}{\frac{d}{2}}$$

$$\frac{136.324 \times 10^6 \times 12}{150 \times d^3} = \frac{120 \times 2}{d}$$

$\therefore$ $$\text{Depth of beam } d = 213.17 \text{ mm}$$

$$\therefore \quad \text{Moment of inertia } I = \frac{bd^3}{12} = \frac{150 \times 213.17^3}{12} = 121.08 \times 10^6 \text{ mm}^4$$

Bending moment at the mid-span,

$$y_C = y_{C1} + y_{C2}$$

$$y_{C1} = \text{Deflection at } C \text{ due to UDL}$$

$$= \frac{5}{384}\frac{w_1 l^4}{EI}$$

$$y_{C2} = \text{Deflection at } C \text{ due to UVL} = \frac{5}{768}\frac{w_2 l^4}{EI}$$

(Deflection at the middle due to UVL $= \frac{1}{2}$ of deflection at the middle due to some UDL)

$$y_C = \frac{5}{384} \times \frac{12 \times 6^4}{EI} + \frac{5}{768} \times \frac{36 \times 6^4}{EI} = \frac{506.25}{EI}$$

$$y_C = \frac{506.25 \times 10^{12}}{210 \times 10^3 \times 121.08 \times 10^6} = 19.91 \text{ mm.}$$

Problem 22 A simply supported 3 m long beam is subjected to a clockwise couple of 150 kN-m at a distance of 1 m from the right side support. Find the deflection at the point of application of the couple and the maximum deflection. The modulus of elasticity is 200 kN/mm^2 and moment of inertia of the section is 1 × 10^8 mm^2.

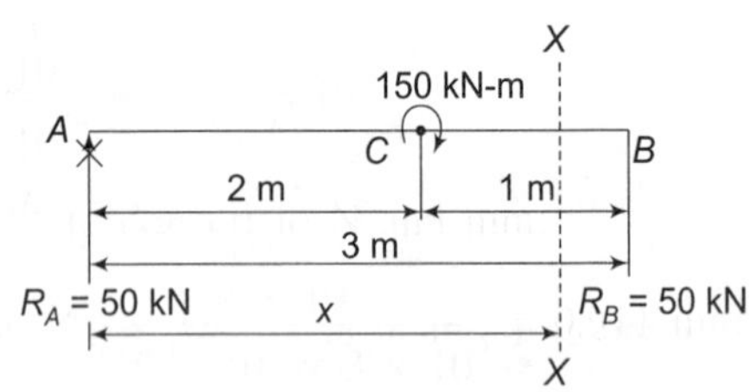

Fig. 7.38

Solution: Figure 7.38 shows the simply supported beam of 3 m long subjected to a clockwise couple of 150 kN-m at C, 1 m from right side support

$$E = 200 \text{ kN/mm}^2 = 2 \times 10^5 \text{ N/mm}^2$$

$$I = 10^8 \text{ mm}^4$$

To find reactions, $R_A + R_B = 0$

Taking moments about A, $150 - R_B \times 3 = 0$

$$\therefore \quad R_B = \frac{150}{3} = 50 \text{ kN} \uparrow$$

$$\therefore \quad R_A = 0 - 50 = -50 \text{ kN, i.e., } 50 \text{ kN} \downarrow$$

Bending moment at any section XX is at a distance x from A.

$$M_x = -R_A\, x + 150$$

$$M_x = -R_A\, x + 150\,(x - 2)^\circ$$

Note: Consider the distance from XX to the couple with power of zero. This term is required for integration as per Macaulay's method.

$$\therefore \qquad EI\frac{d^2y}{dx^2} = -50x + 150(x-2)^0$$

Integrating

$$EI\frac{dy}{dx} = -50\frac{x^2}{2} + C_1 + 150(x-2)$$

$$\text{Integrating again } EIy = -50\frac{x^3}{6} + C_1x + C_2 + \frac{150(x-2)^2}{2}$$

$$= \frac{-25}{3}x^3 + C_1x + C_2 + 75(x-2)^2$$

Applying boundary conditions,

At the support A, $x = 0$, deflection $y = 0$

$$0 = 0 + 0 + C_2 + 0$$

$$\therefore \qquad C_2 = 0$$

At the right support B, $x = 3$, deflection $y = 0$

$$0 = \frac{-25}{3} \times 3^3 + C_1 \times 3 + 75(3-2)^2$$

$$\therefore \qquad C_1 = 50$$

$$\therefore \qquad \text{The deflection equation, } EIy = \frac{-25}{3}x^3 + 50x + 75(x-2)^2$$

$$\text{The slope equation, } EI\frac{dy}{dx} = -25x^2 + 50 + 150(x-2)$$

For the deflection under load at C, $x = 2$

$$EIy_C = \frac{-25}{3} \times 2^3 + 50 \times 2 + 0$$

$$EIy_C = 33.333$$

$$y_C = \frac{33.333}{EI} = \frac{33.333 \times 10^{12}}{2 \times 10^5 \times 1 \times 10^8} = 1.667 \text{ mm}$$

For maximum deflection, slope $\frac{dy}{dx} = 0$

Maximum deflection occurs in the larger segment, i.e., AC

$$0 = -25x^2 + 50$$

$$x = \sqrt{2} = 1.4142 \text{ m from } A$$

Substituting $x = 1.4142$ m in the deflection equation,

$$EIy = \frac{25}{3}x^3 + 50x + 75(x-2)^2$$

$$EIy_{max} = \frac{25}{3}1.4142^3 + 50 \times 1.4142 + 0$$

$$y_{max} = \frac{94.281}{EI} = \frac{94.281 \times 10^{12}}{2 \times 10^5 \times 1 \times 10^8} = 4.714 \text{ mm.}$$

Problem 23 A simply supported 4.8 m long beam is subjected to a point load of 30 kN at 1 m from left end and two, couples 120 kN-m and 60 kN-m at distance of 2 m and 3.5 m respectively from the left side support as shown in Fig. 7.39. Find the deflection at the points of application of the point load and two couples. Flexural rigidity may be taken as 3×10^4 kN m^2.

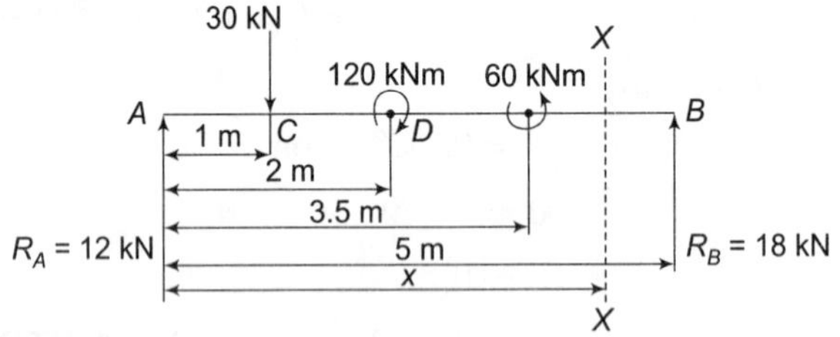

Fig. 7.39

Solution: Flexural rigidity, $EI = 3 \times 10^4 \text{ kN m}^2 = 3 \times 10^4 \times 10^9 = 3 \times 10^{13}$ N-mm^2

Reactions: $R_A + R_B = 30$ kN

Taking moments about A, $30 \times 1 + 120 - 60 - R_B \times 5 = 0$

$\therefore \quad R_B = 18$ kN

$\therefore \quad R_A = 30 - 18 = 12$ kN

Bending moment at any section at a distance x from the left end A,

$$M_x = 12x - 30\,(x - 1) + 120\,(x - 2)^\circ - 60(x - 3.5)^\circ$$

Substituting in the deflection equation,

$$EI\frac{d^2y}{dx^2} = 12\,x - 30\,(x - 1) + 120\,(x - 2)^\circ - 60\,(x - 3.5)^\circ$$

Integrating, $$EI\frac{dy}{dx} = 12\frac{x^2}{2} + C_1 - 30\frac{(x-1)^2}{2} + 120\,(x - 2) - 60\,(x - 3.5)$$

$$= 6x^2 + C_1 - 15\,(x - 1)^2 + 120\,(x - 2) - 60(x - 3.5)$$

Integrating again, $$EIy = \frac{6x^3}{3} + C_1x + C_2 - 15\frac{(x-1)^3}{3} + 120\frac{(x-2)^2}{2} - 60\frac{(x-3.5)^2}{2}$$

$$EIy = 2x^3 + C_1x + C_2 - 5(x - 1)^3 + 60\,(x - 2)^2 - 30\,(x - 3.5)^2$$

Applying boundary conditions,

At support A, $x = 0$, deflection $y = 0$

$$0 = 0 + 0 + C_2$$

$\therefore \quad C_2 = 0$

At support B, $x = 5$, deflection $y = 0$

$$0 = 2 \times 5^3 + C_1 \times 5 - 5(5-1)^3 + 60\,(5-2)^2 - 30(5-3.5)^2$$

$$\therefore \quad C_1 = -402.5$$

$\therefore$ The deflection equation,

$$EIy = 2x^3 - 402.5x - 5\,(x-1)^3 + 60\,(x-2)^2 - 30\,(x-3.5)^2$$

For deflection under point load at C, $x = 1$

$$EIy_C = 2 \times 1^3 - 402.5 \times 1$$

$$\therefore \quad y_C = \frac{-400.5}{EI} = \frac{-400.5 \times 10^{12}}{3 \times 10^3} = -13.35 \text{ mm (Downward)}$$

Deflection under couple 120 kN-m at D, $x = 2$ m

$$EIy_D = 2 \times 2^3 - 402.5 \times 2 - 5(x-1)^3$$

$$y_D = \frac{-794}{EI} = \frac{-794 \times 10^{12}}{3 \times 10^{13}} = -26.467 \text{ mm (Downward)}$$

Deflection under couple 60 kN-m at E, $x = 3.5$ m

$$EIy_E = 2 \times 3.5^3 - 402.5 \times 3.5 - 5(3.5-1)^3$$

$$y_E = \frac{-1401.125}{EI} = \frac{-1401.125}{3 \times 10^{13}} \times 10^{12}$$

$$= -46.704 \text{ mm (Downward)}$$

Problem 24 Figure 7.40 shows a simply supported beam of length 4 m carrying load of 60 kN. Find the deflection at point C. Modulus of elasticity is 200 GPa and moment of inertia = 10^9 N/mm^2.

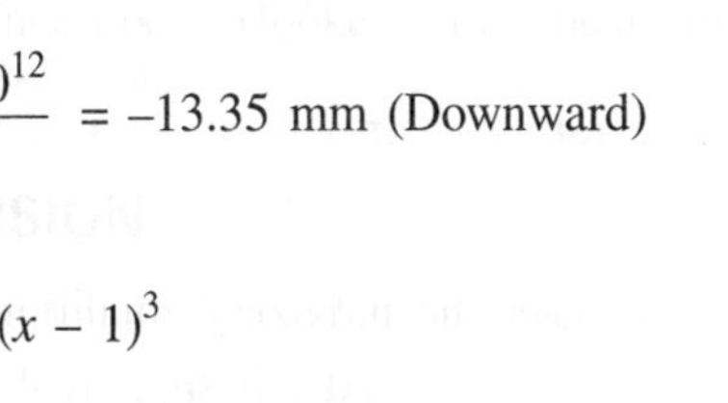

Fig. 7.40

Solution: The given load 60 kN is an eccentric load that will have a combined effect of a direct load of 60 kN at C and a clockwise couple of $60 \times 0.2 = 12$ kNm acting at C as shown in Fig. 7.41.

$$E = 2 \times 10^5 \text{ N/mm}^2 \text{ and } I = 10^9 \text{ mm}^4$$

To find reactions, $R_A + R_B = 60$ kN

Taking moments about A,

$60 \times 3 + 12 - R_B \times 4 = 0$

$$R_B = \frac{192}{4} = 48 \text{ kN}$$

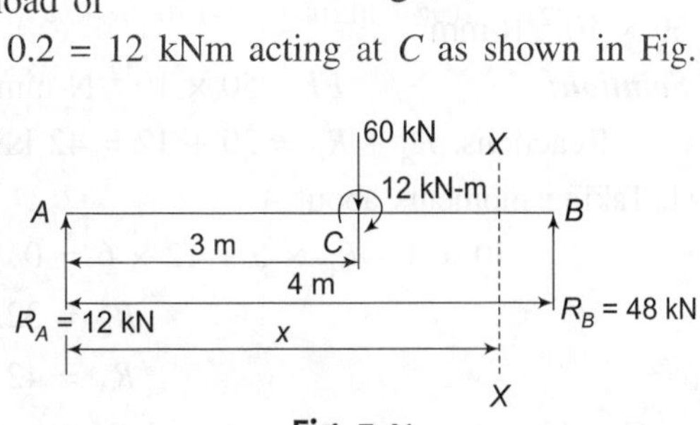

Fig. 7.41

$$R_A = 60 - R_B = 12 \text{ kN}$$

Consider a section XX at a distance x from left end A,

$$\text{Bending moment } M_x = 12\,x - 60\,(x - 3) + 12$$

Substituting in the deflection equation,

$$EI\frac{d^2y}{dx^2} = 12x - 60(x - 3) + 12 = 12x - 60(x - 3) + 12\,(x - 3)^\circ$$

$$\text{Integrating, } EI\frac{dy}{dx} = 12\frac{x^2}{2} + C_1 - 60\frac{(x-3)^2}{2} + 12\,(x - 3)$$

$$= 6x^2 + C_1 - 30\,(x - 3)^2 + 12\,(x - 3)$$

$$\text{Integrating again } EIy = \frac{6x^3}{3} + C_1x + C_2 - \frac{30(x-3)^3}{3} + \frac{12(x-3)^2}{2}$$

$$EIy = 2x^3 + C_1\,x + C_2 - 10(x - 3)^3 + 6(x - 3)^2$$

Applying boundary conditions,

At left support A, $x = 0$, deflection $y = 0$

$$0 = 0 + 0 + C_2$$

$$\therefore \quad C_2 = 0$$

At right support B, $x = 4$, deflection $y = 0$

$$0 = 2 \times 4^3 + C_1 \times 4 - 10(4 - 3)^3 + 6(4 - 3)^2$$

$$C_1 = -31$$

$\therefore$ The deflection equation,

$$EIy = 2x^3 - 31\,x - 10(x - 3) + 6\,(x - 3)^2$$

For deflection at C, $x = 3$ m

$$EIy_C = 2 \times 3^3 - 31 \times 3 - 0 + 0$$

$$y_C = \frac{-39}{EI} = \frac{-39 \times 10^{12}}{2 \times 10^5 \times 10^9} = -1.95 \text{ mm (Downward)}$$

Problem 25 Determine the deflection under the loads for a beam loaded as shown in Fig. 7.42. Flexural rigidity may be taken as 50×10^{12} N-mm^2.

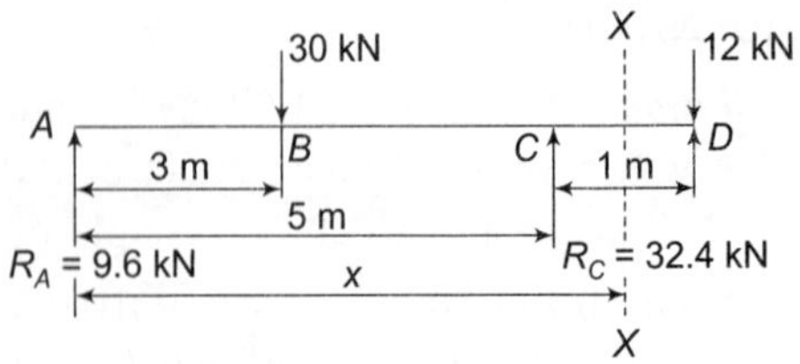

Fig. 7.42

Solution:

$$EI = 50 \times 10^{12} \text{ N-mm}^2$$

Reactions: $R_A + R_C = 30 + 12 = 42$ kN

Taking moments about A

$$30 \times 3 - R_C \times 5 + 12 \times 6 = 0$$

$$R_C = 32.4 \text{ kN}$$

$$R_A = 42 - R_C = 42 - 32.4 = 9.6 \text{ kN}$$

Consider section between C and D at a distance x from A,

Bending moment $M_x = 9.6\ x - 30\ (x - 3) + 32.4\ (x - 5)$

Substituting in deflection equation

$$EI\,\frac{d^2y}{dx^2} = 9.6\ x - 30(x - 3) + 32.4\ (x - 5)$$

$$\text{Integrating, } EI\,\frac{dy}{dx} = 9.6\,\frac{x^2}{2} + C_1 - \frac{30\ (x-3)^2}{2} + 32.4\,\frac{(x-5)^2}{2}$$

$$EI\,\frac{dy}{dx} = 4.8.x^2 + C_1 - 15(x - 3)^2 + 16.2\ (x - 5)^2$$

(Slope equation)

$$\text{Integrating again, } EIy = 4.8\,\frac{x^3}{3} + C_1x + C_2 - \frac{15(x-3)^3}{3} + \frac{16.2\ (x-5)^3}{3}$$

$$EIy = 1.6x^3 + C_1x + C_2 - 5\ (x - 3)^3 + 5.4\ (x - 5)^3$$

(Deflection equation)

Applying boundary conditions

At support A, $x = 0$, deflection $y = 0$

$$0 = 0 + C_1 \times 0 + C_2 - 0$$

$$\therefore \qquad C_2 = 0$$

At support C, $x = 5$ m, deflection $y = 0$

$$0 = 1.6 \times 5^3 + C_1 \times 5 - 5(5 - 3)^3 + 0$$

$$C_1 = -32$$

$\therefore$ The deflection equation,

$$EIy = 1.6\ x^3 - 32x - 5(x - 3)^3 + 5.4(x - 5)^3$$

To find deflection at B, $x = 3$ m

$$EIy_B = 1.6 \times 3^3 - 32 \times 3 - 0 + 0$$

$$\therefore \qquad y_B = \frac{-52.8}{EI} = \frac{-52.8 \times 10^{12}}{50 \times 10^{12}} = -1.056 \text{ mm (Downward)}$$

To find deflection at D, $x = 6$ m

$$EIy_D = 1.6 \times 6^3 - 32 \times 6 - 5(6 - 3)^3 + 5.4(6-5)^3$$

$$y_D = \frac{24}{EI} = \frac{24}{50 \times 10^{12}} \times 10^{12}$$

$$= +\ 0.48 \text{ mm (Upward deflection)}$$

Problem 26 An overhanging beam *ABCD* is loaded as shown in Fig. 7.43. Modulus of elasticity is 200 GPa and moment of inertia is 6×10^8 mm^4. Find the deflection under *B* and *D* and maximum deflection. Also find the slopes at points *A*, *B* and *D*.

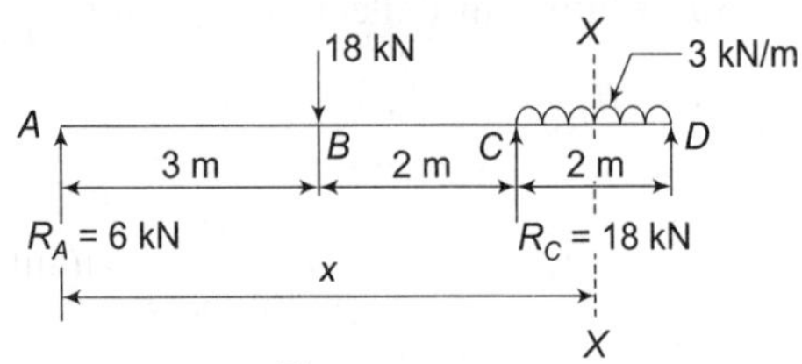

Fig. 7.43

Solution: Since the UDL is touching the right end, we can take section directly between *C* and

$E = 2 \times 10^5$ N/mm^2, $I = 6 \times 10^8$ mm^4

Reactions: $R_A + R_C = 18 + 3 \times 2 = 24$ kN

Taking moments about *A*

$$18 \times 3 - R_C \times 5 + (3 \times 2)\left(5 + \frac{2}{2}\right) = 0$$

$$R_C = 18 \text{ kN}$$

$$R_A = 24 - 18 = 6 \text{ kN}$$

Consider any section *XX* at a distance *x* from left and *A*.

Bending moment at *XX*, $M_x = R_A.x - 18 \times (x - 3) + R_B\,(x - 5) - 3(x - 5)\,\dfrac{(x-5)}{2}$

$$M_x = 6x - 18(x - 3) + 18\,(x - 5) - 1.5\,(x - 5)^2$$

Substituting in the deflection equation

$$EI\frac{d^2y}{dx^2} = M_x = 6x - 18(x - 3) + 18(x - 5) - 1.5\,(x - 5)^2$$

$$\text{Integrating, } EI\frac{dy}{dx} = 6\frac{x^2}{2} + C_1 - \frac{18(x-3)^2}{2} + \frac{18(x-5)^2}{2} - \frac{1.5(x-5)^3}{3}$$

$$EI\frac{dy}{dx} = 3x^2 + C_1 - 9(x - 3)^2 + 9\,(x - 5)^2 - 0.5\,(x - 5)^3$$

Integrating again

$$EIy = \frac{3x^3}{3} + C_1x + C_2 \frac{-9(x-3)^3}{3} + \frac{9(x-5)^3}{3} - \frac{0.5(x-5)^4}{4}$$

$$EIy = x^3 + C_1x + C_2 - 3(x - 3)^3 + 3(x - 5)^3 - 0.125\,(x - 5)^4$$

Applying boundaries conditions,

At support *A*, $x = 0$ deflection $y = 0$

$$0 = 0 + 0 + C_2$$

$$\therefore \quad C_2 = 0$$

At support *C*, $x = 5$ m, deflection $y = 0$

$$0 = 5^3 + C_1 \times 5 - 3(5 - 3)^3 + 0 - 0$$

$$C_1 = -20.2$$

∴ Slope equation,

$$EI\frac{dy}{dx} = 3x^2 - 20.2 - 9(x-3)^2 + 9(x-5)^2 - 0.5\,(x-5)^3 \text{ kN-m}^2$$

and deflection equation,

$$EIy = x^3 - 20.2x - 3(x-3)^3 + 3(x-5)^3 - 0.125\,(x-5)^4 \text{ kN-m}^3$$

To find deflection under load B, $x = 3$,

$$EIy_B = 3^3 - 20.2 \times 3 - 0$$

$$y_B = \frac{-33.6}{EI} = \frac{-33.6 \times 10^{12}}{2 \times 10^5 \times 6 \times 10^8} = -0.28 \text{ mm (Downward)}$$

For deflection at D, $x = 7$ m

$$EIy_D = 7^3 - 20.2 \times 7 - 3(7-3)^3 + 3(7-5)^3 - 0.125(7-5)^4$$

$$y_D = \frac{31.6}{EI} = \frac{31.6 \times 10^{12}}{2 \times 10^5 \times 6 \times 10^8} = 0.2633 \text{ mm (Upward)}$$

For maximum deflection, slope $\frac{dy}{dx} = 0$

It can be seen that maximum deflection occurs in the bigger span AB, considering slope equation up to B, and equate to zero,

$$0 = 3x^2 - 20.2 - 0$$

$$\therefore \quad x^2 = \frac{20.2}{3} \text{ or } x = 2.595 \text{ m from } A$$

Substituting in the deflection equation,

$$EIy_{max} = 2.595^3 - 20.2 \times 2.595 - 0 \text{ kNm}^3$$

$$y_{max} = \frac{-34.944}{EI} = \frac{-34.944 \times 10^{12}}{2 \times 10^5 \times 6 \times 10^8} = -0.2912 \text{ mm (Downward)}$$

Slope:

For slope at A, $x = 0$

Substituting in the slope equation,

$$EI\left(\frac{dy}{dx}\right)_A = -20.2 \text{ kN-m}^2 = -20.2 \times 10^9 \text{ N-mm}^2$$

or $$\text{Slope at } A\left(\frac{dy}{dx}\right)_A = \frac{-20.2 \times 10^9}{EI} = \frac{-20.2 \times 10^9}{2 \times 10^5 \times 6 \times 10^8} = -1.683 \times 10^{-4} \text{ radians}$$

$$\left(\frac{dy}{dx}\right)_A = -1.683 \times 10^{-4} \times \frac{180}{\pi} = 9.645 \times 10^{-3} \text{ degrees}$$

For slope at B, $x = 3$ m

$$EI\left(\frac{dy}{dx}\right)_B = 3(3)^2 - 20.2 - 0 = 6.8 \text{ kN-m}^2 = 6.8 \times 10^9 \text{ N-mm}^2$$

$$\therefore \quad \text{Slope at } B \left(\frac{dy}{dx}\right)_B = \frac{6.8}{EI} \times 10^9 = \frac{6.8 \times 10^9}{2 \times 10^5 \times 6 \times 10^8} = 5.667 \times 10^{-5} \text{ radians}$$

$$\left(\frac{dy}{dx}\right)_B = 5.667 \times 10^{-5} \times \frac{180}{\pi} = 3.247 \times 10^{-3} \text{ degrees}$$

For slope at D, $x = 7$ m

$$EI\left(\frac{dy}{dx}\right)_D = 3(7)^2 - 20.2 - 9(7-3)^2 + 9(7-5)^2 - 0.5\,(7-5)^3 \text{ kNm}^2$$

$$\text{Slope at } D, \left(\frac{dy}{dx}\right)_D = \frac{14.8}{EI} = \frac{14.8 \times 10^9}{2 \times 10^5 \times 6 \times 10^8} = 1.233 \times 10^{-4} \text{ radians.}$$

$$\left(\frac{dy}{dx}\right)_D = 1.233 \times 10^{-4} \times \frac{180}{\pi} = 7.0665 \times 10^{-3} \text{ degrees.}$$

Problem 27 Figure 7.44 shows a overhanging beam carrying a UDL and a point load. Determine the deflection at point B, C and free and E and the maximum deflection. Take flexural rigidity as 2000 kN-m^2.

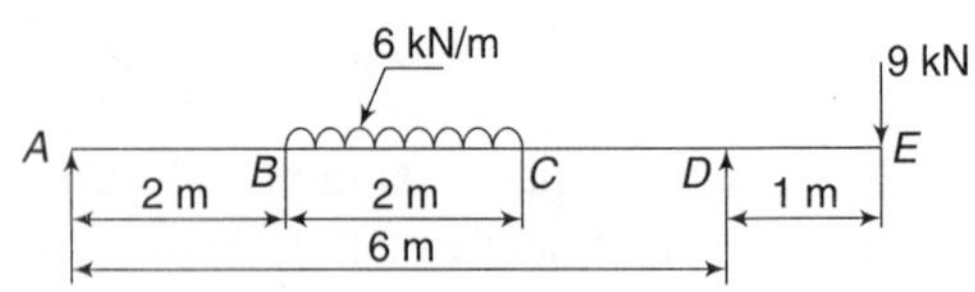

Fig. 7.44

Solution: Flexural rigidity $EI = 2000$ kN-m^2

$$= 2000 \times 10^9 \text{ N-mm}^2$$

$$\text{Reactions: } R_A + R_D = 6 \times 2 + 9 = 21 \text{ kN}$$

Taking moments about A

$$6 \times 2\left(2 + \frac{2}{2}\right) - R_D \times 6 + 9 \times 7 = 0$$

$$R_D = 16.5 \text{ kN}$$

$$R_A = 21 . R_D = 4.5 \text{ kN}$$

Since the UDL is not extending up to the right side and E, extend UDL up to E and apply upward UDL between C and E as shown in Fig. 7.45.

Consider section XX between DE at a distance x from A.

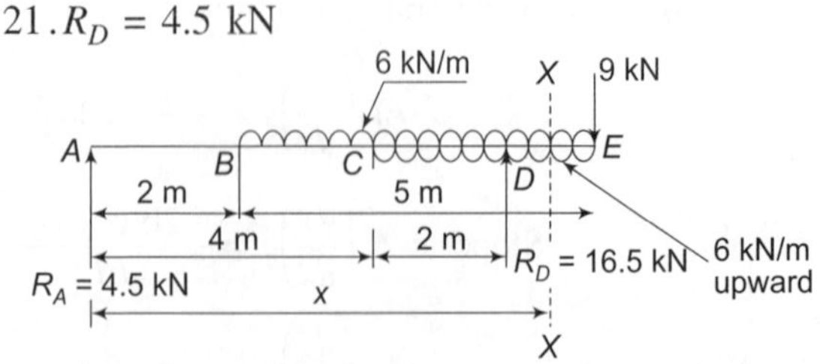

Fig. 7.45

Bending moment at XX,

$$M_x = R_A.x - 6(x-2)\frac{(x-2)}{2} + 6(x-4)\frac{(x-4)}{2} + R_D(x-6)$$

$$M_x = 4.5x - 3(x-2)^2 + 3(x-4)^2 + 16.5(x-6)$$

Substituting in the deflection equation.

$$EI\frac{d^2y}{dx^2} = M_x = 4.5x - 3(x-2)^2 + 3(x-4) + 16.5(x-6)$$

$$\text{Integrating } EI\frac{dy}{dx} = 4.5\frac{x^2}{2} + C_1 - \frac{3(x-2)^3}{3} + \frac{3(x-4)^3}{3} + \frac{16.5(x-6)^2}{2}$$

$$\text{Integrating again } EIy = 4.5\frac{x^3}{6} + C_1x + C_2 - \frac{(x-2)^4}{4} + \frac{(x-4)^4}{4} + \frac{16.5(x-6)^3}{6}$$

Applying boundary conditions,

At left support A, $x = 0$, deflection $y = 0$

$$0 = 0 + 0 + C_2$$

$$\therefore \quad C_2 = 0$$

At support D, $x = 6$, deflection $y = 0$

$$0 = 4.5 \times \frac{6^3}{6} + C_1 \times 6 - \frac{(6-2)^4}{4} + \frac{(6-4)^4}{4} + 0$$

$$\therefore \quad C_1 = -17$$

$$\therefore \quad EIy = 4.5\frac{x^3}{6} - 17x - \frac{(x-2)^4}{4} + \frac{(x-4)^4}{4} + 16.5\frac{(x-6)^3}{6}$$

(Deflection equation)

For deflection at B, $x = 2$ m,

$$EIy_B = 4.5 \times \frac{2^3}{6} - 17 \times 2 \text{ kN-m}^3$$

$$\therefore \quad y_B = \frac{-28}{EI} = \frac{-28 \times 10^{12}}{2000 \times 10^9} = -14 \text{ mm (Downward)}$$

For deflection at C, $x = 4$ m

$$EIy_C = 4.5 \times \frac{4^3}{6} - 17 \times 4 - \frac{(4-2)^4}{4} + 0 \text{ kN-m}^3$$

$$y_C = \frac{-24}{EI} = \frac{-24 \times 10^{12}}{2000 \times 10^9} = -12 \text{ mm (Downward)}$$

For deflection at E, $x = 7$ m,

$$EIy_E = 4.5 \times \frac{7^3}{6} - 17 \times 7 - \frac{(7-2)^4}{4} + \frac{(7-4)^4}{4} + 16.5\frac{(7-6)^3}{6}$$

$$y_E = \frac{4.95}{EI} = \frac{4.95 \times 10^{12}}{2000 \times 10^9} = +\ 2.475 \text{ mm (Upward)}$$

For maximum deflection, slope $\frac{dy}{dx} = 0$. Let us assume that the maximum deflection occurs in the portion BC.

$$0 = 4.5\frac{x^2}{2} + C_1 - \frac{3(x-2)^3}{3}$$

$$= 2.25\,x^2 - 17 - (x^3 - 8 + 12x - 6x^2)$$

$$0 = 2.25\,x^2 - 17 - x^3 + 8 - 12x + 6x^2$$

$$= -x^3 + 8.25\,x^2 - 12x - 9$$

$$x^3 - 8.25x^2 + 12x + 9 = 0$$

Solving $x = 6,\ -0.538$, and 2.788

Taking the feasible value, $x = 2.788$ m which lies between B and C

Substituting in deflection equation

$$EIy_{max} = 4.5 \times \frac{2.788^3}{6} - 17 \times 2.788 - \frac{(2.788-2)^4}{4} + 0 \text{ kN-m}^3$$

$$y_{max} = \frac{-31.239}{EI} = \frac{-31.239 \times 10^{12}}{2000 \times 10^9} = -15.62 \text{ mm (Downward)}$$

Problem 28 Determine the deflection at B and D for an overhanging beam shown in Fig. 7.46. Modulus of elasticity = 210 GPa and moment of inertia of the section may be taken as 1×10^8 mm^4.

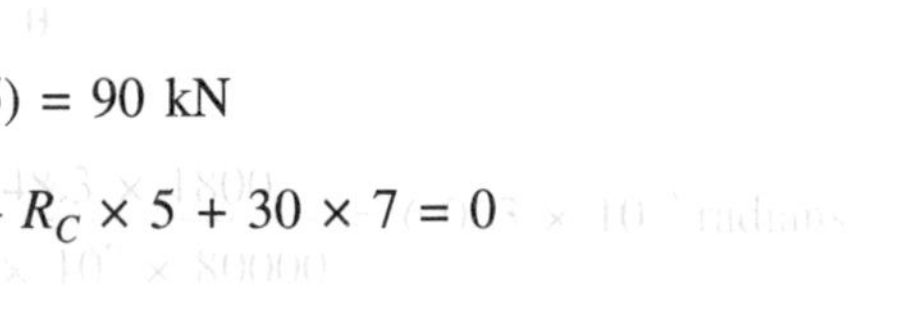

Fig. 7.46

Solution: $E = 210$ GPa $= 210 \times 10^3$ MPa

$I = 1 \times 10^8$ mm^4

To find reactions, $R_A + R_C = 30 + 30 + (6 \times 5) = 90$ kN

Taking moment about A, $30 \times 3 + (6 \times 5)\frac{5}{2} - R_C \times 5 + 30 \times 7 = 0$

$\therefore$

$$R_C = 75 \text{ kN}$$

$$R_A = 90 - R_C = 15 \text{ kN}$$

Since the UDL is not extending upto right end, as per Macaulay's method, extend the UDL from C to D and compensate by same UDL acting upward from C to D as shown in Fig. 7.47.

Consider a section XX at a distance x from A between C and D.

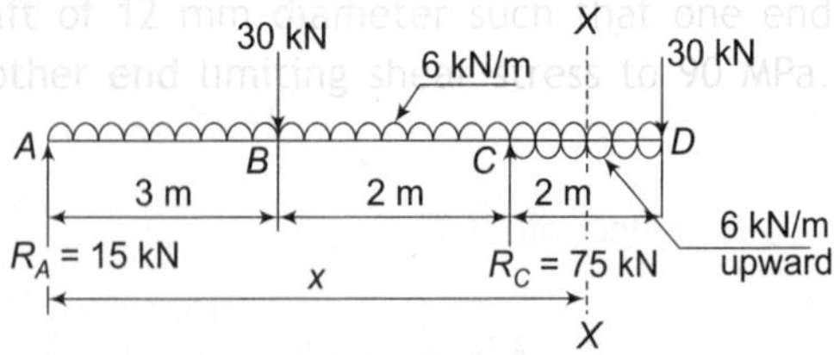

Fig. 7.47

Bending moment at X, $M_x = R_A.x - 30(x-3) + R_C(x-5) - \frac{6x^2}{2} + \frac{6(x-5)^2}{2}$

$$M_x = 15x - 30(x-3) + 75(x-5) - 3x^2 + 3(x-5)^2$$

Substituting in the deflection equation

$$EI = \frac{d^2y}{dx} = M_x = 15x - 30(x-3) + 75(x-5) - 3x^2 + 3(x-5)^2$$

$$\text{Integrating, } EI\frac{dy}{dx} = 15\frac{x^2}{2} + C_1 - \frac{30(x-3)^2}{2} + \frac{75(x-5)^2}{2} - \frac{3x^3}{3} + \frac{3(x-5)^3}{3}$$

$$\text{Integrating again } EIy = \frac{15x^3}{6} + C_1x + C_2 - \frac{30(x-3)^3}{6} + \frac{75(x-5)^3}{6} - \frac{x^4}{4} + \frac{(x-5)^4}{4}$$

$$EIy = 2.5x^3 + C_1x + C_2 - 5(x-3)^3 + 12.5(x-5)^3 - \frac{x^4}{4} + \frac{(x-5)^4}{4}$$

Applying boundary conditions

At the support A, $x = 0$, deflection $y = 0$

$$0 = 0 + 0 + C_2$$

$\therefore$ $C_2 = 0$

At the support C, $x = 5$ m, deflection $y = 0$

$$0 = 2.5 \times 5^3 + C_1 \times 5\ (5-3)^3 + 0$$

$\therefore$ $C_1 = -54.5$

$$EIy = 2.5x^3 - 54.5x - 5(x-3)^3 + 12.5(x-5)^3 - \frac{x^4}{4} + \frac{(x-5)^4}{4} \text{ (Deflection equation)}$$

$$EI\frac{dy}{dx} = 7.5x^2 + C_1 - 15(x-3)^2 + 37.5(x-5)^2 - x^3 + (x-5)^3 \text{ (Slope equation)}$$

For deflection at B, $x = 3$ m, substituting in the deflection equation,

$$EI\,y_B = 2.5 \times 3^3 - 54.5 \times 3 - 0 \text{ kN-m}^3$$

$$y_B = \frac{-96}{EI} = \frac{-96 \times 10^{12}}{210 \times 10^3 \times 1 \times 10^8} = -4.5714 \text{ mm (Downward)}$$

For deflection at the free end at D, $x = 7$,

$$EIy_D = 2.5 \times 7^3 - 54.5 \times 7 - 5(7-3)^3 + 12.5(7-5)^3 - \frac{7^4}{4} + \frac{(7-5)^4}{4} \text{ kN-m}^3$$

$$\therefore \quad y_D = \frac{-340.25}{EI} = \frac{-340.25 \times 10^{12}}{210 \times 10^3 \times 1 \times 10^8} = -16.202 \text{ mm (Downward)}$$

Problem 29 Determine the deflection at B and D for the beam shown in Fig. 7.48. Assume E = 210 GPa, $I = 1.6 \times 10^7$ mm^4.

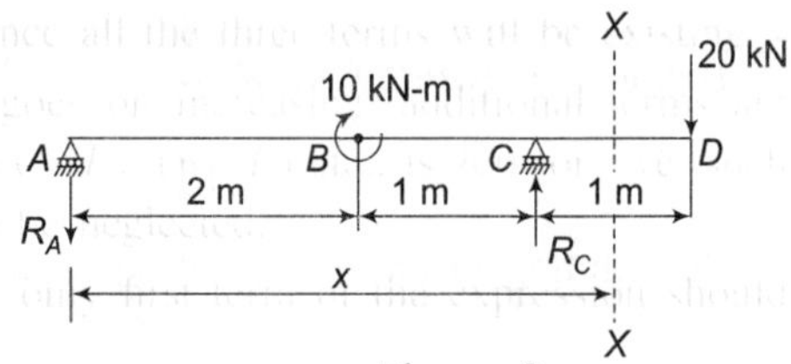

Fig. 7.48

Solution: $E = 210 \text{ GPa} = 210 \times 10^3$ MPa

$I = 1.6 \times 10^7$ mm^4.

Reactions: $R_A + R_C = 20$ kN

Taking moments about A, $10 + 20 \times 4 - R_C \times 3 = 0$

$R_C = 30$ kN

$\therefore \quad R_A = 20 - 30 = -10 \text{ kN} = 10$ kN (Downward)

Consider any section XX between C and D,

Moment of XX, $M_x = -R_A x + 10 + R_C (x-3) = 10 \times x + 10(x-2)^\circ + 30(x-3)$

Substituting in the deflection equation

$$EI \frac{d^2y}{dx^2} = M_x = -10x + 10(x-2)^\circ + 30(x-3)$$

$$\text{Integrating, } EI \frac{dy}{dx} = -10 \frac{x^2}{2} + C_1 + 10(x-2) + \frac{30(x-3)^2}{2}$$

$$\text{Integrating again, } EIy = -10 \frac{x^3}{6} + C_1 x + C_2 + \frac{10(x-2)^2}{2} + \frac{30(x-3)^3}{6}$$

Applying boundary conditions,

At support A, $x = 0$, deflection $y = 0$

$$0 = 0 + 0 + C_2$$

$$\therefore \quad C_2 = 0$$

At support C, $x = 3$m deflection $y = 0$

$$0 = \frac{-10}{6} \times 3^3 + C_1 \times 3 + \frac{10}{2}(3-2)^2 + 0$$

$$C_1 = 13.333$$

$$\therefore \quad EIy = \frac{-10}{6} x^3 - 13.333x + 5(x-2)^2 + 5(x-3)^3 \text{ kN-m}^3$$

For deflection at B, $x = 2$ m

$$EIy_B = \frac{-10}{6} \times 2^3 + 13.333 \times 2 + 0 \text{ kN-m}^3$$

$$y_B = \frac{13.333}{EI} = \frac{13.333 \times 10^{12}}{210 \times 10^3 \times 1.6 \times 10^7} = +3.97 \text{ mm (Upward)}$$

For deflection at D, $x = 4$ m,

$$EIy_D = \frac{-10}{6} \times 4^3 + 13.333 \times 4 + 5(4-2)^2 + 5(4-3)^3$$

$$y_D = \frac{-24.498}{EI} \text{ kN-m}^3$$

$$y_D = \frac{-24.498 \times 10^{12}}{210 \times 10^3 \times 1.6 \times 10^7} = -7.291 \text{ mm (downward).}$$

Problem 30 Determine the deflection under loads in the beam shown in Fig. 7.49. Take flexural rigidity as EI throughout.

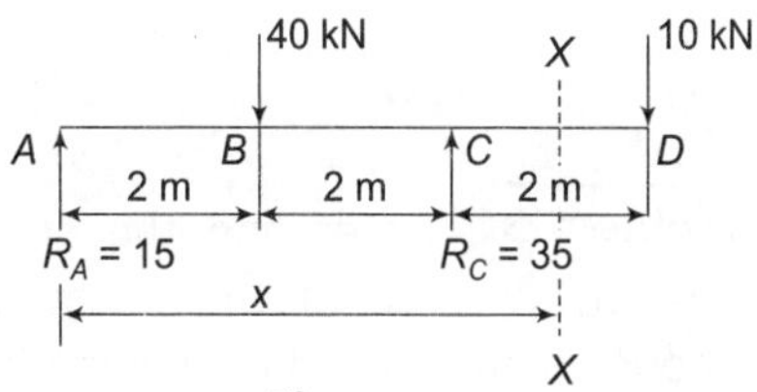

Fig. 7.49

Solution: Flexural rigidity = EI

Reactions: $R_A + R_C = 40 + 10 = 50$ kN

Taking moments about A, $40 \times 2 - R_C \times 4 + 10 \times 6 = 0$

$\therefore$ $R_C = 35$ kN

$R_A = 50 - 35 = 15$ kN

Consider a section between C and D at a distance x from A

Bending moment $M_x = R_A \cdot x - 40(x-2) + R_C(x-4)$

Substituting in the deflection equation,

$$EI\frac{d^2y}{dx^2} = 15x - 40(x-2) + 35(x-4)$$

Integrating, $$EI\frac{dy}{dx} = 15\frac{x^2}{2} + C_1 - \frac{40(x-2)^2}{2} + \frac{35(x-4)^2}{2}$$

Integrating again, $$EIy = 15\frac{x^3}{6} + C_1 x + C_2 - \frac{40(x-2)^3}{6} + \frac{35(x-4)^3}{6}$$

Applying boundary conditions

At support A, $x = 0$, deflection $y = 0$

$0 = 0 + 0 + C_2$

$\therefore$ $C_2 = 0$

At support C, $x = 4$, deflection $y = 0$

$$0 = \frac{15}{6}.4^3 + C_1 \times 4 - \frac{40}{6}(4-2)^3 + 0$$

$\therefore$ $C_1 = -26.667$

To find deflection at B, $x = 2$ m

$$EIy_B = 15 \times \frac{2^3}{6} - 26.667 \times 2 \text{ kN-m}^3$$

$$\therefore \quad y_B = \frac{-33.333}{EI}$$

$$y_B = \frac{-33.333 \times 10^{12}}{EI} \text{ mm}$$

(For E in N/mm^2 and I in mm^4)

For deflection at D, $x = 6$ m

$$EIy_D = 15 \times \frac{6^3}{6} - 26.667 \times 6 - \frac{40(6-2)^3}{6} + \frac{35(6-4)^3}{6}$$

$$\therefore \quad y_D = \frac{-2 \times 10^{-3}}{EI} \times 10^{12} \text{ mm.}$$

Problem 31 Determine the equation of the elastic curve of the prismatic beam supported and loaded as shown (Fig. 7.50). Also find the slope at the left end and deflection at the right end of the beam in terms of uniform flexural rigidity EI.

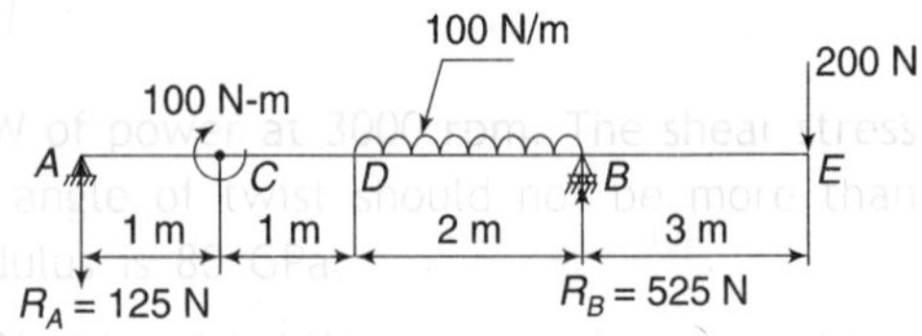

Fig. 7.50

Solution: Flexural rigidity $= EI$

To find reactions: $R_A + R_B = 100 \times 2 + 200 = 400$ N

Taking moment about A

$$100 + 100 \times 2\left(2 + \frac{2}{2}\right) - R_B \times 4 + 200 \times 7 = 0$$

$$R_A = 525 \text{ N}$$

$$\therefore \quad R_A = 400 - R_B = -125 \text{ N (Downward).}$$

For using Macaulay's method, since the UDL is not touching right side and E, extend the UDL up to E and apply same UDL from BE acting upward. As shown in Fig. 7.51.

Consider a section XX at a distance x from A between B and E. Bending moment at this section,

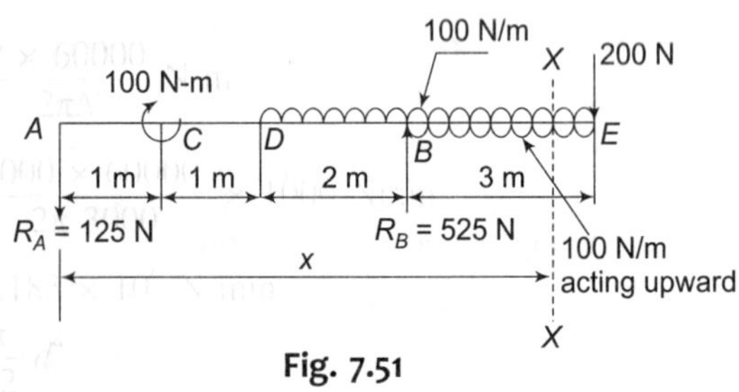

Fig. 7.51

$$M_x = -125\,x + 100 - 100\,\frac{(x-2)^2}{2} + 100\,\frac{(x-4)^2}{2} + 525(x-4)$$

Equating this to deflection equation.

$$EI\frac{d^2y}{dx^2} = -125\,x + 100 - 100\,\frac{(x-2)^2}{2} + 100\,\frac{(x-4)^2}{2} + 525(x-4)$$

For the moment, take the distance from X as $(x-2)^\circ$

$$EI\frac{d^2y}{dx^2} = -125\,x + 100\,(x-1)^0 - 100\,\frac{(x-2)^2}{2} + 100\,\frac{(x-4)^2}{2} + 525\,(x-4)$$

Integrating once

$$EI\frac{dy}{dx} = -\,125\,\frac{x^2}{2} + C_1 + 100(x-1) - 100\,\frac{(x-2)^3}{6} + \frac{100(x-4)^3}{6} + \frac{525(x-4)^2}{2}$$

Integrating again

$$EIy = -\,125\,\frac{x^3}{6} + C_1\,x + C_2 + \frac{100(x-1)^2}{2} - \frac{100(x-2)^4}{24} + \frac{100\,(x-4)^4}{24} + \frac{525(x-4)^3}{6}$$

Applying boundary conditions,

At A, at $x = 0$, deflection $y = 0$

$$0 = 0 + 0 + C_2$$

$\therefore$ $C_2 = 0$

At B, $x = 4$, deflection $y = 0$

$$0 = -\,125 \times \frac{4^3}{6} + C_1 \times 4 + \frac{100}{2}(4-1)^3 - \frac{100(4-2)^4}{24} + 0$$

$\therefore$ $C_1 = 12.5$

$\therefore$ The slope equation,

$$EI\frac{dy}{dx} = -\,125\frac{x^2}{2} + 12.5 + 100\,(x-1) - 100\,\frac{(x-2)^3}{6} + \frac{100(x-4)^3}{6} + \frac{525(x-4)^2}{2}$$

and the deflection equation or equation for elastic curve,

$$EIy = 125\,\frac{x^3}{6} + 12.5x + \frac{100(x-1)^2}{2} - \frac{100(x-2)^4}{24} + \frac{100(x-4)^4}{24} + \frac{525(x-4)^3}{6}$$

To find slope at end A, $x = 0$

$$EI\left(\frac{dy}{dx}\right)_A = 12.5 \text{ N-m}^2$$

or

$$\left(\frac{dy}{dx}\right)_A = \frac{-12.5}{EI}$$

To find deflection at and D, $x = 7$ m

$$EIy_D = -\,125 \times \frac{7^3}{6} + 12.5 \times 7 + \frac{100(7-1)^2}{2} - \frac{100(7-2)^4}{24} + \frac{100(7-4)^4}{24} + \frac{525(7-4)^3}{6}$$

$$EIy_D = 5162.5 \text{ Nm}^3$$

$$\therefore \quad y_D = \frac{-5162.5}{EI} \times 10^9 \text{ mm} \quad (E \text{ in N/mm}^2, I \text{ in mm}^4)$$

Problem 32 For the overhanging beam *ABC* shown in Fig. 7.52, find the slopes at *A* and *C* and the deflection at *A*. Take *EI* as constant. Adopt Macaulay's method.

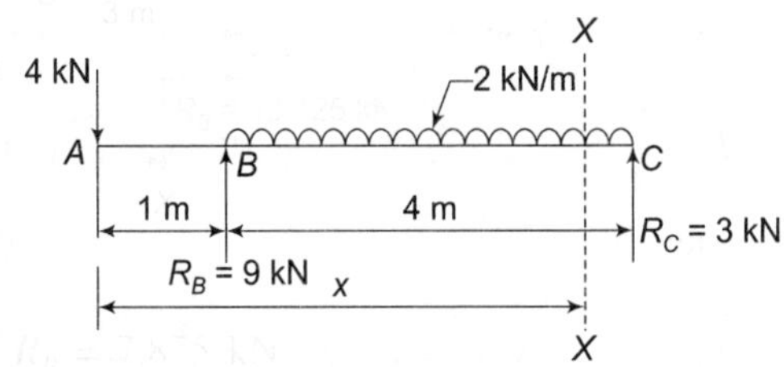

Fig. 7.52

Solution: Flexural rigidity = *EI*

To find reactions,

$$R_B + R_C = 4 + 2 \times 4 = 12 \text{ kN}$$

Taking moment about *B*, $-4 \times 1 + (2 \times 4)\frac{4}{2} - R_C \times 4 = 0$

$$\therefore \quad R_C = 3 \text{ kN}$$

$$R_B = 12 - 3 = 9 \text{ kN}$$

Consider a section *XX* at a distance *x* from left end *A*

Bending moment at this section $M_x = -4x + R_B(x-1) - \frac{2(x-1)^2}{2}$

$$= -4x + 9(x-1) - (x-1)^2$$

Substituting in the deflection equation

$$EI\frac{d^2y}{dx^2} = M_x - 4x + 9(x-1) - (x-1)^2$$

Integrating once, $EI\frac{dy}{dx} = \frac{-4x^2}{2} + C_1 + \frac{9(x-1)^2}{2} - \frac{(x-1)^3}{3}$

Integrating again, $EIy = \frac{-4x^3}{6} + C_1x + C_2 + \frac{9(x-1)^3}{6} - \frac{(x-1)^4}{12}$

Applying boundary conditions

At *B*, *x* = 1m, deflection *y* = 0

$$0 = \frac{-4}{6} + C_1 + C_2$$

$$\therefore \quad C_1 + C_2 = \frac{4}{6} = 0.667 \qquad \text{...(i)}$$

At *C*, *x* = 5 m, deflection *y* = 0

$$0 = \frac{-4}{6} \times 5^3 + C_1 \times 5 + C_2 + \frac{9(5-1)^3}{6} - \frac{(5-1)^4}{12}$$

$$5C_1 + C_2 = 8.667 \qquad \text{...(ii)}$$

$$C_1 + C_2 = 0.667$$

$$\therefore \quad 4C_1 = 8 \text{ or } C_1 = 2$$

$\therefore \qquad C_2 = 0.667 - C_1 = 0.667 - 2 = -1.333$

The slope equation, $EI \dfrac{dy}{dx} = -2x^2 + 2 + 4.5(x - 1)^2 - \dfrac{(x-1)^3}{3}$

The deflection equation,

$$EIy = -\frac{2}{3}x^3 + 2x - 1.333 + 1.5(x-1)^3 - \frac{(x-1)^4}{12}$$

The deflection at free end A. At $x = 0$

$$EIy_A = -1.333 \text{ kNm}^3$$

$$y_A = \frac{-1.333}{EI}$$

Slope at A, $x = 0$

$$EI\left(\frac{dy}{dx}\right)_A = -2 \times 0 + 2 + 0 \text{ kN-m}^2$$

$$\therefore \qquad \left(\frac{dy}{dx}\right)_A = \frac{2}{EI}$$

Slope at C, $x = 5$ m

$$EI\left(\frac{dy}{dx}\right)_C = -2 \times 5^2 + 2 + 4.5(5-1)^2 - \frac{(5-1)^3}{3}$$

$$\left(\frac{dy}{dx}\right)_C = \frac{2.667}{EI}$$

For maximum deflection the slope $\dfrac{dy}{dx} = 0$

Assuming the maximum deflection occurs between B and C,

$$0 = -2x^2 + 2 + 4.5(x-1)^2 - \frac{(x-1)^3}{3}$$

$$= -2x^2 + 2 + 4.5(x^2 + 1 - 2x) - \frac{(x^3 - 1 + 3x - 3x^2)}{3}$$

$$0 = \frac{-6x^2 + 6 + 13.5(x^2 + 1 - 2x) - x^3 + 1 - 3x + 3x^2}{3}$$

$$0 = -6x^2 + 6 + 13.5x^2 + 13.5 - 27x - x^3 + 1 - 3x + 3x^2$$

$$-x^3 + 10.5x^2 - 30x + 20.5 = 0$$

Solving, $x = 6.186$, 1 and 3.314 m

Considering the feasible value, $x = 3.314$ that lies between B and C.

$$EIy_{max} = \frac{-2}{3} \times 3.314^3 + 2 \times 3.314 - 1.333 + 1.5(3.314 - 1)^3 - \frac{(3.314 - 1)^4}{12}$$

$$EIy_{max} = -4.69265 \text{ kNm}^3$$

$$\therefore \quad y_{max} = \frac{-4.69265}{EI}.$$

Problem 33 A horizontal beam AB is simply supported at A and B which are 10 m apart. It carries a UDL of 20 kN/m. A couple of 100 kN-m clockwise is also applied at C which is at a distance of 4 m from A. Determine the deflection at C using Macaulay's method. Take EI = 15 × 10⁴ kN-m².

Solution: $$EI = 15 \times 10^4 \text{ kN-m}^2$$

The given beam with the loads is shown in Fig. 7.53.

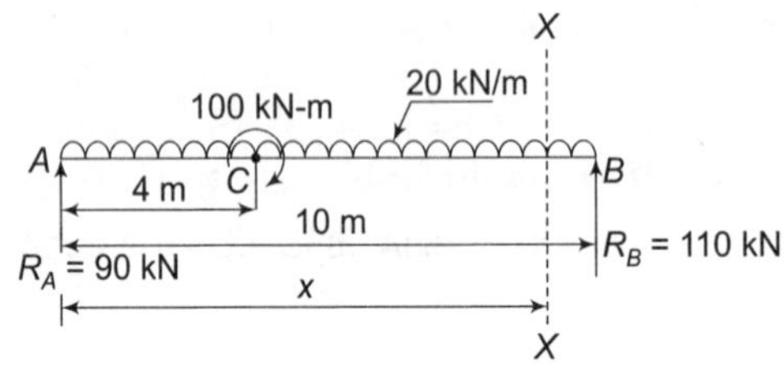

Fig. 7.53

Reactions: $$R_A + R_B = 20 \times 10 = 200 \text{ kN}$$

Taking moments about A, $100 + (200 \times 10) \times \dfrac{10}{2} - R_B \times 10 = 0$

$$\therefore \quad R_B = 110 \text{ kN}$$

$$\therefore \quad R_A = 200 - R_B = 90 \text{ kN}$$

Consider a section XX at a distance x from A

Bending moment at section XX,

$$M_x = 90x + 100 - 20.\frac{x^2}{2}$$

$$= 90x + 100\,(x-4)^0 - 10x^2$$

Substituting in the deflection equation,

$$EI\frac{d^2y}{dx^2} = 90x + 100\,(x-4)^0 - 10x^2$$

$$\text{Integrating, } EI\frac{dy}{dx} = 90\frac{x^2}{2} + C_1 + 100\,(x-4) - 10\frac{x^3}{3}$$

$$\text{Integrating again, } EIy = 90\cdot\frac{x^3}{6} + C_1x + C_2 + \frac{100\,(x-4)^2}{2} - \frac{10\,x^4}{12}$$

Applying boundary conditions, at A, at $x = 0$, deflection $y = 0$

$$0 = 0 + 0 + C_2$$

$$\therefore \quad C_2 = 0$$

At B, at $x = 10$, deflection $y = 0$

$$0 = 90 \times \frac{10^3}{6} + C_1 \times 10 + \frac{100(10-4)^2}{2} - \frac{10 \times 10^4}{12}$$

$$C_1 = -846.667$$

∴ The deflection equation is

$$EIy = \frac{90}{6}x^3 - 846.667\,x + \frac{100\,(x-4)^2}{2} - \frac{10 \cdot x^4}{12}$$

For deflection at C, $x = 4$

$$EIy_C = \frac{90}{6} \times 4^3 - 846.667 \times 4 + 0 \text{ kN-m}^3$$

∴ $$y_C = \frac{-2426.67}{EI} = -\frac{-2426.67}{15 \times 10^4} \text{ m}$$

$$y_C = -0.01618 \text{ m} = -16.18 \text{ mm.}$$

Problem 34 A prismatic cantilever 2 m long beam, carries a uniformly distributed load of 10 kN/m over the entire span. What is the value of point load *W* that is to be applied upward at the free end for the deflection to be zero at the free end?

Solution: Let EI = Flexural rigidity of the beam

UDL, w = 10 kN/m

Length l = 2m

For cantilever with UDL over entire span deflection at the free end $y_B = \frac{wl^4}{8EI} = \frac{10 \times 2^4}{8EI}$

$$y_B = \frac{20}{EI}$$

For cantilever with point load at the free end, The deflection at the free end

$$y_B = \frac{Wl^3}{3EI} = \frac{W \times 2^3}{3EI}$$

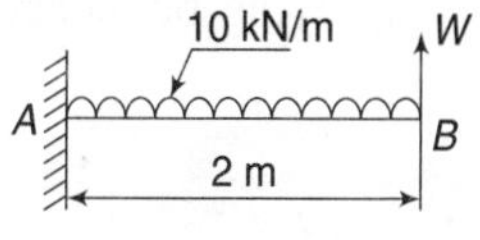

Fig. 7.54

Equating these two, for the deflection to be zero.

$$\frac{20}{EI} = \frac{W \times 2^3}{3EI}$$

∴ $$\text{Point load } W = \frac{20 \times 3}{2^3} = 7.5 \text{ kN.}$$

Problem 35 A beam *ABC* of length *l* has one support at left end *A* and another at *B* at a distance *a* from right and *C*. A point load *W* is applied at the over hanging point *C*. Find the slopes at the supports *A* and *B* and the point *C*. Also find the deflection under load at *C* and maximum deflection between the supports *A* and *B*.

Solution: E = Modulus of elasticity

I = Moment of inertia of the section

W = Point load at the free end

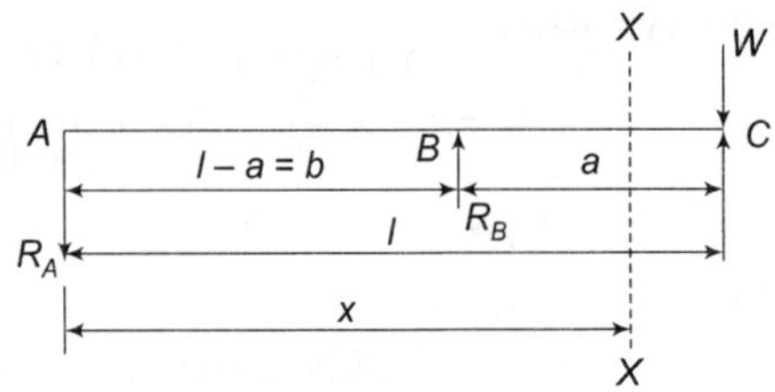

Fig. 7.55

Let $l - a = b$

$$\text{Reactions: } R_A = \frac{W \cdot a}{b} \text{ and } R_B = \frac{Wl}{b}$$

Consider a section between B and C at a distance x from the left support A.

Bending moment at XX, $M_x - R_A.x + R_B\,(x - b)$

$$M_x = \frac{-W \cdot a}{b}\,x + \frac{Wl}{b}\,(x - b)$$

Substituting in the deflection equation

$$EI\,\frac{d^2y}{dx^2} = -\frac{Wa}{b}\,x + \frac{Wl}{b}\,(x - b)$$

$$\text{Integrating, } EI\,\frac{dy}{dx} = -\frac{Wa}{b}\cdot\frac{x^2}{2} + C_1 + \frac{Wl}{b}\,\frac{(x-b)^2}{2}$$

Integrating again

$$EIy = \frac{-Wa}{b}\,\frac{x^3}{6} + C_1x + C_2 + \frac{Wl}{b}\,\frac{(x-b)^3}{6}$$

Applying boundary conditions,

$$\text{At } A\ x = 0, \text{ deflection } y = 0$$

$$0 = 0 + 0 + C_2$$

$\therefore$

$$C_2 = 0$$

$$\text{At } B,\ x = b, \text{ deflection } y = 0$$

$$0 = -\frac{Wa}{b}\cdot\frac{b^3}{6} + C_1b$$

$$0 = -\frac{Wab^2}{6} + C_1b$$

$\therefore$

$$C_1 = \frac{Wab}{6}$$

For slope at support A, $x = 0$, in the slope equation

$$EI\left(\frac{dy}{dx}\right)_A = -\frac{Wa}{b} \times 0 + \frac{Wab}{6} + 0$$

$$\therefore \qquad \left(\frac{dy}{dx}\right)_A = \frac{Wab}{6EI}$$

For slope at support B, $x = b$,

$$EI\left(\frac{dy}{dx}\right)_B = \frac{-Wa}{b}\cdot\frac{b^2}{2} + \frac{Wab}{6} + 0$$

$$= \frac{-Wab}{2} + \frac{Wab}{6} = \frac{-3Wab + Wab}{6} = \frac{-Wab}{3}$$

$$\therefore \qquad \left(\frac{dy}{dx}\right)_B = \frac{-Wab}{3EI}$$

For slope at end C, $x = l$

$$EI\left(\frac{dy}{dx}\right)_C = \frac{-Wa}{b}\frac{l^2}{2} + \frac{Wab}{6} + \frac{Wl}{b}\frac{(l-b)^2}{2}$$

$$= \frac{-Wal^2}{2b} + \frac{Wab}{6} + \frac{Wl}{b}\frac{a^2}{2} \quad (\because l - b = a)$$

$$EI\left(\frac{dy}{dx}\right)_C = \frac{-3Wal^2 + Wab^2 + 3Wla^2}{6b}$$

$$= \frac{Wa(-3l^2 + b^2 + 3al)}{6b}$$

$$= \frac{Wa[-3(a+b)^2 + b^2 + 3a(a+b)]}{6b}$$

$$= \frac{Wa\left[-3(a^2 + b^2 + 2ab) + b^2 + 3a^2 + 3ab\right]}{6b}$$

$$= \frac{Wa\left[-3a^2 - 3b^2 - 6ab + b^2 + 3a^2 + 3ab\right]}{6b}$$

$$= \frac{Wa(-3ab - 2b^2)}{6b} = \frac{-Wab(3a + 2b)}{6b}$$

$$\therefore \qquad \left(\frac{dy}{dx}\right)_C = \frac{-Wa(3a + 2b)}{6}$$

Deflection under the load C

Put $x = l$ in the deflection equation,

$$EIy_C = \frac{-Wa}{b}\frac{l^3}{6} + \frac{Wab}{6}.l + \frac{Wl}{b}\frac{(l-b)^3}{6}$$

$$= \frac{-Wal^3}{6b} + \frac{Wabl}{6} + \frac{Wl}{6b}(a)^3 \quad (\because l - b = a)$$

$$= \frac{-Wal}{6b}[l^2 - b^2 - a^2]$$

$$= \frac{-Wal}{6b}[(a+b)^2 - b^2 - a^2]$$

$$= \frac{-Wal}{6b}[a^2 + b^2 + 2ab - b^2 - a^2]$$

$$EIy_C = \frac{-Wal}{6b}[2ab] = \frac{-Wa^2l}{3}$$

$$\therefore \quad y_C = \frac{-Wa^2l}{3EI}$$

For maximum deflection between supports A and B, the slope $\frac{dy}{dx} = 0$

$$0 = \frac{-Wa}{b}\frac{x^2}{2} + \frac{Wab}{6}$$

$$\frac{Wa}{b}\frac{x^2}{2} = \frac{Wab}{6}$$

$$x = \sqrt{\frac{b^2}{3}} = \frac{b}{\sqrt{3}}$$

Substituting in the deflection equation

$$EIy_{max} = -\frac{Wa}{6b}\left(\frac{b}{\sqrt{3}}\right)^3 + \frac{Wab}{6}\left(\frac{b}{\sqrt{3}}\right) + 0$$

$$EIy_{max} = -\frac{Wa}{6b}\frac{b^3}{3\sqrt{3}} + \frac{Wab^2}{6\sqrt{3}}$$

$$= +\frac{Wab^2}{6\sqrt{3}}\left(-\frac{1}{3} + 1\right) = \frac{Wab^2}{9\sqrt{3}}$$

$$\therefore \quad y_{max} = \frac{Wab^2}{9\sqrt{3}EI} \text{ (Upward).}$$

Problem 36 A beam *ABCD* is 6 m long and is overhanging on both sides, supports *B* and *C* are at a distance of 1.5 m from each end. A load of 30 kN acts at free end *A* and 15 kN at the free end *D*. Calculate the deflection at ends *A* and *D* and at the mid-span of *BC*. Take E = 200 kN/mm^2 and $I = 9 \times 10^7$ mm^4.

Solution:

$$E = 200 \times 10^3 \text{ N/mm}^2$$

$$I = 9 \times 10^7 \text{ mm}^4$$

The given beam with loads is shown in Fig. 7.55.

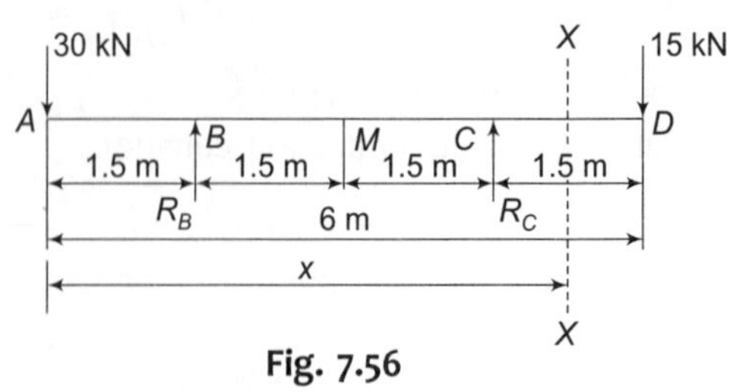

Fig. 7.56

Reactions: $R_B + R_C = 30 + 15 = 45$ kN

Taking moments about *B*.

$$-30 \times 1.5 - R_C \times 3 + 15 \times 4.5 = 0$$

$$R_C = 7.5 \text{ kN}$$

$$\therefore \quad R_B = 45 - 7.5 = 37.5 \text{ kN}$$

Consider a section XX at a distance x from A between C and D.

Bending moment at section XX, $M_x = -30x + R_B(x - 1.5) + R_C(x - 4.5)$

$$M_x = -30x + 37.5(x - 1.5) + 7.5(x - 4.5)$$

Substituting in deflection equation

$$EI\frac{d^2y}{dx^2} = -30x + 37.5(x - 1.5) + 7.5(x - 4.5)$$

$$\text{Integrating, } EI\frac{dy}{dx} = \frac{-30x^2}{2} + C_1 + 37.5\frac{(x - 1.5)^2}{2} + 7.5\frac{(x - 4.5)^2}{2}$$

$$\text{Integrating again, } EIy = \frac{-30}{6}x^3 + C_1x + C_2 + \frac{37.5(x - 1.5)^3}{6} + \frac{7.5(x-4.5)^3}{6}$$

At the support B, at $x = 1.5$, deflection $y = 0$

$$0 = \frac{-30}{6}(1.5)^3 + C_1 \times 1.5 + C_2$$

$$1.5\,C_1 + C_2 = 16.875 \qquad \text{...(i)}$$

At the support C, at $x = 4.5$, deflection $y = 0$

$$0 = \frac{-30}{6}(4.5)^3 + C_1 \times 4.5 + C_2 + \frac{37.5}{6}(4.5 - 1.5)^3 + 0$$

$$4.5\,C_1 + C_2 = 288.875 \qquad \text{...(ii)}$$

Subtracting Eqs. (i) and (ii)

$$3C_1 = 288.875 - 16.875$$

$$\therefore \quad C_1 = 90$$

$$\text{and } C_2 = -118.125$$

$$\therefore \text{ The deflection equation, } EIy = -\frac{30}{6}x^3 + 90x - 118.125 + \frac{37.5(x - 1.5)^3}{6} + \frac{7.5(x - 4.5)^3}{6}$$

At the free end A, $x = 0$

$$EIy_A = -118.125 \text{ kN-m}^3$$

$$\text{Deflection } y_A = \frac{-118.125}{EI} = \frac{-118.125 \times 10^{12}}{200 \times 10^3 \times 9 \times 10^7} = -6.5625 \text{ mm (Downward)}$$

At the right end at D, $x = 6$ m

$$EIy_D = -\frac{30}{6} \times 6^3 + 90 \times 6 - 118.125 + \frac{37.5}{6}(6 - 1.5)^3 + \frac{7.5}{6}(6 - 4.5)^3$$

$$EIy_D = -84.375 \text{ kN-m}^3$$

$$\therefore \quad y_D = -\frac{84.375}{EI} = -\frac{84.375 \times 10^{12}}{200 \times 10^3 \times 9 \times 10^7} = -4.6875 \text{ mm (Downward)}$$

At the mid span, at M, $x = 3$ m

$$EIy_M = \frac{-30}{6} \times 3^3 + 90 \times 3 - 118.125 + \frac{37.5}{6}(3 - 1.5)^3 + 0$$

$$EIy_M = 37.96875 \text{ kN-m}^3$$

$$\therefore \quad y_M = \frac{37.96875}{EI} = \frac{37.96875 \times 10^{12}}{200 \times 10^3 \times 9 \times 10^7} = 2.109 \text{ mm (Upward)}$$

Problem 37 Determine the deflection at points A, C and E for an overhanging beam loaded as shown in Fig. 7.57. Take flexural rigidity $EI = 15 \times 10^{12}$.

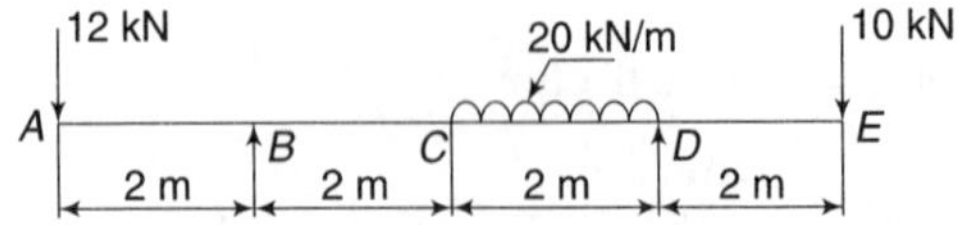

Fig. 7.57

Solution: Since the UDL is not extending up to the right end of the beam. For considering moments at XX between D and E as per Macaulay's method. Extend the UDL up to end E and compensate by upward UDL of same intensity between D and E as shown in Fig. 7.58.

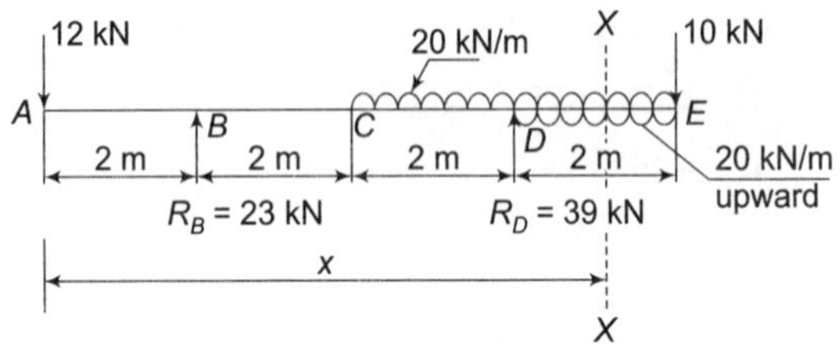

Fig. 7.58

$$EI = 15 \times 10^{12} \text{ N-mm}^2$$

Reactions: $R_B + R_D = 12 + 20 \times 2 + 10 = 62$ kN

Taking moments about B, $-12 \times 2 + 20 \times 2 \times (2 + 1) - R_D \times 4 + 10 \times 6 = 0$

$$\therefore \quad R_D = 39 \text{ kN}$$

$$\therefore \quad R_B = 62 - 39 = 23 \text{ kN}$$

Consider a section XX at a distance x from end A, between D and E.

Bending moment at section XX,

$$M_x = -12.x + 23(x-2) - \frac{20(x-4)^2}{2} + \frac{20(x-6)^2}{2}$$

Substituting in the deflection equation,

$$EI\frac{d^2y}{dx^2} = -12x + 23(x-2) - 10(x-4)^2 + 10(x-6)^2$$

Integrating, $EI\dfrac{dy}{dx} = -12\dfrac{x^2}{2} + C_1 + \dfrac{23(x-2)^2}{2} - \dfrac{10(x-4)^3}{3} + \dfrac{10(x-6)^3}{3}$

Integrating, $EIy = -12\dfrac{x^3}{6} + C_1x + C_2 + \dfrac{23(x-2)^3}{6} - \dfrac{10(x-4)^4}{12} + \dfrac{10(x-6)^4}{12}$

At support B, $x = 2$ m deflection $y = 0$

$$0 = -12 \times \frac{2^3}{6} + C_1 \times 2 + C_2$$

$$2C_1 + C_2 = 16 \quad \text{...(i)}$$

At support D, $x = 6$ m, deflection $y = 0$

$$0 = -12 \times \frac{6^3}{6} + C_1 \times 6 + C_2 + \frac{23(6-2)^3}{6} - \frac{10(6-4)^4}{12} + 0$$

$$6C_1 + C_2 = 200 \quad \text{...(ii)}$$

Subtracting Eq. (i) from Eq. (ii)

$$4C_1 = 200 - 16 = 184$$

$$C_1 = 46$$

$\therefore$ $C_2 = 200 - 6C_1 = 200 - 6 \times 46 = -76$

$\therefore$ The deflection equation is

$$EIy = -12\frac{x^3}{6} + 46x - 76 + \frac{23}{6}(x-2)^3 - \frac{10}{12}(x-4)^4 + \frac{10}{12}(x-6)^4$$

For the deflection at free end A, $x = 0$

$$EIy_A = -76 \text{ kN-m}^3$$

$\therefore$ $y_A = \dfrac{-76}{EI} = \dfrac{-76 \times 10^{12}}{15 \times 10^{12}} = -5.067$ mm (Downward)

For deflection at C, $x = 4$ m

$$EIy_C = -12 \times \frac{4^3}{6} + 46 \times 4 - 76 + \frac{23}{6}(4-2)^3 + 0 \text{ kN-m}^3$$

$$y_C = \frac{74.667}{EI} = \frac{74.667 \times 10^{12}}{15 \times 10^{12}} = 4.978 \text{ mm (Upward)}$$

For deflection at right and E, $x = 8$ m

$$EIy_E = -12 \times \frac{8^3}{6} + 46 \times 8 - 76 + \frac{23}{6}(8-2)^3 - \frac{10}{12}(8-4)^4 + \frac{10}{12}(8-6)^4 \text{ kN-m}^3$$

$$\therefore \quad y_E = \frac{-104}{EI} = \frac{-104 \times 10^{12}}{15 \times 10^{12}} = -6.933 \text{ mm (Downward).}$$

THINGS TO REMEMBER

- The differential equation for deflection, $EI\dfrac{d^2y}{dx^2} = M$

Cantilevers:

- EI is called flexural rigidity
- The maximum deflection at free end of a cantilever of length l subjected to point load W at the free end, $y = \dfrac{Wl^3}{3EI}$
- The deflection of a cantilever of length l under the point load W acting at a distance a from fixed end, $y = \dfrac{Wa^3}{3EI}$
- The maximum slope of a cantilever of length l subjected to a point load W at the free end, $\theta = \dfrac{Wl^2}{2EI}$
- The maximum slope of a cantilever of length l subjected to a point load W at a distance a from the fixed end, $\theta = \dfrac{Wa^2}{2EI}$
- The maximum deflection at the free end of a cantilever of length l subjected to a point load W at a distance a from the fixed end, $y = \dfrac{Wa^3}{3EI} + \dfrac{Wa^2}{2EI}(l - a)$
- The maximum deflection at the free end of a cantilever of length l subjected to a uniformly distributed load w over entire span, $y = \dfrac{wl^4}{8EI}$
- The slope at the free end of a cantilever of length l subjected to a uniformly distributed load w over entire span, $\theta = \dfrac{wl^3}{6EI}$
- The maximum deflection in a cantilever of length l subjected to a moment M at the free end, $y = \dfrac{Ml^2}{2EI}$

- The slope in a cantilever of length l subjected to a moment M at the free end $\theta = \dfrac{Ml}{EI}$
- The maximum deflection at the free end of a cantilever of length l carrying a uniformly varying load, varying from zero at free end to w/unit length at fixed end, $y = \dfrac{wl^4}{30EI}$
- The slope at the free end of a cantilever of length l carrying a uniformly varying load, varying from zero and free end to w/unit length at fixed end, $\theta = \dfrac{wl^3}{24EI}$.
- The maximum deflection at free end of a cantilever of length l carrying a uniformly varying load, varying from zero at fixed end to w/unit length at free and $y = \dfrac{11wl^4}{120EI}$.

Simply Supported Beams

- The maximum deflection at the middle of a simply supported beam of span l subjected to a central load of W, $y = \dfrac{Wl^3}{48EI}$
- The maximum slope at the supports for a simply supported beam of span l subjected to a central load of W,

$$\theta = \frac{Wl^2}{16EI}$$

- The maximum deflection at the middle of a simply supported beam of span l subjected to a uniformly distributed load of w/unit length over entire span, $y = \dfrac{5wl^4}{384EI}$
- The maximum slope at the supports of a simply supported beam of span l subjected to a uniformly distributed of w/unit length over entire span, $\theta = \dfrac{wl^3}{24EI}$
- Maximum deflection of a simply supported beam subjected to a uniformly varying load that varies from O at one end to w/unit length at the other end, $y = 0.00652\dfrac{wl^4}{EI}$ at $x = 0.5193l$ from the end of zero intensity load.
- The deflection at the mid-span of a simply supported beam subjected to a uniformly varying load, that varies from O at one to w/unit length at the other end,

$$y = \frac{5}{768}\frac{wl^4}{EI} \text{ (i.e., half of } y_{max} \text{ for same UDL)}$$

Macaulay's Method of Double Integration

- The integration of terms like $(x - a)$ is to be $\frac{(x-a)^2}{2}$, $(x - a)^2$ is to be $\frac{(x-a)^3}{3}$, etc.
- When writing bending moment equation, for a couple or moment M_0 at a distance a from left end, the equation is written as $M_0 (x - a)^0$ instead of simply M_0, which will be required for integration.
- If the uniformly distributed load is not extending up to the right end, UDL is to considered full up to right end and is to be cancelled by an upward UDL over the extended length.
- The constants of integration C_1 and C_2 are written in the beginning or soon after first term since they are common for all sections.

Exercises

1. A beam of length 4 m is simply supported at the ends and carries two concentrated loads 20 kN and 30 kN at a distance of 1.5 m from each end. Find the deflection under loads and at mid span. Take E = 200 GPa and $I = 3 \times 10^8$ mm^4.
2. Find the maximum deflection and maximum slope for a beam simply supported at ends over a span of 6 m. It carries a point load of 60 kN at 2 m from end and UDL of 30 kN/m over a span of 2 m from the right end. Take flexural rigidity $EI = 3 \times 10^{10}$ kN-mm^2.
3. Find the deflection under the point loads and maximum deflection for a simply supported beam 10 m long carrying two point loads 50 kN and 40 kN at distance 2 m and 6 m from the left end and a UDL of 10 kN/m over the 4 m span between the two point loads. Take $E = 2 \times 10^5$ N/mm^2 and $I = 12 \times 10^{10}$ mm^4.
4. A simply supported beam of span 6 m is subjected to a concentrated load of 18 kN at 4 m from left support. Determine the magnitude and position of maximum deflection. Take E = 200 GPa and $I = 15 \times 10^6$ mm^4.
5. Determine the deflection at B, C and E for an overhanging beam shown in Fig. 7.59. Take $EI = 4 \times 10^{12}$ Nmm^2.

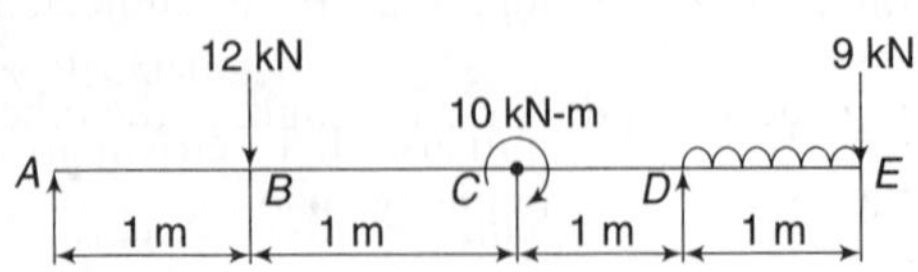

Fig. 7.59

6. A beam $ABCD$ 6 m long rests on two supports B and C which are at 3 m apart. Both overhangs on either side are of equal length of each 1.5 m. It carries a UDL of 30 kN/m over the entire length. Find the deflection at free ends and maximum deflection. E = 210 GPa and $I = 16 \times 10^9$ mm^4.

7. Find the maximum slope at the fixed end and maximum deflection for a cantilever beam loaded as shown in Fig. 7.60. Take $EI = 50{,}000$ kN-m^2

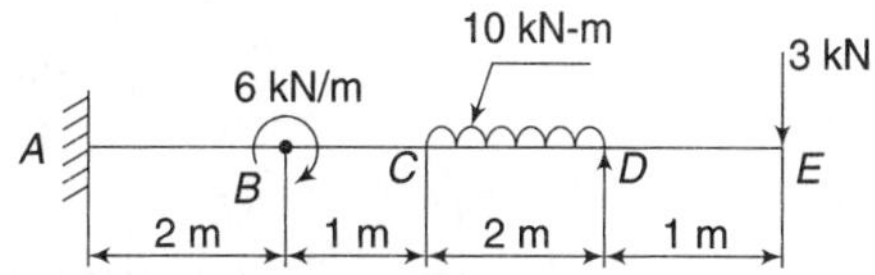

Fig. 7.60

8. A cantilever beam of rectangular cross section 200 mm wide 300 mm deep is 6 m long. It is subjected to a UDL of 10 kN/m over the entire span and a point load of 20 kN at a distance of 4 m from fixed end. Find the deflection under the point load and at the free end. Take $E = 200$ GPa.
9. A cantilever beam of length 3 m carries a UDL of 3 kN/m for a length of 1.5 m from its fixed end and a point load of 1.5 kN at the free end. If it is of circular cross section of diameter 225 mm. Find the deflection of the beam at its free end. Take $E = 1 \times 10^5$ N/mm^2.
10. A overhanging beam *ABC* is 9 m long. It is supported at *A* and *B* at 6 m apart and carries a point load of 12 kN at the free end *C* at a distance of 3 m from support *B*. It also carries a UDL of 4 kN/m over a span of 3 m to the left of support *B*. Determine the slope and deflection at its free end *C* taking *EI* as 100×10^{12} N-mm^2.
11. A overhanging beam *ABC* is 6 m long. It is supported as *A* and *B* at 6 m apart *BC* is over hanging. It carries a UDL of 15 kN/m over the entire span and two point loads 12 kN at *B* and 5 kN at *C*. Find the deflection at the free end *C*. $E = 200$ GPa and $I = 8 \times 10^8$ mm^4.
12. 6 m long beam *ABCD* consists of overhang $AB = 2$ m, $BD = 4$ m carries concentrated loads at free end *A* of magnitude 30 kN and at *C*, the centre of *BD* of 60 kN. Determine the maximum deflection and state where it occurs. Consider uniform *EI*.
13. A simply supported beam is subjected to the loads as shown in Fig. 7.61. Determine the maximum deflection taking flexural rigidity as 1×10^5 N-m^2.

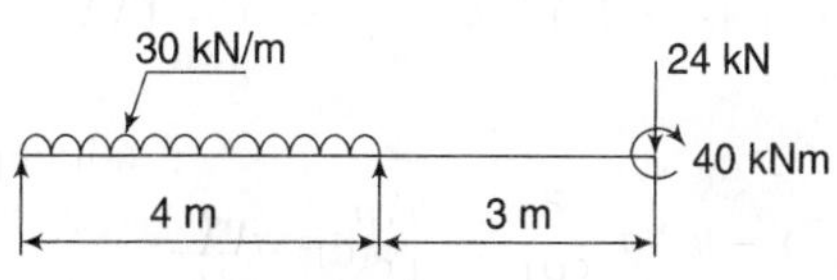

Fig. 7.61

14. A simply supported beam of 6 m span, carries a single point load of 60 kN at a distance of 2 m from left end. Obtain the ratio of maximum deflection and deflection at point load.

15. Figure 7.62 shows a cantilever beam subjected to a UDL and two point loads. Taking $EI = 90$ kN-m^2, determine the deflection at point B, C, D and E.

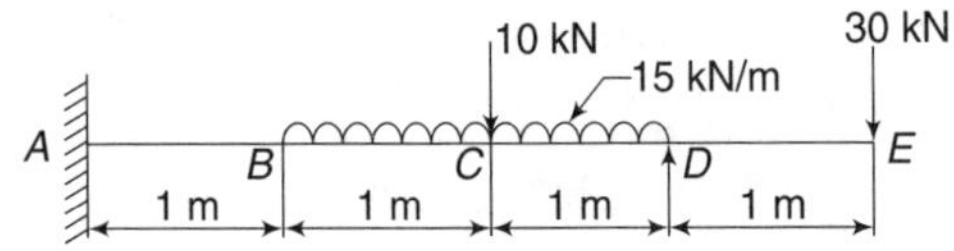

Fig. 7.62

Objective Type Questions

1. The differential equation for deflection is

(a) $EI\dfrac{d^2y}{dx^2} = M$ (b) $EI\dfrac{dy}{dx} = M$

(c) $EI\dfrac{d^2y}{dx^2} = \dfrac{1}{M}$ (d) $EI\dfrac{d^2y}{dx^2} = W$

2. For a simply supported beam of length l with central load W, the maximum deflection equation is

(a) $\dfrac{Wl^3}{3EI}$ (b) $\dfrac{Wl^3}{48EI}$

(c) $\dfrac{5Wl^4}{34EI}$ (d) $\dfrac{Wl^2}{2EI}$

3. For a simply supported beam of length l with central load W, the maximum slope at fixed end is

(a) $\dfrac{Wl^3}{48EI}$ (b) $\dfrac{Wl^2}{24EI}$

(c) $\dfrac{Wl^2}{16EI}$ (d) $\dfrac{Wl^3}{16EI}$

4. For a simply supported beam of length l subjected to a UDL of w over full span, the maximum deflection is

(a) $\dfrac{wl^3}{8EI}$ (b) $\dfrac{wl^3}{3EI}$

(c) $\dfrac{wl^3}{48EI}$ (d) $\dfrac{5wl^4}{384EI}$

5. For a simply supported beam of length l subjected to a UDL of w over full span, the maximum slope at the supports is

(a) $\dfrac{wl^3}{24EI}$ (b) $\dfrac{wl^4}{24EI}$

(c) $\dfrac{wl^2}{16EI}$ (d) $\dfrac{5wl^4}{384EI}$

6. For a cantilever beam of length l subjected to a moment M at the free end, the maximum deflection is

(a) $\frac{Ml^2}{3EI}$ (b) $\frac{Ml^2}{2EI}$

(c) $\frac{Ml}{EI}$ (d) $\frac{Ml^2}{EI}$

7. For a cantilever beam of length l subjected to a moment M at the free end, the maximum slope is

(a) $\frac{Ml^2}{3EI}$ (b) $\frac{Ml^2}{2EI}$

(c) $\frac{Ml}{EI}$ (d) $\frac{Ml^2}{EI}$

8. For a cantilever beam of length l with a point load W at its free end, the maximum deflection is

(a) $\frac{Wl^2}{2EI}$ (b) $\frac{Wl^3}{8EI}$

(c) $\frac{Wl^4}{8EI}$ (d) $\frac{Wl^3}{3EI}$

9. For a cantilever beam of length l with a point load W at its free end, the maximum slope is

(a) $\frac{Wl^2}{2EI}$ (b) $\frac{Wl^3}{8EI}$

(c) $\frac{Wl^4}{8EI}$ (d) $\frac{Wl^3}{3EI}$

10. For a cantilever beam of length l subjected to a UDL of w over the entire span, the maximum deflection is

(a) $\frac{wl^3}{8EI}$ (b) $\frac{wl^4}{8EI}$

(c) $\frac{wl^3}{6EI}$ (d) $\frac{wl^4}{6EI}$

11. For a cantilever beam of length l subjected to a UDL of w over the entire span, the maximum slope is

(a) $\frac{wl^3}{8EI}$ (b) $\frac{wl^4}{8EI}$

(c) $\frac{wl^3}{6EI}$ (d) $\frac{wl^4}{6EI}$

12. For a cantilever beam of length l subjected to a load uniformly varying from 0 at free end to w/unit length at fixed end, the maximum deflection is

(a) $\frac{wl^3}{24EI}$ (b) $\frac{wl^4}{24EI}$

(c) $\frac{wl^3}{30EI}$ (d) $\frac{wl^4}{30EI}$

13. For a cantilever beam of length l subjected to a load uniformly varying from 0 at free end to w/unit length at fixed end, the maximum slope is

(a) $\frac{wl^3}{24EI}$ (b) $\frac{wl^4}{24EI}$

(c) $\frac{wl^3}{30EI}$ (d) $\frac{wl^4}{30EI}$

14. The deflection at the fixed end of a cantilever of length l subjected to a point load W at the free is

(a) $\frac{Wl^3}{3EI}$ (b) zero

(c) $\frac{Wl^2}{EI}$ (d) $\frac{Wl^3}{8EI}$

15. The deflection at the supports for a simply supported beam of length l subjected to a point load at the mid span is

(a) $\frac{Wl^3}{48EI}$ (b) $\frac{Wl^3}{8EI}$

(c) zero (d) $\frac{Wl^2}{16EI}$

16. The slope of the elastic curve for deflection at the position of maximum deflection is

(a) maximum (b) minimum

(c) infinity (d) zero

17. The flexural rigidity is

(a) EI (b) $\frac{E}{I}$

(c) MI (d) $\frac{M}{I}$

18. For a cantilever of length l, subjected to a UDL over one half of its length from the fixed end, the maximum deflection occurs at

(a) middle (b) free end

(c) $\frac{l}{4}$ from fixed end (d) $\frac{l}{4}$ from free end

19. For a simply supported beam of length l carrying a UDL over one half of its span from one end the deflection of the mid span is

(a) $\frac{wl^4}{8EI}$ (b) $\frac{wl^4}{96EI}$

(c) $\frac{5wl^4}{768EI}$ (d) $\frac{5wl^4}{384EI}$

20. A cantilever beam of length 1 m carries a point load of 1 kN at its free end. The maximum deflection is

(a) $\frac{1}{2EI}$ (b) $\frac{1}{4EI}$

(c) $\frac{1}{8EI}$ (d) $\frac{1}{3EI}$

21. A cantilever beam of length 1 m carries a point load of 1 kN at its free end. The maximum slope is

(a) $\dfrac{1}{2EI}$ (b) $\dfrac{1}{48EI}$

(c) $\dfrac{1}{8EI}$ (d) $\dfrac{1}{3EI}$

22. A simply supported beam of length 1 m carries a central load of 1 kN. The maximum deflection is

(a) $\dfrac{1}{2EI}$ (b) $\dfrac{1}{4EI}$

(c) $\dfrac{1}{16EI}$ (d) $\dfrac{1}{24EI}$

23. A simply supported beam of length 1 m carries a central load of 1 kN. The maximum slope is

(a) $\dfrac{1}{2EI}$ (b) $\dfrac{1}{48EI}$

(c) $\dfrac{1}{16EI}$ (d) $\dfrac{1}{24EI}$

24. A cantilever length l carries a UDL of w/unit length. The equivalent point load at the free end to be applied for the same deflection is

(a) $\dfrac{8wl}{3}$ (b) $\dfrac{8w}{3l}$

(c) $\dfrac{3w}{8l}$ (d) $\dfrac{wl}{8}$

Key to Answers

1. (a)	2. (b)	3. (c)	4. (d)	5. (a)	6. (b)	7. (c)	8. (d)
9. (a)	10. (b)	11. (c)	12. (d)	13. (a)	14. (b)	15. (c)	16. (d)
17. (a)	18. (b)	19. (c)	20. (d)	21. (a)	22. (b)	23. (c)	24. (d)

CHAPTER 8

Torsion of Shafts

8.1 INTRODUCTION

A member is said to be in torsion if it is subjected to twisting moments that tend to produce rotation about the longitudinal axis of the member. Figure 8.1 shows a member in torsion.

The moments that produce twisting of the member are called turning moments, twisting moments or torques. Cylindrical members that are subjected to torque and transmit power are called shafts. Some examples are, drive shaft in trucks and buses (a long round rod that connects the engine gearbox to rear axle differential), spindles of lathe machine, drilling machines, propeller shaft of a ship and aeroplane, the rotor shaft of a helicopter, etc.

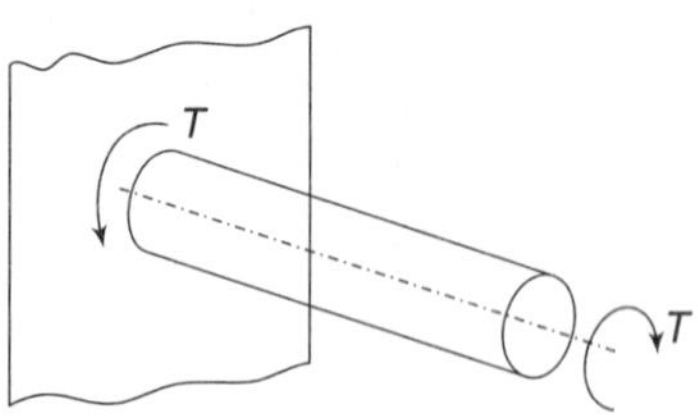

Fig. 8.1

The simple examples for understanding torque is the tightening a bolts with spanners or turning of a screw with screw driver.

In this chapter we will try to understand how the torsional moments are resisted by the members.

8.2 PURE TORSION

A member is said to be in pure torsion when it is subjected to equal and opposite couples at the end and not accompanied by axial forces or bending moment. Figure 8.1 shows a shaft under pure torsion.

Like bending moment causes bending in a member, torque causes twist in the member. The internal resistance developed to oppose the twist is called shear force and the corresponding stress induced is called shear stress.

8.3 ASSUMPTIONS IN THE THEORY OF PURE TORSION

The following assumptions are made in the theory of pure torsion:

(i) The material of the member is homogeneous and isotropic.
(ii) The twist along the shaft is uniform.
(iii) The shaft is of uniform circular section for the entire length.
(iv) Cross sections of shaft which are plane before twist, remain plane after twist.
(v) All radii which were straight before twist, remain straight after twist.
(vi) Torsion is uniform along length.
(vii) The shear stress is within elastic limit and hence obeys Hooke's law (shear stress is proportional to shear strain).

8.4 DERIVATION OF EQUATION OF TORSION

Figure 8.2 shows a solid shaft of uniform radius R and length l, fixed at one end and subjected to a twisting moment T at the free end.

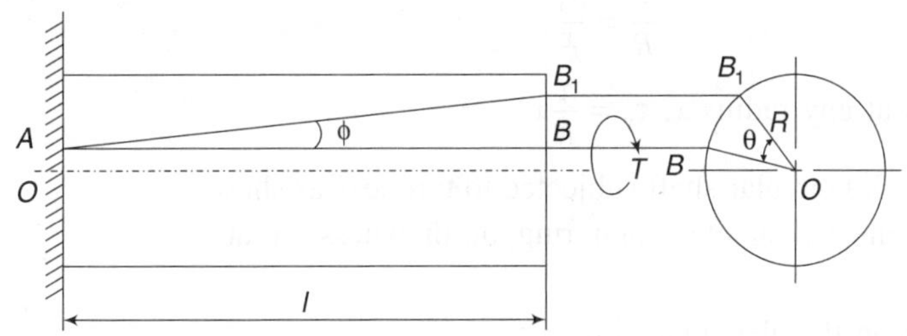

Fig. 8.2

Due to the effect of torque, the straight line AB deforms to AB_1. The angle BAB_1, is the shear strain and is denoted by ϕ. Similarly, the angle BOB_1, is the angle of twist of radius OB and the angle of twist is denoted by θ.

From Fig. 8.2, $$\tan\phi = \frac{BB_1}{AB} = \frac{BB_1}{l}$$

(BB_1 is very small compared to AB and hence treated as a straight line)

For small angles, $$\tan\phi = \phi = \frac{BB_1}{l}$$

$$\therefore \quad BB_1 = l\phi \qquad \ldots(1)$$

From Fig. 8.2 (right side) sector length $BB_1 = R\theta$ ...(2)

Equating Eqs. (1) and (2), $$l\phi = R\theta$$

$$\therefore \quad \text{Shear strain } \phi = \frac{R\theta}{l} \qquad \ldots(3)$$

From the definition of rigidity modulus,

Rigidity modulus, $$G = \frac{\text{Shear stress}}{\text{Shear strain}} = \frac{\tau}{\phi}$$

$$\therefore \quad \phi = \frac{\tau}{G}$$

Substituting this in Eq. (3), we get

$$\frac{\tau}{G} = \frac{R\theta}{l}$$

$$\therefore \quad \frac{\tau}{R} = \frac{G\theta}{l} \quad \ldots(4)$$

From the above equation, it is seen that $\frac{\tau}{r}$ is constant since for a given loading and material. Rigidity modulus G, angle of twist θ and length l are constants. Hence, shear stress is proportional to the radial distance from its axis.

Hence, at a radius x, if the shear stress is τ_x, then $\frac{\tau}{R} = \frac{\tau_x}{x} = \frac{G\theta}{l}$ which is constant

$$\therefore \quad \frac{\tau}{R} = \frac{\tau_x}{x} \quad \ldots(5)$$

Shear stress at any radius x, $\tau_x = \frac{\tau}{R}x$

Consider a solid circular shaft subjected to torque T as shown in Fig. 8.3. Consider an elemental ring of thickness dx at a radius x.

Shear stress on the elemental ring = τ_x

It is seen above that the ratio of shear stress to the radius is constant.

$$\frac{\tau_x}{x} = \frac{\tau}{R} \qquad \therefore \tau_x = \frac{\tau}{R}x$$

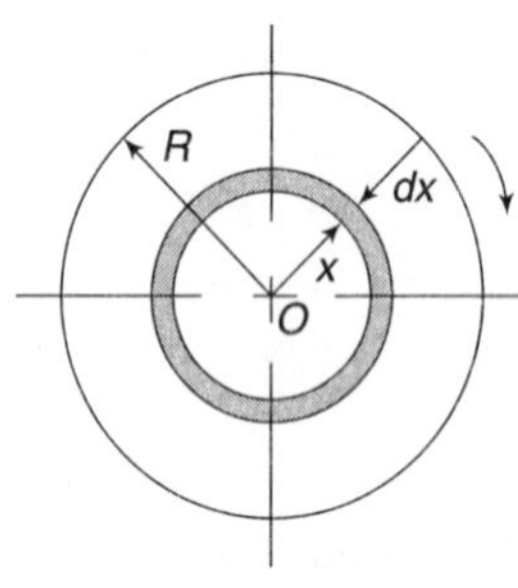

Fig. 8.3 Solid shaft

Since the thickness of elemental ring is very small, considering the ring as a strip of length = $2\pi x$, area of the element ring = $A_x = (2\pi x)dx$

Shear force at the ring, F_x = Shear stress × Area

$$F_x = \tau_x \cdot A_x = \left(\frac{\tau}{R}\cdot x\right)(2\pi x \cdot dx)$$

Torque at the ring $\quad T_x$ = Force at ring × Radius of ring = $F_x \cdot x$

$$T_x = \left(\frac{\tau}{R}\cdot x\right)(2\pi x\, dx)x$$

$$T_x = \frac{\tau}{R}2\pi x^3\, dx \quad \ldots(6)$$

Total torque can be obtained by integrating this w.r.t. x between limits $x = 0$ to R.

$$\text{Total torque, } T = \int_o^R T_x = \int_o^R \frac{\tau}{R} \cdot 2\pi x^3 dx$$

$$= \frac{2\pi\tau}{R} \cdot \left(\frac{x^4}{4}\right)_o^R = \frac{2\pi\tau}{R} \cdot \frac{R^4}{4}$$

$$= \frac{\tau}{R} \cdot \left(\frac{\pi R^4}{2}\right)$$

Polar moment of inertia for a solid circular shaft

$$J = \frac{\pi D^4}{32} = \frac{\pi (2R)^4}{32} = \frac{16}{32}\pi R^4 = \frac{\pi R^4}{2}$$

Substituting $\frac{\pi R^4}{2} = J$ in the above equation

$$T = \frac{\tau}{R} . J$$

or

$$\frac{T}{J} = \frac{\tau}{R} \qquad \text{...(7)}$$

Equating Eqs. (5) and (6), we get

$$\frac{T}{J} = \frac{\tau}{R} = \frac{G\theta}{l} \qquad \text{...(8)}$$

This is called equation of torsion

T = Torque or twisting moment or turning moment, N-mm

J = Polar moment of inertia, mm^4

τ = Shear stress, N/mm^2

R = Radius, mm

G = Rigidity modulus, N/mm^2 (It is also denoted by N or C)

θ = Angle of twist in radians = θ in degrees $\times \frac{\pi}{180}$

l = Length of shaft, mm.

8.5 DERIVATION OF TORSION EQUATION FOR HOLLOW SHAFT

Consider a hollow shaft of outer radius R_o and inner radius R_i as shown in Fig. 8.4. Consider an elemental ring of thickness dx at a radius x.

As discussed in the previous derivation

$$\text{Shear stress, } \tau_x = \frac{\tau}{Ro} \cdot x$$

Area of elemental ring, considering it as a rectangular strip

$$Ax = 2\pi x \cdot dx$$

$$\text{Shear force on the ring, } F_x = A_x \cdot \tau_x = (2\pi x \cdot dx)\left(\frac{\tau}{R} \cdot x\right)$$

Torque on the ring $T_x = F_x \cdot \text{Radius} = F_x \cdot x$

$$= (2\pi x \cdot dx)\left(\frac{\tau}{Ro} \cdot x\right)x$$

$$= \frac{\tau}{R_o} 2\pi x^3 \cdot dx.$$

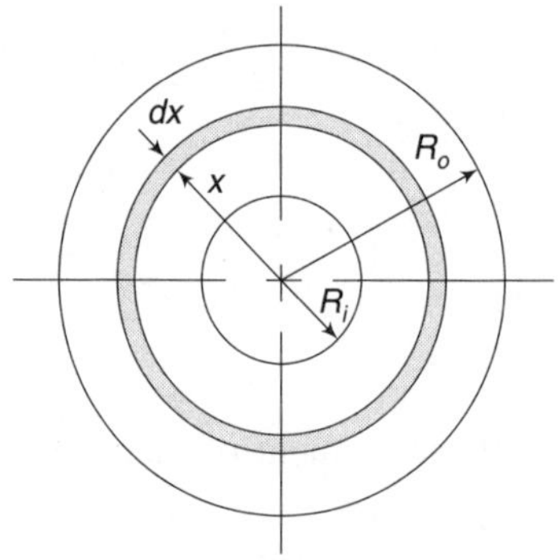

Fig. 8.4 Hollow Shaft

For total torque integrate this w.r.t. x between limits $x = R_i$ to R_o

$$T = \int_{R_i}^{R_o} T_x = \int_{R_i}^{R_o} \frac{\tau}{R_o} \cdot 2\pi x^3 \, dx$$

$$= \frac{\tau}{R_o} \cdot 2\pi \left[\frac{x^4}{4}\right]_{R_i}^{R_o}$$

$$= \frac{\tau}{R_o} 2\pi \left(\frac{R_o^4 - R_i^4}{4}\right)$$

$$T = \frac{\tau}{R_o} \cdot \pi \left(\frac{R_o^4 - R_i^4}{2}\right)$$

But $\dfrac{\pi(R_o^4 - R_i^4)}{2} = J$ for hollow shaft

$\therefore$ $$T = \frac{\tau}{r} \cdot J$$

$\therefore$ $$\frac{T}{J} = \frac{\tau}{R_o} \quad \text{...(9)}$$

Equating Eqs. (5) and (9)

$$\frac{T}{J} = \frac{\tau}{R_o} = \frac{G\theta}{l} \quad \text{...(10)}$$

where, $J = \frac{\pi}{2}(R_o{}^4 - R_i{}^4) = \frac{\pi}{32}(D_o{}^4 - D_i{}^4)$ is polar moment of inertia for hollow shaft

$$R = R_o = \frac{D_o}{2}$$

To find shear stress at inner radius, $R = R_i$

$$\frac{T}{J} = \frac{\tau_i}{R_i} = \frac{G\theta}{l}$$

To find shear stress at outer radius, $R = R_o$

$$\frac{T}{J} = \frac{\tau_o}{R_o} = \frac{G\theta}{l}$$

Since the shear stress is proportional to radius, the shear stress is maximum at the outer radius and minimum at inner radius.

Allowable shear stress is to be taken as shear stress at outer radius since it is maximum.

For solid shaft, $J = \frac{\pi R^4}{2} = \frac{\pi D^4}{32}$ and $R = \frac{D}{2}$.

8.6 POLAR SECTION MODULUS

The ratio of polar moment of inertia to its radius is defined as polar section modulus and is denoted by z_p.

$$Z_p = \frac{J}{R}$$

For solid shaft, $$Z_p = \frac{J}{R} = \left(\frac{\pi D^4}{32}\right)\bigg/\frac{D}{2} = \frac{\pi D^3}{16}$$

For hollow shaft, $$Z_p = \frac{J}{R} = \frac{\pi(D_o{}^4 - D_i{}^4}{\left(\frac{D_o}{2}\right)} = \frac{\pi}{16}\left(\frac{D_o^4 - D_i^4}{D_o}\right)$$

8.7 TORSIONAL RIGIDITY

We had seen earlier in bending stresses that *EI* is known as torsional rigidity, similarly *GJ* is known as torsional rigidity. Torsional rigidity can be defined as the torque required to cause a unit angle of twist in a unit length.

We know that $$\frac{T}{J} = \frac{G\theta}{l}$$

$\therefore$ Torsional rigidity $$JG = \frac{T.l}{\theta}$$

Torsional rigidity is also known as torsional stiffness.

Torsional flexibility is defined as angle of twist per unit torque per unit length.

Torsional and flexibility $$= \frac{1}{GJ} = \frac{\theta}{T}$$

It is the reciprocal of torsional rigidity.

8.8 MOMENT OF RESISTANCE

It is the torsional moment or torque that can be resisted by a shaft

From $$\frac{T}{J} = \frac{\tau}{R}$$

Moment of resistance, $$T = \frac{\tau \cdot J}{R}$$

or $$T = \tau \cdot Z_p$$

For solid shaft moment of resistance, $$T = \frac{\tau \cdot \left(\frac{\pi d^4}{32}\right)}{\frac{d}{2}} = \tau \cdot \frac{\pi}{16} d^3$$

For hollow shaft, moment of resistance, $T = \dfrac{\tau}{\left(\dfrac{d_o}{2}\right)} \dfrac{\pi}{32} (d_o^{\,4} - d_i^{\,4}) = \tau.\dfrac{\pi}{16} \dfrac{(d_o^{\,4} - d_i^{\,4})}{d_o}$

8.9 POWER TRANSMITTED BY SHAFT

Consider a shaft transmitting a torque of T N-m at a speed of N revolutions per minute (rpm)

$$\text{Speed} = N \text{ rpm}$$

$$\therefore \quad \text{Speed per second} = \frac{N}{60} \text{ rev/sec}$$

$$1 \text{ revolution} = 360° = 2\pi \text{ radians}$$

We know that by definition, power P = Work done/sec

P = Work done/sec = Mean torque × Angle rotated/sec

$$= T.\theta$$

$$P = T\cdot\left(\frac{N}{60} \times 2\pi\right) \text{ Nm/s}$$

$$= \frac{2\pi NT}{60} \text{ J/s}$$

$$P = \frac{2\pi NT}{60} \text{ watts}$$

$$P = \frac{2\pi NT}{60000} \text{ kW}$$

Here P = Power in kW

N = Speed, rpm

T = Torque, N-m

Torque transmitted $T = \dfrac{60000 \times P}{2\pi N}$ N-m

$$= \frac{60000\text{P}}{2\pi N} \times 1000 \text{ N-mm}$$

$$T = \frac{60}{2\pi} \times 10^6 \frac{P}{N} \text{ N-mm.}$$

Power P in kW and speed N in rpm

SOLVED PROBLEMS

Problem 1 A solid shaft of 60 mm diameter and 3 m long transmits 45 kW at 1500 rpm. Find the maximum intensity of shear stress induced and angle of twist taking rigidity modulus as 81 GPa.

Solution: **Given data:** Diameter $d = 60$ mm

Length $l = 3$ mm $= 3000$ mm

Power $P = 45$ kW

Speed $N = 1500$ rpm

Rigidity modulus $G = 81$ GPa $= 81 \times 10^3$ MPa

$$\text{Radius } R = \frac{d}{2} = \frac{60}{2} = 30 \text{ mm}$$

$$\text{Polar moment of inertia } J = \frac{\pi}{32} d^4 = \frac{\pi}{32} \times 60^4 = 1.272 \times 10^6 \text{ mm}^4$$

$$\text{Power } P = \frac{2\pi NT}{60000} \text{ kW}$$

$$\therefore \quad \text{Torque } T = \frac{60000}{2\pi} \times \frac{P}{N} \text{ N-m}$$

$$T = \frac{60000}{2\pi} \times \frac{45}{1500} = 286.479 \text{ Nm}$$

$$T = 286.479 \times 1000 \text{ Nmm}$$

$$\text{The equation of torsion, } \frac{T}{J} = \frac{\tau}{R} = \frac{G\theta}{l}$$

$$\therefore \quad \text{Consider } \frac{T}{J} = \frac{\tau}{R}$$

$$\text{Shear stress } \tau = \frac{T \cdot R}{J} = \frac{286.479 \times 1000 \times 30}{1.272 \times 10^6}$$

$$\tau = 6.754 \text{ N/mm}^2$$

$$\text{Consider } \frac{T}{J} = \frac{G\theta}{l}$$

$$\therefore \quad \text{Angle of twist } \theta = \frac{Tl}{JG} = \frac{286.479 \times 1000 \times 3000}{1.272 \times 10^6 \times 81 \times 10^3}$$

$$= 8.34146 \times 10^{-3} \text{ radians}$$

$$\theta = 8.34146 \times 10^{-3} \times \frac{180}{\pi} = 0.478 \text{ degrees.}$$

Problem 2 A hollow circular shaft of length 1.8 m having outer and inner diameters as 120 mm and 72 mm is transmitting power of 600 kW at 1200 rpm. Determine the maximum shear stress induced and the angle of twist. Also find the shear stress at the inner radius. Take rigidity modulus as 80 kN/mm^2.

Solution: **Given data:** Length of shaft $l = 1.8$ m $= 1800$ mm

Outside diameter $d_o = 120$ mm

Inside diameter $d_i = 72$ mm

$$\text{Power transmitted } P = 600 \text{ kW}$$

$$\text{Speed } N = 1200 \text{ rpm}$$

$$\text{Rigidity modulus } G = 80 \text{ kN/mm}^2 = 80 \times 10^3 \text{ N/mm}^2$$

$$\text{Power } P = \frac{2\pi NT}{60000} \text{ kW}$$

$$\therefore \quad \text{Torque } T = \frac{P \times 60000}{2\pi N} \times 1000 \text{ N-mm}$$

$$T = \frac{600 \times 60000 \times 1000}{2\pi \times 1200} = 4774648.3 \text{ N-mm}$$

$$\text{Polar moment of inertia } J = \frac{\pi}{32}(d_o^4 - d_i^4)$$

$$J = \frac{\pi}{32}(120^4 - 72^4) = 17.719 \times 10^6 \text{ mm}^4$$

$$\text{Consider } \frac{T}{J} = \frac{\tau}{R}$$

Maximum shear stress occurs at outer radius = R_o

$$\tau = \frac{T \cdot R_o}{J} = \frac{4774648.3 \times \frac{120}{2}}{17.719 \times 10^6} = 16.166 \text{ N/mm}^2$$

$$\text{Shear stress at inner radius } \tau_i = \frac{T \cdot R_i}{J}$$

$$\tau_i = \frac{4774648.3 \times \left(\frac{72}{2}\right)}{17.719 \times 10^6} = 9.7 \text{ N/mm}^2$$

$$\text{To find angle of twist, consider } \frac{T}{J} = \frac{G\theta}{l}$$

$$\frac{4774648.3}{17.719 \times 10^6} = \frac{80000 \times \theta}{1800}$$

$$\therefore \quad \theta = \frac{4774648.3 \times 1800}{17.719 \times 10^6 \times 80000} = 6.063 \times 10^{-3} \text{ radians}$$

$$\theta = 6.063 \times 10^{-3} \times \frac{180}{\pi} = 0.3474°.$$

Problem 3 What length of solid steel shaft of 12 mm diameter such that one end can be rotated by 90° with respect to the other end limiting shear stress to 90 MPa. Take modulus of rigidity as 84 GPa.

***Solution:* Given data:** Diameter $d = 12$ mm

$$\therefore \quad R = \frac{d}{2} = \frac{12}{2} = 6 \text{ mm}$$

$$\text{Angle of twist } \theta = 90^\circ = 90 \times \frac{\pi}{180} = \frac{\pi}{2} \text{ radians}$$

$$\text{Shear stress } \tau = 90 \text{ MPa}$$

$$\text{Rigidity modulus } G = 84 \text{ GPa} = 84 \times 10^3 \text{ MPa}$$

$$\text{Consider } \frac{\tau}{R} = \frac{G\theta}{l}$$

$$\therefore \quad l = \frac{G\theta R}{\tau} = \frac{84000 \times \left(\frac{\pi}{2}\right) \times 6}{90} = 8796.46 \text{ mm}$$

$$\therefore \quad \text{Length } l = 8.79646 \text{ metres.}$$

Problem 4 Determine the diameter of a solid shaft required to transmit 440 kW at 280 rpm. The angle of twist must not exceed one degree per metre length and maximum torsional stress is to be limited to 40 N/mm^2. Take G = 84 kN/mm^2.

***Solution:* Given data:**

$$\text{Power } P = 440 \text{ kW}$$

$$\text{Speed } N = 280 \text{ rpm}$$

$$\text{Angle of twist } \theta = 1^\circ = 1 \times \frac{\pi}{180} \text{ radians}$$

$$\text{Length of shaft } l = 1 \text{ m} = 1000 \text{ mm}$$

$$\text{Allowable shear stress } \tau = 40 \text{ N/mm}^2$$

$$\text{Rigidity modulus } G = 84 \text{ kN/mm}^2 = 84 \times 10^3 \text{ N/mm}^2$$

$$\text{Torque transmitted } T = \frac{60000P}{2\pi N} \times 1000 \text{ N-mm}$$

$$= \frac{60000 \times 440}{2\pi \times 280} \times 1000$$

$$= 15.006 \times 10^6 \text{ N-mm}$$

$$\text{For solid shaft } J = \frac{\pi d^4}{32} \text{ and } R = \frac{d}{2}$$

$$\text{Consider } \frac{T}{J} = \frac{\tau}{r}$$

$$\frac{15.006 \times 10^6}{\left(\frac{\pi d^4}{32}\right)} = \frac{40}{\left(\frac{d}{2}\right)}$$

$$\therefore \quad \text{Diameter } d = 124.087 \text{ mm}$$

$$\text{Consider } \frac{T}{J} = \frac{G\theta}{l}$$

$$\frac{15.006 \times 10^6}{\left(\frac{\pi}{32} d^4\right)} = \frac{84000 \times 1 \times \frac{\pi}{180}}{1000}$$

$\therefore$ Diameter $d = 101.047$ mm

Recommended diameter $d = 124.087$ mm

(Bigger of the above two values)

Problem 5 A hollow steel shaft 3 m long must transmit 41 kW of power at 100 rpm. The maximum torque to be transmitted is 25% more than the mean torque. The total angle of twist is not to exceed 3° and allowable shear stress for the material is 100 N/mm². Determine the diameter of the shaft if their ratio is 2. The modulus of rigidity may be taken as 85 GPa.

Solution: **Given data:** Length $l = 3\text{ m} = 3000\text{ mm}$

Power $P = 41$ kW

Speed $N = 100$ rpm

Maximum torque = 25% greater than mean torque

$$\therefore \quad T = T_m + 25\%\ T_m = 1.25\ T_m$$

Angle of twist $\theta = 3° = 3 \times \dfrac{\pi}{180}$ radians

Allowable shear stress $\tau = 100\text{ N/mm}^2$

Diameters ratio $= 2 = \dfrac{d_o}{d_i}$

Rigidity modulus $G = 85\text{ GPa} = 85 \times 10^3\text{ MPa}$

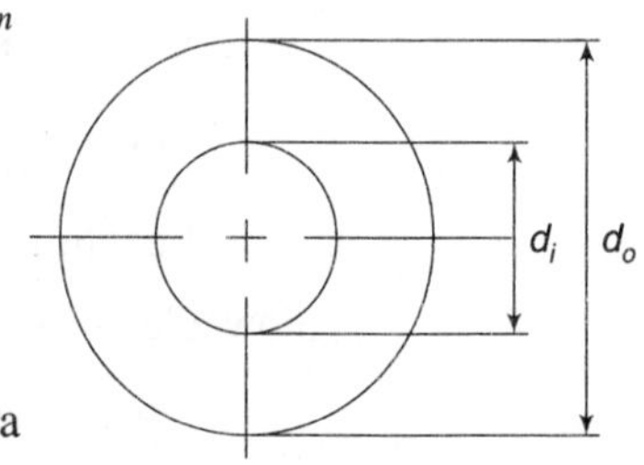

Fig. 8.5

Mean Torque:

$$\text{Power } P = \frac{2\pi N T_m}{60000}\text{ kW}$$

$$\therefore \quad T_m = \frac{P \times 60000}{2\pi N}\text{ N-m}$$

$$= \frac{41 \times 60000}{2\pi \times 100} \times 1000\text{ Nmm}$$

$$T_m = 3.915 \times 10^6\text{ N-mm}$$

$$\therefore \quad T_{max} = 1.25.\ T_m = 1.25 \times 3.915 \times 10^6$$

$$= 4.894 \times 10^6\text{ N-mm}$$

For hollow shaft polar moment of inertia $= J = \dfrac{\pi}{32}(d_o^4 - d_i^4)$

But $d_o = 2\ d_i$

$$\therefore \quad J = \frac{\pi}{32}[(2d_i)^4 - d_i^4] = \frac{\pi}{32}(15d_i^4)$$

$$R = \frac{d_o}{2} = \frac{2d_i}{2} = d_i$$

Consider $\dfrac{T}{J} = \dfrac{\tau}{R}$

$$\frac{4.894 \times 10^6}{\left(\frac{\pi}{32} d_i^4 \times 15\right)} = \frac{100}{d_i}$$

$$\therefore \quad d_i = 32.15 \text{ mm}$$

$$\text{Consider } \frac{T}{J} = \frac{G\theta}{l}$$

$$\frac{4.894 \times 10^6}{\left(\frac{\pi}{32} \times 15 d_i^4\right)} = \frac{85000 \times \left(\frac{3\pi}{180}\right)}{3000}$$

$$d_i = 38.69 \text{ mm}$$

Recommended diameter $d_i = 38.69$ mm (bigger one)

$$\therefore \quad d_o = 2d_i = 77.38 \text{ mm.}$$

Problem 6 Determine the size of solid shafts to transmit 200 kW each without exceeding an allowable shear stress of 70 MPa. One of the shafts operates at 20 rpm and the other at 20000 rpm. Comment on the results.

Solution: **Given data:** Power $P = 200$ kW

Speed $N = 20$ rpm and 20000 rpm

Allowable shear stress $\tau = 70$ MPa

(i) Speed $N = 20$ rpm. Power $P = \frac{2\pi NT}{60000}$ kW

$$\therefore \quad \text{Torque } T = \frac{P \times 60000}{2\pi N} \times 1000 \text{ N-mm.}$$

$$T = \frac{200 \times 60000}{2\pi \times 20} \times 1000 = 95.493 \times 10^6 \text{ N-mm}$$

$$\text{Consider } \frac{T}{J} = \frac{\tau}{R}$$

$$J = \frac{\pi d^4}{32} \text{ and } R = \frac{d}{2}$$

$$\frac{95.493 \times 10^6}{\left(\frac{\pi}{32} d^4\right)} = \frac{70}{\left(\frac{d}{2}\right)}$$

Diameter of shaft $d = 190.82$ mm

(ii) Speed $N = 20000$ rpm

$$\therefore \quad \text{Torque } T = \frac{P \times 60000}{2\pi N} \times 1000 \text{ N-mm}$$

$$= \frac{200 \times 60000}{2\pi \times 20000} \times 1000 = 95.493 \times 10^3 \text{ Nmm}$$

Substituting in torque equation

$$\frac{95.493 \times 10^6}{\left(\frac{\pi}{32} d^4\right)} = \frac{70}{\left(\frac{d}{2}\right)}$$

$$\text{Diameter of shaft } d = 19.082 \text{ mm.}$$

Comment:

$$\text{Diameter of shaft at } N = 20 \text{ rpm} = 190.82 \text{ mm}$$

$$\text{at } N = 20000 \text{ rpm} = 19.082 \text{ mm}$$

The requirement of diameter is inversely proportional to the cube root times the speed.

$$d \,\alpha \left(\frac{1}{N}\right)^{1/3}.$$

Problem 7 Design a shaft to transmit 1 MW of power at 3000 rpm. The shear stress in the shaft should not exceed 60 MPa and angle of twist should not be more than 1° in a length of 10 diameters. Rigidity modulus is 80 GPa.

Solution: **Given data:** Power $P = 1 \text{ MW} = 1 \times 10^3 \text{ kW}$

$$\text{Speed } N = 3000 \text{ rpm}$$

$$\text{Allowable shear stress } \tau = 60 \text{ MPa}$$

$$\text{Angle of twist } \theta = 1° = 1 \times \frac{\pi}{180} \text{ radians}$$

$$\text{Length of shaft } l = 10\, d$$

$$\text{Rigidity modulus } G = 80 \text{ GPa} = 80 \times 10^3 \text{ MPa}$$

$$\text{Power } P = \frac{2\pi NT}{60000} \text{ kW}$$

$$\therefore \quad \text{Torque } T = \frac{P \times 60000}{2\pi N} \text{ N-m}$$

$$= \frac{1000 \times 60000}{2\pi\, 3000} \times 1000 \text{ Nmm}$$

$$T = 3.183 \times 10^6 \text{ N-mm}$$

$$\text{Polar moment of inertia } J = \frac{\pi}{32} d^4$$

and

$$R = \frac{d}{2}$$

$$\text{Consider } \frac{T}{J} = \frac{\tau}{R}$$

$$\frac{3.183 \times 10^6}{\left(\frac{\pi}{32} d^4\right)} = \frac{60}{\left(\frac{d}{2}\right)}$$

$\therefore$ Diameter of shaft $d = 64.65$ mm

$$\text{Consider } \frac{T}{J} = \frac{G\theta}{l}$$

$$\frac{3.183 \times 10^6}{\left(\frac{\pi}{32} d^4\right)} = \frac{80 \times 10^3 \times 1 \times \frac{\pi}{180}}{10\, d}$$

Diameter $d = 61.46$ mm

Recommended diameter $d = 64.65$ mm (bigger one).

Problem 8 A hollow shaft having the internal diameter as 3/4 of its external diameter transmits 600 kW at 150 rpm. Determine the diameter of the shaft if the shear stress is limited to 66 MPa and the angle of twist is to be limited to 0.75° in 20 diameters. Maximum torque to be transmitted is 30% more than the mean torque. Modulus of rigidity is 81 GPa.

Solution: **Given data:** Inside dia $d_i = \frac{3}{4} d_o$

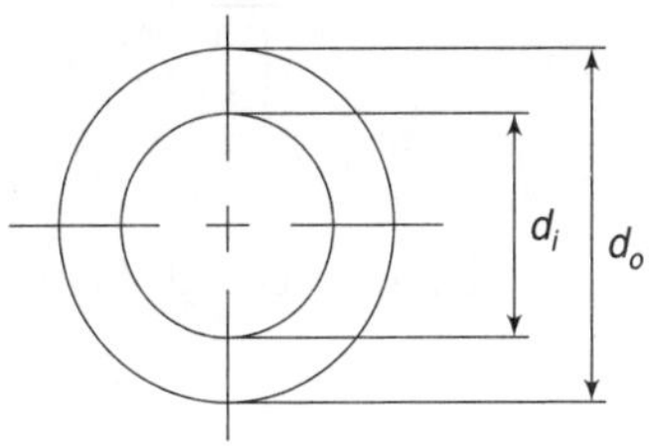

Fig. 8.6

d_o = External diameter

Power $P = 600$ kW

Speed $N = 150$ rpm

Shear stress $\tau = 66$ MPa

Angle of twist $\theta = 0.75° = 0.75 \times \frac{\pi}{180}$ radians

Length $l = 20$ diameters $= 20\, d_o$

Maximum torque $= 1.3 \times T_{mean}$ (30% greater)

Rigidity modulus $G = 81$ GPa $= 81 \times 10^3$ MPa

$$\text{Power } P = \frac{\pi N T_m}{60000} \text{ kW}$$

$$\therefore \quad T_m = \frac{P \times 60000}{2\pi N} \text{ N-m}$$

$$T_m = \frac{P \times 60000 \times 1000}{2\pi N} \text{ N-mm}$$

$$\therefore \quad T_{max} = 1.3\, T_m = 1.3 \times \frac{P \times 60000 \times 1000}{2\pi N}$$

$$= \frac{1.3 \times 600 \times 60000 \times 1000}{2\pi \times 150} = 49.456 \times 10^6 \text{ N-mm}$$

$$\text{For hollow shaft } J = \frac{\pi}{32}(d_o^4 - d_i^4) = \frac{\pi}{32}\left[d_o^4 - \left(\frac{3}{4} d_o\right)^4\right]$$

$$= \frac{\pi}{32} d_o^4 \left(1 - \left(\frac{3}{4}\right)^4\right) = 0.684 \times \frac{\pi}{32} d_o^4$$

$$R = \frac{d_o}{2}$$

Consider $$\frac{T}{J} = \frac{\tau}{R} \quad (T = T_{max})$$

$$\frac{49.456 \times 10^6}{\left(\frac{\pi}{32} d_o^4 \times 0.684\right)} = \frac{66}{\left(\frac{d_o}{2}\right)}$$

$\therefore$ $$d_o = 177.398 \text{ mm}$$

Consider $$\frac{T}{J} = \frac{G\theta}{l}$$

$$\frac{49.456 \times 10^6}{\left(\frac{\pi}{32} d_o^4 \times 0.684\right)} = \frac{80000 \times \left(\frac{0.75 \times \pi}{180}\right)}{20\, d_o}$$

$$d_o = 241.39 \text{ mm}$$

Recommended diameter $d_o = 241.39$ mm (Bigger one)

$\therefore$ $$d_i = \frac{3}{4} d_o = 181.04 \text{ mm.}$$

Problem 9 A solid shaft is required to transmit 210 kW at 2400 rpm. Find the diameter of the shaft if the shear stress is not to exceed 72 MPa. Also find the angle of twist per metre length taking G = 80 GPa.

Solution:

Given data: Power to be transmitted P = 210 kW

Speed N = 2400 rpm

Shear stresses τ = 72 MPa

Rigidity modulus G = 80 GPa = 80×10^3 MPa

Length l = 1 m = 1000 mm

$$\text{Torque } T = \frac{P \times 60000}{2\pi N} \times 1000 \text{ N-mm}$$

$$= \frac{210 \times 60000}{2\pi \times 2400} \times 1000 = 835.56 \times 10^3 \text{ Nmm}$$

For solid shaft, $J = \frac{\pi d^4}{32}$ and $R = \frac{d}{2}$

To find diameter, consider $\frac{T}{J} = \frac{\tau}{R}$

$$\frac{835.56 \times 10^3}{\left(\frac{\pi}{32} d^4\right)} = \frac{72}{\left(\frac{d}{2}\right)}$$

$\therefore$ Diameter $d = 38.95$ mm

To find angle of twist, consider $\frac{T}{J} = \frac{G\theta}{l}$

$$\frac{835.56 \times 10^3}{\left(\frac{\pi}{32} d^4\right)} = \frac{80 \times 10^3 \times \theta}{1000}$$

Substitute $d = 38.95$ mm

$$\frac{835.56 \times 10^3}{\left(\frac{\pi}{32} \times 38.95^4\right)} = \frac{80 \times 10^3 \times \theta}{1000}$$

$\therefore$ Angle of twist $\theta = 0.04622$ radians $= 0.04622 \times \frac{180}{\pi}$

$\theta = 2.6484$ degrees.

Problem 10 A hollow propeller shaft of a turbine of a ship is to transmit 1200 kW at 360 rpm. If the internal diameter is 0.6 times its external diameter and allowable shear stresses for the shaft material is 90 N/mm^2, find the diameters of the shaft and angle of twist in the shaft per metre length. Take G = 84 kN/mm^2.

Solution: **Given data:** Power P =120 kW

Speed N = 360 rpm

Inside diameter $d_i = 0.6\ d_o$

Allowable shear stress τ = 90 MPa

Rigidity modulus G = 84 kN/mm^2 = 84×10^3 N/mm^2

Length l = 1 m = 1000 mm

$$\text{Torque } T = \frac{P \times 60000}{2\pi N} \times 1000 \text{ N-mm}$$

$$= \frac{1200 \times 60000}{2\pi \times 360} \times 1000 = 31.831 \times 10^6 \text{ N-mm}$$

$$\text{For hollow shaft } J = \frac{\pi}{32}(d_o^4 - d_i^4)$$

$$= \frac{\pi}{32}[d_o^4 - (0.6\ d_o)^4] = \frac{\pi}{32} d_o^4\ (0.8704)$$

and $R = \frac{d_o}{2}$

To find diameters, consider $\frac{T}{J} = \frac{\tau}{R}$

$$\frac{31.831 \times 10^6}{\left[\frac{\pi}{32} d_o^4 (0.8704)\right]} = \frac{90}{\left(\frac{d_o}{2}\right)}$$

$\therefore$ Outside diameter $d_o = 127.43$ mm

Inside diameter $d_i = 0.6 \ d_o = 0.6 \times 127.43 = 76.46$ mm

To find angle of twist, consider $\frac{T}{J} = \frac{G\theta}{l}$

$$\frac{31.831 \times 10^6}{\left[\frac{\pi}{32}(127.43^4 - 76.46^4)\right]} = \frac{84000 \times \theta}{1000}$$

$\therefore$ Angle of twist $\theta = 0.0168$ radians $= 0.0168 \times \frac{180}{\pi} = 0.964°$.

Problem 11 A solid circular shaft is required to transmit 60 kW at 1200 rpm. Find suitable diameter of the shaft if the angle of twist is not to exceed 1° per metre length. Take rigidity modulus as 80 GPa. Also find the shear stress induced.

Solution:

Given data: Power to be transmitted $P = 60$ kW

Speed $N = 1200$ rpm

Angle of twist $\theta = 1° = 1 \times \frac{\pi}{180}$ radians.

Length $l = 1$ m $= 1000$ mm

Rigidity modulus $G = 80$ GPa $= 80 \times 10^3$ MPa

For solid shaft $J = \frac{\pi}{32} d^4$ and $R = \frac{d}{2}$

To find diameter, since angle of twist is given as constraint, consider $\frac{T}{J} = \frac{G\theta}{l}$.

$$\text{Torque } T = \frac{P \times 60000}{2\pi N} \times 1000 = \frac{60 \times 60000}{2\pi \times 1200} \times 1000$$

$$T = 47.75 \times 10^3 \text{ N-mm}$$

$$\frac{47.75 \times 10^3}{\left(\frac{\pi}{32} d^4\right)} = \frac{80000}{1000} \times \frac{1 \times \pi}{180}$$

$\therefore$ Required diameter $d = 24.29$ mm

To find shear stress induced, consider $\frac{T}{J} = \frac{\tau}{R}$

$$\frac{47.75 \times 10^3}{\frac{\pi}{32} \times 24.29^4} = \frac{\tau}{\left(\frac{24.29}{2}\right)}$$

Shear stress induced $\tau = 16.97$ N/mm^2.

Problem 12 A hollow machine shaft is required to transmit 30 kW at 600 rpm. The inside diameter is half of the external diameter. The angle of twist is not to exceed 0.6°/m length. Rigidity modulus is 82 GPa. Find suitable diameter of the shaft. Also find the maximum shear stress induced in the shaft.

Solution: **Given data:** Power $P = 30$ kW

Speed $N = 600$ rpm

Inside diameter $d_i = \frac{1}{2}\, d_o = 0.5\, d_o$

Angle of twist $\theta = 0.6° = 0.6 \times \frac{\pi}{180}$ radians

Length $l = 1$ m $= 1000$ mm

Rigidity modulus $G = 82$ GPa $= 82 \times 10^3$ MPa

$$\text{Polar moment of inertia } J = \frac{\pi}{32}\,(d_o^{\,4} - d_i^{4}) = \frac{\pi}{32}\,[d_o^{\,4} - (0.5 d_o)^4]$$

$$J = \frac{\pi}{32}\, d_o^{\,4}\,(0.9375)$$

$$R = R_o = \frac{d_o}{2} \text{ (For maximum shear stress)}$$

$$\text{Torque transmitted } T = \frac{P \times 60000}{2\pi N} \times 1000 \text{ N-mm}$$

$$= \frac{30 \times 60000}{2\pi \times 600} \times 1000 = 477.46 \times 10^3 \text{ N-mm}$$

To find diameters limiting $\theta = 0.6°$, consider

$$\frac{T}{J} = \frac{G\theta}{l}$$

$$\frac{477.46 \times 10^3}{\frac{\pi}{32}\, d_o^{\,4}\,(0.9375)} = \frac{82000 \times 0.6}{1000} \times \frac{\pi}{180}$$

∴ Outside diameter $d_o = 49.58$ mm

∴ Inside diameter $d_i = 0.5\, d_o = 24.79$ mm

To find max shear stress induced, consider $\frac{T}{J} = \frac{\tau}{R_o}$

$$\frac{477.46 \times 10^3}{\frac{\pi}{32}\,(49.58^4 - 24.79^4)} = \frac{\tau}{\left(\frac{49.58}{2}\right)}$$

∴ Shear stress $\tau = 21.28$ N/mm^2.

Problem 13 Find the maximum torque that can be applied on a shaft of diameter 90 mm. The allowable shear stress is 45 N/mm^2 and angle of twist in a length of 3 m is to be limited to 1.2°. Modulus of rigidity = 81 GPa.

Solution: **Given data:** Diameter $d = 90$ mm

$$\text{Allowable shear stress } \tau = 45 \text{ N/mm}^2$$

$$\text{Angle of twist } \theta = 1.2° = 1.2 \times \frac{\pi}{180} \text{ radians}$$

$$\text{Length } l = 3 \text{ m} = 3000 \text{ mm}$$

$$\text{Rigidity modulus } G = 81 \text{ GPa} = 81 \times 10^3 \text{ MPa}$$

$$\text{Polar moment of inertia } J = \frac{\pi}{32} d^4 = \frac{\pi}{32} \times 90^4$$

$$\text{Radius } R = \frac{d}{2} = \frac{90}{2} = 45 \text{ mm}$$

(i) Consider the shear stress limitation $\frac{T}{J} = \frac{\tau}{R}$

$$\therefore \quad \text{Torque } T = \frac{\tau}{R}.J = \frac{45}{45} \times \left(\frac{\pi}{32} \times 90^4\right) = 6.441 \times 10^6 \text{ N-mm}$$

(ii) Consider the angle of twist limitation, $\frac{T}{J} = \frac{G\theta}{l}$

$$\text{Torque } T = \frac{G\theta}{l}.J = \frac{81 \times 10^3 \times 1.2}{3000} \times \frac{\pi}{150} \times \frac{\pi}{32} \times 90^4$$

$$= 3.642 \times 10^6 \text{ N-mm}$$

$\therefore$ Maximum torque that can be transmitted, satisfying both conditions is

$$T = 3.642 \times 10^6 \text{ N-mm.}$$

(Minimum of the above two values)

Problem 14 Find the safe power that can be transmitted by a solid shaft of diameter 50 mm rotating at 900 rpm limiting the shear stress to 54 N/mm^2. The angle of twist is not to exceed 0.9°/m length. Take rigidity modulus as 80 GPa.

Solution: **Given data:** Diameter $d = 50$ mm

$$\text{Speed } N = 900 \text{ rpm}$$

$$\text{Allowable shear stress } \tau = 54 \text{ N/mm}^2$$

$$\text{Angle of twist } \theta = 0.9° = 0.9 \times \frac{\pi}{180} \text{ radians}$$

$$\text{Length } l = 1 \text{ m} = 1000 \text{ mm}$$

$$\text{Rigidity modulus } G = 80 \text{ GPa} = 80 \times 10^3 \text{ MPa}$$

$$\text{Polar moment of inertia } J = \frac{\pi}{32} d^4 = \frac{\pi}{32} 50^4 \text{ mm}^4$$

$$R = \frac{d}{2} = \frac{50}{2} = 25 \text{ mm}$$

From shear stress point of view, consider $\frac{T}{J} = \frac{\tau}{R}$

$$\therefore \quad \text{Torque capacity } T = \frac{\tau}{R} \cdot J = \frac{54}{25} \times \frac{\pi}{32} \times 50^4 = 1.325 \times 10^6 \text{ N-mm}$$

From the consideration of angle of twist, $\frac{T}{J} = \frac{G\theta}{l}$

$$\therefore \quad \text{Torque capacity } T = \frac{G\theta}{l} \cdot J = \frac{80 \times 10^3 \times 0.9}{1000} \times \frac{\pi}{180} \times \frac{\pi}{32} 50^4$$

$$T = 771.06 \times 10^3 \text{ N-mm}$$

$$\therefore \quad \text{Torque capacity of the shaft} = T = 771.06 \times 10^3 \text{ N-mm}$$

(Minimum of the above two values)

$$\text{Power } P = \frac{2\pi NT}{60000} \text{ kW, } T \text{ in N-m}$$

$$T = 771.06 \times 10^3 \text{ Nmm} = 771.06 \text{ N-m}$$

$$\therefore \quad P = \frac{2\pi \times 900 \times 771.06}{6000} = 72.67 \text{ kW.}$$

Problem 15 A hollow shaft has outside diameter of 180 mm and inside diameter of 120 mm and rotates at 1500 rpm. The allowable shear stress is not to exceed 54 MN/m^2 and angle of twist is to be limited to 0.7°/m length. Taking G = 84 GPa determine the power in kW that can be safely transmitted by this shaft.

Solution:

Given data: Outside diameter $d_o = 180$ mm

Inside diameter $d_i = 120$ mm

Speed $N = 1500$ rpm

Allowable shear stress $\tau = 54 \text{ MN/mm}^2 = 54 \text{ MPa} = 54 \text{ N/mm}^2$

Angle of twist $\theta = 0.7° = 0.7 \times \frac{\pi}{180}$ radians

Rigidity modulus $G = 84 \text{ GPa} = 84 \times 10^3 \text{ MPa}$

$$\text{Polar moment of inertia } J = \frac{\pi}{32} (d_o^4 - d_i^4)$$

$$= \frac{\pi}{32} (180^4 - 120^4) = 82.70 \times 10^6 \text{ mm}^4$$

$$\text{Outer radius } R = R_o = \frac{d_o}{2} = \frac{180}{2} = 90 \text{ mm}$$

(i) Considering shear stresses limitation, $\frac{T}{J} = \frac{\tau}{R}$

$$\therefore \quad \text{Torque capacity } T = \frac{\tau}{R} \cdot J = \frac{54}{90} \times 82.7 \times 10^6 = 49.62 \times 10^6 \text{ N-mm}$$

(ii) Considering angle of twist, $\frac{T}{J} = \frac{G\theta}{l}$

$$\therefore \quad \text{Torque capacity } T = \frac{G\theta}{l} \cdot J = \frac{84 \times 10^3}{1000} \times \frac{0.7\pi}{180} \times 82.7 \times 10^6$$

$$T = 84.87 \times 10^6 \text{ N-mm}$$

$$\therefore \quad \text{Maximum torque capacity } T = 49.62 \times 10^6 \text{ N-mm} = 49.62 \times 10^3 \text{ Nm}$$

(Minimum of the above two values)

$$\therefore \quad \text{Power capacity } P = \frac{2\pi NT}{60000} \text{ kW, } T \text{ in N-m}$$

$$P = \frac{2\pi \times 1500 \times 49.62 \times 10^3}{60000} = 7794.29 \text{ kW.}$$

Problem 16 A hollow shaft has an external diameter of 360 mm and internal diameter of 220 mm. While transmitting power, it is observed that the angle of twist was 0.5° in a length of 3.5 m. If $C = 8 \times 10^4$ N/mm^2 and $T_{max} = 1.4\ T_{mean}$, calculate the power transmitted at 5 rps and the maximum shear stress induced in the shaft.

Solution: **Given data:**

$$\text{Outside } d_o = 360 \text{ mm}$$

$$\text{Inside diameter } d_i = 220 \text{ mm}$$

$$\text{Angle of twist } \theta = 0.5° = 0.5 \times \frac{\pi}{180} \text{ radians}$$

$$\text{Length } l = 3.5 \text{ m} = 3500 \text{ mm}$$

$$\text{Rigidity modulus } C = G = 8 \times 10^4 \text{ N/mm}^2$$

$$\text{Speed } N = 5 \text{ rps} = 5 \times 60 = 300 \text{ rpm}$$

$$\text{Polar moment of inertia } J = \frac{\pi}{32}(d_o^4 - d_i^4) = \frac{\pi}{32}(360^4 - 220^4)$$

$$J = 1.419 \times 10^9 \text{ mm}^4$$

(i) From the point of angle of twist, $\frac{T}{J} = \frac{G\theta}{l}$

$$\therefore \quad \text{Maximum torque } T_{max} = \frac{G\theta}{l} \cdot J = \frac{8 \times 10^4}{3500} \times \frac{0.5\pi}{180} \times 1.419 \times 10^9$$

$$T_{max} = 283.038 \times 10^6 \text{ N-mm}$$

$$\text{Given } T_{max} = 1.4\, T_{mean} \quad \therefore \quad T_{mean} = \frac{T_{max}}{1.4} = \frac{283.038 \times 10^6}{1.4}$$

$$T_{mean} = 202.17 \times 10^6 \text{ N-mm} = 202.17 \times 10^3 \text{ N-m}$$

$$\therefore \quad \text{Power transmitted } P = \frac{2\pi NT}{60000} = \frac{2\pi \times 300 \times 202.17 \times 10^3}{60000}$$

$$P = 6351.36 \text{ kW}$$

(ii) Shear stresses induced

$$\text{Consider } \frac{T_{max}}{J} = \frac{\tau}{R}$$

$$\therefore \quad \text{Maximum shear stress } \tau = \frac{T_{max} \cdot R}{J} = \frac{T_{max} \cdot R_o}{J}$$

$$\tau = \frac{283.038 \times 10^6}{1.419 \times 10^9} \times \left(\frac{360}{2}\right) = 35.90 \text{ N/mm}^2.$$

Problem 17 A hollow shaft of 200 mm outer diameter has the same area as that of a solid shaft of diameter 100 mm. Compare

(i) the power transmitted by hollow shaft with that of solid shaft at same speed; and

(ii) the angle of twist of them made of same length and material.

Solution: **Given data:** Outside diameter of hollow shaft d_o = 200 mm

Inside diameter = d_i

Dia of solid shaft d = 100 mm

Given material, length and speed are same.

Hence, l, shear stress, τ, rigidity modulus G and speed N are same.

Area of hollow shaft = Area of solid shaft

$$\frac{\pi}{4}(d_o^2 - d_i^2) = \frac{\pi}{4} d^2$$

$$(200^2 - d_i^2) = 100^2$$

$$\therefore \quad d_i = 173.205 \text{ mm}$$

$$\text{(i) Power transmitted } P = \frac{2\pi NT}{60000} \text{ kW}$$

$$\text{Consider } \frac{T}{J} = \frac{\tau}{R}$$

$$\therefore \quad T = \frac{\tau J}{R}$$

$$\text{For hollow shaft } T_H = \frac{\tau\cdot\frac{\pi}{32}\,(d_o^4 - d_i^4)}{\frac{d_o}{2}}$$

$$= \tau\cdot\frac{\pi}{16}\left(\frac{200^4 - 173.205^4}{200}\right)$$

$$= 687.22 \times 10^3\ \tau \text{ Nmm}$$

$$T_H = 687.22\ \tau \text{ N-mm}$$

$$\text{For solid shaft } T_S = \frac{\tau\cdot\frac{\pi}{32}\,d^4}{\left(\frac{d}{2}\right)}$$

$$= \tau\cdot\frac{\pi}{16}\,(100^3) = 196.35 \times 10^3\ \tau \text{ N-mm} = 196.35\tau \text{ N-m}$$

$$\text{Comparing the powers } \frac{P_H}{P_S} = \frac{\left(\frac{2\pi\, NT_H}{60000}\right)}{\left(\frac{2\pi\, NT_S}{60000}\right)} = \frac{T_H}{T_S}$$

$$\frac{P_H}{P_S} = \frac{687.22\tau}{196.35\tau} = 3.5$$

The hollow shaft can transmit 3.5 time the power that can be transmitted by solid shaft of same area.

(ii) Angle of twist

$$\text{Consider } \frac{T}{J} = \frac{G\theta}{l}$$

$$\therefore \qquad \theta = \frac{Tl}{GJ}$$

$$\text{For hollow shaft } \theta_H = \frac{T_H l}{GJ_H} = \frac{687.22\ \tau\cdot l}{G\cdot\frac{\pi}{32}\,(200^4 - 173.205^4)}$$

$$\theta_H = 10 \times 10^{-6}\,\frac{l\tau}{G}.$$

$$\text{For solid shaft } \theta_S = \frac{T_S\cdot l}{GJ_H} = \frac{196.35\tau\cdot l}{G\cdot\left(\frac{\pi}{32} \times 100^4\right)} = 20 \times 10^{-6}\,\frac{l\tau}{G}$$

$$\text{Comparing } \frac{\theta_H}{\theta_S} = \frac{\left(10 \times 10^{-6}\,\frac{l\tau}{G}\right)}{\left(20 \times 10^{-6}\,\frac{l\tau}{G}\right)} = 0.5$$

∴ The angle of twist of hollow shaft is half of that of the solid shaft.

Problem 18 A solid circular shaft has to transmit 100 kW at 1200 rpm. Find the diameter of the shaft, if the shear stress of the material is not to exceed 72 MPa. The starting torque is 1.25 times the mean torque. What percentage of saving in material would be obtained if the shaft is replaced by a hollow one whose internal diameter is 0.6 times the external diameter. The length, material and allowable shear stress are same.

Solution: **Given data:** Power to be transmitted $P = 100$ kW

$$\text{Speed } N = 1200 \text{ rpm}$$

$$\text{Allowable shear stress } \tau = 72 \text{ MPa}$$

$$\text{Maximum torque } T_{max} = 1.25\ T_m$$

$$\text{For hollow shaft } d_i = 0.6\ d_o$$

$$\text{Power } P = \frac{2\pi N T_m}{60000}$$

$$\therefore \quad T_{mean} = \frac{P \times 60000}{2\pi N} \text{ N-m}$$

$$\therefore \quad T = T_{max} = \frac{P \times 60000}{2\pi N} \times 1.25 \text{ Nm}$$

$$= \frac{100 \times 60000}{2\pi \times 1200} \times 1.25 \times 1000 \text{ N mm}$$

$$T = 994.718 \times 10^3 \text{ Nmm}$$

(i) Consider solid shaft

Torsion equation, $\dfrac{T}{J} = \dfrac{\tau}{R}$

$$J = \frac{\pi}{32} d^4 \text{ and } R = \frac{d}{2}$$

$$\frac{994.718 \times 10^3}{\left(\frac{\pi d^4}{32}\right)} = \frac{72}{\left(\frac{d}{2}\right)}$$

$\therefore$ Diameter of the solid shaft $d = 41.28$ mm

(ii) Consider hollow shaft $\quad d_i = 0.6\ d_o$

$$J = \frac{\pi}{32}(d_o^4 - d_i^4) = \frac{\pi}{32}[d_o^4 - (0.6\ d_o)^4]$$

$$= \frac{\pi}{32} d_o^4 (0.8704)$$

$$R = \frac{d_o}{2}$$

Substituting in torsion equation

$$\frac{994.718 \times 10^3}{\frac{\pi}{32} d_o^4 (0.8704)} = \frac{72}{\left(\frac{d_o}{2}\right)}$$

∴ Outside diameter of hollow shaft d_o = 43.24 mm

∴ Inside diameter $d_i = 0.6\ d_o$ = 25.94 mm

Percentage saving in material

$$\% \text{ saving in} = \frac{\text{Weight of solid shaft} - \text{Weight of hollow shaft}}{\text{Weight of solid shaft}} \times 100$$

$$\text{Weight} = \text{Volume} \times \text{Specific weight} = v \times w$$

$$= \text{Area} \times \text{Length} \times w$$

$$\text{For solid shape } W_S = \frac{\pi}{4} d^2 lw$$

$$\text{For hollow shaft } W_H = \frac{\pi}{4} (do^2 - di^2)\ lw$$

$$\% \text{ saving in material} = \frac{W_S - W_H}{W_S} \times 1000$$

$$= \frac{A_S lw - A_H lw}{A_S \cdot lw} \times 100 = \frac{A_S - A_H}{A_S} \times 100$$

Note: Hence, % saving in weight, is same as % saving in area, since the material and length are same.

$$= \frac{\frac{\pi}{4} \times 41.28^2 - \frac{\pi}{4} (43.24^2 - 25.94^2)}{\frac{\pi}{4} \times 41.28^2} \times 100$$

% saving in material = 29.766%.

Problem 19 Prove that a hollow shaft is stronger than a solid shaft of the same material length and weight.

Solution: **Given data:** Material is same. Hence, allowable shear stress τ, specific weight w and rigidity modulus G are same, length l is same. Weight W is same

$$\text{Shear stresses } \tau_S = \tau_H$$

$$\text{Specific weight } w_S = w_H$$

$$\text{Rigidity modulus } G_S = G_H$$

$$\text{Length } l_S = l_H \text{ and weight } W_S = W_H$$

Let d = Diameter of solid shaft

d_i = Inside diameter of hollow shaft

$$d_o = \text{Outside diameter of hollow shaft}$$

$$\text{Weight} = \text{Volume} \times \text{Specific weight}$$

$$= \text{Area} \times \text{Length} \times \text{Specific weight}$$

$$W = A l w$$

$$\text{For solid shaft } W_S = \frac{\pi}{4} d^2 \, lw$$

$$\text{For hollow shaft } W_H = \frac{\pi}{4} (d_o^2 - d_i^2) \, lw$$

$$\text{Given weight are same, } W_S = W_H$$

$$\frac{\pi}{4} d^2 \, lw = \frac{\pi}{4} (d_o^2 - d_i^2) \, lw$$

$$\therefore \quad d^2 = (d_o^2 - d_i^2)$$

$$d = \sqrt{d_o^2 - d_i^2}$$

$$\text{Assume ratio of diameters } \frac{d_i}{d_o} = k.$$

$$\therefore \quad d_i = k \cdot d_o.$$

Substituting in the above equation

$$d = \sqrt{d_o^2 - (k \, d_o)^2} = \sqrt{d_o^2 (1 - k^2)}$$

$$\therefore \quad d = d_o \sqrt{1 - k^2}.$$

(i) Compression based on torque capacity

$$\text{Equation of torsion } \frac{T}{J} = \frac{\tau}{R}$$

$$\therefore \quad \text{Torque capacity } T = \frac{\tau \cdot J}{R}$$

$$\text{For solid shaft } T_S = \frac{\tau \cdot J_S}{R_S} = \frac{\tau \cdot \left(\frac{\pi}{32} d^4\right)}{\left(\frac{d}{2}\right)}$$

$$T_S = \frac{\pi d^3}{16} \tau$$

$$\text{For hollow shaft, } T_H = \frac{\tau \cdot J_H}{R_H} = \frac{\tau \cdot \frac{\pi}{32} (d_o^4 - d_i^4)}{\left(\frac{d_o}{2}\right)}$$

$$= \tau \cdot \frac{\pi}{16} \left(\frac{d_o^4 - d_i^4}{d_o}\right)$$

$$= \tau \cdot \frac{\pi}{16}\left(\frac{d_o^{\,4} - (k\,d_o)^4}{d_o}\right)$$

$$T_H = \tau \cdot \frac{\pi}{16}\,\frac{d_o^{\,4}(1 - k^4)}{d_o} = \tau \cdot \frac{\pi}{16}\,d_o^{\,3}\,(1 - k^4)$$

$$\text{Comparing, } \frac{T_H}{T_S} = \frac{\tau \cdot \frac{\pi}{16}\,d_o^{\,3}(1 - k^4)}{\tau \cdot \frac{\pi}{16}\,d^3} = \frac{d_o^{\,3}(1 - k^4)}{d^3}$$

Substituting $d = d_o\sqrt{1 - k^2}$ from above.

$$\frac{T_H}{T_S} = \frac{d_o^{\,3}\,(1 - k^4)}{(d_o\sqrt{1 - k^2})^3} = \frac{d_o^{\,3}\,(1 - k^4)}{d_o^{\,3}(\sqrt{1 - k^2})^3}$$

$$= \frac{1 - k^4}{(\sqrt{(1 - k^2})^3} = \frac{(1 + k^2)\,(1 - k^2)}{(1 - k^2)\,\sqrt{1 - k^2}} = \frac{1 + k^2}{\sqrt{1 - k^2}}$$

Since $1 + k^2$ will be $> \sqrt{1 - k^2}$, this ratio $\frac{T_H}{T_S} > 1$.

Hence, a hollow shaft is stronger than a solid shaft for same material, weight and length.

(ii) Compression based on torsional shiftness

$$\text{Torsional shiftness } \frac{T}{\theta} = \frac{GJ}{l}$$

$$\text{For solid shaft } \left(\frac{T}{\theta}\right)_S = \frac{G}{l} \cdot J_S = \frac{G}{l}\,\frac{\pi}{32}\,d^4$$

$$\text{For hollow shaft } \left(\frac{T}{\theta}\right)_H = \frac{G}{l} \cdot J_H = \frac{G}{l} \cdot \frac{\pi}{32}\,(d_o^{\,4} - d_i^{\,4})$$

$$\left(\frac{T}{\theta}\right)_H = \frac{G}{l} \cdot \frac{\pi}{32}\,d_o^{\,4}\,(1 - k^4)$$

$$\text{Comparing } \frac{(T/\theta)_H}{(T/\theta)_S} = \frac{\frac{G}{l} \cdot \frac{\pi}{32}\,d_o^{\,4}(1 - k^4)}{\frac{G}{l} \cdot \frac{\pi}{32}\,d^4} = \frac{d_o^{\,4}\,(1 - k^4)}{d^4}$$

Substituting $d = d_o\sqrt{1 - k^2}$

$$\frac{(T/\theta)_H}{(T/\theta)_S} = \frac{d_o^{\,4}(1 - k^4)}{(d_o\sqrt{1 - k^2})^4} = \frac{d_o^{\,4}\,(1 - k^4)}{d_o^{\,4}\,(1 - k^2)^2}$$

$$= \frac{(1 + k^2)(1 - k^2)}{(1 - k^2)^2} = \frac{1 + k^2}{1 - k^2}$$

Since numerator $1 + k^2 >$ denominator $1 - k^2$,

$$\frac{(T/\theta)_H}{(T/\theta)_S} > 1$$

Hence, hollow shaft is more stronger than solid shaft of same material, weight and length.

Problem 20 Compare the weight of hollow and solid shafts when their torsional strengths are same.

Solution: **Given data:** Torsional strengths are same

$$\text{Torsional strength} = \text{Torque capacity}$$

$$\therefore \quad T_S = T_H$$

Let

d = Diameter of solid shaft

d_o, d_i = Outside and inside diameters of hollow shafts

τ = Allowable shear stress for the material

$= \tau_S = \tau_H$

$w_S = w_H$ = Specific weight of material

W_S = Weight of solid shaft

W_H = Weight of hollow shaft

J_S = Polar *MI* of solid shaft

J_H = Polar *MI* of hollow shaft

$$\text{Consider } \frac{T}{J} = \frac{\tau}{R}$$

$$\therefore \quad \text{Torsional strength } T = \frac{\tau \cdot J}{R}$$

$$\text{For solid shaft } T_S = \tau \cdot \frac{J_S}{R} = \frac{\tau \cdot \frac{\pi}{32} d^4}{\left(\frac{d}{2}\right)} = \frac{\pi}{16} d^3 \tau$$

$$\text{For hollow shaft } T_H = \tau \cdot \frac{J_H}{R} = \frac{\tau \cdot \left[\frac{\pi}{32} (d_o^4 - d_i^4)\right]}{\left(\frac{d_o}{2}\right)} = \frac{\pi}{16} \frac{(d_o^4 - d_i^4)}{d_o} \tau.$$

Equating these two

$$\frac{\pi}{16} d^3 \tau = \frac{\pi}{16} \frac{(d_o^4 - d_i^4)}{d_o} \tau$$

$$d^3 = \frac{d_o^4 \left[1 - \left(\frac{d_i}{d_o}\right)^4\right]}{d_o} = d_o^3 (1 - k^4)$$

where $$k = \frac{d_i}{d_o}$$

$\therefore$ $$d_i = d_o (1 - k^4)^{1/3}$$

$$\text{Weight} = \text{Volume} \times \text{Specific weight}$$
$$= \text{Area} \times \text{Length} \times \text{Specific weight}$$

$\therefore$ $$W_H = A_H \cdot lw$$
$$W_S = A_S lw$$

Comparing $$\frac{W_H}{W_S} = \frac{A_H lw}{A_S lw} = \frac{\frac{\pi}{4}(d_o^2 - d_i^2)}{\frac{\pi}{4}(d^2)}$$

$$= \frac{d_o^2\left[1 - \left(\frac{d_i}{d_o}\right)^2\right]}{d^2} = \frac{d_o^2(1 - k^2)}{[d_o(1 - k^4)^{1/3}]^2}$$

$$= \frac{d_o^2(1 - k^2)}{d_o^2(1 - k^4)^{2/3}} = \frac{1 - k^2}{(1 - k^4)^{2/3}}$$

Since k is less than 1, the numerator is smaller than the denominator. Hence, the ratio

$$\frac{W_H}{W_S} = \frac{1 - k^2}{(1 - k^4)^{2/3}} < 1$$

Hence, the weight of a hollow shaft is less than of a solid shaft for a given length, material and torsional strength.

Problem 21 Compare the weight, strength and shiffness of a hollow shaft with that of a solid shaft made of same material and length having same external diameters. The internal diameter of the hollow shaft is half of its external diameter.

Solution: **Given data:**

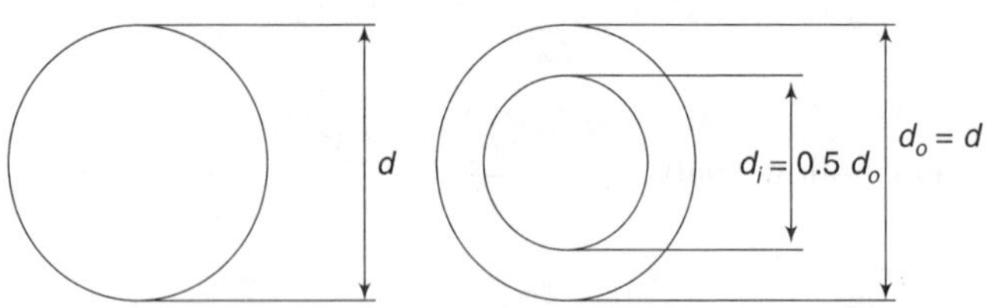

(a) Solid shaft (b) Hollow shaft

Fig. 8.7

Diameter $= d$ Outside diameter

Area $A_S = \frac{\pi}{4} d^2$

Ouside diameter $d_o = d$

Inside diameter $d_i = 0.5\, d_o = 0.5\, d$

$$R = \frac{d}{2}$$

$$J_S = \frac{\pi}{32} d^4$$

$$\text{Area } A_H = \frac{\pi}{4} (d_o^2 - d_i^2)$$

$$= \frac{\pi}{4} [d_o^2 - (0.5\, d_o)^2]$$

$$= \frac{\pi}{4} d_o^2 (1 - 0.5^2)$$

$$= \frac{\pi}{4} d_o^2 (0.75)$$

$$A_H = \frac{\pi}{4} d^2 (0.75)$$

$$R = \frac{d_o}{2} = \frac{d}{2}$$

$$J_H = \frac{\pi}{32} (d_o^4 - d_i^4)$$

$$= \frac{\pi}{32} [d_o^4 - (0.5\, d_o)^4]$$

$$= \frac{\pi}{32} d_o^4 (1 - 0.5^4)$$

$$= \frac{\pi}{32} d^4 (0.9375)$$

(i) Weight W = Volume × Specific weight = $A\,l\,w$

$$\text{For solid shaft } W_S = \frac{\pi}{4} d^2 lw$$

$$\text{For hollow shaft } W_H = \frac{\pi}{4} d^2 (0.75)\, lw$$

$$\text{Comparing, } \frac{\text{Weight of hollow shaft}}{\text{Weight of solid shaft}} = \frac{W_H}{W_S} = \frac{\frac{\pi}{4} d^2\, 0.75 lw}{\frac{\pi}{4} d^2 lw} = 0.75$$

The weight of hollow shaft is 0.75 times the weight of hollow shaft.

(ii) Strength $T = \frac{\tau \cdot J}{R}$

Strength is the torque capacity of a shaft

$$\text{For solid shaft } T_S = \frac{\tau \cdot J}{R} = \frac{\tau \cdot \frac{\pi d^4}{32}}{\left(\frac{d}{2}\right)} = \tau \cdot \frac{\pi d^3}{16}$$

$$\text{For hollow shaft } T_H = \frac{\tau \cdot J_H}{R} = \frac{\tau \cdot \frac{\pi}{32} d^4 (0.9375)}{\left(\frac{d}{2}\right)}$$

$$T_H = \tau \cdot \frac{\pi}{16} d^3 (0.9375)$$

$$\text{Comparing } \frac{T_H}{T_S} = \frac{\tau \cdot \frac{\pi}{16} d^3 (0.9375)}{\tau \cdot \frac{\pi}{16} d^3} = 0.9375$$

The strength of a hollow shaft is 0.9375 times the strength of solid shaft.

(iii) Stiffness: Stiffness for a shaft is $\frac{T}{\theta}$.

$$\frac{T}{J} = \frac{G\theta}{l} \text{ or } \left(\frac{T}{\theta}\right) = \frac{G.J}{l}.$$

$$\text{For solid shaft } \left(\frac{T}{\theta}\right)_S = \frac{G \cdot J_S}{l} = \frac{G}{l}\frac{\pi d^4}{32}$$

$$\text{For hollow shaft } \left(\frac{T}{\theta}\right)_H = \frac{G \cdot J_H}{l} = \frac{G.\frac{\pi}{32} d^4 (0.9375)}{l}$$

$$\text{Comparing } \frac{(T/\theta)_H}{(T/\theta)_S} = \frac{\left(\dfrac{G \frac{\pi}{32} d^4 (0.9375)}{l}\right)}{\left(\frac{G}{l}\frac{\pi}{32} d^4\right)} = 0.9375$$

The stiffness of the hollow shaft is 0.9375 times that of solid shaft.

Problem 22 A solid circular shaft of diameter 50 mm transmits a torque of 1.2 kNm. Determine the shear stress induced at the centre, at a distance of 12 mm from the centre and at the outer surface. Plot a graph for the shear stress distribution.

Solution: **Given data:**

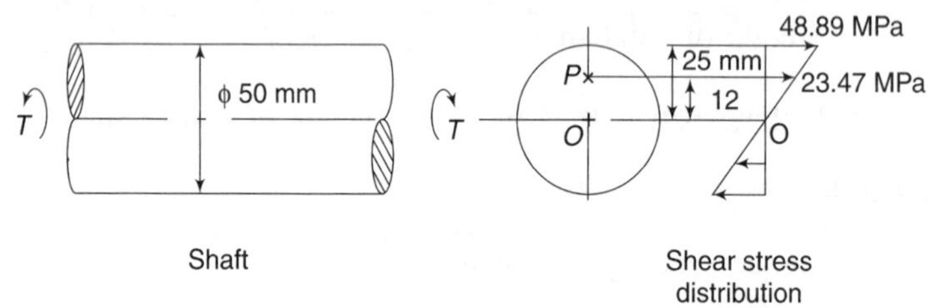

Fig. 8.8

$$\text{Diameter } d = 50 \text{ mm}$$

$$\therefore \quad R = \frac{d}{2} = \frac{50}{2} = 25 \text{ mm}$$

$$\text{Torque transmitted } T = 1.2 \text{ kNm} = 1.2 \times 10^6 \text{ N-mm}$$

$$\text{Polar moment of inertia } J = \frac{\pi}{32} d^4 = \frac{\pi}{32} 50^4 \text{ mm}^4$$

$$\text{Consider } \frac{T}{J} = \frac{\tau}{R}$$

$$\therefore \quad \text{Shear stress } \tau = \frac{T}{J} \cdot R$$

(i) At the centre O, radius $R = 0$

$$\therefore \quad \tau_o = \frac{T}{J} \times 0 = 0$$

Shear stress is zero at the centre.

(ii) At point P at a distance $R = 12$ mm from centre

$$\text{Shear stress } \tau_P = \frac{T}{J} \cdot R = \frac{1.2 \times 10^6}{\left(\frac{\pi}{32} \times 50^4\right)} \times 12 = 23.47 \text{ N/mm}^2$$

(iii) At the outer surface, $R = \frac{d}{2} = 25$ mm

$$\text{Shear stress } \tau = \frac{T}{J} \cdot R = \frac{1.2 \times 10^6}{\left(\frac{\pi}{32} \times 50^4\right)} \times 25 = 48.89 \text{ N/mm}^2$$

The stress variation is linear and is as shown in Fig. 8.8.

Problem 23 What percentage strength of a solid steel shaft of diameter d is lost by boring a hole of 50% of its diameter. Also compare the ratio of weights of the two shafts (or compute the percentage of material saved).

Solution: **Given data:** Dia of solid shaft $= d$

Outside dia of hollow shaft $= d_o = d$

Inside dia of hollow shaft $d_i = 0.5\, d$

$$\text{Equation of torsion } \frac{T}{J} = \frac{\tau}{R}$$

$$\therefore \quad \text{Torque } T = \frac{\tau \cdot J}{R}$$

Strength of a shaft is its torque capacity

$$\text{For solid shaft } T_S = \frac{\tau \cdot \left(\frac{\pi}{32} d^4\right)}{\left(\frac{d}{2}\right)} = \tau \cdot \frac{\pi}{10} d^3$$

$$\text{For hollow shaft } T_H = \frac{\tau \cdot \frac{\pi}{32}(d_o^4 - d_i^4)}{\left(\frac{d_o}{2}\right)}$$

$$= \tau \cdot \frac{\frac{\pi}{32}[d^4 - (0.5d)^4]}{\left(\frac{d}{2}\right)} = \tau \cdot \frac{\pi}{16} d^3 (1 - 0.5^4)$$

$$T_H = \tau \cdot \frac{\pi}{16} d^3 (0.9375)$$

$$\% \text{ loss in strength} = \frac{T_S - T_H}{T_S} \times 100$$

$$= \frac{\tau \frac{\pi}{16} d^3 - \tau \frac{\pi}{16} d^3 (0.9375)}{\left(\tau \frac{\pi}{16} d^3\right)} \times 100 = 6.25\%$$

$$\text{Weight } W = Alw, \quad \text{where } w = \text{specific weight}$$

$$\therefore \quad \text{For solid shaft } W_S = A_S \, lw = \frac{\pi}{4} d^2 lw$$

$$\text{For hollow shaft } W_H = A_H \, lw = \frac{\pi}{4} (d_o^2 - d_i^2) lw$$

$$= \frac{\pi}{4} [d^2 - (0.5d)^2] lw$$

$$= \frac{\pi}{4} d^2 (1 - 0.5^2) lw$$

$$= \frac{\pi}{4} d^2 lw \, (0.75)$$

$$\text{Weight ratio } \frac{W_H}{W_S} = \frac{\frac{\pi}{4} d^2 lw \, (0.75)}{\frac{\pi}{4} d^2 \, lw} = 0.75$$

$$\% \text{ saving in material} = \frac{W_S - W_H}{W_S} \times 100$$

$$= \frac{\frac{\pi}{4} d^2 lw - \frac{\pi}{4} d^2 lw \, (0.75)}{\frac{\pi}{4} d^2 lw} \times 100 = 25\%.$$

Problem 24 A hollow shaft of 120 mm external diameter and thickness of wall of 15 mm transmits power at 1200 rpm. The angle of twist over a length of 1.2 m is 0.6 degrees. The shear stress is not to axceed 60 MPa. The modulus of rigidity is 80 GPa. Find the safe power that can be transmitted by this shaft.

Solution:

Given data: External diameter $d_o = 120$ mm

Wall thickness $t = 15$ mm

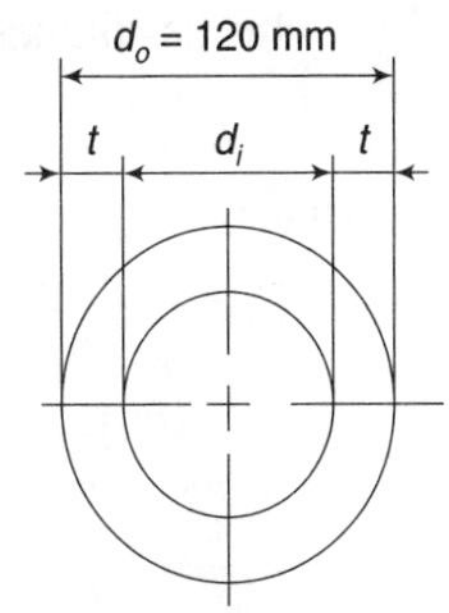

Fig. 8.9

$\therefore$ Internal diameter $d_i = d_o - 2t$

$$d_i = 120 - 2 \times 15 = 90 \text{ mm}$$

Speed $N = 1200$ rpm

Angle of twist $\theta = 0.6° = \dfrac{0.6 \times \pi}{180}$ radians

Shear stress $\tau = 60$ MPa

Length $l = 1.2$ m $= 1200$ mm

Rigidity modulus $G = 80$ GPa $= 80 \times 10^3$ MPa

For hollow shaft $J = \dfrac{\pi}{32}(d_o^4 - d_i^4) = \dfrac{\pi}{32}(120^4 - 90^4)$

$$= 13.916 \times 10^6 \text{ mm}^4$$

Radius $R = \dfrac{d_o}{2} = \dfrac{120}{2} = 60$ mm

To find power transmitted, we have to first find torque transmiting capacity

(i) Consider $\dfrac{T}{J} = \dfrac{\tau}{R}$

$$\therefore \quad T = \frac{\tau \cdot J}{R} = \frac{60 \times 13.916 \times 10^6}{60} = 13.916 \times 10^6 \text{ Nmm}$$

(ii) Again consider $\dfrac{T}{J} = \dfrac{G\theta}{l}$

$$\therefore \quad T = \frac{G\theta J}{l} = 80 \times 10^3 \times \frac{0.6\,\pi}{180} \times \frac{13.916 \times 10^6}{1200} = 9.175 \times 10^6 \text{ N-mm}$$

$\therefore$ Maximum torque capacity $= 9.715 \times 10^6$ N-mm
(Minimum of the above)

$$T = 9.715 \times 10^6 \text{ N-mm} = 9.715 \times 10^3 \text{ N-m}$$

We know that power $P = \dfrac{2\pi NT}{60000}$ kW

$$P = \frac{2 \times \pi \times 1200 \times 9.715 \times 10^3}{60000} = 1220.82 \text{ kW.}$$

Problem 25 Compare the weight of a solid shaft with that of a hollow shaft to transmit a given power at a given speed made of the same material. The inside diameter of the hollow shaft is 2/3 of the outside diameter.

Solution:

Given data: Diameter of solid shaft $= d$

Outside diameter of hollow shaft $= d_o$

Inside diameter of hollow shaft $= d_i = \frac{2}{3} d_o$

Because of the same material shear stress (τ) and specific weight (w) are same for both the shafts.

(i) Consider solid shaft $\frac{T}{J} = \frac{\tau}{R}$

$\therefore$ Torque capacity of solid shaft $T_S = \tau \cdot \frac{J_S}{R}$

$$T_S = \frac{\tau \cdot \frac{\pi}{32} d^4}{\left(\frac{d}{2}\right)} = \tau \cdot \frac{\pi}{16} d^3 \quad \ldots(1)$$

(ii) Consider hollow shaft $T_H = \frac{\tau \cdot J_H}{R_o} = \frac{\tau\left[\frac{\pi}{32}(d_o^4 - d_i^4)\right]}{\frac{d_o}{2}}$

But $\quad d_i = \frac{2}{3} d_o$

$$\therefore \quad T_H = \tau \cdot \frac{\pi}{16} \frac{\left(d_o^4 - \left(\frac{2}{3} d_o\right)^4\right)}{d_o} = \tau \cdot \frac{\pi}{16} \frac{d_o^4\left(1 - \left(\frac{2}{3}\right)^4\right)}{d_o}$$

$$T_H = \tau \cdot \frac{\pi}{16} d_o^3 \,(0.8025) \quad \ldots(2)$$

Since power transmitted and speed are same, torques will be same

$$\therefore \quad T_S = T_H$$

$$\tau \cdot \frac{\pi}{16} d^3 = \tau \cdot \frac{\pi}{16} d_o^3 \,(0.8025)$$

$$\therefore \quad d = d_o \,(0.9293)$$

(iii) Weight = Volume × Specific weight

$$W = Alw$$

For solid shaft $W_S = A_S lw = \frac{\pi}{4} d^2 lw$

For hollow shaft $W_H = A_H \, l \, w = \frac{\pi}{4} (d_o^2 \; d_i^2) lw$

$$= \frac{\pi}{4}\left(d_o^2 - \left(\frac{2}{3} d_o\right)^2\right) lw$$

$$= \frac{\pi}{4} d_o^2\left(1 - \left(\frac{2}{3}\right)^2\right) lw$$

$$= \frac{\pi}{4} d_o^2 \, lw \,(0.5556)$$

Comparing, $\quad \frac{W_S}{W_H} = \frac{\left(\frac{\pi}{4} d^2 lw\right)}{\left(\frac{\pi}{4} d_o^2 lw \times 0.5556\right)} = \frac{d^2}{d_o^2 \,(0.5556)}$

But $$d = d_o\ (0.9293)$$

$$\frac{W_S}{W_H} = \frac{(d_o \times 0.9293)^2}{d_o^2(0.5556)} = 1.5544$$

$\therefore$ Weight of solid shaft is 1.5544 times that of the hollow shaft.

Problem 26 The internal diameter of a hollow shaft is two thirds of its external diameter. Compare its torque capacity (or power capacity at a given speed) with that of a solid shaft of same weight and material.

Solution:

Given data: External diameter of the hollow shaft = d_o

Internal diameter, $d_i = \frac{2}{3} d_o$

Diameter of solid shaft = d

Material and weight are same.

Hence, shear stress τ and specific weight w are same.

$$W_S = W_H$$

$$(Alw)_S = (Alw)_H$$

$$\therefore \quad A_S = A_H$$

$$\frac{\pi}{4} d^2 = \frac{\pi}{4} (d_o^2 - d_i^2)$$

$$d^2 = d_o^2 - \left(\frac{2}{3} d_o\right)^2 = d_o^2 \left[1 - \left(\frac{2}{3}\right)^2\right]$$

$$d = d_o\ (0.7454)$$

(i) For solid shaft $J_S = \frac{\pi}{32} d^4$

$$J_S = \frac{\pi}{32} [0.7454\ d_o]^4 = 0.03031 \times d_o^4$$

and $$R = \frac{d}{2} = \frac{0.7454\ d_o}{2} = 03727\ d_o$$

$$\therefore \quad \text{Torque capacity } T_S = \frac{\tau \cdot J_S}{R} = \frac{\tau \times 0.03031 d_o^4}{0.3727\ d_o}$$

$$T_S = \tau\ (0.08133) d_o^3$$

(ii) For hollow shaft $J_H = \frac{\pi}{32} (d_o^4 - d_i^4)$

$$= \frac{\pi}{32} \left[d_o^4 - \left(\frac{2}{3} d_o\right)^4\right] = \frac{\pi}{32} d_o^4 \left(1 - \left(\frac{2}{3}\right)^4\right)$$

$$J_H = 0.07878\ d_o^4$$

$$R = R_o = \frac{d_o}{2}$$

$\therefore$ Torque capacity of hollow shaft $T_H = \tau.\frac{J_H}{R}$

$$T_H = \tau.\frac{0.07878\ d_o^4}{\frac{d_o}{2}} = 0.15756\ \tau.d_o^3$$

$$\text{Comparing } \frac{T_H}{T_S} = \frac{0.15756\ \tau.d_o^3}{0.08133\ \tau.d_o^3} = 1.937$$

For the same weight, hollow shaft can transmit 1.937 times the torque (power) capacity of a solid shaft.

Problem 27 A test specimen of diameter 12 mm and a gauge length of 50 mm stretched by 0.1325 mm under a tensile load of 60 kN. In a torsion test, the same rod is twisted by 0.9° under a torque of 50 N-m. Evaluate the Poisson's ratio and the three elastic modulii for the material.

Solution: **Given data:** Diameter d = 12 mm

Gauge length l = 50 mm

Elongation δl = 0.1325 mm

Tensile load F = 60 kN = 60×10^3 N

Angle of twist $\theta = 0.9° = 0.9 \times \frac{\pi}{180}$ radius

Torque T = 50 N-m = 50×10^3 N-mm

For tensile load, elongation $\delta l = \frac{Fl}{AE}$

$\therefore$ Young's modulus $E = \frac{Fl}{A\delta l}$

$$E = \frac{60 \times 10^3 \times 50}{\left(\frac{\pi}{4} \times 12^2\right) \times 0.1325} = 200.195 \times 10^3 \text{ N/mm}^2$$

For torsion test, $\frac{T}{J} = \frac{G\theta}{l}$

$$\frac{50 \times 10^3}{\left(\frac{\pi}{32} \times 12^4\right)} = \frac{G \times 0}{50} \cdot \frac{9 \times \pi}{180}$$

$\therefore$ Rigidity modulus $G = 78.18 \times 10^3$ N/mm^2

We know that $E = 2G\,(1 + \mu)$

$\therefore$ Poisson's ratio $\mu = \frac{E}{2G} - 1$

$$\mu = \frac{200.195 \times 10^3}{2 \times 78.18 \times 10^3} - 1 = 0.2803$$

$$\text{Again } E = 3K(1 - 2\mu)$$

$$\therefore \quad \text{Bulk modulus } K = \frac{E}{3(1 - 2\mu)}$$

$$K = \frac{200.195 \times 10^3}{3(1 - 2 \times 0.2803)} = 151.903 \times 10^3 \text{ N/mm}^2$$

Problem 28 A solid aluminium shaft 60 mm diameter is to be replaced by a hollow steel shaft of same outside diameter. Both shafts should have same stiffness. Find the required inner diameter of the steel shaft. Take $G_S = 3\ G_A$.

Solution: **Given data:** Diameter of solid aluminium shaft d = 60 mm

Outside diameter of hollow steel shaft d_o = 60 mm

Inside diameter of steel hollow shaft = d_i

$$G_S = 3\ G_A$$

$$\text{Torsional shiftness is same } \left(\frac{T}{\theta}\right)_A = \left(\frac{T}{\theta}\right)_S$$

$$\text{We know that } \frac{T}{J} = \frac{G\theta}{l}$$

$$\therefore \quad \left(\frac{T}{\theta}\right) = \frac{G \cdot J}{l}$$

$$\text{For aluminium shaft } \left(\frac{T}{\theta}\right)_A = \frac{G_A \cdot \frac{\pi}{32} d^4}{l}$$

$$\left(\frac{T}{\theta}\right)_A = \frac{G_A}{l} \cdot \frac{\pi}{32} \times 60^4 \qquad \text{...(i)}$$

$$\text{For steel shaft, } \left(\frac{T}{\theta}\right)_S = \frac{G_S \cdot \frac{\pi}{32}(d_o^4 - d_i^4)}{l}$$

$$\text{Put} \quad G_S = 3G_A$$

$$\left(\frac{T}{\theta}\right)_s = \frac{3G_A}{l} \cdot \frac{\pi}{32} (60^4 - d_i^4) \qquad \text{...(ii)}$$

Equating Eqs. (i) and (ii) for same stiffness

$$\frac{G_A}{l}\frac{\pi}{32}60^4 = \frac{3G_A}{l}\cdot\frac{\pi}{32}(60^4 - d_i^4)$$

$$60^4 = 3(60^4 - di^4)$$

$$\therefore \quad \frac{60^4}{3} = 60^4 - di^4$$

$$\therefore \quad di^4 = 60^4 - \frac{60^4}{3} = 60^4\left(1 - \frac{1}{3}\right)$$

$$d_i = 54.216 \text{ mm}$$

∴ The inside diameter of the hollow shaft = 54.216 mm

THINGS TO REMEMBER

- Equation of torsion $\frac{T}{J} = \frac{\tau}{R} = \frac{G\theta}{l}$
- Polar moment of inertia for solid shaft $J = \frac{\pi d^4}{32}$
- Polar moment of inertia for hollow shaft $J = \frac{\pi}{32}(d_o^4 - d_i^4)$
- Area of cross section of solid shaft $A = \frac{\pi}{4}d^2$
- Area of cross section of hollow shaft $A = \frac{\pi}{4}(d_o^2 - d_i^2)$
- Weight of a shaft $W = A.l.w$

  where w = Specific weight of material
- Material and length being same, comparison of weight is comparison of areas itself

$$\frac{W_H}{W_S} = \frac{A_H}{A_S}$$

- The weight being same, a hollow shaft is always stronger than a solid shaft
- Polar modulus of solid shaft $Z_P = \frac{J}{R} = \frac{\pi d^3}{16}$
- Polar modulus of hollow shaft $Z_P = \frac{J}{R} = \frac{\pi(d_o^4 - d_i^4)}{16\, d_o}$
- Shear stress is proportional to the radial distance (radius) from the centre or axis of the shaft $\frac{\tau}{r}$ is constant for a given problem
- Shear stress is maximum at the outer surface
- Shear stress at the centre is zero

- If the speed is same, then comparison of power transmission is same as comparison of torque transmission
- Equation of power transmitted by a shaft is $P = \frac{2\pi NT}{60000}$ kW, T in N-m
- A hollow shaft is stronger and stiffer than the solid shaft of same material, length and weight
- GJ is called torsional rigidity
- $\left(\frac{T}{\theta}\right)$ is known as torsional stiffness
- Strength for a shaft is its torque capacity

EXERCISES

1. A steel rod of 60 mm diameter is subjected to a torque of 3 kN-m. Determine the shear stress induced and angle of twist in a length of 1 m. Take G = 80 GPa.

(***Ans.*** τ = 70.74 MPa, θ = 1.689°)

2. A solid shaft of 30 mm diameter is made of steel having allowable shear stress of 63 MPa. Determine the power that can be transmitted by this shaft at 1000 rpm. Also find the angle of twist in a length of 0.9 m taking G = 81 GPa.

(***Ans.*** P = 34.97 kW, θ = 2.674°)

3. A solid shaft made of steel is required to transmit 30 kW at 1200 rpm. Determine the diameter of the shaft if allowable shear stress is 45 MPa and angle of twist is to be limited to 1°/metre length. Rigidity modulus is 84 kN/mm^2.

(***Ans.*** d = 35.89 mm)

4. A hollow shaft made of steel having diameter's ratio as 2 is required to transmit 60 kW at 3000 rpm. Determine the diameters of the shaft if the shear stress is not to exceed 54 MPa and angle of twist is to be 0.9°/m length, assume G = 81 GPa. Maximum torque to be transmitted is 30% greater than the mean torque.

(***Ans.*** d_o = 37.55 mm, d_i = 18.77 mm)

5. Find the safe power that can be transmitted by a solid shaft 30 mm diameter rotating at 1500 rpm limiting the shear stress to 72 MPa. The angle of twist is not to exceed 1.2°/m length. Take G = 80 GPa.

(***Ans.*** P = 20.93 kW)

6. Find the safe power that can be transmitted by a hollow shaft of external diameter 120 mm and internal diameter 75 mm rotating at 2400 rpm. The allowable shear stress is 42 MPa and angle of twist is not to exceed 0.6°/m length. Rigidity modulus is 80 GPa.

(***Ans.*** P = 3034.8 kW)

7. A solid shaft of diameter 180 mm is to be replaced by a hollow shaft of same material and length. The inside diameter of the hollow shaft is 0.5 times its external diameter. Determine the diameters of the hollow shaft and percentage saving in material.

(***Ans.*** d_o = 183.914, d_i 91.957 mm; 21.7%)

8. A solid shaft is to transmit 18 kW at 450 rpm. Taking allowable shear stress as 72 MPa find the diameter of the solid shaft. What percentage of saving in weight is obtained if this shaft is replaced by a hollow shaft whose internal diameter is 0.8 times the external diameter. The length, power to be transmitted and speed are equal in both cases.

(***Ans.*** d = 30 mm, d_o = 35.77 mm, d_i = 28.62 mm, 48.85%)

9. Determine the ratio of power transmitted by a hollow shaft and a solid shaft when both have the same material, weight and length and run at same speed. The diameter of the solid shaft is 150 mm and the outside diameter of the hollow shaft is 250 mm.

(***Ans.*** $P_H/P_S = 2.73$)

10. A solid shaft of 60 mm diameter has same area of cross section as that of a hollow shaft having 30 mm inside diameter. If both the shafts are made of same material and length calculate:

    (i) the ratio of torque transmitting capacities, and

    (ii) the ratio of angle of twist in solid and hollow shafts.

(***Ans.*** $T_H/T_S = 1.342$, $\theta_S/\theta_H = 1.118$)

11. A solid shaft of 120 mm diameter is proposed to be replaced by a hollow shaft of external and internal diameters ratio of 3 d_o = 4 d_i. The material and length are same. Find the diameters of the hollow shaft.

(***Ans.*** d_o = 181.423 mm, d_i = 136.067 mm)

12. A solid circular shaft of diameter 25 mm is 2 m long. Determine the torsional rigidity and the torsional strength taking modulus of rigidity as 50 GPa and allowable shear stress as 40 MPa.

(***Ans.*** GJ = 19175 Nm/rod/m. T = 0.1227 kN-m)

13. A shaft is subjected to a torque of 25 kN-m. Find the suitable diameter of the solid shaft if allowable shear stress and angle of twist are limited to 80 MPa and *i/a* length of 20 diameter respectively.

(***Ans.*** d = 116.75 and 153.93 and d = 153.93 mm)

14. Compare the weight, strength and stiffness of a hollow shaft with that of solid shaft having same external diameter. The inside diameter of the hollow shaft is 0.75 times to external diameter. Both shafts have same length and are made of same material.

$$\left(\textbf{\textit{Ans.}}\ W_H/W_S = 0.4375,\ T_H/T_S = 0.6836,\ \frac{(T/\theta)_H}{(T/\theta)_S} = 0.6836\right)$$

15. Show that hollow shaft whose diameter's ratio as 0.5 has a torsional strength equal to $\frac{15}{16}$ of that of a solid shaft of same outside diameter.

Objective Type Questions

1. Equation of torsion is

    (a) $\frac{T}{J} = \frac{\tau}{R} = \frac{G\theta}{l}$

    (b) $\frac{T}{R} = \frac{\tau}{J} = \frac{G\theta}{l}$

    (c) $\frac{T}{J} = \frac{G}{R} = \frac{\tau\theta}{l}$

    (d) $\frac{T}{J} = \frac{\tau}{l} = \frac{G\theta}{R}$

2. For same material, length and weight, hollow shaft is _______ compared to solid shaft
(a) of same strength (b) stronger
(c) weaker (d) None of the above

3. For same torque capacity, material and length, weights ratio of hollow shaft to solid shaft is
(a) 1 (b) > 1
(c) < 1 (d) None

4. Torsional rigidity is
(a) $G\theta$ (b) T/θ
(c) $\frac{T}{\tau}$ (d) GJ

5. Polar modulus of solid shaft of diameter d is
(a) $\frac{\pi d^3}{16}$ (b) $\frac{\pi d^3}{32}$
(c) $\frac{\pi d^4}{32}$ (d) $\frac{\pi d^4}{64}$

6. Polar modulus of hollow shaft of diameters d_o and d_i is
(a) $\frac{\pi(d_o^4 - d_i^4)}{32}$ (b) $\frac{\pi(d_o^4 - d_i^4)}{16\, d_o}$
(c) $\frac{\pi(d_o^3 - d_i^3)}{16}$ (d) $\frac{\pi(d_o^3 - d_i^3)}{32}$

7. For same material and length of solid and hollow shafts, comparison of weights is comparison of
(a) torque capacities (b) shear stress
(c) area of cross section (d) power transmission

8. For a given torque and cross section, the shear stress is proportional to
(a) polar modulus (b) rigidity modulus
(c) length (d) distance from its axis

9. Rigidity modulus is defined as the ratio of
(a) shear stress to shear strain (b) shear stress to radius
(c) shear stress to torque (d) shear stress to angle of twist

10. Shear stress in a circular shaft is maximum at
(a) centre (b) outer surface
(c) inner surface (d) none of these

11. Shear stress at the axis of a shaft is
(a) maximum (b) average
(c) zero (d) negative

12. For a given torque capacity and material, the angle of twist is proportional to
(a) radius (b) length
(c) both radius and length (d) none

13. A solid shaft of 150 mm diameter is replaced with a hollow shaft of external diameter 250 mm. The internal diameter of the hollow shaft for same weight is
    (a) 200 mm
    (b) 100 mm
    (c) 50 mm
    (d) 180 mm
14. The equation for power transmission by a shaft is $P =$
    (a) $\frac{J}{T} \times \frac{\tau}{r}$
    (b) $\frac{2\pi NT}{60}$
    (c) $\frac{G\theta}{l}$
    (d) $\frac{\pi}{16} d^3 \tau$
15. For a given power, torque is ______ proportional to speed
    (a) directly
    (b) not
    (c) inversly
    (d) cannot say

Key to Answers

1. (a) 2. (b) 3. (c) 4. (d) 5. (a) 6. (b) 7. (c) 8. (d)
9. (a) 10. (b) 11. (c) 12. (b) 13. (a) 14. (b) 15. (c)

CHAPTER 9

Columns and Struts

9.1 INTRODUCTION

A vertical structural member subjected to axial compressive load is known as a column. Columns are relatively very long compared to their cross-sectional dimensions. Columns and stanchions are the vertical members used in the building frames.

The term 'strut' is used for the members of roof truss which is in compression. It may be in vertical, horizontal or inclined position.

The principal compression member in a crane is called 'boom'. The compression members in a bridge girder is known as 'post'. Columns and struts mainly fail due to buckling. Let us consider the buckling of columns which are long and slender and loaded axially in compression. If a compressive member is relatively slender, it may deflect laterally and fail by bending rather than failing in direct compression. For example, if we take a steel ruler or hacksaw blade and subject it to axial compression it deflects laterally and it is called buckling. If the axial compressive load is increased then the lateral deflection will also increase and finally the column will collapse. Buckling can be easily seen when you step on the top of an empty aluminium coke tin, the thin cylindrical walls buckle under the weight and collapse. Buckling is one of the major causes of failure in structures and therefore the possibility of buckling should always be considered in design.

9.2 STABILITY OF COLUMNS

If the compressive force is small, the buckled column will return to its initial position when the load is removed. Under such condition, the column is said to be 'stable'. But if the compressive load is large, lateral displacement will increase and ultimately the structure will collapse. Under such conditions the structure is said to be 'unstable'.

9.3 CRITICAL LOAD

The load on a column up to which the column is stable and beyond which it becomes unstable is known as critical load.

If the axial load is less than the critical load, the structure is stable and if the axial load is greater than the critical load, it is said to be unstable. However, when the load applied is equal to the critical load, then it is called neutral equilbrium, which is the boundary between stability and unstability. But in reality, the column does not behave in the idealized manner because of imperfections like non-homogeneity in material and cross section, not perfectly straight or not having end conditions as presumed in calculations. The load also may not be exactly axial (acting at centroid). This critical load is also called buckling load or crippling load. The value of this critical load depends on the length of the column, least lateral dimension and the end conditions. This buckling load will be less than the crushing load.

The column for which the load acts along centroidal axis are known as centrally loaded columns. The columns for which the load acts away from centroidal axis are called eccentrically loaded centrally columns. Centrally loaded columns experience only axial stress and eccentrically loaded columns experience both axial stress and bending stress due to eccentricity.

9.4 THEORIES OF COLUMNS

Most of the structural materials are elastic in nature, and steel is one commonly used material for columns. Since the columns are generally slender in nature, when compressive load is applied on to them, lateral displacement also may take place due to some unavoidable eccentricity.

Leonhard Euler (1707–1783) a Swiss mathematician derived theoretical expressions for buckling load of columns. The other major theoretical expressions for the buckling load of columns are Rankine's formula, straight line formula, Johnson's parabolic formula and secant formula.

9.5 RADIUS OF GYRATION AND SLENDERNESS RATIO

Radius of gyration of a body or area is defined as the distance from an axis of rotation where whole mass is assumed to be concentrated.

By definition of moment of inertia, it is the moment of moments or second moments of area.

Moment of inertia $I = A \cdot k^2$

where A = Area of cross section and k is the distance at which the mass (Area) is assumed to be concentrated, called radius of gyration.

$$\therefore \quad \text{Radius of gyration } k = \sqrt{\frac{I}{A}}$$

Slenderness ratio is the ratio of effective length of column to its least radius of gyration.

$$\text{Slenderness ratio} = \frac{l_e}{k}$$

when l_e = Equivalent length of column

k = Least radius of gyration

9.6 EULER'S THEORY FOR AXIALLY LOADED LONG COLUMNS

A famous mathematician Leonhard Euler was the first to investigate the buckling of slender columns and determine the critical load (published in 1744). He first dealt with a column which was fixed at the bottom end and free at the top. Later he extended his work to other columns. He derived an expression for the buckling load of long slender coulmns based only on bluckling or lateral bending. The effect of direct compression is not considered. Hence, Euler's formulae are valid or suitable for only long columns and not for short columns.

9.6.1 Assumptions in Euler's Theory of Columns

The following are the assumptions made in Euler's theory of columns:

1. The material of the columns is homogeneous, elastic and isotropic.
2. The columns fail only by buckling.
3. The cross section of the columns is uniform throughout its length.
4. The column is initially straight and is loaded axially.
5. Self weight of a column is negligible.
6. Length of a column is very large compared to its lateral dimension.

9.6.2 Types of End Conditions for Columns

The commonly used end conditions of columns are:

1. Both ends pinned
2. One end fixed, other end free
3. Both ends fixed
4. One end fixed, other end pinned.

Pinned ends are also known as hinged ends, guided ends or supported ends.

Euler's equations are derived for these 4 end conditions.

9.6.3 Derivation of Euler's Buckling Load for Column with Both Ends Pinned

Figure 9.1 shows a column of length l hinged at both ends. It is subjected to an axial compressive load P at end B. Due to this load, the column will bend as shown in Fig. 9.1.

Consider any section X-X at a distance x from end A.

P = Load

l = Length of column

E = Modulus of elasticity

I = Moment of inertia

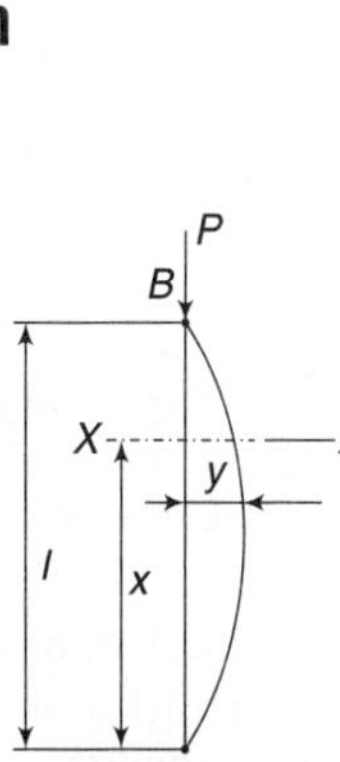

Fig. 9.1

Let y = deflection of the column at section XX. Since the deflection of columns is of concave shape from its initial position, it is treated as –ve.

Bending moment at section XX, $M_x = P.(-y) = -Py$

Equating this to differential equation of deflection

$$EI\frac{d^2y}{dx^2} = M_x = -Py.$$

$$EI\frac{d^2y}{dx^2} + Py = 0$$

$$\frac{d^2y}{dx^2} + \frac{P}{EI}y = 0$$

The general solution of this differential equation is

$$y = C_1 \cos\left(x\sqrt{\frac{P}{EI}}\right) + C_2 \sin\left(x\sqrt{\frac{P}{EI}}\right).$$

where C_1 and C_2 are constants of integration and can be found using boundary conditions.

Boundary conditions.

(i) At $x = 0$, deflection $y = 0$ at end A.

$0 = C_1 \cos 0 + C_2 \sin 0$

$0 = C_1 + 0 \quad \therefore C_1 = 0.$

(ii) At B, at $x = l$, deflection $y = 0$.

Substituting in the above equation of y.

$$0 = 0 + C_2 \sin\left(l\sqrt{\frac{P}{EI}}\right) = 0$$

$$\therefore \quad C_2 \sin\left(l\sqrt{\frac{P}{EI}}\right) = 0$$

$$\therefore \quad \text{Either } C_2 = 0 \text{ or } \sin\left(l\sqrt{\frac{P}{EI}}\right) = 0$$

Since deflection y cannot be zero, $C_2 \neq 0$.

$$\therefore \quad \sin\left(l\sqrt{\frac{P}{EI}}\right) = 0$$

The solution of this is, $l\sqrt{\frac{P}{EI}} = 0, \pi, 2\pi, \ldots,$ etc.

Considering the least significant value

$$l\sqrt{\frac{P}{EI}} = \pi$$

Squaring on both sides

$$l^2\frac{P}{EI} = \pi^2$$

$\therefore$ Critical or buckling load $P = \dfrac{\pi^2 EI}{l^2}$

Euler's critical load for a column with both ends pinned, $P = \dfrac{\pi^2 EI}{l^2}$.

9.6.4 Derivation of Euler's Buckling Load for a Column with Both Ends Fixed

Figure 9.2 shows a column AB of length l with ends fixed and subjected to a axial load of P at B. Due to this load, the column bends are as shown in Fig. 9.2.

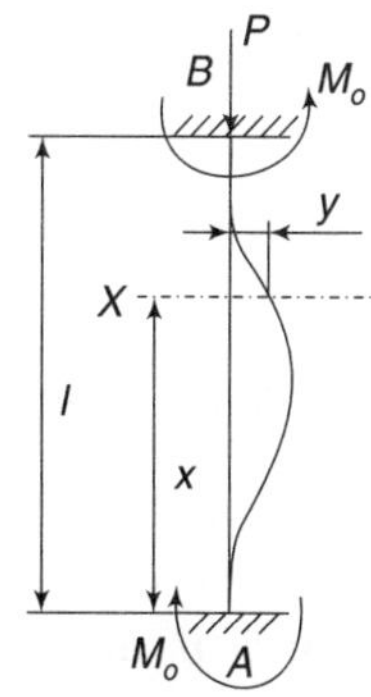

Fig. 9.2

Consider any section XX at a distance x from lower end A. Let y be the deflection at this section.

P = Load

l = Length of column

I = Moment of inertia

E = Modulus of elasticity

Due to the load, a reactive moment will be developed at the fixed ends.

Let M_o = Reactive moment at the fixed ends A and B. Bending moment at section XX, $M_x = M_o - Py$.

Since the deflection of the column is concave from its initial position, deflection y is –ve.

Equating the moment to differential equation for deflection,

$$EI\frac{d^2y}{dx^2} = M_x = M_o - Py$$

$$EI\frac{d^2y}{dx^2} = M_o - Py$$

$$EI\frac{d^2y}{dx^2} = + Py = M_o$$

$$\frac{d^2y}{dx^2} + \left(\frac{P}{EI}\right)y = \frac{M_o}{EI}$$

The general solution of this differential equation is

$$y = C_1 \cos\left(x\sqrt{\frac{P}{EI}}\right) + C_2 \sin\left(x\sqrt{\frac{\pi}{EI}}\right) + \frac{M_o}{P}$$

where C_1 and C_2 are constants of integration and can be found from boundary conditions.

Boundary conditions:

(i) At A, at $x = 0$, deflection $y = 0$

$$0 = C_1 \cos 0 + C_2 \sin 0 + \frac{M_o}{P}$$

$$\therefore \qquad C_1 = -\frac{M_o}{P}$$

(ii) At A, at the fixed end, the slope $\frac{dy}{dx} = 0$

Differentiating the y equation,

$$\frac{dy}{dx} = -C_1 \cdot \sqrt{\frac{P}{EI}} \cdot \sin\left(x\sqrt{\frac{P}{EI}}\right) + C_2\sqrt{\frac{P}{EI}} \cdot \cos\left(x\sqrt{\frac{P}{EI}}\right)$$

At A, $x = 0$, $\frac{dy}{dx} = 0$

$$0 = -C_1\sqrt{\frac{P}{EI}} \cdot \sin 0 + C_2\sqrt{\frac{P}{EI}} \cdot \cos 0$$

$$0 = 0 + C_2\sqrt{\frac{P}{EI}}.$$

Since load $P \neq 0$, $C_2 = 0$

$$\therefore \qquad y = -\frac{M_o}{P} \cdot \cos x\sqrt{\frac{P}{EI}} + \frac{M_o}{P}$$

(iii) At B, at $x = l$, deflection $y = 0$

$$0 = \frac{-M_o}{P} \cos\left(l\sqrt{\frac{P}{EI}}\right) + \frac{M_o}{P}$$

$$\frac{M_o}{P} \cos\left(l\sqrt{\frac{P}{EI}}\right) = \frac{M_o}{P}$$

$$\therefore \qquad \cos\left(l\sqrt{\frac{P}{EI}}\right) = 1$$

The general solution of this equation is

$$l\sqrt{\frac{P}{EI}} = 0,\ 2\pi,\ 4\pi,\ 6\pi, \text{ etc.}$$

Considering the least significant value,

$$l\sqrt{\frac{P}{EI}} = 2\pi$$

Squaring on both sides,

$$l^2 \frac{P}{EI} = 4\pi^2$$

or

$$P = \frac{4\pi^2 EI}{l^2}$$

$$P = \frac{\pi^2 EI}{\left(\frac{l^2}{4}\right)} = \frac{\pi^2 EI}{\left(\frac{l}{2}\right)^2} = \frac{\pi^2 EI}{l_e^2}$$

where, l_e = Equivalent length $= \frac{l}{2}$

Euler's critical load for column with both ends fixed, $P = \frac{4\pi^2 EI}{l^2}$.

9.6.5 Derivation of Euler's Buckling Load for a Column with One End Fixed, Other End Free

Figure 9.3 shows a column AB of length l, with one end A fixed and the other end B is free. It is subjected to a load P at its free end B. Due to this load, the column deflects as shown in Fig. 9.3

Consider any section XX at a distance x from the fixed end A. Let y be the deflection at this section, XX.

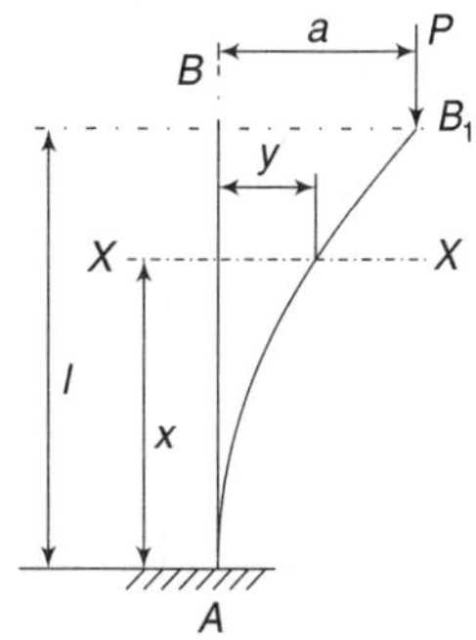

Fig. 9.3

P = Load

I = Moment of inertia

l = Length of column

a = Maximum deflection of column at the free end B

$y_{max} = a$.

Bending moment at section XX, $M_x = P(a - y)$

Here a and y are positive as the beam takes a shape of convex form from its initial position.

Equating this to deflection equation

$$EI\frac{d^2y}{dx^2} = M_x = P(a - y)$$

$$EI\frac{d^2y}{dx^2} = Pa - Py$$

$$EI\frac{d^2y}{dx^2} + Py = Pa$$

$$\therefore \quad \frac{d^2y}{dx^2} + \left(\frac{P}{EI}\right)y = \left(\frac{P}{EI}\right)a$$

The solution of this differential equation is

$$y = C_1 \cos\left(x\sqrt{\frac{P}{EI}}\right) + C_2 \sin\left(x\sqrt{\frac{P}{EI}}\right) + a$$

where, C_1 and C_2 are constants of integration and can be found by applying boundary condition.

Boundary conditions:

(i) At A, at fixed end $x = 0$, deflection $y = 0$

$$0 = C_1 \cos 0 + C_2 \sin 0 + a$$

$$0 = C_1 + a$$

or $$C_1 = -a$$

(ii) At A, at fixed end, $x = 0$ the slope $\frac{dy}{dx} = 0$

Differentiating the above deflection equation.

$$\frac{dy}{dx} = -C_1 \cdot \sqrt{\frac{P}{EI}} \cdot \sin\left(x\sqrt{\frac{P}{EI}}\right) + C_2\sqrt{\frac{P}{EI}} \cdot \cos\left(x\sqrt{\frac{P}{EI}}\right)$$

Put $x = 0, \text{ and } \dfrac{dy}{dx} = 0$

$$0 = -C_1\sqrt{\frac{P}{EI}} \cdot \sin 0 + C_2\sqrt{\frac{P}{EI}} \cdot \cos 0$$

$$0 = 0 + C_2\sqrt{\frac{P}{EI}}$$

Since $\dfrac{P}{EI} \neq 0$, $C_2 = 0$.

Substituting C_1 and C_2 values in y equation

$$y = -a \cos x\sqrt{\frac{P}{EI}} + a$$

(iii) At free end B, $x = l$, deflection $y = y_{max} = a$

$$a = -a \cos\left(l\sqrt{\frac{P}{EI}}\right) + a$$

$$\therefore \quad a - a = a \cos\left(l\sqrt{\frac{P}{EI}}\right)$$

$$a \cos\left(l\sqrt{\frac{P}{EI}}\right) = 0$$

Since deflection $a \neq 0$, $\cos\left(l\sqrt{\dfrac{P}{EI}}\right) = 0$

The general solution is $l\sqrt{\dfrac{P}{EI}} = \dfrac{\pi}{2}, \dfrac{3\pi}{2}, \dfrac{5\pi}{2}$, etc.

Considering the least significant value.

$$l\sqrt{\frac{P}{EI}} = \frac{\pi}{2}$$

Squaring on both sides

$$l^2\frac{P}{EI} = \frac{\pi^2}{4}$$

$$\therefore \quad \text{Critical load } P = \frac{\pi^2 EI}{4l^2}$$

$$P = \frac{\pi^2 EI}{(2l)^2} = \frac{\pi^2 EI}{l_e^2}$$

where, l_e = Equivalent length = $2l$

Euler's critical load for a column with end fixed, other end free is $P = \dfrac{\pi^2 EI}{4l^2}$.

9.6.6 Derivation of Euler's Buckling Load for a Column with One End Fixed and the Other End Pinned

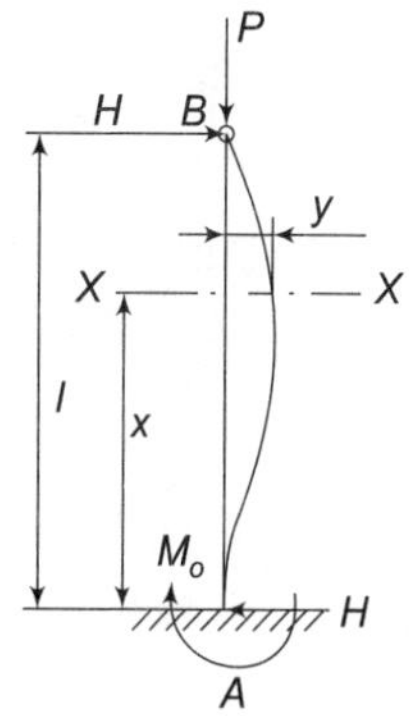

Fig. 9.4

Figure 9.4 shows a column AB of length l, fixed at bottom end A and pinned at top end B. The column is subjected to an axial load of P at the top end B. Under the load, the column deflects as shown in Fig. 9.4.

Consider any section XX at a distance x from the fixed end A. Let the deflection at this section be y. Since the column is deflecting into concave from its initial position, the deflection y is –ve.

At the fixed end, there will be a reactive moment developed equal to M_o. Then for the equilibrium of the column, horizontal forces H are developed at A and B as shown in Fig. 9.4.

$$\text{Moment } M_o = H \cdot l$$

or
$$\text{Horizontal force } H = \frac{M_o}{l}$$

$$\text{Moment at any section } XX, M_x = H(l - x) - Py$$

Equating this to deflection equation, $EI \dfrac{d^2y}{dx^2} = M_x = H(l - x) - Py$.

$$EI \frac{d^2y}{dx^2} + Py = H(l - x)$$

$$\therefore \quad \frac{d^2y}{dx^2} + \left(\frac{P}{EI}\right)y = \frac{H(l - x)}{EI}$$

$$\frac{d^2y}{dx^2} + \left(\frac{P}{EI}\right)y = \left(\frac{P}{EI}\right)\frac{H(l - x)}{P}$$

The solution for this differential equation is

$$y = C_1 \cos\left(x\sqrt{\frac{P}{EI}}\right) + C_2 \sin\left(x\sqrt{\frac{P}{EI}}\right) + \frac{H(l - x)}{P}$$

where C_1 and C_2 are contants of integration and can be found by applying boundary conditions.

Boundary conditions:

(i) At fixed end A, $x = 0$, deflection $y = 0$,

$$0 = C_1 \cos 0 + C_2 \sin 0 + \frac{Hl}{P}$$

$$0 = C_1 + 0 + \frac{Hl}{P}$$

$$C_1 = -\frac{Hl}{P}$$

(ii) At fixed end A, $x = 0$, slope $\frac{dy}{dx} = 0$

Differentiating the deflection equation,

$$\frac{dy}{dx} = -C_1\sqrt{\frac{P}{EI}}\sin\left(x\sqrt{\frac{P}{EI}}\right) + C_2\sqrt{\frac{P}{EI}}\cdot\cos\left(x\sqrt{\frac{P}{EI}}\right) - \frac{H}{P}$$

Put $x = 0$, slope $\frac{dy}{dx} = 0$ and $C_1 = -\frac{Hl}{P}$

$$0 = -\left(\frac{-Hl}{P}\right)\sqrt{\frac{P}{EI}}\cdot\sin 0 + C_2\sqrt{\frac{P}{EI}}\cos 0 - \frac{H}{P}$$

$$0 = 0 + C_2\sqrt{\frac{P}{EI}} - \frac{H}{P}$$

$$\therefore \quad C_2 = \frac{H}{P}\cdot\sqrt{\frac{EI}{P}}$$

$$\therefore \quad y = -\frac{Hl}{P}\cos\left(x\sqrt{\frac{P}{EI}}\right) + \frac{H}{P}\sqrt{\frac{EI}{P}}\cdot\sin\left(x\sqrt{\frac{P}{EI}}\right) - \frac{H(l-x)}{P}$$

(iii) At pinned end B, $x = l$, deflection $y = 0$

$$0 = -\frac{Hl}{P}\cos\left(l\sqrt{\frac{P}{EI}}\right) + \frac{H}{P}\sqrt{\frac{EI}{P}}\cdot\sin\left(l\sqrt{\frac{P}{EI}}\right) - 0$$

$$\frac{Hl}{P}\cdot\cos\left(l\sqrt{\frac{P}{EI}}\right) = \frac{H}{P}\sqrt{\frac{EI}{P}}\cdot\sin\left(l\sqrt{\frac{P}{EI}}\right)$$

$$\therefore \quad \frac{\sin\left(l\sqrt{\frac{P}{EI}}\right)}{\cos\left(l\sqrt{\frac{P}{EI}}\right)} = \frac{\left(\frac{Hl}{P}\right)}{\left(\frac{H}{P}\sqrt{\frac{EI}{P}}\right)} = l\sqrt{\frac{P}{EI}}$$

$$\therefore \quad \tan\left(l\sqrt{\frac{P}{EI}}\right) = l\sqrt{\frac{P}{EI}}$$

This is of the form $\tan\theta = \theta$

The solution of such equation is $\theta = 4.49341$ radians.

$$\therefore \quad l\sqrt{\frac{P}{EI}} = 4.49341$$

Squaring on both sides, $l^2\frac{P}{EI} = 20.191$

$$P = \frac{20.191\ EI}{l^2}$$

The constant 20.191 can be written as $\left(\frac{20.191}{\pi^2}\right) \cdot \pi^2 = 2.046\ \pi^2$

$$\simeq 2\pi^2 \text{ (for simplicity)}$$

$$\therefore \qquad P = \frac{2\pi^2 EI}{l^2}$$

$$P = \frac{\pi^2 EI}{\left(\frac{l^2}{2}\right)} = \frac{\pi^2 EI}{\left(\frac{l}{\sqrt{2}}\right)^2} = \frac{\pi^2 EI}{l_e^2}$$

where equivalent length $l_e = \frac{l}{\sqrt{2}}$

∴ The Euler's critical load for a column with one end fixed and the other end pinned is

$$P = \frac{2\pi^2 EI}{l^2}.$$

9.6.7 Effective Length or Equivalent Length of Column

In the above section, we derived equations for Euler's critical load for columns with four different end conditions. It can be observed that there is lot of similarity in these four equations. All the cases may be represented by one single equation like

$$P = \frac{\pi^2 EI}{l_e^2}$$

where l_e = Equivalent length, which will be different for different end conditions.

Table 9.1 gives the expression for effective length for various end conditions.

Table 9.1 Effective length (Equivalent length)

S.No.	*End condition*	*Euler's crippling load (p)*	*Equivalent length*
1.	Both ends pinned	$P = \frac{\pi^2 EI}{l^2}$	$l_e = l$
2.	Both ends fixed	$P = \frac{4\pi^2 EI}{l^2}$	$l_e = \frac{l}{2}$
3.	One end fixed, other end free	$P = \frac{\pi^2 EI}{4l^2}$	$l_e = 2l$
4.	One end fixed, other end pinned.	$P = \frac{2\pi^2 EI}{l^2}$	$l_e = \frac{l}{\sqrt{2}}$

l = Actual length of column, l_e effective length of column.

9.6.8 Limitations of Euler's Theory of Columns

Euler's critical load is $P_{cr} = \dfrac{\pi^2 EI}{l_e^2}$. The validity of Euler's theory is subject to condition that failure occurs due to buckling.

$$I = \text{moment of inertia} = Ak^2$$

where A = Area of cross section

k = Radius of gyration

$$\therefore \quad P_{cr} = \frac{\pi^2 E\, Ak^2}{l_e^2}$$

or
$$\frac{P_{cr}}{A} = \frac{\pi^2 E}{\left(\dfrac{l_e}{k}\right)^2}$$

We know that $\dfrac{P_{cr}}{A} = \dfrac{\text{Load}}{\text{Area}}$ = Stress σ_{cr}

$$\therefore \quad \sigma_{cr} = \frac{\pi^2 E}{\left(\dfrac{l_e}{k}\right)^2}$$

Here $\dfrac{l_e}{k}$ is known as slenderness ratio, and σ_{cr} is the critical stress.

From the critical stress equation, if the slenderness ratio $\dfrac{l_e}{k}$ approaches zero, the critical stress will be tending to infinity. But it cannot happen since the material will get crushed before this.

Hence, Euler's formula is valid only if the stress induced is less than crippling stress or crushing stress of that material. The limiting value of this crippling stress is the stress corresponding to its elastic limit.

$$\text{Slenderness ratio } \frac{l_e}{k} = \sqrt{\frac{\pi^2 E}{\sigma_{cr}}}$$

Hence, the value of slenderness ratio can be found for various materials so that one can assess the limiting value of the length of column for which Euler's equation is valid.

For example, consider mild steel Modulus of elasticity E = 200 GPa and yield stress σ_{cr} = 320 N/mm^2, then

$$\text{Slenderness ratio } \frac{l_e}{k} = \sqrt{\frac{\pi^2 \times 200 \times 10^3}{320}} = 78.54$$

$$\text{say } \frac{l_e}{k} \simeq 80$$

Hence, when the slenderness ratio is less than 80 for mild steels columns the Euler's formula will not be valid and does not give good results.

Figure 9.5 shows the Euler's curve for critical stress versus slenderness ratio.

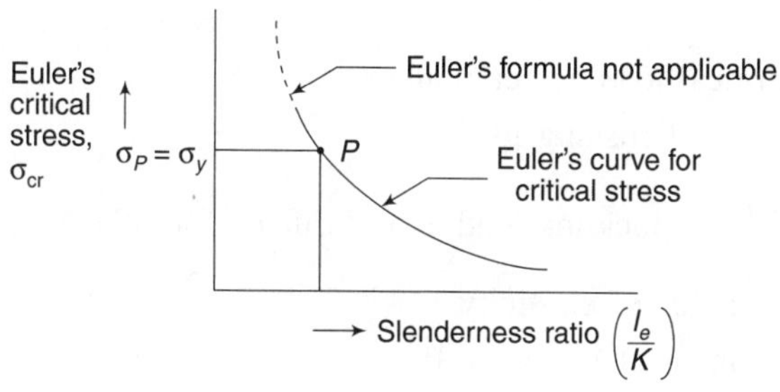

Fig. 9.5

9.7 GRAPHICAL INTERPRETATION OF EQUIVALENT LENGTH

The equivalent length is the distance between the points of inflection on the elastic curves. This is shown in Fig. 9.6 for the four types of end supports.

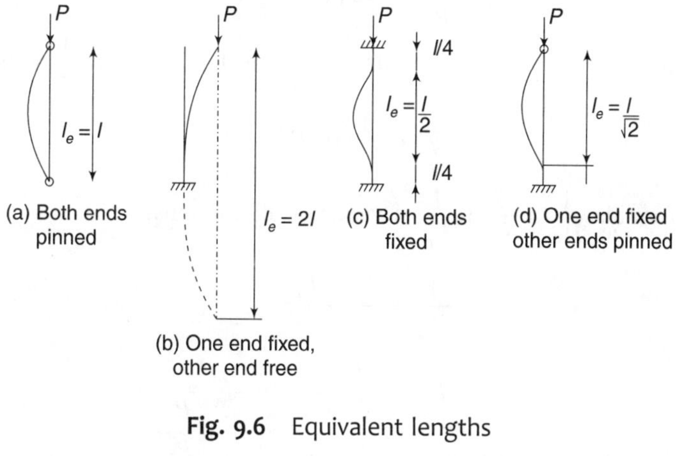

Fig. 9.6 Equivalent lengths

9.8 RANKINE'S FORMULA FOR COLUMNS

It is seen that Euler's formula holds good for long columns and does not yield good results for short columns. The short columns generally fail due to crushing and long columns by buckling. Medium columns fail due to combined effect of crushing and buckling. Considering all this, Rankine and Gorden developed a formula which is widely accepted and found valid for all types of columns—long, medium and short.

Based on his experimental investigations, Rankine devised an empirical formula for the critical load P_{cr}.

$$\frac{1}{P_{cr}} = \frac{1}{P_c} + \frac{1}{P_E}$$

where, P_{cr} = Critical load as per Rankine

$P_c = \sigma_c A$ = Crushing load

$P_E = \frac{\pi^2 EI}{l_e}$ = Buckling load as per Euler's formula

E = Modulus of elasticity

I = Moment of inertia = Ak^2

l_e = Equivalent length of column

A = Area of cross section

k = Radius of gyration = $\sqrt{\frac{I}{A}}$

If the column is short, P_E will be large and $\frac{1}{P_E}$ will be small compared to P_c and can be neglected.

Hence,
$$\frac{1}{P_{cr}} = \frac{1}{P_c} \text{ or } P_{cr} = P_c$$

If the column is long, P_E will be small and $\frac{1}{P_E}$ will be large and $\frac{1}{P_c}$ will be neglected.

$$\frac{1}{P_{cr}} = \frac{1}{P_E} \text{ or } P_{cr} = P_E$$

Hence, this formula $\frac{1}{P_{cr}} = \frac{1}{P_c} + \frac{1}{P_E}$ is valid for all columns.

$$\frac{1}{P_{cr}} = \frac{P_c + P_E}{P_c P_E}$$

$$P_{cr} = \frac{P_c P_E}{P_c + P_E} = \frac{P_c}{1 + \frac{P_c}{P_E}}$$

But $P_c = \sigma_c A$ and $P_E = \frac{\pi^2 EI}{l_e^2}$

$$P_{cr} = \frac{\sigma_c A}{1 + \frac{\sigma_c A}{\left(\frac{\pi^2 EI}{l_e^2}\right)}} = \frac{\sigma_c A}{1 + \frac{\sigma_c A}{\left(\frac{\pi^2 EAk^2}{l_e^2}\right)}}$$

$$= \frac{\sigma_c A}{\left(1 + \frac{\sigma_c Ale^2}{\pi^2 EAk^2}\right)} = \frac{\sigma_c A}{1 + \frac{\sigma_c}{\pi^2 E}\left(\frac{l_e}{k}\right)^2}$$

Let $\alpha = \dfrac{\sigma_c}{\pi^2 E}$, a constant for given material since σ_c and E are constants for a material.

$$P_{cr} = \frac{\sigma_c A}{1 + \alpha\left(\frac{l_e}{k}\right)^2}$$

where,

$$\alpha = \frac{\sigma_c}{\pi^2 E}$$

Table 9.2 gives values of α for various materials.

Table 9.2 Value of Rankine's constant α.

S. No.	*Material*	*critical stress,* $\sigma_c \dfrac{N}{mm^2}$	$\alpha = \dfrac{\sigma_c}{\pi^2 E}$
1.	Mild Steel	320	$\dfrac{1}{7500}$
2.	Cost Iron	550	$\dfrac{1}{1600}$
3.	Wrought Iron	250	$\dfrac{1}{9000}$
4.	Timber	50	$\dfrac{1}{750}$

9.9 JOHNSON'S PARABOLIC FORMULA FOR COLUMN

Johnson suggested the following equation for critical load for columns

$$P_{cr} = \sigma_c A\left[1 - \frac{\sigma_c}{4\pi^2 E}\left(\frac{l_e}{k}\right)^2\right]$$

9.10 STRAIGHT LINE FORMULA FOR COLUMNS

The straight line for critical load for columns is

$$P_{cr} = A\left[\sigma_c - \frac{2\sigma_c}{3\pi\sqrt{\sigma_c/3E}}\left(\frac{l_e}{k}\right)\right]$$

9.11 FACTOR OF SAFETY AND SAFE LOAD

The load carried safely by a column is safe load

$$P = \frac{\text{Crtical load}}{\text{Factor of safety}} = \frac{P_{cr}}{\text{FoS}}$$

SOLVED PROBLEMS

Problem 1 A solid round bar of 60 mm diameter and 2.5 m is used as a strut. Find the safe compressive load for the strut if (i) both ends are hinged and (ii) both ends are fixed. Take $E = 2 \times 10^5$ N/mm^2 and factor of safety = 3.

Solution:

$$\text{Diagram of bar } d = 60 \text{ mm}$$

$$\text{Length of bar } l = 2.5 \text{ m} = 2500 \text{ mm}$$

$$\text{Modulus of elasticity } E = 2 \times 10^5 \text{ N/mm}^2$$

$$\text{Factor of safety, FoS} = 3$$

(i) Both ends are hinged

$$\therefore \quad \text{Equivalent length } l_e = l = 2500 \text{ mm.}$$

$$\text{Euler's critical load } P_{cr} = \frac{\pi^2 EI}{l_e^2}$$

where

$$I = \text{Moment of inertia} = \frac{\pi d^4}{64}$$

$$= \frac{\pi \times 60^4}{64} = 636.17 \times 10^3 \text{ mm}^4$$

$$\therefore \quad P_{cr} = \frac{\pi^2 \times 2 \times 10^5 \times 636.17 \times 10^3}{2500^2} = 200.92 \times 10^3 \text{ N}$$

$$\therefore \quad \text{Safe load } P = \frac{\text{Critical load}}{\text{Factor of safety}} = \frac{P_{cr}}{\text{FoS}}$$

$$P = \frac{200.92 \times 10^3}{3} = 66973.5 \text{ N}$$

(ii) Both ends are fixed

$$\therefore \text{ Effective length of column } l_e = \frac{l}{2} = \frac{2500}{2} = 1250 \text{ m}$$

$$\therefore \quad \text{Euler's critical load } P_{cr} = \frac{\pi^2 EI}{le^2} = \frac{\pi^2 \times 2 \times 10^5 \times 636.17 \times 10^3}{1250^2}$$

$$P_{cr} = 803.68 \times 10^3 \text{ N}$$

$$\therefore \quad \text{Safe load } P = \frac{P_{cr}}{\text{FoS}} = \frac{803.68 \times 10^3}{3} = 267.89 \times 10^3 \text{ N.}$$

Problem 2 Find the diameter of a round rod 3 m long with one end fixed and other end free carrying an axial compressive load of 90 kN. Modulus of elasticity is 200 kN/mm^2. Factor of safety (FoS) = 2.

Solution: Length of rod = 3 m = 3000 mm

One end fixed, other end free.

$\therefore$ Equivalent length $l_e = 2l = 2 \times 3000 = 6000$ mm

Load $P = 90$ kN = 90000 N.

Factor of safety = 2

$\therefore$ Critical load P_{cr} = Safe load × FoS

$P_{cr} = 90000 \times 2 = 180 \times 10^3$ N

Modulus of elasticity $E = 200$ kN/mm^2 = 200×10^3 N/mm^2

Moment of Inentia $I = \dfrac{\pi d^4}{64}$

Euler's formula for columns $P_{cr} = \dfrac{\pi^2 EI}{l_e^2}$

$$180 \times 10^3 = \frac{\pi^2 \times 200 \times 10^3 \times \left(\frac{\pi d^4}{64}\right)}{6000^2}$$

$\therefore$ Diameter $d = 90.43$ mm.

Problem 3 A hollow cylindrical column of 120 mm outside diameter, 60 mm inside diameter 2 m long with one end fixed and the other pinned is made of steel having modulus of elasticity as 200×10^3 N/mm^2. Find the critical load that can be carried by this column.

Solution: Outside diameter $d_o = 120$ mm

Inside diameter $d_i = 60$ mm

Length of column $l = 2$ m = 2000 mm

One end fixed, other end pinned,

$\therefore$ Equivalent length of column $l_e = \dfrac{l}{\sqrt{2}}$

$$l_e = \frac{2000}{\sqrt{2}} = 1414.21 \text{ mm}$$

Modulus of elasticity $E = 200 \times 10^3$ N/mm^2

Euler's critical load $P_{cr} = \dfrac{\pi^2 EI}{l_e^2}$

I = Moment of inertia = $\dfrac{\pi}{64}(d_o^4 - d_i^4)$

$$I = \frac{\pi}{64}(120^4 - 60^4) = 650.80 \times 10^6 \text{ mm}^4$$

$$\therefore \quad P_{cr} = \frac{\pi^2 \times 200 \times 10^3 \times 650.80 \times 10^6}{1414.21^2} = 642.32 \times 10^3 \text{ N}$$

Problem 4 A hollow circular column whose diameters ratio as 0.6, is 3 m on long with both ends pinned. It is subjected to an axial compressive load of 150 kN. Find the suitable diameters of the column. What will be diameters if both end are fixed. Take E = 110 GPa.

Solution:

Outside diameter $d_o = d_o$

Inside diameter $d_i = 0.6\, d_o$

Length of column $l = 3 \text{ m} = 3000 \text{ mm}$

Load $P_{cr} = 150 \text{ kN} = 150 \times 10^3 \text{ N}.$

Modulus of elasticity $E = 110 \text{ GPa} = 110 \times 10^3 \text{ N/mm}^2$

$$\text{Moment of inertia } I = \frac{\pi}{64}(d_o^4 - d_i^4) = \frac{\pi}{64}\left[d_o^4 - (0.6\, d_o)^4\right]$$

$$I = \frac{\pi}{64} d_o^4 (0.8704)$$

(i) Both ends pinned:

$\therefore$ Equivalent length of column $l_e = l = 3000$ mm

$$\text{Euler's initial load } P_{cr} = \frac{\pi^2 EI}{l_e^2}$$

$$150 \times 10^3 = \frac{\pi^2 \times 110 \times 10^3 \times \frac{\pi}{64} d_o^4(0.8704)}{3000^2}$$

$\therefore$ Outside diameter $d_o = 73.45$ mm

Inside diameter $d_i = 0.6\, d_o = 44.07$ mm

(ii) Both ends fixed.

$$\therefore \quad \text{Equivalent length } l_e = \frac{l}{2} = \frac{3000}{2} = 1500 \text{ mm}$$

$$\therefore \quad \text{Euler's critical load } P_{cr} = \frac{\pi^2 EI}{l_e^2}$$

$$150 \times 10^3 = \frac{\pi^2 \times 110 \times 10^3 \times \frac{\pi}{64} d_o^4(0.8704)}{1500^2}$$

Outside diameter $d_o = 51.94$ mm

Inside diameter $d_i = 0.6\, d_o = 0.6 \times 51.94$

$d_i = 31.16$ mm.

Problem 5 A 1.5 long column with square cross section of sides 200 mm is made of cast iron and subjected to a compressive load. Taking modulus of elasticity as 106 GPa, find the safe load that can be applied at a factor of safety of 1.5. The column has one end fixed the other pinned.

Solution: Length of column l = 1500 mm

One end fixed, other end pinned

$\therefore$ Equivalent length of column $l_e = \dfrac{l}{\sqrt{2}} = \dfrac{1500}{\sqrt{2}} = 1060.66$ mm

Side of square section a = 200 mm

$$\text{Moment of inertia } I = \frac{a^4}{12} = \frac{200^4}{12} = 133.33 \times 10^6 \text{ mm}^4$$

Modulus of elasticity E = 106 GPa = 106 × 10^3 MPa.

Factor of safety = 1.5

$$\text{Euler's critical load } P_{cr} = \frac{\pi^2 EI}{l_e^2}$$

$$P_{cr} = \frac{\pi^2 \times 106 \times 10^3 \times 133.33 \times 10^6}{1060.66^2} = 123.988 \times 10^6 \text{N}$$

$$\therefore \quad \text{Safe load } P = \frac{\text{Critical load}}{\text{Factory of safety}} = \frac{P_{cr}}{\text{FoS}}$$

$$P = \frac{123.988 \times 10^6}{1.5} = 82.66 \times 10^6 \text{ N.}$$

Problem 6 A column of rectangular cross section 40 mm wide and 30 mm depth, 1.8 m long is one end fixed other end free. Find the load than can be applied on it taking E = 200 GPa.

Solution: Width b = 40 mm

Depth d = 30 mm

Length l = 1.8 m = 1800 mm

One end fixed, other end free.

$\therefore$ Equivalent length of column $l_e = 2l = 2 \times 1800 = 3600$ mm

Modulus of elasticity E = 200 GPa = 200 × 10^3 MPa

I = Least moment of inertia of I_{XX} or I_{YY}

$$I_{XX} = \frac{bd^3}{12} = = \frac{40 \times 30^3}{12} = 90000 \text{ mm}^4$$

$$I_{YY} = \frac{db^3}{12} = \frac{30 \times 40^3}{12} = 160000 \text{ mm}^4$$

$$\therefore \quad \text{Least } I = 90000 \text{ mm}^4$$

$$\text{Euler's critical load } P_{cr} = \frac{\pi^2 EI}{l_e^2}$$

$$P_{cr} = \frac{\pi^2 \times 200 \times 10^3 \times 90000}{3600^2} = 1370.78 \text{ N}$$

Problem 7 A column of rectangular cross section whose sides ratio is 2 is 2.4 m long with one fixed and the other end hinged. It is subjected to an axial load of 90 kN. Taking modulus of elasticity as 206×10^3 N/mm^2, find the width and thickness of the cross section.

Solution: Rectangular cross section.

$$\text{Width} = b$$

$$\text{Thickness} = d$$

$$\text{Sides ratio } \frac{b}{d} = 2$$

$$\text{Length } l = 2.4 \text{ m} = 2400 \text{ mm}$$

One end fixed, other end hinged.

$$\therefore \quad \text{Equivalent length } l_e = \frac{l}{\sqrt{2}} = \frac{2400}{\sqrt{2}} = 1697.06 \text{ mm}$$

Moment of inertia

$$I_{XX} = \frac{bd^3}{12} = \frac{(2d)\,d^3}{12} = \frac{d^4}{6} = 0.167\ d^4$$

$$\therefore \quad I_{YY} = \frac{db^3}{12} = \frac{d(2b)^3}{12} = \frac{8d^4}{12} = 0.667\ d^4$$

$$I = \text{Least } MI = 0.167\ d^4$$

$$\text{Modulus of elasticity } E = 206 \times 10^3 \text{ MPa}$$

$$\text{Load } P_{cr} = 90 \text{ kN} = 90 \times 10^3 \text{ N}$$

$$\text{Euler's critical load } P_{cr} = \frac{\pi^2 EI}{l_e^2}$$

$$90000 = \frac{\pi^2 \times 206 \times 10^3 \times 0.167 d^4}{1697.06^2}$$

$$\therefore \quad \text{Thickness } d = 29.56 \text{ mm}$$

$$\text{Width } b = 2d = 51.12 \text{ mm.}$$

Problem 8 It is proposed to replace a solid column of diameter 25 with a hollow column of diameters ratio 0.5, made of same material, length and same end conditions subjected to same compressive load. Find the diameters of the hollow column.

Solution: Diameter of solid column d = 25 mm

Inside diameter of hollow column $d_i = 0.5\ d_o$

Material, length, end conditions and load are same,

$\therefore$ E, l_e and P_{cr} are same.

$$\text{Euler's critical load } P_{cr} = \frac{\pi^2 EI}{l_e^2}$$

$$\text{For solid column, } P_{cr} = \frac{\pi^2 E\left(\frac{\pi}{64} d^4\right)}{l_e^2} = \frac{\pi^2 E}{l_e^2} \frac{\pi}{64} \times 25^4$$

$$\text{For hollow column, } P_{cr} = \frac{\pi^2 E}{le^2} \left[\frac{\pi}{64}(d_o^4 - d_i^4)\right]$$

Equating

$$\frac{\pi^2 E}{le^2} \frac{\pi}{64} \times 25^4 = \frac{\pi^2 E}{le^2} \frac{\pi}{64} (d_o^4 - d_i^4)$$

$$25^4 = d_o^4 - d_i^4$$

But

$$d_i = (0.6\ d_o)$$

$$25^4 = d_o^4 - (0.6\ d_o)^4$$

$$= d_o^4 (1 - 0.6^4)$$

$\therefore$

$$d_o = 25.88 \text{ mm}$$

$$d_i = 0.6\ d_o = 0.6 \times 25.88 = 15.53 \text{ mm}$$

The outside and inside diameters of the hollow column are d_o = 25.88 mm and d_i = 15.88 mm.

Problem 9 Compare the ratio of the strength of solid steel column to that of the hollow steel column of the same area of cross section. The internal diameter of the hollow column is $\frac{3}{4}$ of the external diameter. The columns have the same length and same end condition. Use Euler's theory.

Solution: Dia of solid column = d

Outside dia of hollow column = d_o

Inside dia of hollow column = $d_i = \frac{3}{4} d_o$

$$d_i = 0.75\ d_o$$

Areas, length, end conditions and material are same,

Hence, $A_s = A_H$

l_e and E are same.

From

$$A_s = A_H$$

$$\frac{\pi}{4} d^2 = \frac{\pi}{4} (d_o^2 - d_i^2)$$

$$d^2 = d_o^2 - (0.75\ d_o)^2 = d_o^2 (1 - 0.75^2)$$

$$d = d_o (0.6614)$$

$$\text{Euler's critical load } P_{cr} = \frac{\pi^2 EI}{l_e^2}$$

$$\text{For solid column } P_s = \frac{\pi^2 E}{le^2} I_s = \frac{\pi^2 E}{l_e^2} \cdot \frac{\pi}{64} (d^4)$$

$$P_S = \frac{\pi^2 E}{le^2}\frac{\pi}{64}(d_o \times 0.6614)^4 = \frac{\pi^2 E}{l_e^2}.\frac{\pi}{64} d_o^4 \times 0.1914$$

For hollow column, $P_H = \frac{\pi^2 E}{l_e^2} I_H = \frac{\pi^2 E}{l_e^2}\frac{\pi}{64}(d_o^4 - d_i^4)$

$$P_H = \frac{\pi^2 E}{l_e^2}\cdot\frac{\pi}{64}\left[d_o^4 - (0.75\ d_o)^4\right] = \frac{\pi^2 E}{l_e^2}\cdot\frac{\pi}{64} d_o^4 (1 - 0.75^4)$$

$$P_H = \frac{\pi^2 E}{l_e^2}\cdot\frac{\pi}{64} d_o^4 (0.6836)$$

Comparing the critical loads, $\frac{P_H}{P_S} = \frac{\left(\frac{\pi^2 E}{l_e^2}\right)\frac{\pi}{64} d_o^4 (0.6836)}{\left(\frac{\pi^2 E}{l_e^2}\right)\frac{\pi}{64} d_o^4 (0.1914)}$

$$\frac{P_H}{P_S} = \frac{0.6834}{0.1914} = 3.5715$$

Hollow column of same area can carry 3.5715 times the load capacity of solid column.

Problem 10 Find the Euler's crippling load for a hollow cylindrical steel column 40 mm external diameter and 4 mm thick. The length of column is 2.5 m and is hinged at both ends. Also compute the Rankine's crippling load using constants 335 MPa and $\frac{1}{7500}$. Take E = 205 GPa.

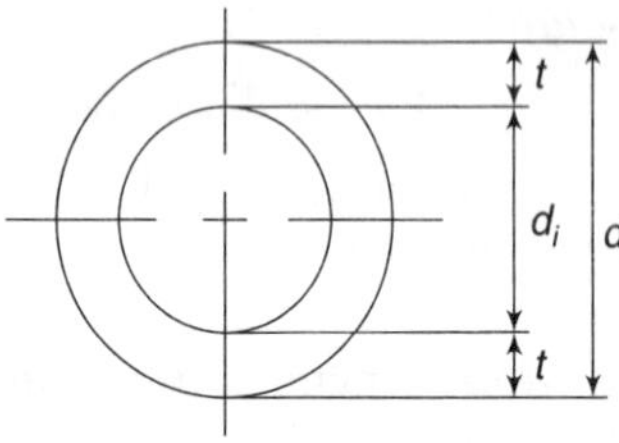

Fig. 9.7

Solution: Outside diameter d_o = 40 mm

Thickness t = 4 mm

∴ Inside diameter $d_i = d_o - 2t$

$di = 40 - 2 \times 4 = 32$ mm

Length l = 2.5 m = 2500 mm

Both ends hinged.

∴ Equivalent length $l_e = l = 2500$ mm

Rankine's constant, σ_{cr} = 335 MPa and

$$\alpha = \frac{1}{7500}$$

Modulus of elasticity E = 205 GPa = 205×10^3 MPa.

Moment of inertia $I = \frac{\pi}{64}(d_o^4 - d_i^4)$

$$I = \frac{\pi}{64}(40^4 - 32^4) = 74191.85\ \text{mm}^4$$

$$\text{Area of cross section } A = \frac{\pi}{4}(d_o^2 - d_i^2)$$

$$A = \frac{\pi}{4}(40^2 - 32^2) = 452.29 \text{ mm}^2$$

$$\text{Radius of gyration } k = \sqrt{\frac{I}{A}} = \sqrt{\frac{74191.85}{452.29}} = 12.806 \text{ mm}$$

(i) Euler's equation for crippling load

$$P_{Ecr} = \frac{\pi^2 EI}{l_e^2} = \frac{\pi^2 \times 205 \times 10^3 \times 74191.85}{2500^2}$$

$$P_{Ecr} = 24017.61 \text{ N}$$

(ii) Rankine's equation for crippling load

$$P_{Rcr} = \frac{\alpha_{cr} A}{1 + \alpha\left(\frac{l_e}{k}\right)^2} = \frac{335 \times 452.29}{1 + \frac{1}{7500}\left(\frac{2500}{12.806}\right)^2}$$

$$P_{Rcr} = 24914.45 \text{ N.}$$

Problem 11 Determine the crippling load for hollow rectangular cast iron column of outer dimensions 200 mm × 150 mm, thickness of metal 25 mm, actual height of column 6 m with one end fixed, the other end free. Consider E = 120 GPa.

Solution: Outer width $B = 200$ mm

Outer height $H = 150$ mm

Thickness $t = 25$ mm

$\therefore$ Inner width $b = B - 2t$

$$b = 200 - 2 \times 25 = 150 \text{ mm}$$

Inner height $h = H - 2t = 150 - 2 \times 25 = 100$ mm,

Height of column = Length l = 6 m = 6000 mm.

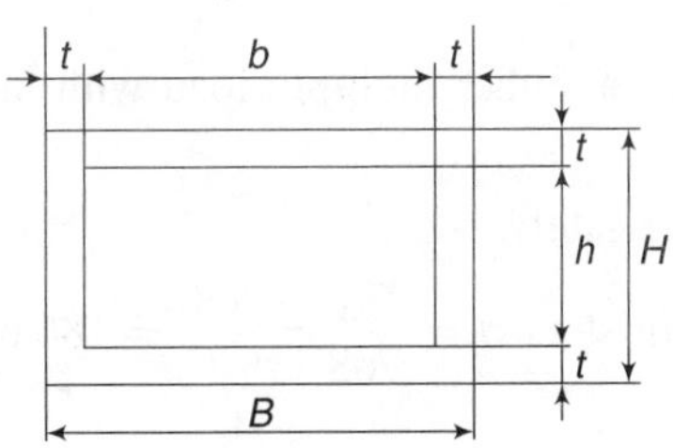

Fig. 9.8

One end fixed, other end free.

$\therefore$ Equivalent of column $l_e = 2l = 2 \times 6000 = 12000$ mm

$$E = 120 \text{ GPa} = 120 \times 10^3 \text{ MPa.}$$

$$\text{Moment of inertia } I_{XX} = \frac{BH^3 - bh^3}{12}$$

$$I_{XX} = \frac{200 \times 150^3 - 150 \times 100^3}{12} = 43.75 \times 10^6 \text{ mm}^4$$

$$I_{YY} = \frac{HB^3 - hb^3}{12} = \frac{150 \times 200^3 - 100 \times 150^3}{12}$$

$$= 71.875 \times 10^6 \text{ mm}^4$$

$$\therefore \quad I = \text{Minimum of } I_{XX} \text{ or } I_{YY} = 43.75 \times 10^6 \text{ mm}^4$$

$$\text{Euler's equation for crippling load } P_{cr} = \frac{\pi^2 EI}{l_e^2}$$

$$P_{cr} = \frac{\pi^2 \times 120 \times 10^3 \times 43.75 \times 10^6}{12000^2} = 359.83 \times 10^3 \text{ N.}$$

Problem 12 Design the section of a circular cast iron column that can safely carry a load of 1000 kN. The length of the column is 5 m. Rankine's constants are $\frac{1}{1600}$ and 560 MPa. Factor of safety = 3. The column has one end fixed, the other pinned.

Solution: Safe load P = 1000 kN

Length of column = 5 m = 5000 mm

One end fixed, other end pinned.

$$\therefore \quad \text{Equivalent length of column } l_e = \frac{l}{\sqrt{2}} = \frac{5000}{\sqrt{2}} = 3535.53 \text{ mm}$$

Factor of safety = 3

$$\text{Rankine's constants, } \alpha = \frac{1}{1600}$$

$$\sigma_{cr} = 560 \text{ MPa.}$$

Since factor of safety is given, either multiply load with factor of safety or divide stress with factor of safety.

Dividing stress by factor of safety,

$$\text{Safe stres } \sigma = \frac{\sigma_{cr}}{\text{FoS}} = \frac{560}{3} = 186.67 \text{ MPa.}$$

For circular section, moment of inertia $I = \frac{\pi d^4}{64}$ and Area of cross section $A = \frac{\pi}{4} d^2$.

$$\text{Radius of gyration } k = \sqrt{\frac{I}{A}} = \sqrt{\frac{\left(\frac{\pi}{64} d^4\right)}{\left(\frac{\pi}{4} d^2\right)}} = \frac{d}{4}$$

$$k = \frac{d}{4} = 0.25\ d$$

Rankine's equation for load capacity

$$P = \frac{\sigma A}{1 + \alpha \left(\frac{l_e}{k}\right)^2}$$

$$1000 \times 10^3 = \frac{186.87 \times \left(\frac{\pi}{4} d^2\right)}{1 + \frac{1}{1600}\left(\frac{3535.53}{0.25\ d}\right)^2}$$

$$1000 \times 10^3 = \frac{146.6\ d^2}{1 + \frac{125 \times 10^3}{d^2}}$$

$$= \frac{146.6\ d^2}{\left(\frac{d^2 + 125 \times 10^3}{d^2}\right)} = \frac{146.6\ d^2 \times d^2}{d^2 + 125 \times 10^3}$$

$$d^2 + 125 \times 10^3 = \frac{146.6\ d^4}{1000 \times 10^3}$$

$$d^2 + 125 \times 10^3 = 1.466 \times 10^{-4}\ d^4$$

$$1.466 \times 10^{-4}\ d^4 - d^2 - 125 \times 10^3 = 0$$

Put $d^2 = x$

$$1.466 \times 10^{-4}\ x^2 - x - 125 \times 10^3 = 0$$

Solving $x = d^2 = 32.809.5$

or $d = 181.13$ mm.

Problem 13 A 1.5 long column of circular cross section 50 mm diameter has both ends fixed. Taking factor of safety as 3, calculate the safe load using

(i) Rankine's formula taking yield stress as 560 N/mm^2 and $\alpha = \frac{1}{1600}$

(ii) Euler's formula taking $E = 1.2 \times 10^5$ N/mm^2

Solution: Length $l = 1.5$ m = 1500 mm

Diameter $d = 50$ mm

Both ends fixed, hence equivalent length $l_e = \frac{l}{2} = \frac{1500}{2} = 750$ mm

Yield stress $\sigma_y = \sigma_{cr} = 560$ N/mm^2

$$\alpha = \frac{1}{1600}$$

$$E = 1.2 \times 10^5 \text{ N/mm}^2$$

$$\text{Factor of safety} = 3$$

$$\text{Moment of inertia } I = \frac{\pi}{64} d^4 = \frac{\pi}{64} \times 50^4 = 306.796 \times 10^3 \text{ mm}^4$$

$$\text{Area of cross section } A = \frac{\pi}{4} d^2 = \frac{\pi}{4} \times 50^2 = 1963.5 \text{ mm}^4$$

$$\text{Radius of gyration } k = \sqrt{\frac{I}{A}} = \frac{d}{4} = \frac{50}{4} = 12.5 \text{ mm}$$

$$\text{(i) Rankine's critical load } P_{Rcr} = \frac{\sigma_{cr} A}{1 + \alpha \left(\frac{l_e}{k}\right)^2}$$

$$P_{Rcr} = \frac{560 \times 1963.5}{1 + \frac{1}{1600}\left[\frac{750}{12.5}\right]^2} = 338.326 \times 10^3 \text{ N}$$

$$\text{Safe load as per Rankine } P_{Rsafe} = \frac{P_{Rcr}}{\text{Factor of safety}}$$

$$P_{Rsafe} = \frac{338.326 \times 10^3}{3} = 112.775 \times 10^3 \text{ N}$$

$$\text{(ii) Euler's critical load } P_{Ecr} = \frac{\pi^2 EI}{l_e^2}$$

$$P_{Ecr} = \frac{\pi^2 \times 1.2 \times 10^5 \times 306.796 \times 10^3}{750^2} = 645.96 \times 10^3 \text{ N}$$

$$\text{Safe load } P_{Esafe} = \frac{P_{Ecr}}{\text{Factor of safety}} = \frac{645.96 \times 10^3}{3} = 215.32 \times 10^3 \text{ N}$$

Problem 14 A steel bar of rectangular cross section 60 mm × 20 mm of length 1.25 m is to be used as a column with both ends pinned. Determine the safe load using Rankine's formula. Adopt a safely factor of 4 and Rankine's constants of 550 MPa and $\frac{1}{7500}$. Using Euler's formula find the length of this column for same critical load. Take $E = 2 \times 10^5$ N/mm^2.

Solution:

$$\text{Width } b = 60 \text{ mm}$$

$$\text{Thickness } d = 20 \text{ mm}$$

$$\text{Length } l = 1.25 \text{ m} = 1250 \text{ mm.}$$

Both ends pinned. Hence, equivalent length of column $l_e = l = 1250$ mm.

$$\text{Factor of safety} = 4$$

$$\text{Rankine's constants, } \sigma_{cr} = 550 \text{ MPa}$$

$$\alpha = \frac{1}{7500}$$

$$E = 2 \times 10^5 \text{ MPa}$$

Moment of inertia

$$I_{XX} = \frac{bd^3}{12} = \frac{60 \times 20^3}{12} = 40000 \text{ mm}^4$$

$$I_{YY} = \frac{db^3}{12} = \frac{20 \times 60^3}{12} = 360000 \text{ mm}^4$$

$\therefore$
$$I = \text{Least of } I_{XX} \text{ or } I_{YY} = 40000 \text{ mm}^4$$

$$A = bd = 60 \times 20 = 1200 \text{ mm}^2$$

$$\text{Radius of gyration } k = \sqrt{\frac{I}{A}} = \sqrt{\frac{40000}{1200}} = 5.774 \text{ mm}$$

$$\text{Rankine's critical load } P_{R\text{cr}} = \frac{\sigma_{cr} A}{1 + \alpha\left(\frac{l_e}{k}\right)^2}$$

$$P_{R\text{cr}} = \frac{550 \times 1200}{1 + \frac{1}{7500}\left(\frac{1250}{5.774}\right)^2} = 91{,}048 \text{ N}$$

$\therefore$
$$\text{Safe load as per Rankine} = P_s = \frac{P_{R\text{cr}}}{\text{Factor of safety}}$$

$$P_S = \frac{91048}{4} = 22762 \text{ N}$$

(i) To find length of column

Euler's equation for critical load

$$P_{E\text{cr}} = \frac{\pi^2 EI}{l_e^2}$$

$$91048 = \frac{\pi^2 \times 2 \times 10^5 \times 40000}{l_e^2}$$

$\therefore$
$$l_e = 931.235 = l. \text{ (Both ends pinned)}$$

$$\text{Length of column } l = 931.235 \text{ mm.}$$

Problem 15 A 28 mm diameter column fixed at its ends is acted upon by an axial load. Determine the length of column taking allowable compressive stress as 320 MPa and modulus of elasticity as of 200 GPa.

Solution: Diameter $d = 28$ mm

Both ends fixed.

$$\therefore \quad \text{Equivalength length } l_e = \frac{l}{2}$$

$$\text{Allowable compressive stress } \sigma_{cr} = 320 \text{ MPa}$$

$$E = 200 \text{ GPa} = 200 \times 10^3 \text{ MPa}$$

We know that Euler's equation for critical load

$$P_{cr} = \frac{\pi^2 EI}{le^2} = \frac{\pi^2 E(Ak^2)}{l_e^2}$$

$$\therefore \quad \text{Stress } \frac{P_{cr}}{A} = \sigma_{cr} = \frac{\pi^2 Ek^2}{l_e^2}$$

where $I = Ak^2$

$$k = \text{Radius of gyration} = \sqrt{\frac{I}{A}} = \sqrt{\frac{\left(\frac{\pi}{64} d^4\right)}{\left(\frac{\pi}{4} d^2\right)}} = \frac{d}{4}$$

$$k = \frac{d}{4} = \frac{28}{4} = 7 \text{ mm}$$

$$\therefore \quad 320 = \frac{\pi^2 \times 200 \times 10^3 \times 7^2}{l_e^2}$$

$$\text{Equivalent length } l_e = 549.789 \text{ mm} = \frac{l}{2}$$

$$\therefore \quad \text{Length of column } l = 2 \times l_e = 1099.558 \text{ mm.}$$

Problem 16 A hollow cylindrical column has one end fixed, other end free and is 4.2 m long, and carries an axial load of 230 kN. Design the column by Rankine's formula. Adopt a factor of safety of 2. Internal diameter may be taken as 0.82 times the external diameter. Take σ_{cr} = 550 N/mm^2 and $\alpha = \frac{1}{1600}$.

Solution: One end fixed, other end free.

$$\therefore \quad l_e = 2l \text{ (one end fixed, other and free)}$$

$$\text{Length } l = 4.2 \text{ m} = 4200 \text{ mm}$$

$$\therefore \quad l_e = 2 \times l = 2 \times 4200 = 8400 \text{ mm}$$

$$\text{Load } P_{safe} = 230 \text{ kN} = 230 \times 10^3 \text{ N}$$

$$\text{Factory of safety} = 2$$

$$\therefore \quad \text{Critical load } P_{cr} = P_{safe} \times \text{Factor of safety}$$

$$= 230 \times 10^3 \times 2 = 460 \times 10^3 \text{ N}$$

$$\text{Inner dia } d_i = 0.82\, d_o.$$

$$\sigma_{cr} = 550 \text{ MPa.}$$

$$\alpha = \frac{1}{1600}$$

$$\text{Moment of inertia } I = \frac{\pi}{64}(d_o^4 - d_i^4)$$

$$= \frac{\pi}{64}\left[(d_o^4 - (0.82\, d_o)^4\right] = \frac{\pi}{64} d_o^4 (1 - 0.82^4)$$

$$I = \frac{\pi}{64} d_o^4 (0.5479)$$

$$\text{Area of cross section } A = \frac{\pi}{4}(d_o^2 - d_i^2)$$

$$= \frac{\pi}{4}\left[d_o^2 - (0.82\, d_o)^2\right] = \frac{\pi}{4} d_o^2 (0.3276)$$

$$\therefore \quad \text{Radius of gyration } k = \sqrt{\frac{I}{A}} = \sqrt{\frac{\frac{\pi}{64} d_o^4 (0.5479)}{\frac{\pi}{4} d_o^2 (0.3276)}}$$

$$k = 0.323\, d_o$$

Rankine's equation for critical load of columns

$$P_{cr} = \frac{\sigma_c A}{1 + \alpha\left(\frac{l_e}{k}\right)^2}$$

$$460 \times 10^3 = \frac{550 \times \left[0.3276 \times \frac{\pi}{4} d_o^2\right]}{1 + \frac{1}{1600}\left(\frac{8400}{0.323\, d_o}\right)^2}$$

$$460 \times 10^3 = \frac{141.51\, d_o^2}{\left(1 + \frac{422.\,7 \times 10^3}{d_o^2}\right)} = \frac{141.51\, d_o^2 \times d_o^2}{d_o^2 + 422.7 \times 10^3}$$

$$141.51\, d_o^4 - 460 \times 10^3\, d_o^2 - 1.944 \times 10^{11} = 0$$

Put $\quad d_o^2 = x$

$$141.51 x^2 - 460 \times 10^3 x - 1.944 \times 10^{11} = 0$$

$$\text{Solving } x = d_o^2 = 38725.14$$

$$\therefore \quad d_o = 196.79 \text{ mm}$$

$$d_i = 0.82\, d_o = 161.37 \text{ mm.}$$

Problem 17 Compute the ratio of crippling loads by Euler's and Rankine–Gordon's formulae for an axially loaded tubular column 3 m long with both ends hinged. The inner diameter of the cross section is 50 mm and it is 10 mm thick. Take yield stress as 415 MPa, Young's and modulus as 200 GPa and Rankine's constant = $\frac{1}{7500}$.

Solution:

$$\text{Length } l = 3 \text{ m} = 3000 \text{ mm}$$

$$\text{Both ends hinged, Equivalent } l_e = l = 300 \text{ mm}$$

$$\text{Inner diameter } d_i = 50 \text{ mm}$$

$$\text{Thickness } t = 10 \text{ mm}$$

$$\therefore \quad \text{Outer diameter } d_o = d_i + 2t = 50 + 2 \times 10 = 70 \text{ mm}$$

$$\text{Yield stess } \sigma_y = \sigma_{cr} = 415 \text{ MPa}$$

$$\text{Modulus of elasticity } E = 200 \text{ GPa} = 200 \times 10^3 \text{ MPa}$$

$$\alpha = \frac{1}{7500}$$

$$\text{Moment of inertia } I = \frac{\pi}{64}(d_o^4 - d_i^4) = \frac{\pi}{4}(70^4 - 50^4) = 871.79 \times 10^3 \text{ mm}^4$$

$$\text{Area of cross section } A = \frac{\pi}{4}(d_o^2 - d_i^2) = \frac{\pi}{4}(70^2 - 50^2) = 1884.96 \text{ mm}^2$$

$$\text{Radius of gyration } k = \sqrt{\frac{I}{A}} = \sqrt{\frac{871 - 79 \times 10^3}{1884.96}} = 21.506$$

$$\text{(i) Euler's formula for critical load } P_E = \frac{\pi^2 EI}{l_e^2}$$

$$P_E = \frac{\pi^2 \times 200 \times 10^3 \times 871.79 \times 10^3}{3000^2} = 191.205 \times 10^3 \text{ N}$$

$$\text{(ii) Rankine formula for critical load } P_R = \frac{\sigma_{cr} A}{1 + \alpha\left(\frac{l_e}{k}\right)^2}$$

$$P_R = \frac{415 \times 1884.96}{1 + \frac{1}{7500}\left(\frac{3000}{21.506}\right)^2} = 217.623 \times 10^3 \text{ N}$$

$$\text{Comparing } \frac{P_E}{P_R} = \frac{191.205 \times 10^3}{217.623 \times 10^3} = 0.8786$$

Euler's load capacity is 0.8786 times the Rankies's load capacity.

Problem 18 A 2 m long pin ended column of square section is to be made of wood. Assuming E = 12 GPa. Determine size of the column to support a load of 32 kN. Use factor of safety 3. Use Euler's equation for buckling.

Solution: Length $l = 2 \text{ m} = 2000 \text{ mm}$

Both ends pinned, equivalent length $l_e = l = 2000 \text{ mm}$

Square section, side $= a$

Modulus of elasticity $E = 12 \text{ GPa} = 12 \times 10^3 \text{ MPa}$

Safe load $P = 32 \text{ kN} = 32000 \text{ N}$

FoS = 3.

$\therefore$ critical load $P_{cr} = P \times \text{FoS} = 32000 \times 3 = 96000 \text{ N}$

$$\text{Moment of inertia } I = \frac{a^4}{12}$$

$$\text{Euler's equation for critical loads } P_{cr} = \frac{\pi^2 EI}{l_e^2}$$

$$96000 = \frac{\pi^2 \times 12 \times 10^3 \times \left(\frac{a^4}{12}\right)}{2000^2}$$

$\therefore$ Side of square $a = 78.98$ mm.

Problem 19 Calculate the critical load of a column which is circular in cross section 3 m long with both end guided. The same bar when used a simply supported beam gives a deflection of 6 mm at the middle under a load of 18 N.

Solution: Length $l = 3 \text{ m} = 3000 \text{ mm}$

Both ends guided (pinned), equivalent $l_e = l = 3000 \text{ mm}$

Deflection of simply supported beam $y = 6$ mm

Load $W = 18$ N

For simply supported beam with central load, deflection

$$y = \frac{Wl^3}{48EI}$$

$$\therefore \quad 6 = \frac{18 \times 3000^3}{48EI}$$

$$\therefore \quad \text{Flexural rigidity } EI = \frac{18 \times 3000^3}{48 \times 6} = 1.687 \times 10^9$$

$$\text{Euler's load } P_{cr} = \frac{\pi^2 EI}{le^2} = \frac{\pi^2 \times 1.687 \times 10^9}{3000^2} = 1850.55 \text{ N.}$$

Problem 20 From the experiment of a steel column conducted in a laboratory, the following results are obtained.

Test No.	*Slenderness ratio*	*Stress at failure*
1	75	180 N/mn^2
2	150	80 N/mm^2

Considering these values are valid with Rankine's theory, find the two Rankine's constants in the formula.

Solution:

$$\text{Test (1)}, \frac{l_e}{k} = 75 \text{ and } \sigma = 180 \text{ N/mm}^2$$

$$\text{Test (2)}, \frac{l_e}{k} = 150 \text{ and } \sigma = 80 \text{ N/mm}^2$$

Rankine's formula for column, $P = \dfrac{\sigma_{cr} A}{1 + \alpha \left(\dfrac{l_e}{k}\right)^2}$

$$\therefore \quad \frac{P}{A} = \frac{\sigma_{cr}}{1 + \alpha \left(\dfrac{l_e}{k}\right)^2}$$

$$\sigma = \frac{\sigma_{cr}}{1 + \alpha \left(\dfrac{l_e}{k}\right)^2}$$

For test (1) $$180 = \frac{\sigma_{cr}}{1 + \alpha (75)^2}$$

$$\therefore \quad \sigma_{cr} = 180 (1 + \alpha \times 75^2) \quad \text{...(i)}$$

For test (2) $$80 = \frac{\sigma_{cr}}{1 + \alpha (150)^2}$$

$$\sigma_{cr} = 80 (1 + \alpha \times 150^2) \quad \text{...(ii)}$$

Equating Eqs. (i) and (ii) $180 (1 + 75^2 \alpha) = 80 (1 + 150^2 \alpha)$

$$180 + 180 \times 75^2 \alpha = 80 + 80 \times 150^2 \alpha$$

$$80 \times 150^2 \alpha - 180 \times 75^2 \alpha = 180 - 80 = 100$$

$$\alpha = \frac{100}{80 \times 150^2 - 180 \times 75^2} = 1.2698 \times 10^{-4} = \frac{1}{7875}$$

Substituting in Eq. (i)

$$\sigma_{cr} = 180 \left(1 + \frac{1}{7875} \times 75^2\right) = 308.57 \text{ N/mm}^2$$

The Rankine's constants are $\sigma_{cr} = 308.57$ N/mm^2 and $\alpha = \dfrac{1}{7875}$.

Problem 21 A hollow circular column 3 m long has one end fixed and the other end pinned. Outside diameter of column is 120 mm and wall thickness is 15 mm. Rankine's constants are 420 MPa and $\frac{1}{7500}$. Compare the critical loads obtained by Rankine's and Euler's formulae. Also find the length of column for which both Rankine and Euler's equation give same critical load. Take E = 200 GPa.

Solution: Length of column $l = 3 \text{ m} = 3000 \text{ mm}$

One end fixed and the other end pinned.

$$\therefore \quad \text{Effective length } l_e = \frac{l}{\sqrt{2}} = \frac{3000}{\sqrt{2}} = 2121.32 \text{ mm}$$

$$\text{Outside diameter } d_o = 120 \text{ mm}$$

$$\text{Wall thickness } t = 15 \text{ mm}$$

$$\therefore \quad \text{Inside diameter } d_i = d_o - 2t = 120 - 2 \times 15 = 90 \text{ mm}$$

$$\text{Rankine's constants } \alpha = \frac{1}{7500}$$

$$\sigma_{cr} = 420 \text{ MPa}$$

$$\text{Modulus of elasticity } E = 200 \text{ GPa} = 200 \times 10^3 \text{ MPa}$$

$$\text{Moment of inertia } I = \frac{\pi}{64}(d_o^4 - d_i^4) = \frac{\pi}{64}(120^4 - 90^4) = 6.958 \times 10^6 \text{ mm}^4$$

$$\text{Area of cross section } A = \frac{\pi}{4}(d_o^2 - d_i^4) = \frac{\pi}{4}(120^2 - 90^2) = 4948 \text{ mm}^2$$

$$\therefore \quad \text{Radius of gyration } k = \sqrt{\frac{I}{A}} = \sqrt{\frac{6.958 \times 10^6}{4948}} = 37.5 \text{ mm}$$

(i) Rankine's equation for critical load

$$P_R = \frac{\sigma_{cr} \cdot A}{1 + \alpha\left(\frac{l_e}{k}\right)^2} = \frac{420 \times 4948}{1 + \frac{1}{7500}\left(\frac{2121.32}{37.5}\right)^2} = 1.4566 \times 10^6 \text{ N}$$

(ii) Euler's equation for critical load

$$P_E = \frac{\pi^2 EI}{l_e^2} = \frac{\pi^2 \times 200 \times 10^3 \times 6.958 \times 10^6}{2121.32^2} = 3.052 \times 10^6 \text{ N}$$

$$\text{Comparing } \frac{P_R}{P_E} = \frac{1.4566 \times 10^6}{3.052 \times 10^6} = 0.477$$

The Rankine's critical load is 0.477 times that of Euler's.

(iii) To find length of column for the same loads.

Given the critical loads are equal

$$P_E = P_R$$

$$\frac{\pi^2 EI}{le^2} = \frac{\sigma_{cr} A}{1 + \alpha\left(\frac{l_e}{k}\right)^2}$$

$$\frac{\pi^2 \times 200 \times 10^3 \times 6.958 \times 10^6}{l_e^2} = \frac{420 \times 4948}{1 + \frac{1}{7500}\left(\frac{l_e}{37.5}\right)^2}$$

$$\frac{1.3735 \times 10^{13}}{l_e^2} = \frac{2.078 \times 10^6}{1 + 9.482 \times 10^{-8}\, l_e^2}$$

$$1 + 9.482 \times 10^{-8}\, l_e^2 = \frac{2.078 \times 10^6}{1.3735 \times 10^{13}} l_e^2 = 1.5129 \times 10^{-7}\, l_e^2$$

$$(1.5129 \times 10^{-7} - 9.482 \times 10^{-8})\, l_e^2 = 1$$

$\therefore$ Equivalent length $l_e = 4208.06$ mm

But $$l_e = \frac{l}{\sqrt{2}}$$

Length of column $l = l_e\sqrt{2} = 5951.1$ mm

The loads obtained by Euler's and Rankine's formulae will be equal for a length of 5951.1 mm for the given conditions.

Problem 22 Determine the Euler's crippling load for an I-section column 180 mm × 360 mm × 10 mm, 6 m long with ends hinged. Take modulus of elasticity as 206 GPa. Also find Rankine's crippling load taking constants as 480 MPa and $\frac{1}{7500}$.

Solution: The given I-Section is shown in Fig. 9.9.

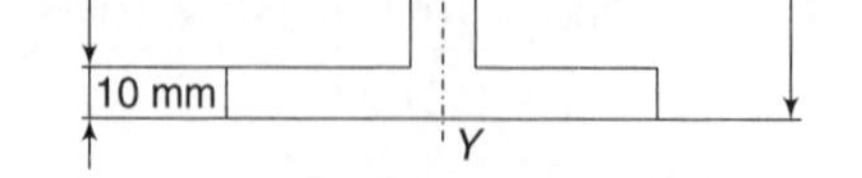

Fig. 9.9

Total width $B = 180$ mm

Total height $H = 360$ mm

Thickness $t = 10$ mm

Inner width $b = B - t = 180 - 10 = 170$ mm

Inner height $h = H - 2t = 360 - 2 \times 10 = 340$ mm

Length $l = 6$ m $= 6000$ mm

Both ends hinged, $\therefore$ Equivalent length $l_e = l = 6000$ mm

Modulus of Elasticity $E = 206$ GPa $= 206 \times 10^3$ MPa.

Moment of inertia:

$$I_{XX} = \frac{BH^3 - bh^3}{12} = \frac{180 \times 360^3 - 170 \times 340^3}{12} = 143.03 \times 10^6 \text{ mm}^4$$

(Considering full rectangle $B \times H$ – The inside rectangle $b \times h$)

$$I_{YY} = I_1 + I_2 + I_3 = \frac{tB^3}{12} + \frac{h \times t^3}{12} + \frac{tB^3}{12}$$

(Consider three rectangles)

$$I_{YY} = \frac{10 \times 180^3}{12} \times 2 + \frac{340 \times 10^3}{12} = 9.748 \times 10^6 \text{ mm}^4$$

Area of cross section $A = BH - bh = 180 \times 360 - 170 \times 340$

$$A = 7000 \text{ mm}^2$$

$\therefore$ $I =$ Least of I_{XX} or $I_{YY} = 9.748 \times 10^6$ mm^4

$\therefore$ Radius of gyration $k = \sqrt{\frac{I}{A}} = \sqrt{\frac{9.748 \times 10^6}{7000}} = 37.317$ mm

Rankine's constants,

Stress $\sigma_{cr} = 480$ MPa.

$$\alpha = \frac{1}{7500}$$

(i) Euler's critical load

$$P_E = \frac{\pi^2 EI}{l_e^2} = \frac{\pi^2 \times 206 \times 10^3 \times 9.748 \times 10^6}{6000^2}$$

$$P_E = 550.53 \times 10^3 \text{ N}$$

(ii) Rankine's critical load

$$P_R = \frac{\sigma_{cr} \cdot A}{1 + \alpha \left(\frac{l_e}{k}\right)^2} = \frac{480 \times 7000}{1 + \frac{1}{7500}\left(\frac{6000}{37.317}\right)^2}$$

$$= 755.58 \times 10^3 \text{ N.}$$

Problem 23 The cross section of a connecting rod is I-section with each flange width of $4t$, total height of $5t$ and thickness of flange and web as t is 300 mm long. It is hinged at both ends in XX plane and both ends fixed in YY plane. It is subjected to an axial load of 30 kN. Taking critical stress as 320 MPa, $\alpha = \frac{1}{7500}$, and factor of safety of 2.5, determine the cross-sectional dimensions of the connecting rod.

Solution: Figure 9.10 shows the I-section.

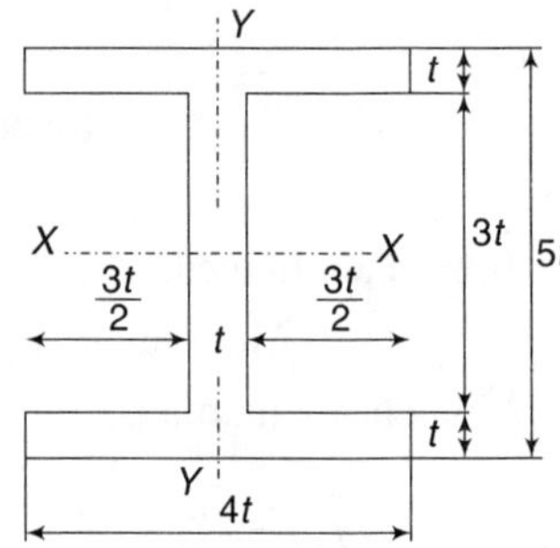

Fig. 9.10

$$\text{Total width } B = 4t$$
$$\text{Total height } H = 5t$$
$$\text{Thickness} = t$$
$$\text{Inner width } b = B - t = 3t$$
$$\text{Inner height } h = H - 2t = 3t$$
$$\text{Load } P = 30 \text{ kN}$$
$$\text{Factor of safely} = 2.5$$
$$\text{Critical stress} = 320 \text{ MPa}$$

∴ Allowable or safe stress

$$\sigma = \frac{\sigma_{cr}}{\text{Factor of safety}} = \frac{320}{2.5} = 128 \text{ MPa}$$

$$\alpha = \frac{1}{7500}$$

$$\text{Length of column, } l = 300 \text{ mm}$$

Both ends pinned in *XX* plane,

$$\therefore \quad l_{e\,XX} = l = 300 \text{ mm}$$

Both ends fixed in *YY* plane,

$$l_{e\,YY} = \frac{l}{2} = \frac{300}{2} = 150 \text{ mm}$$

Moment of inertia

$$I_{XX} = \frac{BH^3 - bh^3}{12} = \frac{4t(5t)^3 - 3t\,(3t)^3}{12} = 34.917\,t^4$$

$$I_{YY} = I_1 + I_2 + I_3 = \frac{t(4t)^3 + 3t \times t^3 + t\,(4t)^3}{12} = 10.917\,t^4$$

$$\text{Area of cross section } A = 4t \times t + 3t \times t + 4t \times t = 11\,t^2$$

$$\text{Radius of gyration } k_{XX} = \sqrt{\frac{I_{XX}}{A}}\sqrt{\frac{34.917\,t^4}{11\,t^2}} = 1.7816\,t$$

$$k_{YY} = \sqrt{\frac{I_{YY}}{A}} = \sqrt{\frac{10.917\,t^4}{11\,t^2}} = 0.996\,t$$

Rankine's formula for columns

$$P = \frac{\sigma A}{1 + \alpha\left(\frac{l_e}{k}\right)^2}$$

Consider *XX* plane

$$30000 = \frac{128 \times 11\,t^2}{1 + \frac{1}{7500}\left(\frac{300}{1.7816\,t}\right)^2}$$

$$30000 = \frac{1408\ t^2}{\left(\dfrac{1 + 3.781}{t^2}\right)} = \frac{1408\ t^2}{\left(\dfrac{t^2 + 3.781}{t^2}\right)}$$

$$1408\ t^4 = 30000\ (t^2 + 3.781)$$

$$t^4 = \frac{30000}{1408}\ (t^2 + 3.781)$$

$$t^4 = 21.31\ (t^2 + 3.781)$$

$$t^4 - 21.31t^2 - 80.56 = 0$$

Putting $t^2 = x$

$$x^2 - 21.31\ x - 80.56 = 0$$

Solving $x = 24.587 = t^2$

$\therefore$ $t = 4.96$ mm

Consider *YY* plane

$$30000 = \frac{128 \times 11\ t^2}{1 + \dfrac{1}{7500}\left(\dfrac{150}{0.996\ t}\right)^2}$$

$$30000 = \frac{1408\ t^2}{1 + \left(\dfrac{3.024}{t^2}\right)} = \frac{1408\ t^2}{\left(\dfrac{t^2 + 3.024}{t^2}\right)}$$

$$30000 = \frac{1408\ t^4}{t^2 + 3.204}$$

$$t^4 = \frac{30000}{1408}\ (t^2 + 3.204) = 21.31\ (t^2 + 3.204)$$

$$t^4 - 21.31\ t^2 - 68.277 = 0$$

Put $t^2 = x$ and solve for x

$$x = t^2 = 24.139$$

$\therefore$ $t = 4.913$ mm

Recommended $t = 4.96$ (bigger of the two)

Flange $4t = 4 \times 4.96 = 19.84$ mm

Height $5t = 5 \times 4.96 = 24.8$ mm

Thickness $t = 4.96$ mm.

Problem 24 Determine the Euler's crippling load for a T-section dimensioned in mm as shown in Fig. 9.11. The length of column is 9 m with one end fixed and the other end free. Modulus of elasticity is 2×10^5 N/mm^2. Also find the crippling load of this column using Rankine-Gordan formula. The critical stress is 360 MPa and $\alpha = \dfrac{1}{7500}$.

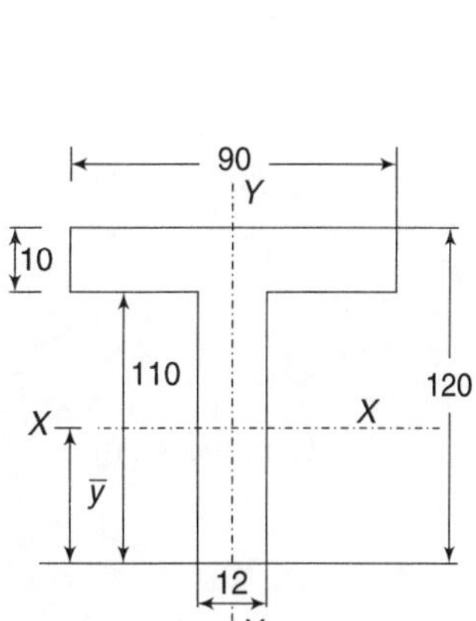

Fig. 9.11

Solution: Length $l = 9 \text{ m} = 9000 \text{ mm}$

One end fixed, other end free, $l_e = 2l = 2 \times 9000 = 18000 \text{ mm}$

Modulus of elasticity $E = 2 \times 10^5 \text{ MPa}$.

Critical stress $\sigma_{cr} = 360 \text{ MPa}$.

$$\alpha = \frac{1}{7500}$$

To find the centre of gravity $\bar{y}$

$$\bar{y} = \frac{a_1 y_1 + a_2 y_2}{a_1 + a_2}$$

$$= \frac{90 \times 10 \times \left(120 - \frac{10}{2}\right) + 110 \times 12 \times \left(\frac{110}{2}\right)}{90 \times 10 + 110 \times 12}$$

$$\bar{y} = 79.324 \text{ mm.}$$

$$\text{Area } A = A_1 + A_2 + = 90 \times 10 + 110 \times 12 = 2220 \text{ mm}^2$$

Moment of Inertia: $I_{XX} = I_1 + I_2$

$$I_{XX} = \left(\frac{bd^3}{12} + A_1 r_1^2\right) + \left(\frac{bd^3}{12} + A_2 r_2^2\right)$$

(From parallel axis theorem)

$$I_{XX} = \frac{90 \times 10^3}{12} + 90 \times 10\,(110 - 79.324)^2 + \frac{12 \times 110^3}{12} + 12 \times 110\,(79.324 - 55)^2$$

$$I_{XX} = 10.589 \times 10^6 \text{ mm}^4$$

$$I_{YY} = I_1 + I_2$$

$$= \frac{10 \times 90^3}{12} + \frac{110 \times 12^3}{12} = 0.623 \times 10^6 \text{ mm}^4$$

$$I = \text{Least of } I_{XX} \text{ or } I_{YY} = 0.623 \times 10^6 \text{ mm}^4$$

$$\text{Radius of gyration } k = \sqrt{\frac{I}{A}} = \sqrt{\frac{0.623 \times 10^6}{2220}} = 16.752 \text{ mm}$$

$\therefore$ Euler's critical load $P_E = \dfrac{\pi^2 EI}{l_e^2}$

$$P_E = \frac{\pi^2 \times 2 \times 10^5 \times 0.623 \times 10^6}{18000^2} = 3797.6 \text{ N}$$

$$\text{Rankine's critical load } P_R = \frac{\sigma_{cr} A}{1 + \alpha\left(\frac{l_e}{k}\right)^2}$$

$$P_R = \frac{360 \times 2220}{1 + \frac{1}{7500}\left(\frac{18000}{16.752}\right)^2} = 5161.81 \text{ N.}$$

Problem 25 A short length of tube 50 mm outside diameter and 5 mm wall thickness failed in compression at a load of 240 kN. When a 2 m long tube of same material was tested as a column with both end fixed, the failure occurred at a load of 160 kN. Taking that the critical stress is correct from first test, find the value of Rankine's constant α. What will be the crippling load of this tube of length 4 m with both ends hinged.

Solution:

$$\text{Outside diameter } d_o = 50 \text{ mm}$$

$$\text{Wall thickness } t = 5 \text{ mm}$$

$$\text{Inside diameter } d_i = d_o - 2t = 40 \text{ mm}$$

$$\text{Length } l = 2 \text{ m} = 2000 \text{ mm}$$

$$\text{Both ends fixed } l_e = \frac{l}{2} = \frac{2000}{2} = 1000 \text{ mm}$$

$$\text{Compressive load } P_c = 240 \text{ kN} = 240 \times 10^3 \text{ N}$$

$$\text{Failure load of column } P_{cr} = 160 \times 10^3 \text{ N}$$

$$\text{Area of cross section } A = \frac{\pi}{4}(d_o^2 - d_i^2) = \frac{\pi}{4}(50^2 - 40^2) = 706.86 \text{ mm}^2$$

$$\therefore \quad \text{Critical stress } \sigma_{cr} = \frac{P_c}{A} = \frac{240 \times 10^3}{706.86} = 339.53 \text{ N/mm}^2$$

$$\text{Moment of inertia } I = \frac{\pi}{64}(d_o^4 - d_i^4) = \frac{\pi}{64}(50^4 - 40^4)$$

$$= 181.13 \times 10^3 \text{ mm}^4$$

$$\therefore \quad \text{Radius of gyration } k = \sqrt{\frac{I}{A}} = \sqrt{\frac{181.13 \times 10^3}{706.86}} = 16 \text{ mm}$$

$$\therefore \text{ Rankine's formula for critical load } P_{cr} = \frac{\sigma_{cr} A}{1 + \alpha\left(\frac{l_e}{k}\right)^2}$$

$$160 \times 10^3 = \frac{339.53 \times 706.86}{1 + \alpha\left(\frac{1000}{16}\right)^2}$$

$$\therefore \quad \text{Rankine's constant } \alpha = 1.28 \times 10^{-4} = \frac{1}{7812.48}$$

(ii) For length $l = 4$ m $= 4000$ mm

$$\text{Both ends hinged, } l_e = l = 4000 \text{ mm,}$$

$$\sigma_{cr} = 339.53 \text{ MPa}$$

$$\alpha = 1.28 \times 10^{-4} \text{ (Same as above)}$$

$$\text{Rankine critical load } P_{cr} = \frac{\sigma_{cr} A}{1 + \alpha\left(\frac{l_e}{k}\right)^2}$$

$$P_{cr} = \frac{339.53 \times 706.86}{1 + 1.28 \times 10^{-4}\left(\frac{4000}{16}\right)^2} = 26666.69 \text{ N.}$$

Problem 26 A built up I-section has a overall depth of 360 mm, width of flanges each 250 mm and thickness of flange 40 mm, thickness of web 30 mm. When it is used as a simply supported beam and carrying a uniformly distributed load of 30 kN/m length, the deflection at the mid-span was found to be 12 mm. Find the safe load of this I-Section when used as a column with one end fixed and the other end pinned taking factor of safety = 2 and E = 200 kN/mm^2.

Solution: The given I-section is shown in Fig. 9.12.

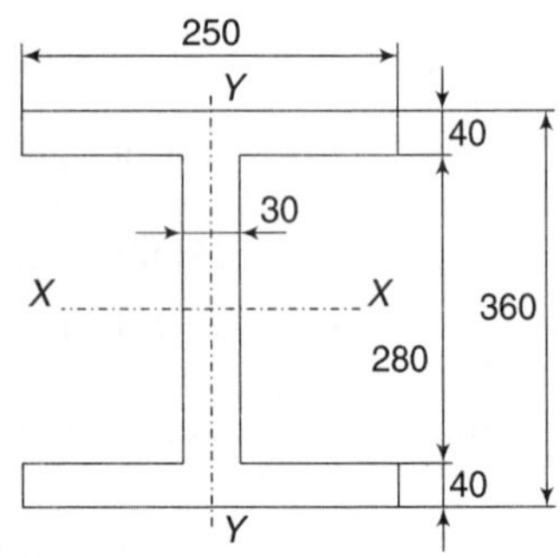

Fig. 9.12

Deflection y = 12 mm

UDL = 30 kN/m = 30000 N/m

Let l = Length of beam; mm

$$E = 200 \text{ kN/mm}^2 = 200 \times 10^3 \text{ N/mm}^2$$

$$I_{XX} = \frac{BH^3 - bh^3}{12}$$

$$= \frac{250 \times 360^3 - (250 - 30) \times 280^3}{12}$$

$$= 569.547 \times 10^6 \text{ mm}^4$$

$$I_{YY} = I_1 + I_2 + I_3$$

$$= \frac{2 \times 40 \times 250^3}{12} + \frac{280 \times 30^3}{12} = 104.797 \times 10^6 \text{ mm}^4$$

(i) Consider the simply supported beam with UDL, deflection at the middle

$$y = \frac{5}{384}\frac{wl^4}{EI}$$

$$w = \text{UDL} = 30 \text{ kN/m} = \frac{30000}{1000} = 30 \text{ N/mm}$$

(Since l is in mm, w should be N/mm)

$I = I_{XX}$ since the simply supported beam bends about XX axis

$$12 = \frac{5}{384}\,\frac{30 \times l^4}{200 \times 10^3 \times 569.547 \times 10^6}$$

$\therefore$ Length of beam l = 7691.22 mm

(ii) Euler's crippling load $P_{cr} = \dfrac{\pi^2 EI}{l_e^2}$

I = Least moment of inertia = 104.797×10^6 mm^4

For one end fixed, other end pinned, $l_e = \frac{l}{\sqrt{2}} = \frac{7691.2}{\sqrt{2}} = 5438.51$ mm

$$P_{cr} = \frac{\pi^2 \times 200 \times 10^3 \times 104.797 \times 10^6}{5438.51^2} = 6.994 \times 10^6 \text{ N}$$

$$\therefore \text{ Safe load} = P = \frac{\text{Critical load}}{\text{Factor of safety}} = \frac{6.994 \times 10^6}{2} = 3.497 \times 10^6 \text{ N.}$$

Problem 27 A beam of length 3 m when used as a simply supported beam subjected to a central load of 120 kN at the middle deflects by 10 mm at the middle. Determine the Euler's crippling load when it is used as a column with

(i) both ends fixed,
(ii) both ends hinged,
(iii) one end fixed, other end free.

Solution: Length between supports l = 3 m = 3000 mm

Central load W = 120 kN = 120×10^3 N

Deflection at the middle y = 10 mm.

Considering simply supported beam with central load,

$$\text{Deflection at the middle } y = \frac{Wl^3}{48EI}$$

$$\therefore \quad \text{Flexural rigidity } EI = \frac{Wl^3}{48y} = \frac{120 \times 10^3 \times 4000^3}{48 \times 10}$$

$$EI = 16 \times 10^{12}$$

$$\text{Euler's crippling load } P_{cr} = \frac{\pi^2 EI}{l_e^2}$$

(i) When both ends fixed, $l_e = \frac{l}{2} = \frac{4000}{2} = 2000$ mm

$$\therefore \quad P_{cr} = \frac{\pi^2 \times 16 \times 10^{12}}{2000^2} = 39478.4 \text{ N}$$

(ii) When both ends pinned, $l_e = l = 4000$ mm

$$\therefore \quad P_{cr} = \frac{\pi^2 \times 16 \times 10^{12}}{4000^2} = 9869.6 \text{ N.}$$

(iii) When one end fixed, other end free, $l_e = 2l = 8000$ mm

$$P_{cr} = \frac{\pi^2 \times 16 \times 10^{12}}{8000^2} = 2467.4 \text{ N.}$$

Problem 28 Compare the strength of two columns 3 m long having same weight and made of same material and both ends pinned. One is I-section of each flange 200 mm × 20 mm and the web of 250 mm × 20 mm. The other is a hollow cylindrical column of wall thickness 20 mm. Use Rankine's equation with σ_{cr} = 330 MPa and $\alpha = \frac{1}{7500}$.

Solution: Length $l = 3 \text{ m} = 3000 \text{ mm}$

Both ends pinned, $\therefore\ l_e = l = 3000$ mm

Same material, weight and length

Hence, areas are same $\sigma_{cr} = 330$ MPa

$$\alpha = \frac{1}{7500}$$

The I-section and hollow circular section are shown in Fig. 9.13.

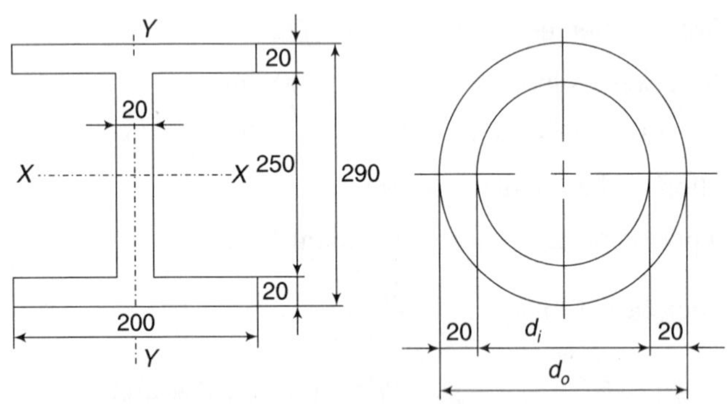

(a) I-section (b) Hollow circular section

Fig. 9.13

(i) Consider I-section:

$$I_{XX} = \frac{BH^3 - bh^3}{12} = \frac{200 \times 290^3 - (200 - 20) \times 250^3}{12} = 172.1 \times 10^6 \text{ mm}^4$$

$$I_{YY} = I_1 + I_2 + I_3 = 2 \times \frac{20 \times 200^3}{12} + \frac{250 \times 20^3}{12} = 26.833 \times 10^6 \text{ mm}^4$$

Area of cross section $A = BH - bh = 200 \times 290 - (200 - 20) \times 250$

$= 13000 \text{ mm}^2$

$\therefore$ I = Least moment of inertia = 26.833×10^6 mm^4

$$\text{Radius of gyration } k = \sqrt{\frac{I}{A}} = \sqrt{\frac{26.833 \times 10^6}{13000}} = 45.432 \text{ mm}$$

Critical load given by Rankine's equation

$$P_{cr} = \frac{\sigma_{cr} A}{1 + \alpha \left(\frac{l_e}{k}\right)^2} = \frac{330 \times 13000}{1 + \frac{1}{7500}\left(\frac{3000}{45.432}\right)^2}$$

$$P_{cr} = 2.713 \times 10^6 \text{ N}$$

(ii) Consider hollow circular column

Given: Wall thickness $t = 20$ mm

Inside diameter $d_i = d_i$

Outside diameter $d_o = d_i + 2t = d_i + 40$ mm

For the same material, weight and length, Area of I section = Area of circular section.

$$13000 = \frac{\pi}{4}(d_o^2 - d_i^2)$$

$$13000 = \frac{\pi}{4}\left((d_i + 40)^2 - d_i^2\right)$$

$$(d_i + 40)^2 - d_i^2 = \frac{4 \times 13000}{\pi} = 16552.114$$

$$d_i^2 + 1600 + 80\, d_i - d_i^2 = 16552.114$$

$$\therefore \quad d_i^2 = 186.9 \text{ mm}$$

$$d_o = d_i + 40 = 226.9 \text{ mm}$$

$$\therefore \quad \text{Moment of inertia } I = \frac{\pi}{64}(d_o^4 - d_i^4)$$

$$= \frac{\pi}{64}(226.9^4 - 186.9^4) = 70.212 \times 10^6 \text{ mm}^4$$

$$\text{Area } A = 13000 \text{ mm}^2 \text{ (Same)}$$

$$\therefore \quad \text{Radius of gyration } k = \sqrt{\frac{I}{A}} = \sqrt{\frac{70.212 \times 10^6}{13000}} = 73.49 \text{ mm}$$

$$\therefore \quad \text{Rankine's critical load } P_{cr} = \frac{\sigma_{cr} A}{1 + \alpha \left(\frac{l_e}{k}\right)^2}$$

$$= \frac{330 \times 13000}{1 + \frac{1}{7500}\left(\frac{3000}{73.49}\right)^2} = 3.51 \times 10^6 \text{ N}$$

$$\text{Comparing, } \frac{P_{cr} \text{ Circular}}{P_{cr} \text{ I-section}} = \frac{3.51 \times 10^6}{2.713 \times 10^6} = 1.2938$$

The hollow circular column of same weight can carry 1.2938 times the load capacity of I-section.

Problem 29 A column is fabricated using two identical I-joists as shown in Fig. 9.14. Both ends of column are hinged. The actual length of column is 6 m. What maximum axial compressive load can be allowed on the column if the load factor is 3 as per Euler's equation. For what length of column, Rankine's and Euler's formulae give the same results. The properties of I-joist are given below.

$$I_{XX} = 0.6 \times 10^8 \text{ mm}^4,\ I_{YY} = 0.1 \times 10^8 \text{ mm}^4$$

$$E = 200 \times 10^3 \text{ N/mm}^2.\ \sigma_{cr} = 330 \text{ N/mm}^2,\ \alpha = \frac{1}{7500}.$$

Area of cross section $A = 5200 \text{ mm}^2$.

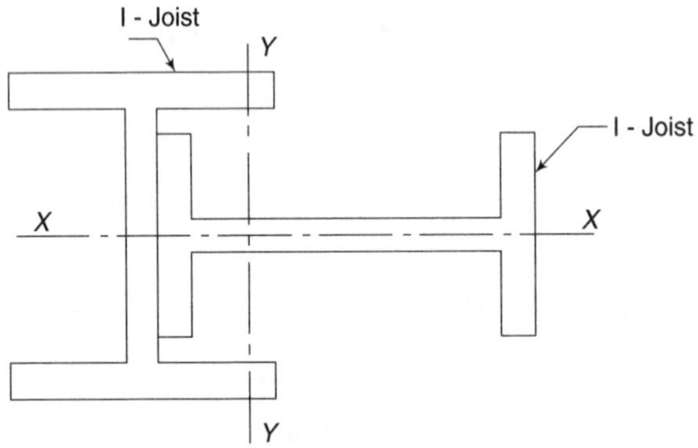

Fig. 9.14

Solution: Length of column = 6 m = 6000 mm

Both ends hinged ∴ Equivalent length $l_e = l = 6000$ mm

Factor of safety = 3

Moment of inertia of each joist, $I_{XX} = 0.6 \times 10^8 \text{ mm}^4$

$I_{YY} = 0.1 \times 10^8 \text{ mm}^4$

Area of cross section $A = 5200 \text{ mm}^2$

Critical stress $\sigma_{cr} = 330 \text{ N/mm}^2$

$$\alpha = \frac{1}{7500}$$

Modulus of elasticity $E = 200 \times 10^3 \text{ N/mm}^2$

For combined section:

$$I_{XX} = I_{XX} \text{ of the I-joist} + I_{YY} \text{ of the joist}$$

$$= 0.6 \times 10^8 + 0.1 \times 10^8 = 0.7 \times 10^8 \text{ mm}^4$$

$$I_{YY} = (I_{YY} \text{ of the I-joist} + Ak_1^2) + (I_{XX} \text{ of the joist} + Ak_2^2)$$

(Parallel axis theorem)

$$= 0.6 \times 10^8 + Ak_1^2 + 0.1 \times 10^8 + Ak_2^2$$

$$= 0.7 \times 10^8 + Ak_1^2 + Ak_2^2$$

I_{XX} is less than I_{YY}

Hence, $$I = I_{XX} = 0.7 \times 10^8 \text{ mm}^4$$

$$\text{Radius of gyration } k = \sqrt{\frac{I}{2A}} = \sqrt{\frac{0.7 \times 10^8}{2 \times 5200}} = 82.04 \text{ mm}$$

(Area = $2A$ for 2 sections)

(i) From Euler's equation critical load $P_E = \dfrac{\pi^2 EI}{l_e^2}$

$$= \frac{\pi^2 \times 200 \times 10^3 \times 0.7 \times 10^8}{6000^2} = 3.8382 \times 10^6 \text{ N}$$

$$\text{Safe load } P = \frac{\text{Critical load}}{\text{Factor of safety}} = \frac{3.8382 \times 10^6}{3}$$

$$P_s = 1.2794 \times 10^6 \text{ N}$$

(ii) For the critical loads obtained by Euler's and Rankine's formulae to be same

$$P_E = P_R$$

$$\frac{\pi^2 EI}{l_e^2} = \frac{\sigma_{cr} A}{1 + \alpha \left(\dfrac{l_e}{k}\right)^2}$$

$$\frac{\pi^2 \times 200 \times 10^3 \times 0.7 \times 10^8}{l_e^2} = \frac{330 \times (2 \times 5200)}{1 + \dfrac{1}{7500}\left(\dfrac{l_e}{82.04}\right)^2}$$

$$\frac{1.3817 \times 10^{14}}{l_e^2} = \frac{343200\ 0}{1 + 1.981 \times 10^{-8}\ l_e^2}$$

$$3432000\ l_e^2 = 1.3817 \times 10^{14}\ (1 + 1.981 \times 10^{-8}\ l_e^2)$$

$$3432000\ l_e^2 - 2612794.7\ l_e^2 = 1.3817 \times 10^{14}$$

$$\therefore \quad l_e = 12987.05 \text{ mm}$$

But given both ends pinned, $l_e = l = 12987.05$ mm.

Problem 30 Compare the load capacities of two identical columns, one with both ends fixed and the other one end fixed, other end pinned.

Solution: Identical columns, $\therefore$ I and E are same.

(i) Both ends fixed, $l_e = \dfrac{l}{2}$

$$\text{Euler's critical load } P_1 = \frac{\pi^2 EI}{l_e^2} = \frac{\pi^2 EI}{\left(\frac{l}{2}\right)^2} = \frac{4\pi^2 EI}{l^2}$$

(ii) Both ends pinned, $l = l$

$$\text{Euler's critical load } P_2 = \frac{\pi^2 EI}{l_e^2} = \frac{\pi^2 EI}{l^2}$$

$$\text{Comparing, } \frac{P_1}{P_2} = 4$$

The load capacity of column with both ends fixed is 4 times that of column with both ends pinned.

Problem 31 The section of a column consists of two channels 350 mm × 100 mm connected by two plates of 400 mm × 20 mm on both top and bottom sides of the flanges as shown in Fig. 9.15. The length of the column is 9 m and both ends are fixed. Determine safe load on the member by Rankine's formula taking factor of safety as 3. σ_{cr} = 320 MPa, $\alpha = \frac{1}{7500}$. The properties of each channel are $I_{XX} = 93.126 \times 10^6$ mm^4, $I_{YY} = 3.946 \times 10^6$ mm^4. Area A = 4947 mm^4. Centroid from back of web C_{YY} = 24.1 mm.

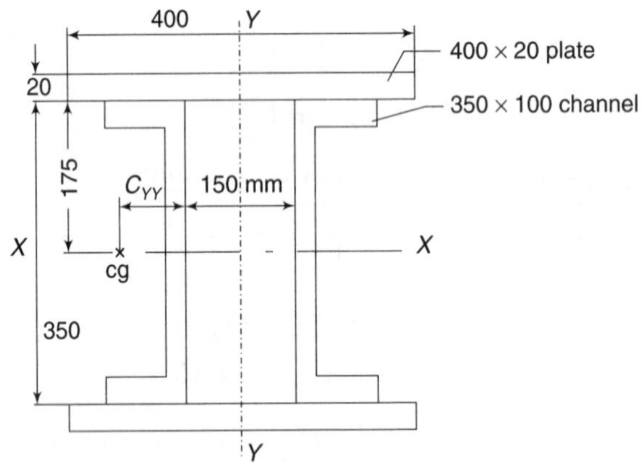

Fig. 9.15

Solution:

$$\text{Plate} = 400 \text{ mm} \times 20 \text{ mm thick}$$

$$\text{Channel} = 350 \text{ mm} \times 100 \text{ mm}$$

$$\text{Length } l = 9 \text{ m} = 9000 \text{ mm}$$

$$\text{Both ends fixed, } \therefore l_e = \frac{l}{2} = 4500 \text{ mm}$$

$$\text{Factor of safety} = 3$$

$$\sigma_{cr} = 320 \text{ MPa}, \alpha = \frac{1}{7500}$$

For each channel $I_{XX} = 93.126 \times 10^6 \text{ mm}^4$, $I_{YY} = 3.946 \times 10^6 \text{ mm}^4$

$A = 4947 \text{ mm}^4$, $C_{YY} = 24.1 \text{ mm}$

Due to symmetry, cg lies in the middle.

$$\text{Total Area } A_{\text{Total}} = 2\, A_{\text{channel}} + 2\, A_{\text{Plate}}$$

$$A_{\text{Total}} = 2 \times 4947 + 2 \times (400 \times 20) = 25894 \text{ mm}^4$$

$$I_{XX} = 2\, I_{XX} + 2\left(\frac{bd^3}{12} + Ak^2\right) \text{ for plate}$$

$$= 2 \times 93.126 \times 10^6 + 2\left[\frac{400 \times 20^3}{12} + 400 \times 20\,(175 + 10)^2\right]$$

$$I_{XX} = 734.385 \times 10^6 \text{ mm}^4$$

$$I_{YY} = 2\left[I_{YY} + A_{ch}k^2\right] + 2\left(\frac{db^3}{12}\right) \text{ for plates}$$

$$\text{For channel } k = \frac{150}{2} + C_{YY} = 75 + 24.1$$

$$= 2\left[3.946 \times 10^6 + 4947\,(75 + 24.1)^2\right] + \frac{20 \times 400^3}{12} \times 2$$

$$I_{YY} = 318.39 \times 10^6 \text{ mm}^4$$

$$\therefore \quad I = \text{Least value of } I_{XX} \text{ or } I_{YY} = 318.39 \times 10^6 \text{ mm}^4$$

$$\text{Radius of gyration } k = \sqrt{\frac{I}{A_{\text{Total}}}} = \sqrt{\frac{318.39 \times 10^6}{25894}} = 110.887 \text{ mm}$$

Rankine's formula for critical load

$$P_{\text{cr}} = \frac{\sigma_{\text{cr}} A}{1 + \alpha\left(\frac{l_e}{k}\right)^2}$$

$$P_{\text{cr}} = \frac{320 \times 25894}{1 + \frac{1}{7500}\left(\frac{4500}{110.887}\right)^2}$$

$$P_{\text{cr}} = 6.7942 \times 10^6 \text{ N}$$

$$\text{Safe load } P_s = \frac{P_{\text{cr}}}{\text{Factor of safety}} = \frac{6.7942 \times 10^6}{3}$$

$$P_s = 2.2647 \times 10^6 \text{ N.}$$

THINGS TO REMEMBER

- Euler's formula for critical load of columns

$$P_{cr} = \frac{\pi^2 EI}{l_e^2}$$

- Rankine's formula for critical load of columns

$$P_{cr} = \frac{\sigma_{cr} A}{1 + \alpha \left(\frac{l_e}{k}\right)^2}$$

- I = Least moment of inertia of I_{XX} or I_{YY}
- Radius of gyration $k = \sqrt{\frac{I}{A}}$
- l_e = Equivalent of column

  $= l$ for both ends pinned, hinged or supported

  $= \frac{l}{2}$ for both ends fixed

  $= \frac{l}{\sqrt{2}}$ for one end fixed, other end pinned

  $= 2l$ for one end fixed, other end free.

- Slenderness ratio $= \frac{l_e}{k} = \frac{\text{Effective length}}{\text{Radius of gyration}}$
- Limitation of Euler's formula: Euler's formula is valid for slenderness ratio $\frac{l_e}{k} \geq \sqrt{\frac{\pi^2 E}{\sigma_c}}$

  where, E = Modulus of elasticity, σ_c = Crushing stress.

- Columns with $\frac{l_e}{k} < \sqrt{\frac{\pi^2 E}{\sigma_c}}$ are known as short columns
- Columns with $\frac{l_e}{k} > \sqrt{\frac{\pi^2 E}{\sigma_c}}$ are long columns
- Equivalent length or effective length of a column is one that gives the same critical load given by column with both ends pinned.

EXERCISES

1. Find the Euler's crushing load for a hollow cast iron column of circular cross section, 110 mm inside diameter and 20 mm thick, 6 m long with both ends pinned. Take E = 100 GPa. Compare this with crushing load given by Rankine's equation taking σ_{cr} 550 MPa and $\alpha = \frac{1}{1600}$. For what length of column Euler's formula fail to apply.

   (***Ans.*** $P_E = 484.16 \times 10^3$ N, $P_R = 393.59 \times 10^3$ N, $\frac{P_E}{P_R} = 1.23$ and $l = 1969.91$ mm)

2. Find the Euler's crippling load for a hollow cylindrical steel column 40 mm outside diameter, 4 mm thick. The length of column is 2.5 m and is hinged at both ends. Also compute the Rankine's crippling load using constants 335 MPa and $\frac{1}{7500}$. Take E = 205 GPa.

   (***Ans.*** $P_E = 24.02 \times 10^3$ N, $P_R = 24.92 \times 10^3$ N)

3. A hollow circular section columns is 7.5 long and is pinned at both ends. The inner diameter of the section is 160 mm and the wall thickness is 20 mm. Find the safe load by Rankine's formula using a factor of safety of 5. Also find the slenderness ratio and the ratio of Euler's and Rankine's critical loads. For cast iron, take σ_c = 550 N/mm^2, $\alpha = \frac{1}{1600}$ and $E = 80 \times 10^3$ N/mm^2.

   (***Ans.*** $P_{Rsafe} = 129.928 \times 10^3$ N, $\frac{l_e}{k} = 117.133$, $P_E = 650.885 \times 10^3$ N, $P_E/P_R = 1.002$)

4. A 1.8 m long column has a circular cross section of 50 mm. One end of column is fixed and the other end is free. Take factor of safety as 3, calculate the safe load using:

   (i) Rankine's formula, take yield stress = 560 MPa and $\alpha = \frac{1}{1600}$

   (ii) Euler's formula, take $E = 1.2 \times 10^5$ N/mm^2

   (***Ans.*** P_{RS} = 30,393.32 N, P_{ES} = 9345.55 N)

5. A simply supported beam of I-section carries a UDL of 30 kN/m on entire span of 10 m. The deflection at the mid point was observed to be 6 mm. Determine the axial compressive load that can be applied on this column if both ends are fixed. Use Euler's criterion.

   (***Ans.*** $P_E = 257.02 \times 10^6$ N)

6. Determine the buckling load for a T-section of dimensions of flange 100 mm × 20 mm and web of 80 mm × 20 mm, length of column 12 m with one end fixed and the other end pinned. Take E = 200 GPa.

   (***Ans.*** $P_E = 47.155 \times 10^3$ N)

7. A timber column of section 150 mm × 120 mm × 6 m long is fixed at one end and the other end free. Young's modulus for timber is 18 kN/mm^2. Determine the

   (i) Euler's crippling load

   (ii) Rankine's crippling load taking σ_c = 50 MPa, $\alpha = \frac{1}{750}$.

   (***Ans.*** $P_E = 26.65 \times 10^3$ N, $P_R = 5.5897 \times 10^3$ N)

8. Determine the Rankine's critical load for a steel column of I section, each flange as 200 mm × 10 mm and web of 380 mm × 10 mm, 9 m long with both ends fixed. Take $\sigma_c = 320$ MPa and $\alpha = \frac{1}{7500}$. Also find Euler's critical load taking $E = 210$ GPa.

(*Ans.* $P_R = 969.035 \times 10^3$ N, $P_E = 1.3679 \times 10^6$ N)

9. Find the Euler's critical load for a hollow cylindrical steel column of 45 mm external diameter and 2.5 mm wall thinckness. Length of column is 1.5 m and hinged at both ends. Also estimate the Rankine's load for this column. Take $E = 207$ GPa. Rankine's parameters as 360 N/mm^2 and $\frac{1}{7500}$.

(*Ans.* $P_E = 68.668 \times 10^3$ N, $P_R = 51.703 \times 10^3$ N)

10. Determine the section of a hollow cylindrical column 3 m long with both ends hinged. It is subjected to an axial compressive load of 90 kN. The external diameter is 1.6 times the internal diameter. Take factor of safety as 5. Rankine's constants are 320 N/mm^2 and $\alpha = \frac{1}{7500}$.

(*Ans.* $d_i = 52.26$ mm, $d_o = 83.62$ mm)

11. Determine the size of a square section column 2 m long with both ends fixed. The column is subjected to an axial compressive load of 45 kN. Take $E = 2 \times 10^5$ N/mm^2.

(*Ans.* = a = 22.87 mm)

12. The piston rod of a hydraulic cylinder 600 mm long with one end fixed and the other end guided is subjected to an axial compressive load of 120 kN. Taking Rankine's constants as 320 MPa and $\frac{1}{7500}$, find the diameter of the rod.

(*Ans.* d = 27 mm)

13. Compare the strength of a hollow steel column to that of a solid column of same cross sectional area, made of same material, length and end conditions. The internal diameter is $\frac{3}{4}$ of the external diameter.

(*Ans.* 3.571)

14. A rectangular cross section, 3 m long and sides ratio as 2 is one end fixed and the other end free. Find the cross sectional dimensions of the rectangle when subjected to an axial load of 72 kN. Take factor of safety as 2. $E = 2 \times 10^5$ N/mm^2.

(*Ans.* b = 52.95 mm d = 105.91 mm)

15. A built up I-section having overall depth of 200 mm and width of flanges 140 mm and thickness of flange and web of 10 mm is used as a cantilever beam and the deflection at the free end is found to be 3 mm when subjected to a UDL of 68 kN/m. Find the safe load if this I-section is used as a column with both ends pinned. Use Euler's formula. Take factor of safety = 3 and E = 200 GPa.

(*Ans.* l = 1207.86, $P_E = 6.21 \times 10^6$ N)

Objective Type Questions

1. For both ends pinned column, the equivalent length of column is

(a) l (b) $2l$

(c) $\frac{l}{2}$ (d) $\frac{l}{\sqrt{2}}$

2. For both ends fixed column, the effective length of column is

(a) l (b) $\frac{l}{2}$

(c) $2l$ (d) $\frac{l}{\sqrt{2}}$

3. For a column with one end fixed, other end free equivalent length of column is

(a) l (b) $\frac{l}{2}$

(c) $2l$ (d) $\frac{l}{\sqrt{2}}$

4. For a column with one end fixed, other end pinned, equivalent length of column is

(a) l (b) $\frac{l}{2}$

(c) $2l$ (d) $\frac{l}{\sqrt{2}}$

5. Euler's equation for critical load of column

(a) $\frac{\pi^2 EI}{l_e^2}$ (b) $\frac{\pi E^2 I}{l_e}$

(c) $\frac{\pi EI}{l_e}$ (d) $\frac{\pi EI}{l_e^2}$

6. Euler's equation for critical load of column of length l, and both ends pinned

(a) $\frac{2\pi^2 EI}{l^2}$ (b) $\frac{\pi^2 EI}{l^2}$

(c) $\frac{4\pi^2 EI}{l^2}$ (d) $\frac{\pi^2 EI}{4l^2}$

7. Euler's equation for critical load of column of length l, with ends fixed

(a) $\frac{2\pi^2 EI}{l^2}$ (b) $\frac{\pi^2 EI}{l^2}$

(c) $\frac{4\pi^2 EI}{l^2}$ (d) $\frac{\pi^2 EI}{4l^2}$

8. Euler's equation for critical load of column of length l, with one end fixed and the other end free.

(a) $\frac{2\pi^2 EI}{l^2}$ (b) $\frac{\pi^2 EI}{l^2}$

(c) $\frac{4\pi^2 EI}{l^2}$ (d) $\frac{\pi^2 EI}{4l^2}$

9. Euler's equation for critical load of column of length l with one end fixed, other end pinned.

(a) $\frac{2\pi^2 EI}{l^2}$ (b) $\frac{\pi^2 EI}{l^2}$

(c) $\frac{4\pi^2 EI}{l^2}$ (d) $\frac{\pi^2 EI}{4l^2}$

10. Euler's equation of critical load gives better result for

(a) short columns (b) long columns

(c) medium columns (d) all the above

11. Rankine's equation of critical load is better valid for

(a) short columns (b) long columns

(c) both (a) and (b) (d) medium columns

12. Limiting condition for Euler's critical load for columns is

(a) $\frac{l_e}{k} = \sqrt{\frac{\pi E}{\sigma_c}}$ (b) $\frac{l_e}{k} = \sqrt{\frac{\pi \sigma_c}{E}}$

(c) $\frac{l_e}{k} = \sqrt{\frac{\pi^2 \sigma_c}{E}}$ (d) $\sqrt{\frac{\pi^2 E}{\sigma_c}}$

13. Slenderness ratio of a column is the ratio of

(a) $\frac{l_e}{k}$ (b) $\frac{l_e^2}{k}$

(c) $\frac{l_e}{k^2}$ (d) $\frac{k}{l_e}$

14. Rankine's equation for critical load of a columns is

(a) $\frac{\sigma_c A}{1 - \alpha \left(\frac{l_e}{k}\right)^2}$ (b) $\frac{\sigma_c A}{1 + \alpha \left(\frac{l_e}{k}\right)^2}$

(c) $\frac{\sigma_c A}{1 - \alpha \left(\frac{k}{l_e}\right)^2}$ (d) $\frac{\sigma_c A}{1 + \alpha \left(\frac{k}{l_e}\right)^2}$

15. The value of constant α in the Rankine's equation for columns is

(a) $\frac{\pi^2 E}{\sigma_c}$ (b) $\frac{\pi^2 \sigma_c}{E}$

(c) $\frac{\sigma_c}{\pi^2 E}$ (d) $\frac{\sigma_c E}{\pi^2}$

16. For the same weight, material and length, a hollow column is ______ stronger than solid column.

(a) less (b) equally

(c) not (d) more

Key To Answers

1. (a)	2. (b)	3. (c)	4. (d)	5. (a)	6. (b)	7. (c)	8. (d)
9. (a)	10. (b)	11. (c)	12. (d)	13. (a)	14. (b)	15. (c)	16. (d)

CHAPTER 10

Thin and Thick Cylinders and Spherical Shells

10.1 INTRODUCTION

Cylinders and spherical shells are commonly used for storage of gases and fluids under pressure. Some examples are, petrol tanker, oxygen cylinder, domestic gas cylinder, water tanks, etc. In this chapter the analysis of such cylinders and spheres will be done. When the wall thickness of a cylinder is less than about $\frac{1}{10}$ of its radius, then such cylinders are treated as thin cylinders. The radical stress in such cases in negligible and the analysis becomes simpler. If the ratio of thickness of the cylinder to the internal radius is more than $\frac{1}{10}$, they are generally called thick cylinders. The analysis of thick cylinders is taken up in the later part of this chapter.

10.2 STRESSES IN THIN CYLINDER SUBJECTED TO INTERNAL PRESSURE

The following assumptions are made in the analysis of thin cylinders

(i) Radical stress in small and hence neglected

(ii) Material is homogeneous and isotropic

(iii) The cylinder is stressed within elastic limit.

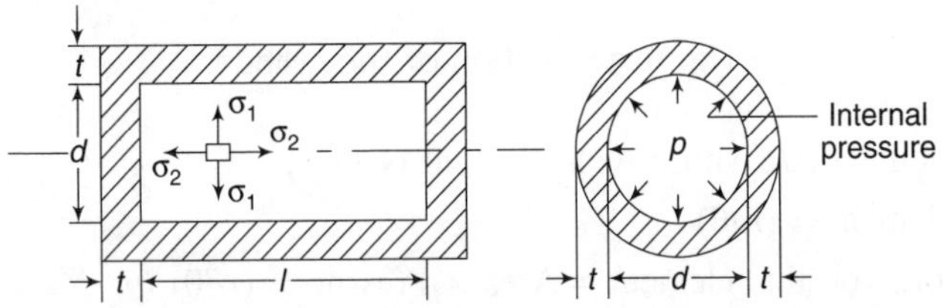

Fig. 10.1 Thin cylinder subjected to internal pressure

Consider a thin cylindrical vessel of diameter d, length l, wall thickness t subjected to an internal pressure p.

This internal pressure will induce two types of stresses in the cylinder, viz., circumferential stress or hoop stress σ_1 and longitudinal stress σ_2.

The stress acting along the circumference of a cylinder is called circumferential stress and the stress lengthwise is called longitudinal stress. The circumferential stress is also known as hoop stress. This circumferential stress causes the cylinder to burst into two troughs as shown in Fig. 10.2 (a). The longitudinal stress causes the cylinder to split into two cylinders as shown in Fig. 10.2 (b).

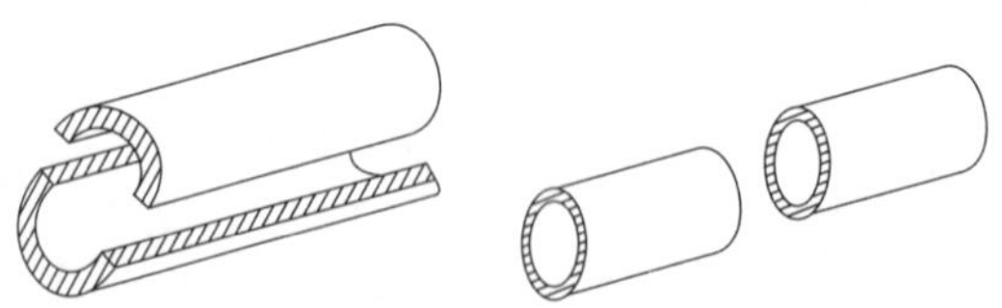

(a) Split into two troughs (b) Split into two cylinders

Fig. 10.2 Failure of thin cylinders

10.3 CIRCUMFERENTIAL STRESS OR HOOP STRESS (σ_1)

Consider a thin cylindrical shell of diameter d, length l and wall thickness t subjected to an internal pressure p.

The cylinder will burst if the force due to fluid pressure is more than the resisting force.

$$r \text{ (radius)} = \frac{d}{2}$$

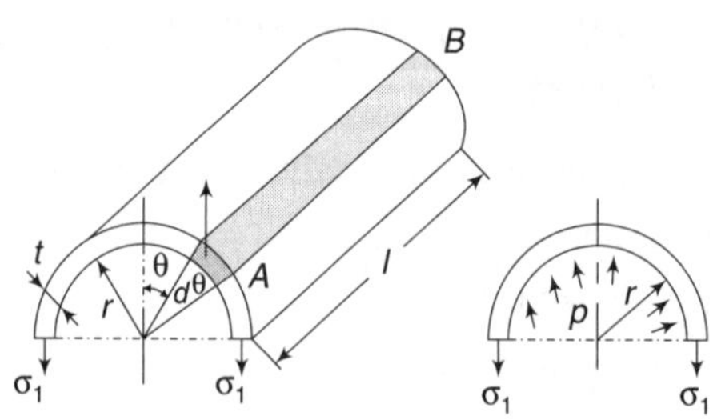

Fig. 10.3 Section of the cylinder

Consider an element AB whose width = $r \cdot d\theta$ (sector)

Area of this element = $(rd\theta) \cdot l$

$\therefore$ Normal force on this element = Area × Pressue = $(rd\theta)\ lp$

Bursting force normal to the section = $rd\theta lp \cdot \cos\theta$

Total bursting force acting normal to the section

$$F_B = \int_{-\pi/2}^{\pi/2} rlp \cos\theta \, d\theta = rlp \, [\sin\theta]_{-\pi/2}^{\pi/2}$$

$$= rlp \, (2) = 2rlp = dlp \quad \text{...(1)}$$

Resisting force offered by the cylinder wall is

$$F_R = \sigma_1 \cdot (l \cdot t) \times 2 \quad \text{(2 for 2 thicknesses)} \quad \text{...(2)}$$

Equating the bursting force to the resisting force,

$$2\sigma_1 \, lt = dlp$$

$$\therefore \quad \text{Circumferential stress } \sigma_1 = \frac{pd}{2t}$$

10.4 LONGITUDINAL STRESS (σ_2)

Consider a thin cylindrical vessel subjected to internal pressure as shown in Fig. 10.4. Due to bursting, the cylinder will be cut longitudinally. Consider a transverse section *AB*.

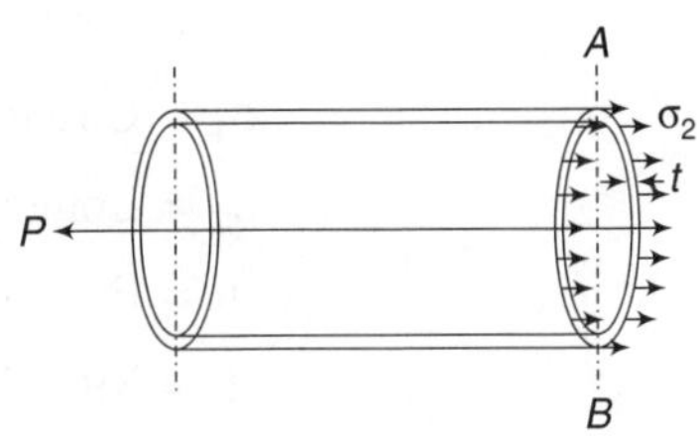

Fig. 10.4 Longitudinal stress in a thin cylinder

p = Internal pressure

t = Wall thickness of cylinder

d = Diameter of cylinder

σ_2 = Longitudinal stress.

Bursting force acting on the cylinder in the longitudinal direction

$$F_B = \frac{\pi}{4} d^2 \cdot p \quad \text{...(3)}$$

$$\text{Resisting force} = \text{Longitudinal stress} \times \text{Area of metal}$$

$$= \sigma_2 \cdot (\pi d \cdot t) \quad \text{...(4)}$$

$$\text{Equating Eqs. (3) and (4) } \sigma_2 \, \pi dt = \frac{\pi}{4} d^2 p$$

$$\therefore \quad \text{Longitudinal stress } \sigma_2 = \frac{pd}{4t}$$

It is seen that longitudinal stress is half of the circumferential stress. Hence, while finding the wall thickness, maximum allowable stress is to be taken as circumferential stress (σ_1) and t is found from equation $\sigma_1 = \frac{pd}{2t}$.

$$\text{Maximum shear stress } \tau_{max} = \frac{\sigma_1 - \sigma_2}{2}$$

$$\tau_{max} = \frac{\left(\frac{pd}{2t} - \frac{pd}{4t}\right)}{2} = \frac{pd}{8t}$$

10.5 CHANGE IN DIMENSIONS OF THIN CYLINDER (CHANGE IN DIAMETER, LENGTH AND VOLUME)

As discussed earlier, for a thin cylinder subjected to internal pressure, there are two types of stresses act:

(i) Circumferential or hoop stress $\sigma_1 = \dfrac{pd}{2t}$ and

(ii) Longitudinal stress $\sigma_2 = \dfrac{pd}{4t}$

These stresses act at principal planes, which are perpendicular to each other.

p = Internal pressue

l = Length of cylinder

d = Diameter of cylinder

t = Wall thickness

σ_1 = Circumferential stress

σ_2 = Longitudinal stress

μ = Poisson's ratio

E = Modulus of elasticity

V = Original volume $= \dfrac{\pi}{4} d^2 l$

δd = Change in diameter

δl = Change in length

$\in_d = \dfrac{\delta d}{d}$ = Circumferential strain $= \in_1$

$\in_l = \dfrac{\delta l}{l}$ = Longitudinal strain $= \in_2$

δV = Change in volume.

$\in_V$ = Volumetric strain $= \dfrac{\delta V}{V}$

10.5.1 Change in Diameter

From generalised Hooke's law

$$\text{Circumferential strain } \in_1 = \in_d = \frac{\sigma_1}{E} - \frac{\mu\sigma_2}{E}$$

$$\in_d = \frac{\delta d}{d} = \frac{pd}{2tE} - \frac{\mu.pd}{4tE} = \frac{pd}{4tE}[2 - \mu]$$

$$\therefore \quad \text{Change in diameter } \delta d = \in_d \cdot d = \frac{pd^2}{4tE}[2 - \mu]$$

It may be noted that circumferential strain is

$$\in_1 = \frac{\text{Final circumference} - \text{Original circumference}}{\text{Original circumference}}$$

$$\in_1 = \frac{\pi(d + \delta d) - \pi d}{\pi d} = \frac{\delta d}{d}$$

10.5.2 Longitudinal Strain and Change in Length

Again from Hooke's Law,

$$\text{Longitudinal strain } \in_2 = \frac{\delta l}{l} = \frac{\sigma_2}{E} - \frac{\mu\,\sigma_1}{E}$$

$$\in_2 = \frac{\delta l}{l} = \frac{pd}{4tE} - \mu\,\frac{pd}{2tE} = \frac{pd}{4tE}(1 - 2\mu)$$

$$\therefore \quad \text{Change in length } \delta l = \in_2 \cdot l = \frac{pdl}{4tE}(1 - 2\mu)$$

10.5.3 Volumetric Strain and Change in Volume

$$\text{Volume } V = \frac{\pi}{4}\,d^2 l$$

$$\text{Differentiating } \delta V = \frac{\pi}{4}\,d^2\,\delta l + \frac{\pi}{4}\,(2d)\,\delta d \cdot l$$

$$= \frac{\pi}{4}\,d^2 \cdot l\left(\frac{\delta l}{l}\right) + \frac{\pi}{4}\,d^2 l \cdot \left(\frac{2\delta d}{d}\right)$$

$$\delta V = \frac{\pi}{4}\,d^2 l\left(\frac{\delta l}{l} + \frac{2\delta d}{d}\right)$$

$$\therefore \quad \text{Volumetric strain } \in_V = \frac{\delta V}{V} = \frac{\delta l}{l} + \frac{2\delta d}{d}$$

Hence,

$$\in_V = 2\in_1 + \in_2$$

$$\in_v = \frac{pd}{4tE}(1 - 2\mu) + 2\,\frac{pd}{4tE}(2 - \mu)$$

$$\in_v = \frac{\delta v}{v} = \frac{pd}{4tE}[1 - 2\mu + 4 - 2\mu] = \frac{pd}{4tE}(5 - 4\,\mu)$$

$$\text{Change in volume } \delta V = \frac{pd}{4tE}(5 - 4\,\mu) . V$$

SOLVED PROBLEMS

Problem 1 A thin cylindrical shell 1.2 m in diameter, 2 m long has a metal thickness of 12 mm. It is subjected to an internal fluid pressure of 3 MPa. Find the nature and magnitude of stresses induced.

Solution:

$$\text{Diameter } d = 1.2 \text{ m} = 1200 \text{ mm}$$

$$\text{Length } l = 2 \text{ m} = 2000 \text{ mm}$$

$$\text{Thickness } t = 12 \text{ mm}$$

$$\text{Internal pressure } p = 2 \text{ MPa.}$$

(i) Circumferential or hoop stress $\sigma_1 = \dfrac{pd}{2t} = \dfrac{3 \times 1200}{2 \times 12} = 150$ MPa

(ii) Longitudinal stress $\sigma_2 = \dfrac{pd}{4t} = \dfrac{3 \times 1200}{4 \times 12} = 75$ MPa

(iii) Shear stress $\tau_{max} = \dfrac{\sigma_1 - \sigma_2}{2} = \dfrac{150 - 75}{2} = 37.5$ MPa.

Problem 2 A steel penstock of diameter 1.5 m and wall thickness of 15 mm is subjected to a water head of 100 m. Calculate the hoop stress and longitudinal stress at the bottom of the penstock.

Solution:

$$\text{Diameter } d = 1.5 \text{ m} = 1500 \text{ mm}$$

$$\text{Wall thickness } t = 15 \text{ mm}$$

$$\text{Waterhead } H = 100 \text{ m}$$

$$\text{We know that 10 m of waterhead} = 1 \text{ kgf/cm}^2$$

$$1 \text{ kgf} = 9.81 \text{ N and cm}^2 = 100 \text{ mm}^2$$

$$\therefore \quad 100 \text{ m} = 10 \text{ kgf/cm}^2 = \frac{10 \times 9.81}{10 \times 10} = 0.981 \text{ N/mm}^2$$

$$\therefore \quad \text{Pressure } p = 0.981 \text{ N/mm}^2$$

(i) Hoop stress $= \sigma_1 = \dfrac{pd}{2t} = \dfrac{0.981 \times 1500}{2 \times 15} = 49.05$ N/mm^2

(ii) Longitudinal stress $\sigma_2 = \dfrac{pd}{4t} = \dfrac{0.981 \times 1500}{4 \times 15} = 24.525$ N/mm^2.

Problem 3 A thin cylindrical vessel of 3 m internal diameter carries a fluid under pressure of 1.8 MPa. Determine the wall thickness of the cylinder if the hoop stress and longitudinal stress are to be limited to 60 MPa and 42 MPa, respectively.

Solution:

$$\text{Diameter } d = 3 \text{ m} = 3000 \text{ mm}$$

$$\text{Fluid pressure } p = 1.8 \text{ MPa}$$

$$\text{Hoop stress } \sigma_1 = 60 \text{ MPa}$$

$$\text{Longitudinal stress } \sigma_2 = 42 \text{ MPa}$$

(i) Hoop stress $\sigma_1 = \dfrac{pd}{2t}$

$$60 = \frac{1.8 \times 3000}{2 \times t}$$

$\therefore$ Thickness $t = 45$ mm.

(ii) Longitudinal stress $\sigma_2 = \dfrac{pd}{4t}$

$$42 = \frac{1.8 \times 3000}{4 \times t}$$

$\therefore$ Thickness $t = 32.14$ mm

$\therefore$ Required thickness $t = 45$ mm (bigger one)

Problem 4 A thin cylindrical sheet 3 m long and 1 m diameter has a metal thickness of 10 mm. It is subjected to an internal fluid pressure of 3 N/mm^2. Determine

(i) Circumferential and longitudinal stresses

(ii) Circumferential, longitudinal and volumetric strains

(iii) Change in length, diameter and volume

Take Poisson's ratio as 0.3 and E = 210 GPa.

Solution:

Diameter $d = 1$ m $= 1000$ mm

Length $l = 3$m $= 3000$ mm

Wall thickness $t = 10$ mm

Internal pressure $p = 3$ N/mm^2

Poisson's ratio $\mu = 0.3$

$E = 210 \times 10^3$ N/mm^2

(i) Circumferential and longitudinal stresses

(a) Circumferential stress $\sigma_1 = \dfrac{pd}{2t} = \dfrac{3 \times 1000}{2 \times 10} = 150$ N/mm^2

(b) Longitudinal stress $\sigma_2 = \dfrac{pd}{4t} = \dfrac{3 \times 1000}{4 \times 10} = 75$ N/mm^2

Shear stress $\tau_{max} = \dfrac{\sigma_1 - \sigma_2}{2} = \dfrac{150 - 75}{2} = 37.5$ N/mm^2

(ii) Strains

(a) Circumferential strain $\in_a = \dfrac{\delta d}{d} = \dfrac{pd}{4tE}(2 - \mu)$

$$= \frac{3 \times 1000}{4 \times 10 \times 210 \times 10^3}(2 - 0.3) = 0.607 \times 10^{-3}$$

(b) Longitudinal strain $\in_l = \dfrac{\delta l}{l} = \dfrac{pd}{4tE}(1 - 2\mu)$

$$= \frac{3 \times 1000}{4 \times 10 \times 210 \times 10^3}(1 - 2 \times 0.3) = 0.1429 \times 10^{-3}$$

(c) Volumetric strain $\in_V = \dfrac{\delta l}{l} + \dfrac{2\delta d}{d}$

$$= \in_l + 2 \in_d$$

$$= 0.1429 \times 10^{-3} + 2 \times 0.607 \times 10^{-3}$$

$$= 1.3569 \times 10^{-3}$$

(iii) Change in length, diameter and volume.

(a) Change in length,

$$\text{Longitudinal strain } \in_l = \frac{\delta l}{l} = 0.1429 \times 10^{-3}$$

$\therefore$ Change in length $\delta l = \in_l \times l = 0.1429 \times 10^3 \times 3000 = 0.4287$ mm

(b) Change in diameter

$$\text{Circumferential strain } \in_d = \frac{\delta d}{d} = 0.607 \times 10^{-3}$$

$\therefore$ Change in diameter $\delta d = \in_d \cdot d = 0.607 \times 10^{-3} \times 1000 = 0.607$ mm

(c) Change in volume

$$\text{Volumetric strain } \in_V = \frac{\delta V}{V} = 1.3569 \times 10^{-3}$$

$\therefore$ Change in volume $\delta V = \in_V \cdot V = 1.3569 \times 10^{-3} \times \left(\frac{\pi}{4} d^2 l\right)$

$$\delta V = 1.3569 \times 10^{-3} \times \frac{\pi}{4} \times 1000^2 \times 3000$$

$$= 3.1971 \times 10^6 \text{ mm}^3.$$

Problem 5 A 1.5 m long thin cylindrical pressure vessel of 600 mm inner diameter undergoes a change in volume of 900×10^3 mm^3. Find the internal pressure and the stresses developed taking E = 207 GPa and Poisson's ratio of 0.3. Wall thickness is 15 mm.

Solution:

$$\text{Length } l = 1.5 \text{ m} = 1500 \text{ mm}$$

$$\text{Diameter } d = 600 \text{ mm}$$

$$\text{Change in volume } \delta V = 900 \times 10^3 \text{ mm}^3$$

$$E = 207 \times 10^3 \text{ N/mm}^2$$

$$\text{Poisson's ratio } \mu = 0.3$$

$$\text{Wall thickness } t = 15 \text{ mm}$$

(i) We know that volumetric strain

$$\in_V = \frac{\delta V}{V} = \frac{pd}{4tE} [5 - 4\mu]$$

$\therefore$ Change in volume $\delta V = \dfrac{pd}{4tE} (5 - 4\mu).V$

$$900 \times 10^3 = \frac{p \times 600}{4 \times 15 \times 207 \times 10^3} (5 - 4 \times 0.3) \times \frac{\pi}{4} d^2 l$$

$$900 \times 10^3 = \frac{p \times 600}{4 \times 15 \times 207 \times 10^3}(5 - 4 \times 0.3) \times \frac{\pi}{4} \times 600^2 \times 1500$$

$\therefore$ Pressure $p = 11.56$ N/mm^2

(ii) Hoop stress $\sigma_1 = \frac{pd}{2t} = \frac{11.56 \times 600}{2 \times 15} = 231.2$ MPa

(iii) Longitudinal stress $\sigma_2 = \frac{pd}{4t} = \frac{11.56 \times 600}{4 \times 15} = 115.6$ MPa

(iv) Shear stress $\tau_{max} = \frac{\sigma_1 - \sigma_2}{2} = \frac{231.2 - 115.6}{2} = 57.8$ MPa.

Problem 6 A 1 m long tube of internal diameter 80 mm and 2 mm thick has closed ends, and is subjected to internal pressure of 3 MPa. If the Young's modulus is 200 GPa and Poisson's ratio of 0.3. Compute the change in volume of the tube.

Solution: Length l = 1 m = 1000 mm

Internal diameter d = 80 mm

Wall thickness t = 2 mm

Internal pressure p = 3 MPa

Young's modulus E = 200 GPa = 200×10^3 MPa

Poisson's ratio μ = 0.3

Since $\frac{t}{r}$ ratio is small, it is thin cylinder

$$\text{Volumetric strain } \in_V = \frac{\delta V}{V} = \frac{pd}{4tE}[5 - 4\mu]$$

$$\therefore \quad \text{Change in volume } \in_V \cdot V = \frac{pd}{4tE}(5 - 4\mu).V$$

$$\delta V = \frac{3 \times 80}{4 \times 2 \times 200 \times 10^3}(5 - 4 \times 0.3) \times \frac{\pi}{4} \times 80^2 \times 1000 = 2865.13 \text{ mm}^3.$$

Problem 7 A cylindrical shell has an external diameter of 500 mm and a wall thickness of 10 mm and is 1.7 m long. Determine the increase in length and internal diameter when inside pressure is 1 N/mm^2. Take E = 210 GPa Poisson's ratio = 0.3.

Solution: External diameter d_o = 500 mm

Wall thickness t = 10 mm

$\therefore$ Inside diameter $d = d_o - 2t = 500 - 2 \times 10 = 480$ mm

Length l = 1.7 m = 1700 mm

Internal pressure p = 1 N/mm^2

Modulus of elasticity $E = 210 \times 10^3$ MPa

Poisson ratio $\mu = 0.3$

Since $\frac{t}{r} = \frac{10}{240} = \frac{1}{24}$, it is thin cylinder

(i) Increase in diameter.

$$\text{Circumferential strain } \frac{\delta d}{d} = \frac{\sigma_1}{E} - \frac{\mu\sigma_2}{E} = \frac{pd}{4tE}(2 - \mu)$$

$$\therefore \quad \text{Change in dia } \delta d = \frac{pd^2}{4tE}(2 - \mu) = \frac{1 \times 480^2}{4 \times 10 \times 210 \times 10^3}(2 - 0.3)$$

Increase in diameter $\delta d = 0.04663$ mm

(ii) Increase in length

$$\text{Longitudinal strain } \frac{\delta l}{l} = \frac{\sigma_2}{E} - \frac{\mu\sigma_1}{E}$$

$$\frac{\delta l}{l} = \frac{pd}{4tE}(1 - 2\mu)$$

$$\therefore \quad \text{Increase in length } \delta l = \frac{pdl}{4tE}(1 - 2\mu)$$

$$= \frac{1 \times 480 \times 1700}{4 \times 10 \times 210 \times 10^3}(1 - 2 \times 0.3) = 0.03886 \text{ mm.}$$

Problem 8 A cylindrical shell 2 m long 180 mm in diameter has a metal thickness of 9 mm. It is completely filled with fluid at atmospheric pressure. If an additional fluid of 20000 mm^3 is pumped in, find the pressure developed. Also find the hoop stress, change in diameter and length. Take modulus of elasticity as 206 GPa and Poisson's ratio as 0.3.

Solution: Length of cylinder $l = 2$ m $= 2000$ mm

Diameter $d = 180$ mm

Wall thickness $t = 9$ mm

Additional volume $\delta V = 20000$ mm^3

Modulus of elasticity $E = 206$ GPa $= 206 \times 10^3$ MPa

Poisson's ratio $\mu = 0.3$

Let the internal $PV = p$

We know that volumetric strain

$$\in_V = \frac{\delta V}{V} = \frac{pd}{4tE}[5 - 4\mu]$$

$$\text{Volume } V = A.l = \frac{\pi d^2}{4} \times l = \frac{\pi}{4} 180^2 \times 2000$$

$$V = 50.8938 \times 10^3 \text{ mm}^3$$

$\therefore \qquad \in_V = \dfrac{20000}{50.8938 \times 10^6} = \dfrac{p \times 180}{4 \times 9 \times 206 \times 10^3}[5 - 4 \times 0.3]$

$\therefore \qquad$ Internal pressure $p = 4.261\ \text{N/mm}^2$

Change in diameter

$$\text{Lateral strain } \in_d = \frac{\delta d}{d} = \frac{Pd}{4tE}[2 - \mu]$$

$$\frac{\delta d}{d} = \frac{4.261 \times 180}{4 \times 9 \times 206 \times 10^3}(2 - 0.3) = 1.75818 \times 10^{-4}$$

$\therefore \qquad$ Change in dia $\delta d = \in_d \times d = 1.75818 \times 10^{-4} \times 180 = 0.0316$ mm

Change in length

$$\text{Longitudinal strain } \in_l = \frac{\delta l}{l} = \frac{pd}{4tE}(1 - 2\mu)$$

$$\frac{\delta l}{l} = \frac{4.261 \times 180\,(1 - 2 \times 0.3)}{4 \times 9 \times 206 \times 10^3} = 4.1369 \times 10^{-5}$$

$\therefore \qquad$ Change in length $\delta l = \in_l \times l = 4.1369 \times 10^{-5} \times 2000 = 0.0827$ mm.

Problem 9 The diameter of a penstock is 1 m and the waterhead is 120 m. Find the thickness of seamless pipe required if the permissible stress is 30 MPa.

Solution: Diameter of pipe $d = 1\ \text{m} = 1000$ mm

Head of water $H = 120$ m

Allowable stress $\sigma_1 = 30$ MPa

(Permissible stress is always σ_1 since $\sigma_1 > \sigma_2$)

Pressure of water $p = w \cdot H$

where w = Specific weight of water = 9810 N/m^3

$p = 9810 \times 120 = 1.1772 \times 10^6\ \text{N/m}^2$

$p = 1.1772\ \text{N/mm}^2$

We know that hoop stress $\sigma_1 = \dfrac{pd}{2t}$

$$30 = \frac{1.1772 \times 1000}{2 \times t}$$

$\therefore \qquad$ Thickness of pipe $t = 19.62$ mm.

Problem 10 A cylindrical pressure vessel is subjected to an internal pressure of 3 MPa. Thickness of plate is 25 mm. If the hoop stress and longitudinal stresses for the material are limited to 120 MPa and 75 MPa, find the diameter of the pressure vessel required.

Solution:

$$\text{Internal pressure } p = 3 \text{ MPa}$$

$$\text{Thickness } t = 25 \text{ mm}$$

$$\text{Hoop stress } \sigma_1 = 120 \text{ MPa}$$

$$\text{Longitudinal stress } \sigma_2 = 75 \text{ MPa}$$

$$\text{We know that, hoop stress } \sigma_1 = \frac{pd}{2t}$$

$$120 = \frac{3 \times d}{2 \times 25}$$

$$\therefore \quad \text{Diameter } d = 2000 \text{ mm}$$

$$\text{Longitudinal stress } \sigma_2 = \frac{pd}{4t}$$

$$75 = \frac{3 \times d}{4 \times 25}$$

$$\text{Diameter } d = 2500 \text{ mm}$$

∴ Recommended maximum diameter d = 2000 mm (minimum of the above two values).

Problem 11 A thin cylinder 1 m long and 500 mm outside diameter has a wall thickness of 15 mm. Its both ends are closed and is completely filled with fluid at atmospheric pressure. Calculate the additional volume of fluid to be pumped into the cylinder so as to raise the pressure to 3 N/mm^2. Take E = 120 GPa and Poisson's ratio as 0.33. Bulk modulus of oil is 2.4 GPa.

Solution:

$$\text{Length } l = 1 \text{ m} = 1000 \text{ mm}$$

$$\text{Outside diameter } d_o = 500 \text{ mm}$$

$$\text{Thickness } t = 15 \text{ mm}$$

$$\therefore \quad \text{Inside diameter } d = d_o - 2t = 500 - 2 \times 15 = 470 \text{ mm}$$

$$\text{Internal pressure } p = 3 \text{ N/mm}^2$$

$$\text{Modulus of elasticity } E = 120 \text{ GPa} = 120 \times 10^3 \text{ MPa}$$

$$\text{Poisson's ratio } \mu = 0.33$$

$$\text{Bulk modulus of fluid } K = 2.4 \text{ GPa} = 2.4 \times 10^3 \text{ MPa}$$

Amount of fluid to be pumped in is the sum of increase in volume of the cylinder and decrease in volume of fluid due to compression of fluid

$$\delta V = \delta V_1 + \delta V_2$$

$$\delta V_1 = \text{Increase in volume due to internal pressure.}$$

$$= \frac{pd}{4tE}(5 - 4\mu) \times V$$

$$= \frac{3 \times 470}{4 \times 15 \times 120 \times 10^3} (5 - 4 \times 0.33) \times \left[\frac{\pi}{4} \times 470^2 \times 1000\right]$$

$$= 125.032 \times 10^3 \text{ mm}^3.$$

δV_2 = Decrease in volume of fluid due to compression

WKT bulk modulus $K = \frac{p}{\in_{V2}}$

where, $\in_{V2}$ = Volumetric strain $= \frac{\delta V_2}{V}$

$$V = \text{Volume} = \frac{\pi}{4} d^2 l = \frac{\pi}{4} \times 470^2 \times 1000$$

$$= 173.494 \times 10^6 \text{ mm}^3$$

$$2.4 \times 10^3 = \frac{3}{\left(\frac{\delta V_2}{173.494 \times 10^6}\right)}$$

$\therefore$ $\delta V_2 = 216.868 \times 10^3 \text{ mm}^3$

$\therefore$ Total volume $\delta V = \delta V_1 + \delta V_2 = 125.0.32 \times 10^3 + 216.868 \times 10^3$

$$= 341.9 \times 10^3 \text{ mm}^3$$

Additional volume of fluid to be pumped to raise the pressure to 3 N/mm^2 = 341.9×10^3 mm^3.

Problem 12 A steel tube is plugged at the ends and is fitted with an extensometer to measure its elongation. When an axial pull of 20 kN is applied to the plugs, an extension of 0.172 mm is observed in a gauge length of 500 mm. When the axial pull is removed and an internal pressure of 3 N/mm^2 is applied, the elongation is found to be 0.03 mm. The tube is 60 mm inside diameter with a wall thickness of 1.5 mm. Find the value of the Poisson's ratio of the material of the tube.

Solution: Inside diameter d = 60 mm

Wall thickness t = 1.5 mm

Axial force F = 20 kN = 20000 N

Extension δl_1 = 0.172 mm

Gauge length l = 500 mm

Internal pressure p = 3 N/mm^2

Elongation under pressure δl_2 = 0.03 mm

Outside diameter $d_o = d + 2t = 60 + 2 \times 1.5 = 63$ mm

Area of cross section $A = \frac{\pi}{4} (d_o^2 - d^2)$

(i) Axial load

$$\text{WKT modulus of elasticity } E = \frac{F \cdot l}{A\delta l_1}$$

$$= \frac{20000 \times 500}{\frac{\pi}{4}(63^2 - 60^2) \times 0.172}$$

$$E = 200.61 \times 10^3 \text{ N/mm}^2$$

(ii) Internal pressure

$$\text{Hoop stress } \sigma_1 = \frac{pd}{2t} = \frac{3 \times 60}{2 \times 1.5} = 60 \text{ N/mm}^2$$

$$\text{Longitudinal stress } \sigma_2 = \frac{pd}{4t} = \frac{3 \times 60}{4 \times 1.5} = 30 \text{ N/mm}^2$$

$$\text{Longitudinal strain } \in_1 = \frac{\delta l_2}{l} = \left(\frac{\sigma_2}{E} - \frac{\mu\sigma_1}{E}\right)$$

$$= \frac{0.03}{500} = \left(\frac{30}{200.61 \times 10^3} - \frac{\mu \times 60}{200.61 \times 10^3}\right)$$

$$30 - \mu \times 60 = \frac{0.03 \times 200.61 \times 10^3}{500}$$

$\therefore$ Poisson's ratio $\mu = 0.29939.$

10.6 THIN SPHERICAL SHELLS

Thin spherical shells are similar to thin cylinders but have spherical shape. Hence, for the analysis of thin spherical shells, the following assumptions are made

(i) Radial stress is small and hence neglected.

(ii) Hoop stress distribution over the cross section is uniform.

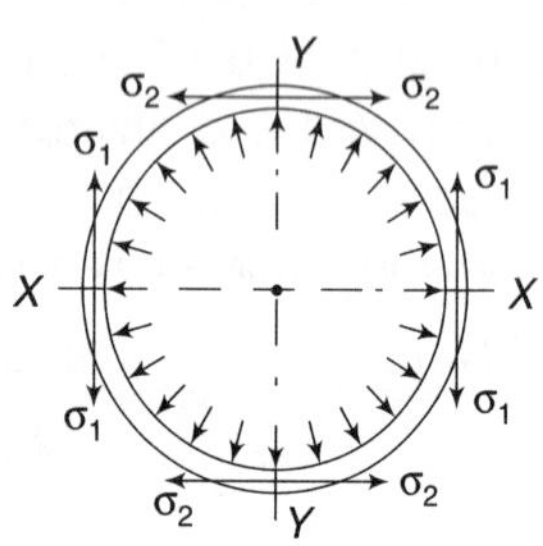

Fig. 10.5 Spherical shell

Figure 10.5 shows a spherical shell of internal diameter d and thickness t, subjected to an internal pressure of p.

Consider a section XX at the centre of the shell. Each section will be a hemisphere subjected to internal pressure p.

$$\text{Bursting force } F_B = \text{Area} \times \text{Pressure}$$

$$= \frac{\pi}{4} d^2 \times p \qquad \text{...(i)}$$

$$\text{Resisting force } F_R = \text{Area of shell} \times \text{Stress}$$

$$= (\pi dt)\, \sigma_1 \qquad \text{...(ii)}$$

Equating these two, $\pi dt\, \sigma_1 = \frac{\pi}{4} d^2 p$

$\therefore$ $$\text{Stress } \sigma_1 = \frac{pd}{4t} \quad \text{...(iii)}$$

Similarly, if a section is taken along *YY*, perpendicular to *XX*, it can be shown that stress

$$\sigma_2 = \frac{pd}{4t} \quad \text{...(iv)}$$

Hence, in the case of this spherical shell, the two principal stresses σ_1 and σ_2 at any point are equal and alike. Hence, every plane through a point is a principal plane having a stress of $\frac{pd}{4t}$.

Since $\sigma_1 = \sigma_2 = \frac{pd}{4t}$, there is no shear stress in the shell.

$$\text{Shear stress } \tau = \frac{\sigma_1 - \sigma_2}{2} = 0$$

The strain in any direction in the shell,

$$\in = \frac{\sigma_1}{E} - \frac{\mu\sigma_2}{E}$$

$$\in = \frac{\sigma_1}{E} - \frac{\mu\sigma_1}{E} = \frac{\sigma_1}{E}(1 - \mu)$$

$$\text{Substituting for } \sigma_1 = \frac{pd}{4t}$$

$$\text{Strain } \in = \frac{pd}{4tE}(1 - \mu)$$

But strain in a spherical shell will be only diametrical strain

$$\in = \frac{\delta d}{d} = \frac{Pd^2}{4tE}(1 - \mu)$$

$$\text{Change in diameter } \delta d = \frac{Pd^2}{4tE}(1 - \mu)$$

$$\text{For spherical shell, volume } V = \frac{4}{3}\pi r^3$$

$$V = \frac{4}{3}\pi\left(\frac{d}{2}\right)^3 = \frac{\pi d^3}{6}$$

$$\text{Differentiating, } \delta V = \frac{\pi}{6}\cdot 3d^2\,\delta d$$

$$\delta V = \frac{\pi}{2}d^2\,\delta d$$

Multiply and divide by $3d$

$$\delta V = \frac{\pi}{6}d^3\cdot\frac{\delta d}{d}\cdot 3 = V\cdot 3\,\frac{\delta d}{d}$$

$\therefore$ $$\text{Volumetric strain } \in_V = \frac{\delta V}{V} = 3\cdot\frac{\delta d}{d}$$

$$\in_V = 3.\in d$$

$$\in_V = \frac{\delta V}{V} = 3\frac{pd}{4tE}(1-\mu)$$

$\therefore$ Increase in volume $\delta V = \dfrac{3pd}{4tE}(1-\mu)V$

Note: If compressibility of the liquid is also considered, then volumetric strain

$$\in_V = \frac{\delta V}{V} = \left(\frac{\delta V_1}{V} + \frac{p}{K}\right)$$

where K is the bulk modulus of the liquid.

δV_1 = Change in volume due to increase in diameter $= \dfrac{3pd}{4tE}(1-\mu).V$

$$V = \text{Volume} = \frac{\pi d^3}{6}.$$

SOLVED PROBLEMS

Problem 13 A thin spherical sheel of 1.2 m diameter, 6 mm thick carries fluid under a pressure of 3 N/mm^2. Determine stress induced. Also find the increase in diameter and volume taking modulus of elasticity as 2×10^5 N/mm^2 and Poisson's ratio as 0.3.

Solution: Diameter $d = 1.2$ m $= 1200$ mm

Thickness $t = 6$ mm

Internal pressure $p = 3$ N/mm^2

Modulus of elasticity $E = 2 \times 10^5$ N/mm^2

Poisson's ratio $\mu = 0.3$

For spherical shell, diametral strain

$$\in_d = \frac{\delta d}{d} = \frac{pd}{4tE}(1-\mu)$$

$$\delta d = \frac{pd^2}{4tE}(1-\mu) = \frac{3 \times 1200^2}{4 \times 6 \times 2 \times 10^5}(1-0.3)$$

$\therefore$ Increase in diameter $\delta d = 0.63$ mm

Increase in volume $= \delta V = \dfrac{3pd}{4tE}(1-\mu).V$

where, V = volume of sphere $\dfrac{\pi d^3}{6}$

$$\delta V = \frac{3 \times 1200}{4 \times 6 \times 2 \times 10^5}(1-0.3) \times \left(\frac{\pi \times 1200^3}{6}\right)$$

$$\delta V = 475008.81 \text{ mm}^3.$$

Problem 14 A spherical shell of 420 mm diameter, wall thickness 7 mm is filled with water at atmospheric pressure. If an additional 7500 mm^3 of water is pumped into the cylinder, find the pressure exerted by the water on the walls of the shell. Also find the hoop stress induced. Take $E = 2.1 \times 10^5$ N/mm^2 and Poisson's ratio = 0.3.

Solution: Spherical shell

$$\text{Diameter } d = 420 \text{ mm}$$
$$\text{Wall thickness } t = 7 \text{ mm}$$
$$\text{Additional volume } \delta V = 7500 \text{ mm}^3$$
$$\text{Modulus of elasticity } E = 2.1 \times 10^5 \text{ MPa}$$
$$\text{Poisson ratio } \mu = 0.3$$

WKT for spherical shell subjected to internal pressure, change in volume

$$\delta V = \frac{3pd}{4tE}(1-\mu)V$$

where,

$$V = \text{Volume} = \frac{4}{3}\pi r^3 = \frac{\pi d^3}{6}$$

$$7500 = \frac{3 \times p \times 420}{4 \times 7 \times 2.1 \times 10^6}(1-0.3) \times \left(\frac{\pi \times 420^3}{6}\right)$$

$\therefore$ Internal pressure exerted $p = 1.289$ MPa

$$\text{Hoop stress } \sigma_1 = \frac{pd}{4t} = \frac{1.289 \times 420}{4 \times 7} = 19.33 \text{ MPa.}$$

Problem 15 A spherical shell 1 m diameter, 12 mm wall thickness is completely filled with fluid at atmospheric pressure. Additional fluid is then pumped in till the pressure rises to 1 N/mm^2. Find the volume of this additional fluid. The modulus of elasticity of the material is 206 GPa and Poisson's ratio is 0.3.

Solution:

$$\text{Diameter } d = 1 \text{ m} = 1000 \text{ mm}$$
$$\text{Thickness } t = 12 \text{ mm}$$
$$\text{Pressure } p = 1 \text{ N/mm}^2$$
$$\text{Modulus of elasticity } E = 206 \text{ GPa} = 206 \times 10^3 \text{ MPa}$$
$$\text{Poisson's ratio } \mu = 0.3$$

WKT for thin spherical shell subjected to internal pressure, volumetric strain

$$\epsilon_V = \frac{\delta V}{V} = \frac{3pd}{4tE}(1-\mu)$$

$$V = \text{Volume of the sphere} = \frac{\pi d^3}{6}$$

$$\therefore \quad \text{Change in volume } \delta V = \frac{3pd}{4tE}(1-\mu)V$$

$$= \frac{3 \times 1000 \times 1}{4 \times 12 \times 206 \times 10^3}(-0.3) \times \left(\frac{\pi \times 1000^3}{6}\right)$$

$$= 111.201 \times 10^3 \text{ mm}^3$$

To raise the pressure to 3 N/mm^2, additional volume of 111.201×10^3 mm^3 of fluid is to be pumped in.

Problem 16 A thin walled spherical shell of diameter 1.2 m 5 mm wall thickness is filled with a water until the pressure is 2.7 MPa. The modulus of elasticity of the shell material is 200 kN/mm^2 and Poisson's ratio is 0.3. Bulk modulus of water may be taken as 2 GPa. Find the additional amount of water to be filled in the shell.

Solution:

$$\text{Diameter } d = 1.2 \text{ m} = 1200 \text{ mm}$$

$$\text{Wall thickness } t = 5 \text{ mm}$$

$$\text{Pressure } p = 2.7 \text{ MPa}$$

$$\text{Modulus of elasticity } E = 200 \text{ kN/mm}^2 = 200 \times 10^3 \text{ N/mm}^2$$

$$\text{Poisson's ratio } \mu = 0.3$$

$$\text{Bulk modulus of water } K = 2 \text{ GPa} = 2 \times 10^3 \text{ N/mm}^2$$

Considering the compressibility of the water

Additional volume of the fluid required,

δV = Additional volume due to increase in volume of the shell (δV_1) + Additional volume due to compression of fluid (δV_2)

$$\text{Volumetric strain in the shell } \in_{v1} = \frac{\delta V_1}{V} = \frac{3pd}{4tE}(1 - \mu)$$

$$\text{and bulk modulus } K = \frac{p}{\in_{v2}}$$

$$\therefore \quad \in_{v2} = \frac{\delta_{v2}}{V} = \frac{p}{K}$$

$$\therefore \quad \delta V = \delta V_1 + \delta V_2 = \left[\frac{3pd}{4tE}(1 - \mu) + \frac{p}{K}\right] V$$

$$= \left[\frac{3 \times 2.7 \times 1200}{4 \times 5 \times 200 \times 10^3}(1 - 0.3) + \frac{2.7}{2000}\right] \times \left(\frac{\pi \times 1200^3}{6}\right)$$

$$\delta V = 2.7605 \times 10^6 \text{ mm}^3$$

Additional volume of 2.7605×10^6 mm^3 of water is pumped in to raise the pressure to 2.7 MPa.

Problem 17 A cylindrical shell shown in Fig. 10.6 is 1.2 m long and the diameter of 600 mm. The wall thickness is 9 mm. Both ends are spherical in shape. Find the change in volume if the internal pressure is 2.4 N/mm^2. Take E = 200 GPa and Poisson's ratio as 0.285.

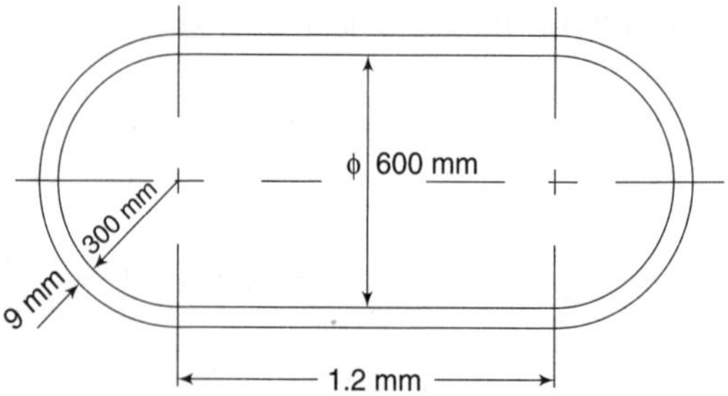

Fig. 10.6

Solution:

$$\text{Length of cylinder } l = 1.2 \text{ m} = 1200 \text{ mm}$$
$$\text{Diameter } d = 600 \text{ mm}$$
$$\text{Wall thickness } t = 9 \text{ mm}$$
$$\text{Internal pressure } p = 2.4 \text{ N/mm}^2$$
$$\text{Modulus of elasticity } E = 200 \text{ GPa} = 200 \times 10^3 \text{ MPa}$$
$$\text{Poisson's ratio } \mu = 0.285$$

Total change in volume δV = Change in volume of cylindrical portion (δV_c) + Change in volume of spherical shell (δV_s)

(Two hemispheres and the ends is one full spherical shell.)

WKT, for cylindrical shell, change in volume

$$\delta V_c = \frac{pd}{4tE}(5 - 4\mu)\, V_C$$

where, V_C = volume of the cylinder = $\frac{\pi}{4} d^2 l$

$$\therefore \quad \delta V_c = \frac{2.4 \times 600}{4 \times 9 \times 200 \times 10^3}(5 - 4 \times 0.285) \times \left(\frac{\pi}{4} \times 600^2 \times 1200\right)$$
$$= 261.93 \times 10^3 \text{ mm}^3$$

For spherical shell, change in volume $\delta_{VS} = \frac{3pd}{4tE}(1 - \mu) \cdot V_S$

where, V_S = Volume of the sphere = $\frac{\pi}{6} d^3$

$$\therefore \quad \delta V_S = \frac{3 \times 2.4 \times 600}{4 \times 9 \times 200 \times 10^3}(1 - 0.285) \times \left(\frac{\pi}{6} \times 600^3\right)$$
$$\delta V_S = 48.519 \times 10^3 \text{ mm}^3$$

$$\therefore \quad \text{Total change in volume } \delta V = \delta V_C + \delta V_S$$
$$= 261.9 \times 10^3 + 48.519 \times 10^3$$
$$\delta V = 310.419 \times 10^3 \text{ mm}^3.$$

Problem 18 Find the ratio of thicknesses of the hemispherical part to the cylindrical part of a cylindrical shell of 200 mm diameter 8 mm thick having hemispherical ends and subjected to internal pressure when

(i) The maximum stress in the both is same

(ii) There should not be distortion at the junction. Take Poisson's ratio as 0.3.

Solution:

$$\text{Diameter } d = 200 \text{ mm}$$
$$\text{Wall thickness of cylinder } t_c = 8 \text{ mm}$$
$$\text{Poisson's ratio } \mu = 0.3$$

Let p be the internal pressure

$$t_s = \text{Thickness of spherical portion}$$

$$E = \text{Modulus of elasticity.}$$

(i) For the maximum stress to be same

$$\sigma_C = \sigma_S$$

$$\frac{pd}{2t_C} = \frac{pd}{4t_S}$$

$$\therefore \quad \frac{t_S}{t_C} = \frac{1}{2} \text{ or } t_S = \frac{1}{2} t_C$$

$$\therefore \text{ Thickness of spherical shell} = \frac{1}{2} \text{ thickness of cylinder}$$

$$t_S = \frac{1}{2} \times 8 = 4 \text{ mm}$$

(ii) For no distortion at the junction:

For no distortion, the volumetric strain in cylinder and volumetric strain in the spherical shell should be same.

$$\in_{Vc} = \in_{VS}$$

$$\frac{pd}{4t_C E}(5 - 4\mu) = \frac{3pd}{4t_S E}(1 - \mu)$$

$$\therefore \quad \frac{t_S}{t_C} = \frac{3(1 - \mu)}{(5 - 4\mu)}$$

$$\frac{t_S}{t_C} = \frac{3(1 - 0.3)}{(5 - 4 \times 0.3)} = 0.5526$$

$$\therefore \quad t_S = 0.5526\, t_C = 0.5526 \times 8 = 4.421 \text{ mm}$$

10.7 THICK CYLINDERS

When the wall thickness of a cylinder is more than $\frac{1}{10}$th of its radius, such cylinders are generally called thick cylinders. For such cylinders, the radial stresses are considerably large and are not negligible and the hoop stress is not uniform across the cross section. Hence, in thick cylinders both radial stress and hoop stress across the cross section are to be considered.

10.8 ASSUMPTIONS IN THE ANALYSIS OF THICK CYLINDERS

(i) The material is homogeneous and isotropic

(ii) The cylinder is stressed within the elastic limit

(iii) The longitudinal strain is uniform across the cross section of the cylinders

This assumption is based on Lame's theory and is almost true for large portions except near the ends.

In thin cylinders, the hoop stress was assumed to be same across the cross section but in thick cylinder the hoop stress will not be uniform but will be maximum at the inner circumference and varies to a minimum value near outer circumference.

10.9 LAME'S EQUATIONS FOR RADIAL PRESSURE AND HOOP STRESS IN THICK CYLINDER

Figure 10.7 shows a thick cylinder of external radius r_o, internal radius r_i and length l. Let p_o be the external pressure and p_i be the internal pressure acting at radii r_o and r_i

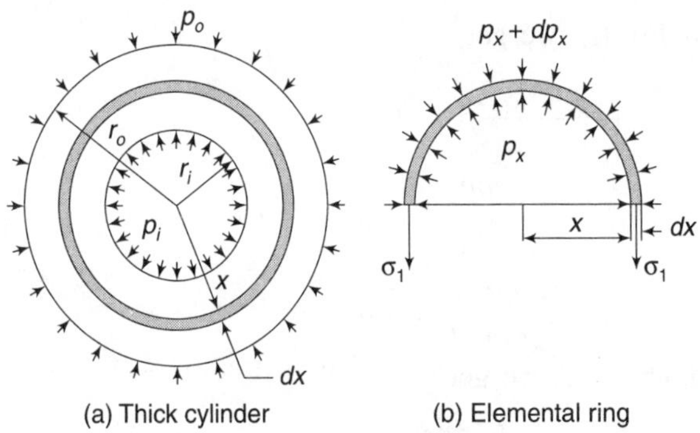

Fig. 10.7 Thick cylinder

Consider an elemental ring of thickness dx and inner radius x as shown in Fig. 10.7(a). For analysis, this can be treated as a thin cylinder of inner radius x, subjected to inner pressure p_x and externl pressure $p_x + dp_x$ as shown in Fig. 10.7(b). The outside radius is $x + dx$.

The bursting force $\quad F_B = p_x\,(2xl) - (P_x + dp_x)\,[2(x + d_x)l]$

The resisting force $\quad F_R = \sigma_1\,(2d_x.l)$

where, $\quad \sigma_1$ = hoop stress, d_x = Thickness

Equating resisting force to the bursting force,

$$\sigma_1.(2\,d_x.l) = p_x.(2xl) - (p_x + d_{Px})\,[2(x + dx)l]$$

$$\sigma_1\,\delta_x = p_x.x - (p_x + d_{Px})\,(x + d_x)$$

$$= p_x \cdot x - (p_x \cdot x + d_{Px} \cdot x + p_x \cdot dx + dp_x \cdot d_x)$$

Since d_x and dp_x are small, their product will be very small and neglected.

$$\sigma_1 d_x = -\,x \cdot dp_x - p_x dx$$

$$\sigma_1 + x \cdot \frac{d_{Px}}{d_x} + P_x = 0 \qquad \ldots(i)$$

$$\text{Longitudinal strain } \in_2 = \frac{\sigma_2}{E} - \frac{\mu\,\sigma_1}{E} - \frac{\mu(-\,p_x)}{E}$$

(p_x is –ve because, it is compressive)

$$\in_2 = \frac{\sigma_2}{E} - \frac{\mu}{E}(\sigma_1 - p_x)$$

As per Lame's theorem, this longitudinal strain is constant at any point. Since longitudinal stress σ_2, Poisson's ratio μ and E are constants, $(\sigma_1 - P_x)$ is also constant.

Let it be equal to $2a$

$$\therefore \quad \sigma_1 - p_x = 2a$$

or

$$\sigma_1 = p_x + 2a \quad \text{...(ii)}$$

Substituting this is Eq. (i), we get

$$(p_x + 2a) + x \cdot \frac{dp_x}{d_x} + p_x = 0$$

$$x \frac{dp_x}{d_x} = -2(p_x + a)$$

$$\frac{dp_x}{p_x + a} = \frac{-2d_x}{x}$$

Integrating on both sides, we get

$$\int \frac{dp_x}{(Px + a)} = \int \frac{-2d_x}{x}$$

$$\log_e (p_x + a) = -2 \log_e x + \text{Constant}$$

Let the constant of integration be $= \log_e b$

$$\therefore \quad \log (p_x + a) = -\log x^2 + \log b$$

$$\log (p_x + a) = \log \left(\frac{b}{x^2}\right)$$

or

$$(p_x + a) = \frac{b}{x^2}$$

$$p_x = \frac{b}{x^2} - a \quad \text{...(iii)}$$

Substituting in Eq. (ii) $\sigma_1 = \left(\frac{b}{x^2} - a\right) + 2a$

$$\sigma_1 = \frac{b}{x^2} + a \quad \text{...(iv)}$$

Equations (iii) and (iv) are known as Lame's equations for thick cylinders.

Lame's equations for thick cylinder are

$$\text{Radial pressure } p_x = \frac{b}{x^2} - a$$

$$\text{Hoop stress } \sigma_x = \frac{b}{x^2} + a$$

where, a and b are constant and x is the radial distance to the element under consideration from the centre of the cylinder (it is the radius).

Thus, at the inner radius of cylinder, $x = r_i$

$$p_i = \frac{b}{r_i^2} - a \quad \text{and} \quad \sigma_i = \frac{b}{r_i^2} + a$$

at outer radius of cylinder, $x = r_o$

$$p_o = \frac{b}{r_o^2} - a \quad \text{and} \quad \sigma_o = \frac{b}{r_o^2} + a$$

The values of the constants a and b can be found applying boundary conditions like p_i, P_o, σ_i and σ_o. After determining the values of constant a and b from σ and p equations, hoop stress and radial pressure at any radius x can be easily determined by substituting the radius x in Eqs. (iii) and (iv).

Figure 10.8 shows the distribution of hoop stress and radial pressure across the wall thickness of the cylinder.

The hoop stress and radial pressure equations are of second degree of the radius x, the distribution is parabolic as shown in Fig. 10.8.

It may be noted that, the difference in the values of hoop stress σ and radial pressure p at any point in the wall of a thick cylinder is constant and equal to $2a$. At any radius x,

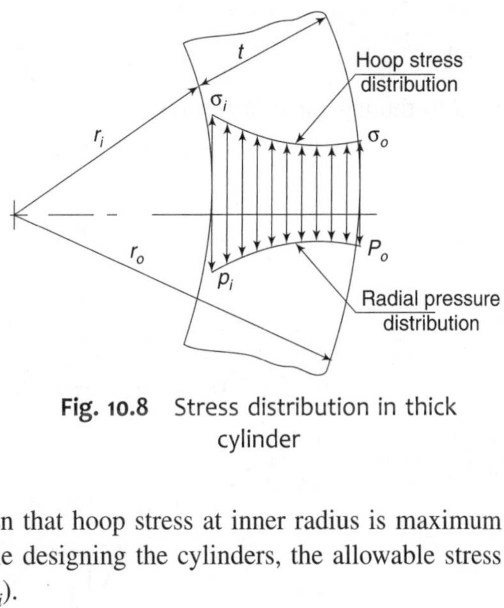

Fig. 10.8 Stress distribution in thick cylinder

$$\sigma_x - p_x = 2a$$

From the equation of hoop stress, it is seen that hoop stress at inner radius is maximum and at outer radius is minimum. Hence, while designing the cylinders, the allowable stress is be taken as hoop stress at inner radius (σ_i).

SOLVED PROBLEMS

Problem 19 A thick cylindrical pipe of diameters 600 mm and 500 mm is subjected to an internal pressure of 9 MPa. Find the maximum and minimum hoop stresses across the section. Sketch the stress distribution across the wall.

Solution: Outside diameter $d_o = 600$ mm

$\therefore \quad r_o = 300$ mm

Inside diameter $d_i = 500$ mm

$$\therefore \quad r_i = 250 \text{ mm}$$

$$\text{Internal pressure } p_i = 9 \text{ MPa}$$

$$\text{External pressure } p_o = 0 \text{ (not given)}$$

For thick cylinders, radial pressure $p = \dfrac{b}{r^2} - a$

$$\text{At inner radius } p_i = \frac{b}{r^2} - a$$

$$9 = \frac{b}{250^2} - a \quad \text{...(i)}$$

$$\text{At outer radius } p_o = \frac{b}{r_o^2} - a$$

$$0 = \frac{b}{300^2} - a$$

or

$$a = \frac{b}{300^2} \quad \text{...(ii)}$$

Substituting for a in Eq. (i)

$$9 = \frac{b}{250^2} - \frac{b}{300^2} = b\left(\frac{1}{250^2} - \frac{1}{300^2}\right)$$

$$\therefore \quad b = 1840909.09$$

$$\therefore \quad a = \frac{b}{300^2} = \frac{1840909.09}{300^2} = 20.4545$$

$$\text{Hoop stress } \sigma = \frac{b}{r^2} + a$$

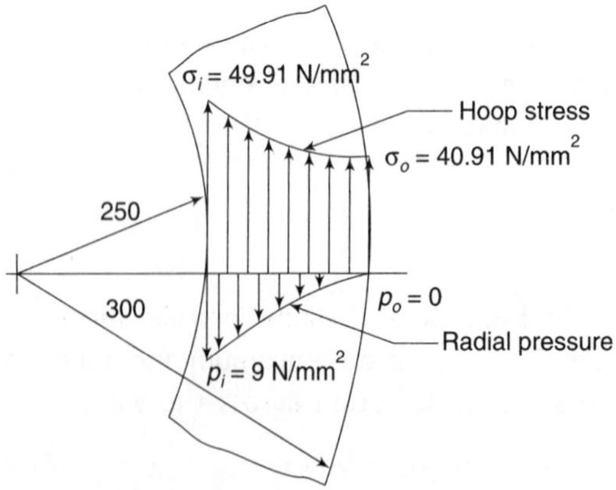

Fig. 10.9 Hoop stress and radial pressure distribution across section

$\therefore$ Hoop stress at inner radius $\sigma_i = \dfrac{b}{ri^2} + a$

$$= \frac{1840909.09}{250^2} + 20.4545 = 49.91 \text{ MPa}$$

Hoop stress at outer radius $\sigma_o = \dfrac{b}{r_o^2} + a = \dfrac{1840909.09}{300^2} + 20.4545 = 40.91$ MPa

The stress distribution is shown in Fig. 10.9.

Problem 20 A pipe of 300 mm internal diameter and 100 mm wall thickness is subjected to an external pressure of 6 N/mm^2. Find the maximum and minimum hoop stresses across the section. Plot the stress distribution across the wall.

Solution:

Inner diameter d_i = 300 mm

Inner radius r_i = 150 mm

Wall thickness t = 100 mm

$\therefore$ Outer radius $r_o = r_i + t = 150 + 100 = 250$ mm

External pressure p_o = 6 N/mm^2

Internal pressure p_i = 0 (not given)

For thick cylinder, radial pressure $p = \dfrac{b}{r^2} - a$

$$\therefore \quad p_i = \frac{b}{ri^2} - a$$

$$0 = \frac{b}{150^2} - a$$

$$a = \frac{b}{150^2} \quad \text{...(i)}$$

$$p_o = \frac{b}{r_o^2} - a$$

$$6 = \frac{b}{250^2} - a \quad \text{...(ii)}$$

Substituting for a from Eq. (i) in Eq. (ii),

$$6 = \frac{b}{250^2} - \frac{b}{150^2}$$

$$b = -210937.5$$

$$a = \frac{b}{150^2} = \frac{-210937.5}{150^2} = -9.375$$

$$\therefore \qquad \text{Hoop stress } \sigma_i = \frac{b}{r_i^2} + a = \frac{-210937.5}{150^2} - 9.375 = -18.75 \text{ N/mm}^2$$

$$\sigma_o = \frac{b}{r_o^2} + a = \frac{-210937.5}{250^2} - 9.375 - 12.75 \text{ N/mm}^2$$

(–ve indicates that hoop stress is compressive)

Figure 10.10 shows the stress distribution

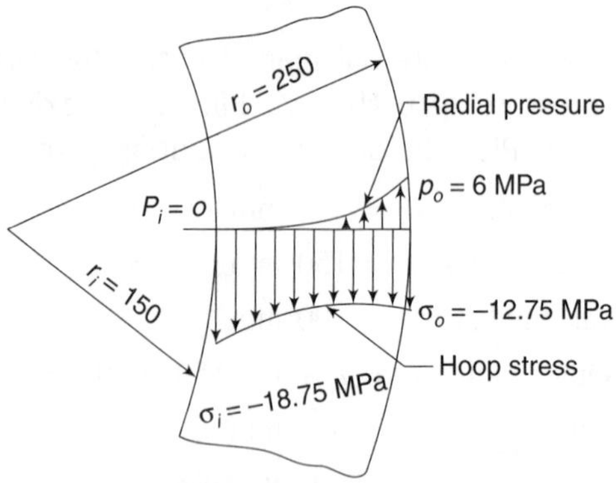

Fig. 10.10 Stress distribution

Problem 21 A thick cylindrical shell of external diameter 420 mm and internal diameter 300 mm is subjected to an internal pressure of 36 MPa and external pressure of 6 MPa. Determine the hoop stresses developed at the inner and outer radii of the cylinder and draw the stress distribution. If the external pressure is doubled, what will be internal pressure to maintain the same maximum stress found above.

Solution: External diameter d_o = 420 mm

External radius r_o = 210 mm

Internal diameter d_i = 300 mm

Internal radius r_i = 150 mm

Internal pressure p_i = 36 MPa

External pressure p_o = 6 MPa

For thick cylinders, pressure $p = \dfrac{b}{r^2} - a$

$$p_i = \frac{b}{r_i^2} - a$$

$$36 = \frac{b}{150^2} - a$$

$$\therefore \qquad a = \frac{b}{150^2} - 36 \qquad \ldots\text{(i)}$$

$$p_o = \frac{b}{r_o^2} - a$$

$$6 = \frac{b}{210^2} - a \qquad \text{...(ii)}$$

Substituting for a from Eq. (i) in Eq. (ii)

$$6 = \frac{b}{210^2} - \left(\frac{b}{150^2} - 36\right)$$

$$6 = \frac{b}{210^2} - \frac{b}{150^2} + 36$$

$$\frac{b}{150^2} - \frac{b}{210^2} = 36 - 6$$

$$\therefore \quad b = 1378125 \text{ and}$$

$$a = \frac{b}{150^2} - 36 = \frac{1378125}{150^2} - 36 = 25.25$$

$$\therefore \quad \text{Hoop stress } \sigma = \frac{b}{r^2} + a$$

$$\sigma_i = \frac{b}{r_i^2} + a = \frac{1378125}{150^2} + 25.25 = 86.5 \text{ MPa}$$

$$\sigma_o = \frac{b}{r_o^2} + a = \frac{1378125}{210^2} + 25.25 = 56.5 \text{ MPa}$$

Figure 10.11 shows the stress distribution

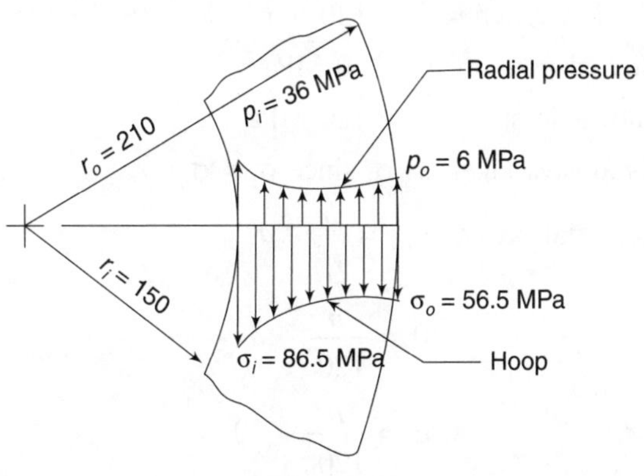

Fig. 10.11

To find internal pressure

$$\sigma_{max} = \sigma_i = 86.5 \text{ MPa}$$

$$p_o = 2 \times 6 \text{ MPa} = 12 \text{ MPa} \quad \text{(doubled)}$$

$$\sigma_i = \frac{b}{r_i^2} + a$$

$$86.5 = \frac{b}{150^2} + a$$

$$\therefore \quad a = 86.5 - \frac{b}{150^2} \qquad \text{...(iii)}$$

$$p_o = \frac{b}{r_o^2} - a$$

$$12 = \frac{b}{210^2} - a \qquad \text{...(iv)}$$

Substitute for a from Eq. (iii) in Eq. (iv)

$$12 = \frac{b}{210^2} - \left(86.5 - \frac{b}{150^2}\right)$$

$$\therefore \quad b = 1467516.89 \text{ and } a = 21.277$$

$$\therefore \quad \text{Internal pressure } p_i = \frac{b}{r_i^2} - a = \frac{1467516.89}{150^2} - 21.277 = 43.95 \text{ MPa.}$$

Problem 22 Find the thickness of metal necessary for a steel cylindrical shell of internal diameter 240 mm required to withstand an internal pressure of 30 MPa. The allowable stress in the section is not to exceed 120 MPa.

Solution: Inner diameter d_i = 240 mm

Inner radius r_i = 120 mm

Internal pressure p_i = 30 MPa

Allowable stress σ_i = 120 MPa

(Allowable stress is always taken as σ_i since $\sigma_i > \sigma_o$)

For thick cylinders, radial pressure $p_i = \dfrac{b}{r_i^2} - a$

$$30 = \frac{b}{120^2} - a$$

$$\therefore \quad a = \frac{b}{120^2} - 30 \qquad \text{...(i)}$$

$$\text{Hoop stress } \sigma_i = \frac{b}{r_i^2} + a$$

$$120 = \frac{b}{120^2} + a \qquad \text{...(ii)}$$

Substituting for a from Eq. (i) in Eq. (ii)

$$120 = \frac{b}{120^2} + \left(\frac{b}{120^2} - 30\right)$$

$\therefore$ $$b = 1080000$$

$$a = \frac{108000}{120^2} - 30 = 45$$

Radial pressure at outer radius $p_o = \dfrac{b}{ro^2} - a$

$p_o = o$ since not given,

$$0 = \frac{1080000}{r_o^2} - 45$$

$\therefore$ Outer radius $r_o = 154.92$ mm

$\therefore$ Wall thickness $t = r_o - r_i = 154.92 - 120 = 34.92$ mm.

Problem 23 A thick cylindrical vessel 250 mm internal diameter and 50 mm wall thickness is subjected to an internal pressure of 10 MPa. Find the maximum hoop stress developed in the cylinder. Also calculate the radial and hoop stresses at a point 20 mm from the inside surface. Sketch the stress distribution.

Solution: Internal radius $d_i = 250$ mm

Inner radius $r_i = 125$ mm

Wall thickness $t = 50$ mm

Outside radius $r_o = r_i + t = 175$ mm

Internal pressure $p_i = 10$ MPa

External pressure $p_o = 0$ (not given)

For thick cylinder, radial pressure $p_i = \dfrac{b}{ri^2} - a$

$$10 = \frac{b}{125^2} - a$$

$\therefore$ $$a = \frac{b}{125^2} - 10 \quad \text{...(i)}$$

External pressure $p_o = \dfrac{b}{r_o^2} - a$

$$0 = \frac{b}{175^2} - \text{a} \quad \text{...(ii)}$$

Substituting for a in Eq. (ii) $0 = \dfrac{b}{175^2} - \left(\dfrac{b}{125^2} - 10\right)$

$$\frac{b}{125^2} - \frac{b}{175^2} = 10$$

$$b\left[\frac{1}{125^2} - \frac{1}{175^2}\right] = 10$$

$$b = 319010.42$$

$$a = \frac{319010.42}{125^2} - 10 = 10.417$$

$$\text{Hoop stress } \sigma = \frac{b}{r^2} + a$$

$$\text{Maximum hoop stress } \sigma_i = \frac{b}{r_i^2} + a = \frac{319010.42}{125^2} + 10.417 = 30.83 \text{ MPa}$$

$$\text{Minimum hoop stress } \sigma_o = \frac{b}{r_o^2} + a = \frac{319010.42}{175^2} + 10.417 = 20.83 \text{ MPa}$$

Radial and hoop stresses at a point 20 mm from the inside surface

$$r = r_i + 20 = 125 + 20 = 145 \text{ mm}$$

$$p_{145} = \frac{b}{r^2} - a = \frac{319010.42}{145^2} - 10.417 = 4.76 \text{ MPa}$$

$$\sigma_{145} = \frac{b}{r^2} + a = \frac{319010.42}{145^2} + 10.417 = 25.59 \text{ MPa}$$

Stress distribution across the wall

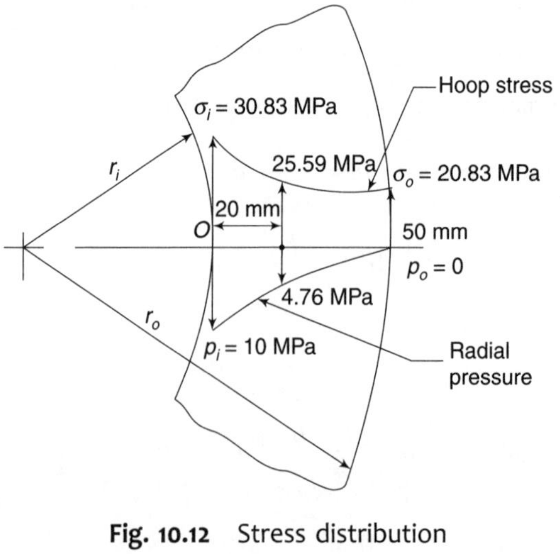

Fig. 10.12 Stress distribution

Problem 24 A thick cylindrical pipe of outside diameter 360 mm and internal diameter 240 mm is subjected to an external fluid pressure of 5 N/mm^2 and internal fluid pressure of 20 N/mm^2. Determine the maximum hoop stress developed. Draw the variation of hoop stress and radial stress across the thickness indicating values at every 20 mm interval.

Solution:

$$\text{Outside diameter } d_o = 360 \text{ mm}$$
$$\text{Outside radius } r_o = 180 \text{ mm}$$
$$\text{Inner diameter } d_i = 240 \text{ mm}$$
$$\text{Inner radius } r_i = 120 \text{ mm}$$
$$\text{External pressure } p_o = 5 \text{ N/mm}^2$$
$$\text{Internal pressure } p_i = 20 \text{ N/mm}^2$$

Consider radius for a thick cylinder,

$$p_i = \frac{b}{r_i^2} - a$$

$$20 = \frac{b}{120^2} - a$$

$$a = \frac{b}{120^2} - 20 \qquad \text{...(i)}$$

and

$$p_o = \frac{b}{r_o^2} - a$$

$$5 = \frac{b}{180^2} - a \qquad \text{...(ii)}$$

Substituting for a from Eq. (i) in Eq. (ii)

$$5 = \frac{b}{180^2} - \left(\frac{b}{120^2} - 20\right)$$

$$\frac{b}{120^2} - \frac{b}{180^2} = 20.5$$

$$b = 388800$$

and

$$a = \frac{388800}{120^2} - 20 = 7$$

$$\text{Maximum hoop stress } \sigma_i = \frac{b}{r_i^2} + a$$

$$\sigma_{max} = \sigma_i = \frac{388800}{120^2} + 7 = 34 \text{ N/mm}^2$$

Variation of stress at 20 mm interval:

(a) At inner radius at $r = r_i = 120$

$$\text{Hoop stress } \sigma_i = 34 \text{ N/mm}^2$$

$$\text{Radial stress } p_i = 20 \text{ N/mm}^2$$

(b) At 20 mm from inside $r = r_i + 20 = 120 + 20 = 140$ mm

$$\text{Hoop stress } \sigma_{140} = \frac{b}{r^2} + a = \frac{388800}{140^2} + 7 = 26.84 \text{ N/mm}^2$$

$$\text{Radial stress } p_{140} = \frac{b}{r^2} - a \; \frac{388800}{140^2} - 7 = 12.84 \text{ N/mm}^2$$

(c) At 40 mm from inside, $r = r_i + 20 + 20 = 160$ mm

$$\text{Hoop stress } \sigma_{160} = \frac{388800}{160^2} + 7 = 22.19 \text{ N/mm}^2$$

$$\text{Radial stress } p_{160} = \frac{388800}{160^2} - 7 = 8.19 \text{ N/mm}^2$$

(d) At 60 mm from inside, $r = r_i + 20 + 20 + 20 = 180 = r_o$

$$\text{Hoop stress } \sigma_{180} = \sigma_o = \frac{388800}{180^2} + 7 = 19 \text{ N/mm}^2$$

$$\text{Radial pressure } p_{180} = p_o = 5 \text{ N/mm}^2$$

The stress distribution is shown in Fig. 10.13.

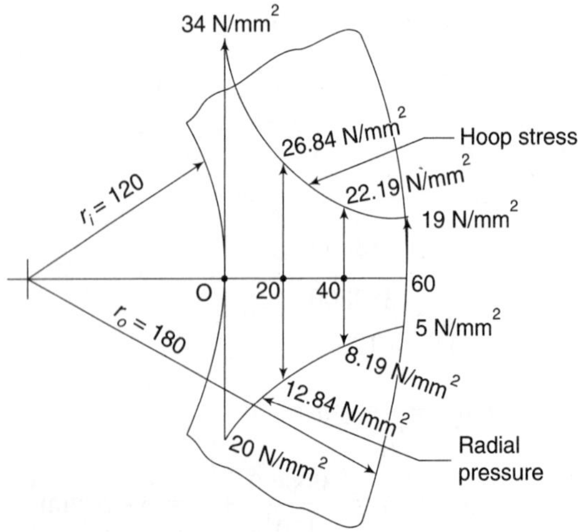

Fig. 10.13

Problem 25 A thick cylinder 500 mm external diameter and 400 mm internal diameter is subjected to both internal and external pressures. The internal pressure is 25 MPa and the hoop stress along the inner face of the cylinder is 45 MPa. Determine the intensity of external pressure and hoop stress along the outer face of the cylinder. Also sketch the stress distribution across the section.

Solution: External diameter $d_o = 500$ mm

External radius $r_o = 250$ mm

Internal diameter $d_i = 400$ mm

Internal radius $r_i = 200$ mm

Internal pressure $p_i = 25$ MPa

Hoop stress at inner radius $\sigma_i = 45$ MPa

For thick cyinders, radial pressure $p_i = \dfrac{b}{r_i^2} - a$

$$25 = \frac{b}{200^2} - a$$

$$a = \frac{b}{200^2} - 25 \qquad \text{...(i)}$$

Hoop stress $\sigma_i = \dfrac{b}{r_i^2} + a$

$$45 = \frac{b}{200} + a \qquad \text{...(ii)}$$

Substituting Eq. (i) in Eq. (ii), $45 = \dfrac{b}{200^2} + \dfrac{b}{200^2} - 25$

$$\therefore \qquad \frac{2b}{200^2} = 45 + 25 = 70$$

$$b = 1400000$$

and $\qquad a = 10$

External pressure $p_o = \dfrac{b}{r_o^2} - a$

$$p_o = \frac{1400000}{250^2} - 10 = 12.4 \text{ MPa}$$

Hoop stress at outer face $\sigma_o = \dfrac{b}{r_o^2} + a = \dfrac{1400000}{250^2} + 10 = 32.4$ MPa

The stress distribution is shown in Fig. 10.14.

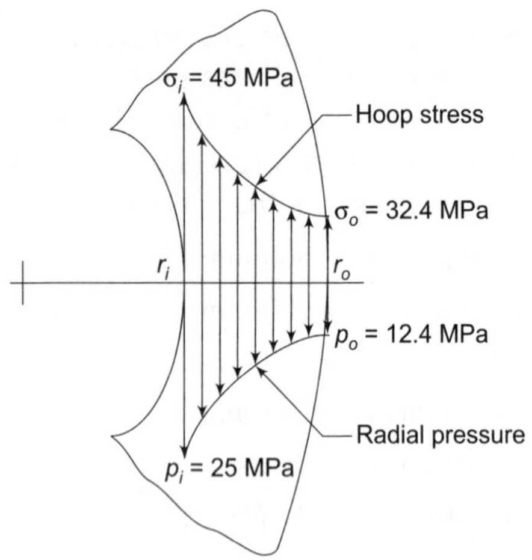

Fig. 10.14

Problem 26 A thick cylindrical pipe of internal radius 250 mm and external radius 300 mm is subjected to an internal pressure of 15 N/mm^2. Calculate the hoop stress across the section. Which is the percentage error if the maximum stress is calculated from the thin cylinder theory?

Solution:

$$\text{Inner radius } r_i = 250 \text{ mm}$$

$$\text{Outer radius } r_o = 300 \text{ mm}$$

$$\text{Internal pressure } p_i = 15 \text{ MPa}$$

$$\text{External pressure } p_o = 0 \text{ MPa (not given)}$$

(a) Considering as thick cylinder

$$\text{Lame's equations } p_i = \frac{b}{r_i^2} - a$$

$$15 = \frac{b}{250^2} - a$$

$$a = \frac{b}{250^2} - 15 \qquad \text{...(i)}$$

$$p_o = \frac{b}{r_o^2} - a$$

$$0 = \frac{b}{300^2} - a \qquad \text{...(ii)}$$

Substituting for a from Eq. (i) in Eq. (ii)

$$0 = \frac{b}{300^2} - \left(\frac{b}{250^2} - 15\right)$$

$$\frac{b}{250^2} - \frac{b}{300^2} = 15$$

$$b = 3068181.82$$

$$a = \frac{b}{300^2} = \frac{3068181.82}{300^2} = 34.09$$

$$\text{Hoop stress } \sigma_i = \frac{b}{r_i^2} + a = \frac{306818.82}{250^2} + 34.09 = 83.18 \text{ MPa}$$

$$\sigma_o = \frac{b}{r_o^2} + a = \frac{306818.81}{300^2} + 34.09 = 6818 \text{ MPa}$$

(b) Considering as thin cylinder

$$\text{Hoop stress } \sigma_1 = \frac{pd}{2t}$$

$$p = p_i = 15 \text{ MPa}$$

$$d = d_i = 500 \text{ mm}$$

$$t = r_o - r_i = 50 \text{ mm}$$

$$\therefore \quad \sigma_1 = \frac{15 \times 500}{2 \times 50} = 75 \text{ MPa}$$

Percentage error in calculating the maximum

$$\text{Hoop stress} = \frac{\sigma_i - \sigma_1}{\sigma_i} \times 100 = \frac{83.18 - 75}{83.18} \times 100 = 9.834\%.$$

Problem 27 A cylindrical pressure vessel has inner and outer radii of 200 mm and 250 mm respectively. The material of the cylinder has an allowable normal stress of 75 MPa, find the maximum internal pressure that can be applied and draw a sketch of radial pressure and circumferential stress distribution across the wall.

Solution:

$$\text{Inner radius } r_i = 200 \text{ mm}$$

$$\text{Outer radius } r_o = 250 \text{ mm}$$

$$\text{Allowable normal stress } \sigma_i = 75 \text{ MPa}$$

$$\text{External pressure } p_o = 0$$

From Lame's equation for thick cylinder

$$\text{Radial pressure } p_o = \frac{b}{r_o^2} - a$$

$$0 = \frac{b}{250^2} - a$$

$$a = \frac{b}{250^2} \qquad \text{...(i)}$$

$$\text{Hoop stress } \sigma_i = \frac{b}{r_i^2} + a$$

$$75 = \frac{b}{200^2} + a$$

$$\text{Substituting for } a \text{ from Eq. (i) } 75 = \frac{b}{200^2} + \frac{b}{250^2}$$

$$\therefore \quad b = 1.829268.29$$

$$a = \frac{b}{250^2} = \frac{1829268.29}{250^2} = 29.27$$

$$\therefore \quad \text{Internal pressure } p_i = \frac{b}{ri^2} - a = \frac{1829268.29}{200^2} - 29.27 = 16.46 \text{ MPa}$$

$$\text{Hoop stress at outer face } \sigma_o = \frac{b}{r_o^2} + a = \frac{1829268.29}{250^2} + 29.27 = 58.54 \text{ MPa}$$

The stress distribution is shown in Fig. 10.15.

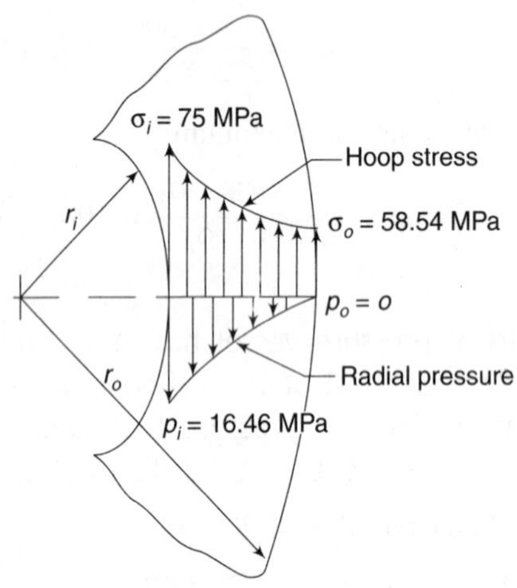

Fig. 10.15

Problem 28 A thick cylinder made of steel with external and internal diameters of 200 mm and 100 mm carries fluid under pressure of 30 N/mm^2 and external pressure of 5 N/mm^2. Find the maximum hoop stress in the cylinder and also find the change in external diameter. Modulus of elasticity is 206 GPa and Poisson's ratio is 0.3.

Solution: External diameter $d_o = 200$ mm

$\therefore \quad r_o = 100$ mm

Internal diameter $d_i = 100$ mm

$\therefore \qquad r_i = 50 \text{ mm}$

Internal pressure $p_i = 30 \text{ N/mm}^2$

External pressure $p_o = 5 \text{ N/mm}^2$

Modulus of elasticity $E = 206 \times 10^3$ MPa

Poisson's ratio $\mu = 0.3$

Lame's equation for radial pressure $p_i = \dfrac{b}{r_i^2} - a$

$$30 = \frac{b}{50^2} - a$$

$$\therefore \qquad a = \frac{b}{50^2} - 30 \qquad \text{...(i)}$$

$$p_o = \frac{b}{r_o^2} - a$$

$$5 = \frac{b}{100^2} - a$$

Substituting for a from Eq. (i), $5 = \dfrac{b}{100^2} - \left(\dfrac{b}{50^2} - 30\right)$

$$\frac{b}{50} - \frac{b}{100^2} = 30 - 5$$

$$\therefore \qquad b = 83333.33$$

$$a = 3.33$$

Maximum hoop stress $\sigma_i = \dfrac{b}{r_i^2} + a = \dfrac{83333.33}{50^2} + 3.33 = 36.67$ MPa

Hoop stress at external diameter $\sigma_o = \dfrac{b}{r_o^2} + a$

$$\sigma_o = \frac{83333.33}{100^2} + 3.33 = 11.67 \text{ MPa}$$

Hoop strain in the outer surface

$$\in_o = \frac{\sigma_o}{E} - \frac{\mu . p_o}{E}$$

$$\in_o = \frac{\delta d_o}{d_o} = \frac{11.67}{206 \times 10^3} - \frac{0.3 \times (-5)}{206 \times 10^3}$$

(P_o is –ve because it is compressive on wall)

$$\in_o = \frac{\delta d_o}{d_o} = 6.38996 \times 10^{-5}$$

$\therefore$ Increase in outer diameter $\delta d_o = 6.38996 \times 10^{-5} \times d_o$

$$\delta d_o = 6.39 \times 10^{-3} \text{ mm.}$$

Problem 29 A cylinder of inner diameter d_i and wall thickness t, subjected to only internal pressure p_i, is assumed to be thin cylinder. What is the greatest value of the ratio $\frac{t}{d_i}$ if the error in the calculation of hoop stress is not to exceed 5 per cent?

Solution:

$$\text{Inner diameter} = d_i$$

$$\text{Thickness} = t$$

$$\text{Outer diameter} = d_o = d_i + 2t$$

$$\text{Internal pressure} = p_i$$

$$\text{External pressure} = p_o = 0$$

$$\text{Error in hoop stress} = 5\% = \left(\frac{\sigma_i - \sigma_1}{\sigma_i}\right) \times 100$$

$$\text{Lame's equation, } \sigma_i = \frac{b}{r_i^2} + a$$

$$\sigma_o = \frac{b}{r_o^2} + a$$

$$p_i = \frac{b}{r_i^2} - a \text{ and } p_o = \frac{b}{r_o^2} - a$$

$$\text{From } p_o = 0 = \frac{b}{r_o^2} - a$$

$\therefore$

$$b = a r_o^2 \qquad \text{...(i)}$$

$$\sigma_i + p_i = \left(\frac{b}{r_i^2} + a\right) + \frac{b}{r_i^2} - a = \frac{2b}{r_i^2} \qquad \text{...(ii)}$$

and

$$\sigma_i - p_i = \left(\frac{b}{r_i^2} + a\right) + \left(\frac{b}{r_i^2} - a\right) = 2a \qquad \text{...(iii)}$$

Substituting for b from Eq. (i) in Eq. (ii)

$$\sigma_i + p_i = \frac{2a\, r_o^2}{r_i^2}$$

Substituting for a from Eq. (iii)

$$\sigma_i + p_i = \frac{2r_o^2}{r_i^2}\left(\frac{\sigma_1 - p_i}{2}\right)$$

$$\sigma_i = \frac{r_o^2}{r_i^2}(\sigma_i - p_i) - p_i$$

$$\sigma_i = \frac{r_o^2(\sigma_i - p_i) - p_i\, r_i^2}{r_i^2}$$

$$\sigma_i\, r_i^2 = \sigma_i\, r_o^2 - p_i\, r_o^2 - p_i\, r_i^2$$

$\therefore$
$$\sigma_i\, r_o^2 - r_i\, r_i^2 = p_i\,(r_o^2 + r_i^2)$$

$$\sigma_i = \frac{(r_o^2 + r_i^2)}{r_o^2 + r_i^2} \times p_i$$

or
$$\sigma_i = \frac{d_o^2 + d_i^2}{d_o^2 - d_i^2} \cdot p_i$$

Put
$$d_o = d_i + 2t$$

$$\sigma_i = \frac{(d_i + 2t)^2 + d_i^2}{(d_i + 2t)^2 - d_i^2} \cdot p_i = \frac{d_i^2 + 4d_i\, t + 4t^2 + d_i^2}{d_i^2 + 4d_i\, t + 4t^2 - d_i} \cdot p_i$$

$$\sigma_i = \frac{2(d_i^2 + 2d_i\, t + 2t^2)}{4t\,(d_i + t)} \cdot p_i = \frac{d_i^2 + 2d_i\, t + 2t^2}{2(d_i + t)} \cdot p_i$$

Let
$$\frac{t}{d_i} = C \quad \text{or } t = d_i \cdot C$$

$$\sigma_i = \frac{d_i^2 + 2d_i^2 \cdot C + 2d_i^2 \cdot C^2}{2(d_i^2\, C + 2 \cdot d_i^2 \cdot C^2)} \cdot p_i$$

$$\sigma_i = \frac{d_i^2(1 + 2\,C + 2\,C^2)}{2d_i^2\,(C + 2\,C^2)} \cdot p_i = \frac{1 + 2C + 2C^2}{2(C + C^2)} \cdot p_i \qquad \text{...(iv)}$$

Considering a thin cylinder

$$\text{Hoop stress } \sigma_1 = \frac{p_i d_i}{2t} = \frac{p_i\, d_i}{2 \cdot d_i C} = \frac{p_i}{2C} \qquad \text{...(v)}$$

$\therefore$
$$\%\text{ error} = \frac{\sigma_i - \sigma_1}{\sigma_i} \times 100 = 5$$

$$1 - \frac{\sigma_1}{\sigma_i} = 0.05$$

$\therefore$
$$\frac{\sigma_1}{\sigma_i} = 1 - 0.05 - 0.95$$

$$\sigma_1 = 0.95\ \sigma_i$$

$$\frac{p_i}{2C} = 0.95\left[\frac{1 + 2C + 2C^2}{2(C + C^2)} \cdot p_i\right]$$

$$\frac{2(C + C^2)}{2C} = 0.95\ (1 + 2C + 2C^2)$$

$$1 + C = 0.95 + 1.9C + 19.C^2$$

$$1.9C^2 + 0.9C - 0.05 = 0$$

$$\text{Solving, } C = \frac{t}{d_i} = 0.05023.$$

Problem 30 A hydraulic cylinder in an earthmoving equipment has internal diameter of 200 mm. Find the thickness required to sustain internal pressure of 50 MPa. The maximum shear stress and tenside stresses and limited to 75 MPa and 120 MPa respectively.

Solution: Internal diameter d_i = 200 mm

Internal radius r_i = 100 mm

Internal pressure p_i = 50 MPa

Maximum shear stress τ_{max} = 75 MPa

Maximum tensile stress σ_i = 120 MPa

External pressure p_o = 0 (not given)

Lame's equations,

$$p_i = \frac{b}{r_i^2} - a$$

$$50 = \frac{b}{100^2} - a$$

$$\therefore \quad a = \frac{b}{100^2} - 50 \quad \text{...(i)}$$

$$\sigma_i = \frac{b}{r_i^2} + a$$

$$120 = \frac{b}{100^2} + a$$

Substitute for a from Eq. (i)

$$120 = \frac{b}{100^2} + \frac{b}{100^2} - 50$$

$$\frac{2b}{100^2} = 120 + 50$$

$$b = 850000 \text{ and } a = \frac{850000}{100^2} - 50 = 35$$

$$\text{Again } \sigma_o = \frac{b}{r_o^2} - a$$

$$0 = \frac{850000}{r_o^2} - 35$$

$$\therefore \quad \text{Outer radius } r_o = 155.839 \text{ mm} \quad \text{...(ii)}$$

Max shear stress will be at inner radius $\tau_{max} = \left(\frac{\sigma_i + p_i}{2}\right)$

$$\text{For } p_o = 0, \quad \sigma_i = \left(\frac{r_o^2 + r_i^2}{r_o^2 - r_i^2}\right) p_i \qquad \text{...(Refer Problem 29)}$$

$$\sigma_i = \frac{r_o^2 + 100^2}{r_o^2 - 100^2} \times 50$$

$$\therefore \quad \tau_{max} = 75 = \frac{\left[\left(\frac{r_o^2 + 100^2}{r_o^2 - 100^2}\right) 50 + 50\right]}{2}$$

$$75 = \frac{50}{2}\left[\frac{r_o^2 + 100^2}{r_o^2 - 100^2} + 1\right]$$

$$75 = 25\left(\frac{r_o^2 + 100^2 + r_o^2 - 100^2}{r_o^2 - 100^2}\right)$$

$$\frac{75}{25}(r_o^2 - 100^2) = 2r_o^2$$

$$3r_o^2 - 3 \times 100^2 = 2r_o^2$$

$$\therefore \quad r_o^2 = 3 \times 100^2$$

$$r_o = 173.2 \text{ mm}$$

$$\therefore \quad \text{Required outer radius } r_o = 173.2 \text{ mm}$$

$$\text{Thickness } t = r_o - r_i = 173.2 - 100$$

$$t = 73.2 \text{ mm.}$$

10.10 THICK SPHERICAL SHELLS

Consider a thick spherical shell of inner radius r_i, outer radius r_o subjected to internal pressure p_i as shown in Fig. 10.16 (a).

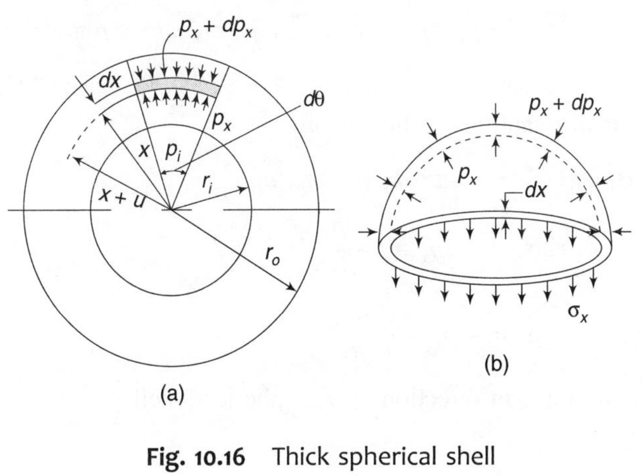

Fig. 10.16 Thick spherical shell

Consider a thin element at radius x and of thickness dx, subtended by an angle $d\theta$ at the centre.

Due to internal pressure, the radius x will increase by a small value, say u. The new radius will be $x + u$.

$$\text{Length of segment } l_x = x \cdot d\theta$$

$$\text{Length of segment after increase in size } l_{(x+u)} = (x + u)\, d\theta$$

$$\therefore \quad \text{Circumferential strain } \in_c = \frac{l_{(x+u)} - l_x}{l_x}$$

$$\in_c = \frac{(x + u)\, d\theta - x d\theta}{x d\theta} = \frac{u}{x} \qquad \ldots\text{(i)}$$

$$\text{Radial strain } \in_r = \frac{d(x + u) - dx}{dx} = \frac{du}{dx} \qquad \ldots\text{(ii)}$$

Let p_x = Pressure at radial distance x

$p_x + dp_x$ = Pressure at radial distance $(x + dx)$

Consider an elemental hemisphere of internal radius x and thickness dx as shown in Fig 10.16 (b)

Let σ_x = Circumferential stress.

Bursting force $\quad F_B = p_x \cdot \pi x^2 - (p_x + dP_x)\, \pi (x + dx)^2$

Resisting force $\quad F_R = \sigma_x \cdot 2\pi x \cdot dx$

Equating bursting force to resisting force,

$$\begin{aligned} \sigma_x \cdot 2\pi x \cdot dx &= p_x \cdot \pi x^2 - (p_x + dp_x)\, \pi\, (x + dx)^2 \\ &= p_x \cdot \pi x^2 - (p_x + dp_x)\, \pi\, (x^2 + 2xdx + dx^2) \\ &= p_x \cdot \pi x^2 - (p_x\, \pi x^2 + p_x\, \pi \cdot 2x dx + p_x \pi\, dx^2 + \pi x^2\, dp_x \\ &\qquad + 2\pi x\, dx \cdot dp_x + \pi\, dp_x \cdot dx^2) \end{aligned}$$

Neglecting the small quantities of higher order

$$\sigma_x \cdot 2\pi x \cdot dx = -2\pi x \cdot dx \cdot p_x - \pi x^2 dp_x.$$

$$2\sigma_x = -\, 2p_x - x \cdot \frac{dp_x}{dx}$$

$$\sigma_x = -\, p_x - \frac{x}{2} \cdot \frac{dp_x}{dx} \qquad \ldots\text{(iii)}$$

This stress σ_x is uniform in directions for a spherical shell.

At any point in the shell, there are three principal stresses acting as shown in Fig. 10.17. They are:

- Radial pressure p_x (Compressive)
- Hoop stress σ_x (Tensile)
- Hoop stress σ_x (Tensile), on the planes at right angles to each other.

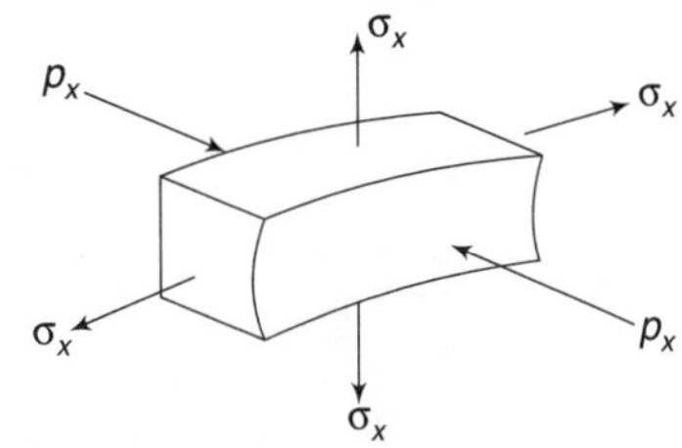

Fig. 10.17

Radial tensile strain

$$\in_r = \frac{-p_x}{E} - \mu\cdot\frac{\sigma_x}{E} - \mu\cdot\frac{\sigma_x}{E}$$

$$= -\frac{1}{E}(p_x + 2\mu\sigma_x) \quad \text{...(iv)}$$

Circumferential tensile strain

$$\in_c = \frac{\sigma_x}{E} - \frac{\mu\sigma_x}{E} - \mu\frac{(-p_x)}{E}$$

$$= \frac{1}{E}[(1-\mu)\,\sigma_x + \mu P_x] \quad \text{...(v)}$$

From Eq. (i) $\quad \in_c = \dfrac{u}{x}$ or $u = x\cdot\in_c$.

From Eq. (ii) $\quad \in_r = \dfrac{du}{dx} = \dfrac{d}{dx}(x\cdot\in_c)$

$$\in_r = \in_c + x\cdot\frac{d\in_c}{dx} \quad \text{...(vi)}$$

Substituting for $\in_r$ and $\in_c$ from Eqs. (iv) and (v)

$$-\frac{1}{E}(p_x + 2\,\mu\sigma_x) = \frac{1}{E}[(1-\mu)\,\sigma_x + \mu p_x] + \frac{x}{E}\cdot\frac{d}{dx}[(1-\mu)\,\sigma_x + \mu p_x]$$

$$= \frac{1}{E}[(1-\mu)\,\sigma_x + \mu p_x] + \frac{x}{E}\left[(1-\mu)d\frac{\sigma_x}{dx} + \mu\cdot\frac{dp_x}{dx}\right]$$

$$0 = (p_x + 2\mu\sigma_x) + (1-\mu)\sigma_x + \mu p_x + x(1-\mu)\frac{d\sigma_x}{dx} + x\cdot\mu\cdot\frac{dp_x}{dx}$$

$$= (p_x + \sigma_x) + (1+\mu) + x(1-\mu)\frac{d\sigma_x}{dx} + x\mu\cdot\frac{dp_x}{dx} = 0 \quad \text{...(vii)}$$

Differentiating Eq. (iii), $\dfrac{d\sigma_x}{dx} = \dfrac{-dp_x}{dx} - \dfrac{1}{2}\dfrac{dp_x}{dx} - \dfrac{x}{2}\cdot\dfrac{d^2p_x}{dx^2}$

$$= -\frac{3}{2}\frac{dp_x}{dx} - \frac{x}{2}\frac{d^2p_x}{dx^2} \quad \text{...(viii)}$$

Substituting for $\dfrac{d\sigma_x}{dx}$ and σ_x in Eq. (vii) and simplifying

We get $\quad x\dfrac{d^2p_x}{dx^2} + 4\dfrac{dp_x}{dx} = 0$

Let $$\frac{dp_x}{dx} = y$$

$$x \cdot \frac{dy}{dx} + 4y = 0$$

$$x \frac{dy}{dx} = -4y$$

$$\frac{dy}{y} = -4 \cdot \frac{dx}{x}$$

Integrating, $\log y = -4 \log x + \log C_1$

where $\log C_1$ is constant of integration

$$\log y = \log C_1 - \log x^4 = \log\left(\frac{C_1}{x^4}\right)$$

or $$y = \frac{C_1}{x^4}$$

But $$y = \frac{dp_x}{dx} = \frac{C_1}{x^4}$$

Integrating, $$p_x = \frac{-C_1}{3x^3} + C_2 \qquad \text{...(ix)}$$

where C_2 is constant of integration.

From Eq. (iii) $\sigma_x = -p_x - \dfrac{x}{2}\dfrac{dp_x}{dx}$

$$\sigma_x = -\left[\frac{-C_1}{3x^3} + C_2\right] - \frac{x}{2}\cdot\left(\frac{C_1}{x^4}\right)$$

$$\sigma_x = \frac{C_1}{3x^3} - C_2 - \frac{C_1}{2x^3} = -\frac{C_1}{6x^3} - C_2$$

Put $C_1 = -6b$ and $C_2 = -a$

$$\sigma_x = \frac{b}{x^3} + a$$

From Eq. (ix) $$p_x = \frac{2b}{x^3} - a$$

These equations are similar to Lame's equation for thick cylinder.

SOLVED PROBLEMS

Problem 31 A spherical shell of diameter 360 mm and 480 mm is subjected to an internal pressure of 60 N/mm^2. Find the maximum hoop stress developed in the shell. Sketch the stress distribution across the wall at intervals of 20 mm.

Solution:

$$\text{Spherical shell of inner diameter } d_i = 360 \text{ mm}$$

$$\text{Inner radius } r_i = 180 \text{ mm}$$

$$\text{Outer diameter } d_o = 480 \text{ mm}$$

$$\text{Outer radius } r_o = 240 \text{ mm}$$

$$\text{Internal pressure } p_o = 60 \text{ MPa}$$

$$\text{External pressure } p_o = 0$$

$$\text{For spherical shell, hoop stress } \sigma = \frac{b}{r^3} + a$$

$$\text{Radial pressure } p = \frac{2b}{r^3} - a$$

$$\text{Consider } p_i = \frac{2b}{r_i^3} - a$$

$$60 = \frac{2b}{180^3} - a \quad \ldots\text{(i)}$$

$$p_o = \frac{2b}{r_o^3} - a$$

$$0 = \frac{2b}{240^3} - a$$

$$a = \frac{2b}{240^3}$$

Substituting for a in Eq. (i)

$$60 = \frac{2b}{180^3} - \frac{2b}{240^3}$$

$$b = 302.63 \times 10^6$$

$$a = 43.78$$

$$\text{Hoop stress } \sigma_i = \frac{b}{r_i^3} + a = \frac{302.63 \times 10^6}{180^3} + 4.378 = 95.67 \text{ MPa}$$

$$\text{At 20 mm } \sigma_{200} = \frac{302.63 \times 10^6}{200^3} + 43.78 = 81.61 \text{ MPa}$$

$$\text{At 40 mm } \sigma_{220} = \frac{302.63 \times 10^6}{220^3} + 43.78 = 72.20 \text{ MPa}$$

$$\text{At 60 mm } \sigma_{240} = \sigma_o = \frac{302.63 \times 10^6}{240^3} + 43.78 = 65.67 \text{ MPa}$$

$$\text{Radial pressure at 20 mm } p_{200} = \frac{2 \times 302.63 \times 10^6}{200^3} - 43.78 = 31.88 \text{ MPa}$$

$$\text{At 40 mm } p_{220} = \frac{2 \times 302.63 \times 10^6}{220^3} - 43.78 = 13.06 \text{ MPa}$$

$$\text{At 60 mm, } p_o = 0$$

The stress variation is shown in Fig. 10.18.

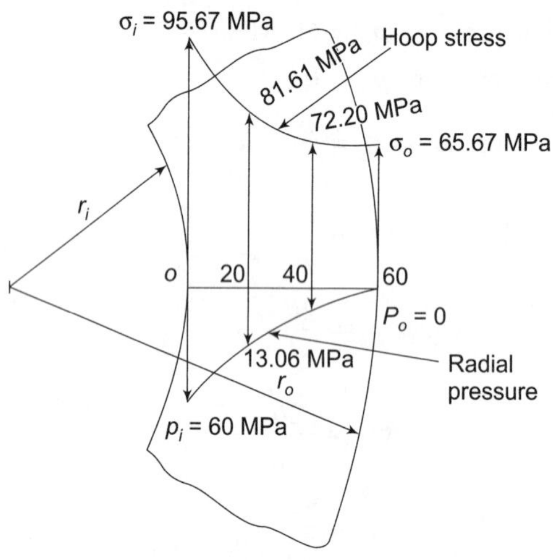

Fig. 10.18

Problem 32 A thick spherical shell 240 mm inner diameter is subjected to an internal pressure of 30 N/mm². If the permissible tensile stress is 90 MPa, find the thickness of the shell.

Solution:

$$\text{Inner diameter } d_i = 240 \text{ mm} \qquad \therefore r_i = 120 \text{ mm}$$

$$\text{Internal pressure } p_i = 30 \text{ N/mm}^2$$

$$\text{Allowable stress } \sigma_i = 90 \text{ MPa}$$

$$\text{External pressure } p_o = 0$$

$$\text{For thick spherical shell, } p_i = \frac{2b}{r_i^3} - a$$

$$30 = \frac{2b}{120^3} - a \quad \text{or } a = \frac{2b}{120^3} - 30 \qquad \text{...(i)}$$

$$\sigma_i = \frac{b}{r_i^3} + a$$

$$90 = \frac{b}{120^3} + a$$

Substituting for a from Eq. (i)

$$90 = \frac{b}{120^3} + \left(\frac{2b}{120^3} - 30\right)$$

$$\therefore \quad \frac{3b}{120^3} = 90 + 30 = 120$$

$$\therefore \quad b = 69.12 \times 10^6 \text{ and } a = \frac{2 \times 69.12 \times 10^6}{120^3} - 30 = 50$$

$$p_o = \frac{2b}{r_o^3} - a$$

$$0 = \frac{2 \times 69.12 \times 10^6}{r_o^3} - 50$$

$\therefore$ Outer radius $r_o = 140.353$ mm

$\therefore$ Wall thickness $t = r_o - r_i = 140.353 - 120 = 20.353$ mm.

Problem 33 A thick spherical shell of 400 mm internal diameter is subjected to a fluid pressure of 15 N/mm². Determine the necessary thickness of the shell is the allowable stress in the shell is 30 N/mm². Also find the stress at outer face.

Solution: Inner diameter $d_i = 400$ mm $\therefore r_i = 200$ mm

Internal pressure $p_i = 15$ N/mm²

Allowable stress $\sigma_i = 30$ N/mm²

External pressure $p_o = 0$

For spherical shell, $p_i = \frac{2b}{ri^3} - a$

$$15 = \frac{2b}{200^3} - a \quad \text{or } a = \frac{2b}{200^3} - 15 \qquad \ldots\text{(i)}$$

$$\sigma_i = \frac{b}{r_i^3} + a$$

$$30 = \frac{b}{200^3} + \left(\frac{2b}{200^3} - 15\right)$$

$$\therefore \quad b = 120 \times 10^6 \text{ and } a = \frac{2 \times 120 \times 10^6}{200^3} - 15 = 15$$

Consider

$$p_o = \frac{2b}{r_o^3} - a$$

$$0 = \frac{2 \times 120 \times 10^6}{r_o^3} - 15$$

$\therefore$ External radius $r_o = 251.98$ mm

$\therefore$ Wall thickness $t = r_o - r_i = 251.98 - 200 = 51.98$ mm

Stress at outer face $\sigma_o = \frac{b}{r_o^3} + a = \frac{120 \times 10^6}{251.98^3} + 15 = 22.5$ MPa.

THINGS TO REMEMBER

A. Thin Cylinder

- For thin cylinders or thin spherical shells, the wall thickness is less than $\frac{1}{10}$ of its radius
- Maximum hoop stress or circumferential stress $\sigma_i = \frac{pd}{2t}$
- Longitudinal stress $\sigma_2 = \frac{pd}{4t}$
- Maximum shear stress $\tau_{max} = \frac{\sigma_1 - \sigma_2}{2} = \frac{pd}{8t}$
- Diameter or circumferential strain $\in_d = \frac{\delta d}{d} = \frac{pd}{4tE}(2 - \mu)$
- Change in diameter $\delta d = \frac{pd^2}{4tE}(2 - \mu)$
- Longitudinal strain $\in_l = \frac{\delta l}{l} = \frac{Pd}{4tE}(1 - 2\mu)$
- Change in length $\delta_l = \frac{pdl}{4tE}(1 - 2\mu)$
- Volumetric strain $\in_V = \frac{\delta V}{V} = 2 \in_d + \in_l = \frac{pd}{4tE}(5 - 4\mu)$
- Change in volume $\delta V = \frac{pd}{4tE}(5 - 4\mu).V$

  where, V = Volume of cylinder $= \frac{\pi d^2}{4} \cdot l$

B. Thin Spherical Shells

- Hoop stress = Longitudinal stress, $\sigma_1 = \sigma_2 = \frac{pd}{4t}$
- Maximum shear stress $= \frac{\sigma_1 - \sigma_2}{2} = 0$
- Diametral strain $\in_d = \frac{\delta d}{d} = \frac{pd}{4tE}(1 - \mu)$
- Change in diameter $\delta d = \frac{pd^2}{4tE}(1 - \mu)$
- Volumetric strain $\in_V = \frac{\delta V}{V} = 3 \in_d = \frac{3pd}{4tE}(1 - \mu)$
- Change in volume $\delta V = \frac{3pd}{4tE}(1 - \mu) \cdot V$

  where, V = Volume of sphere $= \frac{4}{3}\pi r^3 = \frac{\pi d^3}{6}$

- Change in volume due to compressibility of fluid, $\delta V = \frac{p}{K} \cdot V$
  where K = Bulk modulus of the fluid.

C. Thick Cylinders

- Lame's equation for thick cylinders are

$$\text{Hoop stress } \sigma = \frac{b}{r^2} + a$$

$$\text{Radial pressure } p = \frac{b}{r^2} - a$$

- At any point in the cylinder, the difference between hoop stress and radial pressure is constant and is equal to $2a$. That is $(\sigma - p) = 2a$
- Maximum hoop stress occurs at inner surface of the cylinder
- As per Lame's theory, the longitudinal strain at any point is constant.
- The variation of hoop stress and radial pressure across the cross section in a thick cylinder is parabolic.

D. Thick Spherical Shells

- For thick spherical shell,

  Hoop stress $\sigma = \frac{b}{r^3} + a$

  Radial stress $p = \frac{2b}{r^3} - a.$

EXERCISES

1. A cylindrical shell 2 m long, 1.2 m internal diameter with wall thickness of 10 mm is subjected to an internal pressure of 2.1 MPa. Find the stresses developed and changes in diameter, length and volume. Take E = 200 GPa and Poisson's ratio as 0.3.

   (***Ans.*** σ_1 = 126 N/mm^2, σ_2 = 63 N/mm^2, τ_{max} = 31.5 N/mm^2; δd = 0.6426 mm, δl = 0.252 mm, δV = 2.706 × 10^6 mm^3)

2. A cylindrical shell 300 mm diameter and 0.9 m long having wall thickness of 5 mm is filled with fluid at atmospheric pressure. If an additional fluid of volume 20000 mm^3 is pumped in, find the pressure of fluid inside the shell and the hoop stress induced. Take E = 200 kN/mm^2 and Poisson's ratio as 0.27.

   (***Ans.*** p = 1.069 N/mm^2, σ_1 = 32.0796 N/mm^2)

3. A seamless boiler shell is to be made of 12 mm thick plate. The allowable stress is limited to 90 MPa. Find

   (i) the diameter of the boiler to withstand an internal pressure of 3 MPa; and

   (ii) the safe pressure if the diameter is 1.2 metres

   (***Ans.*** d = 720 mm, p = 1.8 N/mm^2)

4. A mild steel cylinder 1.5 m long and 0.5 m diameter having wall thickness of 10 mm has both ends closed by rigid plates. It is initially full of oil at atmospheric pressure. Calculate the additional volume of oil to be pumped into the shell to raise the pressure to 6 MPa. Take E = 200 GPa and Poisson's ratio = 0.32.

(***Ans.*** $\delta V = 410.86 \times 10^3$ mm^3)

5. A cylindrical shell has an external diameter of 500 mm and wall thickness 10 mm and length 1.7 m. Determine the increase in its internal diameter and length when the internal pressure is 1 MPa. Take E = 200 GPa and μ = 0.3.

(***Ans.*** δd = 0.04896 mm, δl = 0.0612 mm)

6. A cylindrical shell 2.7 m long having 0.9 m diameter and wall thickness 12 mm is subjected to internal fluid pressure of 1.2 N/mm^2. Find the maximum shear stress induced and changes in dimensions of the shell. Take $E = 2.1 \times 10^5$ N/mm^2 and Poisson's ratio = 0.33.

(***Ans.*** τ = 11.25 N/mm^2, δd = 0.161 mm, δl = 0.0984 mm)

7. A thin spherical shell of diameter 1 m has metal thickness of 12 mm. Find the change in diameter and the capacity of the shell if it is filled with fluid under pressure of 3 MPa. Take E = 200 GPa and Poisson's ratio = 0.27.

(***Ans.*** δd = 0.228 mm, $\delta V = 358.338 \times 10^3$ mm^3)

8. A thin spherical shell of 1200 mm diameter is subjected to internal fluid pressure of 3 MPa. Find the wall thickness if the hoop stress is to be limited to 90 MPa. Also find the change in diameter and volume taking E = 200 kN/mm^2 and Poisson's ratio = 0.3.

(***Ans.*** t = 10 mm, δd = 0.378 mm, $\delta V = 855.016 \times 10^3$ mm^3)

9. A thin cylindrical shell 0.9 m long and 600 mm inner diameter has a wall thickness of 9 mm. Its both ends are closed and is completely filled with fluid at atmospheric pressure. Determine the additional volume of fluid to be pumped into the cylinder to raise the pressure to 1.8 MPa. Take E = 200 GPa Poisson's ratio = 0.3 and bulk modulus of fluid as 2.1 GPa.

(***Ans.*** 363.16×10^3 mm^3)

10. A thick cylinder of internal diameter 80 mm and external diameter 120 mm is subjected to an external pressure of 4 MPa and internal pressure of 12 MPa. Calculate the circumferential stresses at inner and outer surface. Sketch the stress variation across the section.

(***Ans.*** σ_i = 16.8 MPa, σ_o = 8.8 MPa)

11. A cylindrical pressure vessel has inner and outer radii of 240 mm and 300 mm respectively. The material of the cylinder has an allowable normal stress of 75 N/mm^2. Determine the maximum internal pressure that can be applied and draw sketch of radial pressure and circumferential stress distribution.

(***Ans.*** p_i = 16.465 MPa, σ_o = 58.535 MPa, p_o = 0)

12. A pressure vessel with outer and inner diameter of 400 mm and 300 mm respectively is subjected to an external pressure of 9 MPa. Determine the circumferential stress induced at the inner and outer surfaces.

(***Ans.*** σ_i = 32.14 MPa, σ_o = 23.14 MPa)

13. Determine the wall thickness of a cylindrical shell of internal diameter 200 mm to withstand an internal pressure of 30 MPa. Permissible tensile stress in the material is 120 MPa. Also sketch the hoop stress variation across the wall.

(***Ans.*** t = 29.1 mm, σ_i = 120 MPa, σ_o = 90 MPa, p_i = 30 MPa p_o = 0)

14. The maximum stress permitted in a thick cylinder radii 200 mm and 300 mm is 16 N/mm^2. If internal pressure is 12 MPa, what external pressure can be applied? Plot the curves showing the variation of hoop and radial stresses across the wall.

(***Ans.*** p_o = 4.22 MPa, σ_i = 16 MPa σ_o = 8.22 MPa)

15. A thick cylinder of 160 mm internal diameter is subjected to internal pressure of 60 MPa and no external pressure. Find the thickness of the cylinder if

(i) the maximum tensile stress is limited to 90 N/mm^2

(ii) the maximum shear stress is limited to 80 N/mm^2

(***Ans.*** (i) t = 98.885 mm, (ii) t = 80 mm)

16. A thick spherical shell having internal and external diameter of 240 mm and 360 mm is subjected to an internal fluid pressure of 60 MPa. Find the hoop stress developed at 20 mm internal across the wall.

(***Ans.*** σ_{120} = 110.52 MPa, σ_{140} = 78.95 MPa, σ_{160} = 61.23 MPa σ_{180} = 50.52 MPa)

17. A thick spherical shell of 300 mm dia and 50 mm wall thickness is subjected to internal fluid pressure of 50 MPa. Find the maximum and minimum hoop stress across the wall.

(***Ans.*** σ_i = 61.49 MPa, σ_o = 36.49 MPa)

18. Find the wall thickness of spherical shell of inner radius 150 mm subjected to internal fluid pressure of 50 MPa. The hoop stress is limited to 60 MPa.

(***Ans.*** t = 69.73 mm)

Objective Type Questions

1. Hoop stress in a thin cylinder subjected to internal pressure is

(a) $\frac{pd}{2t}$ (b) $\frac{pd}{4t}$

(c) $\frac{pt}{4d}$ (d) $\frac{2p}{dt}$

2. Longitudinal stress in a thin cylinder subjected to internal pressure is

(a) $\frac{pd}{2t}$ (b) $\frac{pd}{4t}$

(c) $\frac{pt}{4d}$ (d) $\frac{2p}{dt}$

3. Circumferential strain or diametral strain in a thin cylinder subjected to internal pressure

(a) $\frac{pd}{2t}$ (b) $\frac{pd}{4t}$

(c) $\frac{pd}{4tE}(2 - \mu)$ (d) $\frac{pd}{4tE}(1 - 2\mu)$

4. Longitudinal strain in a thin cylinder subjected to internal pressure

(a) $\frac{pd}{2t}$ (b) $\frac{pd}{4t}$

(c) $\frac{pd}{4tE}(2 - \mu)$ (d) $\frac{pd}{4tE}(1 - 2\mu)$

5. Volumetric strain in a thin cylinder subjected to internal pressure

(a) $\frac{pd}{4tE}(2 - \mu)$ (b) $\frac{pd}{4tE}(1 - 2\mu)$

(c) $\frac{pd}{4tE}(5 - 4\mu)$ (d) $\frac{pd}{4tE}(4\mu - 5)$

6. For a thin cylinder subjected to internal pressure, maximum shear stress is

(a) $\frac{pd}{2t}$ (b) $\frac{pd}{4t}$

(c) $\frac{pd}{8t}$ (d) $\frac{pd}{t}$

7. For a thin cylinder subjected to internal pressure, maximum shear stress is

(a) $(\sigma_1 - \sigma_2)$ (b) $(\sigma_2 - \sigma_1)$

(c) $\left(\frac{\sigma_1 - \sigma_2}{2}\right)$ (d) $\left(\frac{\sigma_2 - \sigma_1}{2}\right)$

8. For a thin cylinder subjected to internal pressure, the hoop stress is _____ the longitudinal stress.

(a) thrice (b) equal to

(c) half of (d) twice

9. For this spherical shells subjected to internal pressure the hoop shear is

(a) $\frac{pd}{4t}$ (b) $\frac{pd}{2t}$

(c) $\frac{pd}{3t}$ (d) $\frac{pt}{4d}$

10. In a thin spherical shell subjected to internal pressure the maximum shear stress is

(a) $\frac{pd}{8t}$ (b) zero

(c) $\frac{pd}{4t}$ (d) $\frac{pd}{2t}$

11. The diametral strain for a thin spherical shell subjected to internal pressure is

(a) $\frac{pd}{4tE}(1 - 2\mu)$ (b) $\frac{pd}{4tE}(2 - \mu)$

(c) $\frac{pd}{4tE}(1 - \mu)$ (d) $\frac{pd}{4tE}(\mu - 1)$

12. The volumetric strain for a thin spherical shell subjected to internal pressure is

(a) $\frac{pd}{4tE}(5 - 4\mu)$ (b) $\frac{pd}{4tE}(1 - \mu)$

(c) $\frac{3pd}{2tE}(1 - \mu)$ (d) $\frac{3pd}{4tE}(1 - \mu)$

13. Lame's equation for hoop stress in a thick cylinder is

(a) $\frac{b}{r^2} + a$ (b) $\frac{b}{r^2} - a$

(c) $\frac{b}{r^3} + a$ (d) $\frac{b}{r^3} - a$

14. Lame's equation for radial pressure in a thick cylinder is

(a) $\frac{b}{r^2} + a$ (b) $\frac{b}{r^2} - a$

(c) $\frac{b}{r^3} + a$ (d) $\frac{b}{r^3} - a$

15. The maximum hoop stress in a thick cylinder occurs at

(a) outer face (b) inner face

(c) at the middle of wall (d) none

16. Hoop stress for thick spherical shell is given by

(a) $\frac{b}{r^2} + a$ (b) $\frac{b}{r^2} - a$

(c) $\frac{2b}{r^3} + a$ (d) $\frac{b}{r^3} + a$

17. Radial stress in a thick spherical shell is given by

(a) $\frac{2b}{r^3} - a$ (b) $\frac{2b}{r^3} + a$

(c) $\frac{b}{r^3} - a$ (d) $\frac{b}{r^3} + a$

18. In a thick cylinder, the difference between hoop stress and radial pressure at any given radius across wall is constant and is given by

(a) a (b) $2a$

(c) b (d) $2b$

19. For a thick cylinder, maximum shear stress is given by

(a) $\frac{\sigma_i - p_i}{2}$ (b) $\frac{\sigma_o - p_o}{2}$

(c) $\frac{\sigma_i + p_i}{2}$ (d) $\frac{\sigma_o + p_o}{2}$

20. In a thick cylinder, the stress variation across the wall is

(a) linear (b) cubic

(c) parabolic (d) none

Key to Answers

1. (a)	2. (b)	3. (c)	4. (d)	5. (c)	6. (c)	7. (c)	8. (d)
9. (a)	10. (b)	11. (c)	12. (d)	13. (a)	14. (b)	15. (b)	16. (d)
17. (a)	18. (b)	19. (c)	20. (c)				

Index